Andreas Riedl
Wolfgang Kainz
Gregory A. Elmes

Progress in Spatial Data Handling

Andreas Riedl
Wolfgang Kainz
Gregory A. Elmes
(Editors)

Progress in Spatial Data Handling

12th International Symposium on Spatial Data Handling

With 353 Figures

Springer

Dr. Andreas Riedl
University of Vienna
Department of Geography
and Regional Research
Universitätsstraße 7
1010 Vienna
Austria

e-mail:
andreas.riedl@univie.ac.at

Prof. Wolfgang Kainz
University of Vienna
Department of Geography
and Regional Research
Universitätsstraße 7
1010 Vienna
Austria

e-mail:
wolfgang.kainz@univie.ac.at

Prof. Gregory A. Elmes
West Virginia University
Department of Geology and Geography
215 White Hall
Morgantown WV 26506-6300
USA

e-mail:
gelmes@wvu.edu

Library of Congress Control Number: 2006927798

ISBN-10 3-540-35588-X Springer Berlin Heidelberg New York
ISBN-13 978-3-540-35588-5 Springer Berlin Heidelberg New York

Springer is a part of Springer Science+Business Media
springer.com

Printed in The Netherlands

Cover design: Erich Kirchner
Typesetting: camera-ready by the editors
Production: Christine Adolph
Printing: Krips bv, Meppel
Binding: Stürtz AG, Würzburg

Printed on acid-free paper 30/2133/ca 5 4 3 2 1 0

Foreword

Geographic information is essential to building the global economy, maintaining sustainable environments, and shaping education for the 21st century. There is a continuing need to expand and advocate for research in geographical information science and technology, thereby enhancing numerous applications of geographical information to meet the many challenges of our current era. The International Symposium on Spatial Data Handling is a long-standing expert forum that stands at the leading edge of geographical information developments.

The First International Symposium on Spatial Data Handling (SDH) was organized by the International Geographical Union Commission on Geographical Data Sensing and Processing in Zurich in 1984. The Commission was succeeded by the Commission on Geographic Information Systems, the Study Group on Geographical Information Science and the Commission on Geographical Information Science (http://www.igugis.org/). Over the intervening 26 years, the Symposium has been held in:

1st - Zurich, 1984
2nd - Seattle, 1986
3rd - Sydney 1988
4th - Zurich, 1990
5th - Charleston, 1992
6th - Edinburgh, 1994
7th - Delft, 1996
8th - Vancouver, 1998
9th - Beijing, 2000
10th - Ottawa 2002
11th - Leicester 2004

This volume contains the proceedings of the 12th International Symposium on Spatial Data Handling, which was held in Vienna, Austria, on July 12–14 2006 as part of the conference "Geoinformation Connecting Societies – GICON 2006" in conjunction with the International Cartographic Association (ICA), the International Society for Photogrammetry and Remote Sensing (ISPRS) as well as local associations that focus on geoinformation. The International Symposium on Spatial Data Handling is a fully-refereed conference. All the papers in this book were submitted as full papers and blind reviewed by at least two members of the Program Committee. 125

papers were submitted and the 56 that are included here are clearly of a high standard.

The papers cover the current scope of research challenges in Geographic Information Science, ranging from information sharing (ontology, semantics and participation), through uncertainty and surface modeling to social and policy issues of the use of spatial information. As a result this volume contains valuable contributions to researchers in many specializations.

This publication is the third in a Springer Verlag series, and follows Dr. Peter Fisher's edited volume of the 11th Symposium Proceedings, Developments in Spatial Data Handling. The Commission on Geographical Information Science is pleased to recognize the continuing support of Springer Verlag.

Acknowledgement

Program Committee

I must first acknowledge the very helpful support of my fellow editors, Wolfgang Kainz and Andreas Riedl of the University of Vienna who gave me the freedom to coordinate the call for papers, submission and selection process of the articles in this volume, while graciously accommodating and correcting my missteps. I thank all the members of the Program Committee for their extraordinary efforts to return reviews on time and to Henk Ottens and Peter Fisher for their advice and encouragement from afar.

The final papers would not have been delivered in a timely manner to my colleagues in Vienna without the invaluable assistance of Ms. Christine Titus through the submission, review and revision process. She created and maintained the database, spent hours preparing manuscripts for blind review, and identified and resolved many problems.

I would like to acknowledge the International Cartographic Association (ICA), the International Society for Photogrammetry and Remote Sensing (ISPRS) and the International Geographical Union (IGU) for their commitment to making GICON 2006 and the International Symposium of Spatial Data Handling a first-rate conference.

Thanks also to Mike Phoenix. Through his action and the generosity of ESRI, we were able to award travel scholarships to students and early-career scholars presenting outstanding papers at the conference.

Finally I should also like to thank Jeanne, my wife, and my family for love and support during the long periods when I was focused on the Symposium.

Gregory Elmes
7 May 2006

Local Organizing and Publishing Committee

We acknowledge the support and help of Gregory Elmes and his team at West Virginia University in managing the paper review process and delivering the contributions on time for us to prepare the program and to finish the book of proceedings.

Our special thanks go to Ms. Regina Schneider who spent long hours in editing the contributions according to the template provided by Springer. Without her enormous desktop publishing experience and dedication this book would not have been finished on time. Too often papers deviated substantially from the given template and, moreover, the mixture of papers in Microsoft Word and LaTeX format posed additional challenges to make visual differences in style disappear. Walter Lang assisted in converting and improving difficult figures.

We also would like to acknowledge the support and patience of Springer Verlag who gave us extra time for finishing a book that exceeded the originally planned number of pages by nearly double.

Finally, we thank our wives, Lihui and Doris, and our families for their love and support during the stressful period while preparing the GICON conferences.

Wolfgang Kainz
Andreas Riedl
12 May 2006

Table of Contents

Plenary of Submitted Papers

Spatial Cognition

Data Models

Data Mining

Data Retrieval

Data Quality

Integration and Fusion

Semantics and Ontologies

2D-Visualization

3D-Visualization

Generalization

Uncertainty

Elevation Modeling

Working with Elevation

Spatial Modeling

Author Index

Program Committee

Chair
Gregory Elmes

David Abel
Eliseo Clementini
Leila De Floriani
Peter Fisher
Andrew Frank
W. Randolph Franklin
Mark Gillings
Michael Goodchild
Chris Gold
Francis Harvey
Claire Jarvis
Christopher B. Jones
Brian Klinkenberg
Menno-Jan Kraak
Michael Leitner
(Lily) Chao Li
Martien Molennar
Henk Ottens
Donna Peuquet
Sanjay Rana
Juri Roosaare
Anne Ruas
Monika Sester
Nicholas Tate
Peter van Oosterom
Qihao Weng
Robert Weibel
Mike Worboys
Anthony Yeh

Local Organizing Committee

Chair
Wolfgang Kainz

Michaela Kinberger
Karel Kriz
Alexander Pucher
Andreas Riedl
Regina Schneider

The Devil is *still* in the Data: Persistent Spatial Data Handling Challenges in Grassroots GIS

Sarah Elwood

Department of Geography & Regional Development, University of Arizona, 409 Harvill Building Box 2, Tucson, AZ 85721, USA

1 Introduction[1]

In the past decade we have seen tremendous expansion and diversification of the loosely-knit 'community' of GIS users. Simpler user interfaces, more powerful desktop applications, declining relative costs, and the increasing availability of digital geospatial data have lowered barriers to GIS adoption and use for some new users. The growing involvement of community activists, local-level civic institutions, and non governmental organizations in governance roles has also contributed to this diversification of GIS users. Many of these groups, in urban and rural settings around the world, have begun to use GIS and geospatial data to inform their planning, problem solving and service delivery activities. Participatory GIS (PPGIS) research has devoted nearly a decade of work to understanding the needs, resources and constraints of these grassroots GIS users and examined how their GIS applications are shaped by the social, political, and economic contexts in which they occur.[2] Researchers have developed conceptualiza-

[1] This material is based upon work supported by the National Science Foundation under Grant No. 0443152.

[2] I use the term grassroots GIS users to describe a diverse range of local-level, community-based, or NGO-initiated GIS efforts. In spite of tremendous variation in their political motivations and strategies, relationship to and interaction with structures of government, topical areas of GIS application, and social and political contexts, these GIS users share some common experiences, needs, and barriers.

tions of the social and political impacts of GIS use by grassroots institutions (Elwood 2002), examined the strategies they use to sustain GIS and spatial data access (Leitner et al. 2000), and specified the political, organizational, and technological factors that shape grassroots GIS use (Harris and Weiner 1998; Sieber 2000).

A common argument in PPGIS has been that existing geospatial data from public sources tend to be inappropriate for grassroots GIS users, and researchers have focused on how these users might incorporate their own local knowledge into a GIS (Leitner et al. 2000; Sieber 2004). This focus has meant that while spatial data are clearly an essential variable in grassroots GIS sustainability and effectiveness, PPGIS research has had less engagement with GIScience research on spatial data infrastructures (SDIs). In this chapter I will argue that grassroots users are far more actively seeking and using public spatial data than is reflected in the existing literature and thus, that the accessibility and utility of local SDIs is far more important than has been illustrated previously.

For twenty years or more, GIScience researchers have argued that institutions and their creation and management of digital spatial data matter tremendously in shaping the social and political impacts of GIS (Chrisman 1987). This foundational tenet has shaped a longstanding concern in GIScience with the 'human' side of spatial data representation and analysis, informing research that examines SDIs as socio-technological systems that extend far beyond technologies and policies for spatial data handling. Such work has studied the roles and influence of individuals and institutional actors in producing, sustaining, and transforming SDIs, as well as the significance of governmental and organizational rules and resources (Evans 1999; Harvey 2001; Crompvoets and Bregt 2003; Craig 2005). Researchers have also examined the impact of local and national variability in legal and other regulatory structures that govern accessibility and use of public spatial data (Craglia and Masser 2003; Onsrud and Craglia 2003; Williamson et al. 2003).

Recently researchers and policymakers have become particularly concerned about local data integration in SDIs, coordination between local and national levels, and public participation in spatial data development and policies (Martin 2003; Nedovic-Budic et al. 2004: Van Loenen and Kok 2004; Onsrud et al. 2005). In spite of this notion that highly localized institutions, data, and users matter tremendously, this part of the literature tends to focus primarily on 'traditional' GIS and data users, such as local government, planners, and developers. Grassroots GIS and spatial data users remain relatively absent in these discussions. The growing involvement of grassroots groups in local government in many cities in North America and

Europe makes this omission increasingly problematic. Grassroots groups are making key urban policy recommendations and decisions, and a growing number of them demand GIS and public spatial data to inform these activities (Craig et al. 2002). In spite of the success of PPGIS research in documenting the unique needs of these users, existing local SDIs and modes of access still do not meet their needs. This paper is motivated by these persistent difficulties of spatial data handling at the grassroots.

2 The Humboldt Park GIS Project

Against this backdrop of changing geospatial technologies and data policies, I began a project with two Chicago community development organizations in 2003. The Humboldt Park GIS Project (HPGIS Project) is a participatory action research effort designed to develop strategies for sustaining GIS capacities in community-based organizations, and to better understand the role of GIS and geospatial data in the spatial politics of urban planning and redevelopment.[3] Humboldt Park is an area on the northwest side of Chicago that faces a variety of challenges experienced by many U.S. inner city neighborhoods: higher crime rates, housing and other infrastructural problems, gentrification and problems of residential and small business displacement, and higher levels of unemployment and poverty. Many local non profit agencies are involved in efforts to ameliorate these problems, and the HPGIS Project involves two of these organizations. The participating agencies are broad-based community development groups working on policy advocacy, housing and economic development, and community organizing.

Through the HPGIS project, both groups acquired software, hardware, spatial data, and staff training to begin using GIS to inform and support their activities. In this collaboration, my role and that of project research assistants is to facilitate GIS skill-building in these two groups, but not to shape the substantive direction of their GIS applications and choices about spatial data development and acquisition. Our participant observation of GIS development and use in the HPGIS Project provides a rich opportunity to assess opportunities and barriers of local SDIs for grassroots groups. I incorporate details from this case study not to suggest that they can necessarily be generalized to the experiences of all grassroots GIS initiatives, but rather because they highlight areas in which GIScience theory and

[3] For more detailed discussion of the HPGIS project, see Elwood (2006a) and Elwood (2006b).

practice must be extended to more fully conceptualize and meet the needs of grassroots GIS users.

Given the array of participatory technology and data development strategies emerging from PPGIS research, I anticipated that we would use GIS to include local spatial knowledge in neighborhood planning and revitalization processes, perhaps using multimedia GIS techniques or incorporating qualitative representations of community knowledge. With the expansion of online GIServices, and geoportals for facilitating public access to spatial data infrastructures, I expected data acquisition and development from public sources to be relatively straightforward. The reality has been quite been different. Community participants in the Humboldt Park GIS (HPGIS) Project have articulated a clear preference for fairly traditional applications of GIS and they rely heavily upon public spatial data acquired from local government sources. The biggest challenges to the sustainability and impact of the HPGIS Project stem from persistent problems of data access, appropriateness, and accuracy.

In this paper, I will draw on evidence from the HPGIS Project to suggest that many of the persistent challenges of grassroots GIS are rooted in issues of spatial data and spatial data handling. Heterogeneity between community spatial knowledge and official public data records, data accessibility and compatibility problems, and difficulties in data dissemination continue to hamper grassroots GIS efforts. While a great deal has been gained from the past decade of critical GIS and PPGIS research, the persistence of these difficulties suggests that spatial data handling in grassroots GIS needs further investigation. I suggest that we need to consider the influence of new spatial data handling technologies, policies, and structures; and that we need to draw on a broader range of conceptual and practical resources from across GIScience research. In particular, I will argue that addressing these challenges necessitates building stronger linkages between PPGIS and a diversity of GIScience research on spatial data handling, including work on ontologies, data integration, interoperability, and spatial data infrastructures (SDIs).

3 Geospatial Data in Grassroots GIS

PPGIS researchers have long emphasized the significant role of geospatial data in shaping grassroots GIS efforts, focusing on grassroots groups' difficulties in gaining access to existing data, the inappropriateness of many public and private data sources for grassroots activities and priorities, and the inability of some digital databases and GIS software to fully include and represent community knowledge (Barndt 1998; Elwood and Leitner 2003; Merrick 2003). The HPGIS Project has underscored the extent to which these data problems remain a central challenge of grassroots GIS. In this project, we could rely on a host of alternative practices for incorporating non-traditional representations of spatial knowledge in a GIS, including sketches, photographs, animations, text, or audio files. But the participating community organizations insist that they need a diverse spectrum of GIS practices and forms of geospatial data, ranging from alternative and non-traditional applications and data to more conventional applications and data types commonly used by local government. They argue that this diversity is essential to their political influence. Many of the decisions about urban spaces, structures, and land uses that they seek to influence are negotiated through terminologies, data, and policies set by local government.

In this context, the success of grassroots GIS initiatives rests not just on an ability to provide effective alternative GIS practices, but also upon our ability to ensure that more conventional technologies, infrastructures, data, and policies meet grassroots GIS and geospatial data needs. Evidence from the HPGIS Project clearly indicates our failure to do so. For all our progress in identifying technological, financial, and human resources for sustaining grassroots GIS, a major constraint continues to be consistent access to appropriate and accurate data for community-level GIS applications.

3.1 Data Appropriateness

With respect to the appropriateness of data from existing spatial data infrastructures for grassroots GIS applications, there remain significant gaps between the attribute systems used in public records data and the spatial attributes that grassroots groups use to characterize community needs and conditions. These disjunctures involve semantic or schematic differences in how public spatial data records represent urban spaces and structures and how grassroots groups characterize them. As a result, HPGIS Project participants frequently find that local government data must be significantly revised before being used. One organizer explained:

> '...[The County uses] like 20 different codes just for an empty parcel. At one time there may have been a single family home and it had just been vacant, and there may have been a structure in the back of the lot. So it would have a different code like "vacant structure with detached structure" or whatever, something like that. Or maybe it's just a totally vacant parcel that may have been used for who knows what, and it may just be recorded as something else. There's so many different codes that the County uses...[For us] if it's vacant, it's vacant. That's pretty much the code we use. If it's vacant, we just code it as vacant'. (Alonso 2004)

The organizer went on to describe how he retained the County's designations for future communications with government officials, but added a new field for his organization's own use, populating with a simpler categorization scheme. It comes as no surprise that not all local government data are appropriate for all users. However, I would underscore here the disproportionate challenges that less expert users such as grassroots groups face in transforming these data into more appropriate forms.

Problems with appropriateness of public data for grassroots GIS applications are sometimes also rooted in epistemological differences. For example, in contemporary discourses of urban community development, grassroots groups often try to portray community spaces and needs through "asset-based" strategies that focus on community resources rather than problems. Such assets might include social capital, effective community institutions, or deep experience in effective community activism. HPGIS Project participants note that information that would be useful in assessing, locating, or expanding these assets is not part of local public data records. An ongoing complaint about geospatial data from all levels of government is that these data lend themselves to portraying problems or deficits (such as infrastructural problems or poor property conditions). One community participant offered an example of this data challenge describing multiple perspectives on a single street in his neighborhood:

> '…People talk about "Division Street, that's a horrible place". But then other people talk about, "Paseo Boricua is such a wonderful place". They're the same physical space, but psychologically, they're two different environments, two different realities. And the problem is how do you show that? How do you show the plus and the minus of that'? (Rey 2005)

The critical point here is that these epistemological differences create unique challenges for developing spatial data. The kinds of representations most readily developed from existing spatial data are not always appropriate for grassroots groups or may require significant modification in order to be useful.

3.2 Data Accuracy

Another problematic aspect of spatial data handling in grassroots GIS involves data accuracy. Here, the highly localized perspective of grassroots groups is important. HPGIS Project participants work extensively with spatial and attribute data for individual property parcels, obtained from City of Chicago and Cook County records. Errors in these data, including incorrectly addressed parcels, missing parcels, polygon slivers, and errors in recording owners, taxpayers or zoning designations are a persistent problem for GIS applications (and neighborhood improvement efforts) that rely on these data. Because of their immediate lived connection to the spaces represented in these data, the community organizers readily perceive these errors, as well as potential community development problems that could result from them:

> 'See these parcels here? I don't know if maybe the City has something wrong [in its database] … but if this [address] is 1518, then these two parts should belong to 1518 too. See? But, I'm wondering if he – the person who owns the 1518 property – owns those two little squares in the back of 1506 and 1508. …I'm wondering if those owners know that. My worry as an organizer is that…say a developer is going to be sneaky. They're going to see this and say "Let me buy all the back lots". So they pay the taxes for those little parcels. Do the owners know that those little parcels exist separate from the front section'? (Teresa 2005)

The highly localized scales at which these grassroots institutions intervene in urban life mean that they need highly localized data from the spatial data infrastructure. They are in the position to readily recognize errors in data obtained from the local SDI, because of their intimate knowledge of local conditions. But simultaneously, these errors are extremely problematic for their work, more so than for institutions whose work is not so spatially focused. While HPGIS Project participants express pride in their ability to recognize and correct errors in local government data, they express frustration with the absence of any formalized means of communicating these observations back to the local SDI. In theory, their expertise could position them as important and knowledgeable participants in the local SDI, perhaps resulting in a system better able to meet their data needs. This possibility remains unrealized.

3.3 Data Accessibility

Constraints on data access have been well documented in the PPGIS literature, argued to stem from a range of socio-political, technological, and expertise barriers. Researchers have noted that grassroots users sometimes experience difficulty in navigating complicated bureaucratic structures to obtain public data, occasionally encounter resistance in their attempts to obtain these data, or may be unable to afford cost-recovery fees sometimes charged by public agencies (Laituri 2003). My evidence suggests that access remains a problem, but not always for the reasons articulated in existing research. Certainly we have encountered problems of uncooperative local data stewards, incompatible data formats, and other barriers discussed in the existing literature. But some structures and relationships not previously articulated in PPGIS literature are proving to be important influences upon data accessibility.

While some existing research has focused on the role of data stewards and public agencies in limiting grassroots access to digital spatial data, our early observations suggest that the constraints operate in two directions. Grassroots groups are just as reticent to share their data with local government. In the HPGIS Project, the difficulty of obtaining data from the local SDI seems to create a political economy of scarcity in which individuals and institutions are even more reluctant to share data. As the HPGIS Project has matured, local government officials have themselves begun to request data and maps from the community organizations. Sometimes they ask for basic spatial data that in theory should be readily available from within their own networks of government. As one organizer observed, following such a request by local officials:

> 'Now the maps [showing vacant properties and land use], I kind of didn't want to give it to them because they have that resource already as part of their service through [the City]'. (Teresa 2005).

In a system of scarce resources, this organizer resists sharing data with institutions that may already have access to this information.

New technologies for disseminating and sharing spatial data are also changing issues of data access at the grassroots. The advent of online GIServices or geoportals has dramatically altered public access to geospatial data. The City of Chicago, like many other local governments in the U.S., has created a geoportal that enables download of some basic infrastructural and administrative spatial data, and provides online mapping and other GIServices. HPGIS Project participants suggest that in spite of these myriad new services, data access is still a problem. Participants noted that they are still unable to obtain highly localized data such as parcel boundaries, building footprints, or local streets. Some of the Internet GIS sites they have attempted to use were designed specifically for grassroots groups, but participants found that many of these do not enable spatial data download, only online query and mapping. They noted that online GIServices, while helpful, still do not address their needs for raw data. Raw data access is essential, they argue, so that they can integrate their own information, correct errors, and create their own maps and analysis. Flexibility and autonomy in spatial data handling are essential priorities for grassroots GIS users, and are not necessarily achieved through current trends in making spatial data available to the general public.

In spite of these ongoing challenges of data access, appropriateness, and accuracy, grassroots groups proceed in using these data anyway. They do so with a highly detailed understanding of potential problems. PPGIS research includes many cases in which grassroots groups reject public spatial data and instead strive to include their own spatial information in a GIS (Craig et al. 2002). Evidence from the HPGIS Project does not refute this characterization but underscores the ongoing importance of public spatial data in many grassroots GIS efforts. Grassroots groups use these data with a detailed critical assessment of the potential difficulties and an eye toward inserting their own information and observations. Participants in the HPGIS Project take the position that public data are useful if they are able to add their own information or rework these existing data:

> The nice thing about [the regional planning commission's data] is that they've got a bunch of existing databases. So they have all the licensed health care facilities from the State [of Illinois]. That is probably not going to include everything that is in our area, but if they give us that basic data, then our health committee can say, "Wait, there's that clinic there, there is that source there".' (Kate 2005).

This willingness to proceed with use of public spatial data records, in spite of their problems for grassroots GIS, is in part a recognition that these official data are the currency through which certain kinds of spatial change are negotiated – most especially when the grassroots groups work in concert with local government. As well, it is a reflection of the local expertise that is held by grassroots groups, and enables them to rework and extend public data records.

Perhaps informed by the demonstrated local expertise of grassroots groups, as well as broader governmental restructuring, we are seeing shifts in the roles and interactions of grassroots groups and public institutions in spatial data development and use. In the HPGIS Project, community participants and local government officials alike are calling for more effective means of connecting the intimate local knowledge of grassroots groups to the local SDI. One official, commenting on his own reliance on such local knowledge data argued,

> 'I mean, [local government data] is not as updated. I don't think they have the man-power to update these things at the same level that a community organization can. A community organization is always in that community. They're always in and out of streets and alleys, and so they're able to tell you the most current information'. (Julio 2005).

He went on to echo the calls of community participants for a system that would enable grassroots groups to contribute information to public databases. The community participants emphasized their unique position with respect to local spatial data development, including their nearness to rapidly changing conditions and their knowledge of the broader contextual details:

> 'Most [local officials] are not out in the field. They just ... get the information from the [County] Assessors or whatever ... They don't know is it [operated by] Hispanic Housing, is it this, is it that ... And we can tell them that, because we're out there. So in that way we have an advantage, even with the department of planning ... Even [they] ask us what's for sale, what's for rent'. (Teresa 2005)

These remarks by both local government officials and grassroots groups articulate a shared desire for collaborative interaction around spatial data collection and revision, albeit unrealized. This situation stands in contrast to the dominant framing in the literature of an oppositional, or at least distant, relationship between local government and grassroots in creating and using spatial data.

These brief examples from the Humboldt Park GIS Project illustrate a number of the continuing spatial data handling challenges of grassroots GIS. These data challenges of appropriateness, accessibility, and accuracy have long been observed as part of the unique landscape of grassroots GIS. But simultaneously, they are being altered by new developments such as grassroots groups' growing demand for public spatial data records in spite of their problems, new technologies for gaining access to these data, and calls for stronger interactions of public and grassroots actors in developing, revising and using local geospatial data. An important element in efforts to address these new and persistent data challenges involves broadening our engagement with a diversity of GIScience research, especially work on spatial data handling.

4 Intersecting with GIScience Research on Spatial Data Handling

In their efforts to improve grassroots access to geospatial data and technologies, PPGIS strategies have focused on altering GISystems and geospatial databases, as well as the processes in which they are used. Some researchers have sought ways of embedding GIS in more participatory data development, analysis and decision making processes (Kyem 2004; Williams and Dunn 2003). Others have focused on altering GISystems to expand their epistemological and representational flexibility in handling spatial knowledge, such as Kwan's (2002, 2004) efforts to incorporate time-space concepts, emotion, or ethnographic narrative into GISystems, to Sieber's (2004) proposal to create flexible attribute systems that use new markup languages. Such work has made tremendous contributions to our ability to conceptualize (and intervene to alter) the socio-political, technological, and representational practices of GISystems. But the persistence of data-related challenges in grassroots GIS suggests that we need also to focus more directly on spatial data handling.

Spatial data handling has been an area of tremendous change and new development in GIScience in the past decade. Many pertinent questions in GIScience focus specifically on geospatial data: interoperability and integration of large diverse data sets; challenges posed by semantic, schematic, and ontological heterogeneity in geospatial data; conceptual, technological, and policy issues for spatial data infrastructures (Goodchild et al. 1999; Hunter 2002; Van Loenen and Kok 2004). For the most part, GIScience research on these topics has been conducted without reference to the concerns and empirical contexts of grassroots GIS. Conversely, PPGIS research agendas have not directly engaged these areas of research in GIScience. Yet many of the most intractable challenges of grassroots GIS are rooted in the same spatial data handling issues that underlie GIScience research on data interoperability, integration, or SDIs. Spatial data handling is a critical missing link in our attempts to understand grassroots GIS practices. As grassroots GIS becomes ever more embedded in the spatial decision-making processes that shape our cities, towns, and neighborhoods, addressing these problems of accuracy, access, and appropriateness are imperative. Making stronger links to a diversity of GIScience research related to spatial data handling may be a productive way to begin.

The examples of the previous section, for instance, illustrate that a major issue in grassroots GIS is heterogeneity in geospatial data. Data are representational systems, so it is unsurprising that we would see differences between the data of government and grassroots institutions. Schematic heterogeneity figures highly in the example of a community organizer's difficulty with County property records that use multiple categorizations to represent the status of a vacant lot. His solution of reclassifying the data into a simpler schema is an attempt to address this problem. If these and other public data records are to be appropriate for grassroots users, and relatively easily incorporated into their GIS applications, some of level of scalability is needed so that users can browse or acquire spatial data at multiple levels of complexity and attribute detail. The community organizer accomplished this scalability after the fact, acquiring the data and then reclassifying into a simpler attribute scheme. While he was successful in producing a useful spatial data set, the additional time and skill needed to adapt the data set are important to note as potential limitations within the comparatively resource-poor context of grassroots GIS.

Some of the challenges discussed in the previous section can be read as issues of data interoperability. That is, with respect to exchange of geospatial information between local SDIs and grassroots groups, divergent semantic systems or the variable levels of detail needed by different data users create an interoperability challenge. GIScience research offers a wealth of proven and proposed practices for dealing with data interoperability. Many of these approaches intervene at the database level. Flexible standardization practices, such as Schuurman (2002) proposes, could serve as a means to facilitate this sort of flexibility at the database level, as grassroots GIS users browse and acquire spatial data. Alternatively, metadata offer an existing structure through which we might be able to enhance the appropriateness and accessibility of spatial data for grassroots users. While current metadata standards call for inclusion of technical details about data collection and representation, metadata could certainly also duplicate these details in language accessible to non-expert users.

Strengthening links to spatial data handling research in GIScience is important not just for addressing persistent problems in grassroots GIS, but also for realizing future possibilities. The HPGIS Project illustrates grassroots groups and government officials calling for mutually interactive local spatial data development and revision. Reinforcing this empirical example, the current literature on SDIs also asserts the value of local knowledge, emphasizing the importance of locally-integrated SDIs in fostering good governance and informed decision making (Williamson et al. 2003). But implementing a way for grassroots groups to contribute local knowledge to public databases and then integrating this knowledge into existing data sets is a challenging proposition from technological, administrative, and epistemological standpoints.

At the technological level, enabling grassroots groups to contribute raw data or revisions to existing data would require network architectures to support a two-way flow of data as well as verification and pre-processing of local knowledge contributions, perhaps building on existing geoportal architectures to support these functions. Policies would be needed to establish which individuals and institutions would be able to contribute their observations. Well-documented epistemological and ontological differences between grassroots spatial knowledge and the 'expert' terms of public spatial data records would pose significant data integration challenges. While participants in the HPGIS Project clearly articulate the value of local knowledge contributions for improving the accuracy and relevance of public data, it is also clear that these types of spatial knowledge are often rooted in very different epistemological and ontological systems.

GIScience research on ontologies, linked to the challenges of integrating large diverse data sets, has not been specifically oriented toward grassroots GIS. However, it could make essential contributions in efforts to foster stronger integration of local knowledge into public spatial databases. For instance, Fonseca et al.'s (2002) ontology-based GIS model is an effort to increase interoperability among divergent data sets by building ontologies directly into a GIS and enabling some user control in defining or modifying ontologies. At the conceptual and technological level, their prototype initiative is a long way from widespread use. But it focuses on a fundamental spatial data challenge that cuts across GIScience and affects grassroots GIS practice: the difficulty of integrating and sharing data across diverse ontological systems. These commonalities are precisely why some cutting edge GIScience research has relevance for spatial data challenges in grassroots GIS.

5 Conclusions

In sum, as a grassroots GIS initiative that relies heavily upon a local SDI, the HPGIS Project enables a more specific explication of some of the challenges that local SDIs pose for such users. The PPIGS literature has tended to focus on financial, time, and expertise constraints, or the failure of institutions and policies to include grassroots groups. I have attempted to expand this focus to include critical issues of spatial data, its development and its administration. For instance, problems of the appropriateness of public spatial data for grassroots GIS have been commonly identified by PPGIS researchers but this literature has not as clearly specified underlying causes, such as epistemological differences that foster heterogeneity in spatial data of grassroots groups versus local SDIs. With respect to data accuracy and completeness, it is clear that highly localized actors and their decisions are more strongly affected by these problems. While grassroots groups are well positioned to contribute information about errors and inconsistencies, no widely used formal mechanisms exist to enable such involvement. Finally, with respect to accessibility, this case illustrates that while geoportals, online GIServices, and Internet GIS show great promise for extending SDI access to the general public, they often do not meet the needs of grassroots groups.

How then might these challenges be addressed? In the previous section, I outlined some research directions in GIScience that might be fruitful, as part of a broader effort to strengthen the intellectual engagement between PPGIS and other GIScience research. Additionally however, public institutions and their policies must more consistently recognize grassroots GIS users as an important part of local SDIs, both as users and as potential participants in data and policy development. In practical terms, this means including grassroots groups in focus groups, advisory boards, and user response surveys when new data structures and modes of provision are developed. It also might mean establishing systematic procedures and supporting technologies to enable grassroots groups to contribute and revise highly localized data. Finally, public officials must actively commit to policies that guarantee the ability of grassroots groups to access public data. Too often in the relationships between local government and grassroots groups, information that is legally public is nonetheless made extremely challenging to obtain. Such barriers much be shifted if grassroots groups are to be willing to participate in local SDIs in meaningful ways.

Efforts to address epistemological, technological, and socio-political challenges of grassroots GIS have benefited greatly from PPGIS research and its efforts to modify GISystems and their application. But spatial data handling issues in grassroots GIS has been relatively under-examined and doing so will require strengthening our engagement with spatial data handling research across GIScience. Incorporating ideas, concepts, and practices from contemporary spatial data handling research into PPGIS is a small step within a larger project of continuing to strengthen PPGIS's linkages with other aspects of GIScience research. PPGIS research and practice have been developed with a strong emphasis on alternative practices to those being developed and used in other areas of GIScience. PPGIS has rich basis in social and political theory, but has had relatively less engagement with theories and concepts from areas of GIScience such as spatial data handling, cognition, or decision support. We cannot underestimate the positive impacts of the alternative practices that have emerged from PPGIS, but it is equally important to examine productive intersections of PPGIS research and methods, and 'mainstream' GIScience research. Creating these productive conceptual and technological collisions is challenging, without a doubt, but it is critical to our capacity to build effective, accessible, and flexible geospatial data and technologies.

References

Barndt M (1998) Public participation GIS: Barriers to implementation. Cartography and Geographic Information Systems 25(2):105–112

Chrisman N (1987) Design of geographic information systems based on social and cultural goals. Photogrammetric Engineering and Remote Sensing 53(10): 1367–1370

Craglia M, Masser I (2003) Access to geographic information: A European perspective. The URISA J 15(APAI):51–60

Craig W (2005) The white knights of spatial data infrastructure: The role and motivation of key individuals. The URISA J 16(2):5–13

Craig W, Harris T, Weiner D (eds) (2002) Community Participation and Geographic Information Systems. Taylor and Francis, London

Crompvoets J, Bregt A (2003) World status of national spatial data clearinghouses. The URISA J 15(APAI):43–50

Elwood S (2002) GIS and collaborative urban governance: Understanding their implications for community action and power. Urban Geography 22(8):737–759

Elwood S (2006a) Beyond cooptation or resistance: Urban spatial politics, community organizations, and GIS-based spatial narratives. Annals of the Association of American Geographers 96(2)

Elwood S (2006b) Negotiating knowledge production: The everyday inclusions, exclusions, and contradictions of participatory GIS research. The Professional Geographer 58(2):197–208

Elwood S, Leitner H (2003) GIS and spatial knowledge production for neighborhood revitalization: Negotiating state priorities and neighborhood visions. J of Urban Affairs 25(2):139–157.

Evans J (1999) Organizational and technology interoperability for geographic information infrastructures. In: Goodchild M, Egenhofer M, Fegeas R, Kottman C (eds) Interoperating Geographic Information Systems. Kluwer, Dordrecht, pp 401–411

Fonseca F, Egenhofer M, Agouris P, Camara G (2002) Using ontologies for integrated geographic information systems. Transactions in GIS 6(3):231–257

Goodchild M, Egenhofer M, Fegeas R, Kottman C (eds) (1999) Interoperating Geographic Information Systems. Kluwer, Dordrecht

Harris T, Weiner D (1998) Empowerment, marginalization, and community-integrated GIS. Cartography and Geographic Information Systems 25(2): 67–76

Harvey F (2001) Constructing GIS: Actor networks of collaboration. The URISA J 13(1):29–38

Hunter G (2002) Understanding semantics and ontologies: They're really quite simple if you know what I mean! Transactions in GIS 6(2):83–87

Kwan M (2002) Feminist visualization: Re-envisioning GIS as a method in feminist geography research. Annals of the Association of American Geographers 92(4):645–661

Kwan M (2004) GIS Methods in time-geographic research: Geocomputation and geovisualization of human activity patterns. Geografiska Annaler B 86(4): 205–218

Kyem K (2004) Of intractable conflicts and participatory GIS applications: The search for consensus amidst competing claims and institutional demands. Annals of the Association of American Geographers 94(1):37–57

Laituri M (2003) The issue of access: An assessment guide for evaluating public participation geographic information science case studies. The URISA J 15(APAII):25–32

Leitner H, Elwood S, Sheppard E, McMaster S, McMaster R (2000) Modes of GIS provision and their appropriateness for neighborhood organizations: Examples from Minneapolis and St. Paul, Minnesota. The URISA J 12(4):43–56

Martin D (2003) Reconstructing social GIS. Transactions in GIS 7(3):305–307

Merrick M (2003) Reflections on PPGIS: A view from the trenches. The URISA J 15(APAII):33–40

Nedovic-Budic Z, Feeney M, Rajabifard A, Williamson I (2004) Are SDIs serving the needs of local planning? Case study of Victoria, Australia, and Illinois, USA. Computers, Environment and Urban Systems 28(4):329–351

Onsrud H, Craglia M (2003) Introduction to special issues on access and participatory approaches in using geographic information. The URISA J 15 (APAI): 5–7

Onsrud H, Poore B, Rugg R, Taupier R, Wiggins L (2005) Future of the spatial information insfrastructure. In: McMaster R, Usery L (eds) A Research Agenda For Geographic Information Science. CRC Press, Boca Raton, pp 225–255

Schuurman N (2002) Flexible standardization: Making interoperability accessible to agencies with limited resources. Cartography and Geographic Information Science 29(4):343–353

Sieber R (2000) GIS implementation in the grassroots. URISA J 12(1):5–51

Sieber R (2004) Rewiring for a GIS/2. Cartographica 39(1):25–40

Van Loenen B, Kok B (2004) Spatial Data Infrastructure and Policy Development in Europe and the U.S. DUP Science, Delft

Williams C, Dunn C (2003) GIS in participatory research: Assessing the impacts of landmines on communities in north-west Cambodia. Transactions in GIS 7(3):393–410

Williamson I, Rajabifard A, Feeney M (eds) (2003) Developing Spatial Data Infrastructures: From Concept to Reality. Taylor and Francis, New York

Physical vs. Web Space – Similarities and Differences

Elissavet Pontikakis[1], Gerhard Navratil[2]

[1] Institute for Geoinformation and Cartography, Vienna University of Technology, Vienna, Austria; email: pontikakis@geoinfo.tuwien.ac.at
[2] Institute for Geoinformation and Cartography, Vienna University of Technology, Vienna, Austria; email: navratil@geoinfo.tuwien.ac.at

Abstract

Virtual worlds, and especially the Internet, become increasingly important for advertisement, planning purposes, and simulation. Concepts that proved useful in the real world have been transferred to the Internet. A frequently seen example is the concept of navigation processes. In the real world such processes are based on the properties of space. A transfer to the virtual reality of the Internet is only useful if the properties of real space and virtual space are similar. In this paper we present different concepts of space and discuss their suitability for the Internet.

1 Introduction

The size of the Internet as a virtual world increased dramatically in the last years. The number of hosts advertised in the DNS (Domain Name System) doubled between January 2003 and July 2004 (Internet Software Consortium 2005). This development underlines the growing importance of the Internet in private and business life.

This work has been inspired by concepts of space embedded in the land administration sector and the cadastral system such as boundaries, ownership, and transactions. There are concepts that are used in the real world and are transferred into the Internet. For example, we speak of navigation in the Internet and in the physical world and we also speak of purchasing web space and purchasing land. The concept of navigation has been addressed by Hochmair and Frank in combination with mental maps, to show how we define a destination and navigate towards that destination (Hochmair and Frank 2001). Hochmair and Raubal discussed the transformation of topologic and metric decision criteria from the real world to the Internet (Hochmair and Raubal 2002). This paper discusses the suitability of different concepts of space for the Internet.

2 Concepts of Space

Talking about space has never been straightforward. From the inception of science till now, "place" and "space" lingers within a large range of definitions and points of view (Pontikakis 2005). For Aristotle "place" holds a large list of quantitative and qualitative properties such as the replacement, dimensions, enclosure, inclusion, occupation by a separable physical object, characterized by two universal directions namely up and down, forces resulting in motion or rest and as such connected to time, void, etc. (Aristotle 1941). Couclelis and Gale propose that an algebraic structure of space can assist in transcending the disarray dominating the scientific community when facing the issue of space. They schematize space from the perspectives of Euclidean, physical, sensorimotor, perceptual, cognitive, and symbolic points of view and provide the algebraic axioms and operations that govern each of the above types of space (Couclelis and Gale 1986). Frank proposes algebras as a means for describing spatial features (Frank 1999). The equivocal notion of space is transduced to all its semantically related derivatives, "spatial" being one of them. "Spatial" is defined as pertaining to or involving and having the nature of space (Merriam-Webster 2003).

Montello connects space to locomotion and perception and differentiates between figurative, vista, environmental and geographical space (Montello 1993). Among other things these different scales imply differences in the spatial communication and the validity of simulations.

Frank states that geometries are connected to transformations. He proposes to look at the properties of the different aspects of space that are left invariant when answering questions related to transformations within a certain type of geometry (Frank to appear). He lists four questions to investigate: a) the reason for multiple representations, b) transformations, c) invariants, and d) operations.

3 Physical and Web Space

The Internet consists of two different conceptual layers. The Internet is a network of computers and simultaneously a pool of information. These layers have different types of queries. Whereas in the network, questions on geographic locations or availability of servers are of importance, the Internet as an information pool shall combine data from various sources and places.

The Internet as a computer network consists of nodes and lines. The Internet as used today developed from the ARPANET (Wikipedia 2005), which was a computer network for research sites in the USA. One of the fundamental considerations was the reliability and the availability of the communication paths (Wilkov 1972) leading to network structures. The design of networks lead to optimization processes and design studies for the topology of distributed computer networks, e.g., by minimizing the costs for lines connecting the computers (Gerla and Kleinrock 1977). Later design studies included the geographic concept of landmarks (Siegel and White 1975) for routing large networks (Tsuchiya 1988).

Transfer of data between the nodes requires identification of the receiver and sender. Identification of computers within a network is done by Internet Protocol (IP) addresses. An IP address consists of 4 numbers between 0 and 255. The unique nature of IP addresses allows tracking the computer that has sent a specific message. The assignment is not done arbitrarily. Typically an Internet service provider requests the assignment of a netblock from a registry. Thus, each computer is georeferenced by the IP address because at least the postal address of the Internet service provider is known. Companies like geobytes, MaxMind, or Quova provide services to locate computers based on their IP address. Fraud prevention in business is an application of these locating services (Quova Inc. 2005).

The concept of the computer network is independent of the data carried by it. The network transports the data without knowledge on the data itself or its purpose (Clark and Partridge et al. 2003). Hochmair and Raubal stated that the Internet consists of Web pages and connecting hyperlinks (Hochmair and Raubal 2002). This, however, in not necessarily true for the whole Internet. There may be pages without hyperlinks pointing to these pages. Still, these pages may contain important information. In addition, the navigation process as described by Hochmair and Raubal assumes that there is a step of refinements leading towards the goal. This is true for buying sneakers, but what about other tasks?

Hyperlinks need an identification of the target Web page. The uniform resource locator (URL) provides this identification (Wikipedia 2005). URLs identify the resource by their primary access method, usually http or ftp. A more general structure is the uniform resource identifier (URI). They can also refer to telephone numbers, email addresses, or embedded data. A URL has three main parts: Access method, server name, and path. The URLs pointing to Web pages stored on one server thus only differ in the last part, the path.

4 Geometric Aspects of Interest

Geometries deal with elements situated in space that comprise constructions. Geometries define the ways we conceive space (Frank to appear). Topology is a branch of geometry that defines the spatial relationships among adjacent or neighboring features (Theobald 2001; Frank to appear). Topology is linked to the notion of continuity in space. Frank generalizes from the continuous line to the continuous space that is conceptualized as a set of points (Frank to appear). Topology is linked to the areas neighborhoods, interiors, exteriors, and boundaries. Topological relations remain the same under homeomorphic transformations. Homeomorphic transformations are reflexive, symmetric, and transitive and assume the existence of a reverse.

4.1 The Concept of Boundaries

The concept of boundaries is closely related to continuous space. Although space is continuous, it is the imposed structure that creates discontinuities such as holes. The interior of a feature and its boundary comprise a closed set, namely the feature. Everything else is the exterior. A feature has no hole if for any two points in the interior of the feature there is a path connecting the points, which lies entirely within the interior of the feature. Frank suggests partially open and partially closed objects as in the cases of cadastral and river bed boundaries (Frank to appear).

Smith and Varzi suggest a distinction between bona fide and fiat boundaries. Bona fide boundaries apply to apples and oranges. Fiat boundaries apply to political boundaries (Smith and Varzi 1997).

4.2 Partition

A partition of space shall fulfill two major requirements: The partition shall cover the whole space and there are no overlaps. Gill provides a definition for a partition (Gill 1976: p 15):

Let *pi* be a set of non-empty subsets of *A*. The set *pi* is called a partition of *A* if every element of *A* is in exactly one of the sets in *pi*.

A partition can only contain specific topological relations. A partition consists of different areas. These areas must not overlap. Thus, the relations *contains*, *inside*, *covers*, *is covered by*, and *overlap* are not possible for partitions. The remaining topological relations are thus *equal*, *disjoint*, and *touch*.

4.3 Change within Partitions

Egenhofer relations describe the topological relations between two objects. Egenhofer and Al-Taha showed that in a conceptual neighborhood graph different types of change result in different transitions between relations (Egenhofer and Al-Taha 1992). They limit their discussion on relations between two regions.

Discussion of changes in partitions may not be restricted to two regions. Figure 1 shows an example of a partition. The two situations differ by only one element that changed from being a member of one set to being a member of another set. From a topological point nothing changes because there are still touching relations between each pair of sets. The boundaries, however, changed significantly and all boundary lines are influenced.

Topology can be defined by simplicial complexes. Simplicial complexes are combinations of simplices. A simplex represents a point, line, triangle, or tetrahedron. Changes in the topology occur if points are added or removed. A point moved out of the surrounding triangle changes the topology. However, this cannot be caused by numeric problems since simplicial complexes do not rely on numerical computations.

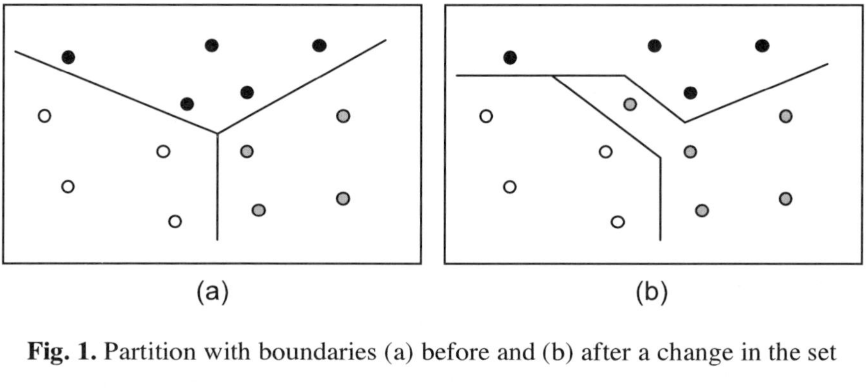

Fig. 1. Partition with boundaries (a) before and (b) after a change in the set

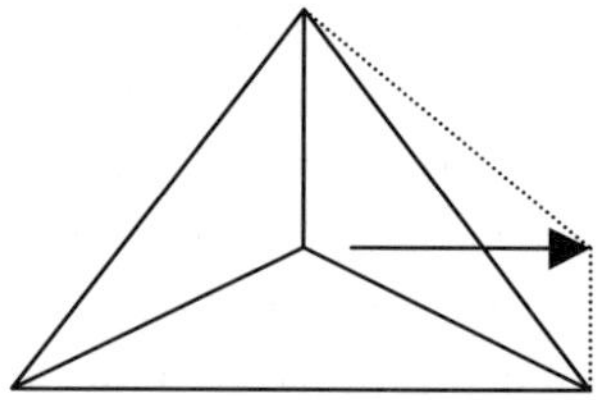

Fig. 2. Point moving from the surrounding triangle

5 Application of Geometry to Physical and Web Space

There are similarities between the physical world and the web. Previous efforts have looked into the way finding and distance issues (Hochmair and Raubal 2002). We suggest that the issues of boundaries, partitions, transactions, and ownership are equally important.

5.1 Boundaries and Partitions

Boundaries in land administration are boundaries as discussed in Section 4. On the web, the servers and connections between them comprise bona fide boundaries. The content however, is subject to semantic limitations – see the semantic web for an example.

A practical example for a partition in the physical world is the cadastre. A cadastre splits the world in small areas and assigns a legal situation to each of these areas. Overlapping areas result in ambiguous legal situations. Uncovered areas result in land without an owner. Cadastral systems shall avoid both situations. Thus, the areas in a cadastre form a partition.

The points in the web space can be clustered in a way that the result is a partition. The IP addresses allow creating a partition. Each address only exists once. Thus, a partition of the IP addresses is easily possible. Using the geographic locations of the identified computers also provides a partition. Simple connected regions in the space of IP addresses can create regions with holes or unconnected regions. We have a region with a hole if not all computers within a geographic region belong to the corresponding IP-address region. The computers outside the IP region will create holes in the geographic region.

As discussed in Section 3, the web pages and the connecting hyperlinks form a network. The Web pages are identified by their URL. Hyperlinks contain the URL of the target page. It is easy to identify hyperlinks to pages located on different servers because the URL shows the server name. A subdivision of the web pages that is based on the server names for separating the different parts is a partition. The Web pages on each server form a subset of the set of Web pages. Each page can only be located on one server. Thus, each page is part of exactly one subset. This is true only if the pages are really stored once. Methods to improve performance include buffering pages recently accessed. This is done in different places like the local machine of users or proxy servers in computer networks. However, these methods still use the URL to identify the pages. These mechanisms can be ignored since they are only a performance issue and the URLs remain unchanged. Thus, the Web pages can be structured in a partition.

The hyperlinks provide connections between the Web pages. The links within a server are irrelevant for the partition because they only define the internal structure of the subset. The links to other servers are directed connections between subsets. A connection between two subsets usually points to the existence of a touch relation between these subsets as defined by Egenhofer and Al-Taha (1992). However, the unidirectional nature of the hyperlink presents a problem. Let us assume there are two servers A and B and there is a hyperlink from a Web page on A to a Web page on B. Thus, A *touches* B. B however does *not touch* A if there is no hyperlink from B pointing to A. We observe therefore a lack of reversibility in this case.

5.2 Change

The example of the cadastre used in the last section also provides an example for change. Change has been discussed intensively for social systems like a cadastre (e.g., Navratil and Frank 2004). The cadastre represents the legal situation. The legal situation changes over time and thus the cadastre changes. Since the cadastre forms a partition, the changes must be performed in a way that keeps the partition. Cadastral systems use documents to reflect the change and the change is either applied completely or not at all. Situations where a change is partially applied do not exist. In the notion of database systems, this parallels the principle of atomicy.

The physical computer network forming the base for the Internet is not stable. Computer networks change whenever a computer is added or removed. Each of these changes influences the network and thus the topology. Typical changes are removing or adding servers or replacing one server by another server with a different IP address. Routing algorithms must adapt to these changes.

The topology of URLs also changes with each new or changed URL. Creation of new Web pages or deletion of existing pages will change the topology as does the setup of new servers or the eliminations of existing servers. Changes in the physical computer network may change the topology of URLs if the added or removed computer is a name server. Moving a server from one IP address to a different IP address may or may not change the physical network and the topology of URLs. These changes are independent from each other and not each change in the physical network will be visible to the URLs.

Removing a Web page may result in broken links. A broken link is a URL pointing to a non-existing Web page. Links to other servers are usually updated by different persons than the target servers and therefore a large number of broken links exits.

Change also applies to the content and involves the semantic boundary. Many Web pages have a "news" section. The content of that section is frequently updated. Links pointing to this section may therefore point to different content at different times and semantic differences may become important. A newspaper may, for example, provide a page with the most important news of the day. This may be the notice that the Olympic games started or a report on a natural disaster. Both reports are news but semantically different.

5.3 Ownership and Access

A part of society deals with legal aspects. One of the most important rights is the right of land ownership. Public infrastructure, such as streets, is situated on land for public use owned by the state. We differentiate here between private and public land ownership. Public infrastructure like streets, sidewalks, and fire hydrants are owned by various levels of government.

The situation is similar with the Internet. The servers and the lines are owned by different persons. The persons may be real persons like the authors of this text or juridical persons like companies or public authorities. Most of the infrastructure is owned by private companies.

The concept of ownership changes when looking at the Web pages and their content. Web pages, like all digital content, can be copied easily. Thus, they must be protected in a different way than physical entities like land or servers. Web pages are protected by copyright law. The creator of the content has the copyright and by making the page publicly available he allows others to read and use the material.

Access to a land parcel is defined by law. We have cases of privately owned land, publicly owned land, and privately owned land with access corridors to other public or private bodies. All these cases define different legal situations. On the web, we observe two types of situations: sites which are open to everybody and those restricted to a specified user group. In this sense, we observe similarities between access rights of physical places and web sites.

6 Conclusions and Future Research

We observe similarities and differences between physical and web space. Navigation and metric issues have been previously addressed. We suggest that the issues of boundaries, partitions, change, and ownership are equally important.

In the web space we observe bona fide and semantic boundaries while the physical space is characterized by bona fide and fiat boundaries. In both, the physical world and the Web, we observe the concept of partition. Change is an important aspect of the real world and the cadastre, for example, reflects changes in the legal situation of land. There are only complete changes in order to keep a valid partition. Changes apply on the Web by changing the bona fide or the semantic boundary. The concept of ownership is similar in the physical and web space.

At various points in this paper, we imply the use of algebraic axioms to describe the above issues. We believe that research will benefit from the use of algebras. These algebras will show possible homomorphisms between the physical and web space. In this paper, the topic of the container scheme that applies to both physical and web space is left for future research. The container scheme can also be examined using algebraic axioms as suggested above.

References

Aristotle (1941) The Basic Works of Aristotle. McKeon R (ed). New York, USA, Random House, Inc.

Clark DD, Partridge C et al. (2003) A Knowledge Plane for the Internet. Proc of the ACM SIGCOMM, Karlsruhe, Germany

Couclelis H, Gale N (1986) Space and Spaces. Geografiske Annaler 68B:1–12

Egenhofer MJ, Al-Taha KK (1992) Reasoning About Gradual Changes of Topological Relationships. In: Frank AU, Camparia I, Formentini U, Theories and Methods of Spatio-Temporal Reasoning in Geographic Space. Heidelberg-Berlin, Springer-Verlag, 639, pp 196–219

Frank AU (1999) One step up the abstraction ladder: Combining algebras – from functional pieces to a whole. In: Freksa C, Mark DM, Spatial Information

Theory – Cognitive and Computational Foundations of Geographic Information Science (Int Conf COSIT'99, Stade, Germany). Berlin, Springer-Verlag, 1661, pp 95–107
Frank AU (to appear) Practical Geometry—Mathematics for Geographic Information Systems
Gerla M, Kleinrock L (1977) On the Topological Design of Distributed Computer Networks. IEEE Transactions on Communications COM-25(1): 48–60
Gill A (1976) Applied Algebra for the Computer Sciences. Englewood Cliffs, NJ, Prentice-Hall
Hochmair H, Frank AU (2001) A Semantic Map as Basis for the Decision Process in the www Navigation. Conf on Spatial Information Theory, Morro Bay, California, USA. Springer
Hochmair H, Raubal M (2002) Topologic and Metric Decision Criteria for Wayfinding in the Real World and the WWW. Spatial Data Handling (SDH'02) (Proc on CD-ROM), Ottawa
Internet Software Consortium (2005) Internet Domain Survey Host Count, Internet Software Consortium (www.isc.org)
Merriam-Webster (2003) Merriam-Webster's Collegiate Dictionary, Merriam-Webster, Inc.
Montello DR (1993) Scale and Multiple Psychologies of Space. In: Frank AU, Campari I, Spatial Information Theory: A Theoretical Basis for GIS. Heidelberg-Berlin, Springer Verlag, 716, pp 312–321
Navratil G, Frank AU (2004) Processes in a Cadastre. Int J on Computers, Environment and Urban Systems 28(5):471–486
Pontikakis E (2005) Contribution of Local Artifacts in Assessing Spatial Experiences What You Keep, What You Throw. CORP 2005 & Geomultimedia05, Vienna, Austria, Selbstverlag des Institutes für EDV-gestützte Methoden in Architektur und Raumplanung
Quova Inc. (2005) Geolocation – New Weapon for E-Retail Fraud Prevention. http://www.quova.com/technology/tech_whitepapers.php. Amsterdam, The Netherlands, p 6
Siegel AW, White SH (1975) The development of spatial representations of large-scale environments. In: Reese HW, Advances in child development and behavior, vol 10. New York, Academic Press, pp 9–55
Smith B, Varzi AC (1997) Fiat and Bona Fide Boundaries: Towards an Ontology of Spatially Extended Objects. In: Hirtle SC, Frank AU, Spatial Information Theory – A Theoretical Basis for GIS (Int Conf COSIT'97). Berlin, Springer-Verlag, 1329, pp 103–119
Theobald DM (2001) Understanding Topology and Shapefiles. ArcUser (April-June)
Tsuchiya PF (1988) The Landmark Hierarchy: A new Hierarchy for Routing in Very Large Networks. ACM SIGCOMM Computer Communication Review 18 (4):35–42
Wikipedia (2005) History of the Internet. Retrieved November 2nd 2005
Wilkov RS (1972) Analysis and Design of Reliable Computer Networks. IEEE Transactions on Communications COM-20 (3):660–678

Utilization of Qualitative Spatial Reasoning in Geographic Information Systems

Carl P.L. Schultz, Timothy R. Clephane, Hans W. Guesgen, Robert Amor

Department of Computer Science, University of Auckland,
Auckland, New Zealand

Abstract

Spatial reasoning is a fundamental part of human cognition, playing an important role in structuring our activities and relationships with the physical world. A substantial body of spatial data is now available. In order to make effective use of this large quantity of data, the focus of GIS tools must shift towards helping a user derive relevant, high quality information from the data available. Standard GIS tools have lacked focus in this area, with querying capabilities being limited, and requiring a user to have specialized knowledge in areas such as set theory, or Structured Query Language (SQL). A fundamental issue in standard GIS is that, by relying entirely on numerical methods when working with spatial data, vagueness and imprecision can not be handled. Alternatively, qualitative methods for working with spatial data have been developed to address some key limitations in other standard numerical systems. TreeSap is a GIS application that applies qualitative reasoning, with a strong emphasis on providing a user with powerful and intuitive query support. TreeSap's query interface is presented, along with visualization strategies that address the issue of conveying complex qualitative information to a user. The notion of a relative feature is introduced as an alternative approach to representing spatial information.

1 Introduction

An immense volume of spatial data is now available [1]. Modern GIS commonly provide powerful tools that allow a user to manipulate, query and view this geographic information, however, many limitations have emerged relating to user interaction and query expressiveness [1,2]. To make effective use of the spatial information available, it is not enough that GIS simply display data to a user [1,2]. People need accessible, intelligent query tools that allow the extraction of specific, relevant information from the raw data provided [1,2]. Standard GIS tools have lacked focus in this area, with querying capabilities being limited, and requiring a user to have specialized knowledge in areas such as set theory, or Structured Query Language (SQL).

A fundamental shortcoming of current GIS is that they rely entirely on numerical approaches when working with spatial data [1,2]. People find numerical methods non-intuitive, for example, a statement such as "The café is at latitude 23 minutes, 8 degrees, and longitude…" is far less natural than "The café is opposite the art gallery on Symonds St" [1,3]. Further to this, numerical approaches cannot handle uncertainty in information, despite uncertainty being an intrinsic property of information about the physical world [3]. For example, it is impossible to define the boundaries of a coastline with absolute numerical accuracy, due to physical limitations of measurement precision, and the issue of information becoming out of date [4,5]. Another example is the inherent vagueness in a statement such as "The Forest is *near* the Pond". Despite this, humans still reason with imprecise and vague spatial information [3].

In everyday situations, humans often reason about spatial information in a qualitative manner, in particular, working with uncertainty [1,5]. A number of formalisms have been developed that apply qualitative techniques to reason about space. These approaches have been strongly influenced by Allen's qualitative temporal logic [3,4,6]. Allen presents a set of thirteen atomic temporal relations that describe relationships between time intervals. He describes key attributes for effective qualitative reasoning, that have been extended to the spatial domain [5]:

- The logic must handle *imprecision* in the data, given that people often express spatial information in a relative manner, with no reference to an absolute coordinate [5].
- *Uncertainty* in the data must be handled, so that a partial relationship between two features is accepted by the calculus, if the exact relationship is not known [5].

Freksa in [7] presents a generalized approach to Allen's temporal logic, by introducing semi-intervals as the basic reasoning unit, along with the notion of conceptual neighbours [7]. This approach supports reasoning with incomplete temporal information, and reduces the computational effort required during the inference process, by compacting the underlying knowledge base [7].

In [8] we introduce a one-dimensional spatial logic directly based on Allen's original temporal logic [8]. The central idea is to represent relative spatial relationships between objects rather than using absolute object positions [8]. This approach is extended to represent spatial relationships of higher dimensions by using an n-tuple of relations between each pair of objects [8]. Each component of the tuple represents a different dimension of the modelled scene [8].

Region Connection Calculus (RCC) proposed by Randell et al. [9] is another approach to qualitative spatial reasoning. RCC describes relationships between different spatial regions based on their topological properties, and is thus independent of any coordinate system [9]. Regions are defined to be the primitive type, with a primitive relation 'X connects with Y': C(X,Y) [4,9]. RCC8 is a system which selects a set of eight of these basic relations, such that the set covers any possible physical relation between two regions, and such that there can be no physical relation which is an instance of two basic relation types [4,9].

While the above approaches address the issue of imprecision, and attempt to provide a more human-friendly system for working with spatial data, they do not typically address the issue of vagueness in spatial relations [3,4]. In order to overcome the limitations of either exclusively numerical or qualitative approaches to spatial reasoning, AI techniques have been applied, such as fuzzy logic, to qualitative formalisms [3,4]. By applying fuzzy logic, the formalisms manage both imprecision and vagueness in spatial relations, and allow qualitative relations to be combined with numerical data [3,4].

Qualitative formalisms can extend a standard GIS by providing more intuitive, sophisticated and powerful querying tools. The primary aim of this work is to show that key issues in GIS (non-intuitiveness, imprecision and vagueness) can be resolved through the use of qualitative spatial reasoning techniques in a software application, and that these formalisms are suitable for practical application to real geographic information.

2 Qualitative Proximity Formalism

Qualitative methods are a coarser, language based approach to working with information, and have been used to specify spatial relationships and properties [3,4,5]. The Qualitative Proximity (QP) formalism is an adapted version of the Fuzzy Proximity formalism described in [10], and is used to reason about **distance relationships** between spatial objects. The possible relationship types, in order of increasing distance, are: touching, very near, near, moderately near, moderately far, far, very far. Figure 1 illustrates two example relationships between a pair of objects, A and B.

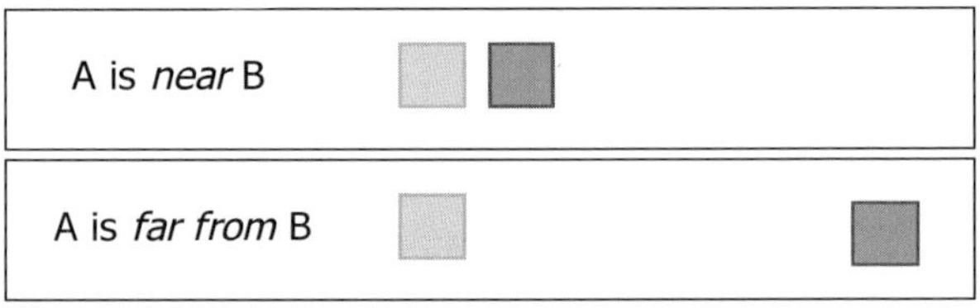

Fig. 1. Subset of the distance relationships defined in QP, where A and B are objects or regions

To address the issue of vagueness in spatial information, we have combined qualitative methods with fuzzy logic [3,4]. To illustrate this, consider the following query: "Find all objects *near* A". As shown in Figure 2, a "near" membership value is assigned to every distance relationship that A shares with some other object, indicating how closely each relationship matches the "near" relationship type. More generally, the standard alpha notation can be used [3,4,10], where $\alpha 0$ indicates the highest possible membership (a value of 100%, where the relationship is definitely considered a "near" relationship), and:

$$\alpha_1 > \alpha_2 > \alpha_3 > \ldots$$

indicating decreasing membership values, where the exact values of α_1, α_2, α_3,... can be determined according to the application [3,4,10].

Fuzzy membership values are thus assigned to relationship types of the QP formalism using the conceptual neighborhood approach proposed in [3,4]. Membership grades are assigned to relations according to the distance the relation is from a reference relation in the conceptual neighborhood graph [3,4]. The further away a relation is from the reference relation, the lower its membership grade [3,4]. Figure 3 illustrates the assigning of membership grades to relations with respect to the "near" relationship type.

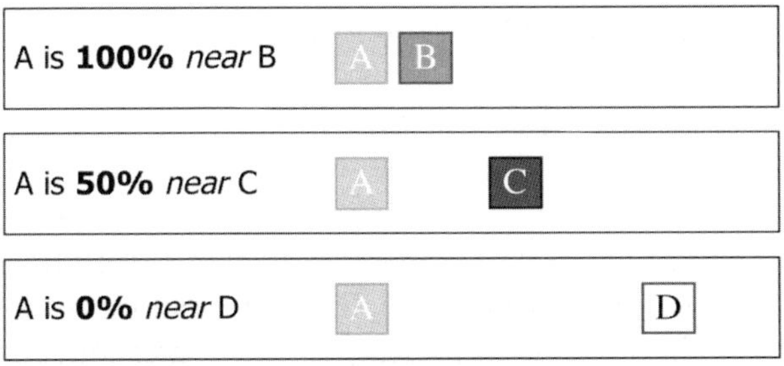

Fig. 2. Results of the query "Find all objects near A". A fuzzy membership value is assigned to every relationship that A shares with another object, representing how well each relationship matches the definition of "near". A is considered definitely "near" B, however A is definitely not "near" D. A is considered partially "near" C

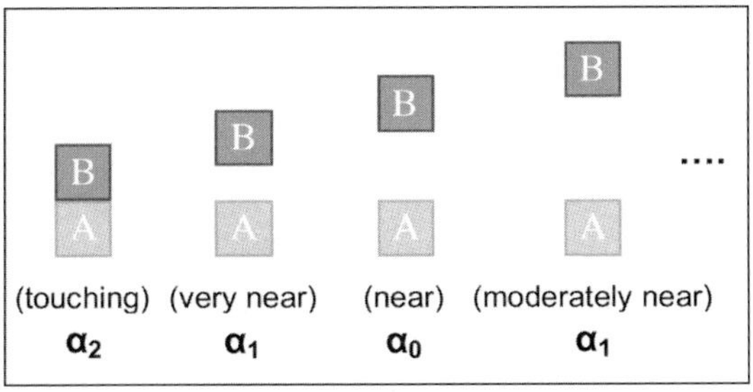

Fig. 3. Extract from the set of QP relations, arranged according to their conceptual neighborhood. Membership values (alpha notation) have been assigned with respect to the "near" relationship type

A complete network of qualitative relationships is constructed [3,4], based on the raw numerical data [10]. The first step is to take the distances between each pair of objects. For example, in a 2D scene, this can be accomplished by computing the minimum distance between each pair of objects, a and b, using (see Eq. 1):

$$\text{distance}(a,b) = \min\left(\sqrt{(x_b - x_a)^2 + (y_b - y_a)^2}\right) \tag{1}$$

The next step is to normalize the distance values to a number between 0 and 1. Thus, a decision needs to be made as to what value is considered a "near" distance. The normalized distance value (d) is then transformed into a fuzzy membership value. If the normalized distance value is 0, then the two objects are touching [10], resulting in a fuzzy value of 1. As the normalized distance value increases, we consider the two objects to be increasingly further apart [10], resulting in a decreasing fuzzy value (between 1 and 0). The following function [10] was used in this case, however, any function with similar properties may also be suitable (see Eq. 2):

$$\text{Fuzzy membership value} = \frac{1}{1+d^2} \tag{2}$$

This fuzzy value is then mapped to one of the seven QP relation types (from "touching", to "very far away"). The set of qualitative relationships (between every pair of objects) makes up the complete relationship network, which can then be referred to, in order to facilitate more advanced query support [3,4].

3 GIS Interface and Usability

Usability and human-computer interaction is a key aspect in introducing qualitative spatial reasoning into GIS [1]. A significant part of the limited usability of current tools in GIS is that they are not intuitive to the predominance of users [1].

To extract any specific piece of information from a large dataset, the user must be able to express arbitrarily complicated queries. That is, if the query tool is not expressive enough, certain users will not be able to present the appropriate criteria to the system. To ensure that **all** users can express their criteria, it is best to not place any restrictions on the complexity of a query. Further to this, the query tool must be intuitive and simple to use, to ensure that it is accessible to all users, not just experts. Many issues arise when a user must provide complicated input:

- A query may be malformed, such as a syntax error in SQL.
- The input may be erroneous or semantically nonsensical, such as incorrect data types being entered into a field.
- A challenge is also in teaching a user how to operate a query interface that allows arbitrarily complex input, at the same time minimizing the potential for a misunderstanding of the system. The user thus requires constant feedback, and reassurance that the intended query is accurately represented by their actual input.
- A user may be required to learn and remember numerous commands and keywords (such as "select", "where", "drop index" from SQL), increasing learning time, along with the chance of a misunderstanding or a syntax error.
- A direct reflection of the underlying formalisms is also desirable, as it allows a user to develop an accurate understanding of how the spatial in-

formation is being managed. A user can then benefit from the full potential of the formalisms.

- A further issue relates to the structuring of a query. If a complicated query has poor structure, or has a format that is too general, then the query may become either ambiguous, or far too difficult to understand. On the other hand, if a query format is too strict, then it may lose the desired expressiveness.

The TreeSap GIS application was produced to address these issues, by demonstrating a qualitative reasoning approach along with different visualization methods. TreeSap's querying and visualization approaches are discussed in the following sections.

4 TreeSap – Qualitative Reasoning GIS

TreeSap (Topographic Reasoning Application) is a desktop GIS application that provides powerful spatial reasoning tools, with a strong emphasis on usability. TreeSap was produced as an example of how the interface into a powerful query tool can be intuitive and simple to use, without requiring an understanding of mathematics, computer science, or artificial intelligence. Effectively conveying a mixture of qualitative and numerical information to a user is also an important issue. Standard geographic information can be extremely detailed. This is further complicated when qualitative and numerical information are combined, as the data can then express uncertainty and imprecision, along with the standard numerical details. TreeSap was also developed to help address this issue, by demonstrating visualization strategies that convey complex qualitative information to a user.

TreeSap provides standard functionality found in current GIS, including presentation and organization of geographic data. In addition to this, two qualitative spatial reasoning tools have been introduced: qualitative querying and relative features. This was accomplished by applying the Qualitative Proximity formalism.

Qualitative reasoning has been applied in two stages of the reasoning process; both in how the query is specified, and in how the system determines which features satisfy a query. Queries are specified in a qualitative manner, that is, the criteria, which the query consists of, are described using qualitative constraints on geographic features. For example a qualitative query might be "Find all Roads *near* all Railways", rather than providing some numerical distance value. Qualitative reasoning is also used in the processing of the qualitative queries. The generated relation networks are used to find solutions to the qualitative queries, while also providing the user with an indication on the viability of the results.

5 Qualitative Querying

TreeSap allows a user to specify a qualitative query with an arbitrary number of conditions. The query interface, illustrated as a screen shot in Figure 4, places a strong emphasis on being intuitive and simple to operate, while providing a user with the full benefits of the underlying qualitative formalisms. The query tool is natural language driven, attempting to reflect the underlying qualitative formalisms as directly as possible. The user consequently does not require specialized knowledge in areas such as set theory, or SQL.

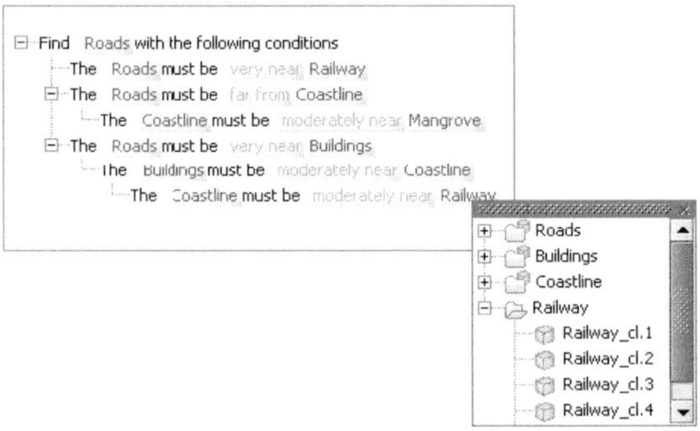

Fig. 4. Screenshot of the interface used to specify a query. The user builds a natural language query tree using the mouse

The user builds their query as a hierarchical tree structure of conditions, where nodes of the tree are search criteria, or constraints. These criteria are described in terms of a subject, and its qualitative spatial relationship with another object. The query tree allows nested relationships to an arbitrary depth, thus allowing for a query of arbitrary complexity. The hierarchical structure allows the complicated queries to be organized in a natural and intuitive manner.

Each nested condition acts to constrain its parent. For example, consider the condition from Figure 4 "The Roads must be *very near* Buildings". The buildings are, in turn, constrained by the nested condition that states: "The Buildings must be *moderately near* Coastline". A further level of nesting now constrains the coastline: "The Coastline must be *moderately near* Railway".

A query is built up in stages, and at each stage the user is presented with the results of the partial query, visualized on a map. This constant feedback is important, as it allows a user to develop a thorough and accurate understanding of how the query tool works.
The query building process is entirely mouse driven, and thus has a number of usability advantages as follows:

- A query can never be malformed, due to the nature in which it is built. This is not the case for other approaches such as SQL, where the query could have syntax errors.
- There is no possibility for erroneous or invalid input, such as incorrect data types being entered into a field. Input is always valid, thus minimizing the opportunity for a misunderstanding to develop.
- The user is given immediate feedback through the mouse-driven interface. When the mouse moves over an interactive component (such as a button), the component lights up, as illustrated in Figure 5. This implicitly suggests to the user that the component is interactive. This is reinforced by a message that explicitly tells the user what interactions the component supports. This feedback encourages a user to explore and familiarize themselves with the interface.

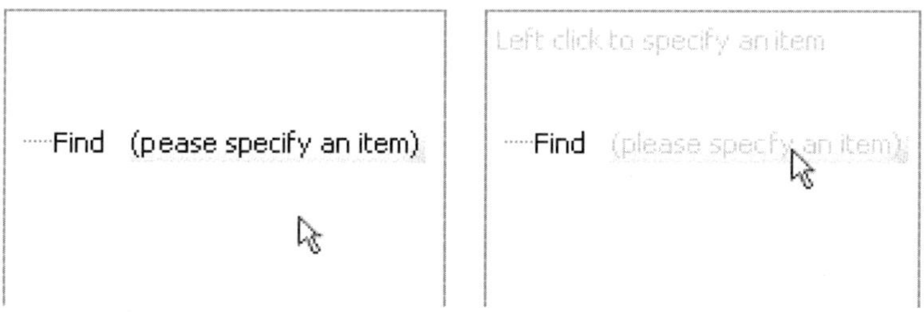

Fig. 5. Screenshots illustrating how interactive components provide the user with feedback. As the mouse moves over an interactive component, the component changes colour (right)

- All user communication and interaction (such as component highlighting, tool-tips, popup selection boxes, and messaging) happens near the mouse pointer. This simplifies the query building process, as it is likely that the user's attention will be focused on the mouse. It also reduces user effort, by avoiding large eye and mouse movements between targets.
- Being mouse driven, the interface is easy to operate, as it does not require a user to learn or remember commands or keywords.

The nature of progressively building a tree, which describes each of the user's desired constraints, ensures that the query building process is simple, while also providing constant feedback (the results of the query are immediately displayed whenever a condition is specified). Further to this, the use of qualitative formalisms to describe the spatial relationships in the query makes it intuitive and easy to learn. Future work will involve allowing a user to combine numerical statements with qualitative statements in a query, for example, "Find all Cafes *near* the Railroad, such that the Railroad is *within 5km* of Downtown". Also, the current query structure uses an implicit AND to join the conditions. This could be extended to include other conditional operators, such as OR, and XOR.

6 Qualitative Visualization

Standard geographic data is often large and detailed. The data that results from a qualitative query is even more complicated, with the introduction of fuzzy values on top of the large, detailed datasets. Further to this, there is a need to reflect the innately ambiguous qualitative notions that are present in the underlying formalisms. The key challenge is to present this information to a user in an intuitive manner, while avoiding an approach that requires knowledge in disciplines such as mathematics, artificial intelligence, or database languages. TreeSap addresses the issue of conveying complex qualitative information to the user by demonstrating two visualization strategies: using transparency, and using a display threshold.

6.1 Transparency

In the first strategy, transparency is used to represent how well a feature fulfills the query criteria, as demonstrated in Figure 6. Features that fulfill the query criteria to a high degree are displayed completely opaquely, while features that are less relevant to the query are displayed transparently. The level of transparency used is proportional to how poorly a feature fulfills the query criteria.

Using this method, all elements displayed to the user directly convey spatial information. Opaque features represent the solution to the query, and are therefore the most important pieces of information being displayed. These features appropriately attract a user's attention, by being displayed more distinctly than non-solution features. By displaying neighboring, non-solution features very faintly, the user is implicitly given some spatial context, to assist in the interpretation of the solution. In this respect, transparency offers an intuitive and visually efficient technique for conveying qualitative information.

Fig. 6. Screenshot of the transparency method used to visualize results of the query: "Find all Roads near a Specific Building (black circle)"

This strategy presents the user with a static snapshot of the solution, with all the information relating to the query result being provided in a single image. This method is thus ideal if it is required that the query solution be ported onto a hardcopy medium, such as a hardcopy report document, or a newsletter.

6.2 Threshold Display

A limitation of the transparency approach is that, while it can provide an instant overview of a query result, it does not effectively convey subtle trends and details. For example, the exact location with the highest solution quality is not always obvious, as subtle differences in transparency can be difficult to recognize. To address this issue, a second approach is proposed that uses a threshold to determine how much information is presented to the user at a given point in time.

Some features fulfill a query's criteria more than others. The notion of a solution quality is thus used to indicate how well each feature meets the given criteria. 100% indicates that a feature meets all the criteria, while 0% indicates that a feature does not meet the criteria in any way. The user can then control a display threshold by dragging a slider. All features that have a solution quality above the threshold are displayed opaquely, and any features with a solution quality that do not meet this threshold are not displayed at all. A scenario is illustrated in Figure 7, where more roads are displayed as the threshold is lowered, revealing an underlying trend.

Fig. 7. Screenshot of the display threshold method used to visualize results of the query: "Find all Roads *near* a Specific Building (black circle)" for differing thresholds

This strategy is a dynamic representation of a query result, with different parts of the solution being revealed at different points in time. A key aspect of this method is that the user has control over the dynamic property by adjusting the threshold, thus revealing trends and patterns in the way that the solution unfolds. For example, consider the scenario that a city council is looking for a site to transform into a reserve for growing native New Zealand kauri trees. After applying the appropriate query, it is observed that one small area meets the criteria by 100%, but a much larger area meets the criteria by around 78%. This second part of the solution will be expressed as a small, independent pocket appearing and growing rapidly, once the threshold has dropped to 78%. This suggests that, with minor adjustments, the larger area may be a more appropriate, long-term solution. Thus, the display threshold approach offers a deeper understanding of the query solution.

7 Relative Features

People often describe spatial features in a relative manner [1,2]. Despite this, standard GIS only allow a user to describe the location of a feature with absolute numerical coordinates [1,2]. In order to support a more intuitive method for expressing the location of features, TreeSap introduces the notion of a relative feature. The difference between a standard feature and a relative feature is that the position of a relative feature is described solely by qualitative relationships that it has with other features.

To describe the location of a relative feature, the user builds a relationship tree. This process is based on the procedure for specifying a query, and as a consequence, is also natural language driven. The user defines relationship constraints, such as "The Café is *near* some Roads". The relationships can be arbitrarily complex, for example "The Café is *near* a Coastline, such that the Coastline is *far away* from a Port". Once a relative feature has been defined, TreeSap can search for a possible numerical location that fulfills the criteria given. The feature is then positioned on the map, and the user is provided with a percentage indicating how well the location meets the relationship criteria. Figure 8 illustrates an example where a user is looking for an appropriate location to build a day care centre.

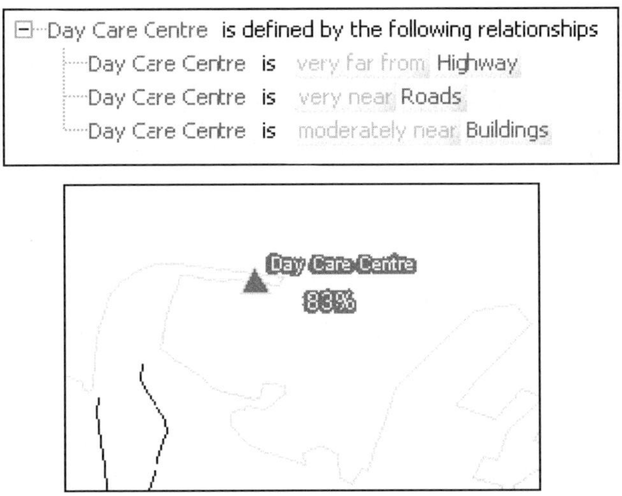

Fig. 8. Defining a day care centre as a relative feature. The relationship constraints that define the location of the day care centre are described using a relationship tree (top). TreeSap can then present a possible instantiation of the relative feature, along with a percentage indicating how well the criteria have been met (bottom)

A key aspect of this method is that it allows a user to determine locations that are consistent with partially defined, or incomplete, qualitative spatial information. For example, consider that a police service uses TreeSap to help isolate the exact position of an emergency. The police station receives a message that an accident has occurred in the Auckland Domain, near the motorway. A user of the application creates a new relative feature to represent this information, and assigns it the label "EMERGENCY_332", along with the partial information relating to its location. TreeSap then attempts to position the relative feature according to the given criteria. As more partial qualitative information is received at the station, the actual position of the emergency is refined. This process would be considerably more complicated if the user could only specify features using numerical coordinates.

This tool provides a user with an intuitive approach to specifying the location of features, compared to standard numerical methods. It allows a user to describe a relative feature with arbitrarily complex relationships, without requiring a user to have specialized knowledge in areas such as SQL or set theory. Further to this, it can handle incomplete spatial information, providing the user with a more powerful system for describing features.

8 Future Works

Currently, TreeSap can only handle small to medium sized datasets. This is due to limited available memory, as the relations network is stored in local RAM. TreeSap could be extended to handle large datasets by storing the network in a database.

Further qualitative spatial formalisms could also be incorporated into TreeSap. For example, the current system generates a relationship between every pair of features in the data. This will cause problems when moving to larger datasets. Formalism could be implemented that groups the data into clusters, and then layers these clusters, producing more sophisticated relation networks. For example, a query such as "is London near New Zealand?" can be translated into two query steps:

1. Where is London? – England
2. Is England near New Zealand?

This would avoid the need to have inter-layer relationships, such as a relationship between London (a city) and New Zealand (a country), vastly reducing the size of the network.

Another example is the Region Connection Calculus (RCC), which specifies containment relationships between regions, such as "The Wharf is *within* Downtown", or "A *overlaps* B" [4]. This would provide a natural extension to TreeSap's querying capabilities.

The notion of relative features could be extended into a form of automated design. Rather than basing relative features on absolute numerical data, a spatial specification could consist entirely of qualitative spatial constraints between different entities. This abstract specification could then be implemented by a computer system, to determine a number of possible numerical configurations that meet all the criteria to some fuzzy degree. A user could then make small modifications to the automatically generated design, and receive feedback on how well the adjusted designs meet the initial qualitative constraints. For example, this could be used in town planning, product design, packaging, bin packing, and other areas that involve spatial reasoning. The concepts demonstrated through TreeSap (particularly qualitative reasoning and more natural human-computer interaction) are not restricted to GIS, or even the spatial domain. Future work could involve researching the applicability of qualitative reasoning in a wide variety of different scenarios and disciplines, particularly areas that work with large amounts of data, such as manufacturing, business services, plan verification, bioinformatics, and others.

9 Conclusions

The amount of stored geographic data has grown significantly [1]. Several key problems exist in standard GIS, due to the reliance on numerical approaches when working with spatial information, including the problems of non-intuitive interfaces (including user controls and data visualization), and the inability to reason under vagueness and imprecision [1].

The area of qualitative spatial reasoning provides powerful formalisms for reasoning about spatial objects, and their relationships, in a 'natural' and intuitive manner. The application of fuzzy logic allow these formalisms to reason under vagueness and imprecision [3,4]. As a consequence, qualitative reasoning formalisms offer benefits to both usability and querying capability in GIS, as demonstrated by TreeSap, thus resolving many of the existing key issues. The notion of the 'relative feature' offers a completely new approach to representing spatial information that can be effectively combined with possibly incomplete, absolute numerical data, to derive further relevant information. Qualitative spatial reasoning formalisms are a promising approach for improved power and flexibility when reasoning about, and interacting with, geographic data.

References

1. Cohn AG, Hazarika SM (2001) Qualitative Spatial Representation and Reasoning: An Overview. Fundamental Informaticae 46(1-2):1–29
2. Frank AU (1996) Qualitative Spatial Reasoning: Cardinal Directions as an Example". Int J GIS 10(3):269–290
3. Guesgen HW (2002) Fuzzifying Spatial Relations. In: Matsakis P, Sztandera L (eds) Applying soft computing in defining spatial relations. Physica-Verlag, Heidelberg, Germany, pp 1–16
4. Guesgen HW (2005) Fuzzy Reasoning About Geographic Regions. In: Petry FE, Robinson VB Cobb MA (eds) Fuzzy Modeling with Spatial Information for Geographic Problems. Springer, Berlin, Germany, pp 1–14
5. Allen JF (1983) Maintaining Knowledge About Temporal Intervals. Communications of the ACM 26(11):832–843
6. Gooday JM, Cohn AG (1994) Conceptual Neighbourhoods in Temporal and Spatial Reasoning. In Pro ECAI-94:57–64
7. Freksa C (1992) Temporal Reasoning Based on Semi-Intervals. Artificial Intelligence 54(1-2):199–227
8. Guesgen HW (1989) Spatial Reasoning Based on Allen's Temporal Logic. Technical Report TR-89-049, ICSI, Berkeley, California
9. Randell DA, Cui Z, Cohn AG (1992) A Spatial Logic Based on Regions and Connection". Pro KR-92:165–176, Cambridge, Massachusetts
10. Guesgen HW (2002) Reasoning About Distance Based on Fuzzy Sets. Applied Intelligence 17:265–270

Identification of the Initial Entity in Granular Route Directions

Martin Tomko[1], Stephan Winter[2]

[1] CRC for Spatial Information, Department of Geomatics, University of Melbourne, Victoria 3010, Australia
email: m.tomko@pgrad.unimelb.edu.au

[2] Department of Geomatics, University of Melbourne, Victoria 3010, Australia; email: winter@unimelb.edu.au

Abstract

Current navigation services assume wayfinders new to an environment. In contrast, our focus is on route directions for wayfinders who are familiar with the environment, such as taxi drivers and couriers. We observe that people communicate route directions to these wayfinders in a hierarchical and granular manner, assuming some shared knowledge of the environment. These route directions do not focus on the route but rather on the destination. In this paper we solve the first problem of automatically generating route directions in a hierarchical and granular manner: finding the initial entity of such a communication. We propose a formal model to determine the initial entity, based on Grice's conversational maxims, and applied to a topological hierarchy of elements of the city. An implementation of the model is tested for districts in a political subdivision hierarchy. The tests show a reasonable behavior for a local expert, and demonstrate the efficiency of granular route directions.

1 Introduction

Current navigation services provide turn-by-turn route directions to human users. In contrast, route directions given by people are significantly different [13, 12, 4]. For example, people *chunk* route segments together, depending on various structural characteristics of the route [14], and they refer to specific salient elements in the environment, such as landmarks, paths, nodes, districts and edges [16].

In particular, people usually adapt route directions to the level of familiarity with the environment shared by the wayfinder. If direction givers can assume that the recipients have some familiarity with the environment, they initiate route directions by reference to an element at a coarse level of granularity replacing the destination, such as 'Ikea ... is in Richmond'. They expect that the recipient shares knowledge of the environment at least at this coarse level, and will be able to find a way to this element autonomously. The direction givers continue then with increasingly detailed directions to the destination starting from this initial element, such as 'In Richmond, at the town hall, take ... '.

The above mentioned elements of the city form a functional hierarchy of levels of granularity, and exploiting this property in route directions produces what we call *granular route directions*. We argue that current navigation services with their detailed turn-by-turn directions serve only a subgroup of wayfinders with low familiarity with the environment. A large group of potential users of navigation services are neglected: people living in a city and having a general idea of its structure. People in this group may perceive turn-by-turn directions patronizing, and hence, they are not satisfied with the quality of service. They are better served by granular route directions.

In this paper we study the initial entity of granular route directions for recipients who are familiar with the environment. We are interested in determining the element of the city that has to be included as the initial entity in such a communication. This element needs to be shared by both agents' mental models. Our hypothesis is that the choice of the element depends on the hierarchical relationship between the start and the destination of the route. The hypothesis builds on the concept of *relevance*, as defined in maxims for communication acts first identified by Grice [8]. These maxims postulate that information conveyed to the recipient should be neither too coarse nor too detailed.

We propose to deduce the relevance of elements of the city to the route from their position in the hierarchy in relation to the start and destination of the route. We formulate conditions that determine the amount of information to be communicated to the recipient, and integrate these conditions into a formalized model. This model is further translated into an executable specification such that its behavior can be demonstrated and tested. Furthermore, specific tests for districts are introduced and explained. The discussion of the test results enable parallels to be drawn for the extension of the model for the remaining elements of the city: paths, landmarks, nodes, and edges.

This paper is structured as follows: Section 2 introduces the basic theoretical foundations from spatial cognition and communication theory upon which we build our proposed model. We then develop our model of granular route directions (Section 3), which is formalized for implementation in Section 4. The algorithm for the identification of an initial entity is formalized in Section 5. The test cases are described and discussed in Section 6, and followed by conclusions in Section 7.

2 Route Communication

2.1 The Structure of Urban Environments

People living in an urban environment learn the spatial layout of that environment through frequent, repetitive interactions, such as wayfinding [17]. The accuracy of the acquired knowledge increases with the continuing interaction with the environment, and so increases the accuracy of the agent's mental model [26]. The evidence that such models contain hierarchically organized knowledge was demonstrated by Hirtle [11, 10]. Hierarchical conceptualization of space enables granular spatial reasoning, for example in wayfinding on a hierarchic path network [25], as well as communication of route knowledge in a granular manner [28, 20]. In comparison to these works, our approach is different by not communicating the full path from the start to the destination, but instead try to describe the destination in a granular manner.

2.2 Communicating Route Directions

Current navigation services provide no direct interaction with the recipient. The communication situation corresponds to indirect human communication: route directions are *read* by the recipients, who then try to realize their understanding of the directions in the physical environment. In this sense route directions form *narratives*, and direction givers are *narrators* [27].

Human communication of route directions has been the object of investigations for decades now [13, 12, 7, 2, 4, 15]. Despite that, there is so far no study that specifically looks into human route communication to wayfinders familiar with the environment. Our observations and examples are, however, consistent with those described in [23, 20].

Past research of direct route communication explored collaboration on references to objects, either mutually known and visually accessible by the recipient [1, 9], or unknown and inaccessible by the recipient [5]. Unlike these studies, the emphasis of this paper is on the selection of the element communicated by a narrator from the set of available elements. The process described is still an indirect one, as the narrator infers which element to refer to from context and builds a narrative from that suggestion.

2.3 Relevance

In his seminal work, Grice [8] made a contribution to the studies of pragmatics by formulating four maxims of conversation, well applicable to effective and efficient information transmission: the maxim of quality, quantity, relevance and clarity. Route directions are a specific type of information communication which is deeply pragmatic and rich in content. This paper explores the impact of the maxims of *quantity* ("make your contribution as informative as required by the purpose of the exchange; do not make your contribution more informative than is required."), *relevance* ("be relevant.") and *clarity* ("avoid obscurity of expression; avoid ambiguity; be brief (avoid unnecessary prolixity); be orderly.") on route directions generation. We assume that the maxim of *quality* ("do not say what you believe to be false; do not say that for which you lack adequate evidence") is respected, as a prerequisite for usable route directions.

3 Route Communication to Familiar Wayfinders

3.1 Relevance in Route Directions

Information that is too coarse is of no value for the recipient. Imagine a passenger entering a taxi at Melbourne University for a trip to the train station. If he says 'To Melbourne, please' the *surprisal value* of this information – being in Melbourne and going to Melbourne – is low [22, 29]. In a different context, for instance in Geelong (next to Melbourne), the same order makes perfect sense to the taxi driver, at least for a first leg of the trip. If a message has the appropriate surprisal value, its *pragmatic information content* [6] is maximal and thus it has high *relevance* for the recipient [24].

In such directions, direction givers refer to an element of the city [16] as small as possible–here, a district. The choice is further limited by the other constraints–the necessity to refer always to shared concepts and to avoid ambiguity. Ambiguity occurs if at a level of granularity several possible references exist. For example, the taxi passenger from Melbourne University should not say 'To the train station, please': there are several of them close by. The surprisal value is too high and the taxi driver is confused.

Thus, a direction giver strives to provide a reference to the finest element presumably shared with the recipient that just avoids ambiguities. This phenomenon was previously observed by Rumelhart in 1974 (as described by Shanon [23]). Shanon then suggests topology as selection criteria for appropriate references.

3.2 Granular Route Directions

Granular route directions represent a specific case of referring expressions as defined by [3]: "[A referring expression is] ... an expression uniquely identifying a specific object". They represent a concatenation of multiple references to entities with an increasingly fine level of detail. This zooming in route directions was previously observed by Plumert et al. [20, 21].

Granular route directions have the potential to be significantly shorter than turn-by-turn directions, with the length measured by the number of references. The difference increases with the length and complexity of the route. For example, a current navigation system needs 18 turn-by-turn directions from Melbourne Airport to 'Turnbull Alley' in the city center. A system using granular route directions could instead deliver a human-like message: "In the city, off Spring Street, opposite the Parliament', consisting of three entities.

Formulating the basic principles of selection of spatial entities from a hierarchical structure of the city enables to construct granular route directions. Human route directions are a mixture of references to various elements of the city, with complex interdependencies. This paper starts the exploration of the subject by considering only a single type of element, districts. Considering a hierarchical partition of the city, we define the conditions for selecting the initial entity of granular route directions, i.

A granular communication process depends on the identification of the context, which is often sufficiently clear in human communication. Consider the taxi scenario again: the passenger concludes from the situation that the driver is familiar with the city. The passenger may also know from experience that taxi drivers typically do not know specific destination addresses. Furthermore, the passenger and driver both share the knowledge of the start of the route: their current location.

In the indirect communication situation of a navigation system the system has to presume that the wayfinder shares the knowledge of some elements of the environment. It can do so by profiles of users, or profiles of standardized application areas. Furthermore, we expect that future systems will be able to conduct dialogs, and thus, can correct wrong assumptions.

3.3 Granular Route Directions in Communication

The quest to keep the amount of information communicated to the necessary minimum provides a means to start the route description at the maximum possible detail and still keep the certainty of the information transmitted, all in accordance with Grice's maxims [8]. If the wayfinder is familiar with the environment, the narrator will refer to this shared knowledge, in order to keep the amount of transmitted information low. The narrator will typically refer to a well-known element of the city in the proximity or containing the destination. This element is at the finest level of shared knowledge, and its communication would be sufficient and most effective.

In some cases, such as those of entities with ambiguous names, the narrator needs to refer to elements at coarser levels in the hierarchy, to uniquely identify the referent. For example, when referring to a destination close to the opera building in Sydney, one could just refer to the *Opera*, and for any wayfinder in Sydney this would be sufficient. If the start of the route is not in Sydney, a disambiguation is necessary, by referring to the *Sydney Opera*. This is a granular route description including two levels of granularity: *Sydney* and *Opera*. A similar process of referring in a document hierarchy was described in [18]. In this work, the authors referred to the effect caused by such ambiguous references as *lack of orientation*. Consider Figure 1, which represents the complete communication process of the narrator and the wayfinder. The spatial relation of the start s and the destination t of the route provides sufficient clues to start the granular route description. An insufficient overlap of the knowledge of the wayfinder and the narrator may require a collaborative identification of the destination in a negotiation segment (N). This process iteratively enables the narrator and the wayfinder to find a shared element at a certain hierarchical level in their respective mental models of the environment.

A similar negotiation process occurs upon reaching the finest shared element, and is followed by turn-by-turn directions. This can also occur in the middle of the route directions, in the case of structural differences in the mental models of the narrator and the wayfinder.

Fig. 1. Negotiations about references and in route direction in the context of the route (see text)

4 A Formal Model for the Initial Element Identification

4.1 Model Constraints and Assumptions

Building on the representation theory of Worboys [29], to reach consensus between the narrator and the wayfinder the representation of the information in the narrator's domain N has to be transmitted through a communication process and matched to the correct representation in the wayfinder's domain W (see Fig. 2). Both N and W are subsets of the domain R, $(N, W \subseteq R)$,

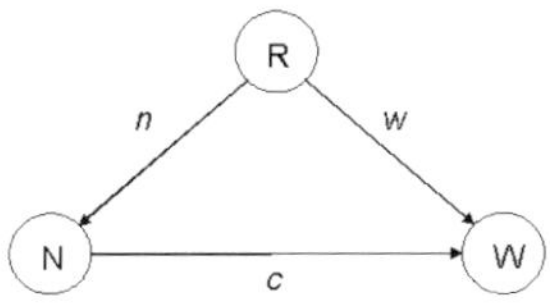

Fig. 2. Domain formation and matching process (see text)

which represents the reality. The domains N and W are mental models of the reality, which can be incomplete and imperfect with regard to R, as they are constructed by learning through interaction with the environment. We call these mapping processes n (reality $\mapsto$ narrator) and w (reality $\mapsto$ wayfinder) respectively.

Let us consider Figure 3. The domain R, and therefore also N and W, consist of elements e, of any of Lynch's five types of elements of the city. These elements are organized hierarchically in levels l. Note that the hierarchies h_N and h_W are not necessarily subtrees of the hierarchy h_R, $(\diamond(h_N, h_W \subseteq h_R))$. As we can see in the hierarchies of the domains N and W, differences in granularity levels between the reality and the mental models occur, including omissions of elements. Therefore, the hierarchies h_N and h_W are not necessarily identical $(\diamond(h_N \neq h_W))$. The intersection of the domains N and W is the subset of elements S that are shared by both agents $(N \cap W = S)$ with identical tree structure h_S.

During the communication, a process c, $(c : e_N \mapsto e_W)$, representing the communication, associates respective elements from the domain N to the domain W. To reach consensus between the narrator and the wayfinder, the process n and the process w have to associate the respective representations e_N and e_W of an element e_R. If these processes are successful, the agents in effect exploit the elements of the domain S in the communication (see Fig. 3).

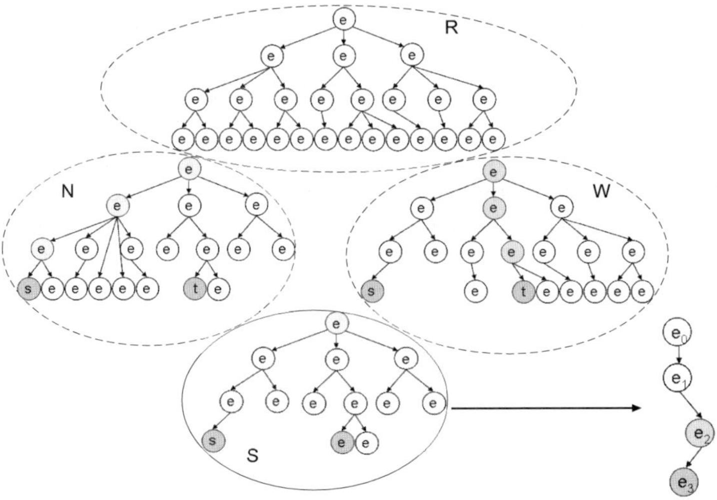

Fig. 3. Domain S and the route hierarchy reconstruction (see text)

In our model, we assume the preservation of relative ordering of elements in the hierarchies (if $l_N(e_1) > l_N(e_2)$ then $l_W(e_1) \geq l_W(e_2)$). If an element e is present in the mental model of the wayfinder and the narrator, its place in the hierarchical structure of the domains N and W is such that the relative ordering between the adjacent elements in the tree is preserved. A violation of the relative ordering condition would result in an incompatible mental model (e.g. cities composed of countries). If this condition is not met, the narrator and the wayfinder have to engage in communication about the reference made.

As an implication, the hierarchies of elements in the two domains may have different depths, e.g., be flatter or deeper. Any elements not shared by both domains N and W can only be found on the finest levels of the respective hierarchies.

In order to describe the whole route, several elements of descending levels of granularity have to be identified and matched through the communication process c. Let us call this set of elements C, $(C \subseteq S)$. This process leads to the reconstruction of a subtree h_C of the hierarchy h_S. The initial entity referred to in granular route directions is a member of the subtree h_C (element e_2 in Figure 3).

Worboys [29] considers the different contexts of two agents engaged in communication. In our case we limit context formally to the knowledge of the agents (narrator and wayfinder), represented by domains with internal structure h_N and h_W. These contexts are, in general, different. A typical example is the situation where the narrator knows the destination, but the wayfinder does not (otherwise route directions are not necessary). The narrator, initiating a wayfinding communication, anticipates the context of the wayfinder. In order to construct successful granular route directions, the conditions on the two domains and their internal structures need to be met, and the context of the wayfinder is correctly anticipated. Otherwise, the agents will enter in a new cycle of negotiation, as discussed in Section 3.3.

4.2 Initial Element Identification

We now apply the basic principles described earlier to the identification of the initial entity i of the granular route directions. These principles are grounded in information relevance and explore the topological relation of the start (s) and target (t) element of the route within the hierarchy h_S (see Fig. 3).

The selection is grounded in a translation of Grice's maxims into the assessment of the information value brought by inclusion or omission of a certain element from a hierarchical partition of space into route directions. Possible topological relations between the start and destination elements of the route in a hierarchical tree structure were analyzed. The topological relationships tested consider only the inside and the boundary of an element (district). Two districts are considered neighbors only if they share a boundary, a one dimensional space. The following conditions can be defined (applies to the selection in a set of *districts*):

1. start and destination must be member of the shared set of elements ($s, t \in S$);
2. start and destination must not be identical (but may meet, be neighbors) ($s \neq t$);
3. the start and the destination should not be neighbors ($\partial s \cap \partial t = \emptyset$));
4. the start and the destination should not have neighboring direct superordinate elements ($\partial Sup_s \cap \partial Sup_t = \emptyset$).

The conditions 2 and 3 must be separated, as they verify a different behavior. If s and t are identical (and thus they have overlapping interiors and boundaries), it is not possible to construct route directions (a route is two dimensional and thus requires a star and a target). If s and t are neighbors, route directions are possible, but the topological distance between the specification of the start and the specification of the target is insufficient to generate *granular* route directions.

Thus, the third and the fourth conditions excludes cases where the start and destination are too close in the hierarchy and turn based directions should be applied.

Consequently, sets $Super_t$ of superordinate elements of t and $Super_s$ of superordinate elements of s can be formed. The set $Super_t$ is a candidate set for the initial element i. Superordinate elements of an element e are elements of coarser granularity than e that have e as descendant. This property is transitive. From now on, the notation Sup_e will be used for a parent element of e. The element i is retrieved from the candidate set $Super_t$ ($i \in Super_t$). Further conditions apply for selection of the element i from the candidate set:

5. element i must not be shared by $Super_s$ and $Super_t$, ($i \notin Super_s$);
   This condition excludes elements that are superior to both the start and the destination, and so do not add information value to the route directions.
6. element i should not be neighbor with an element in $Super_s$ ($\partial i \notin S$);
   This condition excludes elements that are in a neighboring relation with an element of the hierarchy $Super_s$ ($e \cap Sup_i \neq \emptyset, e \in Super_s$). This assures a minimal topological distance between the initial reference and the start s, in order to only add information with sufficient surprisal potential (and thus adhering to the maxim of information quantity). If the condition is not fulfilled, an element one step deeper in the hierarchy should be employed .

These conditions assure that the information communicated through the element i has value for the wayfinder. In condition 5, the wayfinder would get information that was too coarse. For example, *Australia* is not an appropriate initial element for route directions from *Sydney* to *Melbourne Central Business District*, as it is a superordinate element of both.

The second condition assures that the information is detailed enough– mentioning *Victoria* would not provide our wayfinder enough information either. In principle, the granularity level of the resulting element i would be correct, but there is not enough topological distance between the destination and the start element to justify its inclusion. Note that *New South Wales* and *Victoria* are in our hierarchy neighbors, and *Victoria* is a direct superordinate element of *Melbourne* (see Fig. 4). And indeed, telling somebody traveling from *Sydney* to *Melbourne* that she has to go to *Victoria* does not provide any information value. In this case, the starting element of the granular route directions should be *Melbourne*. This element is not a neighbor of any element of the hierarchy of the start element, nor of any of its superordinate elements. Any coarser level of granularity would not satisfy these conditions and would result in a route description violating Grice's maxim of relevance. The conditions mentioned above assume that there is no element with two children with the same identifier. It is, however, possible to have a repetition of identifiers within the domain.

The conditions 1–6 are serialized in an algorithm (Alg. 1), and further implemented in Haskell (Section 5).

Algorithm 1: Initial element i identification (*district hierarchy*)

Data: The urban hierarchical structure of districts: domain S, starting district s and destination district t of the route

Result: The initial element i for route directions from s to t

case $(s, t \notin S) \vee (i \neq s) \vee (i \cap s = 0) \vee (i \neq s)$
- Error: cannot generate granular directions, lack of relevant elements or bad topological context

otherwise
- Construct candidate set $Super_t$ and $Super_s$, where $Super_s = [s, s_1 \dots s_m]$, where $l_{s_m} = 0, l_{s_x} = m - x$; $Super_t = [t, t_1 \dots t_n]$, where $l_{t_n} = 0, l_{t_y} = n - y$;
- Compare hierarchies h_{Super_s}, h_{Super_t} such that $(\forall s_x, t_y \in (Super_s \times Super_t))$:
- **if** $(Sup_{s_x} = Sup_{t_y}) \vee (Sup_{s_x} \cap Sup_{t_y} \neq \emptyset)$ **then**
  - Return list T of t_y;
- Retrieve the element t_y of T with the finest granularity level $(l_{t_y} = max)$;
- Return $i = t_y$;

5 Model Implementation

Algorithm 1 is now implemented in Haskell [19]. Haskell is a purely functional programming language that enables implementation of an executable version of the algorithm, with a focus on the *what* instead on the *how*. Efficiency of our code is neglected, however, some efficiency is gained by the lazy execution paradigm of Haskell.

We call the main analytic function of the program rd (for *route directions*). The function first verifies the four conditions in Section 4.2:

```
rd :: String -> String -> [Object] -> String
rd a b c
 | (!testObj a c || !testObj b c)          = error
 | (a == b)                                = error
```

```
| testShareBounds (obj a c) (obj b c)          = error
| (super (obj a c)) == (super (obj b c))       = error
| otherwise describe a b c
```

The custom data type `Object` contains a name for each object (a string), and other parameters. Names of the objects are identifiers searched for in the above functions. For example, the function `obj` returns the object specified by a name from the list of all objects.

The function `rd` requires the names of the start and destination elements of the route as input parameters (in our case districts, but in principle also other elements of the city), as well as the name of the list of objects, `objects`, on which the selection will be performed. Afterwards, the program reconstructs the superordinate hierarchies of the start and destination elements, bound on the one side by these two elements, and on the other by the root element of the hierarchical tree. These hierarchies are represented as lists of objects, which are then compared according to the conditions 5 and 6 (see Section 4.2). The comparison returns a list of objects – a subset of elements of $Super_t$. The first – coarsest – element of this list, is the element i (Alg. 1).

```
describe :: String -> String -> [Object] -> String
describe a b objlist = fetchName (compareHierarchies
    (findObjects (obj a objlist) objlist)
    (findObjects (obj b objlist) objlist))
```

where

```
compareHierarchies :: [Object] -> [Object] -> Object
compareHierarchies start []      = error
compareHierarchies [] dest       = error
compareHierarchies start dest    = head [y |
    x<-start, y<-target,
    super x == super y || superShareBounds x y]
```

The element i is the initial entity of a sequence of granular route directions. A deeper illustration of the principles summarized in the algorithm and implemented in the program follows.

6 Model Verification and Testing

We have devised a set of tests to verify the behavior of the algorithm. The test data vaguely mirrors the spatial layout of the relations between some *administrative districts* in Victoria and New South Wales, Australia (see Fig. 4). With this data we will assess the results of the algorithm for plausibility.

Our effort is focused on verifying that the conditions for retrieving the elements from the structure of the city provide an amount of information similar to that provided by humans, as drawn from empirical evidence and a small reference corpus from co-workers. More extensive comparison by human subject testing is needed in the future.

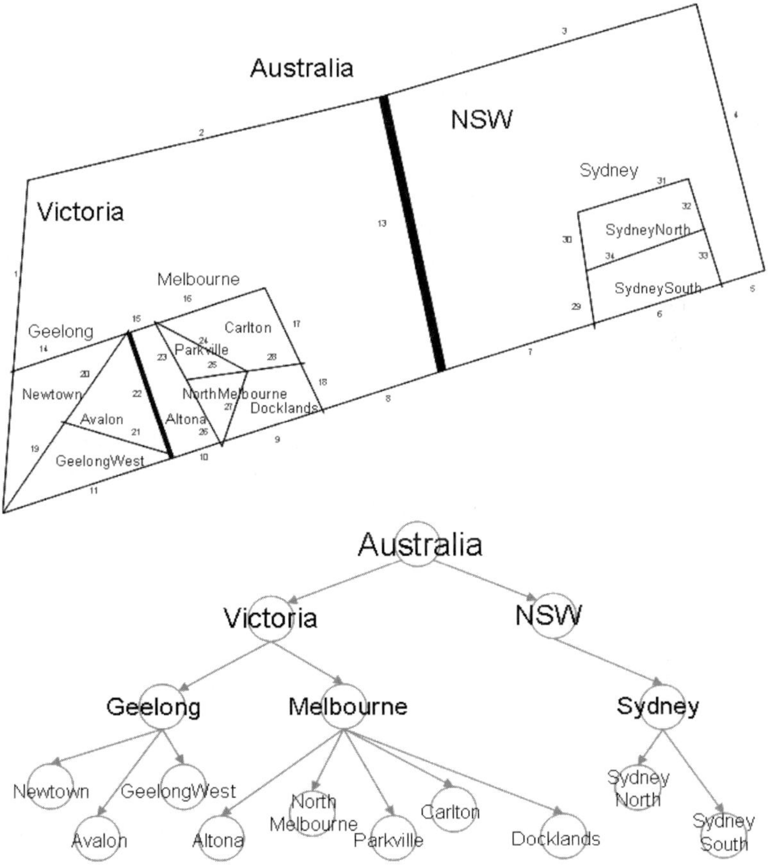

Fig. 4. Test data set

For the test, the `Object` data type for *districts* is structured as follows:

```
data Object = Object Level Super ObjectName Polygon
```

A district *Melbourne* would have the form (cf. Fig. 4):

```
ob4 = Object 2 "VIC" "Melbourne" [22,15,16,17,18,9,10]
```

expressing that Melbourne is of level 2 in the given hierarchy, is part of Victoria and is bound by a polygon of some named edges.

The following eight test cases were devised, to test all the possible eventualities. Each test case consists of a pair of input districts (start and destination) from arbitrary hierarchical levels, and the list of all objects of the hierarchy, `os`. Results of each test case are shown behind the hyphen in each line:

```
rd "SydneyNorth" "Parkville"   os - "Melbourne"
rd "Parkville"   "SydneySouth" os - "Sydney"
rd "Carlton"     "GeelongWest" os - "GeelongWest"
rd "SydneyNorth" "Melbourne"   os - "Melbourne"
rd "SydneyNorth" "Park"        os - "input␣not␣in␣os!"
rd "SydneyNorth" "SydneyNorth" os - "start␣=␣target!"
rd "SydneyNorth" "SydneySouth" os - "neighbors;␣TBT␣dirs"
rd "Parkville"   "Docklands"   os - "same␣super;␣use␣TBT"
```

This set of tests checks the behavior of different possible topological relations between the input objects. Let us have a detailed look at the operation of our test function `rd`, using the first case as an example. The function first tests the inputs against the basic conditions, and if fulfilled, the search for the initial element starts. The hierarchies of the superordinate elements of the start and the destination are then reconstructed, resulting in the following lists:

```
[Object 3 "Sydney" "SydneyNorth" [30,31,32,34],
Object 2 "NSW" "Sydney" [6,29,30,31,32,33],
Object 1 "Australia" "NSW" [3,4,5,6,7,13],
Object 0 "World" "Australia" [1,2,3,4,5,6,7,8,9,10,11,12]]
```

and

```
[Object 3 "Melbourne" "Parkville" [23,24,25],
Object 2 "Victoria" "Melbourne" [22,15,16,17,18,9,10],
Object 1 "Australia" "Victoria" [1,2,13,8,9,10,11,12],
Object 0 "World" "Australia" [1,2,3,4,5,6,7,8,9,10,11,12]]
```

The elements of these lists are then compared on a one–to–one basis through the function `compareHierarchies`, applying the conditions mentioned in Section 4.2. We are looking for a list of objects from the list of the superordinate elements of the destination that satisfies these conditions. The resulting set is:

```
[Object 2 "Victoria" "Melbourne" [22,15,16,17,18,9,10],
Object 1 "Australia" "Victoria" [1,2,13,8,9,10,11,12],
Object 0 "World" "Australia" [1,2,3,4,5,6,7,8,9,10,11,12]]
```

Finally, the first element of this list, *Melbourne*, is returned as the element of finest granularity. This element is proposed as the initial element of granular route directions from *Sydney North* to *Parkville*. And indeed, when asking for route directions from a starting location in Sydney North (a suburb of Sydney, New South Wales), to a location in Parkville (a suburb of Melbourne, Victoria), a familiar wayfinder is likely to expect Melbourne as the element representing the initial entity of route directions. This element provides the optimal trade off of information value and length of the route directions, is not ambiguous and omits irrelevant information. Remaining test cases also produce plausible results, as can be checked with Figure 4 by the interested reader.

7 Conclusions

This paper contributes to bridging the gap between the natural, human way of communicating route directions in a granular manner, using various elements of the urban environment, and the turn-by-turn approach implemented by most of the current navigation services. We focus on the determination of the initial element usable for granular route directions, building on the principles of information content relevance. Topological relations between the member elements of the start and destination element hierarchies are analyzed to identify this element. The conditions for analyzing the hierarchical structures are formalized, and the algorithm is then implemented in Haskell.

The test in this paper focuses on the analysis of district hierarchies, districts being one of the elements of the city most frequently included in human route descriptions, and often the first reference in granular route directions. Our approach enables the use of any type of region, be it with crisp or vague boundaries, as long as they can be organized in a hierarchy. Based on the inputs and the hierarchy, the algorithm returns the initial element. The formalized topological conditions conform with the observations made previously by Rumelhart and explored by Shanon [23], and conform to the findings in the area of hierarchical spatial reasoning. Our further work will strive to integrate the remaining elements, especially paths and landmarks. This will also lead to a more natural definition of topological relationships, depending also on connectivity, and not only on simple relationships between the interiors and boundaries. Connectivity analysis can also provide means to define district hierarchies better reflecting the inherent structure of a city. The analysis of such district hierarchies can provide the basis for identification of the remaining elements, as some dependencies between districts and landmarks [10] or paths were identified.

Acknowledgements

The work has been supported by the Cooperative Research Centre for Spatial Information, whose activities are funded by the Australian Commonwealths Cooperative Research Centres Program.

References

1. Clark HH, Wilkes-Gibbs D (1986) Referring as a Collaborative Process. Cognition 22:1–39
2. Couclelis H (1996) Verbal directions for way-finding: Space, cognition and language. In: Portugali J (ed) The Construction of Cognitive Maps. Kluwer, Dordrecht, pp 133–153
3. Dale R (1992) Generating Referring Expressions: Constructing Descriptions in a Domain of Objects and Processes. ACL-MIT Series in Natural Language Processing. MIT Press
4. Denis M, Pazzaglia F, Cornoldi C, Bertolo L (1999) Spatial Discourse and Navigation: An Analysis of Route Directions in the City of Venice. Applied Cognitive Psychology 13:145–174
5. Edmonds PG (1994) Collaboration on Reference to Objects that are Not Mutually Known. In: 15th Int Conf on Computational Linguistics, COLING-94, Kyoto, pp 1118–1122
6. Frank A (2003) Pragmatic Information Content: How to Measure the Information in a Route Description. In: Duckham M, Goodchild M, Worboys M (eds) Foundations of Geographic Information Science. Taylor & Francis, London and New York
7. Freundschuh SM, Mark DM, Gopal S, Gould MD, Couclelis H (1990) Verbal directions for wayfinding: Implications for navigation and geographic information and analysis systems. In: Brassel K, Kishimoto H (eds) 4th Int Symp on Spatial Data Handling. Department of Geography, University of Zurich, pp 478–487
8. Grice P (1975) Logic and Conversation. In: Cole P, Morgan JL (eds) Speech Acts. Academic Press, New York, pp 41–58
9. Heeman PA, Hirst G (1995) Collaborating on Referring Expressions. Computational Linguistics 21(3):351–382
10. Hirtle S (2003) Neighborhoods and Landmarks. In: Duckham M, Goodchild M, Worboys M (eds) Foundations of Geographic Information Science. Taylor & Francis, London and New York, pp 191–203
11. Hirtle S, Jonides J (1985) Evidence of Hierarchies in Cognitive Maps. Memory and Cognition 13:208–217
12. Jarvella RJ, Klein W (eds) (1982) Speech, Place, and Action. John Wiley & Sons, Chichester, NY
13. Klein W (1979) Wegauskünfte. Zeitschrift für Literaturwissenschaft und Linguistik 33:9–57

14. Klippel A, Tappe H, Habel C (2003) Pictorial Representations of Routes: Chunking Route Segments During Comprehension. In: Freksa C, Brauer W, Habel C, Wender KF (eds) Spatial Cognition III — Routes and Navigation, Human Memory and Learning, Spatial Representation and Spatial Learning (= Lecture Notes in Artificial Intelligence 2685). Springer-Verlag, Berlin, pp 11–33
15. Lovelace KL, Hegarty M, Montello DR (1999) Elements of Good Route Directions in Familiar and Unfamiliar Environments. In: Int Conf on Spatial Information Theory: Cognitive and Computational Foundations of Geographic Information Science. Springer-Verlag, pp 65–82
16. Lynch K (1960) The Image of the City. The MIT Press, Cambridge, Massachusetts, USA
17. Newman EL, Caplan JB, Kirschen MP, Korolev IO, Sekuler R, Kahana MJ (2005) Learning Your Way Around Town: Virtual Taxicab Drivers Reveal the Secrets of Navigational Learning. Cognition, in press
18. Paraboni I, Deemeter K van (2002) Generating Easy References: the Case of Document Deixis. In: Second Int Conf on Natural Language Generation (INLG 2002), New York, USA
19. Peterson J, Chitil O (2005) Haskell.org – The Haskell Home Page. Available http://www.haskell.org (visited on May 15, 2005)
20. Plumert JM, Carswell C, de Vet K, Ihrig D (1995) The Content and Organization of Communication about Object Locations. J of Memory and Language 37:477–498
21. Plumert JM, Spalding TL, Nichols-Whitehead P (2001) Preferences for Ascending and Descending Hierarchical Organization in Spatial Communication. Memory and Cognition 29(2):274–284
22. Shannon CE, Weaver W (1949) The Mathematical Theory of Communication. The University of Illinois Press, Urbana, Illinois
23. Shanon B (1979) Where Questions. In: 17th Annual Meeting of the Association for Computational Linguistics. University of California at San Diego, La Jolla, California, USA, ACL
24. Sperber D, Wilson D (1986) Relevance. Basil Blackwell Ltd, Oxford, UK
25. Timpf S, Volta G, Pollock D, Egenhofer MJ (1992) Conceptual Model of Wayfinding Using Multiple Levels of Abstraction. In: Frank A, Campari I, Formentini U (eds) Theory and Methods of Spatio-Temporal Reasoning in Geographic Space (= Lecture Notes in Computer Science 639). Springer-Verlag, Pisa, Italy, pp 348–367
26. Tversky B (1993) Cognitive Maps, Cognitive Collages, and Spatial Mental Models. In: Frank A, Campari I (eds) Spatial Information Theory: A Theoretical Basis for GIS, COSIT '93 (= Lecture Notes in Computer Science 716). Springer, Berlin, pp 14–24
27. Weissensteiner E, Winter S (2004) Landmarks in the Communication of Route Directions. In: Egenhofer MJ, Miller H, Freksa C (eds) Geographic Information Science 2004 (= Lecture Notes in Computer Science 3234). Springer, Berlin, pp 313–326

28. Wiener JM, Mallot HA (2003) 'Fine-to-Coarse' Route Planning and Navigation in Regionalized Environments. Spatial Cognition and Computation 3(4):331–358
29. Worboys M (2003) Communicating Geographic Information in Context. In: Duckham M, Goodchild M, Worboys M (eds) Foundations of Geographic Information Science. Taylor & Francis, London and New York, pp 33–45

Modeling and Engineering Algorithms for Mobile Data

Henrik Blunck, Klaus H. Hinrichs, Joëlle Sondern, Jan Vahrenhold

Westfälische Wilhelms-Universität Münster, Institut für Informatik,
48149 Münster, Germany
email: {blunck,khh,joellen,jan}@math.uni-muenster.de

Abstract

In this paper, we present an object-oriented approach to modeling mobile data and algorithms operating on such data. Our model is general enough to capture any kind of continuous motion while at the same time allowing for encompassing algorithms optimized for specific types of motion. Such motion may be available in a specific form, e.g., described by polynomials or splines, or implicitly restricted using bounds for speed or acceleration given by the application context.

Key words: spatio-temporal data; object-oriented modeling; algorithm engineering

1 Introduction

Over the past years, mobile data, such as points, lines, and regions whose coordinates change according to some time-variant function, has been the subject of increasing interest in Geographical Information Science and closely related fields. The Spatio-Temporal Databases community has contributed to the modeling and representation of such data, whereas algorithmic aspects have mainly been investigated in the field of Computational Geometry. Two recent surveys [2, 32] not only list the main results but also identify a number of issues that arise when handling mobile data; these issues include the question of numerical robustness, the incorporation of more realistic descriptions of motion, and the question of how to trade off realism and efficiency.

Efficient algorithms for processing and analyzing mobile data usually build on assumptions about the nature of the object's trajectories; commonly, these trajectories are assumed to have a representation based upon piecewise linear or (fixed-degree) polynomial curves – see, e.g., [1, 6, 18, 21, 25]. It has been noted, however, that such an assumption should be avoided wherever possible to allow for more realism [5, 13, 29, 37], especially since these assumptions are crucial only for the analysis of the running time but rather seldom for the correctness of the algorithm. In contrast, real-world scenarios may involve heterogeneous sets of objects whose motion should not be oversimplified by using a single type of motion description such as polynomials. Examples include (air) traffic control, monitoring meteorological phenomena, or tracking wildlife. Furthermore, a complete motion description of the objects may not be known at all or the objects may not be able (or willing) to publish this description, e.g., in the context of mobile services.

The most general (if not: minimalistic) interpretation of the trajectory of a mobile object considers it simply as a continuous function $f(t)$. Hence, a trajectory is modeled as a black box whose interface consists of a single method that allows to obtain the position of the object at any given point in time of its lifespan. Clearly, the trajectory of any "real-world" object can be modeled using this approach; this, however, raises a number of new issues such as the closure of the model and the efficiency of operations on trajectories.

Applications in which the analysis of mobile objects is required can be classified as either *real-time* or *retrospective*; in the first scenario, data can only be accessed at the current point in time (and thus only in increasing chronological order), whereas the second scenario allows random access to data at any point in the past. Application contexts may also provide additional information about mobile objects, e.g., bounds on speed, acceleration, or turning radius, and we need to be able to incorporate such information.

1.1 Related Work

Collision Detection and Motion Restrictions. A frequently performed task in managing mobile data is collision detection and collision warning, and Lin and Manocha [26] review the broad body of literature. Most practical methods are considered in a *real-time* setting, e.g., interaction in virtual-reality environments, and thus cannot make any assumption about the description of motion. To cope with this, most methods use hierarchical decompositions of the objects or temporal coherence for fast pruning of the search space.

A notable exception is the work by Hayward et al. [23] (and the conceptually similar model by Kahan [24]) that relies on a completely different concept: to quickly identify objects that are most likely (or most unlikely) to collide, it uses a restriction-based approach in which, for each mobile object, bounds on the maximal or minimal speed are known. Note, however, that if no such bounds are known (and exploited), the correctness of the approach cannot be guaranteed. Similar concepts have been used in the context of spatio-temporal indexing [31] and in the context of managing uncertainty in spatio-temporal databases [12, 28, 30, 36].

Kinetic Data Structures. A variety of algorithmic problems involving mobile objects has been addressed successfully in the context of *kinetic data structures* [5, 6]. A kinetic data structure (or: *KDS*) for a set $\mathcal{P}$ of mobile objects maintains a time-variant combinatorial description of some property of $\mathcal{P}$, e.g., its extent as given by the convex hull – see Guibas' survey [19] of recent KDS-related results. Even though the objects are moving according to some known, continuous "flight plan" (which may or may not be updated), the combinatorial structure maintained in the KDS will change only at some discrete points in time. To obtain these points in time, the KDS repeatedly identifies roots of so-called certificate functions that guarantee the validity of the combinatorial description.

The main requirement for a meaningful theoretical analysis of the efficiency of a KDS is that the motion, and thus the description of the certificates as well, is given as a polynomial function of time. A major benefit of working with polynomials is that there exists a variety of numerically robust methods for isolating roots – see, e.g., the survey by Schirra [33]. Several of such methods have been integrated in an upcoming extension package [20] for the Computational Geometry Algorithms Library CGAL [14] thus providing efficient support for working with KDSs. The question of whether or not polynomials provide a "good" level of realism for modeling the motion of real-world objects has been raised (and answered) by Basch who concludes that using polynomials may be "too restrictive to be of much use in applications, although it is perfectly adequate for theoretical purposes" [5, p 103].

Modeling Mobile Data. In the context of spatio-temporal databases, a number of approaches to modeling mobile data has been presented [7, 15, 22]. In this paper, we revisit a model that we have proposed earlier [7], and we refer the reader to our original paper for a discussion of related approaches.

Recently, Mount et al. [27] presented a framework that changes the concept of kinetic data structures to make them applicable to the *real-time* setting: In contrast to the original approach, this framework follows an earlier model of Kahan [24] and is based on incremental updates that involve small time steps. Unlike a classical KDS framework, it does not require complete knowledge of the kind of motion but instead estimates future locations revisiting these estimates whenever safety constraints are violated. Again, the correctness depends on the presence of motion restrictions. Kahan's model has also been revisited in the context of the competitive analysis of on-line algorithms – see [11] and the references therein.

1.2 Our Results

The main purpose of this paper is to present a general framework for modeling and engineering algorithms for mobile data. This framework has been successfully implemented in the context of the GOODAC object-oriented geo-database kernel [8], and it makes a first attempt at addressing the issue of engineering robust algorithms for more realistic and possibly heterogeneous motion data. It builds upon a representation scheme we have proposed earlier [7], but whereas our earlier scheme focused on representing and storing moving objects in an object-oriented database management system, we now extend it to also support spatio-temporal (main-memory) algorithms. The main features of our approach are the use of a minimalistic interface (thus allowing for arbitrary continuous motion description) and a strict isolation of algorithmic primitives (thus allowing for better algorithm engineering); it can be seen as extending the concept of kinetic data structures to encompass more general motion description. We use the problem of collision detection for a heterogeneous set of objects as a running example and present corresponding primitives for both the *real-time* and the *retrospective* setting as well as an improved approach that exploits the properties of the *retrospective* setting.

2 Representing Mobile Data

Our approach to representing mobile data [7] is based upon two assumptions: (1) the trajectory of a moving point (and thus also of a segment's endpoint or a polygon's vertex) is a t-monotone, continuous curve $f(t)$, and (2) the representation $f(t)$ can be evaluated at any point in its domain. These two assumptions are the most basic assumptions that can be made about trajectories and do not imply any restrictions for the representation of the motion of real-world objects.

Almost all motion data obtained from real-world moving objects is either available in advance as a complete motion description or is given as a collection of timestamped locations, e.g., obtained through the use of GPS-based devices. In order to convert the latter discrete points to a continuous function, interpolation and approximation techniques are applied. It has been noted frequently that there is no single interpolation technique, e.g., piecewise linear or polynomial, that is optimal over a wide range of scenarios; the trajectory of an airport's ramp-agent or of a plane being towed on ground level may be represented by a piecewise linear function, but the trajectory of an airborne plane with a limited possible turning radius can be represented in a much more realistic way using ν-splines; that is, even if we are working with a single class of objects, e.g., planes, their motion may have completely different characteristics.

These considerations result in a class design (see Fig. 1) whose core class `mpoint`$\langle d \rangle$ represents a time-variant point in d dimensions. Each instance of `mpoint`$\langle d \rangle$ stores a collection of timestamped location data, and the (continuous) representation of the motion restricted to each dimension can be (re-) constructed using a specialization of the `Function` interpolation class. More specifically, an instance of `mpoint`$\langle d \rangle$ aggregates d instances of specializations of `Function` and delegates the evaluation of the trajectory to them.

When working with this framework, a number of issues have to be taken into consideration, most notably the issue of maintaining the model closed under (concatenated) operations such as (time-variant) difference or distance computation. For a more detailed description of the framework and for a discussion of the practical efficiency of its implementation we refer the reader to our previous paper [7].

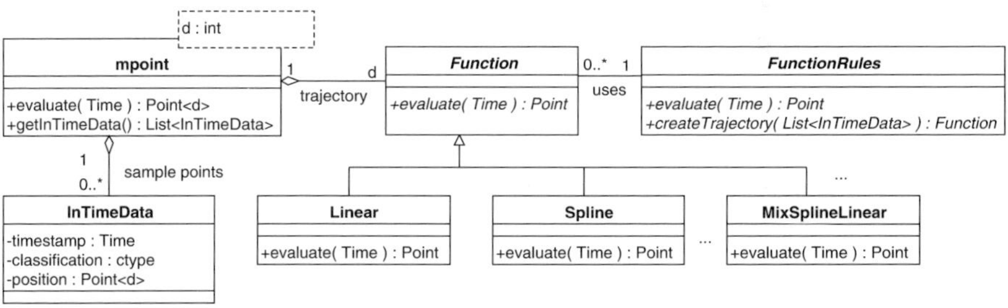

Fig. 1. Class diagram for representing mobile point data (see [7])

3 Modeling Algorithms for Heterogeneous Sets of Mobile Data

As we have mentioned in the introduction, polynomials are a very popular type of curves for modeling motion. A main reason for this is that there exists a variety of numerically robust methods for isolating the roots of a polynomial; this is exploited, e.g., in the context of (certificates for) kinetic data structures. However, in certain applications involving mobile objects (such as in the above-mentioned airport scenario), it is necessary to consider sets of objects that have different motion characteristics.

In this section, we extend the model sketched in the previous section to allow for modeling algorithms for heterogeneous sets of mobile objects. In Section 4, we discuss how to provide means for a more efficient treatment of special instances for which additional information about the objects involved is available.

Example. Our exposition proceeds using the following well-known problem setting as a running example: Given a set of mobile objects that move along the real axis, find all collisions between them. In the two-dimensional (t, y)-parameter space, this setting translates to the problem of finding all intersections induced by a set of t-monotone curves. At first, this seems identical to what templated algorithms for t-monotone curves can handle, e.g., the industrial-strength methods of CGAL's `Arrangement_2` class. A closer look, however, reveals that the implementation of such methods always assumes that the curves belong to the same class of functions, e.g., polynomials, and since we do not make any assumption about the nature of the objects involved, our setting is much more general and thus encompasses a wider range of scenarios. Any (two-dimensional) specialization of our `Function` class, on the other hand, would work fine in the context of the `Arrangement_2` class as long as all primitive operations required by the corresponding algorithm are realized. This issue is addressed as part of the following discussion.

3.1 Isolating Primitive Operations from Algorithms

Our design for modeling algorithms for heterogeneous sets of mobile objects follows a classic "black box"-based approach, that is, we isolate from the general algorithm all operations (*primitives*) that are dependent on the type of trajectory. For each kind of primitive, e.g., intersection-finding, we have a `Decider` class that encapsulates knowledge about how to handle different types of trajectories. Whenever the algorithm needs to process heterogeneous motion descriptions, it polls a Decider-instance which, depending on the types of motion, selects an appropriate specialization of the primitive (see Fig. 2).

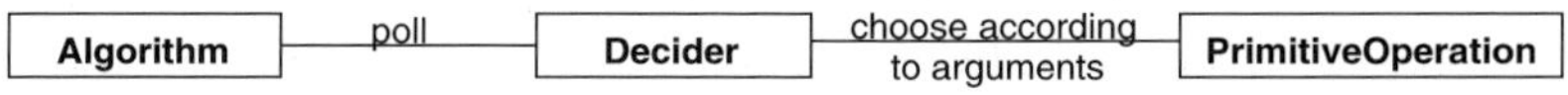

Fig. 2. Handling heterogeneous sets of mobile objects using a Decider-object

Example. In our intersection-finding example, the main operation that needs to be isolated is the test for whether or not two given trajectories intersect in some given (possibly unbounded) time interval $[begin, end]$; Boissonnat and Vigneron [10] declare this predicate as "mandatory". Assuming that we have to check two curves s and t, this test is implemented as follows:

```
if ( s.intersectsWithin(t, begin, end) )  /* ... */
```

In the above situation, the class of which s is an instance needs to provide a polymorphic version of the method intersectsWithin for each additional type of trajectory that is supported by the system.

Using a Decider-instance, we decouple the knowledge about other classes in the system from the class representing a certain trajectory type. This knowledge (and thus the main administrative burden) is encapsulated in the corresponding Decider-class, and the above code fragment then looks as follows:

```
IntersectionPred ip = myIntersectionDecider.poll(s, t);
if ( ip.eval(s, t, begin, end) == true )  /* ... */
```

The "Double Dispatch"-Problem. The problem we have addressed in this section is known as the *double dispatch* problem [16] where the (type of) result of an operation depends on the type of its operands. While some programming languages, e.g. Smalltalk, provide mechanisms to directly address this issue, the generic solution is to employ the so-called *Visitor* design pattern [16] which reduces the problem to type-dependent single-argument dispatching. As we show in Section 4.2, our algorithms not only depend on the *type* of objects but also on (a combination of) their properties. Thus, the *Visitor* pattern cannot be used, and we feel that our solution discussed above is the best-suited approach for the problem at hand.

3.2 Modeling Compound Functions

For our running example of intersection-finding, we observe that finding intersections between two curves s and t is equivalent to determining the zeros of $s - t$. At this point, the minimalistic interface provided by the class mpoint$\langle d \rangle$ (see Section 2) turns out to be a strong design advantage: the concatenation of continuous functions again is a continuous function (with the necessary care taken for the case of division). A modification of the base framework allows to represent compound functions (such as `Difference`) that are composed of other functions, and we have implemented the framework given in Figure 3.

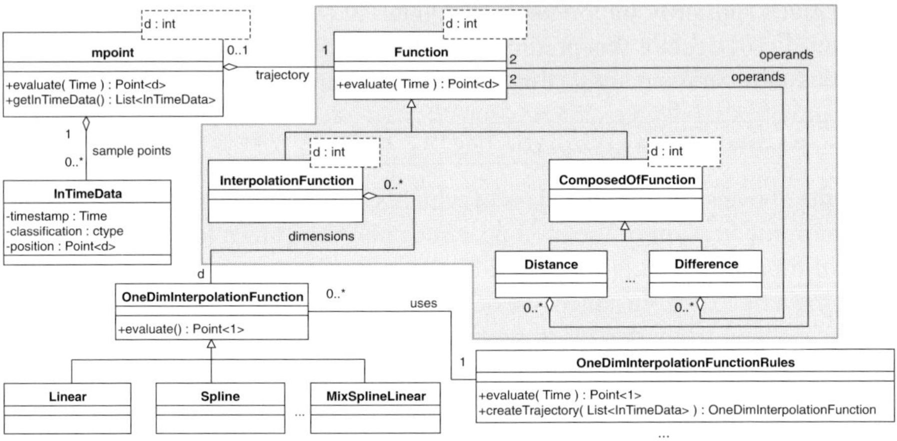

Fig. 3. Class diagram for representing mobile point data (extensions highlighted)

To allow for a nested composition of functions, we need to slightly modify the original framework (cf. Fig. 1): in our extended setting, each instance of class mpoint$\langle d \rangle$ now aggregates a single d-dimensional function instead of d one-dimensional functions.

Example. Assuming that we have a `Decider`-instance for selecting an appropriate specialization of the root-finding primitive, the code for intersection-finding can be rewritten as follows.

```
Difference diff = new Difference(s, t);
ZeroFinderPred zfp = myZeroFinderDecider.poll(diff);
if ( zfp.eval(diff, begin, end) == true )  /* ... */
```

4 Handling Motion with Known Restrictions

The example at the end of the previous section reduces the problem of intersection-finding to the problem of isolating roots, and there exists a number of numerical and algebraic methods for isolating roots of functions whose mathematical description is known. For example, if both trajectories are given by cubic polynomials, the difference between them is a cubic polynomial as well, and we may use an algebraic approach to implementing a root-finding primitive. If, on the other hand, one trajectory is approximated using a wavelet while the other is approximated using a ν-spline, we have to resort to iterative numerical methods. In Section 4.2, we demonstrate that an iterative algorithm can be guaranteed not to miss any root if we can exploit additional information such as upper bounds on the velocity of both objects. Figure 4 illustrates a simple iteration rule: If $f(t_i)$ and $g(t_i)$ are known, no root of $f - g$ can occur prior to time t_{i+1} which is determined by assuming that f and g move towards each other at maximum speed. The time t_{i+1} at which $f - g$ can have its "next" root is the time of the intersection of bounded-slope segments extending the trajectories of f and g from time t_i onwards—or, equivalently, the root of a bounded-slope linear function extending the trajectory of $f - g$. This method is referred to as L_{one} as it involves one linear function.

If no restrictions are known, the only feasible approach to root-finding is to employ "classical" numerical methods such as Newton's Method, the Secant Method, or Bisection. However, as these methods may fail to produce the roots in chronological order, skip roots, or even fail to converge [35], the correctness of algorithms using them as a primitive cannot be guaranteed.

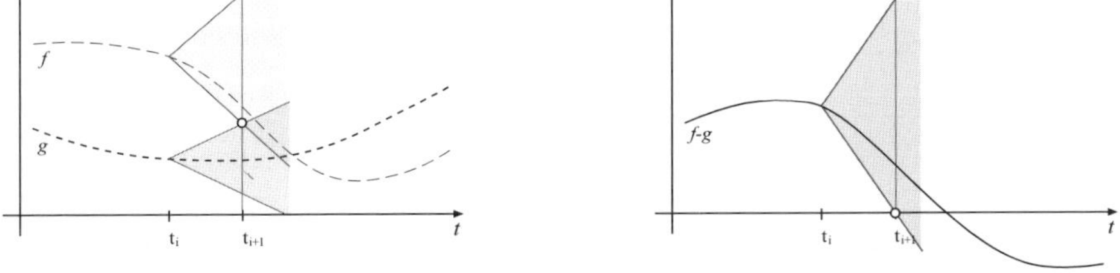

Fig. 4. Intersection-finding by exploiting an upper bound on the velocity

4.1 Modeling Algorithms: The Case of Known Restrictions

By design, our class model does not make any assumption about the type of trajectories (except for assuming continuity). To fully integrate known results for handling classes of trajectories for which additional information is available, we introduce the concept of *restrictions*.[1] A realization of the interface `MotionRestriction` models additional information about a trajectory (or a composition thereof); examples are "real-world" restrictions such as bounds on speed or acceleration given as part of the application context.[2] Such restrictions are defining features for applications involving mobile real-world data and distinguish our setting from related settings involving t-monotone curves.

Figure 5 displays the extension to our class diagram resulting from the inclusion of the concept of restrictions: In addition to the design discussed above, an instance of (a non-abstract specialization of) class `Function` can aggregate any number of instances of realizations of `MotionRestriction`,[3] and different realizations are distinguished by unique identifiers.

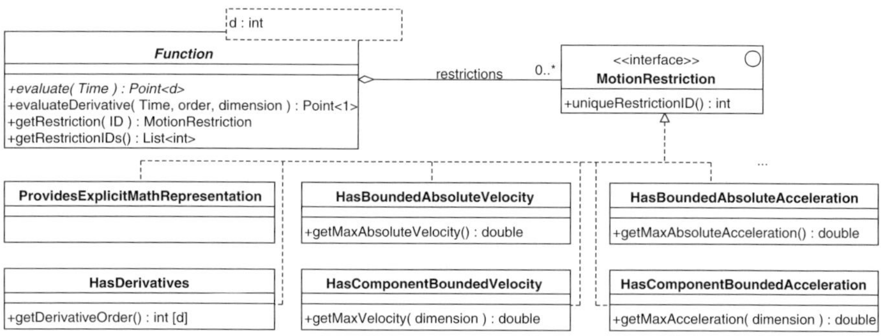

Fig. 5. Additions to the class diagram to model motion with known restrictions

[1] A preliminary version of the approach discussed in this subsection has been presented in our previous work [9].

[2] The European Organization for Safety of Air Navigation maintains a database http://www.eurocontrol.fr/projects/bada of the inflight behavior, e.g., velocity or descent speed, of over 250 aircraft types to support exact modeling. For obtaining a correctness guarantee, it is sufficient to use conservative estimates for velocity or acceleration.

[3] The data type `double` is to be read as a placeholder for the actual numeric data type used in the application.

To allow for easy access to motion status data, e.g., speed and velocity, we enhance the interface of `Function`, such that derivatives at a given time can be evaluated. Whether or not such information is available, is modeled by class `HasDerivatives`, a realization of `MotionRestriction`.

We also use a realization of `MotionRestriction` to indicate whether or not we may explicitly access a mathematical representation such as the coefficients of a fixed-degree polynomial; this allows a `Decider`-instance to also consider (semi-)algebraic methods and thus to encompass the techniques discussed for the polynomial-based KDS-framework of Guibas et al. [20]. The representation can be accessed using the well-known *Factory* design pattern [16] or using a language-dependent construction, such as the `Java` *Reflection* API [4].

4.2 Designing Primitives: The Case of Known Restrictions

In this section, we continue to consider our running example and focus on primitives for root-finding. The iterative approach L_{one} uses a velocity-based restriction to determine a "next" iteration point t_{i+1} such that $[t_i, t_{i+1}]$ is guaranteed not to contain a root. We show that acceleration-based restrictions can be used to obtain a similar result; an important observation is that such a restriction can lead to more efficient algorithms in the *retrospective* setting than in the *real-time* setting.

As a proof-of-concept we present two restriction-based methods for our running example that we have implemented within our framework. Due to space constraints we omit proofs for their correctness and efficiency; the reader may find these proofs in the thesis of one of the authors [34].

Methods for Root-Finding Using Acceleration-Based Restrictions. Let us assume that the objects subject to collision detection have an upper bound b_{acc} on their acceleration. The earliest collision after time t_i can be computed by assuming that both objects move towards each other with maximum possible acceleration. The resulting "exclusion region" for occurrence of the next possible root is induced by a parabola (see Fig. 6 left), and t_{i+1} can be computed – assuming w.l.o.g. that $f(t_i) > 0$ – as follows [34, Sec. 4.3.1]:

$$t_{i+1} = t_i + \frac{1}{b_{\text{acc}}} \left(f'(t_i) + \sqrt{f'^2(t_i) + 2 \cdot b_{\text{acc}} \cdot f(t_i)} \right) \tag{1}$$

This approach, which we refer to as P_{one} since it involves one parabola, has also been used for collision detection. An alternative, more efficient approach, computes the earliest possible point in time t_{i+1} at which a *second* root may occur. The earliest such occurrence coincides with a double root

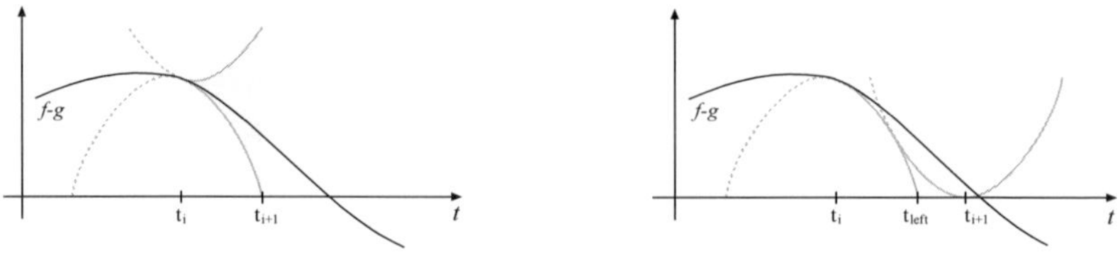

Fig. 6. Intersection-finding by exploiting an upper bound on the acceleration

of the distance function, i.e., the (real-world) objects touch each other. The continuity of motion and speed implies that this point in time t_{i+1} can be computed by first extending $f - g$ by a parabola P as in P_{one}, but then to model the deceleration by an inverted copy $\hat{P}$ of P that continuously extends P such that the vertex of $\hat{P}$ lies on the t-axis (thus inducing a double root) – see Fig. 6 (right). If even a maximal deceleration cannot avoid a collision, the extension of the distance function by $\hat{P}$ from t_i onwards intersects the t-axis, and we choose its first intersection point as t_{i+1}. In both cases, t_{i+1} can be computed in the following, surprisingly simple way [34, Sec. 4.3.1]:

$$t_{i+1} = t_i + \begin{cases} \frac{1}{b_{\text{acc}}} \left(f'(t_i) + \sqrt{2} \cdot \sqrt{f'^2(t_i) + 2 \cdot b_{\text{acc}} \cdot f(t_i)} \right) & \text{if } f'(t_i) \geq \sqrt{2 \cdot b_{\text{acc}} \cdot f(t_i)} \\ \frac{1}{b_{\text{acc}}} \left(-f'(t_i) - \sqrt{f'^2(t_i) - 2 \cdot b_{\text{acc}} \cdot f(t_i)} \right) & \text{else (collision unavoidable)} \end{cases} \tag{2}$$

If $f - g$ has the same sign at time t_i and at time t_{i+1}, no root lies within $[t_i, t_{i+1}]$, and we iterate. Otherwise we can employ Newton's Method to efficiently find the root inside $[t_{\text{left}}, t_{i+1}]$, where t_{left} is determined as in P_{one}.[4] Since this approach involves two parabolas, we refer to it as P_{two}.

Quality and Applicability of the Methods P_{one} and P_{two}. Equations 1 and 2 indicate that the cost of computing the next increment, i.e., the number of arithmetic operations, is almost identical for both methods; the cost is exactly the same if our cost measure is the number of invocations of `evaluate` and `evaluateDerivative`. We analyzed increment and order of convergence:

Fact 1 ([34, Sec. 4.1]) *The increment $\Delta_{P_{\text{two}}}$ is always larger than $\Delta_{P_{\text{one}}}$:*

[4] The correctness of Newton's Method is guaranteed since there is exactly one root inside $[t_{\text{left}}, t_{i+1}]$. Also, due to the bound on the acceleration, the method cannot leave $[t_{\text{left}}, t_{i+1}]$ – for a better understanding of this crucial property, see, e.g., [35].

$$\Delta_{P_{\mathrm{one}}} < \Delta_{P_{\mathrm{two}}} \leq (\sqrt{2}+1) \cdot \Delta_{P_{\mathrm{one}}}.$$

For decelerating compound speed, e.g., when approaching a root, we have:

$$\sqrt{2} \cdot \Delta_{P_{\mathrm{one}}} < \Delta_{P_{\mathrm{two}}} \leq (\sqrt{2}+1) \cdot \Delta_{P_{\mathrm{one}}}.$$

Fact 2 ([34, Sec. 4.2]) *P_{one} and P_{two} both converge quadratically while L_{one} converges linearly.*

However, as both methods use a global bound on the acceleration, their *actual* quality inside some time interval depends on how much the acceleration locally deviates from this global bound. Thus P_{two} is "better" than P_{one}, since it eventually switches to Newton's Method, which then is guaranteed to have quadratic convergence with a (globally) good asymptotic error constant [35]. We conclude that fully exploiting the power of the *retrospective* setting, i.e., being able to move forward and backward in time, can lead to more efficient algorithms than "simply" using known *real-time* algorithmic primitives.

4.3 Generalizations

Combination of Restrictions. We mention in passing that we can also engineer primitives for trajectories that underlie a combination of restrictions [34, Sec. 2.3]; one of these primitives, for example, combines L_{one} and P_{two}. All of these primitives can be implemented using the methods `evaluate` and `evaluateDerivative` provided by class `Function`. All such specializations of a primitive can be incorporated into the framework we have presented: the only necessary modification is the incorporation into the decision process represented by the `Decider`-class associated with the respective kind of primitive.

Applications to Kinetic Data Structures. Many problems in the context of KDSs can be reduced to tracing the relative position of objects and hyperplanes, and algorithms employ root-finding primitives for real-valued functions to check for changes. Our framework thus can be used to extend the concept of KDS to encompass more realistic motion descriptions. Moreover, we can transfer the idea underlying the method P_{two} to the above certificate functions: using a bound on the acceleration we can – in a *retrospective* setting – determine the earliest possible point in time at which the relative position of an object and a hyperplane can have changed twice.

Collision Warning in Multiple Dimensions. The problem of collision warning is to find the time intervals during which the distance between two objects is smaller than some threshold $\varepsilon > 0$. This problem can be solved, e.g., by solving a collision *detection* problem in which one of the objects is extended by a buffer of width ε, see, e.g., [36]. The boundary of this buffer can be seen as a replacement for the hyperplane used in the context of a KDS certificate, and – provided that the description of the buffer is not too complicated – we can again employ a *retrospective* P_{two}-like approach whose efficiency unfortunately diminishes for very small values of ε. A *real-time* P_{one}-like approach is discussed by Hayward et al. [23]. Note that the seemingly more fundamental problem of collision *detection* in a multidimensional spatio-temporal setting reduces to intersection finding of curves in more than two dimensions; for this problem, no (theoretically) efficient algorithms are known – even if the curves are straight lines. This is in contrast to the collision warning setting where (at least for the three-dimensional case) nontrivial algorithms are known [3].

5 Conclusions

We have presented an object-oriented approach to modeling mobile data and algorithms operating on such data. Our model is general enough to capture not only polynomial motion descriptions but also more general (and thus more realistic) descriptions of continuous motion, e.g., of motion restricted only by bounds for the absolute speed or acceleration. In addition to being able to encompass "classical" exact algorithms for polynomials, our approach addresses the problem of numerical robustness and efficiency by modeling and efficiently utilizing motion restrictions. Using algorithmic primitives for collision detection as a proof-of-concept, we have shown how to engineer and to implement efficient algorithmic primitives that exploit such restrictions. A beneficiary side effect of our approach is that these primitives also have a direct applicability in the context of kinetic data structures; thus they extend this concept to encompass more realistic motion descriptions.

References

1. Agarwal PK, Arge LA, Vahrenhold J (2001) Time responsive external data structures for moving points. In: Proc 7th Int Workshop Algorithms and Data Structures (= LNCS 2125), pp 50–61

2. Agarwal PK et al. (2002) Algorithmic issues in modeling motion. ACM Comp Surveys 34(4):550–572
3. Agarwal PK, Sharir M (2000) Pipes, Cigars, and Kreplach: The Union of Minkowski Sums in Three Dimensions. Discrete & Computational Geometry 24(4):645–657
4. Arnold K, Gosling J, Holmes D (2006) The Java^{TM} Programming Language, 4th ed. Addison-Wesley
5. Basch J (1999) Kinetic Data Structures. PhD Thesis, Dept of Computer Science, Stanford University
6. Basch J, Guibas LJ, Hershberger J (1999) Data structures for mobile data. J Algorithms 31(1):1–28
7. Becker L, Blunck H, Hinrichs HK, Vahrenhold J (2004) A framework for representing moving objects. In: Proc 15th Int Conf Database and Expert Systems Applications (= LNCS 3180), pp 854–863
8. Becker L, Voigtmann A, Hinrichs KH (1996) Developing Applications with the Object-Oriented GIS-Kernel GOODAC. In: Proc 7th Int Symp Spatial Data Handling vol I:5A1–5A18
9. Blunck H, Hinrichs KH, Puke I, Vahrenhold J (2004) Verarbeitung von Trajektorien mobiler Objekte (in German). In: Beiträge zu den Münsteraner GI-Tagen, pp 29–41
10. Boissonnat JD, Vigneron A (2002) An elementary algorithm for reporting intersections of red/blue curve segments. Computational Geometry: Theory and Applications 21(3):167–175
11. Bruce R, Hoffmann M, Krizanc D, Raman R (2005) Efficient Update Strategies for Geometric Computing with Uncertainty. Theory of Computing Systems 38:411-423
12. Cheng R, Kalashnikov DV, Prabhakar S (2004) Querying imprecise data in moving object environments. IEEE Trans Knowledge and Data Engineering 16(9):1112–1127
13. Chomicki J, Revesz PZ (1999) A general framework for specifying spatiotemporal objects. In: Proc 6th Int Workshop Temporal Representation and Reasoning, pp 41–46
14. Fabri A, Giezeman GJ, Kettner L, Schirra S, Schönherr S (2000) On the design of CGAL a computational geometry algorithms library. Software – Practice and Experience 30(11):1167–1202
15. Forlizzi L, Güting RH, Nardelli E, Schneider M (2000) A data model and data structures for moving objects databases. In: Proc ACM Int Conf Management of Data, pp 319–330
16. Gamma E, Helm R, Johnson R, Vlissides J (1995) Design Patterns: Elements of Reusable Object-Oriented Software. Addison-Wesley
17. Goodman JE, O'Rourke J (eds) (2004) Handbook of Discrete and Computational Geometry. Discrete Mathematics and its Applications, 2nd ed. CRC Press
18. Gudmundsson J, van Kreveld M, Speckmann B (2004) Efficient Detection of Motion Patterns in Spatio-Temporal Data Sets. In: Proc 12th Symp Geographic Information Systems, pp 250–257

19. Guibas LJ (2004) Modeling motion. In: Goodman JE, O'Rourke J (eds) Handbook of Discrete and Computational Geometry. Discrete Mathematics and its Applications, chapter 50, 2nd ed, pp 1117–1134
20. Guibas LJ, Karavelas MI, Russel D (2004) A computational framework for handling motion. In: Proc 6th Workshop Algorithm Engineering and Experiments, pp 129–141
21. Guibas LJ, Mitchell JSB, Roos T (1992) Voronoi diagrams of moving points in the plane. In: Proc 17th Int Workshop Graph-Theoretic Concepts in Computer Science (= LNCS 570), pp 113–125
22. Güting RH, Böhlen MH, Erwig M, Jensen CS, Lorentzos NA, Schneider M, Vazirgiannis M (2000) A foundation for representing and querying moving objects. ACM Trans Database Systems 25(1):1–42
23. Hayward V, Aubry S, Foisy A, Ghallab Y (1995) Efficient collision prediction among many moving objects. Int J Robotics Research 14(2):129–143
24. Kahan S (1991) A model for data in motion. In: Proc 23rd ACM Symp Theory of Comp, pp 267–277
25. Kollios G, Gunopulos D, Tsotras VJ (1999) On indexing mobile objects. In: Proc 18th ACM Symp Principles of Database Systems, pp 261–272
26. Lin MC, Manocha D (2004) Collision and proximity queries. In: Goodman JE, O'Rourke J (eds) Handbook of Discrete and Computational Geometry. Discrete Mathematics and its Applications, chapter 35, 2nd ed, pp 787–807
27. Mount DM, Netanyahu NS, Piatko CD, Silverman R, Wu AJ (2004) A computational framework for incremental motion. In: Proc 20th ACM Symp Computational Geometry, pp 200–209
28. Pfoser D, Jensen CS (1999) Capturing the uncertainty of moving-object representations. In: Proc 6th Int Symp Spatial Databases, (= LNCS 1651), pp 111–132
29. Pfoser D, Jensen CS (2001) Querying the trajectories of on-line mobile objects. In: Proc 2nd Int ACM Workshop Data Engineering for Wireless and Mobile Access, pp 66–73
30. Pfoser D, Tryfona N (2001) Capturing fuzziness and uncertainty of spatiotemporal objects. In: Proc 5th East European Conf Advances in Databases and Information Systems (= LNCS 2151), pp 112–126
31. Prabhakar S, Xia Y, Kalashnikov DV, Aref WG, Hambrusch SE (2002) Query indexing and velocity constrained indexing: Scalable techniques for continuous queries on moving objects. IEEE Trans Computers 51(10):1124–1140
32. Roddick JF, Egenhofer MJ, Hoel E, Papadias D, Salzberg B (2004) Spatial, temporal and spatio-temporal databases. Hot issues and directions for PhD research. SIGMOD Record 33(2):126–131
33. Schirra S (2000) Robustness and precision issues in geometric computation. In: Sack JR, Urrutia J (eds) Handbook of Computational Geometry, chapter 14. Elsevier, pp 597–632
34. Sondern J (2005) Nutzung von Bewegungsrestriktionen für Algorithmik in Moving-Objects-Datenbanken. Master's Thesis, Department of Computer Science, University of Münster (in German)
35. Suli E, Mayers DF (2003) An Introduction to Numerical Analysis. Cambridge

36. Trajcevski G, Wolfson O, Hinrichs KH, Chamberlain S (2004) Managing uncertainty in moving objects databases. ACM Trans Database Systems 29(3):463–507
37. Yeh TS, De Cambray B (1995) Modeling highly variable spatio-temporal data. In: Proc 6th Australasian Database Conf, pp 221–230

Database Model and Algebra for Complex and Heterogeneous Spatial Entities

Gloria Bordogna[1], Marco Pagani[1], Giuseppe Psaila[2]

[1] CNR-IDPA, Dalmine (BG), Italy
email: firstname.secondname@idpa.cnr.it
[2] Facoltà di Ingegneria, Università di Bergamo, Dalmine (BG), Italy
email: psaila@unibg.it

Abstract

Current Geographic Information Systems (GISs) adopt spatial database models that do not allow an easy interaction with users engaged in spatial analysis operations. In fact, users must be well aware of the representation of the spatial entities and specifically of the way in which the spatial reference is structured in order to query the database. The main reason of this inadequacy is that the current spatial database models violate the independence principle of spatial data. The consequence is that potentially simple queries are difficult to specify and strongly depends on the actual data in the spatial database.

In this contribution we tackle the problem of defining a database model to manage in a unified way *spatial entities* (classes of spatial elements with common properties) with different levels of complexity. Complex spatial entities are defined by aggregation of primitive spatial entities; instances of spatial entities are called *spatial grains*.

The database model is provided with an algebra to perform spatial queries over complex spatial entities; the algebra is defined in such a way it guarantees the independence principle and meets the closure property. By means of the operators provided by the algebra, it is possible to easily perform spatial queries working at the logical level only.

1 Introduction

Current Geographic Information Systems (GISs) are successfully exploited to support applications dealing with complex spatially referenced data. However, spatial databases supporting GISs are not easy to use, in particular as far as query formulation and spatial data analysis are concerned.

In fact, the need to model complex situations leads the database designer to structure the database schema in such a way that does not ease the formulation of queries. As it will be shown in Section 3, the structure of the spatial entities and their interrelationships constrain the formulation of spatial operations that can be performed on the database. For example, many common operations of spatial allocation/localization of resources on the territory need long sequences of spatial operations for identifying the suitable places where to allocate/localize a given resource.

The main reason of this inadequacy of current GIS is that they adopt spatial database models that do not clearly separate the database model, which describes spatial entities with their properties (such as countries, rivers, motor ways with their valued properties), and the representation model of the spatial component, which actually represents spatial references of spatial entities (in terms of points, polylines, regions, etc.). Consequently, the set of operations that can be performed instance of spatial entities strongly depends on the particular representation chosen for their georeference; the result is that the independence principle is violated [8].

Another important limitation of database models for GISs is their inability to flexibly deal with complex spatial entities and environments. This is the case of technological infrastructures, such as communication networks, water supply infrastructures, etc., that are usually composed of heterogeneous components. Although current GISs allow to model these environments, their intrinsic heterogeneity makes very difficult to query the database, requiring to write very complex and unnatural queries.

In this paper, we tackle the problem of defining a spatial database model suitable for complex and heterogeneous spatial entities. It is provided with an algebra that guarantees the independence of queries by the particular representation of spatial references and the closure property. To achieve this goal, we rigorously separate the *database level* and the *representation level*; at the database level, spatial operations can be performed on any type of spatial entity, irrespective of its actual representation modality and structure; at the representation level, the specific spatial operations to perform are chosen by the system, depending on the actual representation of the involved spatial objects. In this paper, we concentrate on the database level.

In Section 2 we will review the related works, in Section 3 we will describe an example of spatial query depending on the structure of the database that is needed in a current GIS to perform spatial analysis. In Sections 4 and 5 the database model and the algebra for the main spatial operations are defined, respectively; we will recall the example in Section 3 and show how the spatial analysis is simplified by adopting the new database model and algebra. Finally Section 6 will draw the conclusions.

2 Related Work

Two main types of approaches were proposed to incorporate in a GIS database functionalities, hybrid approaches and integrated approaches [8]. Hybrid approaches use a DBMS to store and manage non-spatial data, and spatial data are separately managed by either a proprietary file system (e.g., ARC/INFO) [6] or a spatial data manager (e.g., Papyrus [5]). Integrated approaches extend the relational data model by adding new data types (Spatial Abstract Data Types, SADT) and operations to capture spatial semantics [4, 1, 7, 9, 3].

These approaches however showed their shortcomings in real applications as it discussed in the next section.

Some authors pointed out that a suitable approach to solve this deficiency of GISs would be to adopt an object-oriented database model [11], and several GISs based on an object-oriented or an object-relational database model (that adds object capabilities to relational databases) were proposed.

As an example, the publicly available freeware package PostGIS[1] "spatially enables" the PostgreSQL server, allowing it to be used as a spatial database for GISs. PostGIS, used both by the open source GRASS system (http://grass.itc.it/) and by the Open Geospatial Consortium GeoTools GIS[2], follows the "Simple feature specification for SQL" (which aims at becoming the standard extension of SQL to support storage, retrieval, query and update of geospatial features), proposed by the Open Geospatial Consortium.

However, while these systems show that the object-oriented approach is suitable for representing spatial references, they still keep together both the database level and the representation level.

The work which is most closely related to this paper is [10]. The authors propose a database model able to represent thematic maps: a map is a compound object, whose components can be in turn compound; an algebra for querying maps is provided. the model and algebra in [10]. Nevertheless, this

1 http://postgis.refractions.net/
2 http://docs.codehaus.org/display/GEOTOOLS/Tutorials

model presents some limitations: first of all, the components of a thematic map are homogeneous, i.e., they must belong to the same class of spatial entities; consequently, the model is unable to flexibly deal with maps composed of heterogeneous components, such as the case of technological infrastructures. Second, the query algebra deals with nesting mainly as far as the definition of aggregate operations over numerical attributes are concerned, while it seems not to consider the problem of flexibly dealing with spatial references. In formulating our proposal we had in mind the overcoming of these problems as it will be discussed in the next sections.

3 Spatial Querying in Current Spatial Databases

In order to represent a reality in a spatial database, it is necessary to undertake a three level abstraction process, as outlined in [2].

To clarify the dependency of the operations on the representation and structure of spatial entities, we illustrate and discuss a simple example of querying in a GIS based on the relational model [10]. We recall that most of the commercial GISs are still based on the relational model and on spatial extensions of the SQL language for querying the database.

First we consider the schema representing the administrative units in a Country consisting of Regions and Provinces. This schema can be implemented in the relational model by defining a relation for each entity.

```
Country(country_code, country_name)
Region(region_code, region_name, country_code)
Province(province_code, province_name, region_code, population,
        geometry)
```

Underlined attributes identify the primary key of relations. Attribute `geometry` in table `Province` is defined over the spatial data type `polygon`; it is the spatial reference of provinces.

Notice that, in the reported schema, the spatial reference appears only at the level of `Province`. This design choice avoids redundancy; however, to obtain the spatial reference for regions, it is necessary to explicitly perform the union of spatial references of provinces.

Let us now consider a highway network schema, which is a typical and well known application of GISs ([8]). The geometry of a highway, a polyline, does not appear in the `Highway` relation but it appears at the level of sections. Then, we define two relations: `Highway` and `Section`.

```
Highway(highway_code, highway_name, highway_type)
Section(section_code, section_name, number_of_lanes,
         city_start, city_end, highway_code, geometry)
```

Attribute `geometry` in table `Section` is of type `line`. Similarly to the previous case, to obtain the full spatial representation of highways, it is necessary to aggregate the spatial references of all the component sections.

Consider now the request for the highways going through Italian regions. We must specify the query reported in Figure 1.

```
SELECT h.highway_name, s.section_code, t.region_name,
    OverlapsLP(s.geometry, t.geometry) AS geometry
FROM Highway AS h INNER JOIN Section AS s
   ON h.highway_code=s.highway_code
   INNER JOIN
   (SELECT region_code, region_name,
            RegionUnion(p.geometry) AS geometry
   FROM Province AS p INNER JOIN Region AS r
            ON p.region_code=r.region_code
            INNER JOIN Country AS c
            ON r.country_code=c.country_code
   WHERE c.country_name='Italy'
   GROUP BY region_code, region_name) AS t
   ON OverlapsLP(s.geometry, t.geometry) IS NOT  NULL
```

Fig. 1. A Spatial SQL query in current GIS

Let us examine the nested query. It is necessary to obtain the spatial representation for Italian regions: in fact, the `GROUP BY` clause groups together provinces belonging to the same region. Then, for each region, it is possible to obtain the overall spatial representation by means of the aggregate spatial function *RegionUnion*. This function takes a set of *polygons* as arguments, and returns the *Region* that contains all the input polygons. Notice that, even though provinces are modeled as polygons, it is necessary to obtain geometrical regions, since islands are not geographically connected to the rest of their country. This shows that often queries inside GISs based on the relational model can be implemented only being aware of the model of data.

The external query combines the aggregated description of regions with highway sections if (join condition) spatial references of sections and regions overlap. Function `OverlapLP` returns the `line` that results by overlapping a line and a polygon, or the null value in case they do not overlap. This function is used both to join tuples in relation `s` (sections) with tuples in relation `r` (regions), and to produce attribute `geometry`. Thus, the query returns, for each administrative region, the sections of highways that traverse it.

Notice that to write the query, it is necessary to be strongly aware of how spatial objects are represented. Furthermore, spatial aggregations are not implicitly supported by the system, but the user is required to build complex queries (see the nested query) to obtain the desired aggregation level.

4 Model

The spatial database model that we propose borrows some notions on which it is founded from the object oriented database model. It is based on the notion of spatial entity, considered as a class of spatial elements sharing a common set of attributes. This definition is the same as that of class in the context of object oriented databases. Then, rivers, mountains, buildings correspond to spatial entities. In defining a spatial database schema based on our model, one should not be aware of the specific representation of spatial references of the entities. In effect, the spatial reference might be defined among spatial data types, e.g., point, line, polyline, polygon, region, etc. as it occurs in object oriented databases for basic attributes of entities defined among primitive types (integer, real etc.); nevertheless, different representation models can be adopted for physically representing spatial data types, such as the *spaghetti model*, the *network model*, and the *topological model* [8], for which the Open Geospatial Consortium defined several spatial operations.

In our proposal, the *Representation Level* deals with the actual representation of spatial references; the *Database Level* deals with elementary and complex entities, whose instances are associated with a spatial reference. A *Spatial Reference* is referenced by the database level by means of a unique identifier *ID*; it is responsibility of the Representation Level to ensure uniqueness of *ID*s and perform all operations and transformations over spatial references, such as union, intersection, and so on of points, polylines, regions, etc.

4.1 Spatial Entities and Spatial Grains

Let us define the *ingredients* of the database model.

Definition 1: Spatial Entity. A *Spatial Entity* describes a particular class of spatial elements; any two instances of the same spatial entity have the same set of properties (attributes).
Spatial entities can be *named* or *unnamed*; the former ones are uniquely identified by a *Name*, i.e. explicitly defined; the latter ones are not identified by a name and are derived and not explicitly defined. □

Definition 2: Spatial Grain. A *Spatial Grain* sg is an instance of a spatial entity as occurs in the object orirnted database model. It has both values for attributes and a value for the spatial reference. □

In practice, a spatial entity specifies the common characteristic of an homogeneous set of real spatial elements in the world, an intensional definition;

the real spatial elements are named spatial grains and constitute the extensional definition of the spatial entity. In other words (see next definitions) spatial entities define the schema for spatial grains. We hereafter distinguish between *Elementary* and *Complex* spatial entities.

Definition 3: **Elementary Spatial Entity.** An *Elementary Spatial Entity* ESE describes elementary (i.e. not compound) spatial grains. The schema of an elementary spatial entity ESE is the tuple

$$Schema(ESE) : (\mathcal{E} : string, \mathcal{A} : (P_1.A_1, \ldots, P_n.A_n))$$

where $\mathcal{E}$ is the (possibly missing) name of the spatial entity, $\mathcal{A}$ is the set of descriptive (non spatial) attributes; each attribute is denoted as $P_i.A_i$, where P_i is a prefix and A_i is the attribute name; with $Dom(P_i.A_i)$ we denote the domain of the attribute. □

In the following, given an elementary spatial entity E, we use the notations $E.\mathcal{A}$ and $Schema(E).\mathcal{A}$ to denote the list of attributes in its schema; the same notations hold for an entity name.

Note that attributes are identified as $P_i.A.i$, where P_i is a prefix. Although not essential, the prefix is very useful to semantically group attributes somehow correlated: two attributes with the same prefix describes two correlated properties, for instance, `River.ID` and `River.Name` denotes the identifier (`ID`) and the name of a river, respectively.

Definition 4: **Complex Spatial Entity.** A *Complex Spatial Entity* CSE describes spatial grains obtained as aggregation of other spatial grains. The schema of a complex spatial entity CSE is the tuple

$$Schema(CSE) : (\mathcal{E} : String, \mathcal{A} : (P_1.A_1, \ldots, P_n.A_n), \mathcal{CE} : \{N_1, \ldots N_m\})$$

where $\mathcal{E}$ is the (possibly missing) name of the spatial entity, $\mathcal{A}$ is the set of descriptive (non spatial) attributes, defined as in Definition 3. $\mathcal{CE}$ is a set of *entity names* N_j; it denotes that a spatial grain defined over a complex entity aggregates spatial grains defined over named entities in $\mathcal{CE}$. □

Notice that N_j can be either an elementary or complex spatial entity name. In the following, given a complex spatial entity E, we use the notations $E.\mathcal{A}$ and $Schema(E).\mathcal{A}$ to denote the list of attributes in the schema, while we use the notations $E.\mathcal{CE}$ and $Schema(E).\mathcal{CE}$ to denote the set of component entities in its schema; the same notations hold for an entity name.

Definition 5: **Elementary Spatial Grain.** An *Elementary Spatial Grain* esg defined over an elementary entity E is a tuple

$$esg : [\,\texttt{Values} : (v_1, \ldots, v_n),\ \texttt{Sref} : spatial\_reference\,]$$

where `Values` is the tuple with actual attribute values (with $v_i \in Dom(P_i.A.i) \cup \{null\}$ and $P_i.A_I \in E.\mathcal{A}$), while `Sref` is the spatial reference identifier (*ID*) of the grain. □

In the following, given an elementary spatial grain g, we adopt the notation g.`Values` and g.`Sref` to denote the grain's attribute value tuple and the spatial reference, respectively.

Definition 6: **Complex Spatial Grain.** A *Complex Spatial Grain* csg defined over a complex entity E is a tuple

$$esg : [\,\texttt{Values} : (v_1, \ldots, v_n),\ \texttt{Components} : \{g_k\}\,]$$

where `Values` is the tuple with actual attribute values (with $v_i \in Dom(P_i.A.i) \cup \{null\}$ and $P_i.A_I \in E.\mathcal{A}$).

`Components` is the set of spatial grains aggregated by the complex grain; for each $g_k \in$ `Components`, g_k is an instance of an entity $E_k \in E.\mathcal{CE}$. `Sref`, the spatial reference of the grain, is not explicitly specified, but is derived; it is defined as: `Sref` $= \cup_k\, g_k$.`Sref` for each $g_k \in$ `Components` (the union of spatial references is autonomously performed by the representation level). □

In the following, given a complex spatial grain g, we adopt the notation g.`Values` g.`Components` and g.`Sref` to denote the grain's attribute value tuple, set of aggregated grains and spatial reference, respectively.

Example 1: In our model, the concepts of *highway* and *section*, as well as the concepts of *country*, *region* and *province* can be modeled by the following spatial entities. Observe that we introduce an entity named `Connection`, which models connections between highways and external roads; consequently, a highway is composed by sections and connections (thus, a highway is representative of spatial entities composed of heterogeneous components).

Notice that the relationships between regions and provinces, and between countries and regions are not described by means of attributes, but in terms of composition: a `Region` is composed by `Provinces`, and a `Country` is composed by `Regions`.

($\mathcal{E}$: `Province`, $\mathcal{A}$: (`Province.code`, `Province.name`,
`Province.population`))

($\mathcal{E}$: `Region`, $\mathcal{A}$:(`Region.code`, `Region.name`), $\mathcal{CE}$: { `Province` })

($\mathcal{E}$: `Country`, $\mathcal{A}$: (`Country.code`, `Country.name`), $\mathcal{CE}$: { `Region` })

($\mathcal{E}$: `Section`, $\mathcal{A}$: (`Section.code`, `Section.name`, `Section.number_of_lanes`,
`Section.city_start`, `Section.city_end`))

($\mathcal{E}$: `Connection`, $\mathcal{A}$: (`Connection.code`, `Connection.name`))

($\mathcal{E}$: `Highway`, $\mathcal{A}$: (`Highway.code`, `Highway.name`, `Highway.type`),
$\mathcal{CE}$: { `Section`, `Connection` }) □

4.2 Spatial Database

Defined the concepts of spatial entity and spatial grain, we are now ready to define the concept of *Spatial Database*, starting from its basic element, the concept of *Container*.

Definition 7: **Spatial Grain Container.** A *Spatial Grain Container* is a set of spatial grains defined over a given spatial entity. In particular given a container *c*, with $Entity(c)$ we denote the spatial entity over which the container is defined, with $Schema(c) \equiv Schema(Entity(c))$ we denote the schema of the container (which is the schema of the entity).
The *Instance* of the container $Instance(c) \equiv \{g_i\}$ is a possibly empty set of grains g_i for which it must hold that g_i *is-instance-of* $Entity(c)$ (the grain is an instance of the entity over which the container is defined).
Similarly to entities, containers can be named or unnamed. □

Definition 8: **Spatial Database.** A *Spatial Database* SDB is defined by the tuple

$$SDB : (\mathcal{E}, \mathcal{C})$$

where $\mathcal{E}$ is a possibly empty set of *named spatial entities*, while $\mathcal{C}$ is a possibly empty set of *named spatial grain containers*. To be correct, a spatial database must meet the following condition:

$$\forall c \in \mathcal{C} \Rightarrow Entity(c) \in \mathcal{E}$$

i.e. each container in the spatial database must be defined over a named entity defined in the database.
$\mathcal{E}$ is also called the *schema* of the database, while $\mathcal{C}$ is also called the *instance* of the database (denoted as $Schema(SDB) \equiv \mathcal{E}$ and $Instance(SDB) \equiv \mathcal{C}$). □

Example 2: Suppose that the schema of the spatial database is constituted by entities defined in Example 1. The instance might contain two named containers: the first one, named `EU_Countries` represents countries in the European Union; the second one, named `EU_Highways`, represents highways in the EU.

$$\texttt{EU\_Countries} = \{\, g_{Italy} = [(\texttt{"ITA", "Italy"}), \{g_{Italy,1}, \ldots\}],$$
$$g_{France} = [(\texttt{"FR", "France"}), \{g_{France,1}, \ldots\}], \ldots\,\}$$

Grains $g_{Italy,i}$ describe Italian regions, which in turns aggregates grains describing provinces, as follows.

$$g_{Italy,1} = [(\texttt{"LOM", "Lombardia"}), \{g_{Lom,1}, \ldots\}]$$
$$g_{Lom,1} = [(\texttt{"MI", "Milano", xxx}), \texttt{Sref}_{\texttt{MI}}]$$
$$g_{Lom,2} = [(\texttt{"VA", "Varese", yyy}), \texttt{Sref}_{\texttt{VA}}]$$

If we consider container `EU_Highways`, its instance might be the following.

$\texttt{EU\_Highways} =$
$\{ g_{A1} = [(\texttt{"ITA-A1", "A1", "Pay Toll"}), \{g_{A1,s1}, \ldots, g_{A1,c1}, \ldots\}],$
$g_{A4} = [(\texttt{"ITA-A4", "A4", "Pay Toll"}), \{g_{A4,s1}, \ldots, g_{A4,c1}, \ldots\}],$
$g_{A3} = [(\texttt{"ITA-A3", "A3", "Free"}), \{g_{A3,s1}, \ldots, g_{A1,c1}, \ldots\}], \ldots \}$

Grains $g_{A1,si}$ (and similarly for other highways) describes highway sections and connections. Some of them might be the following.

$g_{A1,s1} = [(\texttt{"A1-01", "MI-BO", 4, "Milano", "Bologna"}), \texttt{Sref}_{\texttt{A1-01}}]$
$g_{A4,s1} = [(\texttt{"A4-01", "TO-MI", 3, "Torino", "Milano"}), \texttt{Sref}_{\texttt{A4-01}}]$
$g_{A4,s2} = [(\texttt{"A4-02", "Milano", 3, "West Milano", "East Milano"}),$
$\texttt{Sref}_{\texttt{A4-02}}]$
$g_{A4,s3} = [(\texttt{"A4-03", "MI-BG", 3, "Milano", "Bergamo"}), \texttt{Sref}_{\texttt{A4-03}}]$

The same is for some grains describing connections on highway `"A1"`.

$g_{A1,c1} = [(\texttt{"A1-Parma", "Parma"}), \texttt{Sref}_{\texttt{A1-Parma}}]$
$g_{A1,c2} = [(\texttt{"A1-Modena", "Modena"}), \texttt{Sref}_{\texttt{A1-Modena}}]$ □

5 Algebraic Operators

Once defined the database model, we are now ready to define an algebra to derive containers from containers. The operators that constitutes the algebra are all defined to be independent of the specific representation of grains and to meet the closure property. The operators are inspired to classical relational algebra operators.

5.1 Selection, Projection and Renaming

Let us start with the adaptation of the classical set of operators for selection,, projection and renaming. In particular, we define two distinct types of projection.

Definition 9: **Selection.** The selection operator σ selects a subset of spatial grains in a container. It is defined as follows.

$$\sigma_{pred} : c \longrightarrow c' \text{ (written } c' = \sigma_{pred} \ (\ c \ ))$$

where c and c' are spatial grain containers, while $pred$ is a boolean predicate over attributes in $Schema(c).\mathcal{A}$.

Schema. $Entity(c') \equiv Entity(c)$, both c and c' are defined over the same entity.

Instance. $Instance(c') = \{g_1, \ldots, g_q\}$ is the set of all $g_i \in Instance(c)$ such that $eval(pred, g_i.\texttt{Values}, Entity(c).\mathcal{A}) = true$ (where function $eval : pred, \langle v_1, \ldots, v_k \rangle, \langle P_1.A_1, \ldots, P_k.A_k \rangle \rightarrow \{true, false\}$ evaluates the predicate $pred$ over the values $\langle v_1, \ldots, v_k \rangle$ which instantiate attributes $\langle P_1.A_1, \ldots, P_k.A_k \rangle$). □

Definition 10: **Projection over Attributes** The projection over attributes π projects a container over a subset of its attributes. It is defined as follows.

$$\pi_{attlist} : c \rightarrow c' \text{ (written } c' = \pi_{attlist}\ (\ c\ ))$$

where c and c' are containers, while $attlist =< P_1.A_1, \ldots, P_l.A_l >$ (with $l > 0$) is a list of attributes (with $l \geq 0$) such that $\forall P_i.A_i \in attlist$ it is $P_i.A_i \in Entity(c).\mathcal{A}$.

Schema. c' is defined over a new unnamed spatial entity.
If $Entity(c)$ is an elementary spatial entity, $Entity(c')$ is an elementary spatial entity as well, and $Entity(c').\mathcal{A} = attlist.$
If $Entity(c)$ is a complex spatial entity, $Entity(c')$ is a complex spatial entity as well, with $Entity(c')).\mathcal{A} = attlist$ and $Entity(C').\mathcal{CE} = Entity(c).\mathcal{CE}$.
Instance. Given a grain g'_i, $g'_i \in Instance(c')$ if and only if $\exists g \in Instance(c)$, $\forall v'_k \in g'.\texttt{Values}$ it is $v'_k = v_j$ such that $v_j \in g.\texttt{Values}$ with $j = Pos(P_i.A_i, Schema(c).\mathcal{A})$ ($Pos : P.A, \langle P_1.A_1, \ldots, P_k.A_k \rangle \rightarrow$ *natural-number* is a function that returns the position of an attribute within a list of attributes); furthermore, if $Entity(c)$ and $Entity(c')$ are elementary entities, then it is $g'.\texttt{Sref} = g.\texttt{Sref}$, while if $Entity(c)$ and $Entity(c')$ are complex entities, it is $g'.\texttt{Components} = g.\texttt{Components}$. □

Definition 11: **Projection over Components** The projection over components $\overline{\pi}$ projects a container over grains in the `Components`, w.r.t. entity names. It is defined as follows.

$$\overline{\pi}_{entityset} : c \rightarrow c' \text{ (written } c' = \overline{\pi}_{entityset}\ (\ c\ ))$$

where c and c' are containers defined over complex entities, while $entityset = \{N_1, \ldots, N_l\}$ (with $l > 0$) is a set of entity names such that $\forall N_i \in entityset$ it is $N_i \in Entity(c).\mathcal{CE}$.

Schema. c' is defined over a new unnamed complex spatial entity.
$Entity(c').\mathcal{A} = Entity(c).\mathcal{A}$ and $Entity(c').\mathcal{CE} = entityset.$
Instance. Given a grain g'_i, $g'_i \in Instance(c')$ if and only if $\exists g \in Instance(c)$, such that $g'.\texttt{Values} = g.\texttt{Values}$ and $\forall \overline{g}_k \in g.\texttt{Components}$ it is $\overline{g}_k \in g'.\texttt{Components}$ if and only if $Schema(Entity(\overline{g}_k)).\mathcal{E} \in entityset.$ □

Definition 12: **Attribute Renaming** The operator for attribute renaming ρ renames attributes in the schema of a container. It is defined as follows.

$$\rho_{attlist \rightarrow attlist'} : c \rightarrow c' \text{ (written } c' = \rho_{attlist \rightarrow attlist'}\ (\ c\ ))$$

where c and c' are containers, while $attlist =< P_1.A_1, \ldots, P_l.A_l >$ and $attlist' =< P'_1.A'_1, \ldots, P'_l.A'_l >$ (with $l > 0$) are lists of attributes such that $\forall P_i.A_i \in attlist$ it is $P_i.A_i \in Schema(c).\mathcal{A}$.

Schema. $Entity(c')$ is a newly generated unnamed entity.
If $Entity(c)$ is an elementary spatial entity, $Entity(c')$ is an elementary spatial entity as well, and $Entity(c').\mathcal{A} = \overline{attlist}$).
If $Entity(c)$ is a complex spatial entity, $Entity(c')$ is a complex spatial entity as well, with $Entity(c').\mathcal{A} = \overline{attlist}$ and with $Entity(c').\mathcal{CE} = Entity(c).\mathcal{CE}$).
Let us define $\overline{attlist}$. For each attribute $\overline{P}_i.\overline{A}_i \in \overline{attlist}$ (with $len(\overline{attlist}) = l$), one of the two following situations must hold: $\overline{P}_i.\overline{A}_i = P_i.A_i$, with $P_i.A_i \in Schema(c).\mathcal{A}$, if $P_i.A_i \notin attlist$; $\overline{P}_i.\overline{A}_i = P'_j.A'_j$, with $j = Pos(P_i.A_i, attlist)$ and $P'_j.A'_j \in attlist'$, if $P_i.A_i \in attlist$ ($Pos : P.A, \langle P_1.A_1, \ldots, P_k.A_k \rangle \rightarrow$ *natural-number* is a function that returns the position of an attribute within a list of attributes).
Instance. $Instance(c') = Instance(c)$. □

Example 3: By means of the following expression, the user selects the Italian highway coded `"ITA-A4"`, but is interested only in the name and in highway sections (connections must be discarded); furthermore, she/he changes attribute name `Highway.type` into `HighWay.PayToll`.

4. $c' = \overline{\pi}_{\{\texttt{Section}\}}($
3. $\quad \pi_{(\texttt{Highway.name, Highway.PayToll})}($
2. $\quad\quad \rho_{(\texttt{Highway.type}) \rightarrow (\texttt{Highway.PayToll})}($
1. $\quad\quad\quad \sigma_{(\texttt{Highway.code="ITA-A4"})}(\texttt{EU\_Highways}))))$

First of all (line 1.), only the grain corresponding to the desired highway is selected. Then (line 2.)attribute `Highway.type` is renamed into `HighWay.PayToll`. Line 3. projects over attributes `Highway.name` and `Highway.PayToll`). Finally, line 4 projects the components over entity `Section`. We obtain the following container (with one single grain).

$$c' = \{\, g = [(\texttt{"A4", "Pay Toll"}), \{g_{A4,s1}, \cdots\}] \,\}$$

Here, we report the sequence of schemas produced by each line. The schema generated by line 4. is the schema of the resulting container. Notice that schemas in lines 2., 3., 4. are unnamed.

4. $(\mathcal{E} : , \mathcal{A} : (\texttt{Highway.name, Highway.PayToll}),$
   $\mathcal{CE} : \{\, \texttt{Section} \,\})$
3. $(\mathcal{E} : , \mathcal{A} : (\texttt{Highway.name, Highway.PayToll}),$
   $\mathcal{CE} : \{\, \texttt{Section, Connection} \,\})$
2. $(\mathcal{E} : , \mathcal{A} : (\texttt{Highway.code, Highway.name, Highway.PayToll}),$
   $\mathcal{CE} : \{\, \texttt{Section, Connection} \,\})$
1. $(\mathcal{E} : \texttt{Highway}, \mathcal{A} : (\texttt{Highway.code, Highway.name, Highway.type}),$
   $\mathcal{CE} : \{\, \texttt{Section, Connection} \,\})$ □

5.2 Un-nesting of Grains

An important issue concerning nested grains is to be able to un-nest them. We define a basic operator, named *Push-up* and three derived operators.

Definition 13: **Push-up of Grains** The push-up of grains χ extracts grains from within the `components` of grains in a grain container, composing the external and internal grains' attributes. It is defined as follows.

$$\uparrow: c \rightarrow c' \text{ (written } c' = \uparrow (c))$$

where c and c' are containers, with c defined over complex entities. Given c, either $\forall E \in Schema(c).\mathcal{CE}$ E must be an elementary entity, or $\forall E \in Entity(c).\mathcal{CE}$ E must be a complex entity. Furthermore, $\forall E \in Entity(c).\mathcal{CE}$, it must be $Entity(c).\mathcal{A} \cap E.\mathcal{A} = \emptyset$.

If $\forall E \in Entity(c).\mathcal{CE}$, E is an elementary entity, c' is as follows.
Schema. $Entity(c')$ is a new unnamed elementary entity.
Let us denote with $\mathcal{SCE} = Sort(Entity(c).\mathcal{CE}))$ the list of component entities sorted in alphabetical order;
$Entity(c').\mathcal{A} = Entity(c).\mathcal{A} \bullet (E_1.\mathcal{A}\ comp\ E_2.\mathcal{A}\ comp\ \ldots\ comp\ E_n.\mathcal{A})$ where $attlist_i\ comp\ attlist_{i+1} = attlist_i \bullet (attlist_{i+1} - attlist_i)$ (where $\bullet$ is the classical sequence composition operator).
Instance. Given a grain g_i', $g_i' \in Instance(c')$ if and only if $\exists g \in Instance(c)$ and $\exists \overline{g} \in g$.`Components` such that the following holds. $\forall v_i' \in g'$.`Values`, it is $v_i' = v_i$, with $v_i \in g$.`Values`, if $P_i.A_i \in Schema(c').\mathcal{A}$ and $P_i.A_i \in Schema(c).\mathcal{A}$; $v_i' = \overline{v}_j$, with $\overline{v}_j \in \overline{g}$.`Values` and $j = Pos(P_i.A_i, Schema(\overline{g}).\mathcal{A})$, if $P_i.A_i \in Entity(c').\mathcal{A}$ and $P_i.A_i \in Entity(\overline{g}).\mathcal{A}$ ($Pos : P.A, \langle P_1.A_1, \ldots, P_k.A_k\rangle \rightarrow$ *natural-number* is a function that returns the position of an attribute within a list of attributes); $v_i' = null$ if $P_i.A_i \in Entity(c').\mathcal{A}$ and $P_i.A_i \notin Entity(\overline{g}).\mathcal{A}$.

If $E \in Entity(c).\mathcal{CE}$ is a complex entity, c' is as follows.
Schema. $Entity(c')$ is a new unnamed complex entity.
$Entity(c').\mathcal{A}$ is defined as in the previous case.
$Entity(c').\mathcal{CE} = \cup_E Entity(E).\mathcal{CE}, \forall E \in Entity(c).\mathcal{CE}$.
Instance. Given a grain g_i', $g_i' \in Instance(c')$ if and only if $\exists g \in Instance(c)$ and $\exists \overline{g} \in g$.`Components` such that .`Values` is defined as in the previous case, while g'.`Components` $= \overline{g}$.`Components`. □

Example 4: Consider the following expression.

$$c' = \uparrow (\texttt{EU\_Highways})$$

The expression generates a new spatial grain container whose schema and instance are reported hereafter.

$Schema(c') = (\mathcal{E}:, \mathcal{A}:$ (`Highway.code`, `Highway.name`, `Highway.type`, `Connection.code`, `Connection.name`, `Section.code`, `Section.name`, `Section.number_of_lanes`))

$Instance(c') = \{ g_1 = [($"ITA-A1", "A1", "Pay Toll", null, null, "A1-01", "MI-BO", 4, "Milano", "Bologna"), $\texttt{Sref}_{\texttt{A1-01}}$],
$g_2 = [($"ITA-A1", "A1", "Pay Toll", "A1-Parma", "Parma" null, null, null, null, null), $\texttt{Sref}_{\texttt{A1-Parma}}$],
$g_3 = [($"ITA-A1", "A1", "Pay Toll", "A1-Modena", "Modena" null, null, null, null, null), $\texttt{Sref}_{\texttt{A1-Modena}}$],
$\ldots\}$

In the schema, notice that all attributes in the nested grains are pushed-up; attribute prefix is then useful to distinguish different attributes with the same name. In the instance, from the initial grain of highway "A1" we obtain three grains: g_1 describes a section, while g_2 and g_3 describe connection (obviously, the set is not completed). □

Now three derived operators are defined, called *selective push-up*, *extraction*, *selective extraction*; the later two unnest grains loosing external grains' attributes.

Definition 14: **Selective Push-Up.** The selective push-up of grains $\overline{\uparrow}$ extracts grains from within the `components` of grains in a grain container, in such a way only grains defined over a sub-set of component entities are pushed-up. It is defined as follows.

$$\overline{\uparrow}_{entityset} : c \rightarrow c' \text{ (written } c' = \overline{\uparrow}_{entityset} (c))$$

where c and c' are containers, with c defined over complex entities. Given either c, $\forall E \in (Entity(c).\mathcal{CE} \cap entityset)$ E must be an elementary entity, or $\forall E \in Entity(c).\mathcal{CE}$ E must be a complex entity. Furthermore, $\forall E \in Entity(c) - \mathcal{CE}$, it must be $Entity(c).\mathcal{A} \cap E.\mathcal{A} = \emptyset$.

$$c' = \overline{\uparrow}_{entityset}(c) = \uparrow (\overline{\pi}_{entityset}(c)).$$

□

In practice, this operator allows to select a subset of entities, in order to extract only grains defined over these entities.

Definition 15: **Extraction of Grains** The extraction of grains χ extracts grains from within the `components` of grains in a grain container, loosing external grains' attributes. It is defined as follows.

$$\chi : c \rightarrow c' \text{ (written } c' = \chi (c))$$

where c and c' are containers, with c defined over complex entities. Given either c, $\forall E \in Entity(c).\mathcal{CE}$ E must be an elementary entity, or $\forall E \in Entity(c).\mathcal{CE}$ E must be a complex entity.

$$c' = \chi(c) = \pi_{attlist}(\uparrow (c)).$$

where $attlist = Entity(\uparrow (c)).\mathcal{A} - Entity(c).\mathcal{A}$. □

Definition 16: **Selective Extraction.** The extraction of grains $\overline{\chi}_{entityset}$ extracts grains from within the `components` of grains in a grain container. It is defined as follows.

$$\overline{\chi}_{entityset} : c \rightarrow c' \text{ (written } c' = \overline{\chi}_{entityset} (c))$$

where c and c' are containers, with c defined over complex entities. Given c, either $\forall E \in (Entity(c).\mathcal{CE} \cap entityset)$ E must be an elementary entity, or $\forall E \in Entity(c).\mathcal{CE}$ E must be a complex entity.

$$c' = \overline{\chi}_{entityset}(c) \equiv \chi(\overline{\pi}_{entityset}(c)).$$

Since χ is in turn a derived operator, we obtain $\overline{\chi}_{entityset}(c) = \pi_{attlist}(\uparrow (\overline{\pi}_{entityset}(c)))$, where $attlist = Entity(\uparrow (c)).\mathcal{A} - Entity(c).\mathcal{A}$. □

In practice, this operator allows to select a subset of entities, in order to extract only grains defined over these entities.

Example 5: Let us continue Example 4. The expressions

$c_1 = \overline{\uparrow}_{\{\texttt{Section}\}}(\texttt{EU\_Highways})$,

$c_2 = \chi(\texttt{EU\_Highways})$, $c_3 = \overline{\chi}_{\{\texttt{Section}\}}(\texttt{EU\_Highways})$

generates three containers c_1, c_2 and c_3, hereafter reported. The reader can see that: the selective push-up operator extracts only grains defined over entity `Section` (attributes of entity `Connection` are not reported); the grain extraction and selective grain extraction operators behave similarly to push-up operators, but attributes of entity `Highway` are automatically discarded.

$Schema(c_1)$ = ($\mathcal{E}$:, $\mathcal{A}$: (`Highway.code`, `Highway.name`, `Highway.type`, `Section.code`, `Section.name`, `Section.number_of_lanes`))
$Instance(c')$ = { $g_{1,1}$ = [(`"ITA-A1"`, `"A1"`, `"Pay Toll"`, `"A1-01"`, `"MI-BO"`, `4`, `"Milano"`, `"Bologna"`), $\texttt{Sref}_{\texttt{A1-01}}$], ...}

$Schema(c_2)$ = ($\mathcal{E}$:, $\mathcal{A}$: (`Connection.code`, `Connection.name`, `Section.code`, `Section.name`, `Section.number_of_lanes`))
$Instance(c')$ = { $g_{2,1}$ = [(`null`, `null`, `"A1-01"`, `"MI-BO"`, `4`, `"Milano"`, `"Bologna"`), $\texttt{Sref}_{\texttt{A1-01}}$],
$g_{2,2}$ = [(`"A1-Parma"`, `"Parma"`, `null`, `null`, `null`, `null`, `null`), $\texttt{Sref}_{\texttt{A1-Parma}}$],
$g_{2,3}$ = [(`"A1-Modena"`, `"Modena"`, `null`, `null`, `null`, `null`, `null`), $\texttt{Sref}_{\texttt{A1-Modena}}$], ... }

$Schema(c_2)$ = ($\mathcal{E}$:, $\mathcal{A}$: (`Section.code`, `Section.name`, `Section.number_of_lanes`))
$Instance(c')$ = { $g_{3,1}$ = [(`"A1-01"`, `"MI-BO"`, `4`, `"Milano"`, `"Bologna"`), $\texttt{Sref}_{\texttt{A1-01}}$], ...} □

5.3 Overlay Operator

The *overlay* operator performs the operation also known as *spatial join*; it allows to combine containers based on the spatial intersection of grains.

Definition 17: **Overlay.** The *overlay* operator $\overline{\bowtie}$ combines two spatial grains in two containers, having a common spatial intersection. It is defined as follows.

$$\overline{\bowtie} : c_1, c_2 \rightarrow c' \text{ (written } c' = c_1 \overline{\bowtie} c_2)$$

where c_1 and c_2 are containers such that $Entity(c_1).\mathcal{A} \cap Entity(c_2).\mathcal{A} = \emptyset$

Schema. $Entity(c')$ is a new unnamed elementary entity.
It is $Entity(c').\mathcal{A} = Entity(c_1).\mathcal{A} \bullet Entity(c_2).\mathcal{A}$.
Instance. Given a grain g_1', $g_1' \in Instance(c')$ if and only if there exist two grains $g_1 \in Instance(c_1)$ and $g_2 \in Instance(c_2)$ such that $Overlap(g_1.\texttt{Sref}, g_2.\texttt{Sref}) = true$ ($Overlap$ is a boolean function that returns $true$ if two spatial references overlap).
For each $v_i' \in g'.\texttt{Values}$, the following two alternatives must hold: $v_i' = v_i$, with $v_i \in g_1.\texttt{Values}$, if $i \leq len(Entity(c_1).\mathcal{A})$; $v_i' = v_{(i-L)}$, with $v_{(i-L)} \in g_2.\texttt{Values}$, if $i > len(Entity(c_1).\mathcal{A})$ ($L = len(Entity(c_1).\mathcal{A})$).
$g'.\texttt{Sref} = Intersection(g_1.\texttt{Sref}, g_2.\texttt{Sref})$ (function $Intersection$ generates a new spatial reference corresponding to the intersection of the operand spatial references). □

Example 6: Consider the SQL query in Figure 1. With our operators, we obtain the following equivalent formulation, that extracts the sections of highways traversing Italian regions.

4. $c' = \pi_{(\texttt{Highway.name, Section.code, Region.name})}($
1. $\quad(\pi_{(\texttt{Highway.name, Section.code})}(\uparrow_{\{\texttt{Section}\}}(\texttt{EU\_Highways})))$
3. $\quad\overline{\bowtie}$
2. $\quad(\chi(\sigma_{(\texttt{Country.name = "Italy"})}(\texttt{EU\_Countries}))))$

The query is extremely simple. Line 1. pushes up grains defined over entity `Section` from within highways (see Example 5), and projects the resulting container only on attributes `Highway.name` and `Section.code`. Line 2. selects the grain describing Italy from within container `EU_Countries`, then extracts grains describing Italian regions. The containers (highway sections and Italian regions) are joint by line 3.: a grain in this new container describes a piece of highway section traversing a specific region; each grain has a new generated spatial reference that geographically describes the intersection of the original section with the region area. Observe that it is not necessary to specify how to compute the region area: this is automatically done by the system. Finally, Line 4. projects the container only on the desired attributes. A sketch of container c' is the following, where $\texttt{Sref}'_1$, $\texttt{Sref}'_2$ and $\texttt{Sref}'_3$ are new spatial references resulting from the intersection of joined grains.

$$
\begin{aligned}
Schema(c') &= (\mathcal{E} :, \mathcal{A} : (\,\texttt{Highway.name}, \texttt{Section.code}, \texttt{Region.name}\,))\\
Instance(c') &= \{\, g'_1 = [(\texttt{"A1"},\ \texttt{"A1-01"},\ \texttt{"Lombardia"}), \texttt{Sref}'_1],\\
&\quad\ g'_2 = [(\texttt{"A4"},\ \texttt{"A4-01"},\ \texttt{"Lombardia"}), \texttt{Sref}'_2],\\
&\quad\ g'_2 = [(\texttt{"A4"},\ \texttt{"A4-02"},\ \texttt{"Piemonte"}), \texttt{Sref}'_3], \ldots\}
\end{aligned}
$$

□

5.4 Set-Oriented Operators

Finally, we redefine the classical set-oriented operators.

Definition 18: **Union, Intersection, Difference.** The union, intersection and difference operators combines two containers having the same set of attributes. They are defined as follows.

$$\overline{\cup} : c_1, c_2 \rightarrow c' \qquad \overline{\cap} : c_1, c_2 \rightarrow c' \qquad \overline{/} : c_1, c_2 \rightarrow c'$$

where c_1 and c_2 are containers such that $Schema(c_1).\mathcal{A} = Schema(c_2).\mathcal{A}$

Schema. c' is a container of elementary grains, i.e.
$Schema(c') \equiv Schema(Entity(c')) \equiv (\mathcal{A} : (P_1.A_1, \ldots, P_n.A_n))$.
It is $Schema(c').\mathcal{A} = Schema(c_1).\mathcal{A}$.
Instance. In case of *union*, $Instance(c') = \phi(Instance(c_1) \cup Instance(c_2))$.
In case of *intersection*, $Instance(c') == \phi(\overline{I})$, where $\overline{I} = \{g_i\}$ such that either $g_i \in Instance(c_1)$ and $\exists\, g \in Instance(c_2)$ such that $g_i.\texttt{values} = g.\texttt{values}$, or $g_i \in Instance(c_2)$ and $\exists\, g \in Instance(c_1)$ such that $g_i.\texttt{values} = g.\texttt{values}$.
In case of *difference*, $Instance(c') = \phi(\overline{D})$, where $\overline{D} = \{g_i\}$ such that $g_i \in Instance(c_1)$ and $\nexists\, g \in Instance(c_2)$ such that $g_i.\texttt{values} = g.\texttt{values}$.
ϕ is a function that fuse into one single grain all grains with the same attribute values, i.e. given a set of grains $\{g_1, \ldots, g_k\}$ such that $g_1.\texttt{values} = \ldots = g_k.\texttt{values}$, they are fused into an elementary grain g such that $g.\texttt{values} = g_1.\texttt{values}$ and $g.\texttt{Sref} = fusion(g_1.\texttt{Sref}, \ldots, g_1.\texttt{Sref})$ (with $fusion$ we mean the spatial transformation that generates one single spatial reference obtained fusing the k spatial references). □

In practice, all grains having the same attribute values are fused together into a unique elementary grain; in particular, the union fuse all grains in both operands, the intersection fused together grains if and only if they are in both the operands.

The reader can easily verify that, based on the previous definition, the intersection can be derived as $c_1 \overline{\cap} c_2 = (c_1 \overline{\cup} c_2) \overline{/} [(c_1 \overline{/} c_2) \overline{\cup} (c_2 \overline{/} c_1))]$

Example 7: Suppose we have a database that describes in which European regions languages are spoken. By performing a few queries, we might obtain the following containers, one describing languages spoken in French and one languages spoken in Switzerland.

$\texttt{France} = \{\, g_{F,1} = [(\texttt{"French"}), \texttt{Sref}_1] \,\}$
$\texttt{Switzerland} = \{\, g_{S,1} = [(\texttt{"French"}), \texttt{Sref}_2], g_{S,2} = [(\texttt{"German"}), \texttt{Sref}_3],$
$g_{S,3} = [(\texttt{"Italian"}), \texttt{Sref}_4]\}$

By means of the set-oriented operators, the union obtains where French, German and Italian are spoken in France and Switzerland ($\texttt{Sref}'$ is the fusion of $\texttt{Sref}_1$ and $\texttt{Sref}_2$), the intersection obtains where French is spoken in both countries and the difference obtains where languages not spoken in France are spoken in Switzerland.

$\texttt{France} \,\overline{\cup}\, \texttt{Switzerland} =$
$\{\, g_{U,1} = [(\texttt{"French"}), \texttt{Sref}'], g_{S,2} = [(\texttt{"German"}), \texttt{Sref}_3],$
$g_{S,3} = [(\texttt{"Italian"}), \texttt{Sref}_4]\}$
$\texttt{France} \,\overline{\cap}\, \texttt{Switzerland} = \{\, g_{I,1} = [(\texttt{"French"}), \texttt{Sref}']\}$
$\texttt{Switzerland} \,\overline{/}\, \texttt{France} =$
$\{\, g_{S,2} = [(\texttt{"German"}), \texttt{Sref}_3], g_{S,3} = [(\texttt{"Italian"}), \texttt{Sref}_4]\}$ □

6 Conclusions and Future Work

The proposed database model is able to represent complex spatial entities, composed of possibly heterogeneous entities; this is the case of common environments, such as technological infrastructures. Nevertheless, the rigid separation between the database level, and the representation level (where actual spatial references are represented) allows us to define a query algebra that is independent of spatial reference representations, simple to use and able to flexibly deal with complex entities composed of heterogeneous components.

The algebra is defined in such a way it meets the closure property; furthermore, by writing simple expressions, it is possible to perform complex queries over the database. The algebra of the most common spatial operations have been also formalized so as to meet the closure property: in fact, all the operators works on spatial grain containers, thus remaining in the same data model.

Examples, discussed through the paper, show the use of the various spatial operators; furthermore, they show that the proposed algebra allows the specification of simple queries, even though complex entities are queried.

The next step of the work is the definition of the interface between the database level and the representation level, for which we plan to exploit a current spatial DBMS.

References

1. Batty P (1992) Exploiting relational database technology in a gis. Computers and Geosciences 18(4):453–462
2. Burrough PA, McDonnel RA (2000) Principles of Geographical Information Systems. Oxford
3. DeWitt DJ, Kabra N, Luo J, Patel JM, Yu JB (1994) Client-server paradise. In: Proc of the 20th VLDB Conf, Santiago, Chile, September
4. Hadzilacos T, Tryfona N (1997) An Extended Entity-Relation Model for Geographic Applications. SIGMOD Record 26(3):September
5. Hasan W, Heytens M, Kolovson C, Neimat MA, Potamianos S, Schneider D (1993) Papyrys gis demonstration. In: Proc ACM SIGMOD Conf on Management of Data, Washington DC, May. British Columbia, pp 207–216
6. Morehouse S (1992) The Arc/Info Geographic Information System. Computers and GeoSciences 18(4):435–441
7. Parent C, Spaccapietry S, Zimanyi E (1999) Spatio-temporal conceptual models: Data structures + spaces + time. ACM-GIS:26–33
8. Rigaux P, Scholl M, Voisard A (2002) Spatial Databases with application to GIS. Morgan Kaufmann
9. Tryfona N, Hadzilacos R (1998) Logical data modeling of spatiotemporal applications: Definitions and a model. IEEE IDEAS:14–23
10. Voisard A, David B (2002) A database perspective on geospatial data modeling. IEEE Transactions on Knowledge and Data Engineering 14(2):226–243
11. Worboys M (1995) GIS: a Computing Perspective. Taylor and Francis, London
12. Zhao D, Yu B, Randolph D, Hong B (2003) Relational geographic databases. In: IIIS SCI 2003: Systematics, Cybernetics, and Informatics, vol VI, pp 95–100

QACHE: Query Caching in Location-Based Services

Hui Ding[1], Aravind Yalamanchi[2], Ravi Kothuri[2], Siva Ravada[2], Peter Scheuermann[1]

[1] Department of EECS, Northwestern University, hdi117
email: peters@ece.northwestern.edu
[2] Oracle USA; email: [aravind.yalamanchi]@oracle.com

Abstract

Many emerging applications of location-based services continuously monitor a set of moving objects and answer queries pertaining to their locations. Query processing in such services is critical to ensure high performance of the system. Observing that one predominant cost in query processing is the frequent accesses to the database, in this paper we describe how to reduce the number of moving object to database server round-trips by caching query information on the application server tier. We propose a novel-caching framework, named QACHE, which stores and organizes spatially-relevant queries for selected moving objects. QACHE leverages the spatial indices and other algorithms in the database server for organizing and refreshing relevant cache entries within a configurable area of interest, referred to as the cache-footprint, around a moving object. QACHE contains appropriate refresh policies and prefetching algorithms for efficient cache-based evaluation of queries on moving objects. In experiments comparing QACHE to other proposed mechanisms, QACHE achieves a significant reduction (from 63% to $99%) in database roundtrips thereby improving the throughput of an LBS system.

Key words: location-based services, query processing caching

1 Introduction

Location-based services (LBS) [10] typically operate in a three-tier architecture: a central database server that stores past and current locations of all moving objects, applications that register to the database server their queries that are pertaining to the moving objects locations, and a set of moving objects that continuously change their locations (as shown in Fig. 1). As moving objects report their changing locations periodically, new answers are delivered to the applications when certain criteria are met. These queries on moving objects may contain predicates on the spatial locations as well as any other non-spatial attributes associated with the moving objects.

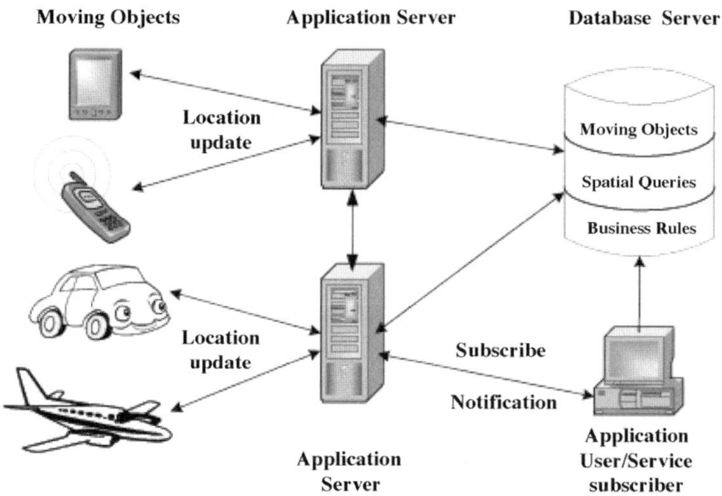

Fig. 1. Location-Based Services

Consider the following motivational scenario: a LBS system for local restaurant promotion sends appropriate restaurant information to nearby tourists. A registered restaurant specifies an area around its location using a spatial predicate (e.g., within-distance operator in commercial spatial databases: see [5] for more details) and restricts promotions only to tourists (identified by checking for "area_code != restaurant_area_code") who are interested in its specific type of food (specified by predicate "user_food_interest == Chinese"). [12] describes how to specify such queries in Oracle database. Upon location updates of all mobile users, the LBS system must quickly decide whether one (or more) user matches all query criteria of a registered restaurant so that the promotion message can be sent before he/she travels out of the target area.

A critical problem about answering such queries in LBS is that any delay of the query response may result in an obsolete answer, due to the dynamic nature of the moving objects (in our example, tourists). This requires highly efficient query evaluation. On the other hand, while moving objects frequently report their location updates to the database server, many of the updates do not result in any new query answer. Take the above scenario as an example, the service system receives location updates from all tourists once every minute; it is too expensive to evaluate all location updates against the query criteria of all registered restaurants in the database server. Yet it is not necessary to do so because each query includes both spatial criteria and non-spatial criteria [8, 12] and an answer update should be delivered *only if both criteria* are met, e.g., location updates of tourists preferring Indian cuisine need not be evaluated even if they are in the area of Chinatown; likewise, location updates of tourists that are too far away from Chinatown need not be evaluated even if they do like Chinese food. In summary, query evaluation against irrelevant updates should be avoided as much as possible to reduce database burden and average response time.

To improve the performance of LBS on the delivery of in-time query answers, we focus on reducing query evaluation cost by minimizing the number of database accesses and the amount of computation required during evaluation. One effective technique toward this goal is to cache relevant data for fast answer delivery. In a three-tier LBS system, caching can be achieved on any of the three tiers.

- **On the mobile devices of end users:** queries are assumed to be issued by mobile users asking about its vicinity; when a user issues a query, the received answers are stored and used for answering future queries since spatial queries issued by the same mobile user usually exhibit high spatial locality. Unfortunately, this approach can only be used to cache objects that are static. Moreover, it highly relies on the tight processing and storage ability of the mobile devices and thus is not widely applicable.

- **On the database server:** most frequently referenced data and most frequently executed query plans can be cached by the database server to improve the performance of query processing. However, this approach increases burden on the already heavily loaded database server with large volume of incoming location updates [13].

- **On the middle-tier, i.e., application server:** relevant data items can be stored in the application server that serves as an external cache. When location updates are received, the application server can frequently use the cached data to process the updates and respond to the application users efficiently; location updates that cannot be evaluated are forwarded to the database for further processing.

In this paper, we adopt the third approach because it has the following advantages: (1) caching on the application server does not rely on the limited processing and storage ability of end users and it does not impose additional burden on the database server; (2) the application server can effectively cache data coming from heterogeneous sources to a single application; (3) the application server can provide caching for each moving object and this granularity is usually desirable in LBS, because a moving object may frequently be monitored for a series of events; and (4) the application server can filter out many of the updates that will not result in any new query answer and thus avoid unnecessary database accesses.

We present QACHE, a dynamic query-caching framework on the application server in LBS. This framework builds and improves on existing research solutions based on safe distance [7]. The main goal of QACHE is to improve the system performance in spatial query monitoring. To achieve this goal, QACHE identifies the most relevant spatial queries for the moving objects (in the sense that the upcoming location updates may result in new answers to these queries), and cache information of these queries in the application server. QACHE has the following characteristics:

- The items cached are not the moving objects but are the pending spatial queries pertaining to the moving objects. Since moving objects update their locations frequently, caching their locations would involve frequent cache replacement and update, and introduce significant overhead. In contrast, pending spatial queries are relatively stable[1] and should be cached to improve query response time.
- The granularity of the cache is per moving object, i.e., session-wise. The cache entry for a moving object stores queries that are interested in the moving object and are close to its current location. In addition, different sessions can share queries in the cache to minimize the storage requirement.

[1] The pending spatial queries may also change due to insertion or deletion, or modification to the query patterns etc. However, these changes occur much less frequently than the location updates.

- For a given moving object, only those queries that match the non-spatial (static) predicates can be cached in the cache entry.
- The queries cached are carefully organized to support efficient access for query answer update. In the cases where database access is necessary after cache access, the number of disk accesses can still be reduced by using the information stored in the cache.
- Our cache is dynamically updated as moving objects change their locations, so that queries that become farther away from a moving object are removed from the cache to make space for queries that get in the vicinity of that object.
- We propose the concept of *cache-footprint* for a cache entry, which is configured in terms of the minimum time interval between consecutive updates of the cache entries. This is represented as a distance D_{max} from the location of the moving object based on its known maximum velocity ($D_{max} = refresh\_interval \times max\_velocity$). For a fixed size of the cache entry, QACHE employs a two-pronged approach of storing the closest queries in *true detail* and the rest of the queries in cache-footprint region as approximations. The queries in true detail provide exact answers for a moving object whereas the approximated query regions reduce the false-positives. This two-level filtering improves the cache-effectiveness thereby increasing the throughput of the LBS system.

The rest of this paper is organized as follows. Section 2 presents the related work. Section 3 describes the main components of QACHE and Section 4 elaborates on its implementation details. Section 5 describes our experimental evaluation results. Finally, Section 6 concludes the paper.

2 Related Work

Various techniques have been proposed to efficiently process spatial queries in LBS. The main approaches can be categorized as follows: (1) reducing the amount of computation when location updates are received by grouping pending queries using grid or similar indexing structures and conducting spatial join between moving objects and pending queries [6]; (2) reducing the number of queries performed by introducing safe distance/region for moving objects [7]; and (3) reducing the number of disk access by building a query index for all pending queries [7]. Unfortunately, the above techniques either focus on optimizing the performance within the database and hence fail to make use of the processing and storage power provided by the middle-tier, or have certain constraints on realistic applications. For example, many of the frequent location updates from the moving objects will not generate any new query answer and it is thus unnecessary to evaluate the pending queries against these updates.

Caching has been extensively studied in the area of operating systems, web information retrieval and content delivery networks. For example, the middle-tier data caching products developed by Oracle [3] was designed to prevent the database from being a bottleneck in content delivery networks. The main idea is to cache data outside of the database server to reduce database access load. QACHE differs from the traditional caches in that the items cached are queries instead of data. More importantly, the cached data is carefully organized for efficient access to minimize overhead. Recently, caching has also been applied to the area of mobile computing. The prevailing approach is to cache received answers at the client side for answering future queries. A Furthest Away Replacement (FAR) cache replacement policy was proposed in [9] where the victim is the answering object furthest away from the moving object's current location. Proactive caching for spatial queries [4] extends the caching granularity to per query object level. The compact R-tree presented in this work facilitates query processing when the cached item cannot answer the query. We use a similar approach in QACHE that treats both the database (query) index and the query data as objects for caching and manage them together to reduce the cache miss penalty.

3 Overview of QACHE

In this section we first briefly state the assumptions held when building the QACHE framework and provide an overview of the architecture and main components of QACHE. We then describe how QACHE handles location updates and maintains correct query answers.

3.1 Assumptions

The basic assumptions of QACHE are as follows:

1. Moving objects have the ability to determine their current location through GPS device. They also have the ability to communicate with the server periodically to report their location updates.
2. The only constraint on the motion of moving objects is that they are subject to a maximum speed.

3. All moving objects report their location updates to the server synchronously. Please note that this assumption simplifies our simulation and performance analysis, but is not necessary for QACHE to function correctly.
4. The queries stored in the database are indexed using spatial indices, such as the R-tree [1, 5].

3.2 System Architecture

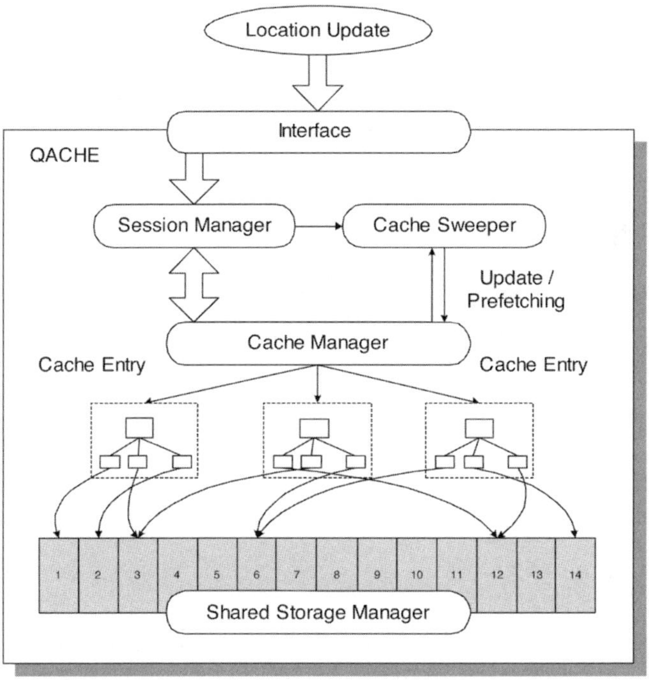

Fig. 2. System Architecture

As illustrated in Figure 2, QACHE has five main components: an interface that accepts location updates from moving objects, a *session manager* that manages the safe distance for each connected session (moving object), a *cache manager* that manages the cached contents for selected sessions, a *shared storage manager* that actually stores the spatial queries loaded from the database, and *a cache sweeper* that evicts invalid entries and prefetches new entries into the cache.

- **Session manager:** The session manager maintains a *look-up directory* that keeps, for each moving object, current location, a reference location called the base location, the safe distance from the reference location which is defined as the distance to the closest query region [7] and meta information about the corresponding cache-entry (such as cache-footprint if relevant). This look-up directory is indexed for efficient access. For objects that do not match any non-spatial predicates in any query the safe distance is set to infinity. The session manager could also maintain a location-logger that records all location updates and has them flush back to the database periodically.
- **Cache manager:** For each selected moving object, its cache entry consists of all relevant query regions in true detail/approximation. The cache manager manages such entries for moving objects whose safe distance does not exceed the cache-footprint. Due to memory constraints, the cache manager may create cache entries only for a subset of such moving objects based on their probability of being relevant to a query as described in Section 4.
- **Shared storage manager:** While the cache manager maintains cache entries for selected moving objects on a per session basis, it does not store the actual cached queries. Instead, all cached queries are managed by the shared storage manager to avoid duplication and thus save memory space. This is because a single query may be interested in multiple moving objects and hence may be cached more than once in QACHE. When a cache entry is accessed from the cache manager, a pointer is provided to visit the shared storage manager where the actual query is stored.
- **Cache sweeper:** The purpose of the cache sweeper is to refresh cache entries, evict invalidated cache entries and prefetch new entries that are not currently in the cache manager. Cache sweeper may refresh a cache entry for a moving object as it approaches boundary of the cache-footprint (refer to Section 4.2) and prefetch those prospective queries into QACHE. The refreshed/prefetched cache entry will center on the latest location of the moving object, i.e., within D_{max} distance from the latest location. Note that although prefetching introduces extra accesses to the database server, the operation is performed asynchronously thus the disk access is not on the critical path for query evaluation. Instead, when the prefetched queries do need to be evaluated against the next location update, no database access is necessary because those queries are already in QACHE thanks to prefetching. The cache sweeper can be implemented as background process that operates cooperatively with the cache manager.

3.3 Processing Location Updates

Figure 3 illustrated how QACHE handles location updates. When a location update from a moving object is received, the session manager first examines its look-up directory and checks whether the moving object is a new session. If so, the moving object is registered to the session manager, and the location and maximum speed of this moving object are used to query the database server for query evaluation and safe distance calculation. The safe distance calculated is then inserted into the look-up directory for future updates. If the calculated safe distance is less than the cache-footprint for the moving object, the corresponding cache-entry is created and inserted into the cache manager. On the other hand, if the location update is from an existing session, the session manager first examines its lookup directory and checks whether the moving object is still in its safe distance. If so, nothing needs to be done. Otherwise, the corresponding cache-entry is accessed to decide if this moving object has entered any query region. Note the cache entry has query regions in true detail or in approximate form.

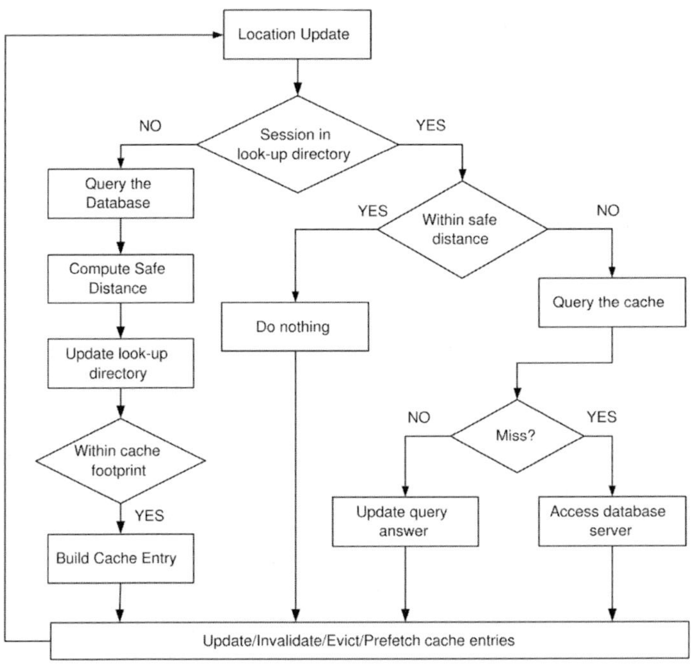

Fig. 3. Handling Location Updates in QACHE

For all true-detail query regions that the moving object matches, the query results are propagated to the application. For the matching approximate query regions, additional processing is performed in the database tier. This database processing is also required when a cache entry is missing (due to memory constraints, or invalidation by the cache sweeper).

In summary, when a new query is registered to the system, it is initially stored in the database and evaluated against all moving objects in the look-up directory of the session manager. A cache entry may be created, or an old cache entry may be replaced by the cache sweeper.

4 Design and Implementation of QACHE

This section elaborates on the design and implementation of three key components of QACHE, i.e., session manager, cache manager, shared storage manager. We describe: (1) how session manager maintains the safe distance for each moving object; (2) how cache manager selects moving objects and maintains a cache entry for each selected object to support efficient evaluation on location updates; and (3) how cached items are managed by shared storage manager and shared across selected moving objects to avoid duplication.

4.1 Maintaining the Safe Distance

The safe distance is the minimum distance within which a moving object will not enter any query region. Location updates of a moving object that are not beyond the safe distance need not be evaluated against any query, which indicates that the safe distance can serve as a filter in query processing.

When a moving object first registers to the application server, an initial safe distance is calculated for it by performing a nearest neighbor search on queries from the database server; the safe distance is then stored in the look-up directory of session manager. When a cache entry is created for this moving object, depending on the cache replacement policy such as LRU, the new safe distance must be recalculated and updated by the cache sweeper that mediate between the session manager and the database.

4.2 Building a cache entry

For each moving object, its corresponding cache entry (if presents) stores selected queries that are interested in the object. The selection of queries is decided by: the *QACHE refresh period* (QRP), i.e., the time interval between two consecutive cache updates, the maximum speed of the moving object V_{max} and the cache entry size B, i.e., the maximum number of items that can be stored in each cache entry.

QACHE attempts to cache queries within the cache-footprint of the moving object. Cache-footprint is described by a maximum distance D_{max} (see Eq. 1):

$$D_{max} = V_{max} \times QRP \qquad (1)$$

Ideally, any query within distance D_{max} to the moving object should be cached since the moving object is very likely to enter the query region before the next cache refreshing. However, if the number of such queries exceeds the maximum size B of each cache entry, QACHE can't possibly cache all queries in full detail and has to aggregate some of them. Based on our assumption 4 in Section 3.1, queries are indexed using an R-tree in the database and hence the internal nodes of the R-tree can be used as an approximation of query aggregation.

As a consequence, each cache entry with a capacity of B stores two categories of items: (1) query regions that are stored in true detail: any moving object that satisfies such query regions is a *true-positive* match. A hit on this cached item indicates the moving object is a query answer; (2) query regions that are stored using *approximations*: any moving object that satisfies any such query approximations could be a *false-positive*. Additional processing needs to be done for such queries in the database. Moving objects not intersecting either category of regions is a *true-negative* and no further processing is required. This multi-category-based filtering serves as the backbone for the performance of QACHE in pending query evaluation.

To efficiently process location updates, QACHE organizes the cached items of each cache entry using an in-memory R-tree, i.e., the content of each cache entry is the internal nodes of the R-tree, while the actual cached items are managed by the share storage manager (please refer to Section 4.3). The algorithm used in QACHE for the construction of a cache entry is presented below. The algorithm starts by descending the query-index tree in the database from root and recursively explores child nodes that may contain eligible objects. A *priority queue* stores all nodes that are within distance $D_{\max}$ of the moving object. When a node is met, its children are enqueued; when a query object is met, it is added to a *query list* given that the non-spatial criteria of the query are also satisfied. This process terminates when the total size of the priority queue and the query list reaches the cache entry capacity B, or when the priority queue becomes empty. The query list stores all queries that are explicitly cached and the priority queue stores all cached nodes that aggregate the rest of eligible queries.

Algorithm 1 Create a cache entry

Input: Query-index tree in the database **R**, location $\{x,y\}$, maximum speed V_{max}, QRP,cache entry size B

Output: Cache tree R

```
Priority queue Q ⇐ ∅, query list L ⇐ ∅
D_max ⇐ V_max × QRP
Enqueue(root(R))
while Q not empty AND (Q.size() + L.size()) < B do
    e ⇐ dequeue(Q)
    if e is a leaf node of R then
        c ⇐ e's closest child to {x, y}
        e ⇐ e − c, // remove the child from the node
        if distance(c, {x,y}) ≤ D_max AND Expression(c) evaluates to 'true' then
            Enqueue(c)
        end if
        if e not empty AND distance(e, {x,y}) ≤ D_max then
            Enqueue(e)
        end if
    else if e is an internal node of R then
        c ⇐ e's closest child to {x, y}
        e ⇐ e − c, // remove the child from the node
        if distance(c, {x,y}) ≤ D_max then
            Enqueue(c)
        end if
        if e not empty AND distance(e, {x,y}) ≤ D_max then
            Enqueue(e)
        end if
    else {// e is a qualifying query object}
        Add e to LC
    end if
end while
Create R-tree R from objects in Q and L
return R
```

For example, in Figure 4, O is the current location of a moving object for which a cache entry is to be constructed. I is the root of the database R-tree with three children: I_1, I_2, and I_3. The circle illustrates the region that is within distance D_{max} to the moving object; queries that intersect this region should be explicitly or implicitly cached. Suppose that the cache entry size B is set to five. I is first dequeued, it's three children are then examined. Only I_1 and I_2 are enqueued because they are within D_{max} (Step 2). I_1 is then dequeued and its three children are added to the query list (Step 3, 4, 5) So far, four items are cached: Q_1, Q_2, Q_3 in the query list and I_2 in the priority queue. Subsequently I_2 is dequeued; its closest child Q_5 is added to the query list, while $Q_4$and $Q_6$are re-aggregated to a new node which is put back to the priority queue (Step 6). At this time we have exactly five items in total: Q_1, Q_2, Q_3 and Q_5 in the query list and $Q_4 + Q_6$ in the priority queue. These five items are then used to build a in-memory R-tree for the cache entry

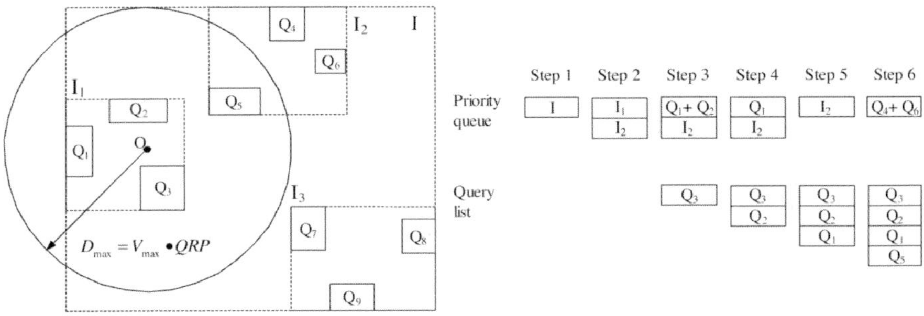

Fig. 4. Example: Create a Cache Entry

4.3 Sharing Cache Contents Among Sessions

One novelty of QACHE is its session-wise granularity. When a location update is received, it need not be evaluated against all queries in the cache because queries that are interested in this particular moving object are already selected into its own cache entry. Unlike conventional approach, this prevents the non-spatial predicate of a query to be evaluated every time: the non-spatial predicate is evaluated exactly once when the cache entry is create, while the spatial predicate may be evaluated on every subsequent location update.

However, this session-wise granularity has its own deficiency: potential waste of memory space. A query may be interested in multiple moving objects, and hence may be cached in multiple cache entries. To solve this problem, we implemented a shared storage manager that actually holds the data cached in memory. Each cache entry only stores pointers to the corresponding slots in the shared storage manager. This guarantees that only one copy of each query/node is kept in memory at any time.

The shared storage manager is implemented as a hash table that is a tuple of index, data, and a reference counter. When a query or an intermediate node is selected for caching, only its index is stored in the cache entry. The actual data, i.e., the geometry of a query or the minimum bounding box (MBB) of an intermediate node, will be stored in an entry in the storage manager based on the index. During a query evaluation, the storage manager identifies the location of the data using a hash function and the ID of the query/node as a hash key. When the storage manager receives a request for a data insert, it first checks whether the data already exists. If so, the storage manager increases the reference counter by one; otherwise, a new entry is created. When a cache entry is evicted, all queries/nodes cached will have their reference counter decreased by one. When a counter becomes zero, the actual data can be safely removed from the shared storage manager.

5 Performance Evaluation

We have built a simulation environment for QACHE with the *Java* programming language. We compare QACHE with two other approaches: (1) the naive approach where location updates are directly sent to the database server and evaluated every time; (2) the *safe distance approach* (SD) where only safe distance is used to reduce number of query evaluation. We examined the number of disk accesses to the database (R-tree) as well as storage requirement of each approach. With the experimental data, we also analyzed the processing time of different approaches to demonstrate the efficiency of QACHE.

5.1 Simulation Setup

Using our own data generator modified from the GSTD tool [11], a data set is generated that simulates a mobile environment where N objects moves following the *Random Waypoint Model* [2], a well accepted model in the mobile computing community. Each object starts at a randomly selected location in the region of [0...1, 0...1], moves for a period randomly generated between [0, *QRP*] at a speed randomly selected between [0, *QRP*], and sends its new location to the application server at time *QRP*; after this the same process repeats. When an object hits the boundary, its moving direction is adjusted to guarantee constant number of moving objects in the simulation space. The query workload contains 1000 queries that are evenly distributed in the simulation space; currently only static range queries are considered.

In our simulation, new location updates from all N objects are collected at the same time and processed before the next round of location updates arrives. Our simulation processes 5000 rounds of location updates. All experiments were performed on a 3.0 GHz Pentium 4, 1 GB memory workstation running Windows XP SP2.

5.2 Disk Access and Memory Requirement

We conducted three sets of experiments where the number of moving objects ($N_{m.o.}$) grows from 1000 to 10000. In each set, we varied the number of cache entries ($N_{c.e.}$) from 5% to 20% of ($N_{m.o.}$). The cache entry capacity B, i.e., the number of cached items in each entry, is set to 10. A fixed number of queries (1000) are organized in the database server as an R-tree, the size of which is 640KB excluding the non-spatial predicates. For the three approaches (in short, naive, SD, and QACHE), we collected the expected number of disk page accesses ($E(dpa)$) to the database index R-tree on every round of location updates. We also recorded the memory requirement and the cache hit ratio when applicable. The performance of QACHE and the other two approaches are presented in Table 1.

Table 1. Disk access and memory requirement of the three different approaches

$N_{m.o.}$		1000			5000			10000		
$N_{c.e.}$		50	100	200	250	500	1000	500	1000	2000
$E(dpa)$	naive	3626	3626	3626	18255	18255	18255	36191	36191	36191
	SD	374	374	374	2051	2051	2051	4009	4009	4009
	QACHE	137	49	2	754	193	7	1310	353	15
Cache hit ratio	naive	-	-	-	-	-	-	-	-	-
	SD	-	-	-	-	-	-	-	-	-
	QACHE	56%	85%	99%	54%	88%	99%	56%	89%	99%
Memory requirement (Byte)	naive	-	-	-	-	-	-	-	-	-
	SD	4000	4000	4000	20000	20000	20000	40000	40000	40000
	QACHE	11283	18035	24200	52074	73773	93805	93850	124740	150021

Compared to the safe distance approach, QACHE reduces by $E(dpa)$ at least 63%. In each set of experiments, $E(dpa)$ for the other two approaches remains constant for a given number of moving objects, but decreases significantly for QACHE when the number of cache entries is decreased. When $N_{c.e.}$ is 20% of $N_{m.o.}$, the expected disk page accesses is almost negligible. This is because almost all query evaluation can be completed by QACHE and only a few disk page accesses are generated from false-positive hits in the cache.

Another major observation from Table 1 is that QACHE is scalable in terms of memory storage requirement. We recorded the total number of bytes required by the look-up directory, cache manager and the shared storage manager; the results indicate that the total memory requirement does not grow in proportion to the number of moving objects. Moreover, considering the total size of query R-tree in the database, QACHE is highly efficient in utilizing memory space and providing a high hit ratio.

5.3 Processing Time

While the number of disk accesses is an important criteria when evaluating the effectiveness of QACHE, a quantitative analysis is necessary to decide the exact performance improvement. In this section we demonstrate the overall speed up that QACHE can achieve in query evaluation over the naive approach and the safe distance approach. In our analysis, the following terms are frequently used: (1) disk page access time T_{disk}; (2) memory access time T_{mem}; (3) query evaluation time T_{eval}; and (4) the height of the query R-tree in the database H_Q. For simplicity, we assume that an access to the query R-tree in the disk reads $0.75xH_Q$ disk pages. We also assume that the cache entry R-tree has a fan out of 2, thus the in-memory cache R-tree has a height of $\log_2 B$. The average response time to a location update can be calculated as follows:

- Naive approach:

$$T_{naive} = 0.75 \times H_Q \times (T_{eval} + T_{disk}) \tag{2}$$

- Safe distance approach: assuming that in each round of location updates, 10% are beyond the safe distance so that database accesses are required, the average response time is:

$$T_{sd} = T_{mem} + 0.1 \times 0.75 \times H_Q \times (T_{eval} + T_{disk}) \tag{3}$$

- QACHE: assuming that $N_{c.e.}$ is 20% of $N_{m.o.}$, then only 0.2% of the location updates will result in database access (see Table 1), the average response time is:

$$T_{qache} = T_{mem} + 0.75 \times \log_2 B \times (T_{eval} + T_{disk}) + 0.002 \times 0.75 \times H_Q \times (T_{eval} + T_{disk}) \quad (4)$$

Based on a reasonable estimation of the relative parameters presented in Table 2, QACHE achieves a 498 times speed up over the naive approach and a 50 times speed up over the safe distance approach.

Table 2. Estimations of the required time for each operation

$T_{mem}(ns)$	$T_{eval}(ns)$	$T_{disk}(ns)$	H_Q	B
100	100	5000000	10	10

6 Conclusions

We have described and evaluated QACHE, a novel query caching framework for LBS systems. By caching spatial queries for appropriate moving objects on the application tier, a significant amount of database accesses can be eliminated, resulting in a dramatic performance improvement of LBS. We examined several important implementation issues and proposed effective solutions to them for QACHE to be deployed in real LBS systems. We compared QACHE with existing solutions based only on safe distance. Our simulation results indicate that with the cache capacity 20% of total number of moving objects, and the memory requirement ranging from 3% to 20% of the query R-tree size in database (depending on the number of moving objects), QACHE is capable of eliminating 99% of the disk accesses. On real LBS systems, this memory requirement is totally affordable. Further more, our quantitative analysis shows that QACHE achieves a 50 times speed up over the safe distance approach and a 498 times speed up over the naive approach where all location updates are directly processed in the database.

References

1. Beckmann N, Kriegel H-P, Schneider R, Seeger B (1990) The R*-Tree: An Efficient and Robust Access Method for Points and Rectangles. In: SIGMOD Conf, pp 322–331
2. Broch J, Maltz DA, Johnson DB, Hu Y-C, Jetcheva J (1998) A performance comparison of multi-hop wireless ad hoc network routing protocols. Mobile Computing and Networking:85–97
3. Greenwald R, Stackowiak R, Stern J (2001) Oracle Essentials. O'Reilly &Associates Inc., CA
4. Hu H, Xu J, Wong WS, Zheng B, Lee DL, Lee WC (2005) Proactive caching for spatial queries in mobile environments. In: ICDE, pp 403–414
5. Kanth KVR, Ravada S, Sharma J, Banerjee J (1999) Indexing medium-dimensionality data in oracle. In: SIGMOD Conf, pp 521–522
6. Mokbel MF, Xiong X, Aref WG (2004) SINA: Scalable incremental processing of continuous queries in spatio-temporal databases. In: SIGMOD Conf, pp 623–634
7. Prabhakar S, Xia Y, Kalashnikov DV, Aref WG, Hambrusch SE (2002) Query indexing and velocity constrained indexing: Scalable techniques for continuous queries on moving objects. IEEE Trans Computers 51(10):1124–1140
8. Kothuri R, Beinat EGA (2004) Pro Oracle Spatial. Apress
9. Ren Q, Dunham MH (2000) Using semantic caching to manage location dependent data in mobile computing. In: MOBICOM:210–221
10. Schiller J, Voisard A (2004) Location-Based Services. Morgan Kaufmann Publishers, CA
11. Theodoridis Y, Silva JRO, Nascimento MA (1999) On the generation of spatiotemporal datasets. In: SSD, pp 147–164
12. Yalamanchi A, Kanth Kothuri VR, Ravada S (2005) Spatial Expressions and Rules for Location-based Services in Oracle. IEEE Data Eng Bull 28(3): 27–34
13. Yalamanchi A, Srinivasan J, Gawlick D (2003) Managing expressions as data in relational database systems. In: CIDR

A Voronoi-Based Map Algebra

Hugo Ledoux, Christopher Gold

GIS Research Centre, School of Computing, University of Glamorgan
Pontypridd CF37 1DL, Wales, UK
email: hledoux@glam.ac.uk; christophergold@voronoi.com

Abstract

Although the map algebra framework is very popular within the GIS community for modelling fields, the fact that it is solely based on raster structures has been severely criticised. Instead of representing fields with a regular tessellation, we propose in this paper using the Voronoi diagram (VD), and argue that it has many advantages over other tessellations. We also present a variant of map algebra where all the operations are performed directly on VDs. Our solution is valid in two and three dimensions, and permits us to circumvent the gridding and resampling processes that must be performed with map algebra.

1 Introduction and Related Work

The representation and modelling of geographical data can be done with two contrasting approaches: the object and the field models (Peuquet 1984; Couclelis 1992; Goodchild 1992). The former model considers the space as being 'empty' and populated with discrete entities (e.g. a house or a road) embedded in space and having their own properties. The latter model considers the space as being continuous, and every location in space has a certain property (there is *something* at every location). The property can be considered as an attribute of the location in space, and the spatial variation of an attribute over a certain spatial extent is referred to as a *field*. This is used to represent continuous phenomena such as the ambient temperature or the humidity of the soil.

To store object-based models in a geographical information system (GIS), a variety of data structures with different properties have been developed and implemented. For instance, several GISs explicitly store the adjacency relationships between objects [e.g. TIGER (Boudriault 1987) and ARC/-INFO (Morehouse 1985)], while some others use non-topological structures (e.g. the so-called spaghetti model) and reconstruct on-the-fly the spatial relationships when needed (Theobald 2001). By contrast, within the GIS community, field models are more or less synonymous with raster structures (Goodchild 1992), i.e. a regular tessellation of the plane into squares (pixels) such that each pixel contains the value of the attribute studied. The tools implemented in most GISs to model and analyse different fields are based on the *map algebra*, which is a framework developed for the analysis of fields stored in a raster format (Tomlin 1983). With this approach, each field is represented by a grid, and a set of primitive GIS operations on and between fields can be used and combined together to extract information and produce new fields. The framework, and its different operations, are further described in Section 3.

Since its conception, several weaknesses and shortcomings of map algebra have been discussed, and many have proposed improvements. Caldwell (2000) introduces a new operator to extend the spatial analysis capabilities, and Eastman et al. (1995) do the same to help the decision-making process. Takeyama (1996) proposes Geo-Algebra, a mathematical formalisation and generalisation of map algebra that integrates the concepts of *templates* and cellular automata under the same framework. The templates, developed for *image algebra* (Ritter et al. 1990), extends the concept of neighbourhood of a location, and the addition of cellular automata permits us to model geographic processes. Pullar (2001) also uses the idea of templates and shows how they can help to solve several practical GIS-related problems. As explained in Section 3, the fact that map algebra was developed for raster structures is problematic, firstly because of the dangers of using pixels for analysis (Fisher 1997), and secondly because complete representations (as in a complete grid) are rarely found in GIS applications, unless datasets come from photogrammetry or remote sensing. Indeed, it is usually impossible to measure geographic phenomena everywhere, and we have to resort to collect samples at some finite locations and reconstruct fields from these samples. Thus, a raster structure implies that some sort of manipulations have already been performed on a field. Kemp (1993) states that "map algebra requires us to enforce a structure on reality rather than allowing reality to suggest a more appropriate structure for our analysis", and shows that alternative representations (e.g. a triangulated irregular network (TIN), or contour lines; the possible representations are listed in Section 2) are a viable solution. She proposes to have operations – similar to map algebra's – for modelling fields, which are not all stored under the same representation. She therefore defines a set of rules to convert the different types of fields to other ones when binary operations are applied. For example, if two fields, one stored as a TIN and the other as contour lines, are analysed then the contours must first be converted to TIN before any manipulation is done. Haklay (2004), also to avoid the drawbacks of raster structures, proposes a system where only the data points (samples) and the spatial interpolation function used to reconstruct the field are stored. Each field is thus defined mathematically, which permits us to manipulate different fields in a formulaic form.

It should be noticed that the concept of field also generalises to three dimensions, for the modelling of such phenomena as the salinity of water bodies or the percentage of gold in the rock. Mennis et al. (2005) have recently extended map algebra to three dimensions, the tessellation they use is regular (the pixels become cubes called *voxels*) and the operations are straightforward generalisations of their two-dimensional counterparts.

As an alternative to using raster structures and to converting back and forth between different representations of a field, we propose in this paper representing fields with the Voronoi diagram (VD), i.e. a tessellation of space into 'proximity' regions. As explained in Section 4, the VD provides a natural way to represent continuous phenomena, and its properties are valid in any dimensions, which makes it ideal for modelling d-dimensional fields. Our proposition is similar to Haklay's (Haklay 2004) – get rid of raster and keep only the samples! – but we argue that the VD has many advantages over other tessellations. We also introduce in Section 5 a variant of the map algebra framework where every field and every operation is based on the VD. Perhaps the main contribution of this paper is that the framework is valid in any dimensions. However, since most GIS-related applications are concerned with two and three dimensions, the description and examples will focus on these two cases.

2 Fields

A field is a model of the spatial variation of an attribute a over a spatial domain, and it can be represented by a function mapping the location to the value of a, thus

$$a = f(location). \tag{1}$$

The function can theoretically have any number of independent variables (i.e. the spatial domain can have any dimensions), but in the context of geographical data it is usually bivariate (x, y) or trivariate (x, y, z). The domain can also incorporate time as a new dimension, and *dynamic* fields, such that $a = f(location, time)$, are thus obtained (Kemp 1993). This notion is useful for modelling phenomena in oceanography or meteorology that continually change over time. Also, notice that in the case of modelling the elevation of a terrain, the function is bivariate as the elevation is assumed to be a property of the surface of the Earth, and no cliffs or overfolds are allowed (as in a so-called 2.5D GIS).

Since fields are continuous functions, they must be *discretised* – broken into finite parts – to be represented in computers. The space covered by a field can be partitioned, or tessellated, into *regular* or *irregular* regions. In a regular tessellation, all the regions will be of the same shape and size, while in an irregular one, elements of any shape and size are allowed. In the plane, each region is a polygon, while in three dimensions it is a polyhedron. Regular tessellations arbitrarily divide the space, while irregular tessellations follow the outline of the data points (the samples that were collected to study the field), albeit this is not a requirement. Subdividing the space based on the samples has the main advantage of producing a tessellation that is adaptive to the sample distribution and to the complexity of the phenomenon studied. It also permits us to preserve the samples, which are the only "ground truth" of the field studied, and have even been referred to as the *meta-field* (Kemp and Vckovski 1998). Converting scattered samples to a grid means that the original data are 'lost'.

Once the space is tessellated, the field function becomes a *piecewise function*: to each region is assigned a function describing the spatial variation in its interior. As Goodchild (1992) points out, this function can be constant, linear, or of a higher order. A constant function means that the value of the attribute is constant within one region. An example of the use of a linear function is a TIN: the spatial variation within each region (a triangle) is described by the linear function (a plane) defined by the three vertices (usually samples) lifted to their respective elevation. Akima (1978) shows the advantages of using higher order functions in each region of a TIN – the main one being that the slope of the terrain is continuous everywhere. For the two-dimensional case, some other representations have also been mentioned and used, notably contour lines and irregularly spaced points (the samples to which attributes are attached). In our opinion, the latter representation is incomplete if the spatial function used to reconstruct the field is not explicitly defined, and therefore should not be considered a valid representation of a field.

While the dependent variable a in the function representing a field can theoretically be a vector (mostly used in physics to model for instance the magnetic field), we assume in this paper that it is always a scalar. Depending on the scale of measurement used for the values of the attribute, different types of fields are possible:

Continuous scale: the value of an attribute can have any value. Temperature, precipitation or salinity are examples because they can be measured precisely. The *interval* and *ratio* scales commonly used in GIS, as defined by Stevens (1946), fall into this category. We refer to this type of field as a *continuous field.*

Discrete scale: the values of an attribute are simply labels. Stevens's *nominal* and *ordinal* scales fall into this category. Nominal values are meaningless: an example is a map of Europe where each location contains the name of the country. Ordinal values are labels that can be ordered, e.g. a certain region can be categorised according to its suitability to agriculture from 1 to 5: 1 being poor, and 5 very good. We refer to this type of field as a *discrete field.*

Notice that here the terms "continuous" and "discrete" refer to the scale of measurement, and not to the spatial continuity of a field. Indeed, both types of fields are spatially continuous, as they are represented by a function. It is also important to notice that not all operations are possible on both types of fields. While many arithmetic operations (addition, subtraction, multiplication, etc.) are possible on continuous fields, they are meaningless for discrete fields.

3 Map Algebra

Map algebra refers to the framework, first developed and formalised by Tomlin (1983), to model and manipulate fields stored in a raster structure. It is called an algebra because each field (also called a map) is treated as a variable, and complex operations on fields are formed by a sequence of primitive operations, like in an equation (Berry 1993). A map algebra operation always takes a field (or many fields) as input and returns a new field as output (the values of the new field are computed location by location). Operations can be unary (input is a single field), binary (two fields) or n-ary (n fields); because n-ary operations can be obtained with a series of binary operations we describe here only the unary and binary cases. Tomlin (1983) describes three categories of operations:

Local operation: (see Fig. 1a) the value of the new field at location x is based on the value(s) of the input field(s) at location x. An unary example is the conversion of a field representing the elevation of a terrain from feet to meters. For the binary case, the operation is based on the

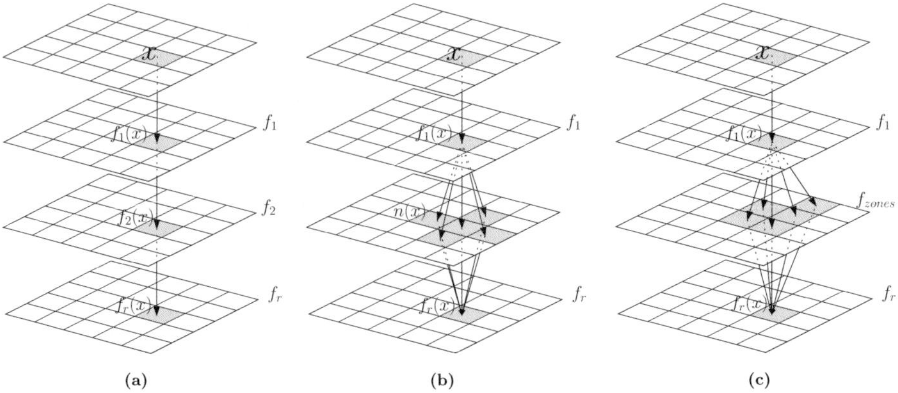

Fig. 1. The map algebra operations with a raster structure. **(a)** A binary local operation. **(b)** An unary focal operation. **(c)** A zonal operation that uses a set of zones (f_{zones}) stored as a grid

overlay in GIS: the two fields f_1 and f_2 are superimposed, and the result field f_r is pointwise constructed. Its value at location x, defined $f_r(x)$, is based on both $f_1(x)$ and $f_2(x)$. An example is when the maximum, the average or the sum of the values at each location x is sought.

Focal operation: (see Fig. 1b) the value of the new field at location x is computed as a function of the values in the input field(s) in the neighbourhood of x. As Worboys and Duckham (2004) describe, the neighbourhood function $n(x)$ at location x associates with each x a set of locations that are "near" to x. The function $n(x)$ can be based on distance and/or direction, and in the case of raster it is usually the four or eight adjacent pixels. An unary example is the derivation of a field representing the slope of a terrain, from an elevation field.

Zonal operation: (see Fig. 1c) given a field f_1 and a set of *zones*, a zonal operation creates a new field f_r for which every location x summarises or aggregates the values in f_1 that are in a given zone. The set of zones is usually also represented as a field, and a zone is a collection of locations that have the same value (e.g. in a grid file, all the adjacent cells having the same attribute). For example, given a field representing the temperature of a given day across Europe and a map of all the countries (each country is a zone), a zonal operation constructs a new field such that each location contains the average temperature for the country.

Although the operations are arguably simple, the combination of many makes map algebra a rather powerful tool. It is indeed being used in many commercial GISs, albeit with slight variations in the implementations and the user interfaces (Bruns and Egenhofer 1997). It should be noticed that the three categories of operations as not restricted to the plane, and are valid in any dimensions (Mennis et al. (2005) have recently implemented them with a voxel structure). Despite its popularity, the biggest handicap to the use of map algebra is arguably that is was developed for regular tessellations only, although the concepts are theoretically valid with any tessellation of space (Takeyama 1996; Worboys and Duckham 2004). Using raster structures has many drawbacks. Firstly, the use of pixels as the main element for storing and analysing geographical data has been criticised (Fisher (1997) summarises the issues). The problems most often cited are: (1) the meaning of a grid is unclear (are the values at the centre of each pixel, or at the intersections of grid lines?), (2) the size of a grid (if a fine resolution is wished, then the size of a grid can become huge), (3) the fact that the space is arbitrarily tessellated without taking into consideration the objects embedded in that space. Secondly, in order to perform binary operations, the two grids must "correspond", i.e. that the spatial extent, the resolution and the orientation of the two grids must be the same, so that when they are overlaid each pixel corresponds to one and only one pixel in the other grid. If the grids do not correspond, then *resampling* of one grid (or both) is needed. This involves the interpolation of values at regularly distributed locations with different methods such as nearest neighbour or bilinear interpolation, and each resampling degrades the information represented by the grid (Gold and Edwards 1992). Thirdly, unless a grid comes from a sensor (remote sensing or photogrammetry), we can assume that it was constructed from a set of samples. Converting samples to grids is dangerous because the original samples, which could be meaningful points such as the summits, valleys or ridges or a terrain, are not present in the resulting grid. Also, when a user only has access to a grid, he often does not know how it was constructed and what interpolation method was used, unless meta-data are available.

4 Voronoi Diagrams

Let S be a set of n points in a d-dimensional Euclidean space $\mathbb{R}^d$. The Voronoi cell of a point $p \in S$, defined $\mathcal{V}_p$, is the set of points $x \in \mathbb{R}^d$ that are closer to p than to any other point in S. The union of the Voronoi cells of all generating points $p \in S$ form the Voronoi diagram of S, defined VD(S). In two dimensions, $\mathcal{V}_p$ is a convex polygon (see Fig. 2a), and in 3D it is a convex polyhedron (see Fig. 2b). It is relatively easy to implement algorithms to construct a VD in two dimensions (Fortune 1987; Guibas and Stolfi 1985) and to delete a single point from one (Devillers 2002). In three dimensions, the algorithms are more complex but still implementable and efficient. The

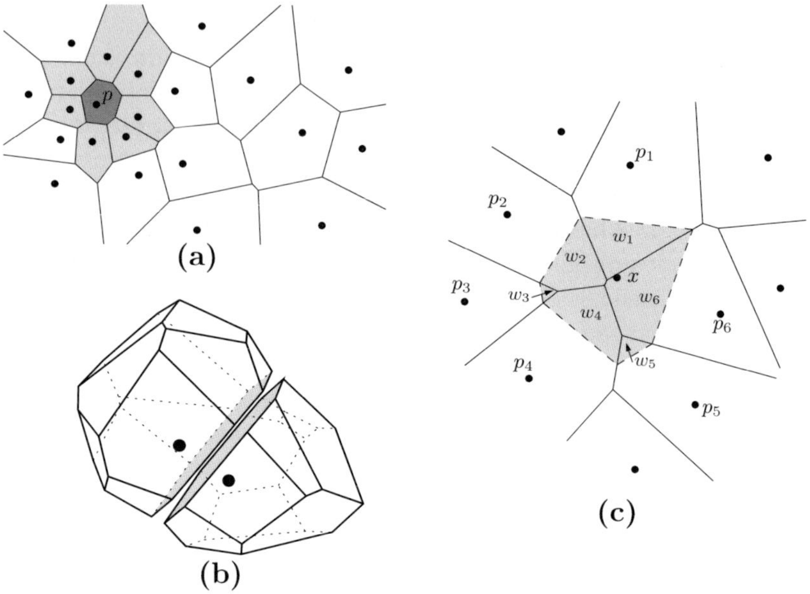

Fig. 2. **(a)** The VD for a set of points in the plane. **(b)** Two Voronoi cells adjacent to each other in 3D (they share the grey face). **(c)** The insertion of point x in a VD creates a new Voronoi cell that steals area to its 'would be' natural neighbours

most popular algorithms to construct a 3D VD are incremental (Edelsbrunner and Shah 1996; Watson 1981), which means that a VD is constructed by adding every point one by one. The deletion of a point is also possible in three dimensions, and it is a local operation (Ledoux et al. 2005). All these algorithms exploit the fact that the VD is the dual structure of the Delaunay triangulation – the knowledge of one structure implies the knowledge of the other – and perform their operations on the dual Delaunay triangulation because it is simpler to manipulate triangles/tetrahedra over arbitrary polygons/polyhedra.

Since most fields in geography must first be sampled to be studied, we argue in this paper that the Voronoi tessellation has many advantages over other tessellations for representing fields. First, it gives a unique and consistent definition of the spatial relationships between unconnected points (the samples). As every point is mapped in a one-to-one way to a Voronoi cell, the relationships are based on the relations of adjacency between the cells. For example in Figure 2a, the point p has seven neighbours (the lighter grey cells). Note that the points generating these cells are called the *natural neighbours* of the point p because they are the points that are naturally both close to p and 'around' p (Sibson 1981). This is particularly interesting for Earth sciences because the datasets collected often have highly anisotropic distribution, especially three-dimensional datasets in oceanography or geology because they are respectively gathered from water columns and boreholes (data are therefore usually abundant vertically but sparse horizontally). Second, the size and the shape of Voronoi cells is determined by the distribution of the samples of the phenomenon studied, thus the VD adapts to the distribution of points. Observe in Figure 2a that where the data distribution is dense the cells are smaller. Third, the properties of the VD are valid in any dimensions. Fourth, it is *dynamically* modifiable, which permits us to reconstruct the field function, and to add or delete samples at will.

If a constant function is assigned to each Voronoi cell, the VD permits us to elegantly represent *discrete fields*. To know the value of a given attribute at a location x, one simply has to find the cell containing x – Mücke et al. (1999) describe an efficient way to achieving that. To reconstruct a *continuous field* from a set of samples, more elaborate techniques are needed since the VD creates discontinuities at the border of each cell. The process by which the values at unsampled locations are estimated is called interpolation, and many methods have been developed over the years. An interesting one in our case is the natural neighbour interpolation method (Sibson 1981), because it has been shown by different researchers to have many advantages over other methods when the distribution of samples is highly anisotropic and is an automatic method that does not require user-defined parameters (Gold 1989; Sambridge et al. 1995; Watson 1992). This is a method entirely based on the VD for both selecting the samples involved in the interpolation process, and to assign them a weight (an importance). It uses two VDs: one for the set of samples, and another one where a point x is inserted at the interpolation location. The method is based on the area (or volume in three dimensions) that a new point inserted at the interpolation location x 'steals' from some of the Voronoi cells already present, as shown in Figure 2c. The resulting function is exact (the samples are honoured), and also smooth and continuous everywhere except at the samples themselves. See Gold (1989) and Watson (1992) for further discussion of the properties of the method, and Ledoux and Gold (2004) for a description of an algorithm to implement it in any dimensions.

5 A Voronoi-based Map Algebra

With a Voronoi-based map algebra, each field is represented by the Voronoi diagram of the samples that were collected to study the field. This eliminates the need to first convert to grids all the datasets involved in an operation

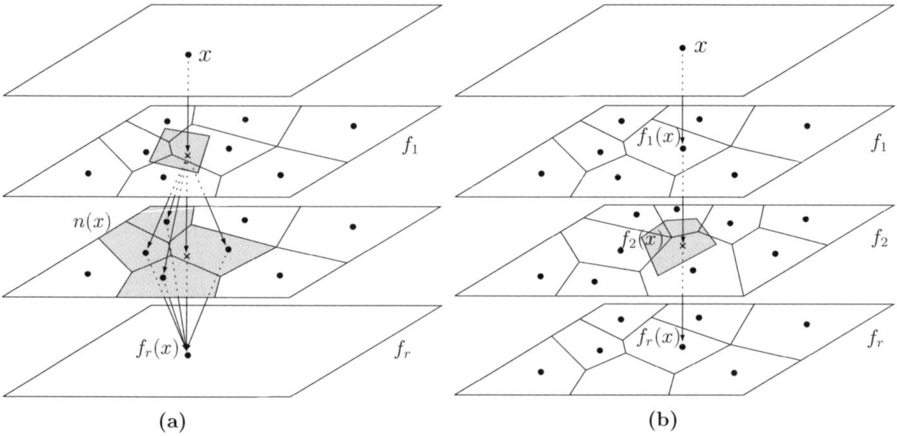

Fig. 3. Two Voronoi-based map algebra operations. The top layer represents the spatial extent of the fields, and x is a location for which the value in the resulting field f_r (bottom layer) is sought. **(a)** A unary focal operation performed on the field f_1. The third layer represents the neighbourhood function $n(x)$. **(b)** A binary local operation performed on the fields f_1 and f_2

(and moreover to grids that have the same orientation and resolution), as the VD can be used directly to reconstruct the fields. The permanent storage of fields is also simplified because only the samples need to be stored in a database, and the VD can be computed efficiently on-the-fly and stored in memory (problems with huge raster files, especially in three dimensions, are thus avoided).

When a field is represented by the VD, unary operations are simple and robust. To obtain the value of the attribute at location x (for a local operation), the two interpolation methods described in the previous section for discrete and continuous fields can be used directly. Also, the neighbouring function needed for focal operations is simply the natural neighbours of every location x, as defined in the previous section. Figure 3a shows a focal operation performed on a field f_1. Since at location x there are no samples, a data point is temporarily inserted in the VD to extract the natural neighbours of x (the generators of the shaded cells). The result, $f_r(x)$, is for example the average of the values of the samples; notice that the value at location x is involved in the process and can be obtained easily with natural neighbour interpolation.

Although Kemp (1993) claims that "in order to manipulate two fields simultaneously (as in addition or multiplication), the locations for which there are simple finite numbers representing the value of the field must correspond", we argue that there is no need for two VDs to correspond in order to perform a binary operation because the value at any locations can be obtained readily with interpolation functions. Moreover, since the VD is rotationally invariant (like a vector map), we are relieved from the burden of resampling datasets to be able to perform operations on them.

When performing a binary operation, if the two VDs do not correspond – and in practice they rarely will do! – the trickiest part is to decide where the 'output' data points will be located. Let two fields f_1 and f_2 be involved in one operation, then several options are possible. First, the output data points can be located at the sampled locations of f_1, or f_2, or even both. An example where the output data points have the same locations as the samples in f_1 is shown in Figure 3b. Since there are no samples at location x in f_2, the value is estimated with natural neighbour interpolation. The result, $f_r(x)$, could for example be the average of the two values $f_1(x)$ and $f_2(x)$. It is also possible to randomly generate a 'normal' distribution of data points in space (e.g. a Poisson distribution) and to output these. But one should keep in mind that in many applications the samples can be meaningful, and we therefore recommend to always keep the original samples and if needed to densify them by randomly adding some data points. The VD also permits us to vary the distribution of data points across space, for example having more data points where the terrain is rugged, and less for flat areas.

As with the other map algebra operations, a zonal operation must also output a field because its result might be used subsequently as the input in another operation. With a Voronoi-based map algebra, the output has to be a VD, and the major difficulty in this case it that we must find a VD that conforms (or approximates) the set of zones. Since zones come from many sources, different cases will arise. The first example is a remote sensing image that was classified into several groups (e.g. land use). Such a dataset can easily be converted to a VD: simply construct the VD of the centre of every pixel. Although this results in a huge VD, it can easily be simplified by deleting all data points whose natural neighbours have the same value. Notice in Figure 4 that the deletion of a single point is a local operation, and the adjacent cells will simply merge and fill up the space taken by the original cell. The second example is with *in situ* data, for instance in oceanography a dataset indicating the presence (or not) of fish in a water body. The VD of such a dataset can obviously be used directly. The third example is a set of arbitrary zones, such as a vector map of Europe. In two dimensions, it is possible to approximate the zones with a VD (Suzuki and Iri 1986), but the algorithm is complex and the results are not always satisfactory. A simpler option is to define a set of "fringe" points on each side of a line segment, and label each point with the value associated to the zone. Gold et al. (1996) show that the boundaries can be reconstructed/approximated automatically

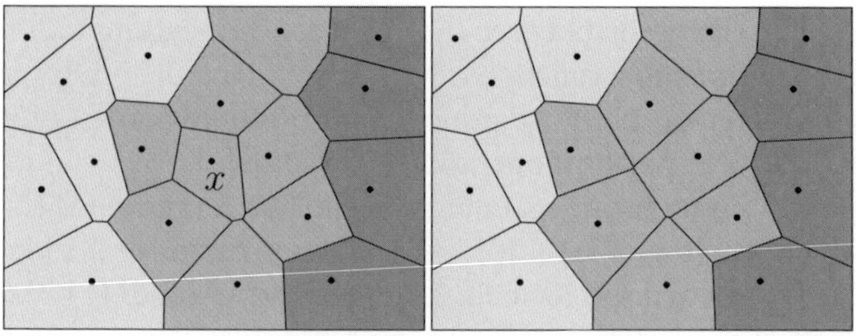

Fig. 4. Simplification of a discrete field represented with the VD. The data point x is completely surrounded by data points having the same value (here the value is defined by the colour), and deleting it does not change the field

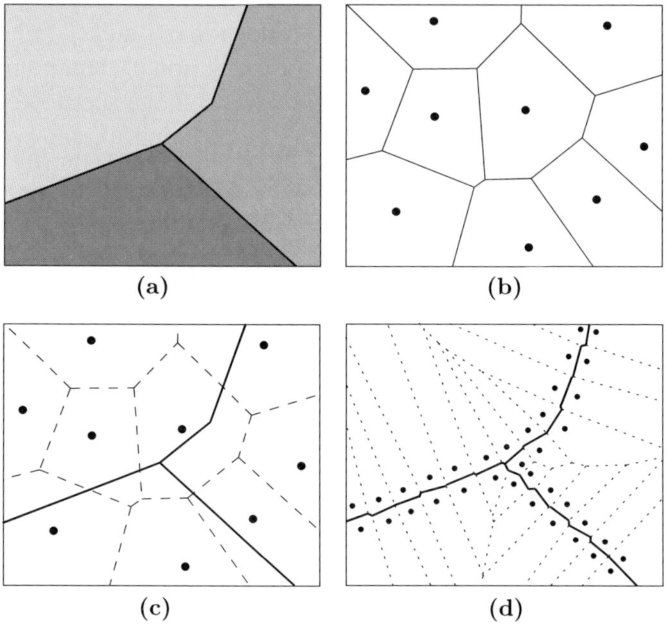

Fig. 5. (a) A vector map of three zones. **(b)** A continuous field represented with the VD. **(c)** When overlaid, notice many Voronoi cells overlap the zones. **(d)** Approximation of the borders of the zones with the VD

with Voronoi edges. An example is shown in Figure 5: a set of three zones appears in Figure 5a, and in Figure 5d the Voronoi edges for which the values on the left and right are different are used to approximate the boundaries of the zones. Since each location x in the output field of a zonal operation summarises the values of a field in a given zone, we must make sure that the locations used for the operation are sufficient and distributed all over the zone. Let us go back to the example of the temperature across Europe to find the average in each country. Figure 5a shows a vector map with three countries, and the temperature field f_1 is represented by a VD in Figure 5b. Notice that when the two datasets are overlaid (see Fig. 5c), many Voronoi cells cover more than one zone. Thus, simply using the original samples (with a point-in-polygon operation) will clearly yield inaccurate results. The output field f_r, which would contain the average temperature for each country, must be a VD, and it can be created with the fringe method (see Fig. 5d). Because the value assigned to each data points correspond to the temperature for the whole zone, we suggest estimating, with the natural neighbour interpolation, the value at many randomly distributed locations all over each zone.

6 Discussion

The wide popularity of map algebra is probably due to its simplicity: simple operations performed on a simple data structure that is easy to store and manipulate. Unfortunately this simplicity has a hefty price. Tomlin's map algebra forces an unnatural discretisation of continuous phenomena and implies a fair amount of preprocessing of datasets, which is usually hidden to the user. As stated in Gold and Edwards (1992), continual reuse and resampling of gridded datasets produce massive degradation of the information conveyed by the data, and can lead to errors and misinterpretations in the analysis.

As we have demonstrated in this paper, the tessellation of the space with the Voronoi diagram has many advantages for modelling fields, and a Voronoi-based map algebra permits us to circumvent the gridding and resampling processes when we want to manipulate several fields. Although the algorithms to manipulate VDs are admittedly more complex than the ones for raster structures, they are readily available and efficient, and that for the two- and three-dimensional cases.

Acknowledgments

We would like to thank the financial support of the EU Marie Curie Chair in GIS at the University of Glamorgan.

References

Akima H (1978) A method of bivariate interpolation and smooth surface fitting for irregularly distributed data points. ACM Transactions on Mathematical Software, 4(2):148–159.

Berry JK (1993) Cartographic Modeling: The Analytical Capabilities of GIS. In M Goodchild, B Parks, and L Steyaert, editors, *Environmental Modeling with GIS*, chapter 7, pages 58–74. Oxford University Press, New York.

Boudriault G (1987) Topology in the TIGER File. In *Proceedings 8th International Symposium on Computer Assisted Cartography*. Baltimore, USA.

Bruns HT and Egenhofer M (1997) Use Interfaces for Map Algebra. Journal of the Urban and Regional Information Systems Association, 9(1):44–54.

Caldwell DR (2000) Extending Map Algebra with Flag Operators. In *Proceedings 5th International Conference on GeoComputation*. University of Greenwich, UK.

Couclelis H (1992) People Manipulate Objects (but Cultivate Fields): Beyond the Raster-Vector Debate in GIS. In A Frank, I Campari, and U Formentini, editors, *Theories and Methods of Spatio-Temporal Reasoning in Geographic Space*, volume 639 of *LNCS*, pages 65–77. Springer-Verlag.

Devillers O (2002) On Deletion in Delaunay Triangulations. International Journal of Computational Geometry and Applications, 12(3):193–205.

Eastman J, Jin W, Kyem A, and Toledano J (1995) Raster procedures for multi-criteria/multi-objective decisions. Photogrammetric Engineering & Remote Sensing, 61(5):539–547.

Edelsbrunner H and Shah NR (1996) Incremental Topological Flipping Works for Regular Triangulations. Algorithmica, 15:223–241.

Fisher PF (1997) The Pixel: A Snare and a Delusion. International Journal of Remote Sensing, 18(3):679–685.

Fortune S (1987) A Sweepline algorithm for Voronoi diagrams. Algorithmica, 2:153–174.

Gold CM (1989) Surface Interpolation, spatial adjacency and GIS. In J Raper, editor, *Three Dimensional Applications in Geographic Information Systems*, chapter 3, pages 21–35. Taylor & Francis.

Gold CM and Edwards G (1992) The Voronoi spatial model: two- and three-dimensional applications in image analysis. ITC Journal, 1:11–19.

Gold CM, Nantel J, and Yang W (1996) Outside-in: An Alternative Approach to Forest Map Digitizing. International Journal of Geographical Information Science, 10(3):291–310.

Goodchild MF (1992) Geographical Data Modeling. Computers & Geosciences, 18(4):401–408.

Guibas LJ and Stolfi J (1985) Primitives for the Manipulation of General Subdivisions and the Computation of Voronoi Diagrams. ACM Transactions on Graphics, 4:74–123.

Haklay M (2004) Map Calculus in GIS: A Proposal and Demonstration. International Journal of Geographical Information Science, 18(2):107–125.

Kemp KK (1993) Environmental Modeling with GIS: A Strategy for Dealing with Spatial Continuity. Technical Report 93-3, National Center for Geographic Information and Analysis, University of California, Santa Barbara, USA.

Kemp KK and Vckovski A (1998) Towards an ontology of fields. In *Proceedings 3rd International Conference on GeoComputation*. Bristol, UK.

Ledoux H and Gold CM (2004) An Efficient Natural Neighbour Interpolation Algorithm for Geoscientific Modelling. In P Fisher, editor, *Developments in Spatial*

Data Handling—11th International Symposium on Spatial Data Handling, pages 97–108. Springer.

Ledoux H, Gold CM, and Baciu G (2005) Flipping to Robustly Delete a Vertex in a Delaunay Tetrahedralization. In *Proceedings International Conference on Computational Science and its Applications – ICCSA 2005*, LNCS 3480, pages 737–747. Springer-Verlag, Singapore.

Mennis J, Viger R, and Tomlin CD (2005) Cubic Map Algebra Functions for Spatio-Temporal Analysis. Cartography and Geographic Information Science, 32(1):17–32.

Morehouse S (1985) ARC/INFO: A Geo-Relational Model for Spatial Information. In *Proceedings 7th International Symposium on Computer Assisted Cartography*. Washington DC, USA.

Mücke EP, Saias I, and Zhu B (1999) Fast randomized point location without preprocessing in two- and three-dimensional Delaunay triangulations. Computational Geometry—Theory and Applications, 12:63–83.

Peuquet DJ (1984) A Conceptual Framework and Comparison of Spatial Data Models. Cartographica, 21(4):66–113.

Pullar D (2001) MapScript: A Map Algebra Programming Language Incorporating Neighborhood Analysis. GeoInformatica, 5(2):145–163.

Ritter G, Wilson J, and Davidson J (1990) Image Algebra: An Overview. Computer Vision, Graphics, and Image Processing, 49(3):297–331.

Sambridge M, Braun J, and McQueen H (1995) Geophysical parameterization and interpolation of irregular data using natural neighbours. Geophysical Journal International, 122:837–857.

Sibson R (1981) A brief description of natural neighbour interpolation. In V Barnett, editor, *Interpreting Multivariate Data*, pages 21–36. Wiley, New York, USA.

Stevens S (1946) On the Theory of Scales and Measurement. Science, 103:677–680.

Suzuki A and Iri M (1986) Approximation of a tesselation of the plane by a Voronoi diagram. Journal of the Operations Research Society of Japan, 29:69–96.

Takeyama M (1996) *Geo-Algebra: A Mathematical Approach to Integrating Spatial Modeling and GIS*. Ph.D. thesis, Department of Geography, University of California at Santa Barbara, USA.

Theobald DM (2001) Topology revisited: Representing spatial relations. International Journal of Geographical Information Science, 15(8):689–705.

Tomlin CD (1983) A Map Algebra. In *Proceedings of the 1983 Harvard Computer Graphics Conference*, pages 127–150. Cambridge, MA, USA.

Watson DF (1981) Computing the n-dimensional Delaunay tessellation with application to Voronoi polytopes. Computer Journal, 24(2):167–172.

Watson DF (1992) *Contouring: A Guide to the Analysis and Display of Spatial Data*. Pergamon Press, Oxford, UK.

Worboys MF and Duckham M (2004) *GIS: A Computing Perspective*. CRC Press, second edition.

Modeling Geometric Rules in Object Based Models: An XML / GML Approach

Trevor Reeves[2], Dan Cornford[1], Michal Konecny[1], Jeremy Ellis[2]

[1] Knowledge Engineering Group, School of Engineering and Applied Science, Aston University, Birmingham B4 7ET, UK
[2] Key Traffic Systems Ltd., Ardencroft Court, Ardens Grafton, Alcester, Warwickshire, B49 6DP, UK

Abstract

Most object-based approaches to Geographical Information Systems (GIS) have concentrated on the representation of geometric properties of objects in terms of fixed geometry. In our road traffic marking application domain we have a requirement to represent the static locations of the road markings but also enforce the associated regulations, which are typically geometric in nature. For example a give way line of a pedestrian crossing in the UK must be within 1100–3000 mm of the edge of the crossing pattern. In previous studies of the application of spatial rules (often called 'business logic') in GIS emphasis has been placed on the representation of topological constraints and data integrity checks. There is very little GIS literature that describes models for geometric rules, although there are some examples in the Computer Aided Design (CAD) literature. This paper introduces some of the ideas from so called variational CAD models to the GIS application domain, and extends these using a Geography Markup Language (GML) based representation. In our application we have an additional requirement; the geometric rules are often changed and vary from country to country so should be represented in a flexible manner. In this paper we describe an elegant solution to the representation of geometric rules, such as requiring lines to be offset from other objects. The method uses a feature-property model embraced in GML 3.1 and extends the possible relationships in feature collections to permit the application of parameterized geometric constraints to sub features. We show the parametric rule model we have developed and discuss the advantage of using simple parametric expressions in the rule base. We discuss the possibilities and limitations of our approach and relate our data model to GML 3.1.

1 Introduction

The use of object based modeling frameworks is well established in GIS (Worboys and Hearnshaw 1990). However, present implementations are largely based on static representations of the application domain model and do not address the relationship between different spatial objects, other than topological constraints. In the application for which we are developing the object based model it is necessary to impose and check constraints (which we will call rules) relating to the geometric relationship between features. For example we need to be able to constrain one straight line to be within a given orthogonal offset range of another line. An additional problem is that we require that the offset range might vary according to certain external parameters, and we show how this form of constraint can be included in the model. In this paper we show the approach we adopt to solving this problem in an object based framework. In particular we show the underlying object based model and show how this can be mapped to an XML schema, and how the schema can be used to store real features that we use in our application domain.

In the following section we review previous approaches to rule representation within GIS systems. We then describe the application domain, road marking modeling, which motivates the solution we have developed. We go on to illustrate the object-based model we have created and show how rules, and in particular parameterized rules can be represented within this. To ensure that the rules are flexible and simple to update or extend we next illustrate their implementation using XML schema and show how this relates to the object based models. We look at how our approach relates to GML 3.1, and conclude with suggestions as to how the work might be extended.

2 Rule Representation in Spatial Data

Methods for modeling and enforcing geometric constraints (rules) between spatial entities currently exist i.e. restrictions and relationships pertaining to the positional attributes of geometric shapes. Since there is a discernible difference in the way these rules are handled in the CAD and GIS environments, both of these domains are reviewed.

2.1 GIS Approach

Work in the GIS field has previously focused on the semantic integrity of data sets, to ensure the logical correctness of geographical data. The uniqueness of integrity constraints for spatial data is identified in Cockcroft (1996), where *spatial integrity constraints* are introduced under the concept of 'business rules' for spatial application domains. In non-spatial application domains, business rules are identified to preserve the integrity of the logical data model; this is no different for spatial application domains, only that spatial business rules refer to the spatial properties of the data or enforce spatial relations between data.

The importance of spatial data integrity is addressed in Borges et al. (1999), where a method for specifying 'integrity rules' within the Object Modeling Technique – G (OMT- G), a geographic applications extension to OMT, at an early stage within the database design sequence is suggested (Borges 1997). It is suggested, as in Cockcroft (1998), that the integrity rules must be enforced at data entry and update; this ensures the integrity of any state of the database. The integrity rules or constraints are essentially enforced by ensuring that certain spatial relations are present between spatial entities within the database at any given time.

The modeling of topological relationships for complex spatial objects are described by Price et al. (2001), which builds on earlier work (Price et al. 2000). They show, at the conceptual level, how the different topological relationships can be defined as constraints to be imposed on spatial entities that comprise a higher composite spatial entity. That is, when modeling a spatial phenomenon that can be described as the composition of a number of other individual spatial phenomena, to ensure the integrity of the data, the composition relationships between the spatial parts and the spatial whole may be required to meet some topological relationship condition. An example of where these *part-whole* relationship constraints could be imposed would be to ensure that the total phone service coverage area contains or equals, the geometric union of the spatial extents of the individual phone service cells, which comprise it.

Spatial integrity constraints are classified in Cockcroft (1997) as: *topological, semantic*, and *user*.

- The topological constraints imposed are based on the topological relationships; Contains, Does not contain, Contains Entire, Intersects, Entirely within, Is not within, Within, Is not entirely within, Does not intersect, Does not contain entire. The topological relationships employed by Price et al. (2001) are based on those specified by Egenhofer and

Herring (1991) and Clementini and Felice (1994), and are given as; Boundary-Overlap, 4 Interior-Overlap, Mixed-Overlap, Disjoint-Separate, Disjoint-Interpenetrating, Contains, Equals, Inside.

- Semantic constraints are concerned with the meaning of geographical features, and apply to the properties of the geographical objects that need to be stored.
- User integrity constraints are much more specific to the application domain and are not necessarily based on semantics.

Different methods for realizing the inclusion of integrity constraints in spatial databases exist, these methods include; constraints as queries, schema based approaches, object-oriented approaches and the business rule approach. Cockcroft (1998) and Cockcroft (2001) go on to specify the means for implementing a spatial business rules system; rules are stored in their own 'rule repository', separate from the data itself. The business rules, or integrity constraints are then enforced at data entry, and each time an update occurs to the data. The integrity rules are stored at the metadata level; that is they are defined for different types of geographical features, not individually for each instance. The system outlined in Cockcroft (2001) provides a means for the application domain modeler to define the integrity constraints on data types present in a spatial database.

2.2 CAD Approach

Parametric models within many CAD systems can be used to enforce geometric constraints between a number of different geometric objects. Parameterization in CAD models is defined by Pratt (1998) as:

> *'the association of named variables, or expressions involving named variables, with certain quantities in a model. In a shape model these quantities are usually dimensions.'*

Pierra et al. (1994) describe a 'parametric instance', i.e. an instance of a feature represented in a parametric model, to consist of a set of parameters (potentially either numeric or Boolean), a set of geometric items (points, curves etc), and for each one of these primitives, a function that is able to compute it from the parameter values and the other geometric items in the model.

These functions are called parametric functions and can contain four main constructs:

- Constraint-Based Definitions: these are typically spatial relationships, which describe one geometric item as some form of constraint between other geometric items. Geometric constraints are used in many parametric CAD models to control the behavior of shape elements in a design, included for the improvement of design functionality (Pratt 1998). The geometric constraints are in the form of explicit spatial relationships, for example a perpendicularity constraint between two planar faces, or a tangency constraint between a line and a circle (Pratt 1998).
- Numeric and Boolean Valued Expressions: these describe numerical or Boolean logic based relationships between parameters in the model and properties of the geometric items in the model.
- Grapho-Numeric Expressions: these allow geometric items within the model to be used as arguments to functions/operators within parametric functions; e.g. distanceBetween(point1, point2).
- Constraint-Based virtual definitions: e.g. projectionOf point1 onto line2.

The explicit geometric constraints included in the parametric functions are very different to the topological constraints that are imposed in much of the GIS literature to maintain spatial integrity.

The term variational is used to denote the type of model that exhibits both parameterization and geometric constraints (Pratt 1998). Pratt (1998) and Pierra et al. (1994) outline the two methods that currently exist for the representation of variational models:

- Explicit Models: Parameters are associated with dimensional elements in the model, and constraints are explicitly specified between particular elements such as faces or edges.
- Implicit or History-Based Models: The primary representation of the model is in terms of the sequence of operations used to construct it. In this case there is no explicit information about the model shape at all; that information does not become available until after the specified operations have been performed, the result of this being an explicit model.

Our work shares some scope and motivation with the GIS approach to the application of 'business rules' in spatial data but the form and representation of our rules draw largely from rule representation within the CAD domain.

3 Road Markings and Rules: The Application Domain

The application domain is that of a Road Traffic design software to aid transport professionals to quickly and easily represent traffic features.

Traffic controls are seen every day on our roads, typically installed by an appropriate governing authority with the purpose of controlling the behavior of traffic in a defined way. These typically fall into three categories:

- Road Markings (stop lines, double yellow lines etc)
- Traffic Lights
- Road Signs

Governing authorities throughout most countries in the world have developed strict and complete regulations governing the selection, location and physical appearance of traffic controls on their public highways. The designing of new highways or updating of existing highways must adhere to these strict regulations on the use of traffic controls.

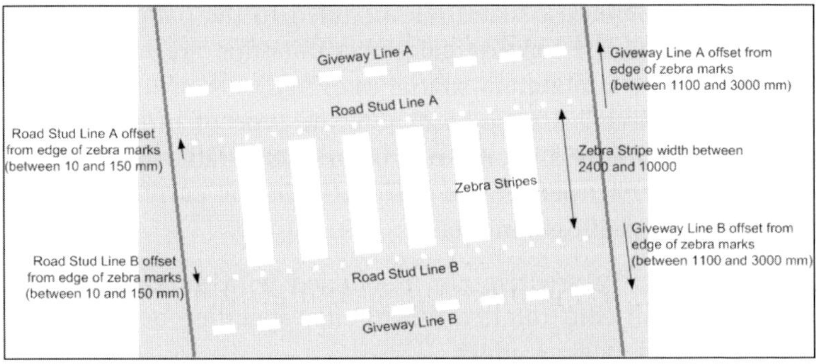

Fig. 1. UK Zebra Crossing Regulations (UK DOT 2003)

Typically a set of traffic control regulations defines many types of traffic control; each traffic control type has a shape or geometry defined by means of one or more geometric shapes that must exist within prescribed geometric constraints. Instances of these traffic control types exist as discrete entities in the real world. Traffic control regulations can be viewed as a set of rules for constructing and placing instances of traffic controls in the real world. Initially our work has focused on the regulations pertaining to road markings within the UK and the USA although it is envisaged that this will be extended to other countries in the near future.

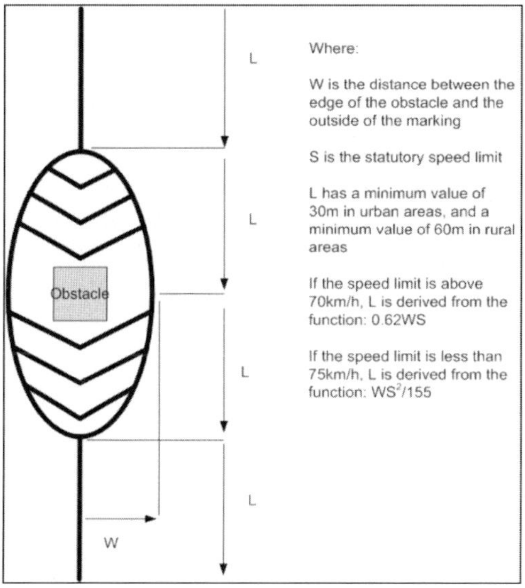

Fig. 2. USA Obstruction Marking (US DOT 2003)

Figures 1 and 2 show examples of the constraints imposed by the traffic regulations on zebra crossings (UK) and obstruction markings (USA) respectively. These and all other Traffic Controls need to be designed by the Transport Professional easily using the software but at the same time the software should enforce the rules defined.

4 Object Based Approaches to Modeling Traffic Features

Traditional GIS layer based vector approaches to modeling road markings would represent each component of the road marking as a point / line / polygon on a road markings layer. As such it would be very difficult to logically group the basic geometric components of individual traffic features, although this could be achieved using attribute information. A more natural representation is to use object based models. Object models represent the world in terms of features (or objects) rather than layers (Worboys and Hearnshaw 1990) and thus are more flexible and richer. In our approach we take advantage of the flexibility of modern object oriented modeling (Shalloway and Trott 2005) exploiting abstraction, inheritance and aggregation to provide a flexible, powerful model for road markings, which shares a number of features with GML 3.1. The high level overview of the model is given in Figure 3, with the geometry model in Figure 4.

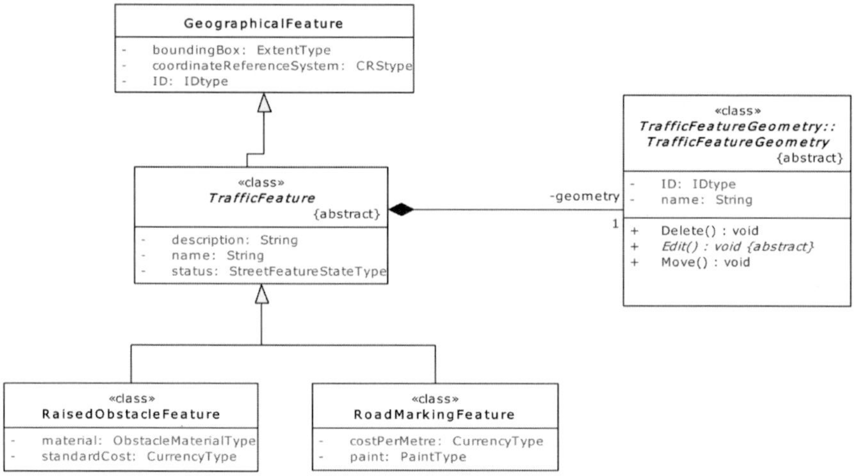

Fig. 3. Traffic Feature Model overview

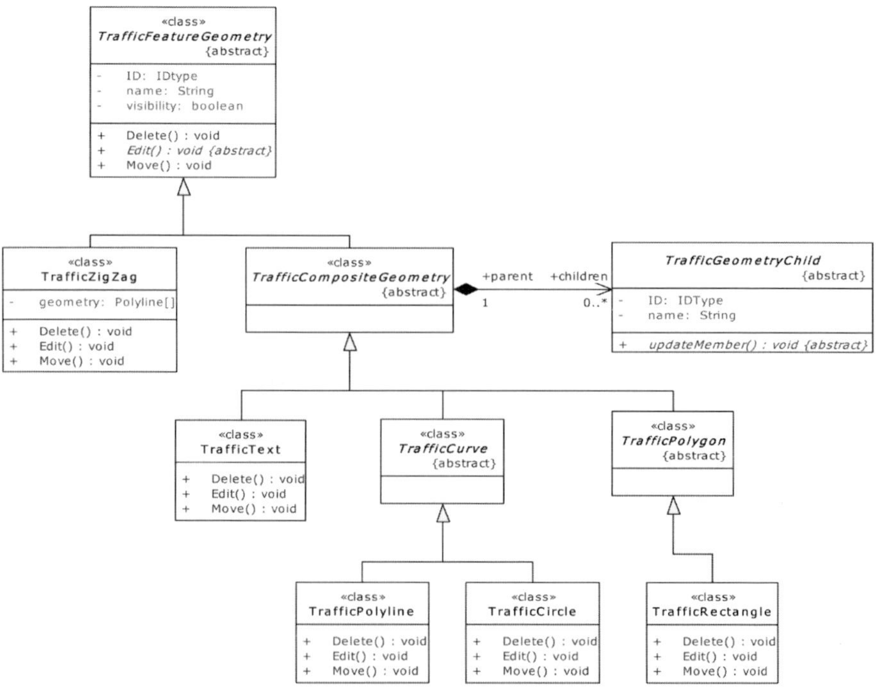

Fig. 4. Traffic Feature Geometry overview

4.1 Modeling Rules between Feature Components

In order to model the rules/constraints between geometric elements within a feature's geometry, using a standard GML-like geometry model is insufficient. The model presented here takes its inspiration from many of the concepts presented earlier for variational CAD models.

Elements in the feature's geometry are described through constraint relationships; these are typically explicit spatial relationships that are used to determine the geometric state of one geometric element based on the geometry of another 'parent' geometry element. Figure 5 provides an overview of the UML model for the relationships between a geometry element and any 'child' geometries it may contain. The offset relationship is only included here for simplicity, though as suggested by Pratt (1998) relationships such as perpendicularity and tangency could be included. Geometry structures can now be viewed to take the form of recursive trees, where a geometry element contains potentially many 'child' geometries (whose relationships are constrained by the child types), and these child geometries themselves can contain children.

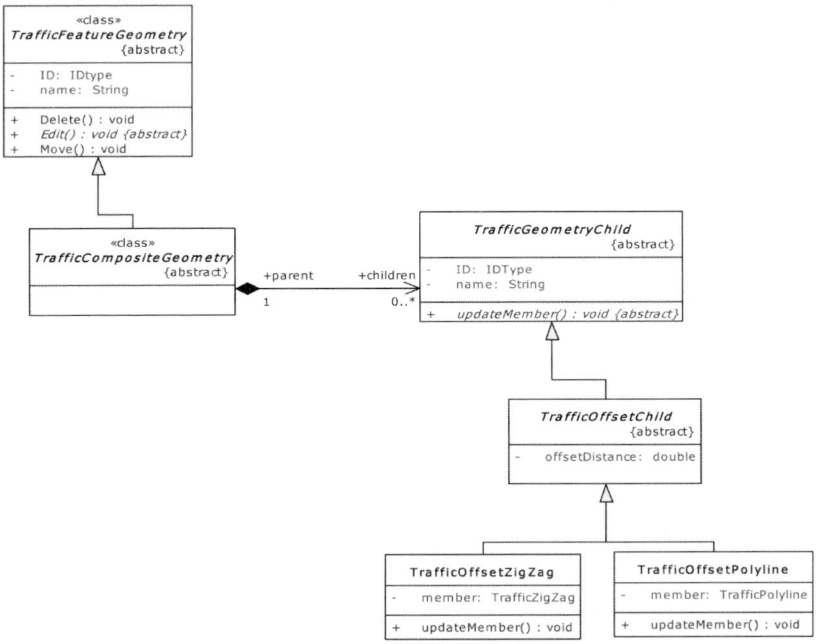

Fig. 5. Traffic Feature Geometry Child overview

This geometry model is essentially an *Explicit Variational Model* as described in Pratt (1998) and Pierra et al. (1994), though the definition and instantiation of these geometry structures through the use of the XML rule bases have more in common with the History-Based Models. The parametric nature of this model will be discussed in the next section.

4.2 Parameterized Rules and Simple Expressions

To enable geometric elements and geometric constraints within geometry definitions to relate to each other in more complex and less structured ways than the constraint-based geometry model permits, a parameter model has been devised. The parameter mechanism allows the rule base designer to specify relationships between certain attributes of the geometry and geometry constraint (child) classes in terms of real number or Boolean-value based expressions. A parameter value is essentially defined as a function of a number of other parameters.

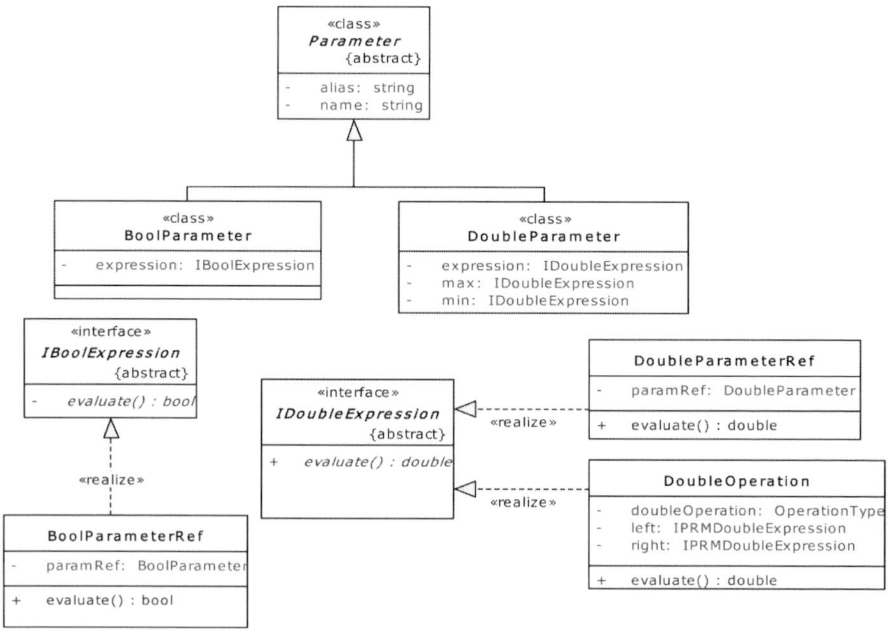

Fig. 6. Basic UML Model of Boolean and Double Parameters

A real number valued parameter (or double parameter), for example, consists of an expression, used to determine a real number value. There are also minimum and maximum bounds that this value is only permitted to lie between. The minimum and maximum bounds provide the means to enforce the upper and lower limits specified in the traffic regulations. These minimum and maximum bounds are themselves defined by expressions, providing increased flexibility for creating feature-geometry definitions.

Figure 6 shows the simplified UML model for parameters, the capability of the expressions can be extended by implementing the IDoubleExpression and IBoolExpression interfaces. When numeric and Boolean based attributes in the various geometry and geometric constraint classes are exposed as parameters they can be referenced by other parameters and have their values defined as functions of other parameters belonging to other elements in the model. Figure 7 shows how the distance attribute of the TrafficOffsetChild class can now be defined as a double parameter, and hence the distance at which the member geometry is offset from its parent is derived through the evaluation of a function that is potentially dependent on other parameter values.

The following example illustrates how an expression might be used in the definition of a UK Zebra Crossing (see Fig. 1). The distance at which a stud line is permitted to be offset from a line representing the centre of the crossing could be expressed as:

OffsetDistance = -($StripeWidth/2 + $StudOffset)

Where $StripeWidth and $StudOffset are references to other parameters within the same feature definition that represent other quantities of the feature or its geometry.

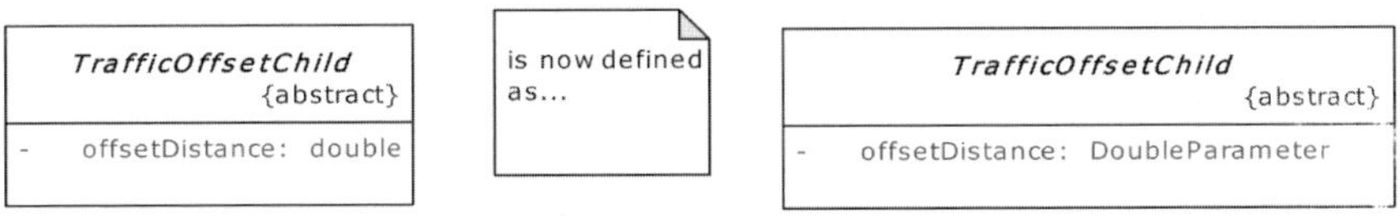

Fig. 7. Exposing attributes of the geometry and child types as parameters

Similarly, the Boolean-valued attribute 'visibility', which controls whether or not a geometric element within a feature's geometry is visible/included in the current instance can be exposed as a Boolean parameter:

Visibility = $IncludeRoadStuds

4.3 Example Road Marking Features

Figure 8 shows a representation of an XML instance describing the structure of a zebra crossing. This encoding aims to encapsulate those constraints specified by the traffic regulations as seen in Figure 1.

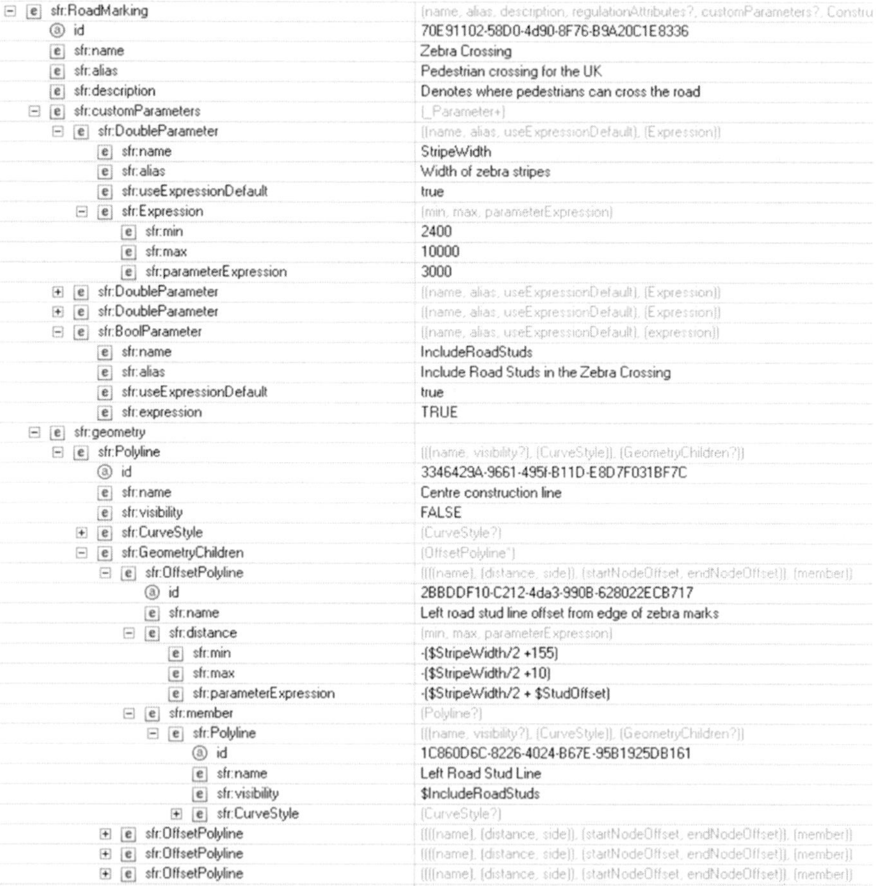

Fig. 8. A representation of the XML instance defining a Zebra Crossing, to encapsulate the constraints imposed by the UK Traffic Regulations

A number of 'custom' feature level parameters are defined for the road-marking feature itself. One of these, 'StripeWidth', is used to model the width of the zebra stripes that occur along the centre of the road marking. The stripes themselves are not included in this definition for simplicity. The parameters 'StudOffset' and 'GivewayOffset' are used to hold the offset distances of the stud lines and giveway lines from the edge of the zebra stripes respectively. A single polyline is used to represent the centre of the zebra crossing, around which the other elements in its geometry are positioned.

Each geometry element has a visibility attribute, represented as a Boolean-valued parameter. The traffic regulations state that the inclusion of the road studs in Zebra Crossings are optional; this is realized in this encoding by the custom parameter 'IncludeRoadStuds'; the visibility parameters of both road stud lines are set equal to 'IncludeRoadStuds', ensuring that both road studs are either included or not included in the zebra crossing.

Figure 9 shows a properties dialog box–type representation of an instance of this zebra crossing. The panel on the left hand side displays the geometry elements along with the child association types that are used to constrain them. The panel on the right provides the means for a user to edit the properties of the geometry and child types. Here we can see access to the offset distance of the right road stud line – represented as a double parameter. The offset distances for the stud lines and give way lines for the Zebra Crossing are derived from feature-level parameters, and so these values will only be modified through the manipulation of the feature properties.

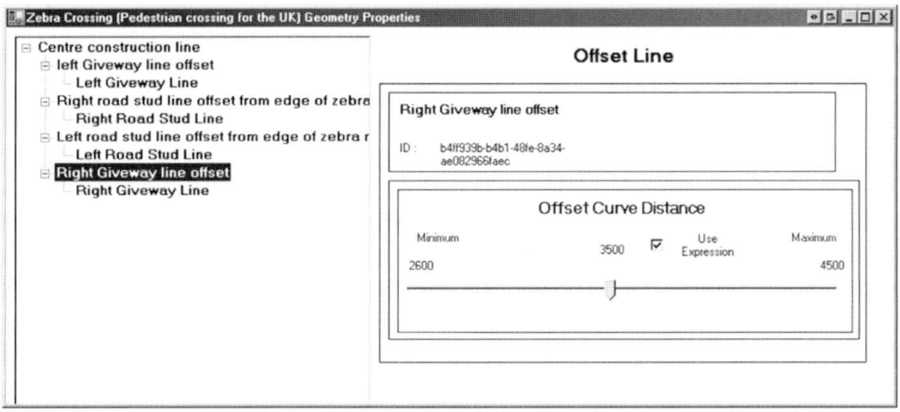

Fig. 9. The structure of the geometry for an instance of a UK Zebra Crossing as defined by the XML encoding shown in Figure 8

4.4 Links with Geography Markup Language

GML 3.1 is a developing set of standards for the encoding and transmission of spatially referenced data (Lake et al. 2004). GML was developed by the Open Geospatial Consortium[1] (OGC) and provides a rich set of XML based schema for describing spatial data, including geometry, topology, coordinate systems, coverages and grids, temporal data and observations. The central pattern used in GML is the feature (or object) – property model. The manner in which GML is used in applications is by extending the base abstract feature (or feature collection) model provided by the GML feature schema in a user defined application schema. This gives GML immense flexibility but also introduces its own semantic problems since each user can in theory develop their own application schema. The key benefit of using GML is that we have an open standard for the transmission of our data, which can be achieved using web services (Graham et al. 2001), making genuine interoperability a real possibility.

In this work we make two contributions; we develop an application schema for road marking features, but because the GML 3.1 model is not rich enough to permit us to represent this model using the standard GML 3.1 geometry schema, we produce a novel geometry schema to permit a range of constrained geometry types, such as an offset line. While this goes somewhat against the GML recommended best practices (Lake et al. 2004), it is the only plausible method to impose such business rules in GML. In developing the schema for the representation of the road marking features we have incorporated the GML feature – property model. The benefit of the feature – property model is that the name of the property conveys weak semantic meaning to that property in a manner similar to the resource description framework, on which GML was initially based.

The schema constrain the creation of instances of the road marking features, for example that shown in Figure 8. These instances maintain knowledge of their own internal associations, however it is possible to resolve all parameterized components, that is evaluate all parameterized rules to represent the feature in terms of static geometry, to produce a pure GML 3.1 compatible feature collection. At some point the resolution of static geometry is necessary to display the features in a GIS or CAD environment in any case. Of course in doing this 'explode' or 'export' we lose the flexible representation of the rules within the feature, however since very few applications, other than those we are developing are likely to support the use of these rules, this is not relevant. Evaluating the model to a pure GML 3.1 representation does however offer the ability to communi-

[1] Open Geospatial Consortium: http://www.opengeospatial.org/

cate the resulting road marking design across a web service to a whole range of GML enabled clients, allowing easy communication with external clients and the public.

5 Summary

In this paper we have reviewed the representation of geometric rules within both GIS and CAD, with emphasis on the types of rules that are relevant to the representation of Traffic Controls. We have shown how we have extended an object based spatial model to permit the representation of parameterized rules within a GIS context. This has united the CAD and GIS approaches to rule representation, and is very flexible. We have defined a set of schema that implement the UML represented object models that are closely tied to the developing GML format. In particular we are able to create a set of data driven object types from our rule instances, which provides great flexibility in modeling a wide range of traffic features and rules, without the need to change any source code. The instances are readily converted to GML for easy display across the web.

An interesting direction to take the work in the future would be to integrate the model more tightly with GML, by further abstracting the geometric rule representation to allow it to be included in a future GML specification, however at present it is not clear that there are a sufficient range of applications requiring this extra complexity to merit its inclusion in GML.

References

Borges K (1997) Geographic Data Modelling – An Extension of the OMT Model for Geographic Applications. Master's Thesis, Joao Pinheiro Foundation (in Portuguese)

Borges K, Laender A, Davis JrC (1999) Spatial Data Integrity Constraints in Object Oriented Geographic Data Modelling. In: Proc 7th ACM GIS, Kansas City, USA

Clementini E, Felice P (1994) A Model for Representing Topological Relationships Between Complex Geometric Features in Spatial Databases. Info Sciences:1–17

Cockcroft S (1996) Towards the Automatic Enforcement of Integrity Rules in Spatial Database Systems. In: Proc of the Spatial Information Research Centres' 8th Colloquium

Cockcroft S (1997) A Taxonomy of Spatial Data Integrity Constraints. GeoInformatica 1(4):327–343

Cockcroft S (1998) User Defined Spatial Business Rules: Storage, Management and Implementation – A Pipe Network Example. In: Proc of the Spatial Information Research Centres' 10th Colloquium

Cockcroft S (2001) Modelling Spatial Data Integrity Rules at the Metadata Level. In: Proc of the Sixth Int Conf on GeoComputation

Egenhofer M, Herring JR (1991) Categorizing Binary Topological Relationships Between Regions, Lines, and Points in Geographic Databases. Technical Report:1–33, Dept of Surveying Engineering, University of Maine, Orono, ME

Graham S, Simeonov S, Boubez T, Daniels G, Davis D, Nakamura Y, Neyama R (2001) Building Web Services with Java: Making Sense of XML, SOAP, WSDL and UDDI, Pearson Education, London

Lake R, Burggraf D, Trninic M, Rae L (2004) Geography Mark-Up Language: Foundation for the Geo-Web. John Wiley and Sons, London

Pierra JC, Potier G, Girard P (1994) Design and Exchange of Parametric Models for Parts Library. In: Proc of the 27th Int Symp on Advanced Transportation Applications, Aachen, Germany, pp 397–404

Pratt M (1998) Extension of the Standard ISO 10303 (STEP) for the Exchange of Parametric and Variational CAD Models. In: Proc of the Tenth Int IFIP WG5.2/5.3 Conf

Price N, Tryfona R, Jensen CS (2000) Modeling Part-Whole Relationships for Spatial Data. In: Proc of the 8th ACM GIS, pp 1–8

Price N, Tryfona R, Jensen C (2001) Modeling Topological Constraints in Spatial Part-whole Relationships. In: Proc of the 20th Int Conf on Conceptual Modeling, Yokohama, Japan.

Shalloway A, Trott JR (2005) Design Patterns Explained: A New Perspective on Object-Oriented Design. Addison-Wesley, London

UK Department for Transport (DOT) (2003) Traffic Sign Manual Chapter 5 – Road Markings, The Stationary Office

U.S. Department of Transportation (DOT) (2003) Federal Highway Administration (FHWA), Manual on Uniform Traffic Control Devices for Streets and Highways [Online]. Available: http://mutcd.fhwa.dot.gov/pdfs/2003/pdf-index.htm.

Worboys MF, Hearnshaw HM (1990) Object-Oriented Data Modelling for Spatial Databases. Int J of Geographical Information Systems 4:369–383

Exploring Geographical Data with Spatio-Visual Data Mining

Urška Demšar[1], Jukka M. Krisp[2], Olga Křemenová[2]

[1] Geoinformatics, Royal Institute of Technology (KTH), Stockholm, Sweden; email: urska.demsar@infra.kth.se
[2] Cartography and Geoinformatics, Helsinki University of Technology, Helsinki, Finland; email: jukka.krisp@hut.fi, olga.kremenova@hut.fi

Abstract

Efficiently exploring a large spatial dataset with the aim of forming a hypothesis is one of the main challenges for information science. This study presents a method for exploring spatial data with a combination of spatial and visual data mining. Spatial relationships are modeled during a data pre-processing step, consisting of the density analysis and vertical view approach, after which an exploration with visual data mining follows. The method has been tried on emergency response data about fire and rescue incidents in Helsinki.

1 Introduction

Digital data generated in recent years often contain geographical references, which makes it possible to integrate diverse types of data into spatial databases. How to find meaningful information in these complex data is a challenge to the information scientists. The discipline that tries to extract as yet unknown, but potentially useful information from data is called data mining (Hand et al. 2001; Fayyad et al. 2002).

Spatial data mining is a branch of data mining, which focuses on the spatial nature of the data. According to Tobler's first law of geography, all objects are related to each other, but closer objects are more related than distant ones. The implication of this is that the standard assumptions of independence and identically distributed random variables in classical data mining are not applicable for mining of spatial data. Spatial data are usually heterogeneous, thus the overall characteristics often do not hold for particular fractions of the dataset. In addition, complex geometrical and topological relationships between the spatial or spatio-temporal objects need to be considered (Miller and Han 2001).

The mining methods that handle these special spatial issues can be separated into two main groups. Methods in the first group apply classical non-spatial data mining algorithms to specifically prepared data where the spatial relationships have been pre-encoded (Koperski and Han 1995; Estivill-Castro and Lee 2001; Malerba et al. 2001), while the second group consists of new techniques where spatial information is processed as a part of the data mining algorithm (Ester et al. 1997; Chawla et al. 2001). Processing of spatial data mining tasks by computational methods is, however, very demanding.

A type of data mining where the user performs the exploration in a series of interactively connected visualizations is called visual data mining. All data mining is a form of pattern recognition. The most formidable pattern recognition apparatus is the human brain and mind. Human ability of perception enables the analyst to analyze complex events in a short time interval, recognize important patterns and make decisions much more effectively than any computer can do. Given that vision is a predominant sense and that computers have been created to communicate with the humans visually, computerized data visualization provides an efficient connection between data and mind. The basic idea of visual data mining is to present the data in some visual form, in order to allow the human to get insight into the data, recognize patterns, draw conclusions and directly interact with the data. The recognized patterns reveal relationships between data attributes or data elements or identify global structure in the data (Ankerst 2000; Keim and Ward 2003). Visual data mining has several advantages over computational data mining. The result is obtained faster and with a higher degree of confidence in the findings, because the exploration is intuitive and doesn't require understanding of complex mathematical algorithms. It is effective when little is known about the data and when the exploration goals are vague. It can be used to explore heterogeneous and noisy data where automatic mining methods fail. It is therefore well suited for exploration of spatial data (Keim et al. 2004).

Visual data mining has in recent years been used for spatial data in a number of cases. Recent examples include pixel-based geovisualizations (Keim et al. 2004), several attempts to explore geographical metadata

(Demšar 2004; Ahonen-Rainio 2005; Klein 2005), visual mining of spatial time series data (Andrienko et al. 2004), exploration of spatially distributed genetic datasets (Joost and the Econogene Consortium 2005) and exploration of geochemical data (Grünfeld 2005).

One disadvantage of applying visual data mining to spatial data is that the spatial component of the data is difficult to visualize in any other way than by using maps, which have several drawbacks from the visualization point of view. For example, since the data are usually not uniformly distributed over space, some display areas might be sparsely populated while in other areas a high degree of overprinting occurs. In order to include the spatial component into visual data mining, efficient strategies how to represent large spatial datasets have to be found. Alternatively, data mining methods focusing on the spatial component of the data could be combined with the visual mining (Keim et al. 2005).

This paper proposes a way to explore spatial data with a combination of spatial and visual methods. Spatial relationships are modeled during a data pre-processing step, after which visual data mining is used for the exploration. The pre-processing step consists of density analysis (Silverman 1986; O'Sullivan and Unwin 2003) combined with a vertical view approach for spatial data mining (Estivill-Castro and Lee 2001; Karasová 2005). The approach is similar to the methods from the first group of spatial data mining methods except that visual data mining is used instead of classical automatic algorithms after the data have been spatially pre-processed. Additionally, a temporal exploration of the data is possible by providing a special visualization for temporal attributes. The approach has been tested on emergency response data from the central part of Helsinki.

The rest of the paper is organized as follows: Section 2 presents the data and the case study area. The spatio-visual exploration method is explained in Section 3. Section 4 presents some of the more interesting exploration results. Finally, Section 5 evaluates the exploration approach and presents some ideas for future research.

2 Exploration Data

The study area covered a 14x14km square area in the central part of the Helsinki city. Data about fire and rescue incidents, population, a business register, topology and infrastructure were used.

The Helsinki Fire & Rescue Services have supplied sample datasets, which contain all the fire alarms, rescue missions and automated fire alarm systems missions within Helsinki city for the years 2001–2003. The material includes selected information, such as mission type codes, dates, addresses and X/Y coordinates. The data is not publicly accessible.

The second data source was a product named "SeutuCD" published by the regional council of Helsinki (YTV) in 2003. SeutuCD is a data collection from different sources and includes population information, a business register and various background information on topology and infrastructure. The population information in SeutuCD originates from the Population register centre. Their Population Information System contains information for the whole Finland on Finnish citizens and foreigners who are permanently residing in Finland. A business register of Helsinki with geocoded and classified data for different business types was used to obtain information on bars and restaurants. This source also supplied different background information on topology and infrastructure, such as data about water areas, roads, railway network and built-up areas.

3 The Spatio-Visual Exploration Method

This section presents the spatio-visual data exploration approach, which consists of a combination of spatial pre-processing and visual data mining.

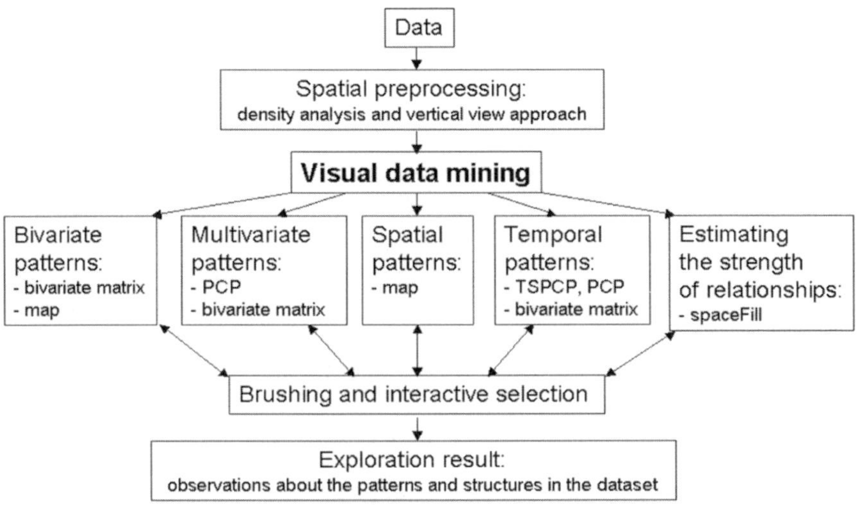

Fig. 1. Framework for the visual data mining of spatially pre-processed data

The framework for exploration of the incidents dataset using the spatio-visual approach is presented in Figure 1. Density analysis and vertical view approach are used to produce the exploration dataset with intention to encode the spatial neighborhood relationships in the architecture of the dataset. Then, different visualizations in combination with interactive selection and brushing (simultaneous highlighting of the selected data in all visualizations) allow discovery of bivariate, multivariate, spatial and temporal patterns in the spatially pre-processed dataset. Additionally, the strength of the bivariate relationships can be visually estimated. Pattern recognition, visual mining and interactive selection form a dynamic iterative process that can be repeated a number of times, until a satisfactory outcome in the form of observations about patterns in the incident dataset has been found.

This section is divided into two parts. The first part presents how the spatial dependencies in the data were modeled in preparation for the visual mining. The second part describes how the visual data mining system used for data exploration was built.

3.1 Spatial Pre-Processing

In order to mine spatial relationships among different geographical layers (incidents, topographical data, locations of various point features, such as population information or bars and restaurants, etc.), all features were converted to rasters with identical extent and resolution by producing either density surfaces or proximity surfaces. All the surfaces covered the same study area, which was divided into a 250m raster containing 3136 (56x56) grid cells.

The incidents were represented as a continuous density surface in order to give an efficient impression of their spatial distribution. Applying a kernel density estimations scheme recognizes this continuity. The kernel density method replaces each point with a kernel and then sums the kernels of all points into a surface (Silverman 1986). The determining factors are the bandwidth, sometimes referred to as the kernel search radius, and the output grid size of the estimation. If the bandwidth is too large, the estimated densities will be similar everywhere and close to the average incident density of the entire study area. When the bandwidth is too small, the surface pattern will be focused on the individual incident records. Experimentation is required to derive the optimal bandwidth setting to acquire a satisfactory density surface (O'Sullivan and Unwin 2003).

To recognize the different types of incidents during the visual data mining process, the incident data were classified into six classes according to incident types shown in Table 1. These incident types were separated into night-time and daytime incidents for each of the years 2001 to 2003. The density was calculated for each type, time and year. This provided 36 different density layers (including night-time incident density and daytime incident density for 2001, 2002, 2003 for each of the six types of accidents). Three additional layers were produced as aggregated densities for all night-time/daytime incidents and all incidents over the whole time period 2001-2003. Figure 2 shows the aggregated 2001-2003 density of the night-time incidents.

Table 1. Incident types

Class no.	Code by fire & rescue services	Description
1	A	automated fire alarms
2	T1, T2, T3, T4	fires
3	P3, P4, P5, P6, P7	other incidents
4	P2	traffic incidents
5	T5	boat incidents
6	P1	people rescue incidents

The population density, the bars & restaurant density and the railway stations density were also produced, based on the data from the SeutuCD.

The kernel search radius was 500m for all densities. The grid size was 250m. The output was classified into six classes using a natural breaks classification to optimize breaks between the individual classes. Values ranged from 1 indicating low density to 6 indicating high density.

In addition to densities, background information was included into the database by producing proximity surfaces based on intersection. A cell in a proximity surface was assigned value 1 if the feature in question intersected the cell and 0 if there was no intersection. These surfaces were produced for each of the following features: water areas (sea, lakes and rivers), built-up areas, railway and six different road types (derived by classification of the road network according to Table 2). The data used in this step came from SeutuCD.

Table 2. Classification of road types

Class no.	Road type
0	no roads in the cell
1	highway
2	major road
3	ring or major road
4	road
5	residential street

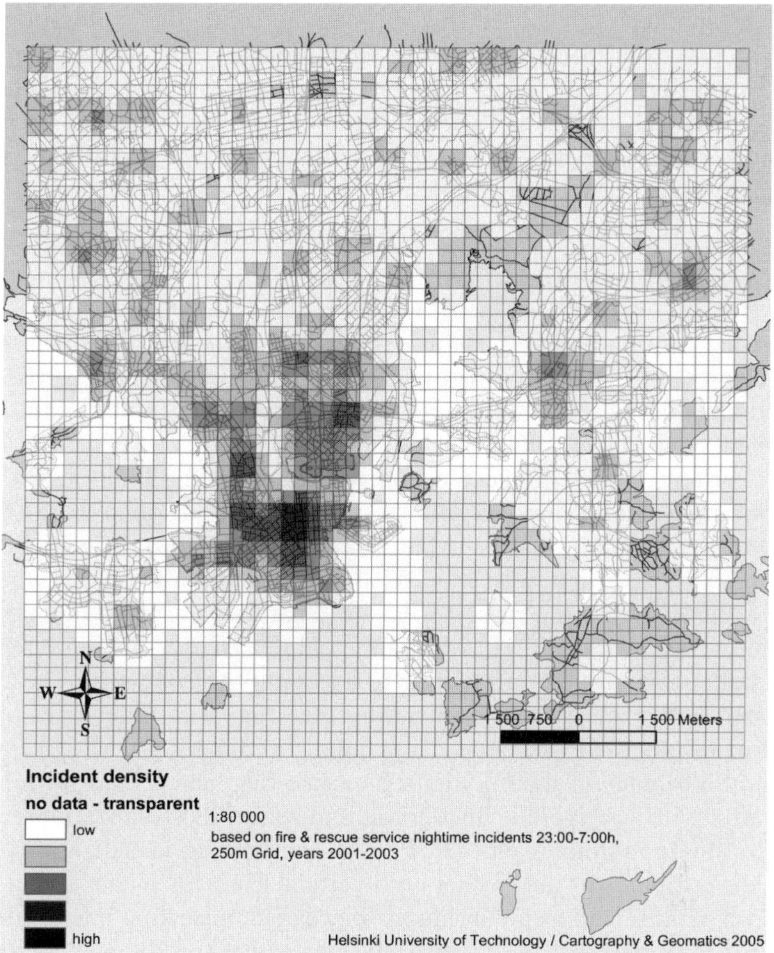

Fig. 2. Density of nighttime incidents in the centre of Helsinki

The densities and the proximity surfaces were integrated in the exploration dataset by the vertical view approach to spatial data mining (Estivill-Castro and Lee 2001; Karasová 2005). Spatial data mining requires comparison of all the attributes for a particular location. Therefore data needs to be arranged into a form similar to the one used in relational databases. The vertical view approach to spatial data mining, which is based on map overlay, offers a solution to this problem. In this approach there is a separate layer of raster cells covering the area of interest for each attribute of the dataset. Information about a particular location is contained in the corresponding cell in all layers. One cell identifies the smallest neighborhood unit. The database of neighborhoods is then created by a virtual overlay of all rasters into a table of cells and their respective raster values.

In our application, we used this approach to create the exploration database for the visual data mining. The final database was in the form of an area shape-file, where each polygon represented one raster cell with 15 non-temporal attributes (aggregated incident densities and proximity surfaces) and 36 temporal attributes (incident densities according to time, type and year) as described above.

3.2 Visual Data Mining

The visual data mining system for exploration of the incidents dataset was built using GeoVISTA Studio, a java-based collection of geovisualizations and computational data mining methods for geoscientific data exploration (Gahegan et al. 2002; Takatsuka and Gahegan 2002).

The system included selected components from ESTAT design – the Exploratory Spatio-Temporal Analysis Toolkit (Robinson et al. 2005) and a multiform bivariate matrix (MacEachren et al. 2003). The decision to build the system in this way was based on the nature of the data: the temporal attributes were the reason to use a separate parallel coordinates plot for time series as in ESTAT, while the goal to discover bivariate relationships in the data was the motivation to use a multiform matrix. The system consisted of the following visualizations (see Fig. 3): a time series parallel coordinates plot (TSPCP) on top, a parallel coordinates plot (PCP) for non-temporal attributes in the middle, a bivariate map and a multiform bivariate matrix with scatterplots, histograms and spaceFill visualizations.

A parallel coordinates plot (PCP) is a multivariate visualization, which maps the attributes of the dataset onto vertical axes. Each data object in the dataset is represented as a continuous piecewise linear line intersecting the axes at appropriate attribute values. It is effective for complex spatiotemporal datasets, but has a disadvantage of overprinting when there are many data elements with similar values, which conceals the strength of the correlations (Inselberg 1999; Edsall 2003).

An element in the row *i* and column *j* in a multiform bivariate matrix is a scatterplot of the variables *i* and j, if it is located above the diagonal, a spaceFill visualization of the same two variables, if it is located below the diagonal and a histogram of variable *i*, if it is on the diagonal. Scatterplots display data elements as (x, y) points, where x and y represent the respective attribute values for each data element. They can be used to detect correlation between the two variables, but when there are too many data elements and overprinting occurs, it is not possible to see the strength of the actual correlation (Hand et al. 2001). Since the range of many of the variables in the accidents dataset consisted of six discrete values, there was significant overprinting in the scatterplots and in the PCP. Therefore it was decided to add a spaceFill visualization in the matrix. In a spaceFill visualization each grid square represents one data element. The color of the square is assigned according to one of the two display variables and the order of the squares in the grid according to the second display variable. The visualization solves the problem of overprinting, as it displays all data elements at non-overlapping positions (MacEachren et al. 2003).

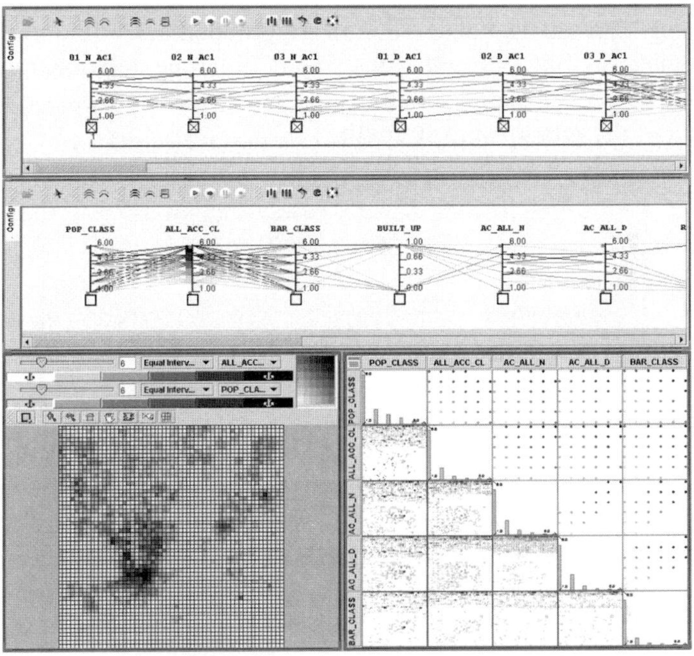

Fig. 3. The visual data mining system for the incidents dataset with a TSPCP, a PCP, a bivariate map and a multiform bivariate matrix

Finally, a geoMap, a component in GeoVISTA which shows the spatial extent of the data, was included in the system. This is a bivariate choropleth map, whose color scheme is interactively transferred to other visualizations that do not have own color schemes. Alternatively the color scheme for the map and all other visualizations can be interactively defined in one of the PCPs according to one of the attributes. The same color scheme in all visualizations makes it easy to visually identify the same data object in different components (Gahegan et al. 2002; Takatsuka and Gahegan 2002).

All visualizations in GeoVISTA-based exploration systems are interactively connected by the principle of brushing. This means that when a subset of data elements is selected in one visualization, the same elements are selected everywhere, providing a better visual impression and easier pattern recognition (Gahegan et al. 2002; Takatsuka and Gahegan 2002).

4 Results of Data Exploration

This section presents a selection of the more interesting observations about the incidents dataset. The observations are grouped according to the exploration framework into observations about bivariate relationships, multivariate relationships, spatial relationships, temporal relationships and about estimation of the relationship strength.

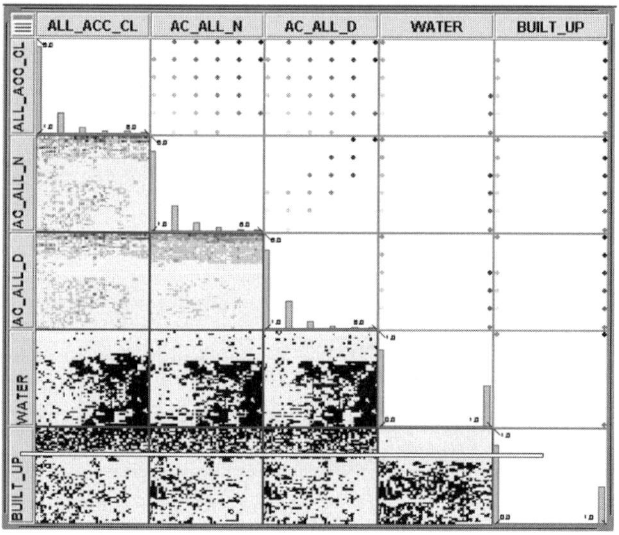

Fig. 4. A bivariate matrix of all incidents, nighttime incidents, daytime incidents, proximity to water and built-up areas

The bivariate relationships were discovered using mainly the spaceFills and the map. The PCP, TSPCP and scatterplots were not very useful for this purpose, because of the high level of overprinting. There is a very good correlation between the location of night and day incidents. This can be seen from the scatterplot between night incidents (AC_ALL_N) and day incidents (AC_ALL_D) in the bivariate matrix in Figure 4. There is a general trend towards the diagonal in this scatterplot. However, the overprinting makes it impossible to estimate the strength of the relationship. The incidents do not correspond very well to the population density. There is a better correlation between the incidents density and the density of bars and restaurants. High incident density corresponds to built-up areas and areas that are not close to water, as can be deduced from the spaceFills in the lower two lines of the bivariate matrix in Figure 4. In the next to last row of this matrix, the color represents vicinity to water and in the last row built-up areas. Dark color means that there is water or built-up area in the cell represented by each grid square. The order of grid squares in these spaceFills is defined according to the density of incidents, starting from the lowest value in the bottom-left corner and proceeding along a scan line towards the highest value in the top-right corner. In order to illustrate how the deduction was made, a thin white horizontal line in the bottom row has been added to this figure (it is not a part of the visual data mining system). The cells above this line have a high density of incidents, but are predominantly dark, which indicates that these are built-up areas. A similar conclusion can be made about vicinity to water and incidents in the three spaceFills in the fourth row. Here the grid squares that correspond to high-density areas are light, indicating that there is no water in these cells.

Multivariate relationships can be identified by using interactive selection in several visualizations. An example is the selection of the built-up areas with high bar and restaurant density. The selection was performed in the scatterplot and transferred interactively to other visualizations. From these, the characterization of these areas could be inferred: the map indicated that these areas correspond fairly well with the centre of Helsinki. From the PCP it could be observed that these areas have a high incidents density, medium to low population density, are far from water and contain roads of types 2–5. The TSPCP (see Fig. 5) indicated that in these areas the incidents of type 1 (automatic fire alarms) and 3 (other incidents) are common, but that these areas have no incidents of type 5 (boat incidents) and lower density of incidents of type 6 (people rescue incidents). The incidents of type 2 (fires) were distributed over all values in these areas and incidents of type 4 (traffic incidents) had a higher density in these areas during the day than at night.

Using the interactive selection for different types of roads in the PCP and the transfer of this selection to the map and the TSPCP it was possible to spatially identify areas where a certain type of accidents occurred frequently. Two such hotspots were identified for traffic accidents: one on the highways and one on the major roads. With the help of the PCP, these two areas were characterized further: they both have low population density, are built-up, have no bars, are not near water or railways or railway stations, but have roads of classes 1, 3 and 5 (highways, ring and residential streets). The TSPCP revealed that these two spots have high values of traffic incidents for all years and at any time. Other accidents were low for these two particular spots, except for fires, which were frequent during both day and night in 2002 for one of the spots.

A selection of the highest densities of particular incidents in the TSPCP and comparison of their locations on the map helped to detect a hot spot of automated fire alarm incidents. It is located around the central railway station, a place with high density of bars and restaurants. During the daytime, these incidents cover a slightly larger area than at night. In the same way, a hot spot of people rescue incidents was identified. The place is again located in the Helsinki centre, and it is further characterized by higher bar and restaurant density and by proximity to water.

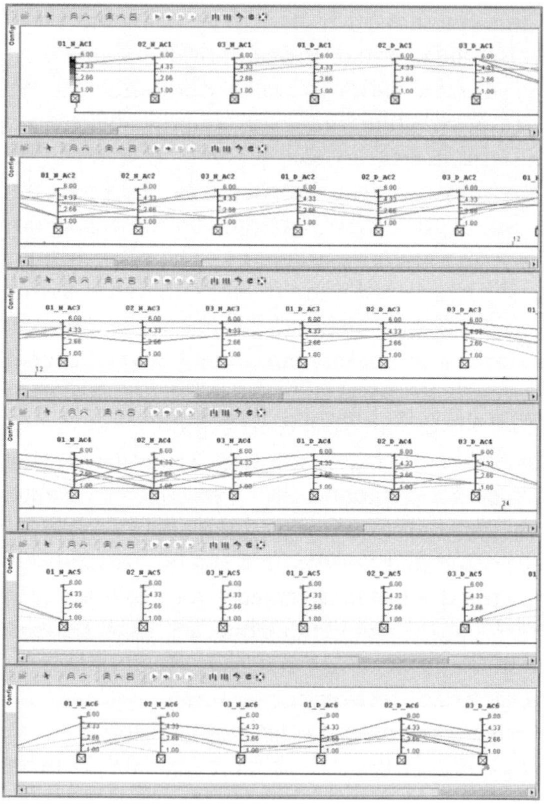

Fig. 5. The complete TSPCP with all 36 temporal attributes displaying the selection of built-up areas with high density of bars. The color of the lines follows the density of all incidents, dark for high density and light for low density

The TSPCP in combination with other visualizations revealed interesting temporal patterns through the years as well as changes over the daytime and the nighttime. The traffic accidents occurred with lower density during the night. In addition, the decrease in their density from 2001 to 2003 is also apparent. Using the map, different hot spots for day and night traffic accidents were discovered. The pattern of traffic accidents is similar for each year, with an exception of a distinct shape for nighttime incidents in 2001, which disappears in the following years. Similarly, there was a decrease in the boat accidents for year 2003, during both day and nighttime. Using the combination of TSPCP, PCP and spaceFills, the difference in characterization of people rescue incidents for day and night-time was discovered: while during the daytime people were often rescued in populated and built-up areas, at night this kind of incidents occurred in less populated areas and in the proximity of bars.

The VDM tools can be used to visually estimate the strength of the relationship between two attributes. An example of this is shown in Figure 6, which shows spaceFills for the density of bars and restaurants and population density versus the aggregated density of night incidents. The color in both SpaceFills is assigned from light to dark according to bar density and population density respectively. The density of the night-time incidents defines the order of the grid cells: the cell with the lowest night-time incidents density is situated in the bottom-left corner, from where the cells then proceed along a scan line towards the cell with the highest night-time incidents density in the top-right corner. If the attribute defining the color of the cells is correlated with the attribute defining the order of the cell, there will be a relatively smooth transition from the lightest to the darkest color from bottom to top (or from top to bottom), as it is the case with the spaceFill showing the bar density versus the night-time incidents density. The spaceFill of population density and density of the nighttime incidents is much more scattered, i.e. the propagation from light towards dark is not as smooth. This suggests that the correlation between the bar density and the nighttime incidents is stronger than the correlation between the population density and the night-time incidents. This is however just a visual estimation, but the observation can be used to form a hypothesis, which could then be explored further using other tools, for example statistics or spatial analysis.

Fig. 6. SpaceFills for bar and restaurant density vs. night-time incidents density on the left and population density vs. night-time incidents density on the right

5 Conclusions and Future Work

The spatio-visual approach for the exploration of the incident dataset proved to be valuable for detecting patterns and visually estimating the strength of observed relationships. From the observed patterns and relationships hypotheses could be made about the incident dataset. The method however has limitations to value the significance of potential relationships. Even though the method is strong to preliminary identify relationships between attributes, further investigation is needed to value the findings. Experts within the Finnish Fire & Rescue Services might evaluate potential results to see if they are significant. The findings should be evaluated and compared with reality. For example, the hot spots of traffic accidents that have been identified as built-up, low populated areas with many different road types could possibly be industrial areas or major road junctions. This would have to be checked on the spot, to see what there really is in these areas. Some of these critical places exist only for one particular year – was there perhaps a temporary obstruction (i.e. road works) that made these areas temporarily more dangerous? Another example are the temporal trends for the traffic accidents: they occurred at different places during the day and the night. This might for example indicate potential problems with illumination at night. In other places there is a decrease in traffic accidents from 2001 to 2003. Was there something done in these places, such as introduction of new lower speed limits or warning signs? Does this correspond to the overall trend of amount of accidents? Have drivers become generally more careful, or do they pay more attention just at these critical places?

Mining spatial data always requires special data pre-processing. This step is application-dependent and therefore cannot be completely automated. In this study, the generation of all densities proved to be particularly time consuming. However, once this is done, various mining methods can be applied to the dataset, either classical automatic algorithms or visual methods.

Representing neighborhood relationships by using the raster format for the vertical view approach might be problematic: two objects in two neighboring cells may be closer to each other than objects in the same cell. This problem was partially overcome by using the density analysis, where a continuous density surface represents the objects. The density analysis, however, is dependent on the bandwidth and the output grid size. Grid resolution also affects the visual exploration: different patterns can be discovered at different resolutions.

Another problematic issue beside the raster resolution is the representation of a neighborhood by a square cell. This method of representing neighborhoods might not define spatial relations sufficiently well. Different basic neighborhood units might therefore be considered, such as buffers around particular features or census tracts.

In this sample case the SpaceFill diagrams seemed to be more useful to find relationships than either the PCP or the TSPCP. This is mainly due to significant overprinting caused by classification. Working with classified densities in the input data is therefore problematic, because of the loss of information during the visual analysis. It would be interesting to look at the same data pre-processed in a different way, so that the density attributes ranged over a continuous interval instead of being classified into discrete values.

Trends were best seen in the TSPCP and PCP in combination with the map. The map in combination with other visualizations was also useful for identification of spatial patterns and hotspots of incidents.

An important feature of the system that enables effective exploration is the interactive selection and brushing. However, this can be confusing for potential explorers. Generally visual data mining needs training and is not a tool for inexperienced users, such as the Fire and Rescue domain experts might be. There is, however, the advantage that the results are easy to show and explain to the general audience, as there is no requirement to understand difficult algorithms or other procedures. In order to evaluate if such a tool could be useful for practical everyday purposes and how difficult it would be for domain experts to learn to use it properly, user-studies and a usability evaluation would be necessary.

Another idea for future research is to combine the visual data mining with spatial data mining algorithms, such as spatial clustering, spatial association rules, a Self-Organizing Map, etc. in order to try to detect patterns in the data in an even more effective way.

Acknowledgements

The authors would like to acknowledge support of the Helge Ax:son Johnson Foundation, the Helsinki Fire & Rescue Services, the Regional Council of Helsinki (YTV) and the Helsinki University of Technology.

Contributors

At the time of writing this paper all three authors are PhD students: Urška Demšar at the Royal Institute of Technology in Stockholm and Jukka Krisp and Olga Křemenová at the Helsinki University of Technology. Urška is working with visual and automatic data mining of spatial data, Jukka's research is dedicated to geovisualization and Olga's to spatial data mining. Urška and Jukka prepared the exploration dataset for this paper and Urška designed the visual data mining system. The exploration of the data and writing of the paper was done by all three as a joint contribution.

References

Ahonen-Rainio P (2005) Visualization of geospatial metadata for selecting geographic datasets. PhD Thesis, Helsinki University of Technology, Helsinki

Andrienko G, Andrienko N, Gatalsky P (2004) Visual Mining of Spatial Time Series Data. In: Proc of the 8th European Conf on Principles and Practice of Knowledge Discovery in Databases, PKDD 2004 (= LNCS 3202). Springer, Berlin Heidelberg New York, pp 524–527

Ankerst M (2000) Visual Data Mining. PhD Thesis, Ludwig-Maximilans-Universität, München

Chawla S, Shekhar S, Wu W, Ozesmi U (2001) Modelling spatial dependencies for mining geospatial data. In: Miller HJ, Han J (eds) Geographic Data Mining and Knowledge Discovery. Taylor & Francis, London, pp 131–159

Demšar U (2004) Exploring geographical metadata by automatic and visual data mining. Licenciate Thesis, Royal Institute of Technology (KTH), Stockholm

Edsall RM (2003) The parallel coordinate plot in action: design and use for geographic visualization. Computational Statistics & Data Analysis 43:605–619

Ester M, Kriegel H-P, Sander J (1997) Spatial Data Mining: A Database Approach. In: Proc of the 5th Int Symp on Large Spatial Databases, SSD 1997, Berlin, Germany

Estivill-Castro V, Lee I (2001) Data Mining Techniques for Autonomous Exploration of Large Volumes of Geo-referenced Crime Data. In: Proc of 6th Int Conf on Geocomputation, Brisbane, Australia

Fayyad U, Grinstein GG, Wierse A (eds) (2002) Information Visualization in Data Mining and Knowledge Discovery. Morgan Kaufmann Publishers, San Francisco

Gahegan M, Takatsuka M, Wheeler M, Hardisty F (2002) Introducing GeoVISTA Studio: an integrated suite of visualization and computational methods for exploration and knowledge construction in geography. Computers, Environment and Urban Systems 26:267–292

Grünfeld K (2005) Visualization, integration and analysis of multi-element geochemical data. PhD Thesis, Royal Institute of Technology (KTH), Stockholm

Hand D, Mannila H, Smyth P (2001) Principles of Data Mining. The MIT Press, Cambridge, Massachusetts

Inselberg A (1999) Multidimensional detective. In: Card SK, MacKinley JD, Shneiderman B (eds) Using Vision to Think, Readings in Information Visualization. Morgan Kaufmann, San Francisco

Joost S, The Econogene Consortium (2005) Combining biotechnologies and GIScience to contribute to sheep and goat genetic resources conservation. In: Proc of FAO Int Congress: The Role of Biotechnology, Turin, Italy, pp 109–116

Karasová V (2005) Spatial data mining as a tool for improving geographical models. Master Thesis, Helsinki University of Technology, Helsinki

Keim DA, Ward M (2003) Visualization. In: Berthold M, Hand DJ (eds) Intelligent Data Analysis, 2nd ed. Springer, Berlin Heidelberg, pp 403–428

Keim DA, Panse C, Sips M, North SC (2004) Pixel based visual data mining of geo-spatial data. Computers & Graphics 28:327–344

Keim DA, Panse C, Sips M (2005) Information Visualization: Scope, Techniques and Opportunities for Geovisualization. In: Dykes J, MacEachren AM, Kraak M-J (eds) Exploring Geovisualization. Elsevier and International Cartographic Association, Amsterdam

Klein P (2005) TheCircleSegmentView: A User Centered, Meta-data Driven Approach for Visual Query and Filtering. PhD Thesis, Universität Konstanz, Konstanz

Koperski K, Han J (1995) Discovery of Spatial Association Rules in Geographic Information Databases. In: Proc of the 4th Int Symp on Large Spatial Databases, Portland, Maine, USA

MacEachren AM, Dai X, Hardisty F, Guo D, Lengerich G (2003) Exploring High-D Spaces with Multiform Matrices and Small Multiples. In: Proc of the Int Symp on Information Visualization, Seattle, pp 31–38

Malerba D, Esposito F, Lisi F A (2001) Mining Spatial Association Rules in Census Data. In: Proc of the Joint Conf on New Techniques and Technologies for Statistics and Exchange of Technology and Know-how, Crete, Greece

Miller JH, Han J (2001) Geographic Data Mining and Knowledge Discovery: an overview. In: Miller HJ, Han J (eds) Geographic Data Mining and Knowledge Discovery. Taylor & Francis, London, pp 3–32

O'Sullivan D, Unwin DJ (2003) Geographic information analysis. John Wiley & Sons Inc., New Jersey

Robinson AC, Chen J, Lengerich EJ, Meyer HG, MacEachren AM (2005) Combining Usability Techniques to Design Geovisualization Tools for Epidemiology. In: Proc of the Auto-Carto 2005, Las Vegas

Silverman BW (1986) Density estimation for statistics and data analysis. Chapman and Hall, London

Takatsuka M, Gahegan M (2002) GeoVISTA Studio: a codeless visual programming environment for geoscientific data analysis and visualization. Computers & Geosciences 28:1131–1144

Continuous Wavelet Transformations for Hyperspectral Feature Detection

Jelle G. Ferwerda, Simon D. Jones

School of Mathematical and Geospatial Sciences, GPO Box 2476V, Melbourne, VIC 3001 Australia; email: Jelle.Ferwerda@rmit.edu.au

Abstract

A novel method for the analysis of spectra and detection of absorption features in hyperspectral signatures is proposed, based on the ability of wavelet transformations to enhance absorption features. Field spectra of wheat grown on different levels of available nitrogen were collected, and compared to the foliar nitrogen content. The spectra were assessed both as absolute reflectances and recalculated into derivative spectra, and their respective wavelet transformed signals. Wavelet transformed signals, transformed using the Daubechies 5 motherwavelet at scaling level 32, performed consistently better than reflectance or derivative spectra when tested in a bootstrapped phased regression against nitrogen.

1 Introduction

Recent advances in remote sensing have resulted in the development of high spectral and spatial resolution sensors. These sensors enable us to measure more objects more accurately. Recent work has shown the utility of hyperspectral data to detect foliar nitrogen (Kokaly 2000; Lilienthal et al. 2000; Ferwerda et al. 2005), phosphorous (Mutanga et al. 2003), chlorophyll (Haboudane et al. 2002; Coops et al. 2003) and phenolic compounds (Soukupova et al. 2002; Ferwerda et al. in press).

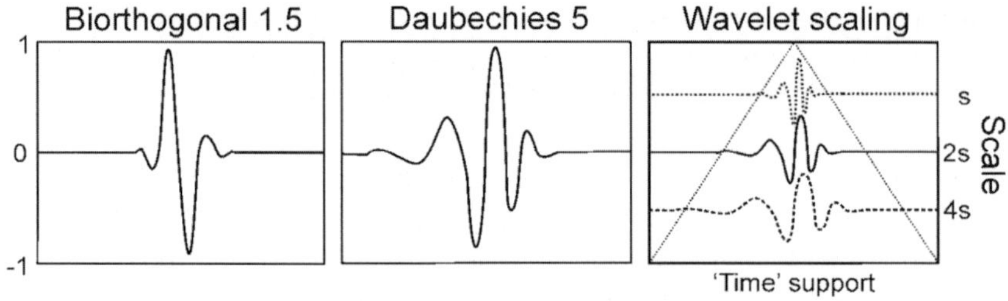

Fig. 1. Wavelets used for decomposition, and an illustration of the effect of scaling wavelets in the time-domain. With increasing scale levels the window of analysis, or time support, increases, and the frequency resolution decreases

With the development of high spectral resolution sensors, reflectance data has become near continuous. This has created an opportunity to treat individual measurements (in the case of field spectra) or the combined band values for individual pixels (in the case of hyperspectral images) as a continuous signal, where electromagnetic radiation reflectance is measured as a function of wavelength. Several methods, which were developed to detect absorption features in hyperspectral data, such as derivative analysis (Demetriades-Shah et al. 1990; Tsai and Philpot 1998) and continuum removal (Clark and Roush 1984), are based on the fact that the data forms a near-continuous signal. It is therefore surprising to note that studies using wavelet transformations for the analysis of absorption features in hyperspectral reflectance data are rare. Wavelet analysis is based on the Fourier transform, developed by Joseph Fourier in 1807. It decomposes a complex signal into component sub-signals. Each of these sub-signals is active at a different signal scale, or frequency. This makes it ideal for the detection of absorption features in complex signals

Wavelets look like small oscillating waves, and they have the ability to analyze a signal according to signal scale (frequency, see Fig. 1). In other words, during analysis the original wavelet, or mother wavelet, is scaled along the time-axis, to match signals at different frequencies. Thus a narrow wavelet (low scaling number) is used to match high frequency signals, irrespective of the underlying low-frequency changes. Low frequency signals are picked up using a wide wavelet (high scaling number), while high frequency signals (e.g., noise) is ignored.

The analysis of a signal is equivalent to computing all the correlations between the wavelet function at a certain scale and the input signal (Abbate et al. 2002, p 26). The results of the wavelet transform are the wavelet coefficients C, which are a function of scale and position of the transform. Multiplying each coefficient by the appropriately scaled and shifted wavelet yields the constituent wavelets of the original signal. This can be repeated at different wavelet scales, to either match against low or high frequency signals in the input signals. A specific form of wavelet transformation, the continuous wavelet transformation, decomposes a signal into an output signal with the same length as the input signal. To provide the reader with a comprehensive technical background of wavelet analysis falls outside the scope of this paper. For more information on the mathematical and historical aspects of wavelet analysis, please see the work by Hernandex (1996), Abbate (2002) or Ogden (1997).

Although wavelet analysis is used in remote sensing, this is mainly for the purpose of data compression (Amato et al. 2000; Bjorke and Nilsen 2002) and edge detection (Gruen and Li 1995). Examples of the use of wavelet transforms to enhance absorption features in hyperspectral data originate predominately from the food industry, where quality control is performed using near infrared lab-based spectroscopy. Chen (2002) for instance improved prediction of oil content in instant noodles by applying a 4-scale Mallat wavelet transform to NIR reflectance spectra. Fu (2005) successfully applied a Daubechies wavelet transform to NIR spectra of vinegar to derive sugar content. In the environmental sciences, Koger (2003) used wavelet-based analysis of hyperspectral reflectance signals for early season detection of 'Pitted morning-glory' in soybean fields. A field where the unambiguous identification of the characteristics of absorption features is of critical importance is that of hyperspectral remote sensing of foliar chemistry. Chemical components in foliage result in distinct absorption features, with a specific spectral location, and a depth and width related to the concentration of that component (Curran 1989). Therefore, during data processing it is crucial to be able to quantify changes in the depth and width of absorption-features. For the project presented here we hypothesized that wavelet analysis might do just that. Since wavelet transformed signals represent a measure of resemblance between the mother wavelet stretched at a certain scale, and the input signal at each specific spectral location, we hypothesize that this match between wavelet and derivative spectra will provide us with an unbiased measure of the shape of the absorption feature at that spectral location. Typically only the information of one or a few bands is included in a model to predict the concentration of specific foliar components. However, since wavelet transformation matches a subset of the input signal against a scaled wavelet, the output signal combines information on the variation over a number of surrounding bands. Therefore it is expected that the wavelet transform is less sensitive to noise and will result in a better relation to the component of interest than pure reflectance or derivative signals.

The ability to predict chemical composition of plants using remote sensing is directly dependent on the selection of appropriate bands to use. When specific absorption features are unknown, most studies have used stepwise regression techniques to determine which wavebands are most appropriate to use. This may however result in over-fitting of prediction models. Principal component analysis is another method, which combines the information from several bands into one predictor variable, and reduces this problem. It however minimizes the understanding of the relation between absorption features and chemical composition, since the effects of individual bands are combined into one factor. Because it reduces the number of variables while maintaining most of the information, principal component analysis has also been used to compress data. Still, selecting the appropriate predictor bands, and managing the volume of this data remains a problem. Ferwerda et al. (in Press) suggested a method they referred to as *bootstrapped phased regression*, that partially overcomes this problem. In short, from the original dataset of n unique samples, a bootstrap dataset of n samples is selected, allowing duplicate samples to occur (Efron and Tibshirani 1993). This is repeated 10 000 times, a number comparable to that suggested by Potvink and Roff (1993) and Pitt and Kreutzweiser (1998) for bootstrapping routines. For each repetition, the waveband with maximum correlation to the component of interest is recorded, which results in a frequency table detailing the number of times that each waveband has maximum correlation with the component of interest. The band with the highest frequency after 10 000 iterations is selected, where the position of a band is defined as the central wavelength of that band.

This step is repeated in order to build a linear regression model with more than one predictor by calculating the regression goodness of fit between the component of interest and the already selected band combined with each of the other bands, again selecting the band with the highest frequency of maximum correlation for 10 000 random datasets. This routine deviates from a full stepwise regression within each bootstrap repetition because the aim is to select the best waveband to use with respect to already selected wavebands. This process is repeated until the required number of bands is reached or until the maximum frequency of maximum correlation drops below 5%.

This study explores the ability of wavelet transformation to enhance absorption features in reflectance signatures. Wavelet transformation was applied to derivative signatures of wheat of which the foliar nitrogen content was known. Subsequently a bootstrapped phased regression was applied to select the best bands for prediction of foliar N. The predictive power of wavelet factors was compared to that of reflectance and derivative spectra.

2 Materials and Methods

2.1 Field Data

Wheat was grown during the 2004-growing season in Horsham, Victoria, Australia (Sowing date: June 17th). As part of a larger experiment on the effects of nitrogen and water availability on the productivity of wheat in semi-arid systems, wheat was grown on a split-plot factorial design. The treatments consisted of irrigated (390 mm; decile 9 for Horsham) and rain-fed (270 mm; decile 5 for Horsham), combined with two plant densities (300 and 150 plants/m^2) and four levels of nitrogen applied as urea (0, 34, 84 and 354 kg urea/ha) in subplots with three replications.

2.2 Hyperspectral Measurements

At the end of the growing season (November 8th, 144 days after sowing) spectral properties were recorded. Approximately 1 m^2 of the canopy was recorded using an ASD Fieldspec FR field spectrometer. The FieldSpec® FR spectrometer combines three spectrometers to cover the 350 to 2500 nm range of the electromagnetic spectrum. This spectrometer uses a photo diode array to cover the 350 to 1000 nm spectral range with 1.4 nm sampling interval and 3 nm bandwidth. Two fast scanning spectrometers provide coverage for the wavelength range from 1000 to 2500 nm with 2 nm sampling interval and 10 nm bandwidth. The optic fiber was placed in a pistol grip and mounted on a steel boom 2.5 m above ground surface pointing downwards in a 900 angle to measure the up-welling radiance of the wheat. Absolute reflectance was calculated using a calibrated Spectralon Reflectance Target (Labsphere, Inc, North Sutton, New Hampshire) as a reference. The centre of the measured area was harvested (0.9 m^2) and a random sub-sampled was chemically analyzed to determine total canopy nitrogen content.

2.3 Data Processing

The reflectance spectra were very noisy between 1850 and 2500 nm. Therefore this part of the spectrum was excluded from further analysis. During wavelet transform, the wavelet used should represent the signal to be detected as closely as possible. In this paper it was decided to use two mother wavelets (e.g., Fig. 1): one which closely represented the derivative of a Gaussian distribution; the second of which represented a combination of the derivative of multiple Gaussian absorption features. Therefore derivative spectra were calculated of the original reflectance spectra. An added advantage is the signal normalizing effect of calculating derivative spectra. Derivative analysis (Tsai and Philpot 1998) assumes that differences in the absolute reflectance do not affect the actual absorption features. Therefore derivative spectra are less affected by sun angle and structural variation than absolute reflectance spectra, and by using the slope of the spectrum instead of the absolute reflectance, the same signal is produced for samples with different absolute reflectance but the same absorption features.

Using the reflectance spectra, derivative spectra were calculated using an adjusted version of the seven band moving spline smoothing technique (Savitzky and Golay 1964; Tsai and Philpot 1998). Instead of smoothing the spectra first and then calculating the derivative spectra from the smoothed spectra, the parameters of the moving polynomial were used to directly calculate the derivative at the centre waveband of the moving spline window.

Reflectance and derivative spectra were averaged by individual treatments, and plotted against wavelength to visually analyze the differences in reflectance between treatments, and the effect of normalizing data through derivative calculation. To better understand the sources of variation in the dataset, A full factorial ANOVA was performed on a subset of wavebands (350 nm to 1250 nm, step 100 nm), with applied nitrogen level (n=4), irrigation (n=2) and planting density (n=2) as interacting factors. Similarly, the effect of individual treatments on the foliar nitrogen content was tested using a factorial analysis of variance with nitrogen level (n=4), irrigation (n=2) and planting density (n=2) as interacting factors. The concentration of nitrogen was recorded as a fraction of the dry weight. Since these values are typically low (< 10%), these concentrations require a log-transformation to meet requirements of normality (Zar 1999). After transformation groups did not deviate from normality (Shapiro wilks' W; $p > 0.05$).

2.4 Wavelet Transform

The Matlab environment was used to perform continuous wavelet transform on the derivative spectra, decomposing the derivative spectra at 7 scale levels, in 2^a step increments ($0 < a < 7$). Two mother wavelets were used for decomposition. The first is the biorthogonal wavelet 1.5 '(Bior.1.5, see Fig. 1), which represents the derivative of a Gaussian curve.

The other is the Daubechies wavelet 5 (DB.5) a basic ripple with 5 convolutions (see Fig. 1).

The 16 resulting datasets (Reflectance spectra, Derivative Spectra and 2 times 7 wavelet transforms) were tested for their relation foliar nitrogen content. To select the most robust wavebands for detecting nitrogen content, a bootstrapped phased regression was applied.

3 Results

3.1 Effects of Treatment

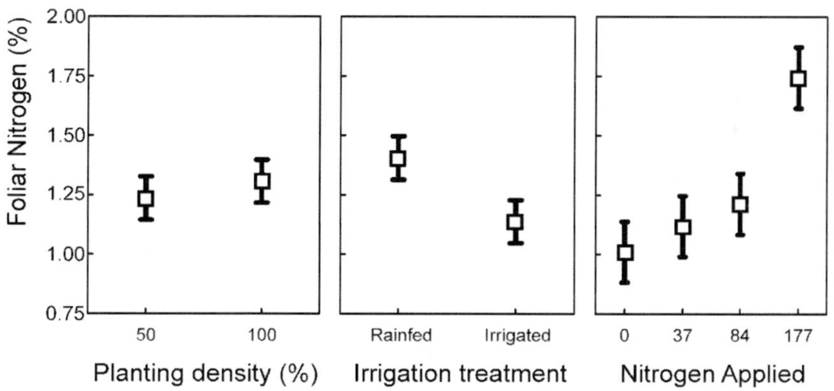

Fig. 2. Foliar nitrogen content in wheat samples, grouped by nitrogen treatment and irrigation treatment. Vertical bars denote 95% confidence interval for the mean

Foliar nitrogen concentration was affected by nitrogen application and irrigation treatment individually (ANOVA; $p \leq 0.001$; see Fig. 2) but not by sowing density (ANOVA; $p \geq 0.1$). Second degree interaction terms were not significant. Group sizes were too small to reliably calculate full factorial interaction terms.

The effects of irrigation and nitrogen treatments on reflectance spectra (see Fig. 3) were significant, but the effect of 'between planting densities' was not (ANOVA; $p \leq 0.001$). The interactions between treatments were not significant.

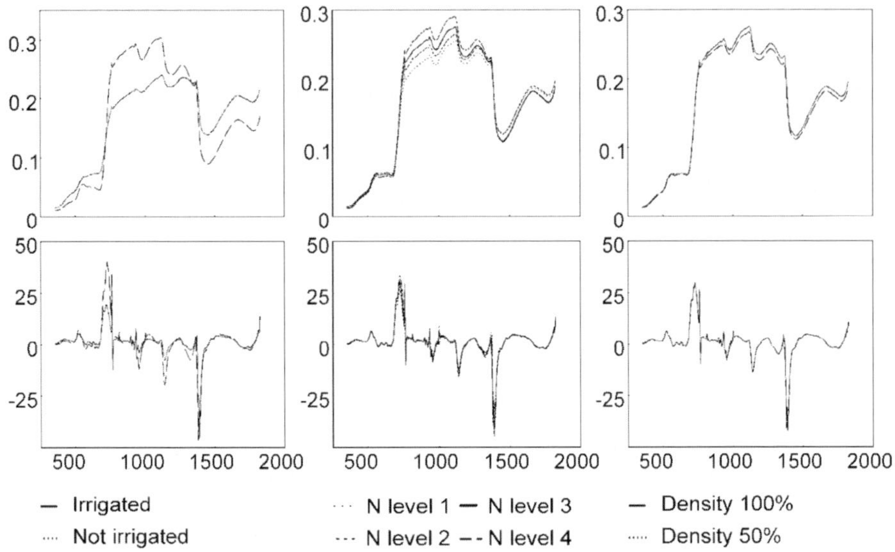

Fig. 3. Comparison of reflectance and derivative spectra averaged by treatments. N levels: 1: 0 kg urea/ha, 2: 34 kg urea/ha, 3: 84 kg urea/ha and 4: 354 kg urea/ha

Decomposition of spectra using a wavelet transform showed localized responses, in particular around 1000 nm and 1400 nm for lower scale transforms (matching higher frequency signals, see Fig. 4). Higher scale levels (Matching lower frequency signals) result as expected, in responses over wider regions of the signal (see Fig. 4).

Figure 5 depicts the regression goodness of fit (r^2) between foliar nitrogen concentration and reflectance spectra, derivative spectra, and wavelet transformed spectra for models with 1 to 6 predictor bands. Derivative spectra of wheat, transformed using a DB.5 wavelet transform at scale 32, show a stronger relation to foliar nitrogen concentration than the original derivative spectra (see Fig. 5). Figure 5 shows a mean regression r^2 of 0.54 between the best band for DB.5 scale level 32 transformed derivative spectra over 10 000 bootstrap iterations, whereas this is only 0.03 and 0.31 for the pure reflectance spectra, and derivative spectra respectively. For all models a maximum mean regression goodness of fit is achieved when derivative spectra are transformed using DB.5 at scale level 32 (see Fig. 5). Derivative spectra have a higher mean regression goodness of fit than reflectance spectra. Derivative spectra transformed using Bior.1.5 perform less good than the input derivative spectra (see Fig. 5).

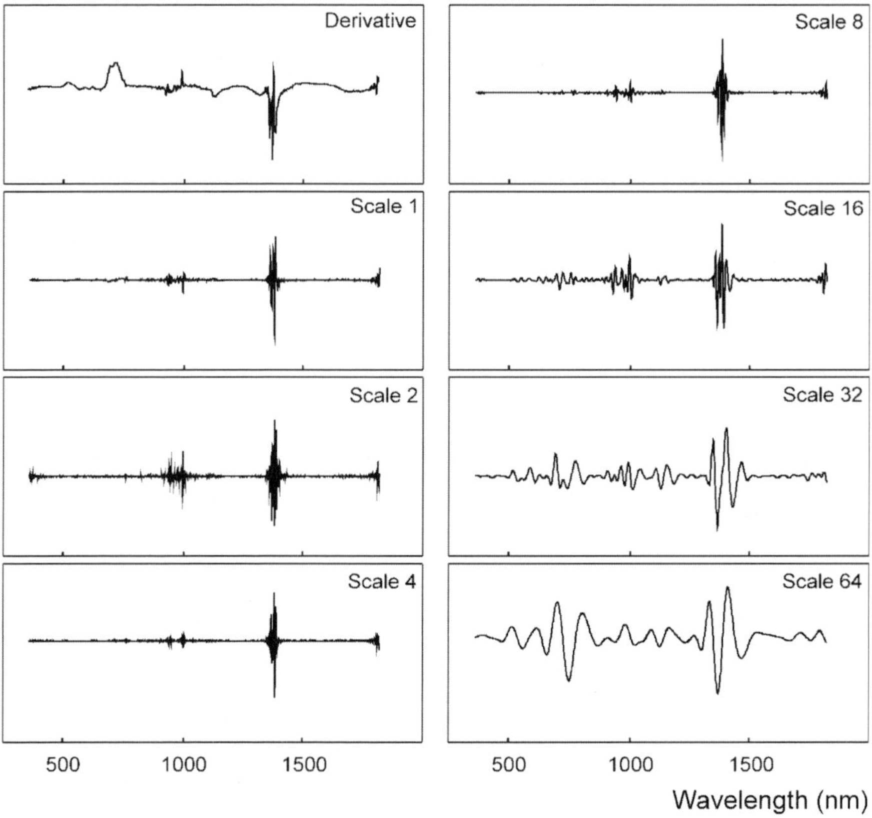

Fig. 4. Derivative spectrum of wheat and the 7 wavelet transforms using a Daubechies 5 continuous wavelet transform

4 Discussion

Transforming derivative spectra in the wavelet transformed signals resulted in an increased correlation with foliar nitrogen concentration. Low scale analysis uses a small window for analysis, in other words: the mother wavelet is stretched across only a small part of the spectrum during the analysis. Therefore it is well suited to detect high-frequency changes in the signal (absorption features). An increase scale results in a wider analysis window and the mother wavelet is scaled along a wider stretch of the spectrum. Wavelet transformations at higher scales are consequently more suitable for detecting changes in large absorption features. The graphs in Figure 5 show a steep increase in the relation between the Daubechies wavelet transformed signal and foliar nitrogen when moving from scale 2 to scale 32, for all regression models. At scale 64 the regression goodness of fit is lower than at scale 32. This suggests that the optimal scale for detection of nitrogen using derivative spectra is located between 16 and 64.

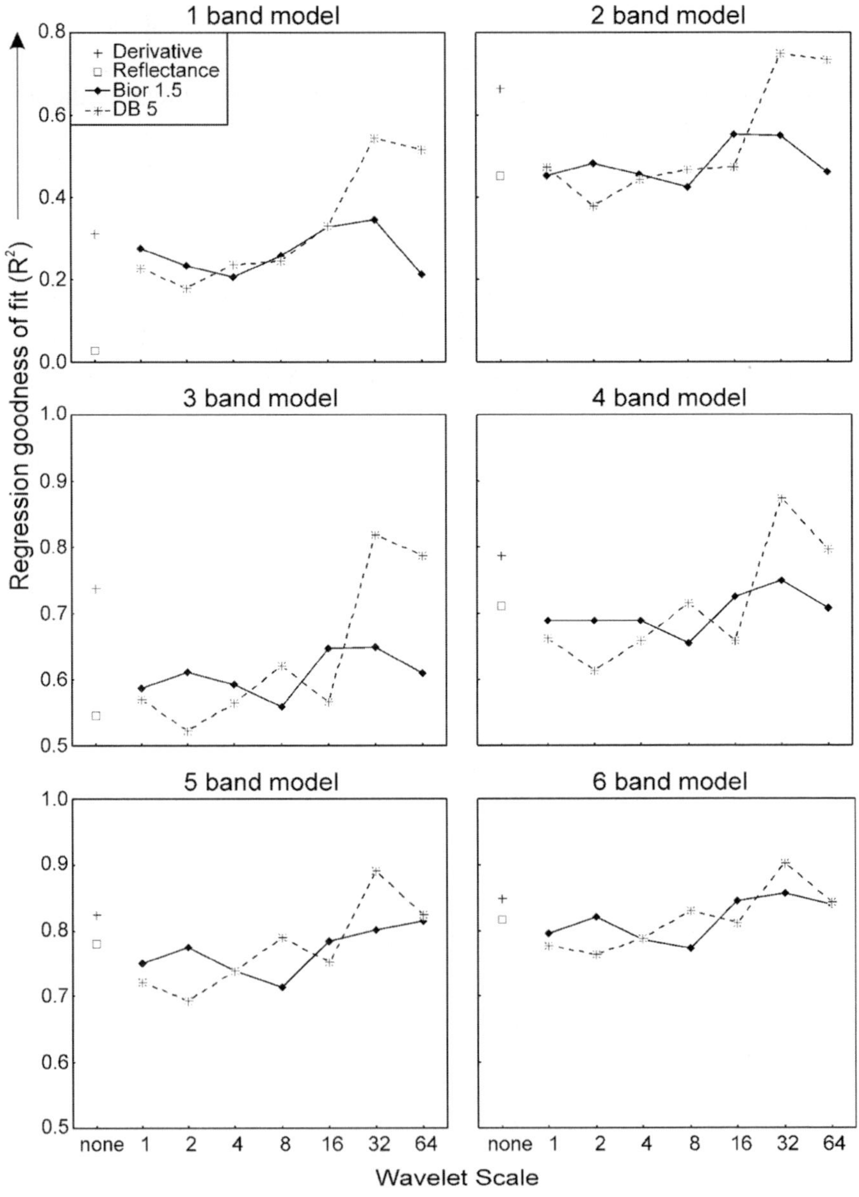

Fig. 5. Mean regression goodness of fit (r^2) over 10 000 bootstrap iterations for regression models with 1 to 6 predictors, in regression between nitrogen and reflectance, derivative and wavelet transformed spectra

The current work does not take into account the various aspects of remote sensing using imaging spectrometers, such as mixed signals, canopy shading and varying canopy architecture. Obviously these factors will affect the outcome of these results, and to fully understand whether wavelet transformations are appropriate signal processing tools for hyperspectral data, a subsequent study is required. This would ideally integrate the ideas presented here in a purpose designed image-based study.

Acknowledgements

The authors would like to thank the Victorian Department of Primary Industries, Australia in Horsham, Australia for generously sharing their data. The field site and experimental data used in the analysis in this paper were derived from an "Our Rural landscape" project. We thank Lene Christensen for acquisition of the data, Russel Argall for conduct of the field experiment, and Garry O'Leary for advice and discussion in relation to the dataset.

References

Abbate A, DeCuusatis CM, Das PK (2002) Wavelets and Subbands. Fundamentals and applications. Birkhauser, Bosten

Amato U, Angelini C, Serio C. (2000) Compression of AVHRR images by wavelet packets. Environmental Modelling and Software 15:127–138

Bjorke JT, Nilsen S (2002) Efficient representation of digital terrain models: compression and spatial decorrelation techniques. Computers & Geosciences 28: 433–445

Chen B, Fu X-G, Lu D-L (2002) Improvement of predicting precision of oil content in instant noodles by using wavelet transforms to treat near-infrared spectroscopy. J of Food Engineering 53:373–376

Clark RN, Roush TL (1984) Reflectance Spectroscopy: Quantative Analysis Techniques for Remote Sensing Applications. J of Geophysical Research 89: 6329–6340

Coops NC, Stone C, Culvenor DS, Chisholm LA, Merton RN (2003) Chlorophyll content in eucalypt vegetation at the leaf and canopy scales as derived from high-resolution spectral data. Tree Physiology 23:23–31

Curran PJ (1989) Remote sensing of Foliar Chemistry. Remote Sensing of Environment 30:271–278

Demetriades-Shah TH, Steven MD, Clark JA (1990) High resolution Derivative Spectra in Remote Sensing. Remote Sensing of Environment 33:55–64

Efron B, Tibshirani RJ (1993) An introduction to the Bootstrap. Chapman & Hall/CRC, London

Ferwerda JG, Skidmore AK, Mutanga O (2005) Nitrogen detection with hyperspectral normalized ratio indices across multiple plant species. Int J of Remote Sensing 26:4083–4095

Ferwerda JG, Skidmore AK, Stein A (in press) A bootstrap procedure to select hyperspectral wavebands related to tannin content. Int J of Remote Sensing

Fu X, Yan G, Chen B, Li H (2005) Application of wavelet transforms to improve prediction precision of near infrared spectra. J of Food Engineering 69: 461–466

Gruen A, Li H (1995) Road extraction from aerial and satellite images by dynamic programming. ISPRS J of Photogrammetry & Remote Sensing 50:11–20

Haboudane D, Miller JR, Tremblay N, Zarco-Tejada PJ, Dextraze L (2002) Integrated narrow-band vegetation indices for prediction of crop chlorophyll content for application to precision agriculture. Remote Sensing of Environment 81:416–426

Hernandez E, Weiss H (1996) A first course on Wavelets. CRC Press, Boca Raton

Koger CH, Bruce LM, Shaw DR, Reddy KN (2003) Wavelet analysis of hyperspectral reflectance data for detecting pitted morningglory (*Ipomoea lacunosa*) in soybean (*Glycine max*). Remote Sensing of Environment 86: 108–119

Kokaly RF (2000) Investigating a physical basis for spectroscopic estimates of leaf nitrogen content. Remote Sensing of Environment 75:153–161

Lilienthal H, Haneklaus S, Schnug E, Haveresch E (2000) Utilisation of hyperspectral data for the evaluation of the spatial variability of the nitrogen status of wheat. Aspects of Applied Biology 60:189–194

Mutanga O, Skidmore AK, Prins HHT (2003) Predicting in situ pasture quality in the Kruger National Park, South Africa, using continuum-removed absorption features. Remote Sensing of Environment 89:393-408

Ogden RT (1997) Essential wavelets for statistical applications and data analysis. Birkhauser, Boston

Pitt DG, Kreutzweiser DP (1998) Applications of computer-intensive statistical methods to environmental research (review). Ecotoxicology and environmental safety 39:78–97

Potvink C, Roff DA (1993) Distribution-Free and Robust Statistical Methods: Viable alternatives to parametric statistics. Ecology 74:1617–1628

Savitzky A, Golay MJE (1964) Smoothing and differentiation of data by simplified least squares procedures. Analytical Chemistry 36:1627–1629

Soukupova J, Rock BN, Albrechtova J (2002) Spectral characteristics of lignin and soluble phenolics in the near infrared – a comparative study. Int J of Remote Sensing 23:3039–3055

Tsai F, Philpot W (1998) Derivative analysis of hyperspectral data. Remote Sensing of Environment 51:66–41

Zar JH (1999) Biostatistical Analysis. Prentice-Hall, London

Measuring Linear Complexity with Wavelets

Geoff J. Lawford

Department of Geomatics, University of Melbourne, and
Geospatial and Earth Monitoring Division, Geoscience Australia
Corner Jerrabomberra Avenue and Hindmarsh Drive
Symonston, ACT 2609, Australia

Abstract

This paper explores wavelets as a measure of linear complexity. An introduction to wavelets is given before applying them to spatial data and testing them as a complexity measure on a vector representation of the Australian coastline. Wavelets are shown to be successful at measuring linear complexity. The technique used breaks a single line into different classes of complexity, and has the advantages that it is objective, automated and fast.

1 Introduction

Linear complexity is the extent to which a line is intricate and convoluted, a highly complex line being one that is angular, inflected, and tightly circuitous, as opposed to straight. The study of linear complexity in cartography waxed in the 1980s and 1990s when it was seen as potentially enhancing automatic feature recognition and the assessment and functionality of line simplification algorithms. Notably, McMaster (1986) defined and assessed a suite of complexity-related measures for evaluating the effect of line simplification algorithms. Also, Buttenfield (1991) proposed five complexity-related indices, together forming a "structural signature", for restraining simplification algorithms from removing important linear detail, and Jasinski (1990) summarized eight of the most promising complexity measures and studied their response to successive levels of line simplification. Sixteen years later interest in the study of linear complexity has waned, yet it still promises to improve automated feature recognition, in a broader sense terrain and morphology recognition, linear simplification and even enhancement, and "scale" determination for digital data. In this paper the focus returns to linear complexity with the introduction, not of a new index of complexity, but of a new method of measuring complexity using an existing index. The new method is based on wavelet technology. The paper begins with a review of wavelets, using the Haar wavelet to explain and illustrate the main concepts. A section in which wavelets are applied to spatial data follows this, with the Australian mainland coastline used for examples. Two techniques are then tested for using wavelets to classify a single line into sections based on complexity.

2 Terminology

In this paper the author uses a new word for the "scale" of digital data. Although digital data are scale free, they are often captured from paper maps or customized for plotting on paper maps at a particular scale. As noted by Chrisman (1982, p 161), scale "is more than the mathematical relationship between the earth and the map". It implies "the degree of detail shown in the geometric form ... the physical size of the smallest entity shown ... [and] the positional accuracy of mapped features" (Goodchild and Proctor 1997, p 11). To accommodate the fact that digital data are scale free, and at the same time retain all the connotations of the word scale, and the name by which they are most commonly known, the term "e-scale" will be used throughout this paper. The "e-scale" of digital data can be taken to mean the scale of the paper map from which the data was originally captured or to which it is customized for output. It is both an equivalent scale, and a "scale" that applies to electronic data.

3 Wavelet History

Wavelets, like Fourier's sinusoidal waves, are simple functions used, through superimposition, to build compound functions. Stollnitz et al. (1995, p 76) describe wavelets as "a mathematical tool for hierarchically decomposing functions. They allow a function to be described in terms of a coarse overall shape, plus details that range from broad to narrow. Regardless of whether the function of interest is an image, a curve, or a surface, wavelets offer an elegant technique for representing the levels of detail present".

Wavelet theory is based on Fourier theory, though as early as 1873 it broke from Fourier theory with the work of the German mathematician Karl Weierstrass on building functions from scaled copies of a simple function (Stollnitz et al. 1996). Further work on wavelet theory was undertaken in 1909 by the Hungarian mathematician Alfred Haar though much of the theory was not developed until the 1980s and early 1990s when numerous scholars, in particular European physicists, geophysicists and mathematicians, broadly expanded the theory. One of these scholars, the French geophysicist Jean Morlet, is generally credited with coining the term "wavelet" in the early 1980s, translating it from the French word *ondelette* for "small wave" (Wikipedia 2005). However, Stollnitz et al. (1996, p 3) states that the term "comes from the field of seismology, where it was coined by Ricker in 1940 to describe the disturbance that proceeds outward from a sharp seismic impulse or explosive charge". Norman Ricker was an American physicist who worked in both industry and academia.

Following work carried out in the late 1980s by the French mathematicians Stephane Mallat and Yves Meyer, which led to the development of the Fast Wavelet Transformation, wavelet theory was increasingly applied to real world problems. Like the Fast Fourier Transformation, the Fast Wavelet Transformation, which also requires that the number of sample points be an integer power of 2, greatly increased the speed of transformation. Indeed, the Fast Wavelet Transformation is faster than the Fast Fourier Transformation. Today wavelets are used in a broad range of applications, in particular in the fields of astronomy, seismology, image processing, and signal analysis.

While wavelet and Fourier theory have the same ancestry they differ in two fundamental ways. Firstly, while Fourier theory is based on waves of sinusoidal shape, wavelet theory is non-prescriptive regarding wave shape. For wavelets, wave shape can be chosen to best suit a particular application. Secondly, while Fourier theory repeats each wave end on end, wavelet theory requires that each wave occur only once, and outside the limits of this single pulse there is no oscillation. This enables wavelet theory to better localize, or position, detail along a compound function, and ensures it can be applied to non-periodic phenomena. Thus, when applied to spatial data, Fourier theory is largely limited to closed polygonal features such as coastlines, while wavelet theory can be applied to both closed and open ended features, including roads and watercourses. The ability to localize detail along a function and work with the full range of linear phenomena makes wavelets appealing as a potential measure of linear complexity.

In the following section the simplest and first wavelet developed, the Haar wavelet is used to illustrate the main characteristics of wavelets and how the wavelet transformation works.

4 Wavelets in One Dimension

The Haar wavelet and its associated scaling function, illustrated in Figure 1, are both box shaped. A scaling function is the base within which a wavelet operates and defines the coarsest level of detail of a wavelet transformation. For the Haar wavelet the scaling function is a horizontal line the start and end of which define the limits of the wavelet operation. The Haar wavelet has the following attributes:

- adjacent wavelets at the same level do not overlap
- small wavelets fall within sections of large wavelets and their scaling functions, where the latter two are constant
- the wavelet is zero outside a narrow domain

The first and second attributes are a result of a property called "orthogonality", which means that adjacent wavelets and wavelets at larger and smaller frequencies or levels, and their scaling functions, are "perpendicular" to each other. This means that their inner vector product is zero. The third characteristic means that the wavelet has "compact support". The Haar wavelet is unique in being orthogonal and compactly supported, as well as symmetrical.

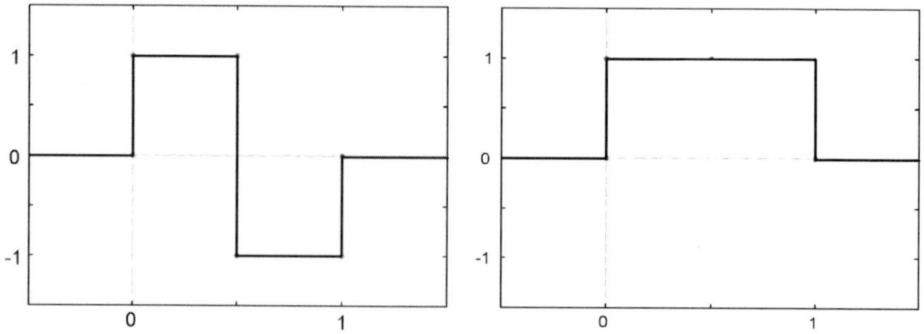

Fig. 1. Left, the Haar wavelet, and right, its associated scaling function, both shown in a thick black line

The following example illustrates how a wavelet transformation using the Haar wavelet can turn the number sequence 4,2,6,8,4,8,7,9 into the wavelet coefficients 6,-1,-2,-1,1,-1,-2,-1. The example is divided into three steps, each with an associated figure.

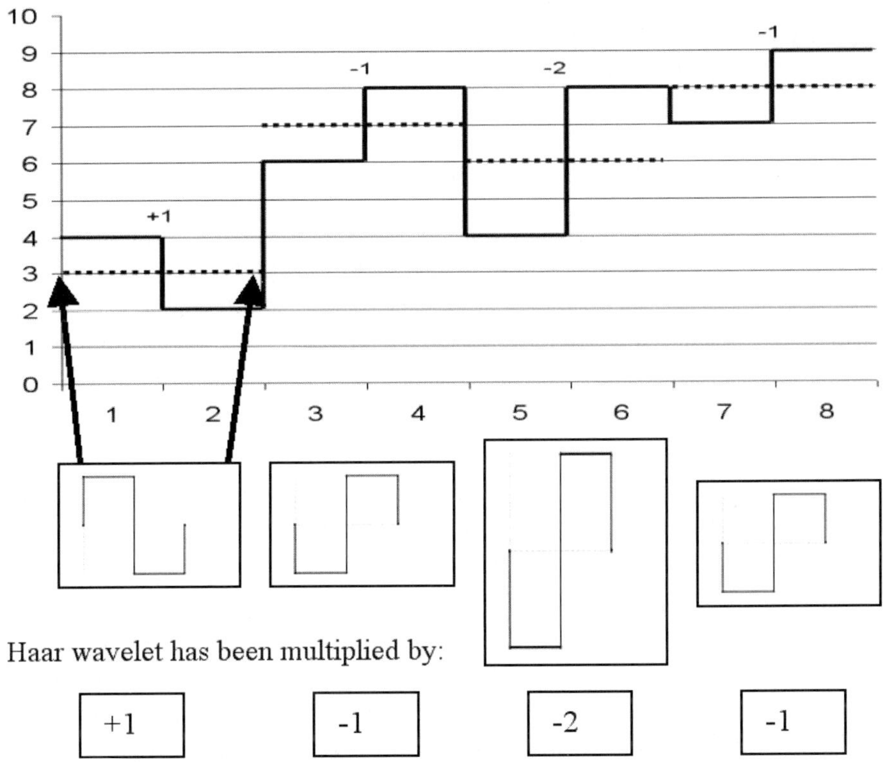

Fig. 2. Step 1. Pairwise averaging of the number sequence 4,2,6,8,4,8,7,9, using the Haar wavelet as a base

Step 1. Do a pairwise averaging of the number sequence 4,2,6,8,4,8,7,9, as shown by the dark black line in the graph in Figure 2, to obtain a lower resolution version of the number sequence, plus four coefficients.

$$\begin{aligned}
&\left.\begin{matrix}4\\2\end{matrix}\right\} = 3, coefficient = +1 \\
&\left.\begin{matrix}6\\8\end{matrix}\right\} = 7, coefficient = -1 \\
&\left.\begin{matrix}4\\8\end{matrix}\right\} = 6, coefficient = -2 \\
&\left.\begin{matrix}7\\9\end{matrix}\right\} = 8, coefficient = -1
\end{aligned} \tag{1}$$

A lower resolution version of the number sequence, represented in the graph by the dotted lines is thus 3,7,6,8. The set of wavelet coefficients 1, -1,-2,-1 represent the factors by which the Haar wavelet must be multiplied when positioned and scaled so that it is coincident with each averaged pair of numbers.

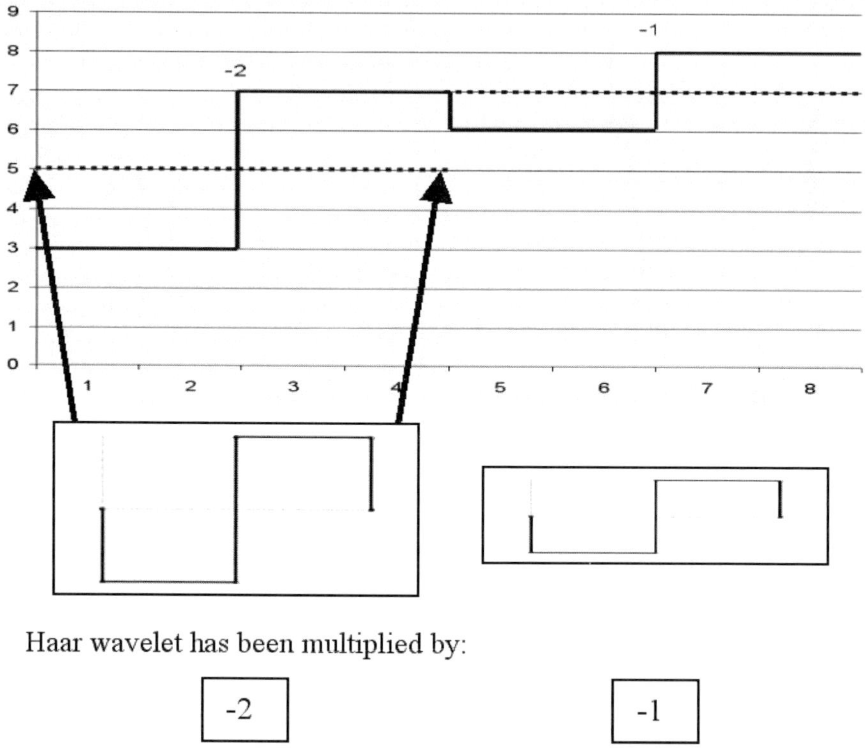

Fig. 3. Step 2. Pair wise averaging of the number sequence 3,7,6,8, using the Haar wavelet as a base.

Step 2. Do a pair wise averaging of the averaged number sequence 3,7,6,8, as shown by the dark black line in the graph in Figure 3, to obtain a lower resolution version of the number sequence, plus two more coefficients:

$$\left.\begin{matrix}3\\7\end{matrix}\right\} = 5, coefficient = -2$$
$$\left.\begin{matrix}6\\8\end{matrix}\right\} = 7, coefficient = -1 \qquad (2)$$

A new even lower resolution version of the number sequence, represented in the graph by the dotted lines is thus 5,7. The additional wavelet coefficients -2,-1 represent the factors by which the Haar wavelet must be multiplied when positioned and scaled so that it is coincident with each new averaged pair of numbers. These coefficients are added to the front of the existing set of coefficients as follows: -2,-1,1,-1,-2,-1

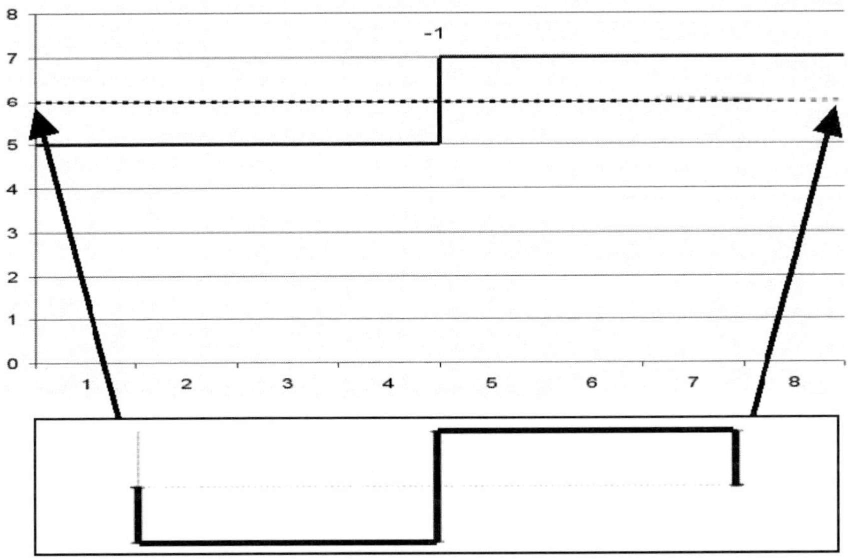

Haar wavelet has been multiplied by:

-1

Fig. 4. Step 3. Pair wise averaging of the number sequence 5,7, using the Haar wavelet as a base.

Step 3. Do a pair wise averaging of the previous averaged number sequence 5,7, as shown by the dark black line in the graph in Figure 4, to obtain a lower resolution version of the number sequence, plus one more coefficient.

$$\left.\begin{matrix}5\\7\end{matrix}\right\} = 6, coefficient = -1 \qquad (3)$$

A new even lower resolution version of the number sequence, represented in the graph by the dotted lines is thus 6. This is the coefficient by which the scaling factor, which is coincident with the entire original number sequence, is multiplied. The additional wavelet coefficient -1 represents the factor by which the Haar wavelet must be multiplied when positioned and scaled so that it is coincident with the entire original number sequence.

The output of the wavelet transformation is the number sequence 6 which is the coarse average value of the entire original number sequence, followed by –1,-2,-1,1,-1,-2,-1 which are the multiplication factors of the progressively re-scaled and re-positioned Haar wavelet such that the original number sequence can be regenerated. In this example there is one scaling factor and three "levels" of wavelets, one level for each described step. In the following section the Haar wavelet is applied to spatial data.

5 Wavelets with Spatial Data

As can be shown with regard to Fourier series, wavelet technology can break down complex linear features into a series of simple functions, or wavelets. This is done by applying a discrete wavelet transformation to a sequence of X and Y axis coordinate values defining the complex features to derive coefficients for wavelets operating along each of the X and Y axes respectively. The coordinate values to be transformed can be for the vertices defining the line, for a set of equally spaced sample points along the line, or for any other set of points defining the line. A complex linear feature broken into wavelets in this way can be regenerated by merging the separate wavelets back together. This is done by executing an inverse discrete wavelet transformation on the unaltered coefficients. This will regenerate perfectly the original set of coordinates as they were prior to the forward transformation. Interestingly, a generalized version of the original complex linear feature can be created by excluding higher resolution wavelets from the inverse transformation. Also, the coefficient data can be compressed, with minimum loss of detail to the original line, by calculating coefficients under a specified threshold to zero.

To illustrate how the wavelet transformation works with spatial data, Figure 5 shows the "centroid" plus the first three wavelets constituting the 1:10 million e-scale vector representation of the Australian mainland coastline. The Haar wavelet is used. The coastline was sampled with 2^{13} (8192) equally spaced points. The X and Y coordinates of the sampled points, in decimal degrees of longitude and latitude respectively, were transformed separately. The resultant wavelet coefficients are given in the first two rows of the table, ignoring the header row.

1st coefficient	2nd coefficient	3rd coefficient	4th coefficient
202.7	-717.8	-86.8	162.5
1870.0	398.1	-678.3	504.9

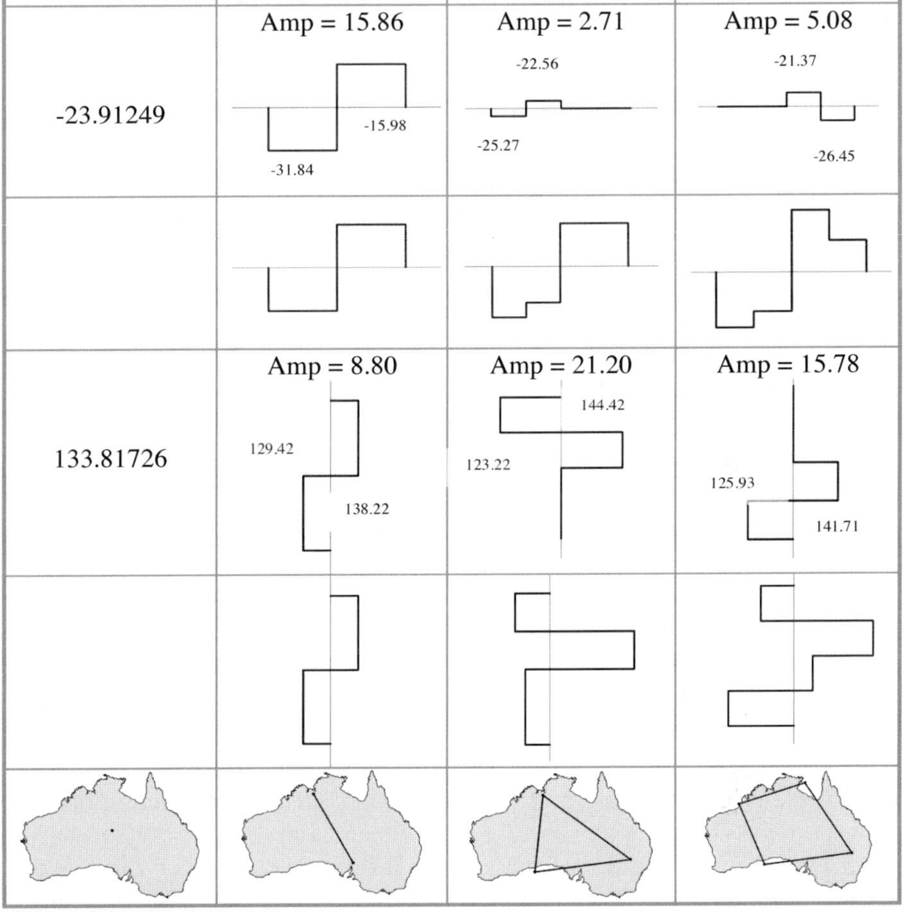

Fig. 5. The "centroid" plus the first three wavelets, and their progressive merger, for both X and Y axes, created from a fast wavelet transformation of the Australian 1:10 million e- scale coastline using the Haar wavelet. The original coastline was sampled with 2^{13} (8192) equally spaced points. Excluding the heading, the rows illustrate, from top: (1) the Y wavelet coefficient, (2) the X wavelet coefficient (3) the "centre" and wavelets for the Y coordinates with amplitude (Amp) and Y dimensions, (4) the progressive merger of the Y axis wavelets, (5) the "centre" and wavelets for the X coordinates with amplitude (Amp) and X dimensions, (6) the progressive merger of the X axis wavelets, and (7) the "centroid" and merger of both the X and Y wavelets relative to the original coastline

The figure shows visually the first three wavelets for both X and Y axes, plus the progressive merger of those wavelets for each axes. The last row of the figure shows the shape generated by combining the merged wavelets in both X and Y directions, against a backdrop of the original coastline. In this example and throughout this research a Fast Wavelet Transformation has been used. The wavelets and merged wavelets shown in the figure have been created by executing an Inverse Fast Wavelet Transformation only on those coefficients of interest. In other words, prior to executing the inverse transformation all other coefficients were set to zero.

Figure 6 shows four of the more advanced stages of regenerating the 1:10 million e-scale Australian mainland coastline shown in Figure 5.

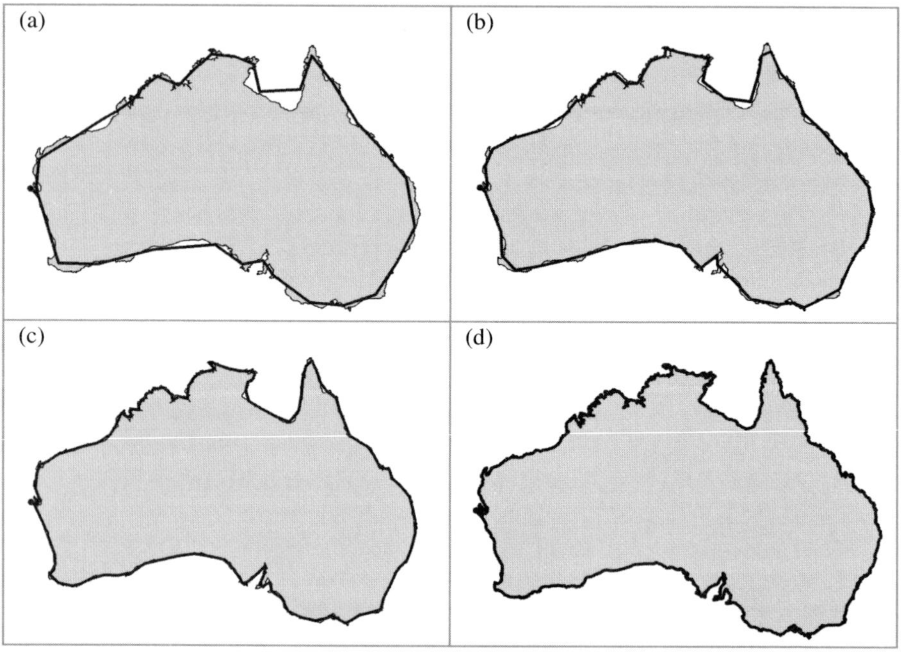

Fig. 6. The Australian mainland coastline from data at 1:10 million e-scale, overlaid with, in a heavy line, the coastlines formed by merging the wavelets defined by the first **(a)** 32, **(b)** 64, **(c)** 128, and **(d)** 2^{13} (8192) coefficients. The latter coastline includes all the wavelets derived when sampling the coastline with 2^{13} (8192) equally spaced points, as was the case here

It shows the coastlines regenerated from the first 32, 64, 128 and 8192 coefficients respectively from a fast wavelet transformation of a 2^{13} (8192) point sampling of the original coastline. The transformation used the Haar wavelet. The latter coastline perfectly matches the original coastline as sampled, as it includes all 2^{13} (8192) coefficients. In regenerating the first three coastlines, all coefficients but those of interest were set to zero prior to executing the inverse transformation. The output from the inverse transformation was a sequence of coordinates that defined the new coastline. As previously, the transformations of the X and Y coordinate values were undertaken independently.

Figure 7 shows more clearly the gain in fine detail as the higher-level coefficients are retained prior to executing the inverse transformation. The figure shows Shark Bay, on the western tip of the Australian coastline, using 1:10 million e-scale data. On top, shown in a thick black line, from left to right, are the coastlines regenerated from the first 128, 512, 2048, and 8192 coefficients respectively from a fast wavelet transformation of a 2^{13} (8192) point sampling of the entire original coastline around the whole continent. The same method and parameters were used as for Figure 6.

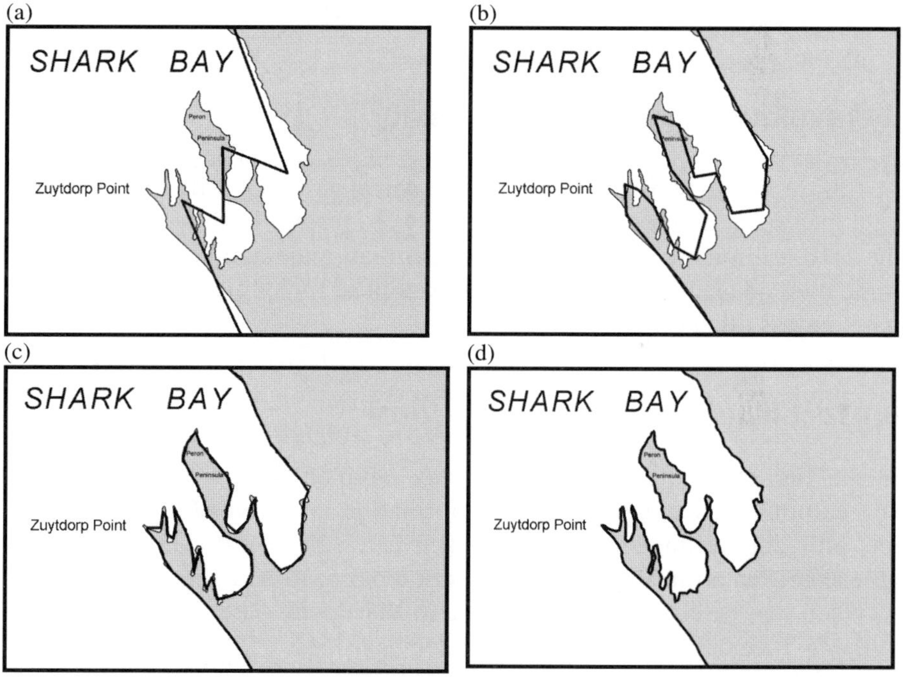

Fig. 7. The western tip of the Australian mainland coastline, at Shark Bay, from data at 1:10 million e-scale, overlaid with, in a heavy line, the coastlines formed by merging the wavelets defined by the first **(a)** 128, **(b)** 512, **(c)** 2048, and **(d)** 2^{13} (8192) coefficients

Figure 8 shows the difference in the regenerated coastline for wavelets other than, and including, the Haar wavelet. The wavelets are from the Daubechies family of wavelets, invented in the 1980s and named after their inventor, the Belgium physicist Ingrid Daubechies. They are nowadays used in many applications and differ from other wavelets by being both compactly supported and orthogonal. The wavelets in the figure are of dimension (a) 2, that is, the Haar wavelet, (b) 4, (c) 6, (d) 8, and (e) 10, with the dimension representing the number of filtering loops executed within the wavelet algorithm. In this example the same method and parameters were used as in the previous examples, except that here all but the first 1024 coefficients were zeroed. Although the higher dimension Daubechies wavelets are considered superior to the Haar wavelet at modeling smooth shaped functions, as can be seen from the figure, they overlap sideways, making them less suitable for isolating singularities and linear complexity.

6 Wavelets and Linear Complexity

In this research two techniques using wavelets to classify linear spatial data into sections based on complexity were tested. Both techniques used the 1:10 million e-scale data of the Australian mainland coastline, and the same method as used in the previous examples, including the Haar wavelet base, unless otherwise specified.

6.1 Technique 1

In the first technique the coastline was sampled with 2^{13} (8192) equally spaced points, meaning that the transformation involved 12 levels of wavelets plus one scaling function. For each level, if a coefficient based variable exceeded a threshold then the section of coastline coincident with that wavelet was categorized as complex to that level. The coefficient based variable, here called C_{var}, was calculated as follows

$$C_{\text{var}} = C_{wav} \times \sqrt{2^j} \div L \tag{4}$$

where C_{wav} was the coefficient returned by the forward transformation, j was an index to the wavelet level, where $j \in 0,1,2\ldots 11$ for the 12 levels of the transformation, and $L = \frac{1}{2} l$ where l is the length of coastline

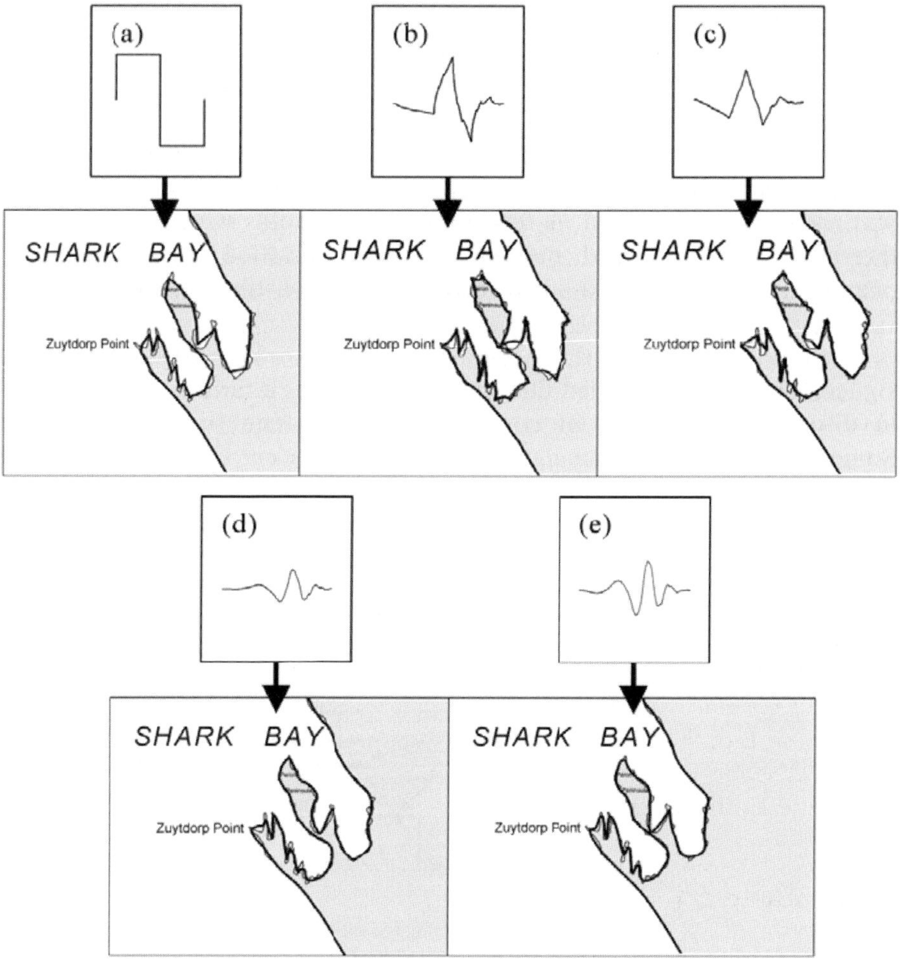

Fig. 8. The western tip of the Australian mainland coastline, at Shark Bay, from data at 1:10 million e-scale. This is overlaid with, in a heavy line, the coastlines formed by merging the wavelets defined by the first 1024 coefficients using Daubechies wavelets of dimension (a) 2, (b) 4, (c) 6, (d) 8, and (e) 10. The original coastline was sampled with 2^{13} (8192) equally spaced points. The shape of each wavelet is shown above each map

associated with the wavelet in question. For the low level, coarse wavelets l was long, and for the high level, fine wavelets it was short. Note that $\sqrt{2^j}$ is the standard wavelet factor used to generate normalized coefficients, and division by L was undertaken so that a constant threshold could be used regardless of the wavelet level.

The finest, or highest, level wavelets were tested first and their associated coastline sections assigned complexity “one” if the threshold was exceeded by either X or Y wavelets. The eleven lower levels were then processed in turn, however coastline sections already classified were not reclassified. As lower level wavelets were longer than higher-level wavelets they had the potential to classify long stretches of coastline, where it was not already classified. A few different thresholds were used. When a threshold of 25 was used, the coastline was classified into complexities one through nine, however some sections of the coastline remained unclassified.

Inspection of plots of the coastline, using symbology based on the assigned complexity, revealed that the technique was a failure. It was unable to discriminate sections of coastline with disparate linear complexity. Some sections of coastline classified as complex were in fact smooth, and vice versa. Figure 9 illustrates why the technique failed, showing that even a straight line, which displays no linear complexity, can reduce to non-trivial wavelets.

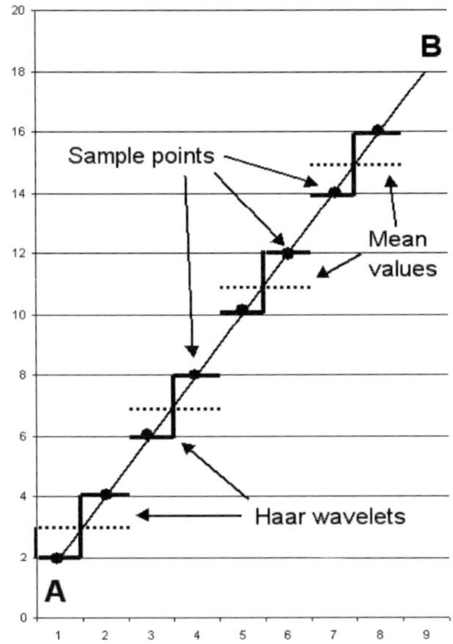

Fig. 9. Straight line $\overline{AB}$ sampled at eight locations, shown with the large black dots, modelled in the Y direction using Haar wavelets. The sampled section is broken into four mean values, shown with dotted lines, and four Haar wavelets, shown with a thick black line. Although the line is straight the wavelets that model it are of non-trivial amplitude, as are their coefficients

6.2 Technique 2

The second technique was similar to the first except that there were 2^{15} (32768) equally spaced sample points rather than 2^{13}. Also, the test was altered to be $Abs(C_{\text{var}_i} - C_{\text{var}_{i+1}}) > threshold$, where $Abs(C_{\text{var}_i} - C_{\text{var}_{i+1}})$ was the absolute value of the difference in the coefficient based variable of adjacent wavelets, C_{var_i} being the coefficient based variable as defined in the previous technique and applying to wavelet i, and $C_{\text{var}_{i+1}}$ being the coefficient based variable applying to its adjacent wavelet, indexed $i+1$. In short, this technique differed from the previous technique by focusing on the difference in the amplitude of adjacent wavelets rather than the wavelet amplitudes themselves. In all other ways the technique was the same as the previous technique. When a threshold of 40 was used, the coastline was classified into complexities one through 12, with only a small section of low complexity coastline remaining unclassified. Note that with 2^{15} equally spaced sample points there were 14 levels of wavelets plus one scaling function.

By plotting the processed coastline, using symbology to discriminate the assigned complexities, and visually inspecting the result, it was apparent that the technique was successful. Figure 10 shows the result for sections of the coastline, though note that the complexities have been reassigned into two classes, those with complexities one to four inclusive, and those with complexities five to 12 inclusive plus those sections unclassified. In the figure those sections of the coastline that appear complex have been coded as such, and vice versa. It can be imagined that had the classification been done manually by eye the result would be similar. Note that while only sections of the coastline are shown in the figure, the entire coastline was processed.

The technique works by responding to changes in line direction. Whereas wavelets associated with a straight line show no coefficient variation, wavelets "either side" of a direction change have markedly different amplitudes and coefficients in X, Y or both axes. Moreover, whereas the smaller, high level, wavelets isolate tight curves and fine detail, the larger wavelets isolate broader sweeping changes in direction, thus enabling the complexity to be broken into different complexity classes.

While it is not suggested here that a new index or measure for linear complexity has been developed, it is suggested a new method of indirectly measuring and utilizing the existing index of "angularity" has been developed. Angularity, as described by McMaster (1986), is based on the angular change between segments of a line, and can be quantified in numerous ways, "average angularity" for an entire line being the one favored and reviewed by Jasinski (1990). While the technique described here does not directly measure angularity, it indirectly measures it because the change in coefficients for adjacent wavelets is a function of angularity.

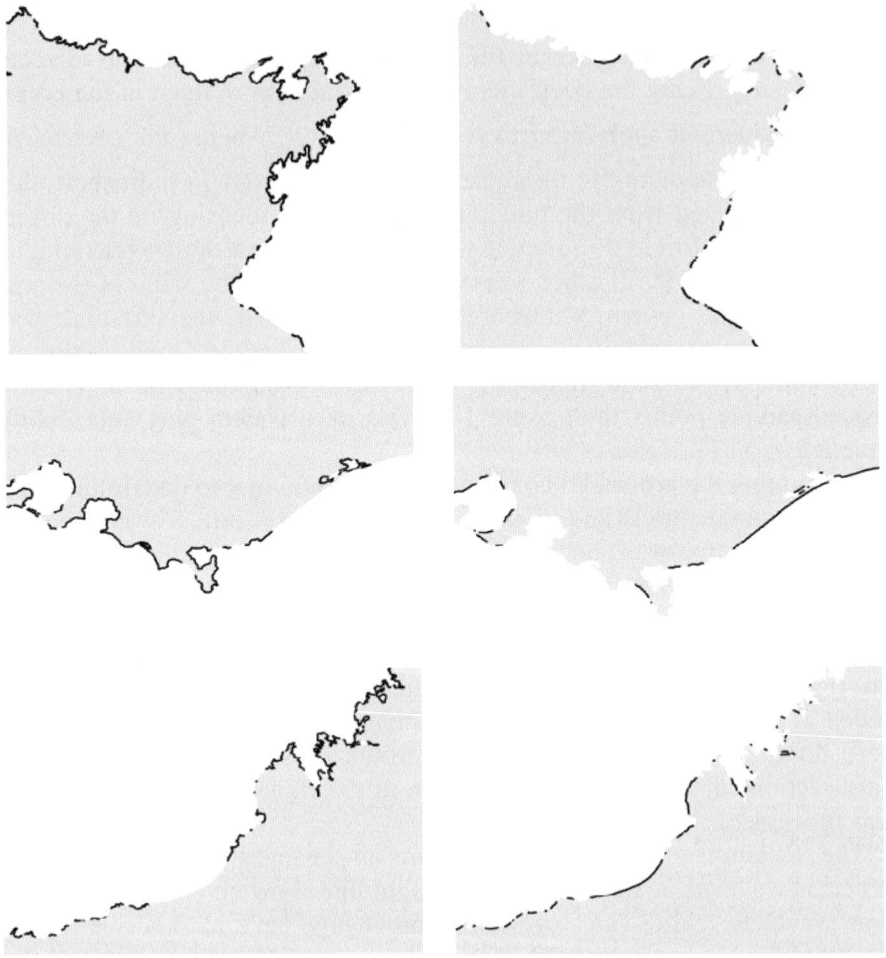

Fig. 10. Sections of the Australian mainland, shaded, from data at 1:10 million e-scale. At left, those sections of the coastline satisfying high complexity criteria are shown with a thick black line. At right, those sections satisfying low complexity criteria are shown in black. At top is the coastline around Gove Peninsula and the western side of the Gulf of Carpentaria, Northern Territory. In the middle is the coastline of central to mid-eastern Victoria. At bottom is the coastline of north-west Western Australia. The coastline was sampled with 2^{15} (32768) equally spaced points. A fast wavelet transformation, using the Haar wavelet and technique 2 described above, was used to determine the complexity of the coastline in the vicinity of each sample point

The new method has the advantages that it is objective, automated, localizes complexity along the different sections of a line, and is fast. Classifying and redrawing the 1:10 million e-scale coastline of mainland Australia, using 2^{13} sample points, took just three seconds on a Pentium 4 personal computer (PC). Table 1 lists the processing and drawing time on the same Pentium 4 PC for the same coastline using 2^{13}, 2^{15}, and 2^{17} sample points. In each case the bulk of this processing time was consumed by point sampling the coastline. It should be noted, however, that the method also has the disadvantage that the number of sample points is restricted to 2^n, where n is a positive integer. The user therefore does not have the freedom to specify an exact distance between sample points. Perhaps further research will overcome this shortcoming. Such research might look at whether it is viable to use an excess of the 2^n samplings to repeatedly sample the end point of the line.

Table 1. Approximate running time on a Pentium 4 PC with a 2.6 gigahertz processor and 480 megabyte of random access memory (RAM) of technique 2 for three different sample point numbers

# of sample points	Time
2^{13} (8192)	3 seconds
2^{15} (32768)	1 minute 30 seconds
2^{17} (131072)	24 minutes 30 seconds

7 Technical Implementation

The techniques described in this paper were developed and run on a Pentium 4 PC with a 2.6 gigahertz processor and 480 megabyte of RAM. The operating system was Microsoft Windows XP. ESRI's ArcGIS, version 9.1, software was used to access and display the digital spatial data, all of which were held in ESRI shape file format. Code was written on top of the ArcGIS application using the Visual Basic for Applications language, and making use of ESRI's ArcObjects development toolkit. Code was written for sampling and redrawing the digital vector coastlines, drawing the individual and compound wavelets, executing the Fast Wavelet Transformation, manipulating the coefficients, and executing the Inverse Fast Wavelet Transformation.

8 Conclusions

It is many years since research into linear complexity was popular even though further research has the potential to advance automated feature recognition, line simplification techniques, and e-scale determination. In this research the focus returns to linear complexity with an investigation of the utility of wavelets as a classifier of linear complexity. A background to wavelet technology has been given and two tests undertaken examining how successfully wavelets break a vector representation of the Australian coastline into classes of different complexity. While the first test failed, the second test, based on measuring the difference between the amplitudes of adjacent wavelets and thereby the line's angularity, succeeded. Plots showed that it successfully classified the Australian coastline in a way expected of manual classification The technique was objective, automated and fast. Further research using wavelets to simplify, enhance, compress, and spatially conflate data, may prove equally fruitful.

Acknowledgements

The author wishes to thank Dr. Ole Nielsen, of Geoscience Australia for kindly providing the author with a wavelet transformation algorithm, and instructing the author in its use and more broadly in wavelet technology.

References

Buttenfield BP (1991) A rule for describing line feature geometry. In: Buttenfield BP, McMaster RB (eds) Map generalization: making rules for knowledge representation. London, Longman Scientific & Technical, pp 150–171

Chrisman NR (1982) A theory of cartographic error and its measurement in digital data bases. Auto-Carto 5 – Proc, Crystal City, Virginia

Goodchild M, Proctor J (1997) Scale in a digital geographic world. Geographical and environmental modelling 1(1):5–23

Jasinski MJ (1990) The comparison of complexity measures for cartographic lines. National Center for GI and Analysis, technical paper 7, Buffalo

McMaster RB (1986) A statistical analysis of mathematical measures for linear simplification. The American Cartographer 13(2):103–116

Stollnitz EJ, DeRose TD, Salesin DH (1995) Wavelets for computer graphics: a primer, part 1. IEEE Computer Graphics and Applications 15(3):76–84

Stollnitz EJ, DeRose TD, Salesin DH (1996) Wavelets for computer graphics: theory and applications. San Francisco, Morgan Kaufmann

Wikipedia (2005). Wavelet. Wikimedia Foundation, Inc. http://en.wikipedia.org/wiki/Wavelet 2005

Expert Knowledge and Embedded Knowledge: Or Why Long Rambling Class Descriptions are Useful

R.A. Wadsworth, A.J. Comber, P.F. Fisher

Lead author Wadsworth at: CEH Monks Wood, Abbots Ripton, Cambridgeshire PE28 2LS, UK; e-mail: rawad@ceh.ac.uk

Abstract

In many natural resource inventories class descriptions have atrophied to little more than simple labels or ciphers; the data producer expects the data user to share a common understanding of the way the world works and how it should be characterized (that is the producer implicitly assumes that their epistemology, ontology and semantics are universal). Activities like the UK e-science programme and the EU INSPIRE initiative mean that it is increasingly difficult for the producer to anticipate who the users of the data are going to be. It is increasingly less likely that producer and user share a common understanding and the interaction between them necessary to clarify any inconsistencies has been reduced. There are still some cases where the data producer provides more than a class label making it possible for a user unfamiliar with the semantics and ontology of the producer to process the text and assess the relationship between classes and between classifications. In this paper we apply computer characterization to the textual descriptions of two land cover maps, LCMGB (land cover map of Great Britain produced in 1990) and LCM2000 (land cover map 2000). Statistical analysis of the text is used to parameterize a look-up table and to evaluate the consistency of the two classification schemes. The results show that automatic processing of the text generates similar relations between classes as that produced by human experts. It also showed that the automatically generated relationships were as useful as the expert derived relationships in identifying change.

Key words: inconsistent data sets, class labels, text mining, land cover

1 Introduction

Over time scientific knowledge advances, policy objectives change and technology develops these three factors all contribute to a common problem in the environmental sciences; almost every survey creates a new "base line" rather than being part of a sequence (Comber et al. 2002; 2003). For some phenomenon, like solid geology, we can be fairly sure that the differences between "maps" represent changes in; understanding, technology or objectives. For other phenomena, like land cover, differences may also be because the phenomenon has changed in significant and interesting ways.

Thirty or forty years ago this problem was not so acute because the map was used to support a description of the phenomena contained in a detailed survey monograph. The EU through initiatives such as INSPIRE (Infrastructure for Spatial Information in Europe)[1] and the UK Research Councils through e-science and GRID initiatives[2] are encouraging seamless access to data over the Internet; a traditional monograph may not be available to the user and may not even have been produced (Fisher 2003). Not only has the "balance of power" between the graphical (map) and textual (monograph) changed, but crucially the user may be led to wrongly treat the map as if it were data (a measurement) and not as information (an interpretation).

Recently a number of methods have been proposed to reconcile inconsistently defined data sets; these include:

- Statistical-semantic, Comber et al. (2004a,b,c)
- Fuzzy-logic, Fritz & See (2005)
- Conceptual overlaps, Ahlqvist (2004, 2005)

Each of these three methods relies on the existence of human experts to characterize the classes and the relationship between all the classes in all the data sets. The statistical-semantic approach of Comber et al. (2004a; 2004b, 2004c) elicited expert knowledge on the relationship between classes in different systems and represents this Expert Knowledge in look-up-tables (LUT). Values in the LUT are restricted to three values, +1 an expected relation, -1 an unexpected relation and 0 an uncertain relation. Different experts produced different LUTs reflecting their academic interests and familiarity with the data products. An alternative way of using expert knowledge to reconcile inconsistent land cover data has recently been

[1] http://eu-geoportal.jrc.it/

[2] http://www.nerc.ac.uk/funding/escience

shown by Fritz & See (2005) who employ a fuzzy logic framework. The idea of calculating a conceptual overlap to compare land cover classes has been explored by Ahlqvist (2004 & 2005) and Wadsworth et al. (2005) with variously continuous, discreet, ordinal and nominal scaled data such as "intensity of use", "crown closure", "tree species" etc. Data primitives are manually extracted from the class descriptions using expert opinion.

There may be situations where a human expert is unavailable or where differences between experts lead to significantly different results. Where the class descriptions are more than labels it should be possible to process the text to estimate the amount of overlap or similarity between classes. Processing natural language is very challenging and text mining is the subject of considerable research at the moment including in 2004 the establishment in the UK of worlds first "National Centre for Text Mining" (Ananiadou et al. 2005). A word can have many meanings that depend on context; for example, a "figure" might refer to a person, a picture, a number, an idea and so on. Words can be synonyms in some contexts but not others; for example "ground" and "land" are usually synonymous, but in the phrases "the traveler spent a few hours on the ground" and "the traveler spent a few hours on land" one suggests travel by an airplane the other by a ship. Sophisticated processing of the context, tense, roots, synonyms, acronyms and abbreviations to determine context and meaning can be employed entailing the construction of thesauri, dictionaries and agreed ontologies.

Kavouras et al. (2005) propose deconstructing geographic class descriptions into a formal semantic structures by defining hypernyms (generic terms), semantic properties (such as; 'purpose', 'cause', 'location' etc.) and semantic relations (such as; 'is-a', 'is-part-of', 'has-part', etc.) for each class similarity is calculated between these formal structures. While the Kavouras et al. (ibid) proposal works well for their example case (wetland terms in a "geographic" context (CORINE, Bossard et al. 2000) and a lexical context (WordNet Princeton University 2005) it contains no discussion of what to do with conflicting or multiple definitions. For example, from WordNet Kavouras (ibid, Table 4) characterize "Ditch" as hypernym: waterway, size: small, nature: natural. However, WordNet (v. 2.1) has two noun definitions of a ditch "a long narrow excavation in the earth" and "any small natural channel". Similarly almost all CORINE definitions have multiple included and excluded cases. Kavouras (ibid) do not explain how their method works where there are potentially many equally valid but different semantic structures. Fortunately, simple methods based on "word lists" work (surprisingly) well in representing or classifying documents and messages provided the text is reasonably long (Lin 1997; Honkela 1997).

This paper illustrates the issues in automatically processing text to quantify relations between classes and classifications with reference to land cover.

2 Data Used

The Countryside Survey (CS) is a sequence of increasingly elaborate assessments of the UK environment conducted in 1978, 1984, 1990 and 2000; CS 2007 is planned. As part of CS90 and CS2000 two land cover maps of Great Britain were produced; the LCMGB (Land Cover Map of Great Britain) (Fuller et al. 1994) and the LCM2000 (Land Cover Map 2000) (Fuller et al. 2002).

LCMGB used summer and winter pairs of Landsat satellite scenes to distinguish 25 target classes using a maximum likelihood classifier on a per-pixel basis. Because of problems with cloud cover images were collected over a three-year period. A description of the LCMGB classes can be found on the Countryside Survey web page[3], where the class descriptions have been grouped into the simplified 17-class system but are annotated in a manner that allows the 25 class descriptions to be disentangled.

LCM2000 also had problems with cloud cover. Although produced by the same organization (CEH) from similar data (Landsat satellite images) as part of the same sequence of assessments (the CS) it is a rather different product from the LCMGB. Changes in UK and EU Government policy led to a change from "Target classes" to "Broad Habitats", changes in science, technology and user feedback from LCMGB led to a change from per-pixel to a parcel based classification. Unusually parcel based metadata is available, detailing spectral heterogeneity and variants and an operating history. The fullest description of the LCM2000 classes is contained in Appendix II of a report to the LCM Steering Committee; a slightly shorter but more accessible description of the classes is[4]. The LCM2000 was issued with a caveat: "NB. The LCM2000 raster dataset is not directly comparable with the LCM1990 dataset, as it has been constructed by different methods. It is **NOT** suitable for estimating change over the 10-year period."[5]

[3] http://192.171.153.203/data/lcm/

[4] http://192.171.153.203/data/lcm/lcm2000_fulldatasetdescription.pdf

[5] http://192.171.153.203/data/lcm/productversionsandformats/pdf

Three hundred and fifty randomly selected locations in the Ordnance Survey SK tile were visited in 2003. The UK is divided by the Ordnance Survey into a regular pattern of 100 by 100 km blocks or "tiles" each tile has a unique two letter identifier[6]. The SK tile is centered approximately 52°03'N 1°15'W in the East Midlands of England. SK contains a range of environments from intensive agriculture and horticulture in Lincolnshire, to intensive urban and industrial activities in Leicester and Nottingham through general farmland to bleak moor lands in the Derbyshire Peak District. At each location a single expert with extensive knowledge of land cover mapping judged whether they considered the LCM2000 to be correct, and whether it was likely that the parcel had changed since 1990. Photographs were taken at each location and the authors reviewed the attributions. Certain classes, particularly acid grassland, are often difficult to definitively distinguish from other Broad Habitats (such as improved grassland and neutral grassland), where there was doubt we accepted the LCM2000.

3 Proposed Method

The first three stages of the proposed method are similar to Lin (1997) and Honkela (1997) who use SOM (self organizing (feature) maps, Kohonen (1982) to classify documents and messages. Here we use a similar approach using the descriptions of classes in the LCMGB and LCM2000. The stages of the process are:

a) For each dataset the class descriptions are converted into a word list (a single column of terms) by converting spaces to line breaks, converting capital letters to lower case and deleting most of the punctuation. A limited number of phrases are gathered into single terms; phrases concern measurements, e.g. "25 m", geographic locations, e.g. "North Yorks Moors" or species names, e.g. "*Nardus stricta*". In the case of the LCM2000 data some of the "Appendix II" descriptions are very short but refer to previous descriptions. For example the description of acid grassland is; "As above, but pH <4.5 denotes 'acid' soils. This range is appropriate". A decision was made to substitute the "as above" with the full text from the previous description.

b) A very sparse matrix of class v. terms is constructed with the cells containing the number of times that particular word appears in a class description. We wrote a "C" programme to do this but there are an increasing

[6] http://edina.ac.uk/digimap/support/gridreference.shtml

number of "off-the-shelf" software solutions such as "text to matrix generator" in Matlab®. The matrix is sparse because word frequency is very skewed; in the case of the land cover descriptions 53% of terms (710 out of 1337) occur once, while the term "the" makes up 4% (239 out of 5461) of the entire corpus.

c) Each term is weighted using the "tf.idf" (total frequency x inverse document frequency) scheme (see Robertson 2004 for a discussion on weighting schemes):

$$W_{ij} = \frac{n_i}{\sum n_i} \ln \frac{D}{n_j} \tag{1}$$

Where
W_{ij} is the weight of the ith word in the jth class
n_i is the number of times the word appears in the jth class
Σn_i is the total length of the jth class description.
D is the total number of classes
n_j is the number of classes containing the ith word.

The weighting has the effect that a word that appears in all class descriptions has a zero weight, but a word appearing frequently in a few short classes has a high weight.

d) The conceptual overlap between every pair of classes is calculated. Measures of conceptual overlap have the advantage over measures of distance in that the relationship between categories can be asymmetric. A conceptual overlap, allows the idea that category A is a (partial) sub-set of category B to be expressed, which is not possible with most measures of distance (where A is as far from B as B is from A). We use the measure of overlap suggested by Bouchon-Meunier et al. (1996) for non-ordered qualitative domains:

$$O(p_a,p_b) = \Sigma \min(p_a,p_b)/\Sigma p_a \tag{2}$$

Where p_a and p_b are the values (properties) of the two categories at each point within that domain. $O(p_a,p_b)$ can vary from 1 when "A" is a perfect sub-set of "B" to zero when there is no overlap (no terms occur in both A and B).

e) The calculated overlaps form an n by n matrix (where n is the number of classes). Patterns between the classes are rather difficult to visualize; we have tested both PCA (principle component analysis) and SOMs [self organizing feature maps, Kohonen (1982)]. SOMs are slow to compute but can produce a more easily interpreted result where there are a large number of classes (say over 100). Note that we use this step to provide qualitative quality control "do the results seem sensible?"

f) For comparison with the expert knowledge (expressed in a LUT) produced by Comber et al. (2004a) it is necessary to convert the degree of overlap into "expected", "uncertain" and "unexpected" relationships. This was done by first converting the overlap values to approximately N(0,1) (a Gaussian or "normal" distribution with a mean of zero and standard deviation of 1) by taking the natural logarithm of the raw values. Transformed values greater than one are "expected", values less than zero are "unexpected" and values between zero and one are "uncertain"; thresholds have not been optimized and produce just over a tenth expected (10.8%) and just under a half unexpected (47.7%).

g) The categorized overlap values were substituted for expert knowledge in the statistical-semantic methodology. Full details of the development of the statistical-semantic method can be found in Comber et al. (2004a,b,c), however, as an *aide-memoir* the process is as follows. Figure 1 illustrates a fragment of LUT capturing expert opinion about the relation between classes in two different classification systems. Within the fragment there are apparent cases of precise equivalence indicated by a single "expected" (+1) relationship, more often the expert believes there is has a positive relationship with several classes; in other cases the expert attributes no more than an uncertain relationship.

Look-up-Table fragment
LCM2000 (rows),
LCMGB (columns)

		Open shrub moor (10)	Dense Shrub Moor (11)	Bracken (12)	Dense Shrub Heath (13)	Scrub / Orchard (14)	
...	...	...	...	...	...	...	...
Bracken (9.1)	...	-1	-1	**1**	-1	-1	...
Dense Dwarf Shrub Heath (10.1)	...	0	**1**	-1	**1**	0	...
Open Dwarf Shrub Heath (10.2)	...	**1**	**1**	0	**1**	0	...
Fen, Marsh, Swamp (11.1)	...	-1	-1	-1	-1	-1	...
Bog (deep peat) (12.1)	...	0	0	-1	0	-1	...
Water (13.1)	...	-1	-1	-1	-1	-1	...
...	...	...	...	...	...	...	...

Fig. 1. Fragment of a LUT; comparing LCM2000 with LCMGB

In the process of producing a land cover map from satellite data, the producer must have an estimate of how "close" in spectral / attribute space every location (pixel) is to all the possible end points (classes). Although the producer has this information it is rarely saved and even less often made available to the user. Fortunately, the statistical-semantic approach can be used without that information by using the diversity or homogeneity of pixels within a parcel (patch, segment, polygon) as a surrogate. The diversity of pixels within the segment is used with the LUT values to estimate the extent to which the co-occurrence of classes are *consistent*. Inconsistency may arise from change or, much more commonly, error.

The semantic-statistical approach can be explained with reference to a hypothetical segment shown in Figure 2 for which we have information on its constituent land cover from two sources. These can be derived by intersecting the segment with alternative land cover data.

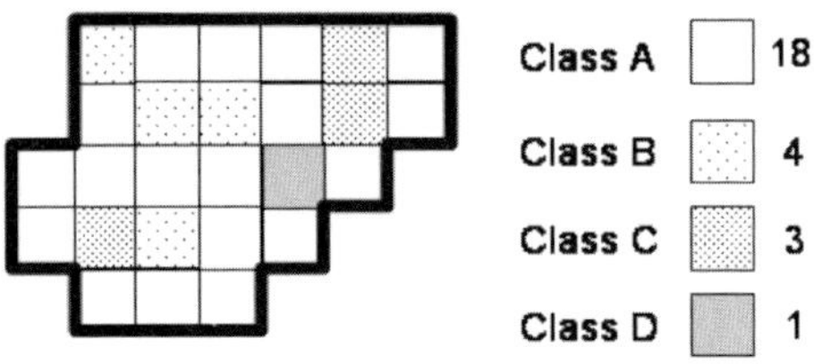

	A	B	C	D
A	1	0	0	-1
B	0	1	0	-1
C	0	0	1	-1
D	-1	-1	-1	1

Fig. 2. A hypothetical segment and the associated look-up-table

The hypothetical segment contains four classes. For each class we can calculate an expected, unexpected and uncertain score from the LUT. For example, for class A:

- expected score = 18 (18 class A),
- uncertain score = 7 (4 class B + 3 class C) and
- unexpected score is 1 (1 class D).

In a similar manner for class B:

- expected score = 4 (4 class B),
- uncertain score = 21 (18 class A + 3 class C) and
- unexpected score = 1 (1 class D).

Figure 3 shows a second "map" of the same segment but produced using a different classification scheme and part of the LUT between the classes.

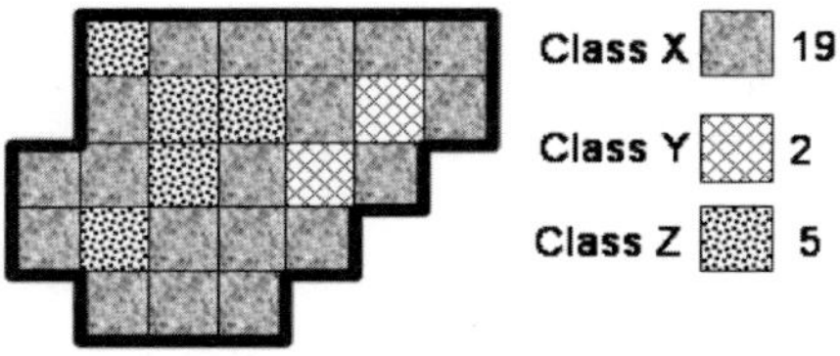

	X	Y	Z
A	1	-1	0

Fig. 3. The same segment but a different classification and part of the associated LUT between the classifications

Using the LUT between the first and second classifications we can again calculate an expected, unexpected and uncertain score for each class; for class A:

- expected score = 19 (class X),
- uncertain score = 5 (class Z)
- unexpected score = 2 (class Y).

Interpreting proportions as probabilities (which strictly speaking they are not), allows access to a number of useful techniques. A "Bayesian" approach would be to determine the extent to which the additional information (from the second classification) allows us to revise our opinion about the attribution supplied by the first classification. Dempster-Shafer (Dempster 1967; Shafer 1976) can be considered as an extension to Bayesian statistics that introduces an explicit description of uncertainty, "belief" + "uncertainty" + "disbelief" = 1 ("belief" + "uncertainty" = "plausibility"). The importance of this is that a weak belief in a proposition does not have to imply a strong belief in its negation.

Combining beliefs is done using Equation 3 and 4 (formulation from Tangestani & Moore 2002).

$$\text{Belief} = (\text{Bel}_1.\text{Bel}_2 + \text{Unc}_1.\text{Bel}_2 + \text{Unc}_2.\text{Bel}_1) \,/\, \beta \qquad (3)$$

$$\beta = (1 - \text{Bel}_1.\text{Dis}_2 - \text{Bel}_2.\text{Dis}_1) \qquad (4)$$

Bel_1 & Bel_2 are the two beliefs,
Unc_1 & Unc_2 are the uncertainties and
Dis_1 & Dis_2 the disbeliefs.

Applying the values from the hypothetic segment, for the "hypothesis" that the segment is class A.
$\text{Bel}_1 = 18/26 = \mathbf{0.692}$, $\text{Unc}_1 = 7/26 = 0.269$, $\text{Dis}_1 = 1/26 = 0.038$
$\text{Bel}_2 = 19/26 = \mathbf{0.731}$, $\text{Unc}_2 = 2/26 = 0.077$, $\text{Dis}_2 = 5/26 = 0.192$

Therefore:
$\beta = 1 - 0.692{*}0.192 - 0.731{*}0.038 = \mathbf{0.839}$
Belief = (0.692*0.731 + 0.693*0.077 + 0.731*0.269) / 0.839 = **0.901**

In this case our belief has *increased* with the addition of the extra information; therefore, we consider that this segment is consistent.

4 Results

The relationship between categories expressed in the matrix of calculated overlaps can be visualized in a number of ways. Figure 4, uses a PCA to show the relationship between classes in the LCMGB. Note that because the PCA is based on text overlap and the overlaps are weighted by absolute and relative frequency it impossible to assign a simple meaning to any axis.

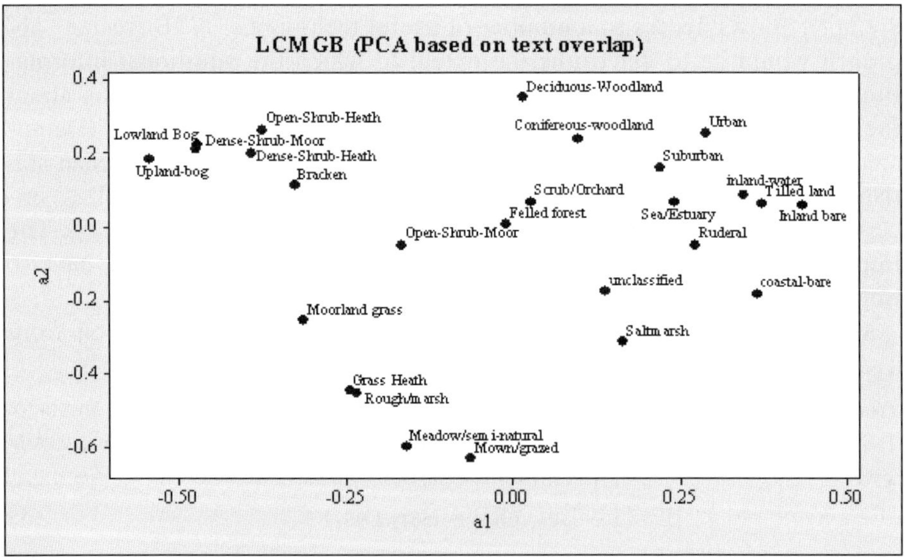

Fig. 4. PCA plot of LCMGB classes based on text overlap

The overall structure of the PCA seems "reasonable" because classes from similar environments are relatively close. Broadly regions in the scatter plot can be associated with woodland, moorland, improved grassland, bare ground, and settlement, however there is some mixing within these regions and some surprises such as sea/estuary between bare land and woodland. These results can be visually compared to the PCA derived from the LUT developed by the expert user, Figure 5.

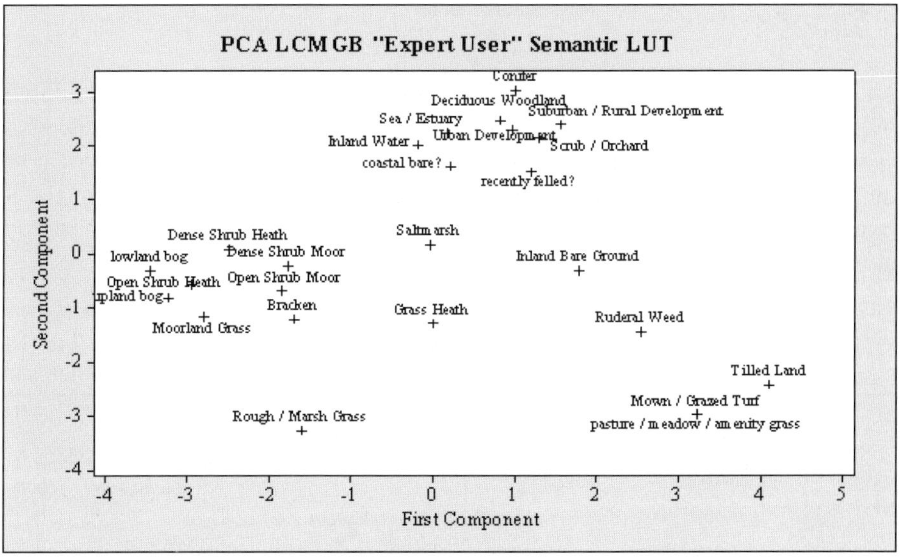

Fig. 5. PCA plot of LCMGB Classes based on the "Expert Users" LUT

When using the text descriptions the LCM2000 classes show a more clustered structure than the LCMGB (see Fig. 6). Note that the text descriptions for LCM2000 are only about half the length of the LCMGB descriptions but that the distinct, "neutral-grass", "acid-grass" and "calcareous-grass" cluster is partly an artefact of the way the classes are described, and the use of "as above" noted in Section 3.

When the "Expert User" LUT is used to create the PCA (see Fig. 7) it shows a slightly more dispersed pattern; interestingly the user seems to draw a strong distinction in the coastal classes between sediment and rocks a distinction that is not evident in the text descriptions.

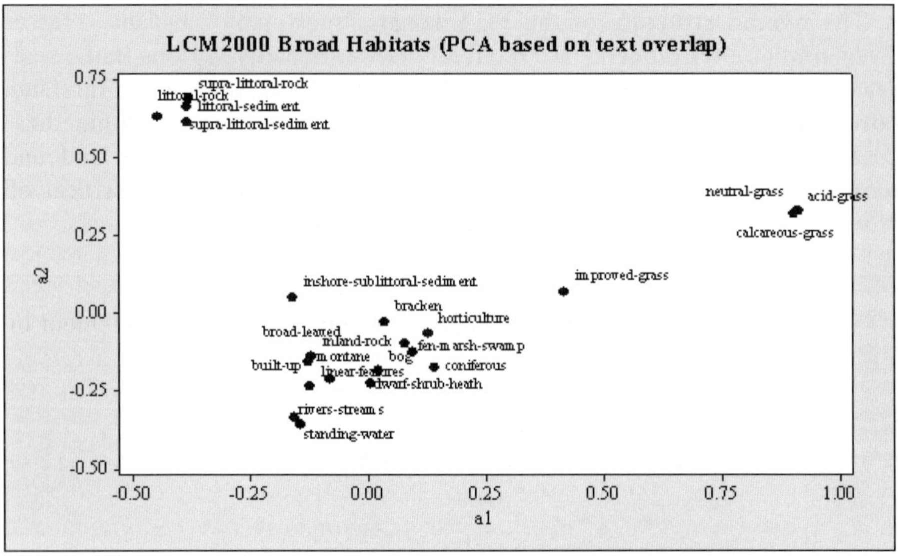

Fig. 6. PCA plot of LCM2000 classes based on text overlap

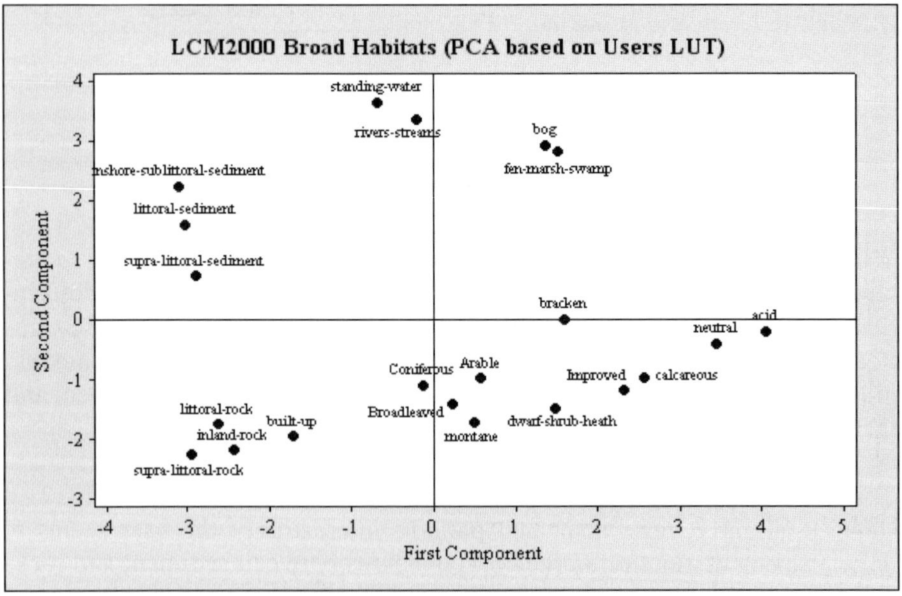

Fig. 7. PCA plot of LCM2000 classes based on "Expert User" LUT

4.1 Success in Predicting Consistency

The information from the 343 randomly visited locations was used to test how well the EK performs compared to the human experts. Following the terminology of Comber et al. the human experts are designated as "distributor", "producer" and "user". Problems in the post-hoc assessment of remotely sensed products are well known and not necessarily relevant to the purpose of this comparison. Comparison between the "Predicted" results (using Eq 3, the LUTs and the two maps) are compared to the field assessment and are shown in Tables 1a to 1d with the overall agreement in Table 1e. The "predicted" results are categorized as consistent, uncertain, undefined and inconsistent. Results are "uncertain" when the neither the belief for or against change exceed 0.5; results are "undefined" when there is perfect disagreement between two strands of evidence as β (Eq. 4) evaluates to zero. "Field Agreement" means that the field visit in 2003 thought the LCM2000 attribution was corrected, "Field Disagreement" that the LCM2000 was wrong.

Table 1a. Results from using the Embedded Knowledge LUT

Predicted	Field Agreement	Field Disagreement	Total
Consistent	186	44	230
Uncertain	44	5	49
Undefined	6	2	8
Inconsistent	50	6	56
Total	286	57	343

Table 1b. Results from using the Distributor's LUT

Predicted	Field Agreement	Field Disagreement	Total
Consistent	138	32	170
Uncertain	29	4	33
Undefined	8	3	11
Inconsistent	111	18	129
Total	286	57	343

Table 1c. Results from using the Producer's LUT

Predicted	Field Agreement	Field Disagreement	Total
Consistent	134	31	165
Uncertain	22	2	24
Undefined	9	3	12
Inconsistent	121	21	142
Total	286	57	343

Table 1d. Results from using the User's LUT

Predicted	Field Agreement	Field Disagreement	Total
Consistent	169	31	200
Uncertain	32	7	39
undefined	10	0	10
Inconsistent	75	19	94
Total	286	57	343

Table 1e. Comparison of all Experts

	Correct	Uncertain	Wrong
EK	56.0%	16.6%	27.4%
Distributor	45.5%	12.8%	41.7%
Producer	45.2%	10.5%	44.3%
User	54.8%	14.3%	30.9%

By expressing the values in Table 1e as counts it is possible to test the differences using the χ^2 test. EK is significantly different from both the Producer ($p<0.001$) and Distributor ($p<0.001$); the User is significantly different from the Producer ($p=0.001$) but not the Distributor ($p=0.013$). The Distributor and Producer are not significantly different ($p=0.583$), nor are the EK and User ($p=0.505$).

Table 2 compares the success of the embedded knowledge against the human experts across the Broad Habitats (grassland and mountain classes have been amalgamated because there were too few cases).

Table 2. Correct predictions summarized by Broad Habitat

Habitat	EK	Producer	Distributor	User
Broadleaved woodland	80.0%	28.6%	28.6%	65.7%
Coniferous woodland	45.0%	35.0%	40.0%	50.0%
Arable	75.6%	70.0%	70.0%	50.0%
Improved grass	77.1%	70.2%	70.2%	44.7%
Neutral, calcareous, acid grass	34.4%	32.8%	31.3%	60.9%
Moorland, bog	12.5%	0.0%	0.0%	66.7%
Built	47.2%	39.6%	41.5%	49.1%

5 Discussion and Conclusion

Although PCA is not a very sophisticated method to visualize highly multivariate data it is sufficient to demonstrate that the overall pattern of "relatedness" is similar whether it is derived from the overlap of terms in a text description (see Fig. 4) or from a LUT produced by a human expert (see Fig. 5). For the LCMGB a notable feature is that both figures have a cluster of moorland/heath classes, the "LUT" group is eight strong: dense shrub heath, dense shrub moor, open shrub heath, open shrub moor, upland bog, lowland bog, moorland grass and bracken. In the "text" PCA the open shrub moor and moorland grass are adjacent but separated with the moorland grass slightly closer to the rough/marshy grass and grass heath classes, which are themselves close to meadow/semi improved and mown/grazed grasslands. In both cases the two main woodland classes (broadleaved and coniferous) are close to each other and to the two built sub-classes (industrial and suburban/rural development). One of the most noticeable differences between the patterns is that the human expert put the arable land close to the agricultural grasses (mown/grazed, pasture/ meadow/amenity) while the "text" puts arable close to sparsely vegetated classes, (inland bare and inland water).

While it is interesting to compare the patterns produced by different experts and by the automatic processing of text descriptions the real issue is whether there is a practical application. A simple conversion from the relative amount of overlap to an expected, uncertain and unexpected attribution allows direct comparison of how good the processing might be. Table 1e reveals the rather startling fact that the Embedded Knowledge is superior or equal to all three human experts. To a great extent the success of the EK is due to the fact that it predicts higher levels of consistency; it does not perform as well as the human experts in predicting inconsistent segments.

Table 2 shows that the Producer and Distributor are inferior to EK in all cases. The User is superior in four cases, coniferous woodland, the combined neutral, calcareous, acid grassland, the built environment and the combined moorland and bog class. With the exception of the combined moorland and bog class differences between the User and the EK are relatively small.

This experiment with land cover class descriptions seems to suggest that reasonably long (> 100 word) *descriptions* of classes provides information that can be processed and used by someone unfamiliar with the epistemology, ontology and semantics of the data set. The full-blown survey memoir is not going to make a come back, but we do need more than labels and ciphers. We therefore urge data producers to express themselves and not restricted themselves to cryptic and gnomic utterances.

Acknowledgements

Many of the ideas on the semantic-statistical approach were developed during the REVIGIS project which was funded by the European Commission, (Project Number IST-1999-14189), Project Coordinator Robert Jeansoulin Laboratoire d'Informatique de Marseille, CMI – Universite de Provence, FRANCE.

References

Ahlqvist O (2004) A parameterized representation of uncertain conceptual spaces. Transactions in GIS 8(4):493–514

Ahlqvist O (2005) Using uncertain conceptual spaces to translate between land cover categories. Int J of Geographical Information Science 19(7):831–857

Ananiadou S, Chruszcz J, Keane J, McNaught J, Watny P (2005) The national centre for text mining: aims and objectives. Ariadne Issue 42 June 2005. http://www.ariadne.ac.uk/issue42

Bossard M, Feranec J, Otahel J (2000) CORINE land cover technical guide – Addendum 2000. http://www.epa.ie/OurEnvironment/Land/CorineLandCover/Technicaldetails/FileUpload,5858,en.pdf

Bouchon-Meunier B, Rifqi M, Bothorel S (1996) Towards general measures of comparison of fuzzy objects. Fuzzy sets and systems 84:143–153

Comber AJ, Fisher PF, Wadsworth RA (2002) Creating Spatial Information: Commissioning the UK Land Cover Map 2000. In: Richardson D, Oosterom P van (eds.) Advances in Spatial Data. Springer-Verlag, Berlin, pp 351–362

Comber AJ, Fisher PF, Wadsworth RA (2003) Actor Network Theory: a suitable framework to understand how land cover mapping projects develop? Land Use Policy 20:299–309

Comber AJ, Fisher PF, Wadsworth RA (2004a) Identifying Land Cover Change Using a Semantic Statistical Approach. In: Atkinson PM, Foody GM, Darby SE, Wu F (eds) Geodynamics. CRC Press, Boca Raton, pp 73–86

Comber AJ, Fisher PF, Wadsworth RA (2004b) Assessment of a Semantic Statistical Approach to Detecting Land Cover Change Using Inconsistent Data Sets. Photogrammetric Engineering and Remote Sensing 70(8):931–938

Comber A, Fisher PF, Wadsworth R (2004c) Integrating land cover data with different ontologies: identifying change from inconsistency. Int J of Geographical Information Science 18(7):691–708

Comber AJ, Fisher PF, Wadsworth RA (in press) Combining expert relations of how land cover ontologies relate. Paper to be published in Int J of Applied Earth Observation and Geoinformation.

Dempster AP (1967) Upper and lower probabilities induced by a multi-valued mapping. Annals Math Star 38:325–339

Fisher PF (2003) Multimedia reporting of the results of natural resource surveys. Transactions in GIS 7:309–324

Fritz S, See L (2005) Comparison of land cover maps using fuzzy agreement. Int J of Geographical Information Science 19(7):787–807

Honkela T (1997) Self-Organising maps in natural language processing. PhD Thesis Helsinki University of Technology, Department of Computer Science and Engineering. http://www.cis.hut.fi/~tho/thesis/

Kavouras M, Kokla M, Tomao E (2005) Comparing categories among geographic ontologies. Computers & geosciences 31:145–154

Kohonen T (1982) Self-organized formation of topologically correct feature maps. Biological Cybernetics 43:59–69

Lin X (1997) Map displays for information retrieval. J of the American Society for Information Science 48:40–54

Princeton University (2005) WordNet a lexical database of the English Language version 2.1 http://www.cogsci.princeton.edu/~wn/

Robertson S (2004) Understanding inverse document frequency: on theoretical arguments for IDF. J of Documentation 60:503–520

Robertson SE, Spärck Jones K (1976) Relevance weighting of search terms. J of the American Society for Information Science 27(3):129–146

Shafer G (1976) A mathematical theory of evidence. Princeton University Press, Princeton NJ. USA

Tangestani MH, Moore F (2002) The use of Dempster-Shafer model and GIS in integration of geoscientific data for porphyry copper potential mapping, north of Shahr-e-Babak, Iran. Int J of Applied Earth Observation and Geoinformation 4:65–74

Wadsworth RA, Fisher PF, Comber A, George C, Gerard F, Baltzer H (2005) Use of Quantified Conceptual Overlaps to Reconcile Inconsistent Data Sets. Session 13 Conceptual and cognitive representation. Proc of GIS Planet 2005, Estoril Portugal 30th May – 2nd June 2005. ISBN 972-97367-5-8. 13pp

Preference Based Retrieval of Information Elements

Claudia Achatschitz

Institute for Geoinformation and Cartography, TU Vienna
email: achatschitz@geoinfo.tuwien.ac.at

Abstract

Spatial information systems like GIS assist a user in making a spatial decision by presenting information that supports the decision process. Planning a holiday via the Internet can be a daunting process. Decision-making is based on finding the relevant data sets in short time among an overwhelming amount of data sources. In many cases the user has to figure out how the presented information can fit the decision he has to make on his own. To address this problem a tourist would need tools to retrieve the spatial information according to his current preferences. The relevant data is determined as data elements describing facts corresponding to the tourist's preferences. The present work suggests a way how a user dealing with a tourist information system can indicate his preferences through the user interface. According to the suggested alternatives the user can change his preferences and generate a new set of alternatives. A feedback loop provides a tool that allows the user to explore the available alternatives. The work was motivated by classical dialog based booking processes between human operators, carried out on the telephone. We introduce the conceptual model of a user interface, which considers the agents preferences and describes the overall interaction process.

1 Introduction

Internet portals offer various possibilities for tourists to plan a vacation. However, the handling of such systems can be frustrating. Sometimes it is impossible to find the relevant data elements for a particular spatial decision, although it would be available. Hence decision making supported by an information system is based on finding the decision relevant information among an overwhelming amount of data sources. The information retrieval process has to be accomplished within a limited time period. A tourist does not want to spend more time than absolutely necessary to find a suitable hotel. The outcome of traditional tourist information systems on the web would present the user with a result regardless of his preferences.

Every hotel has certain attributes. A possibility to choose among them is to select them at the user interface where the order of the attributes is fixed. In the example below the user interface includes an uncountable amount of attributes (see Fig. 1).

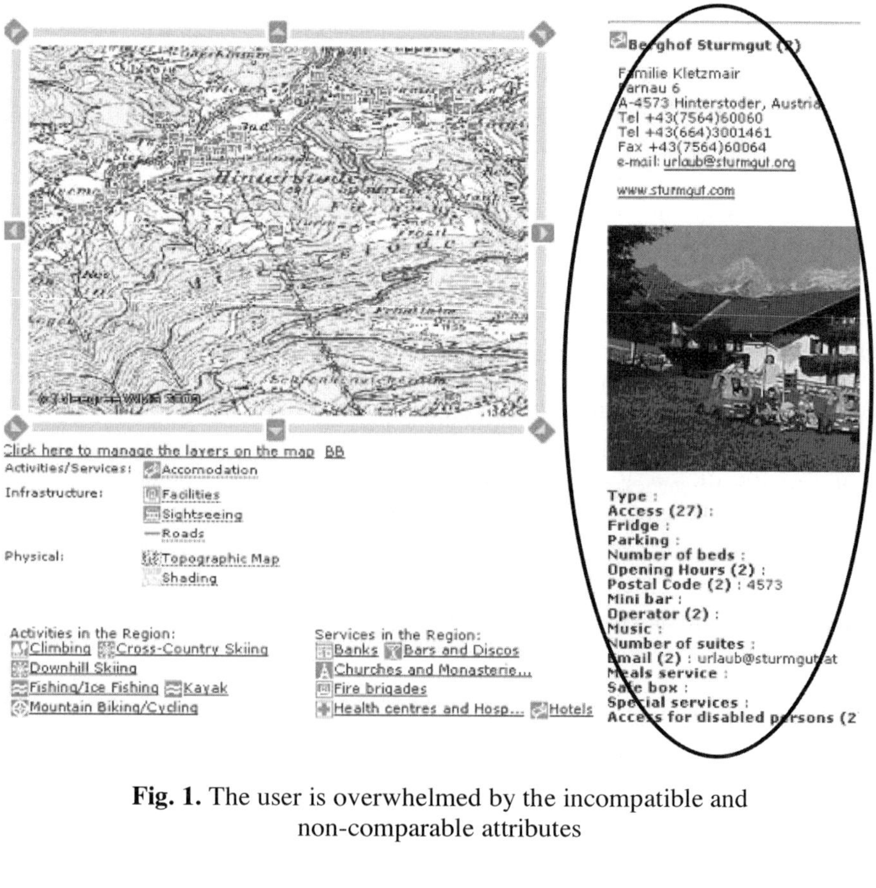

Fig. 1. The user is overwhelmed by the incompatible and non-comparable attributes

The approach cannot be successful for two reasons. First empirical studies show that human users seem to be able to cope with 7+/- 2 elements at the same time and that they can capture around 3–4 elements at a quick (Miller 1994). The amount of the displayed attributes has to be reduced in comparison to traditional systems. Second users have preferences that should be considered when retrieving data from the information system. A way to consider the user's preferences is to let the user decide upon the importance of the attributes.

An example for a hotel selection scenario can be described as following. A user might know exactly that he needs to go to Vienna at a specific date, which he specifies on the available website. Thus he needs an accommodation at a specific location at a specific date. These aspects must be satisfied. We call these obligatory criteria. Other aspects might be convenient but not at all obligatory. We call these aspects soft or compensatory criteria.

Currently in many cases the user could end up with no results, because there is no accommodation available that fulfils the specified criteria at the first attempt. The tourist then would have to start the search again.

We want to stress that when searching a web interface the user should have recurring opportunities to set his preferences. This means that the user after setting the preferences should be able to reconsider them. This would result in a softening or sharpening process of his preferences.

If the priority of finding a hotel in Vienna is higher than the priority of carrying out a certain activity, then this should be considered during the data retrieval process. In any case the web portal should at least display the available hotels in Vienna.

The remainder of the paper is organized as follows. Section 2 introduces findings in spatial decision-making and tourist information systems. Section 3 presents the framework for the user model. Section 4 discusses the direct manipulation interface approach. Section 5 explains the conceptual model of the interaction process and the concluding sixth section gives an outlook to future research

2 Decision Making Processes

Every decision is driven by a goal that a decision maker wants to achieve.

In the literature we are confronted with different ways of describing a decision making process. The general division in risk less and risky decision making processes (Tversky and Kahneman 1981), rational decision making process and decision making with bounded rationality (Edwards

1967) structures the various models of decision making processes. A further differentiation between static approaches and dynamic approaches as well as between linear versus cyclic approaches builds another basis to model decisions (Baron 1988; Mintzberg and Westley 2001).

2.1 Spatial Decisions

Spatial decision-making is part of our every day life. No matter if we decide where to go for lunch, when to do shopping, and how to get to the next gas station as fast as possible, space is always an important factor. Spatial decision-making incorporates every decision that includes space as a factor. Golledge and Stimson divide the process of spatial decision making into two acts. In the first act a human is searching for information and developing a plan to get to a destination. The second act incorporates the choice act and the evaluation phase (Golledge and Stimson 1997).

In the present paper we are primarily interested in supporting the user of a tourist information system during the first act. In this work we consider spatial decision making to be a dynamic process. We understand dynamic in a sense that factors influencing the decision process can change as the process moves on. This affects the user preferences as they might vary depending on the decision maker's situation. In the present paper we propose a dynamic way of supporting the establishment of user preferences through a user interface. The proposed user interface allows the user to return to any previous state of the process.

One approach to model decision-making is using a cyclic model. Thereby decision-making is seen as a cyclic process where a decision maker can go through the applied steps again and again until the final satisfying decision can be made. Mintzberg states that within many decision-making processes we keep cycling back, get influenced by new information until somehow the solution emerges. He questions if the decision-making process really exists. If we consider recent decisions that we made, we realize that we did not always structure the decision-making process as proposed by the different decision models. It is more that we pass through the same process many times until finally we are able to come to a decision. We can also apply this cyclic approach to spatial decision making and in particular to spatial decision-making supported by spatial information systems like GIS.A tourist can go back within the information system until he finally can decide where to go.

2.2 Establishing Preferences

A decision problem is defined by available alternatives (Yntema and Torgerson 1961). The available alternatives determine the preferences a user can specify for a particular decision situation (McClennen 1990). Establishing those preferences is a part of the decision making process. Preferences are factors that considerably influence the decision making process. They support or restrict the achievement of a goal. These factors can be divided in factors that a user can influence, factors that he cannot influence and factors that he is not aware of. In this paper we deal primarily with the factors that we can influence and we call these factors *user preferences*.

A user interface should support the dynamic character of establishing preferences and not change it into a static process. The proposed interaction process allows the user to reformulate the preferences according to the suggested output of his first attempt to retrieve the relevant data elements out of the tourist information system. The interface supports the user with the help of a feedback loop that offers the possibility to redefine the preferences if the outcome does not satisfy the user. We utilize the example of a site selection. A user wants to find a hotel that best fits to his actual requirements.

2.3 Spatial Decision Making and the Web

Personalized user interfaces are supporting the user in retrieving decision relevant information on the Internet (Staab, Werthner et al. 2002). The revision of the user preferences can be a tedious process. Developing a shared knowledge base of the system and its user through a direct feedback approach is one way to agree upon a common understanding of vague spatial concepts (Cai, Wang et al. 2003). User interface agents are also an attempt to facilitate the use of an information system. The user interface agent operates parallel to the user and tries to retrieve information by applying certain strategies (D'Aloisi and Giannini 1995; Li, Zhou et al. 2002).

Though there have been recently a number of approaches that try to consider the user's needs in web-GIS (Raubal and Rinner 2004; Hochmair 2004) they require a lot of user input and some knowledge about algorithms used to retrieve data from a system. A dialog based approach seems promising. The user answers a series of questions comparable to a telephone booking. However it has to be restricted to a limited number of questions. Users have to be very acquainted with a website (like Amazon) to be willing to spend more time on feedback.

3 Framework for a User Model

In the present work a user is seen as a rational decision making agent. Within this framework a user is assumed to always maximize his overall wealth (Edwards 1967). Rational decision-making is a prescriptive way to model a decision making process. The prescriptive approach aims at revealing how decisions should be made to get the optimal output. Thus it is an unrealistic way to deal with decision-making processes. Gigerenzer states that people don't optimize but they pretend as if they do. To reach a better ex post common understanding of decisions, people tend to reason about their taken decisions in a rational way (Gigerenzer 2004). Even though a decision was not taken rationally, the outcome is still presented as if the decision maker had complete knowledge about the available alternatives and was able to state which alternative he preferred. The decision is presented as if it maximizes the decision maker's contentment.

Hence the theory of rational decision making supposes that the decision maker is always able to put the available alternatives into a weak order, has complete knowledge about the facts, and always wants to maximize something. Rationality assumes that a decision maker maximizes the overall utility.

3.1 Preferences and Utilities

As stated above a decision needs always a goal that should be achieved. In every situation we try to achieve different goals. Utility makes it possible to measure the achievement of this goal. The basis for utility theory is the utility of each outcome combined with a probability value. Utility theory follows some basic principles and the underlying philosophy is called utilitarianism developed by Jeremy Bentham and Stuart Mill.

Luce and Raiffa defined weak ordering and independence as the two basic axioms for utility. Weak ordering means that a decision maker either must prefer something or be indifferent about it. The independence principle indicates that the utility of an alternative is independent of other alternatives (Luce and Raiffa 1957). Decision-making that involves utility theory is based on a number of principles.

The first basic principle is that a decision maker knows his goals. Every goal is bounded by consequences. The second principle is the ability of the decision maker to always rank the available alternatives according to his preferences (Keeney and Raiffa 1993). In the present work the alternatives are composed of a vector of attributes describing each alternative.

Jonathan Baron stresses that utility theory consists of three different parts. The expected utility deals with the tradeoffs between the probability of an outcome and its utility. The second part is the multiattribute utility that is concerned with the tradeoffs between different goals and is used to calculate the utility of an outcome. The third part deals with the tradeoffs of goals from different people (Baron 1988). Assigning trade off values to the alternatives suggested by a tourist information system is topic for future research. Questioning to what extend one alternative compensates the absence of another one is a next step into that direction.

4 Direct Manipulation Interface – A Dynamic Approach

The direct manipulation interface approach supports the dynamic approach of a decision-making problem. The concept of an interactive direct manipulation interface offers the user the possibility to modify his preferences for certain attributes directly through the user interface. This interactive display allows the user to experiment with the preferences and thresholds. The effect is a change of offered alternatives in the display.

The term direct manipulation was introduced by Shneiderman (1998). But also Norman and Hutchins did research in that field. They refer to the "psychological distance" between user goals and user actions at the interface. Direct manipulation aims at minimizing distance and maximizing engagement (Frohlich 1993). A continuous representation of the object of interest allows the user to physically perform actions. The rapid reversible operation lets the user recognize the effect on the object. The major benefits of direct manipulation interfaces are the ability to learn basic functionality quickly. A wide range of tasks can be performed also from non-expert users. Users can immediately see if the action performed directly on the user interface brings them closer to their goal, which leads to a greater system comprehension (Hutchins, Hollan et al. 1985). A direct manipulation interface should facilitate the human-computer interaction process and should not get into the users way (Heeter 1991).

5 Modeling the Dynamic Interaction Process

The presented model is a dialogue-based model in extension to the work of (Linden, Hanks et al. 1997). Linden et al. concentrated on booking a flight. The user could ask the system for available flights. According to the direct manipulation approach where directness is among others achieved by maximizing engagement Linden's example corresponds to the *conversation metaphor*. Within this metaphor the user and the interface conduct a conversation about the assumed world (Hutchins, Hollan et al. 1985).

We will take over some of the findings to apply it to the case of searching for a suitable hotel. The desired hotel has to have certain attributes, some are obligatory and we call them obligatory criteria (location, price, …) some are mandatory (sauna, pool, breakfast, …). For the tourist and his decision which hotel to book, only hotels where the attributes correspond to the user's preferences are relevant. User preferences are subdivided into obligatory criteria and weights.

Through directly manipulating the interface the user can specify these preferences, by setting the obligatory criteria. He can set weights and indicate the importance of the criteria. Changing the preferences will result in a revision of the relevant information. By direct manipulation we mean an interface that immediately reacts to the user's settings and manipulations in the interface (Shneiderman 1998). This extended model corresponds to the *model-world metaphor* of the direct manipulation approach, where the interface itself represents a world where the user can perform actions. The interface responds to the action and changes state.

5.1 Conceptual Model

We apply an agent-based approach to simulate the interaction process. An agent can be defined as "Anything that can be viewed as perceiving its environment through sensors and acting upon the environment through effectors" (Russell and Norvig 1995). The present model contains three actors: a data repository, a user-interface, and a hotel seeking agent (see Fig. 2).

Fig. 2. The users' preferences have to be included into the system. The double arrows indicate the feedback loops

Objects and their attributes are stored in the data repository of the information system. The user interface stands at the same time for the information system (Frank 1993). The agent can observe and manipulate the user-interface.

Several assumptions have been made:

- We restrict our model to a site selection problem. A user wants to find an accommodation at a specific location. The agent knows where he wants to go and can state his preferences in a weak order.
- We divide preferences into obligatory criteria and weights.
- The agent weights the attribute types of the data repository
- The agent assigns his preferences to each attribute.
- An agent can change his mind by changing his preferences. Therefore feedback loops are foreseen in the model.

The underlying assumption is that the user knows his preferences and chooses the alternative that best fits his preferences, thus acts rationally.

The following elements have been identified in this interaction process where the user and the information system are involved:

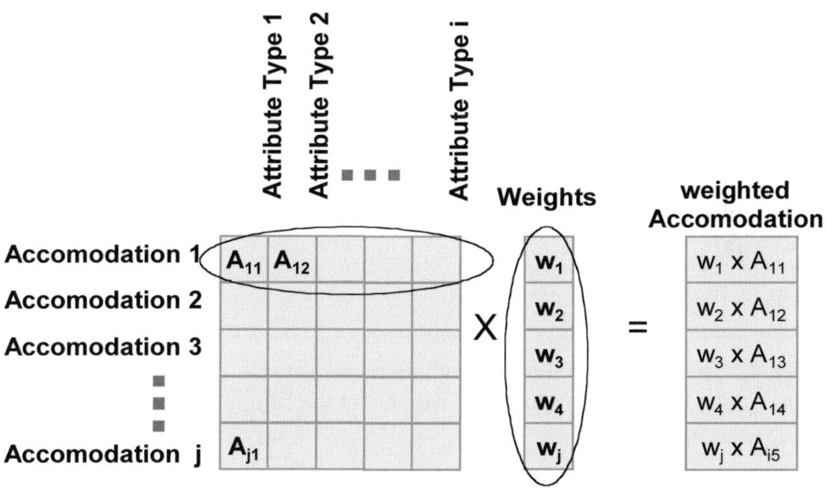

Fig. 3. The obligatory criteria are multiplied by weights. The user sets weights through e.g. sliders in the user interface

1. The user has to state his preferences by setting the obligatory criteria and weighting the attribute types in the interface.
2. According to the user preferences the information system has to produce candidate solutions by searching the data repository.

3. The user has to evaluate the candidate solutions.
4. If the user is not satisfied with the candidate solutions, the user has to select the alternative that fits his preferences better and set the preferences again.

The user would set his preferences in a user interface shown in the figure below. In the left part he sets the obligatory criteria. In the right section he can influence the importance of the attribute type. Here the attribute types are accommodation type, region, room type etc. Another part of the user interface would immediately react to the user's action and list the possible results. A map supports the presentation of the results.

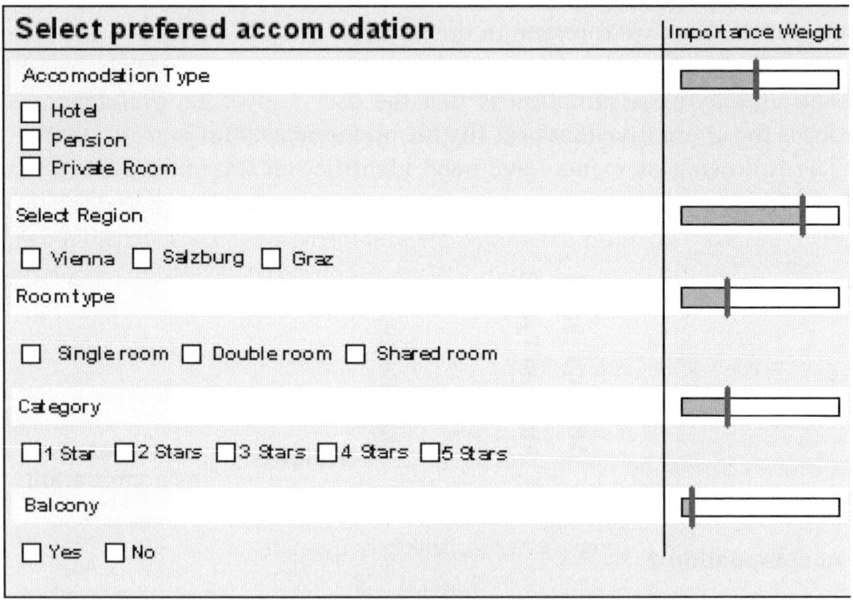

Fig. 4. A part of a user interface where a user sets the obligatory criteria and the importance weights at the same time

5.2 Modelling the Interaction Process

To clarify the conceptual model we use pseudo code. We identified smaller elements of the interaction process.

```
fun:: a -> b -> c        f (a,b) = c
```

The notation above will be used throughout the following paragraphs. The simplicity of the system was at the core of the conceptual model; therefore we kept the possibility of interaction small. All the user can do is, setting the obligatory criteria and the weights for the attribute types. Placing weights allows the user to make tradeoffs. Investigating the tradeoff values for the attributes will be future research work. The attributes as well as the preference values have to be standardized in order to be comparable (Raubal and Rinner 2004).

We introduce data types as simple as possible. Preferences, weights of attribute types, and scores of the utilities are represented as floating numbers. The hotel is an object with the usual attributes such as name, category (string), price (float) …

```
type Pref = (Oblig, Weight)
type Weight = Float
type Oblig = String
type Score = Float
DataAccomodation = Accom Name Category Price Location ...
```

These types will be used in the subsequent functions that formalize the overall interaction process. We represent these functions by their signatures. The agent uses them when manipulating the user interface. "[]" indicates lists of objects. This allows the calculation of weighted preferences.

The user can set the obgligatory criteria.

```
setObligatory:: [Accom]->[Oblig]->[Accom]
```

The user weights the Obligatory criteria. These weights can cause a softening or a sharpening of the set criteria.

```
stateWeight::[Accom]->[Weight]->[(Accom,Weight)]
```

The following step creates candidate solutions. We are not interested in the result but in the overall process. It is important to use standardized values of the attributes for evaluating the candidates.

```
createCandidates::[Pref]->[Accom]->[(Accom,Score)]
```

In any case, the result will be ranked as a list of accommodation possibilities based on a utility score.

```
utility::Accom->[Pref]->Score
```

The resulting list will be sorted and displayed to the user.

```
sortByUtilityScore::[(Accom,Score)]->[(Accom,Score)]
```

The user has two options: If he likes a hotel and its displayed attributes he can continue and select it. Otherwise he can criticize the information system by redefining his preferences. In the present model this is done by changing the weights of the attribute types, and by resetting some of the obligatory criteria. In this feedback loop the function createCandidates will be called again until the user decides to select one of the displayed alternatives. This phase is critical because too many loops could cause the user to give up the search. The selection of one candidate among the displayed solutions is the final step of the decision process.

```
selectCandidate::[(Accom,Float)]->(Accom,Float)
```

The described interaction process iP needs as inputs a data repository with their attributes (hotellist) and the user's preferences (preflist). Optionally the hotel list can be pre-processed by excluding all hotels that do not fulfill obligatory criteria. The values of the attributes and preferences are standardized in order to be comparable.

6 Conclusions and Future Work

We presented a conceptual model for a hotel-seeking agent. The interaction process with the information system has been decomposed into the elementary units. While previous research focused on creating candidate solutions (Raubal and Rinner 2004; Hochmair and Rinner 2005) we focused on the overall process.

Within the interaction process the user sets obligatory criteria and weights the attribute types. The weighting can soften the obligatory criteria. The user can also weight attribute negatively. The direct manipulation approach makes it easier for the user to control his actions.

The investigation of the interaction process is a first step into the direction of establishing a negotiation process between a user and the system. We argue that this makes it easier for the user to establish preferences according to his current decision making task. In a possible future negotiation scenario not only the user can set criteria but also the system could indicate preferences. The system might react to the changed criteria of the user with a more attractive offer. Multi-agent systems that consider hotel recommendations between agents are another goal of future research. Selection strategies will influence the displayed results.

Investigating selection strategies will influence the displayed results. An extension of the model in several directions seems possible. The results motivate especially the investigation of the new paradigm of data retrieval that is based on the user desires rather than on the available data sources.

Acknowledgement

This work has been funded by the E-Ten project GEOCOMPASS N° C517343. I would like to thank all members of the Institute for Geoinformation and Cartography especially Florian Twaroch and Elissávet Pontikakis contributing to this work with valuable discussions.

Reference

Baron J (1988) Thinking and Deciding. Cambridge University Press

Cai G, Wang H et al. (2003) Communicating Vague Spatial Concepts in Human-GIS Interactions: A collaborative Dialogue Approach. COSIT 2003, Springer Verlag, Berlin Heidelberg

D'Aloisi D, Giannini V (1995) The Info Agent: An Interface for Supporting Users in Intelligent Retrieval. ERCIM Workshop Towards Interfaces for all: Current Trend and Future Efforts

Edwards W (1967) The Theory of Decision Making. Decision Making:13–64

Frank AU (1993) The Use of Geographical Information Systems: The User Interface is the System. In: Medyckyj-Scott D, Hearnshaw HM, Human Factors in Geographical Information Systems, pp 3–14

Frohlich D (1993) The History and Future of Direct Manipulation, Hewlett Packard

Gigerenzer G (2004) Striking a Blow for Sanity in Theories of Rationality. Models of a Man: Essays in Memory of Herbert A. Simon

Golledge RG, Stimson RJ (1997). Spatial Behavior: A Geographic Perspective. The Guildford Press, New York

Heeter C (1991) The Look and Feel of Direct Manipulation. HYPERNEXUS: J of Hypermedia and Multimedia Studies

Hochmair HH (2004) Decision Support for Bicycle Route Planning in Urban Environment. 7th AGILE Conf on Geographic Information Science, Heraklion. Crete University Press, Greece

Hutchins EL, Hollan JD et al. (1985) Direct Manipulation Interfaces. Human Computer Interaction 1:311–338

Keeney RL, Raiffa H (1993) Decisions with Multiple Obejctives. Preferences and Value Tradeoffs, Cambridge University Press

Li M, Zhou S et al. (2002) Multi-Agent Systemsfor Web-Based Information Retrieval. GIScience 2002. Springer Verlag, Berlin Heidelberg

Linden G, Hanks S et al. (1997) Interactive Assessment of User Preference Models: The Automated Travel Assistant. Sixth Int User Modeling Conf, UM97, Vienna. Springer, Vienna New York.

Luce RD, Raiffa H (1957) The Axiomatic Treatment of Utility. In: Edwards W, Tversky A, Decision Making. Penguin Books, Harmondsworth, Middlesex, England

McClennen EF (1990) Rationality and Dynamic Choice. Foundational Explorations. Cambridge University Press

Miller GA (1994) The Magical Number Seven, Plus or Minus Two. Some Limits on our Capacity for Processing Information. Psychological Review 101(2): 343–352

Mintzberg H, Westley F (2001). Decision Making: It's not what you think. MIT Sloan Management Review 42(3):89–93

Raubal M, Rinner C (2004) Multi-Criteria Decision Analysis for Location Based Services. 12th Int Conf on GeoInformatics – Geospatial Information Research: Bridging the Pacific and Atlantic, Gävle, Sweden, University of Gävle

Russell SJ, Norvig P (1995) Artificial Intelligence. Prentice Hall, Englewood Cliffs, NJ

Shneiderman B (1998) Designing the User Interface: Strategies for Effective Human Computer Interaction. Addison Wesley Longman, Reading, MA

Staab S, Werthner H et al. (2002) Intelligent Systems for Tourism. IEEE Intelligent Systems, Trends & Controversies 17(6):53–66

Tversky A, Kahneman D (1981) The framing of decisions and the psychology of choice. Science 211:4538

Yntema DB, Torgerson WS (1961) Man-Computer Cooperation in Decisions Requiring Common Sense. In: Edwards W, Tversky A, Decision Making. Penguin Books, Harmondsworth, Middlesex, England, pp 300–314

Spatiotemporal Event Detection and Analysis over Multiple Granularities

Arie Croitoru, Kristin Eickhorst, Anthony Stefandis, Peggy Agouris

Department of Spatial Information Science & Engineering and National Center for Geographic Information and Analysis, The University of Maine, U.S.A; email: {arie, snoox, tony, peggy}@spatial.maine.edu

Abstract

Granularity in time and space has a fundamental role in our perception and understanding of various phenomena. Currently applied analysis methods are based on a single level of granularity that is user-driven, leaving the user with the difficult task of determining the level of spatiotemporal abstraction at which processing will take place. Without a priori knowledge about the nature of the phenomenon at hand this is often a difficult task that may have a substantial impact on the processing results. In light of this, this paper introduces a spatiotemporal data analysis and knowledge discovery framework, which is based on two primary components: the spatiotemporal helix and scale-space analysis. While the spatiotemporal helix offers the ability to model and summarize spatiotemporal data, the scale space analysis offers the ability to simultaneously process the data at multiple scales, thus allowing processing without a priori knowledge. In particular, this paper discusses how scale space representation and the derived deep structure can be used for the detection of events (and processes) in spatiotemporal data, and demonstrates the robustness of our framework in the presence of noise.

1 Introduction

Many phenomena in virtually all areas of natural sciences involve the study of change, and in particular change in spatial data over time. A primary reason for this interest in change is simple: change has a fundamental role in our perception and understanding of the world as it provides a systematic approach to the evolution of things in space and time. The identification and formalization of change patterns allows us to achieve what is often taken for granted: formalize rules, apply reasoning, and predict future behaviors of a given phenomenon. Consequently, the study of change in spatial data over time is essential in various areas, such as meteorology, geophysics, forestry, biology, and epidemiology.

The study of change in all these disciplines is closely related to the study of *events*. The description of change in terms of events is natural to us, primarily because as humans we intuitively tend to perceive an activity as consisting of discrete events (Zacks and Taversky 2001). Yet the term event, which is often used rather loosely in daily life, may have different meanings under different circumstances and different contexts. In view of their wide variability in space and time, there have been various suggestions for a more general model of events. For example, Galton (2000), based on an analysis of change, has identified three classes, namely states (the absence of change), processes (on-going change), and events (a predefined amount of change). Yet, as Galton has indicated, the distinction between states and events is not always straightforward: *"Thus processes seem to have a chameleon-like character, appearing now as states, now as events, depending on the context in which they are considered."* (Galton 2000, p 215).

Zacks and Taversky (2001) have also addressed the nature of events in human perception and conception and have defined events as a segment of time at a given location that is conceived by an observer to have a beginning and an end. Yet, in light of this definition they have also indicated that *"In general, it seems that as we increase the time-scale of our view, events become less physically characterized and more defined by the goals, plans, intentions and traits of their participants"* (Zacks and Taversky 2001, p 7).

These different examples of how events are defined and considered emphasize the intricate nature of events. In particular, a primary factor that contributes to the dual nature of events (processes⇔events and objects⇔events) is *granularity* in time and space, which has a fundamental role in our perception and understanding of various phenomena. Granularity refers to the notion that the world is perceived at different grain sizes

(Hornsby and Egenhofer 2002), and in the context of this work relates to the amount of detail necessary for a data analysis task.

The motivation for this work stems from the effect granularity has on our ability to analyze and understand spatiotemporal events. Currently applied analysis methods are based on a *single* level of granularity that is user-driven, that is, the user has the difficult role of determining a level of spatiotemporal abstraction at which the processing will take place. Yet in many cases, and in particular when analyzing unfamiliar phenomena, it is difficult to determine beforehand the granularity level at which event processing should take place without *a priori* knowledge: a too fine granularity will result in detail overloading while a too coarse granularity will result in over abstraction and loss of detail. In other cases, users may not be interested in events in a single granularity level, but rather in a range of levels in the study of events in given phenomenon (for example, users may be interested in analyzing rapid changes in the cloud mass of a hurricane both at the hourly and daily levels). This often leads to a need to repeat the processing of the data at each granularity level, thus resulting in low efficiency and high computational cost. Furthermore, such an approach makes it difficult to compare between phenomena that have a similar behavior but occur at different temporal scales.

It should be noted that the issue of the effect of granularity on the analysis of events is closely related to the problem of temporal zooming (Hornsby 2001) and the modeling of moving objects over multiple granularities (Hornsby and Egenhofer 2002). Yet, while this previous work focused primarily on the transition between identity states of objects and on adjusting the level-of-detail of object representations as a result of a change in granularity, the focus of this contribution is on the simultaneous analysis of spatiotemporal data over multiple granularities rather than a specific one.

In light of this, the main contribution of this work is the introduction of a novel spatiotemporal data analysis framework, which is based on two primary building blocks: *the spatiotemporal helix* – a formal representation of spatiotemporal data, and a *scale-space analysis* of the data in the temporal domain. Such an analysis offers two distinct advantages, namely, the ability to analyze the data at multiple granularities instantaneously, and the ability to reveal the hierarchical structure of events within the given phenomenon. To illustrate this, we will focus on a hurricane data analysis application, in which it is required to discover similar hurricanes by clustering. Our data source in this case is a time-series of remotely sensed imagery of each of the hurricanes, as provided by the National Oceanic and Atmospheric Administration (NOAA) (NASA 2005).

The remainder of this paper is organized as follows: Section 2 provides an overview of the spatiotemporal helix model that is used in this work to represent and summarize spatiotemporal phenomena. Section 3 provides a review of the scale space representation, followed by an analysis of the utilization of this representation as an event detection and analysis framework in spatiotemporal helixes. Section 4 describes an example application of the proposed framework using real-world data. Finally, conclusions and future work are summarized in Section 5.

2 The Spatiotemporal Helix

In our paradigm the spatiotemporal helix is used as a formal model of a spatiotemporal phenomenon. The spatiotemporal helix is a framework for describing and summarizing a spatiotemporal phenomenon. It is composed of a compact data structure and a set of summarizing tools capable of generalizing and summarizing spatiotemporal data, thus allowing efficient querying and delivering of such data. This framework can be applied to a variety of spatiotemporal data sources, ranging from manual monitoring of an object through time to spatiotemporal data that is collected through a single sensor or a sensor network (Venkataraman et al. 2004).

The idea behind the spatiotemporal helix is based on the observation that as an object moves, two key characteristics change over time: location and deformation. Using an image time series [see Fig. 1(a)] we extract the object using image-based feature extraction, from which we can track the object's location by calculating its center of mass and collecting this information in a database. In addition, we also track the deformation of the object by recording expansion and contraction magnitudes in each of four cardinal directions. This inclusion of deformation in the helix model provides the ability not only to track the changes in the location of the object, but also changes in its morphology. The result of the feature extraction process is depicted in Figure 1(b), which shows that a visualization of the extraction results in a three-dimensional space, consisting of a number of object out-lines stacked one on top of the next, with time as the vertical axis.

While initially all possible location and deformation information is gathered from the image time series, the summarization aspect of the helix is introduced by retaining only the frames that include significant information based on a user-defined change threshold. In this process we collect two types of entities: *nodes* consisting of the coordinates of the object's center of mass for each time instance, and *prongs* that capture information about the object's expansion or contraction. These *prongs* consist of a record of the time instance, magnitude, and direction in which the object has expanded or contracted. Nodes and prongs are therefore the building blocks of the spatiotemporal helix, where nodes construct the spine of the helix and prongs provide an annotation of the spine [see Fig. 1(c)]. An outline of the helix construction process from motion imagery is depicted in Figure 1. The interested reader may find more information about the spatiotemporal helix model in Agouris and Stefanidis (2003) and Stefanidis et al. (2005).

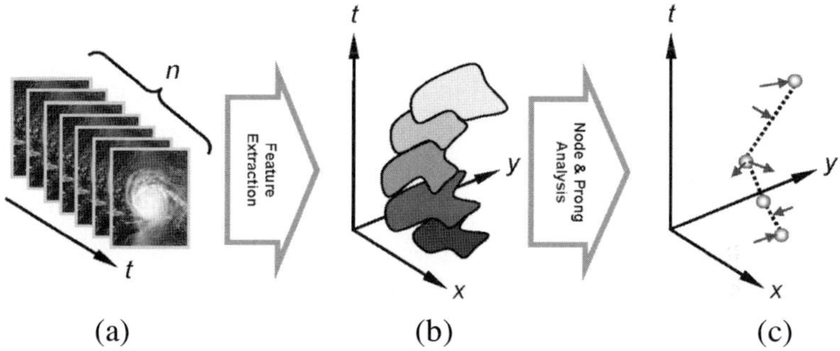

Fig. 1. Spatiotemporal helix visualized as stacking of objects over time. **(a)** An image time series of size n. **(b)** Feature extraction results in space and time. **(c)** The spatiotemporal helix (arrows represent prongs, circles represent nodes)

3 Scale Space Analysis

As was mentioned earlier, scale space analysis has a central role in our framework due to several distinct advantages it offers in the context of spatiotemporal event analysis. In order to demonstrate this we will first provide a short overview of scale space representation, and will then analyze the different characteristics in the context of spatiotemporal event analysis. It should be noted that in view of the ample body of literature on scale space representation, only a brief and non-exhaustive description of the key ideas and results are provided here. The interested reader may refer to Lindeberg (1994a, 1994b) and Sporring et al. (1997) for further details.

3.1 Scale Space Representation and Deep Structure

The development of the scale space representation stemmed from the understanding that scale plays a fundamental and crucial role in the analysis of measurements (signals) of physical phenomena. In order to demonstrate this, consider a signal *f*, which was obtained from a set of real-world measurements. The extraction of information from *f* is based on the application of an operator with a predefined scale. A fundamental question in this process is therefore the determination of the *proper scale* of the operator. Clearly, there is a direct connection between the scale of the operator we chose to apply and the scale of the structures (information) in *f* that we wish to detect (Lindeberg 1994b). If the scale of the operator is too large or too small, our ability to derive information from *f* will be compromised, leading to either high sensitivity to noise or low sensitivity to the structures sought. Consequently, proper scale should be used in order to ensure the optimal extraction of meaningful information from *f*. The determination of the proper scale is straightforward in cases where a priori knowledge about *f* exists, yet in other cases where there is no a priori knowledge the determination of the proper scale becomes a fundamental challenge and all scales should be considered. This notion of considering all possible scales is at the heart of the scale space representation (Lindeberg 1994b).

The construction of a scale space representation is carried out by embedding the signal *f* into a one-parameter family of derived signals, in which the scale is controlled by a scale parameter σ. More formally, given a signal $f(x):\Re\rightarrow\Re\ \forall x\in\Re$, the (linear) *scale space representation* $L(x,\sigma):\Re\times\Re+\rightarrow\Re$ of $f(x)$ is defined such that $L(x,0)=f(x)$, and the representation at coarser scales are given by $L(x,\sigma)=g(x,\sigma)*f(x)$, where $*$ is the convolution operator, and $g(x,\sigma)$ is a Gaussian kernels of increasing width (Lindeberg 1994b). In the case of a one-dimensional signal, $g(x,\sigma)$ is taken as the one-dimensional Gaussian kernel (Witkin 1983; Lindeberg 1990):

$$g(x,\sigma)=\frac{1}{\sqrt{2\pi\sigma}}e^{-x^2/2\sigma} \tag{1}$$

The two-dimensional space formed by (x,σ) is termed *scale space*. The scale space representation of $f(x)$ is therefore comprised from a family of curves in scale space that have been successively smoothed by the kernel.

While the generation of the scale space representation (L) results in a family of signals with an increasing level of smoothing, it is the inner structure of the scale space representation that exhibits distinct inherent behavior. In particular, it was found that the extrema points (the zero-crossings of the nth derivative) in scale space representation form paths in scale space that will not be closed from below and that no new paths will be created as σ increases. Hence, as σ increases new extrema points cannot be created (Witkin 1983; Mokhtarian and Mackworth 1986). We carry out the generation of such paths by computing the location of the zero-crossings for each of the derived signals in L, and then stacking these different locations in scale space. As result of this process a binary image showing these paths is created. Following Florack and Kuijper (2000) and Kuijper et al. (2001), we term the resulting binary image the *deep structure* of the Gaussian scale space, that is the structure of at all levels of granularity simultaneously.

To illustrate how the scale space representation and the deep structure are used for the detection of features in the data consider the example depicted in Figure 2, which shows the analysis of a one-dimensional signal [see Fig. 2(a) top].

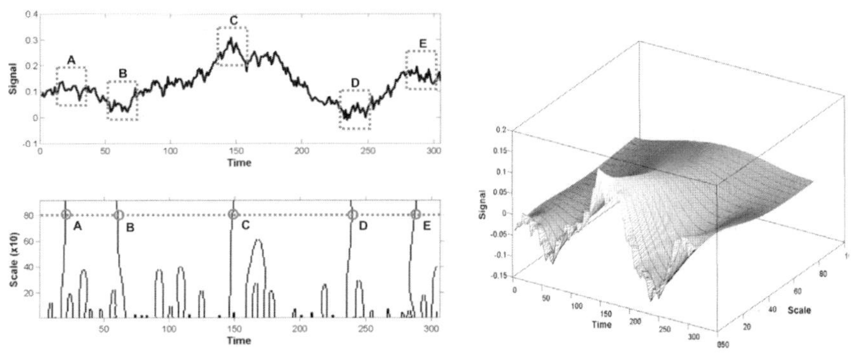

(a) (b)

Fig. 2. An example of a scale space analysis of a one-dimensional signal. **(a)** Top – A one-dimensional signal; Bottom – the deep structure of the signal as de-rived by the zero-crossings of the 1st derivative **(b)** The scale space representation of the signal, which was derived using a Gaussian kernel of increasing size. In all figures the x-axis represents time

By convolving this signal with a Gaussian kernel of an in-creasing size (see Eq. 1) the scale space representation (L) is derived. Figure 2(b) shows the scale space representation as a three-dimensional surface (note how the original signal becomes smoother as the scale of the Gaussian kernel increases). Once the scale space representation is derived, the zero-crossings of the nth derivative are detected for each scale level and the deep structure [see Fig. 2(a), bottom] is recovered by stacking the zero-crossings one on top of the other. Here, for the sake of simplicity, we have chosen to use the zero crossing of the 1st derivative to construct the deep structure due to its direct relation to features in the signal. Consequently, the paths formed in the deep structure describe how extrema points in the signal evolve as scale increases. To demonstrate this, consider the deep structure at a scale level of 8 [marked by the dashed horizontal line in Fig. 2(a), bottom] which shows five points, A through E. Clearly these five points correspond to local extrema in the signal, as marked by the five rectangles in Figure 2(a), top. It is easy to see that as the scale parameter decreases more extrema points appear in the deep structure due to the noisy nature of the signal.

3.2 Analysis of Events Using Scale Space Representation

The adaptation of the scale space representation approach to the analysis of events in spatiotemporal helixes can offer several distinct advantages. In order to demonstrate this we first define events within the helix framework and then analyze the different characteristics of the scale space representation in light of this definition.

Following Galton (2000) and Grenon and Smith (2004), we define events within the context of spatiotemporal helixes as the *transition between states*. As such, we regard events as all entities that exhaust themselves in a single instant of time (Grenon and Smith 2004). Consequently, events are used to define the boundaries of processes and indicate the transition within processes.

Having this definition in mind, let us now analyze how the scale space representation and the deep structure of a physical measurement signal could be used for the detection of events. As noted earlier, the zero crossing of the nth (commonly n=2) derivative of L is used to construct the deep structure. Since the zero-crossing condition ensures a change of sign in the second derivative (note that this is not the same as requiring that the second derivative will be zero), the deep structure serves as an indicator of inflection points[1] in the given signal f. Such inflection points indicate either a minimum or a maximum in the gradient of L. In conclusion, the deep structure can be used for detecting minimum or maximum rates of change (the first derivative) of a process, or changes in the sign of the rate of change (the second derivative) of a process. Note that in our interpretation we view processes as occurring between inflection points (events), which corresponds to our definition of events.

To illustrate this, let us assume that f is a vector of the x coordinate of a hurricane that was tracked in time. The first derivative of f indicates the speed of the hurricane in the x direction, while the second derivative indicates the acceleration of the hurricane in the x direction. Transforming f into a scale space representation and the recovery of the deep structure en-

[1] Given a twice differentiable function $g(x)$, a point $x=c$ on g is an inflection point if $g(c)'$ is an extremum point and $g(c)''$ changes its sign in the neighborhood of c (Binmore 2001).

ables the detection of the following events: *maximum speed events, minimum speed events, acceleration to deceleration events, and deceleration to acceleration events.*

Let us now turn back to the deep structure of spatiotemporal helixes and analyze its characteristics in relation to the analysis of events. Several key observations can be made here:

1. As was mentioned earlier, the derived deep structure is based on the detection of inflection points, which form paths that are guaranteed not be closed from below. It is also guaranteed that no new paths will be created as σ (the scale factor) increases. These characteristics are essential in the analysis of events as it is expected that no new processes will emerge, as the time granularity of a physical process is made coarser. Furthermore, this property ensures that a process (which is defined between inflection points) cannot disappear and then reemerge in coarser time granularities.
2. In general, paths in the deep structure will not cross each other (Mokhtarian and Mackworth 1986). This property of the paths assures that time conflicts will not occur. Consider for instance the example in Figure 2: since the x axis of both the data and the deep structure is time, paths that cross each other will indicate that two events that occurred in one order in one time granularity level will occur in the opposite order in another granularity level. This property therefore ensures that the proper order of events will be maintained at all granularity levels.
3. In general, paths in the deep structure will either form arch-like paths that are closed from above [for example, curve e in Fig. 3(a)], or will form a single path line. In the context of events, these properties indicate that as time granularity is made coarser processes that are defined between two events (inflection points) will converge to a single *event point* (the top point of a path for which the gradient is zero) and eventually disappear. This can be seen in Figure 3(a), where path b converges to a single point (tangent to the dashed horizontal line) at scale σ_i, and disappears at coarser granularities. In addition, single path lines indicate *transition events* [for example, curve a in Fig. 3(a)] that are not defining processes within the framework of the data provided but rather a change in the process.

In summary, the deep structure (and the scale space representation) can be used for the detection of events through which processes can be defined. Furthermore, the Gaussian scale space ensures that as time granularity is made coarser (a) no new processes will emerge, (b) processes can not disappear and reemerge, (c) the proper order of events (and processes) is maintained, (d) processes in lower granularity will tend to converge to a single event point, and eventually disappear.

In addition to these characteristics it is important to note that the deep structure inherently offers the ability to reveal the *hierarchy* of events and processes. To illustrate this, consider the scale space path b in Figure 3(a). As can be seen, this path contains two additional paths, c and d, which can be seen as two *sub-processes*. It should be noted that as granularity increases sub-processes turn to a single point event and eventually disappear. This hierarchy can be further described in a process tree [see Fig. 3(b)].

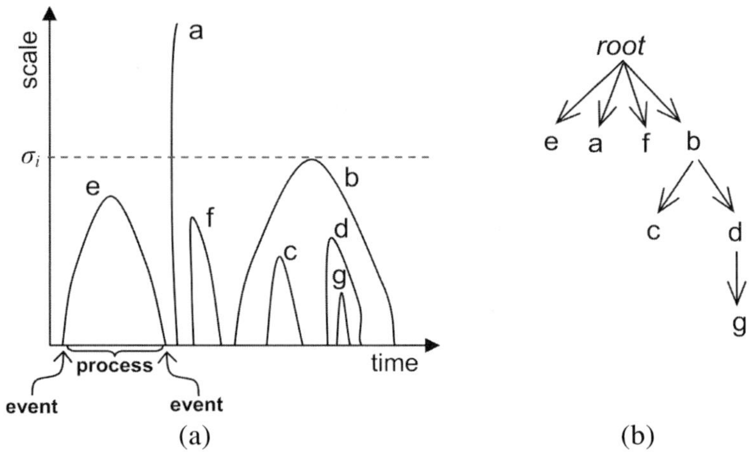

Fig. 3. Event and process hierarchy discovery through scale space representation. **(a)** A sample deep structure. **(b)** The derived process tree

3.3 Scale Space Analysis of Events in Spatiotemporal Helixes

As was described in Section 2, the spatiotemporal helix is a framework for describing and summarizing a spatiotemporal phenomenon. Given an image time series which contains an object that should be analyzed, the spatiotemporal helix collects the following information about the object along its spine (Stefanidis et al. 2005): the x and y location of the center of mass, acceleration, rotation, and expansion/contraction in north, east, south, and west directions.

In order to analyze the helix and detect events (and processes), we treat each of the 8 attributes of the spatiotemporal helix as a one-dimensional signal. Based on this, we can then apply a scale space analysis for each of these signals. This will result in 8 scale space representations and deep structures that could be used in various ways in order to understand the underlying physical phenomenon at hand. In particular, the following applications of the deep structures are considered in the analysis of helixes:

1. *Single helix dimension analysis* – in this case, a single dimension of a single helix can be analyzed (for example, the x or y location of the center of mass). Here, the deep structure can be used as an inspection tool for the detection of events and processes at multiple granularities. Furthermore, the deep structure provides an insight into the evolution of events and to the *hierarchy* of processes [see Fig. 3(b)].
2. *Multiple helix dimension analysis* – in this case, two or more dimensions of a single helix are analyzed by overlaying their deep structures. This would allow, for instance, detecting of processes that occur at the same time interval in different dimensions, from which higher-level inferences about the phenomenon could be derived.
3. *Helix clustering* – in this case, the primary goal is to estimate the similarity between helixes for the purpose of discovering similar physical phenomenon. In this case the similarity function, $S(\cdot,\cdot)$, between helixes H_i and H_j can be computed by:

$$S\left(H_i, H_j\right) = \sum_{k=1}^{n} w_k C\left(DS_k\left(H_i\right), DS_k\left(H_j\right)\right) \tag{2}$$

where k is the number of dimensions (attributes) in each helix, w_k is a weight assigned to each dimension (user defined), $DS_k(\cdot)$ is the deep structure of the k_{th} dimension (a two-dimensional matrix of size $u \times v$), and $C(\cdot,\cdot)$ is the two-dimensional cross correlation coefficient between $DS_k(H_i)$ and $DS_k(H_j)$ that is given by:

$$C\left(DS_k\left(H_i\right), DS_k\left(H_j\right)\right) = \frac{\sum_u \sum_v \left(DS_k\left(H_i\right) - \overline{DS_k}\left(H_i\right)\right)\left(DS_k\left(H_j\right) - \overline{DS_k}\left(H_j\right)\right)}{\sqrt{\sum_u \sum_v \left(DS_k\left(H_i\right) - \overline{DS_k}\left(H_i\right)\right)^2} \sqrt{\sum_u \sum_v \left(DS_k\left(H_j\right) - \overline{DS_k}\left(H_j\right)\right)^2}} \tag{3}$$

If the length of the two helixes is not the same, a template matching approach is used, in which the deep structure of the shorter helix is shifted along the x axis (time) of the deep structure of the longer helix, and the maximum cross correlation coefficient [see Eq. (3)] is taken. It should be emphasized that because the deep structure of the kth dimension is used in Equation (3), the cross correlation is being simultaneously computed in multiple granularities. Thus, S [see Eq. (2)] is a measure of the overall similarity of the two helixes over multiple granularities.

4 An Example: Spatiotemporal Hurricane Data Clustering

In order to demonstrate the capabilities and robustness of our approach we have applied the proposed framework to the following problem: *given a data set of n imagery time series of physical phenomena, partition the data set into subsets of similar phenomena without any a priori knowledge about the granularity of the phenomena.* The primary motivation for selecting this particular task was the centrality that role clustering has in numerous areas, such as data mining and knowledge discovery, search engines, machine learning, and pattern recognition. In all these areas reliance on minimal a priori knowledge and robustness to noise are crucial.

For this work we have collected real-world satellite imagery time series of five different tropical storms and hurricanes. The satellite data was obtained from NASA-GFSC's GOES project (http://goes.gsfc.nasa.gov) that provides GOES-12 imagery. The storms that were collected are Alex (1–5 August, 2004), Allison (4–14 June, 2001), Charley (11–14 August, 2004), Dennis (7–11 July, 2005), and Frances (August 31 – September 7, 2004).

The preliminary processing of each of the five hurricane image time series included the delineation of the contour of the hurricane cloud mass from each image frame. This process, which resulted in a binary image time series, was then used as input to the helix construction process (Stefanidis et al. 2005) from which a spatiotemporal helix was created for each hurricane (see Fig. 4). Then, from each of the five helixes four more permutations were created by corrupting the original helix data with an increasing level of random noise that was added to the center of mass and the expansion/contraction dimensions. This process resulted in a data set consisting of a total of 25 helixes. In order to cluster this data set we have implemented two different methods:

1. *The local extreme event approach* – in this approach we defined extreme events based on the deviation from the average attribute value using a moving window, that is, given a confidence level d all helix attribute values within a window of a user-defined size that deviate more than $d{\cdot}\sigma$ from the average are considered to be extreme events. By changing the window size and the value of d the user can then control the level of granularity in which extreme events are detected. Based on these extreme events we then computed the distance between all possible helix pairs using the technique described by Stefanidis et al. (2005), and constructed a distance matrix from which a dendrogram was derived.
2. *The scale space approach* – in this approach we implemented the pro-posed scale space clustering technique that was described in Sec-

tion 3.3. Similar to the first approach, here we have also computed the distance between all possible helix pairs using Equation (2) and (3), and constructed a distance matrix from which a dendrogram was derived.

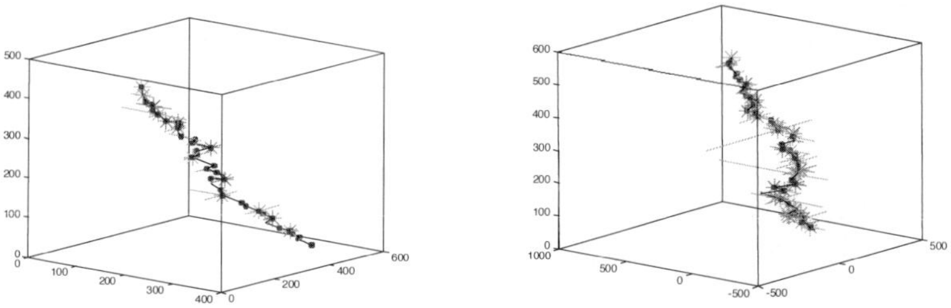

(a) (b)

Fig. 4. Examples of the derived spatiotemporal helixes for the hurricane data sets. **(a)** hurricane Frances, **(b)** hurricane Alison. In both figures the central black line represents the spine of the helix, the black circles represent the nodes, and the gray lines represent prong information

Using these methods two different clustering experiments were conducted. The first experiment included the clustering of data from two hurricanes, Frances and Alison, including their permutations (a total of 10 helixes) using both methods. The second included the clustering of the entire hurricane data set (a total of 25 helixes) using the scale space approach. In both experiments a correct clustering would result in distinct clusters in the dendrogram, where each cluster contains data from only one of the hurricanes. An example of the deep structure (2nd derivative) that was derived and utilized for each helix in both experiments is depicted in Figure 5.

The results of the first experiment are depicted in Figure 6, where Figure 6(a) through (c) show the distance matrices and dendrograms that were obtained from the local extreme event approach.

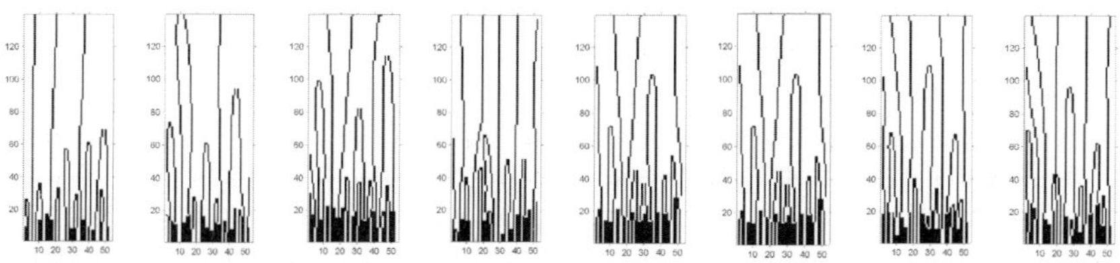

(a) (b) (c) (d) (e) (f) (g) (h)

Fig. 5. The deep structure of the spatiotemporal helix of hurricane Alison. **(a)** *x* location, **(b)** y location, **(c)** acceleration, **(d)** rotation, **(e)** through **(h)** expansion / contraction in the north, east, south, and west direction respectively

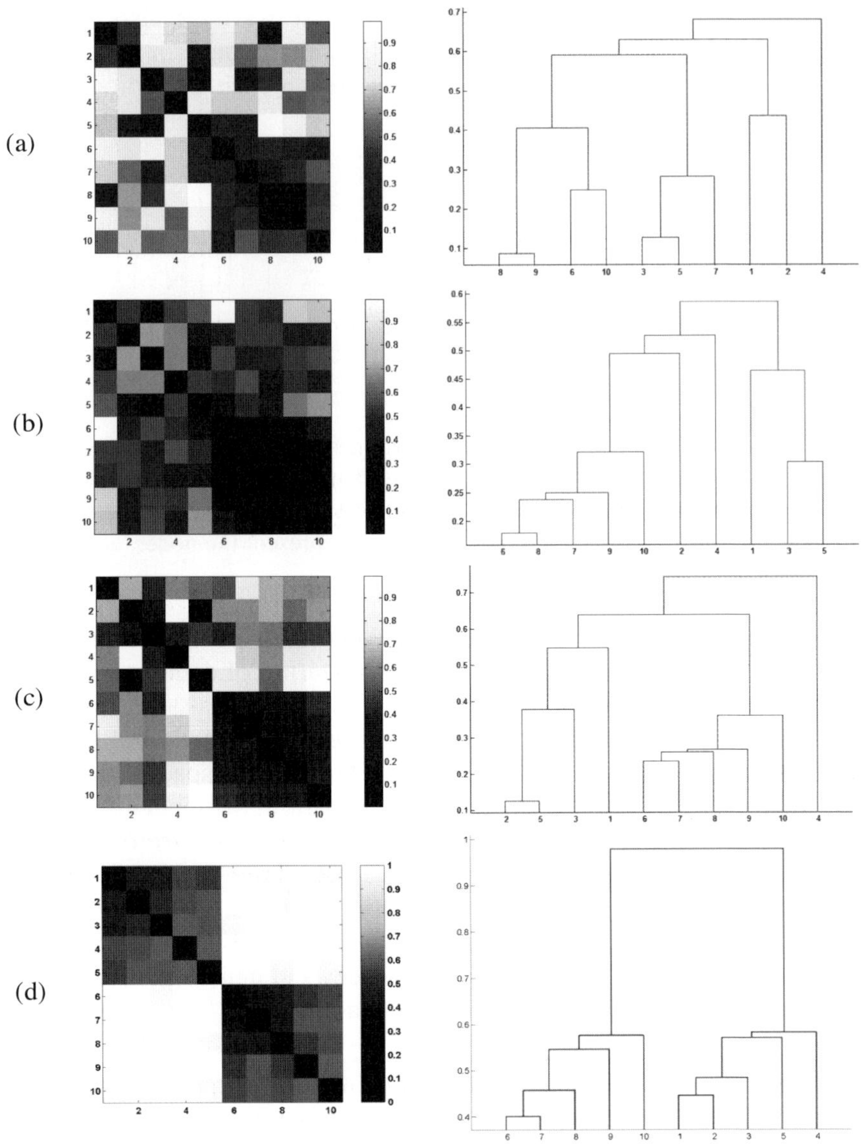

Fig. 6. Results of the first clustering experiment. **(a)** through **(c)** – The distance matrix (left) and dendrogram (right) using the local extreme event approach with a moving window size of 5, 7, and 9 respectively. **(d)** The distance matrix (left) and dendrogram (right) using the scale space approach. In all figures numbers 1–5 correspond to Allison and numbers 6–10 correspond to Frances. Darker shades in the distance matrices correspond to higher similarity

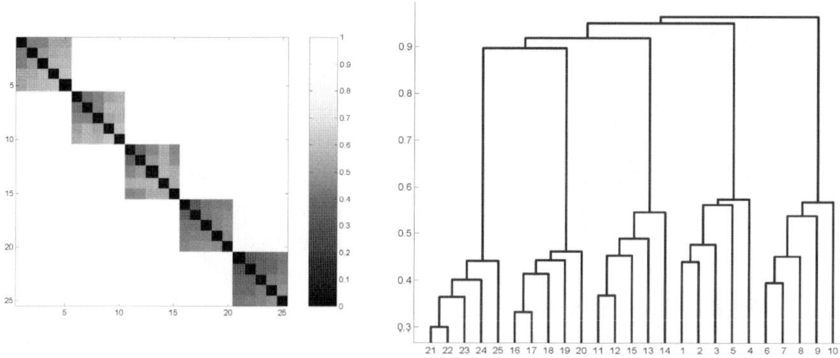

(a) (b)

Fig. 7. Results of the second clustering experiment using the entire hurricane data set. **(a)** The distance matrix. **(b)** The resulting dendrogram. In both figures numbers 1–5 correspond to Allison, 6–10 correspond to Frances, 11–15 correspond to Dennis, 16–20 correspond to Alex, and 21–25 correspond to Charley

The effect of the granularity at which the processing takes place is evident: as the window size increases some clusters do begin to emerge; yet the correct clustering is not obtained. In practical application this demonstrates the difficulty users are likely to face when analyzing such data without proper a priori knowledge. In contrast, the scale space approach produced the correct clustering [see Fig. 6(d)], resulting in two well-defined clusters, one for each set of hurricane data.

The results of the second experiment are depicted in Figure 7. As can be seen, the scale space approach successfully recovered the five clusters of hurricanes in this case as well.

5 Conclusions

Granularity in time and space has a fundamental role in our perception and understanding of various phenomena. Furthermore, since improper granularity may lead to erroneous results, it is essential that proper granularity be used in spatiotemporal data analysis and knowledge discovery. In spite of the importance of granularity, it is often difficult to determine at which granularity data processing should take place without a priori knowledge. This paper addressed this problem by adopting a scale space approach, in which all granularity levels are considered instead of applying a single granularity level. Based on this approach we presented a framework consisting of the spatiotemporal helix as a modeling and summarization tool, and the scale space representation as an analysis and knowledge discovery tool. The primary advantage of our framework is that it does not require a priori knowledge about granularity.

We analyzed how scale space representation and the derived deep structure could be used for the detection and analysis of events and processes and showed that due to its unique characteristics, deep structure can be used for the detection of events through which processes can be defined. Furthermore, we showed that the deep structure ensures that as time granularity is made coarser no new processes will emerge, processes can not disappear and reemerge, the proper order of events (and processes) is maintained, and that processes in lower granularity will tend to converge to a single event point, and eventually disappear. Additionally, we described how the deep structure could be used for the discovery of a hierarchy of events and processes. To demonstrate the capabilities of our approach we applied the proposed framework to the problem of real-world hurricane data clustering, and showed its robustness in the presence of noise.

In the future we plan to further explore and expand our framework. In particular, we are interested in utilizing the proposed approach for determining the proper granularity that should be used in the analysis of a given data set, and in developing additional similarity functions for scale space representations.

Acknowledgements

We would like to thank the anonymous reviewers for their helpful comments and suggestions. This work was supported by the National Geospatial-Intelligence Agency through grant NMA 401-02-1-2008, and by the National Science Foundation through grant ITR-0121269.

References

Agouris P, Stefanidis A (2003) Efficient summarization of spatiotemporal events. Communications of the ACM 46(1):65–66

Binmore KG (2001) Calculus. Cambridge University Press

Florack L, Kuijper A (2000) The topological structure of scale-space images. J of Mathematical Imaging and Vision 12:65–79

Galton A (2000) Qualitative spatial change. Oxford University Press

Grenon P, Smith B (2004) SNAP and SPAN: towards dynamic spatial ontology. Spatial Cognition and Computation 4(1):69–104

Hornsby K (2001) Temporal zooming. Transactions in GIS 5(3):255–272

Hornsby K, Egenhofer M (2002) Modeling moving objects over multiple granularities, Special issue on Spatial and Temporal Granularity. In: Annals of Mathematics and Artificial Intelligence 36. Kluwer Academic Press, Dordrecht, pp 177–194

Kuijper A, Florack LMJ, Veirgever MA (2001) Scale space hierarchy. Technical report UU-CS-2001-19, Utrecht University, Department of Computer Science

Lindeberg T (1990) Scale space for discrete signals. IEEE Transactions on Pattern Analysis and Machine Intelligence 12(3):234–254

Lindeberg T (1994a) Scale-space Theory: A Basic Tool for Analyzing Structures at Different Scales. J of Applied Statistics 21(2):225–270

Lindeberg T (1994b) Scale space theory in computer vision. Kluwer Academic Press, Dordrecht

Mokhtrian F, Macworth A (1986) Scale based description and recognition of planar curves and two-dimensional shapes. IEEE Transactions on Pattern Analysis and Machine Intelligence 8(1):34–43

NASA (2005) The GOES science project. http://rsd.gsfc.nasa.gov/goes/goesproject.html (last visited: December 2005)

Sporring J, Florack L, Nielsen M, Johnsen P (1997) Gaussian scale-space theory. Kluwer Academic Publishers

Stefanidis A, Agouris P, Eickhorst K, Croitoru A (2005) Modeling Object Movements and Changes with Spatiotemporal Helixes. Submitted to the Int J of Geographical Information Science

Venkataraman V, Srinivasan S, Stefanidis A (2004) Object Color Propagation in an Unregistered Distributed Video Sensor Network. IEEE Int Conf on Image Processing (ICIP) 2004, Oct 2004, Singapore

Witkin AP (1983) Scale-space filtering. Proc of the 8th Int Joint Conf on Artificial Intelligence, Karlsruhe, Germany, pp 1019–1022

Zacks JM, Tversky B (2001) Event structure in perception and conception. Psychological Bulletin 127(1):3–21

Reduced Data Model for Storing and Retrieving Geographic Data

Andrew U. Frank

Dept. of Geoinformation and Cartography, Technical University Vienna
email: frank@geoinfo.tuwien.ac.at

Abstract

The 'industry-strength' data models are complex to use and tend to obscure the fundamental issues. Going back to the original proposal of Chen for Entities and Relationships, I describe here a reduced data model with Objects and Relations. It is mathematically well founded in the category of relations and has been implemented to demonstrate that it is viable. An example how this is used to structure data and load data is shown.

1 Introduction

Numerous groups investigate how to structure data for use in Geographic Information Systems (GIS); they use tools prepared for 'real life' applications using the tried and trusted methods of the past – mostly object concepts based on languages like C++ or Java and the relational data model. Design is typically using UML despite the known lack of formal definition. These methods and tools are useful to build GIS applications but they are not appropriate as foundations for GIScience research. Shortcomings have been identified years ago, but convincing solutions are still missing (Dijkstra 1976). This paper addresses the fundamental question of structuring data for permanent storage and proposes a reduction of data models to *Objects* and *Relations* with 6 operations. Building the program to store and retrieve the data for a graph shows that the reduction is viable and has preserved the essence.

Current GIScience research focuses – among other things – on

- adding support for temporal data and processes to the GIS (Langran 1992);
- drive the design of a geographic application from an ontological perspective (Fonseca and Egenhofer 1999) and incorporating the results of the 'GIS ontology discussion' (Mark, Smith et al. 2000; Frank 2003);
- integrating data from different sources (Nyerges 1989; Bishr 1998);
- use an agent oriented design and construct agent based simulations (Ferber 1998; Bittner 2001; Raubal 2001);
- improve usability by connecting the user interface design with the ontological analysis (Achatschitz 2005).

For these and other similar research efforts, the construction of a repository for the data used in the application code is very time consuming. The performance orientation of the 'industry-strength' systems impose restrictions and limitations, and – last but not least – the theoretical bases on which these industry systems are built are in conflict with the research goals. It is hard to build new tools if the machinery to build them is imposing assumptions that we try to overcome!

Over the past years we have built many programs to store the data necessary for experiments in handling spatial data. We learned to reduce the data model to a minimum, which gives maximum flexibility and the least restrictions. This paper describes a usable set of functions for storage and retrieval of data for experimentation with advanced concepts of geographic data handling, especially research focused on temporal data, ontology, and integration. It achieves:

- programs are structured around objects,
- a data structure and the data collection can be changed and extended,
- modules describing an object type can be freely combined.

A number of limitations in programming languages had to be overcome to translate a clean design founded in a mathematical theory into executable code that validates the design. The code is now ready for use by others and made available from our CVS repository (gi07.geoinfo.tuwien.ac.at/CVSroot/RelDB).

Before starting with a new design, I review in Section 2 the achievements in Computer Science that are used for GIS, explain their rationale and the shortcomings relevant for GIS. Section 3 then restricts the various concepts employed to two fundamental ones: objects and binary relations. Section 4 explains the bundling of data description and object code. Section 5 gives examples with code and the concluding section touches on some of the limitations of the current code compared to full object-orientation or database concepts; it describes directions for future work as well.

2 Issues with the Current Technology

The current technology used to build applications that manage large collections of geographic data, gives the stability and the performance necessary for the GI industry, but the resulting systems are complex to install and manage. The flexibility to fulfill novel requirements or to adapt to changing situations is limited and numerous restrictions apply. New theoretical foundations, especially the development of the theory of programming languages, allow new approaches and we must rethink our design choices (Chen 2006). In this section I argue for the specific choices behind the reported solution.

2.1 Database Concept

To consider the data as a resource of an organization was a first step towards the information age in the 1970s. Data management became a central task and generalized database management software developed. The three schema model separates the stable description of the data stored from the often changing application programs. Unfortunately, the ability to describe the data in the schema is limited and many important aspects of data descriptions are dispersed in the application programs.

Computer science research investigates how the data descriptions can be embedded with the data using the XML language (Ceri, Fraternali et al. 2000); ontology languages try to include more from the data description (Dieckmann 2003) but seem to lack abilities to describe processes.

2.2 Relational Data Model

Codd invented the relational data model to facilitate the management of administrative data, specifically data that was ordinary represented on paper as tables. The relational data model was very successful because it gave a formal base to the intuitive and widely used concept of tables; Codd defined a small number of operations on tables that are closed and 'relationally complete' (Codd 1982). For applications, the SQL query language presented a human readable interface. The relational data model is 'value based', which means that all operations compare just values and there is no concept of objects in the formal model.

Härder observed already 20 years ago that geographic data, similar to data from CAD and other applications that relate to space and time, require for their representation multiple tables, connected by common values of their identifiers (Härder 1986). The representation of a graph – which is at the core of most spatial applications, e.g., transportation, cadastral data – can become easily inconsistent by changing the names of nodes and not maintaining the corresponding edge data (see Fig. 1):

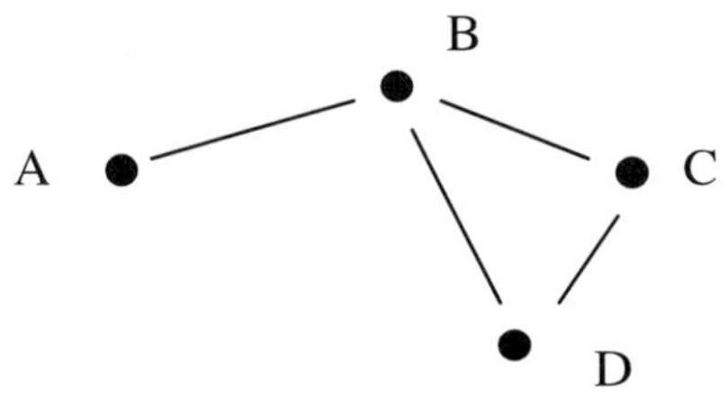

Nodes

Name	X	Y
A	0.7	1.3
B	1.5	2.2
C	2.9	2.0
D	2.8	0.9

Edges

Start	End
A	B
B	C
C	D
B	D
C	D

Fig. 1. A graph with 2 tables for nodes and segments

Codd himself has seen this limitation and suggested the use of substitutes (i.e., identifiers) to establish the relations between tuples (rows) of tables (Codd 1979). It is possible to store geographic data under the relational data model but code to maintain the data consistent is necessary.

It was found that structuring data in relational tables leads to anomalies in updates. Redundancy can be hidden within the tables and leads to inconsistencies when changes are applied. Dependencies between values within a tuple (row of a table) must be avoided and some multi-column tables must be broken in smaller tables to eliminate such dependencies. It was not possible to give a small set of rules to identify all harmful dependencies and to normalize a set of relational tables.

2.3 Object-orientation

The structuring of design and code centering on objects and operations applicable to them is the dominant paradigm of software engineering the past 20 years. Code for objects and their interactions as operations can hide the internals of an object (so-called encapsulation). Inheritance of object behavior into subclasses gives extensibility (Borning 1977). Object-orientation was welcomed in the GIS community to help with the analysis of the complex structure of geographic reality (Egenhofer and Frank 1987; Worboys, Hearnshaw et al. 1990).

Programming languages offer methods to structure data into objects for processing, but different languages have selected slightly different approaches (Cardelli and Wegner 1985); the controversy, regarding contra- vs. co-variance has not yielded a usable and implementable solution (Lämmel and Meijer 2005). Particular difficulties arise with multiple inheritances, i.e., cases where an object is a specialization of two (parent) classes, which is important for geographic data (Frank 1988; Frank and Timpf 1994) (see Fig. 2) but convincing solutions to model the difference in meaning of concepts like *boat-house* and *house-boat* are missing (Goguen and Harrell 2006). I use here a class-bounded (parametric) concept of polymorphism because the difficulties with subtyping polymorphism seem insurmountable (Abadi and Cardelli 1996; Lämmel and Meijer 2005).

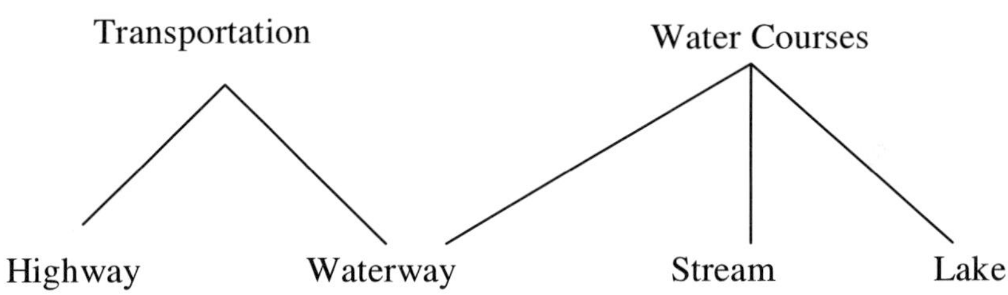

Fig. 2. A Waterway inherits properties from the transportation system and the water bodies

2.4 Object-oriented Databases

An impedance mismatch was observed between data handling in an application program written in an imperative language, which is organized 'a piece of data at a time', and the relational database, which operates on whole relations. Object-oriented databases (OODB) combine database concepts with the object-orientation in data structuring (Atkinson, Bancilhon et al. 1989; Lindsay, Stonebraker et al. 1989).

Practically, the subtle differences between variants of object-oriented concepts in programming languages and databases led to difficulties with structuring application data: only the concepts available in both the OODB and the object-orientation programming languages could be used; most OODB systems are tied to specific object-oriented languages. This makes the integration of data that is organized under different OODBMS very difficult. The tight coupling of the object concept in the application program and the long term view of the database seems to be fundamentally different and impedes evolution of a database over time.

Mapping the data structure from the program view to a simple structure maintained by the data storage system on secondary (disk) storage seems to be the answer. The object-relational approach combines a relational database with an object-oriented programming language (Stonebraker, Rowe et al. 1990), but simpler solutions, going directly to storage emerge (e.g., db4o, objectStore).

2.5 Desired Solution

Current GIS applications use an improved, but not theory based, "object-relational" database for storage of data. For experimentation with programs that handle spatial data I felt the following aspects important:

- combination of object description in the schema with the operations to handle the object instances (the data);
- binary relations to avoid dependencies;
- composability of object definitions;
- object orientation, with multiple inheritance using parametric polymorphism;
- focus on long-term secondary storage and direct connection to object-orientation structure of programming language;
- formal, mathematically sound framework.

The next section describes how these goals were achieved. The suggested solution is designed for experimentation and leaves out a number of questions important for processing large amounts of data.

3 Concepts to Retain

The desired solution should rather contain less than more concepts and the concepts should be simpler and more generally applicable, more oriented towards the user or the world ontology. The basic concepts, *Entities* and *Relationship*, to structure data but also the conceptualization of the world was described by Peter Chen in a landmark paper (Chen 1976).

3.1 Types and Instances

The world contains individuals and we structure our concepts of the world in entities, things that are thought to exist independently. The representation of these entities we will call *instances*. The discussion of data models concentrates on collections of similar instances, which in programming are called types (Cardelli 1997).

3.2 Objects

Objects represent entities that have permanence in time. The difference to the relational data model is that

1. the values and operations applicable to an object may change; but also
2. two objects may coincide in their values but are still distinct objects.

The term 'object' is generally used both to describe object types or classes and object instances, which are specific objects, representing individuals; compare the class 'dog' and my dog 'Fido', which is an individual. Object classes can be seen as algebras, with domains and operations (Ehrich, Gogolla et al. 1989; Loeckx, Ehrich et al. 1996).

3.3 Relations

Object (instances) are related to values. A city has a name, a coordinate to describe its position, and the name of the state it is in. These can be described formally as functions from the object instance to the values (Shipman 1981), but this is not general enough. For example a person can have several children, the mapping from person to children is therefore not a function but a relation. Relations have an advantage over functions as they have always a *converse*: a city is related to the state it is in, the converse relations relate the state to the cities it contains (some functions have inverses, but not all of them!).

Relations can be used to store two different aspects, namely (1) the values associated with an object and (2) a relationship to another object. An example for the first is the name of the city; an example for the second is the state the city is in, which is usually not stored as a name, but as the identifier of an object of type *state*. This use of relations to store relationships between objects solves the problem of maintaining the representation of a graph and other geometric data structures.

4 Description of the Solution

A first step towards simplification and flexibility is to select a modern Functional Programming (FP) language (Peyton Jones, Hughes et al. 1999), because FP languages are closely connected to the mathematics of computers (Asperti and Longo 1991; Walters 1991). The resulting conceptual simplicity is demonstrated by the restriction to two concepts and 6 functions to manipulate the data.

4.1 Objects Map to Identifiers

As already suggested by Codd (Codd 1979) the objects are represented in the long-term data storage as identifiers, which are permanent and never reused. When creating a new object in the database, only a new identifier is assigned to it.

Representing objects by identifiers, not data fields, is arguably the most radical decision here, which is sensible only in an environment where data is primarily stored on secondary (disk) storage. Giving up the combination of data storage and object representation cuts away many of the complexities of object management.

4.2 Materialized (Stored) Relations

The data associated with the objects are stored in binary relation. Breaking all data into binary relations from object identifier to value removes all potential for anomalies and gives automatically the highest level of normalization.

That operations on relational tables could compose is a major strength of the relational data model and must be preserved. Operations on relations must have inputs and results that are sets of values, not just single values. Thus operations can compose, i.e., the result of one operation can be the input for the next one. An isomorphic mapping, called the power transpose (Bird and de Moor 1997: p 108), is necessary because the category of relations cannot be directly implemented. The power transpose maps from relations to functions over the powerset, which can be implemented.

4.3 Bundles of Functionality are Modules

Bundles of functionality, e.g., support for graphs, ownership cadastre, or agents moving, etc. must be designed and coded separately. They are represented by modules, which are independently compliable units and care will be necessary to keep their dependencies minimal. In particular, the data structure and the related operations must be included in the same module, allowing different applications to use different combinations of bundles.

4.4 Operations for Data Handling

To store data, the operation is *assert*, which adds a new entry to a relation. The operation *change* takes a function that is applied to the currently stored value and the result is then stored (this can be used to *set* a value to *v* by passing the function *const v*). The function *delete* will delete the corresponding entries.

To retrieve two base functions are sufficient:

to: given a set of identifiers and a relation label, find the values related to the identifiers.

from: given a set of values and a relation label, find the identifiers related to the values. (Note: *to* and *from* are not inverse to each other but *from* = *to.conv*!) For the special case of searching in functions (which are a special kind of relations) with a single value and expecting a single result, specialized forms of *to* and *from* are given as *to*' and *from*'. *To* and *from* are total functions—they produce always a result, but *to*' and *from*' can fail and produce then a descriptive error message.

5 Example: How to Use RelDB

This section shows code for storing and retrieving data describing a simple graph: the entities are NODE and EDGE; there are relations from NODE to the value of its number and to the value of its coordinate. EDGE has its length as a value and two relations to the points where it starts and where it ends. The ER diagram in Figure 3 shows that there are five relations, 3 to values (ellipses) and 2 between entities (diamonds).

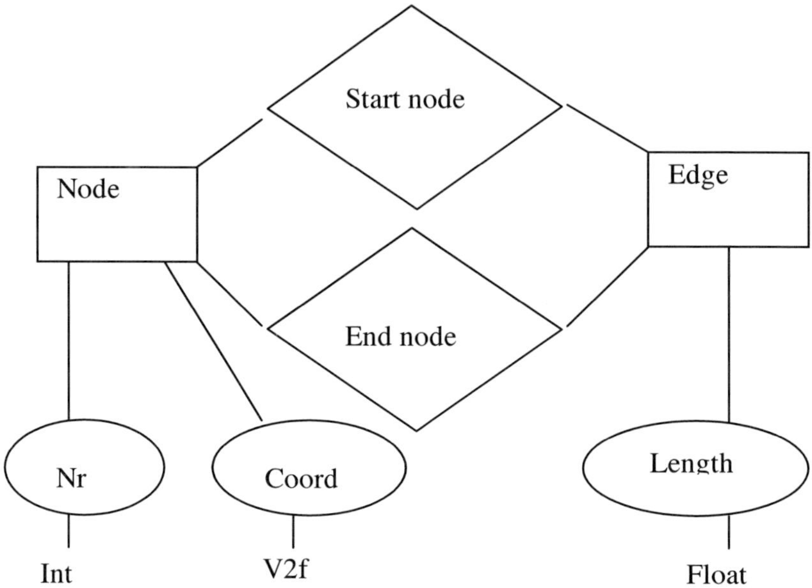

Fig. 3. The ER diagram in the original style (Chen 1976)

The coordinates are formatted as type *V2f* (Vector with *x* and *y* coordinates represented as floats), the number is an integer (*Int*), the length of the edge is a *Float*. There are five functions, which will be represented as relations:

```
coord :: PointID-> V2f
nr :: PointID -> Number
startNode :: EdgeID -> PointID
endNode ::EdgeID -> PointID
length :: EdgeID -> Float.
```

5.1 Definition of Relations

The five relations are defined each in 3 lines of code. Consider the relation from the node to the coordinate value: Define a type for the relation label and define a value of it. This creates two entries in the namespace of the module, for example *ID2Coord* and *id2coord*. These will be used as the relation label; for convenience it is also used to contain a descriptive string that is used when printing the relation.

```
newtype ID2Coord = ID2Coord String
id2coord = ID2Coord "coordinates of points"
```

A function *id2c* to add this relation to the empty database is declaring the type of the values, namely *V2f*:

```
id2c = HCons (id2coord, zero::RelVal V2f)
```

This must be repeated for the 4 other relations: for the number, for the two relation edge to node and finally for the length:

```
newtype ID2Nr = ID2Nr String
id2nr = ID2Nr "identification of nodes"
id2n = HCons (id2nr, zero::RelVal Nr)

newtype ID2StartNode = ID2StartNode String
id2startnode = ID2StartNode "start node of edge"
id2s = HCons (id2startnode, zero:: RelID)

newtype ID2EndNode = ID2EndNode String
id2endnode = ID2EndNode "end node of edge"
id2e = HCons (id2endnode, zero:: RelID)

newtype ID2EdgeLength = ID2EdgeLength String
id2edgelength = ID2EdgeLength "length of edge"
id2d = HCons (id2edgelength,  zero:: RelVal Float)
```

5.2 Construction of Database (Schema)

The database is constructed by adding these relations to the empty database (*emptydb1* exported from the generic database module). A database with support for node with number and coordinates would be:

```
pointdb1 = id2c (id2n emptydb1).
```

Remember: the constructs *id2n*, *id2c* are functions that can be applied or composed (with "."). To construct a database for graphs is:

```
graphdb1 = (id2c . id2n . id2s . id2e) emptydb1
```

or equivalently using the already existing pointdb1, which may already contain data:

```
graphdb1 = (id2s . id2e) pointdb1.
```

5.3 Handling Object Data

5.3.1 Points

Assuming that the data describing the points is in a file as a sequence of pairs consisting of number and coordinate values. A single entry describing a point looks as follows: *(3, V2f 4.1 2.9)*.

The function *loadNode* takes a pair of number and coordinate as input. It creates first a new object and gets the identifier into the variable *i*. Then it asserts that this identifier *i* has for the relation *id2nr* the value of *n* (the node number from the input) and then asserts that this identifier *i* has for the relation *id2coord* the value of *p* (the coordinate from the input).

```
loadNode (n, p) = do        i <- createM     nodeT
                            assertM  id2nr i n
                            assertM  id2coord i p
```

To load a series of points, stored in list *fh*, to the *pointdb1* is achieved with:

```
pointdb2 = (mapM loadNode fh) *** pointdb1.
```

To find for a given point number the corresponding identifier, we use the function *from'* that is specialized for cases where we expect only a single value as a result (as intended, the function from point number to identifier is an isomorphism):

```
identifyByNr db nr =  from' "identifyNr: not found" db id2nr
```

where the message is printed if for this point number no point is found in the data. The function to retrieve the point coordinate from a given identifier is very similar:

```
pos  db i = to' "position i pl in loadFreihaus"   db  id2coord i.
```

5.3.2 Edges

Loading an edge is somewhat more involved: Assume that the external file contains pairs of numbers of the start and end points for each edge. The relations between the edges and the nodes however are based on identifiers; it is necessary to find the identifier with *identifyByNr* to enter in the relation. To compute for an edge the length from start to end, we have to retrieve the position of the two nodes using *pos*. A function *dist'* to compute the distance between to coordinate pairs exists and is extended (overloaded) with a new instance, such that it computes the distance between to points given by their identifier (in the context of the current data):

```
instance  (FromTos a ID2Coord V2f)
        =>  Vec2x ID (State a Float) where
  dist' a b = State $ \s ->
        let    ap  = pos s a
               bp  = pos s b
        in (  (dist' ap bp), s)
```

Combining these support functions to form a single loadEdge function:

```
loadEdge (s, e) =
            do  i <- createM edgeT
                si   <- identifyNrM s
                se   <- identifyNrM e
                assertM id2startnode i si
                assertM id2endnode i se
                (c :: Float) <- dist' si se
                assertM id2edgelength i c.
```

It takes a pair of node numbers as input and creates first an edge with an identifiers. Then it retrieves the identifiers for the start (*si*) and the end node (*se*). It asserts that these are the values for the *id2startnode* and the *id2endnode* relations respectively. Then the distance between the two nodes (given by their identifiers) is computed and the result asserted for the relation *id2edgecost*.

6 Conclusions

The complex issues of designing a database schema have been reduced to a very small number of essential concepts. Additional tools may be necessary to achieve better performance, to install spatial access methods and to connect with a transaction management system, tasks left for future research. The goal was to identify what is essential, and to separate it from the desirable aspects. I have found it necessary to provide more than a 'paper and pencil' analysis but to implement the result and to show how it can be done in a running program.

- The restriction to binary relations simplifies the query language to 2 commands, one to find the related terms to an entry (*to*) and the other to find the identifiers related to a value (*from*), which is using the converse of the relation.
- Application programming in a functional programming language using the monadic style is straightforward.
- Modularization such that the schema information and the code to operate on an object class can be bundled and an application can use multiple of these bundles without interference.

A number of questions remain open for future work:

- Should the identifiers be typed?
- Consistency constraints: if we know that a relation is a function, where is the best place to enforce this restriction?

Acknowledgements

Many of my colleagues and students have helped me to advance to this point. Particular thanks go to Werner Kuhn for many valuable discussions during the past years. John Herring has asked years ago very valuable questions that have pushed me in this direction, but special recognition goes to Peter Chen for his early insight into the fundamental aspects of data modeling. An anonymous reviewer made useful suggestions to improve the paper.

References

Abadi M, Cardelli L (1996) A Theory of Objects. Springer-Verlag, New York

Achatschitz C (2005) Identifying the Necessary Information for a Spatial Decision: Camping for Beginners. CORP 2005 & Geomultimedia05. Selbstverlag des Institutes für EDV-gestützte Methoden in Architektur und Raumplanung, Vienna, Austria

Asperti A, Longo G (1991) Categories, Types and Structures – An Introduction to Category Theory for the Working Computer Scientist. The MIT Press, Cambridge, Mass

Atkinson M, Bancilhon F et al. (1989) The Object-Oriented Database System Manifesto. First Int Conf on Deductive and Object-Oriented Databases, Elsevier

Backus J (1978 Can Programming Be Liberated from the von Neumann Style? A Functional Style and Its Algebra of Programs. CACM 21:613–641

Bird R, Moor O de (1997) Algebra of Programming. Prentice Hall Europe, London

Bishr Y (1998) Overcoming the Semantic and Other Barriers to GIS Interoperability. Int J of Geographical Information Science 12(4):299–314

Bittner S (2001) An Agent-Based Model of Reality in a Cadastre. Department of Geoinformation, Technical University Vienna, Vienna

Borning A (1977) ThingLab – An Object-Oriented System for Building Simulations Using Constraints. IJCAI 1:497–498

Cardelli L (1997) Type Systems. Handbook of Computer Science and Engineering. AB Tucker, CRC Press:2208–2236

Cardelli L, Wegner P (1985) On Understanding Types, Data Abstraction, and Polymorphism. ACM Computing Surveys 17(4):471–522

Ceri S, Fraternali P et al. (2000) XML: Current Development and Future Challenges for the Database Community. In: Zaniolo C, Lockemann PC, Scholl MG, Grust T, Advances in Database Technology – EDBT 2000 (7th Int Conf on Extending Database Technology, Kontanz, Germany). Springer-Verlag, Berlin Heidelberg, 1777, pp 3–17

Chen PP-S (1976) The Entity-Relationship Model – Toward a Unified View of Data. ACM Transactions on Database Systems 1(1):9–36

Chen PP (2006) Entity-Relationship Modeling:Historical Events, Future Trends, and Lessons Learned. Retrieved 01.09.06, 2006

Codd E (1979) Extending the Database Relational Model to Capture More Meaning. ACM TODS 4(4):379–434

Codd EF (1982) Relational Data Base: A Practical Foundation for Productivity. Communications of the ACM 25(2):109–117

Dieckmann J (2003) DAML+OIL und OWL XML-Sprachen für Ontologien. Berlin, p 21

Dijkstra EW (1976) A Discipline of Programming. Prentice Hall, Englewood Cliffs, NJ

Egenhofer MJ, Frank AU (1987) Object-Oriented Databases: Database Requirements for GIS. Int Geographic Information Systems (IGIS) Symp: The Research Agenda, Crystal City, VA, NASA

Ehrich H-D, Gogolla M et al. (1989) Algebraische Spezifikation abstrakter Datentypen. BG Teubner, Stuttgart

Ferber J (ed) (1998) Multi-Agent Systems – An Introduction to Distributed Artificial Intelligence. Addison-Wesley

Fonseca FT, Egenhofer MJ (1999) Ontology-Driven Geographic Information Systems. 7th ACM Symp on Advances in Geographic Information Systems, Kansas City, MO

Frank AU (1988) Multiple Inheritance and Genericity for the Integration of a Database Management System in an Object-Oriented Approach. Advances in Object-Oriented Database Systems – Proc of the 2nd Int Workshop on Object-Oriented Database Systems, Bad Muenster am Stein-Ebernburg, F.R. Germany. Springer-Verlag, New York

Frank AU (1999) One Step up the Abstraction Ladder: Combining Algebras – From Functional Pieces to a Whole. In: Freksa C, Mark DM (eds), Spatial Information Theory – Cognitive and Computational Foundations of Geographic Information Science (Int Conf COSIT'99, Stade, Germany). Springer-Verlag, Berlin, 1661, pp 95–107

Frank AU (2003) Ontology for Spatio-Temporal Databases. In: Koubarakis M, Sellis T et al., Spatiotemporal Databases: The Chorochronos Approach. Springer-Verlag, Berlin, pp 9–78

Frank AU, Timpf S (1994) Multiple Representations for Cartographic Objects in a Multi-Scale Tree – An Intelligent Graphical Zoom. Computers and Graphics Special Issue on Modelling and Visualization of Spatial Data in GIS 18(6): 823–829

Goguen J, Harrell DF (2006) Information Visualization and Semiotic Morphisms

Härder T (1986) New Approaches to Object Processing in Engineering Databases. Proc on the 1986 Int Workshop on Object-Oriented Database Systems, Pacific Grove, California, United States. IEEE Computer Society Press

Lämmel R, Meijer E (2005) Mappings Make Data Processing Go' round. Microsoft Corp., Redmond, USA

Langran G (ed) (1992) Time in Geographic Information Systems. Technical Issues in GIS. Taylor and Francis

Lindsay B, Stonebraker M et al. (1989) The Object-Oriented Counter Manifesto

Loeckx J, Ehrich H-D et al. (1996) Specification of Abstract Data Types. John Wiley and B.G. Teubner, Chichester, UK and Stuttgart

Mark D, Smith B et al. (2000) Ontological Foundations for Geographic Information Science:18

Nyerges T (1989) Schema Integration Analysis for the Development of GIS Databases. Int J of Geographical Information Systems 3(2):153–183

Peyton Jones S, Hughes J et al. (1999) Haskell 98: A Non-Strict, Purely Functional Language

Raubal M (2001) Agent-Based Simulation of Human Wayfinding: A Perceptual Model for Unfamiliar Buildings. Institute for Geoinformation, Vienna University of Technology, Vienna, p 159

Shipman DW (1981) The Functional Data Model and the Data Language DAPLEX. ACM Transactions on Database Systems 6 (March)

Stonebraker M, Rowe LA et al. (1990) Third-generation Data Base System Manifesto. Electronics Research Lab, UC Berkeley

Walters RFC (1991) Categories and Computer Science. Carslaw Publications, Cambridge, UK

Weiss G (1999) Multi-Agent Systems: A Modern Approach to Distributed Artificial Intelligence. The MIT Press, Cambridge, Mass

Worboys MF, Hearnshaw HM et al. (1990) Object-Oriented Data Modelling for Spatial Databases. Int J of Geographical Information Systems 4(4):369–383

Filling the Gaps in Keyword-Based Query Expansion for Geodata Retrieval

Hartwig H. Hochmair

St. Cloud State University, Department of Geography
720 Fourth Avenue South, St. Cloud, MN 56301, USA
hhhochmair@stcloudstate.edu

Abstract

Query expansion describes the automated process of supplementing a user's search with additional terms or geographic locations to make it more appropriate for the user's needs. Such process relies on the system's knowledge about the relation between geographic terms and places. Geodata repositories host spatial data, which can be queried over their metadata, such as keywords. One way to organize the system's knowledge structure for keyword-based query expansion is to use a similarity network. In a complete similarity network the total number of similarity values between keyterms increases with the square of included keywords. Thus, the task of determining all these values becomes time consuming very quickly. One efficient method is to start with a sparse similarity network, and automatically estimate missing similarity values from other values with an algorithm. Hence, this paper introduces and evaluates four such algorithms.

1 Introduction

Geodata repositories contain data that are spatially referenced and made accessible through access points in the Internet. The data sets can be searched in a Spatial Data Infrastructure (SDI) based on their metadata. Metadata are data about data and typically describe thematic, geographic, and data format characteristics of a data set. Query expansion supports the user in finding related data files or documents, which allows the user to specify the direction of the search. Geodata repositories often provide multi-modal search functionality, such as search after a theme or keyword. This research focuses on semantic query expansion of thematic keywords, which derives additional query terms by traversing a knowledge structure.

Most search models for documents or data in the Internet are based on simple keyword matching. More advanced search architectures use ontologies as knowledge structure for query expansion. They describe and formalize the concepts and vocabulary of the domain of interest. Ontologies are traditionally based on a thesaural structure controlling hierarchies, associative relations, and synonymy (Voorhees 1994; Järvelin et al. 2001) or constructed from document collection based on term clustering (Mandala et al. 1999). The FACET Web demonstrator[1] described by Binding and Tudhope (2004) allows query expansion functionality across thesaurus data, which, in this application, are artifacts of an exhibition. One of the functions provided is fuzzy query expansion through the use of a simple coarse-grained control. Examples for geodata repositories that use a structured knowledge base for geodata retrieval are CERES[2] or the GEON[3] portals. These catalogs structure their feature classes in ontologies based on subset-superset relations, part-of relations, synonyms, and associate relations.

The keyterms describing data repositories can stem from many different categories and are difficult to place in purely hierarchical relationships. To handle these diverse keyterms for query expansion, we adopt similarity networks (Dagan et al. 1993) as knowledge structure for the system. The similarity-based approach promotes an "unstructured" paradigm of representing relations between concepts, as opposed to taxonomical structures. The keyterms are connected through weighted edges. Weights are used to express the semantic similarity between two terms in the similarity network. This approach of using similarity links refers to the findings of Shirky (2005) who suggests that a pre-classification of documents into categories may have shortcomings, especially when it comes to search in the WWW. Links between documents without categorization would be a more organic way for organizing information. As a working hypothesis we assume that there is no need for taxonomic structures in the knowledge

[1] http://www.comp.glam.ac.uk/~FACET/webdemo/

[2] California Environmental Resources Evaluation System: http://ceres.ca.gov/catalog/

[3] Geosciences Network: http://www.geongrid.org/

base and that the more general relation of similarity (relatedness) may be sufficient as an organizing principle.

The price that needs to be paid for this relatively simple knowledge structure is a high number of links that connect all keyterms with each other in the similarity network. Similarity values for these links can be stored in a similarity matrix. Creating a complete similarity matrix is time consuming if many terms are included. One way to avoid such costs is to use a *skeleton matrix*, which is a sparsely filled similarity matrix. Missing values would then be automatically estimated from other edge weights using an algorithm. This paper introduces and evaluates a set of four algorithms that estimate the missing values in a skeleton matrix. It further examines the potential impact of the structure of "gaps" on the accuracy of the estimated missing values. The work reported in this article is part of the Andeon Amazon GIS Web Portal Project[4], which has the goal to build a user-friendly geodata warehouse hosting hydrology related data sets about the Andean Amazon region of Bolivia, Colombia, Ecuador, and Peru.

The remainder of this paper is structured as follows: Section 2 explains the purpose of a reference matrix in our approach, Section 3 reviews definitions from graph theory needed to describe various types of incomplete similarity networks which are introduced in section 4. Section 5 explains four different algorithms that estimate missing similarity values. These algorithms are tested in Section 6. Section 7 gives an interpretation of the results and lists some challenges for future work.

2 Creating the Reference Matrix

Both similarity matrices and weighted graphs are interchangeable mathematical models for describing the structure of a similarity network. In order to test the quality of how different algorithms would estimate values for the empty elements in a skeleton matrix, we first created a complete reference matrix, from which we removed arbitrary elements in a second step to get a skeleton matrix. Within the simulation, the difference between the estimated values in the skeleton matrix and the correct reference values from the complete matrix provide a measure for the quality of the algorithms that fill the gaps in the skeleton matrix. The reference matrices in our simulation relate 18 geographical keywords from data sets of the Andean Amazon GIS Web Portal project. The keywords fall into several the-

[4] http://aagwp.fiu.edu/

matic classes, thus it would be difficult to place them in purely hierarchical relationships.

In order to fill the reference matrix, we conducted an empirical study, where a total of 28 volunteers from Florida International University (FIU) participated. 24 participants were undergraduate students in an environmental studies course (age between 20 and 37 years), and four were employees of the GIS-RS center (age between 26 and 62 years). A reference matrix for n terms contains n^2-n relations (excluding self relations), which gives 306 relations for n=18. Each participant was assigned a random subset of pairs of keywords and asked to assign to them similarity values S between 0 (not related at all) and 10 (strongly related). We received a total of six filled similarity matrices (A–F). All computations discussed later in this paper were performed using a similarity matrix Σ that averaged the values over the six matrices. Some computations were also completed with the A-matrix only. Fig. 1 shows a part of Σ, where similarity values were normalized to a range between 0 and 1. The remaining four terms that are not shown in the figure are catchments, water sampling stations, deforestation rates, and land cover.

	rivers	landscape	foothills	hydrological data	topography	meterological data	meterological stations	soils	geology	watersheds	roads	population centers	coast line	digital terrain model
rivers	1.00	0.84	0.22	0.84	0.58	0.24	0.02	0.35	0.48	0.90	0.14	0.40	0.58	...
landscape	0.53	1.00	0.32	0.52	0.70	0.22	0.22	0.60	0.52	0.65	0.38	0.48	0.48	
foothills	0.10	0.48	1.00	0.22	0.85	0.33	0.05	0.28	0.57	0.18	0.27	0.10	0.05	
hydrological data	0.97	0.64	0.20	1.00	0.50	0.63	0.55	0.22	0.23	0.80	0.02	0.36	0.64	
topography	0.74	0.58	0.72	0.44	1.00	0.08	1.80	0.57	0.64	0.40	0.38	0.24	0.74	
meterological data	0.35	0.32	0.32	0.56	0.38	1.00	0.97	0.13	0.12	0.58	0.06	0.38	0.08	
...							...							

Fig. 1. Part of the Σ reference matrix with normalized similarity values

Analysis of Σ revealed expected asymmetries. According to Tversky (1977), a subclass would be more similar to a superclass than the other way round. In Σ, this predictable effect can, for example, be found between the superclass "landscape" and its part meronym "river" where *S(landscape, rivers)* = 0.53 and *S(rivers, landscape)* = 0.84 (see Fig. 1). Another example is the pair *landscape-foothills*. The average of the asymmetries over all pairs (Σ_{ij} - Σ_{ji}) was 0.03, the standard deviation was 1.65.

3 Matrices and Graphs

This section reviews some definitions of matrix and graph theory that are needed to characterize the topology of a similarity network. We use both matrix and graph concepts to explain the algorithms and the network topology. Most definitions in this section are based on work by Sporns (2002). We use G(V, E) to represent a directed graph (digraph) G, with its vertex set V and edge set E. An edge $e_{ij} \in E$ is an ordered pair (v_i, v_j), where $v_i, v_j \in V$. Vertices of the similarity network are mapped to row names and column names in the matrix M, and its weighted edges are mapped to matrix elements. In a *complete* graph, each vertex is adjacent to each other. Such graph would be mapped to a complete matrix, whereas missing edges of G appear as unknown values in the matrix, i.e., as "gaps" in a sparse matrix. A degenerated edge of a graph which joins a vertex to itself is called a *self-loop*. A *simple* graph has no parallel edges or self-loops. *Paths* are ordered sequences of distinct edges and vertices from a source vertex to a target vertex. A graph is said to be *k-connected* if and only if there exist k disjoint paths between any two vertices of G. A directed graph is called *strongly connected* if there is a directed path from any vertex to any other vertex. The algorithms presented in this paper are analyzed for strongly connected simple graphs.

The topologic structure is described by the graph's *adjacency matrix A(G)* with binary entries a_{ij}, where $a_{ij} = 1$ if there exists a path from v_i to v_j and 0 otherwise. The *average connection density (d)* of an adjacency matrix A(G) is the number of all its non-zero entries, divided by the maximal possible number of edges. The sparser the graph, the lower *d* will be. A graph's *diameter* is the largest number of vertices, which must be traversed in order to travel from any vertex to another along the shortest path (the graph geodesic). The *distance matrix D(G)* holds the lengths of the shortest paths between the vertices of a network. It provides information about the "directedness" with which two units in a network interact. For example, if the distance between two vertices v_i and v_j is 2, then v_i can influence v_j through just one unit. If no path exists between v_i and v_j, the distance between v_i and v_j is infinite. The *characteristic path length $\lambda(G)$* of a graph is defined as the global mean of its finite entries of its distance matrix. The path with the shortest graph distance between two vertices can, for example, be found with Dijkstra's algorithm, breadth-first or depth-first search. To find all paths of a certain graph distance, the latter two algorithms are appropriate, whereas Dijkstra's algorithm can be used to search the shortest path for weighted graphs.

4 Matrix Structures

A similarity matrix must be complete in order to function as basis for query expansion. Creating a sparse reference matrix, i.e., a skeleton matrix, and deriving its missing values automatically from other values will save effort. This section explains the topologic parameters that we use to characterize the structure of a skeleton matrix, i.e., its *schema*. The schema impacts the quality of the estimated values for filling the gaps. Some types of skeleton matrices that are used for testing the estimation algorithms are introduced.

4.1 Topologic Parameters

Following parameters are used to characterize the structure of a skeleton matrix: Characteristic path length λ, average connection density d, and connectedness k of the underlying graph. Assuming transitivity for similarity, a small λ is desired as it keeps path lengths short between sources and gaps when estimating similarities. A large d is desirable as it generally yields a large k, which allows the missing value to be derived over several paths from the same source. λ and d are indirect proportional to each other for a given k. Table 1 summarizes the general relations between any two topologic parameters when the third parameter is kept constant. A ↑ means that a parameter value increases, a ↓ means a decrease, and c indicates the parameter held constant.

Table 1. Relation between characteristic path length λ, average connection density d, and connectedness k for a skeleton matrix

λ	**d**	**k**
c	↑	↑
↑	c	↑
↓	↑	c

4.2 Skeleton Matrices

We analyze a few selected schemes of skeleton matrices. Each scheme is visualized through a 6 x 6 adjacency matrix, where all empty elements indicate missing similarity values to be estimated from other matrix elements. In addition, the distance matrix and the graph are shown.

The *simple diagonal* scheme (see Table 2) requires the theoretically smallest possible number of reference values to derive values for all gaps. Even if the assumed transitivity for similarity would strictly hold, derived similarity values for gaps would most likely be computed too small because the graph is only 1-connected: Any edge with a similarity value of 0 along the path causes a 0 similarity between source and target concept.

Table 2. The simple diagonal scheme

Adjacency matrix

	A	B	C	D	E	F
A		1				
B			1			
C				1		
D					1	
E						1
F	1					

Distance matrix

	A	B	C	D	E	F
A	0	1	2	3	4	5
B	5	0	1	2	3	4
C	4	5	0	1	2	3
D	3	4	5	0	1	2
E	2	3	4	5	0	1
F	1	2	3	4	5	0

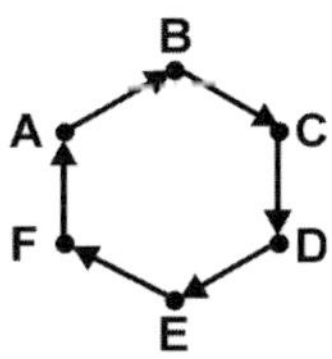

d	*1/(n-1)*
λ	*n/2*
k	*1*

The *horizontal-vertical (HV)* scheme has one central vertex that connects all other vertices (see Table 3). The distance between all vertices is 2 except if the central vertex is involved. Compared to the diagonal scheme this scheme has a twice as high *d*, which reduces λ to approximately 2 for large *n*. The 1-connectivity stays unchanged, which is an undesirable characteristic. The smaller λ however should increase the accuracy of derived gap values. Depending on which vertex is selected as central, *n* variations are possible for the HV scheme.

Table 3. The horizontal-vertical scheme – one of n possible variations

Adjacency matrix

	A	B	C	D	E	F
A		1	1	1	1	1
B	1					
C	1					
D	1					
E	1					
F	1					

Distance matrix

	A	B	C	D	E	F
A	0	1	1	1	1	1
B	1	0	2	2	2	2
C	1	2	0	2	2	2
D	1	2	2	0	2	2
E	1	2	2	2	0	2
F	1	2	2	2	2	0

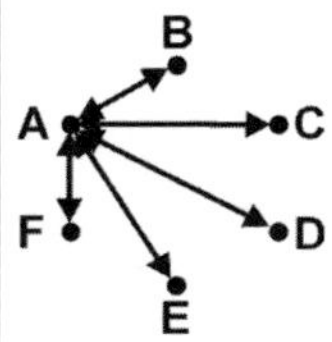

d	*2/(n-1)*
λ	$\frac{2n-4}{n-1}$
k	*1*

The combination of the two previous schemes gives the *horizontal-vertical-diagonal (HV-diag)* scheme (see Table 4). Whereas λ stays close to 2, this scheme has the advantage of a 2-connected graph, however with an increased *d*.

Table 4. The horizontal-vertical-diagonal scheme

Adjacency matrix

	A	B	C	D	E	F
A		1	1	1	1	1
B	1		1			
C	1			1		
D	1				1	
E	1					1
F	1					

Distance matrix

	A	B	C	D	E	F
A	0	1	1	1	1	1
B	1	0	1	2	2	2
C	1	2	0	1	2	2
D	1	2	2	0	1	2
E	1	2	2	2	0	1
F	1	2	2	2	2	0

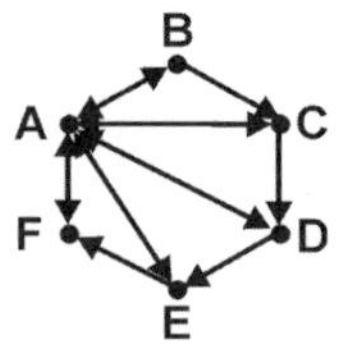

d	*3(n-1)*
λ	$\frac{2n-5}{n-1}$
k	2

Adding all reverse edges to the diagonal scheme gives the *symmetric diagonal (diagSym)* scheme (see Table 5). The advantage of the 2-connectivity and a small d has to be paid with a large λ when compared to the *HV-diag* scheme.

Table 5. The symmetric diagonal scheme

Adjacency matrix

	A	B	C	D	E	F
A		1				1
B	1		1			
C		1		1		
D			1		1	
E				1		1
F	1				1	

Distance matrix

	A	B	C	D	E	F
A	0	1	2	3	2	1
B	1	0	1	2	3	2
C	2	1	0	1	2	3
D	3	2	1	0	1	2
E	2	3	2	1	0	1
F	1	2	3	2	1	0

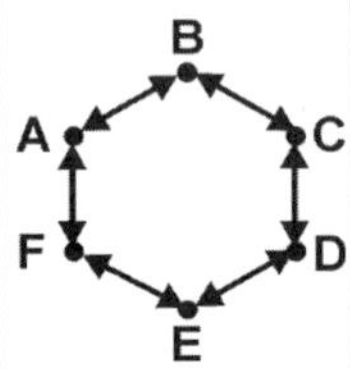

d	*2(n-1)*
λ	*n/4*
k	2

The group of *multidiagonal* schemes contains similarity values on selected diagonal stripes only. The average connection density of these skeleton matrices ranges between $0 < d \leq 0.5$. λ and k can be derived from d. Table 6 shows the connectivity matrices and the graph visualization for 6 x 6 matrices with $d = 1/2$ and $d = 1/3$. Elements on every second respectively third diagonal stripe are filled. $k=n*d$, and $\lambda=0.5/k+0.5$. For $n=6$ and $d=1/2$ (left figure) this gives $k=3$ and $\lambda=1.5$, and for $n=6$ and $d=1/3$ (right figure) this gives $k=2$ and $\lambda=2$. Thus, a smaller average connection density deteriorates the expected quality of the results by decreasing the connectivity of the graph and by increasing the characteristic path length.

Table 6. The multidiagonal scheme for average connection densities 1/2 and 1/3

d=1/2, k=3, λ=1.5

	A	B	C	D	E	F
A		1		1		1
B	1		1		1	
C		1		1		1
D	1		1		1	
E		1		1		1
F	1		1		1	

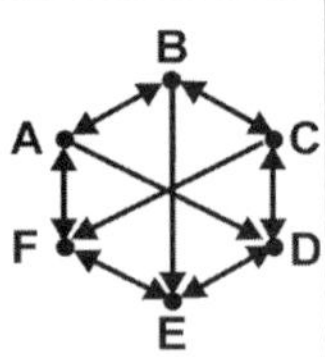

d=1/3, k=2, λ=2

	A	B	C	D	E	F
A		1			1	
B			1			1
C	1			1		
D		1			1	
E			1			1
F	1			1		

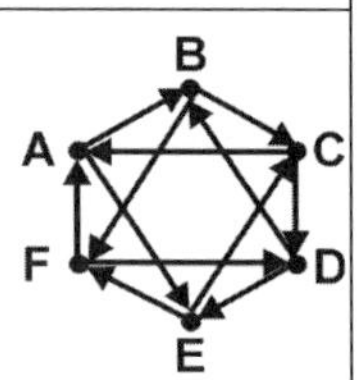

5 Algorithms to Fill the Gaps

This section introduces four algorithms that estimate missing values for gaps. The algorithms are demonstrated along with a 6 x 6 sample skeleton matrix. All non-diagonal empty elements in the sample matrix (see Table 7, left figure) indicate missing values. The "X" marks the missing value that will be computed along with the explanation of the algorithms. This element expresses the unknown similarity between concepts B and E. Numbers in the visualization of the directed graph (right figure) are normalized similarity values S(i,j) between connected concepts.

Whereas equivalence is transitive, most definitions of similarity relations are intransitive (Dagan, Marcus et al. 1993; Widdows 2003). In opposition, we want to examine whether the proposed algorithms compensate for a simplified assumption, which is: Transitivity holds for geographic concepts to a certain extent at least for short paths. We assume therefore that if both concept1 and concept2, and concept2 and concept3 are similar, concept1 and concept3 are also somewhat similar. We use the term *chaining* for deriving similarity values along paths.

Table 7. Sample skeleton similarity matrix with distance matrix and graph

Similarity matrix

	A	B	C	D	E	F
A				6		5
B	4		1	9	**X**	
C					7	
D					9	
E						8
F		8	3			

Distance matrix

	A	B	C	D	E	F
A	0	2	2	1	2	1
B	1	0	1	1	2	2
C	4	3	0	4	1	2
D	4	3	3	0	1	2
E	3	2	2	3	0	1
F	2	1	1	2	2	0

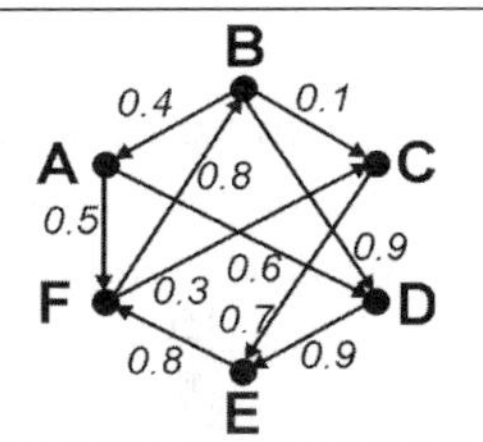

In the presented algorithms, which assume transitivity, the missing similarity *u* between two concepts, when derived from a source along a selected path, decreases with an increasing graph distance, which is the number of edges between two vertices. The computed similarity depends also on the similarity values of edges along the chosen path. One way to derive a similarity value is by multiplication of all S_i along a path together with a weighting factor *p* expressed as exponent of a power function with base S_i:

$$u = \prod_{i=1}^{l} S_i^p \text{, with } l\text{...graph distance, } S_i\text{...normalized similarities on path}$$

With a $p=1/2$, for example, the algorithm uses the root of normalized similarity values. For the demonstration of the algorithms we set $p=1$. The four algorithms may find different paths (chains) for the similarity computation.

The first algorithm, called *topo-avg*, uses all paths of the shortest graph distance *l* between source and target concept, and averages over all estimated similarities values between source and target. A breadth first search algorithm was implemented to retrieve all paths of length *l*. For the given example (see Table 7), the distance matrix shows that the shortest path between B and E has length 2. According to the graph, two of such paths exist, namely BDE and BCE. Using linear similarity values (p=1) gives u_{BDE} = 0.81 and u_{BCE} = 0.07, and an average of 0.44. It is apparent that using this algorithm estimated values may be too small: If one of the paths contains an edge connecting dissimilar concepts (i.e., B-C), this decreases the derived overall similarity value. In general, any method that estimates the missing similarity as a product (or average) of the other two similarities in the triangle cannot work for all of the possible missing pairwise values, as the estimate can never be greater than the larger of the two observed values. As an example let us take the triangle A-B-F from Table 7 and assume symmetric similarity on these edges for now. In this example, if S(A,F) and S(F,B) are observed, multiplicative chaining works, as S(A,F)=0.5, S(F,B)=0.8, and S(A,B)=S(A,F)*S(S,B)=0.4. On the contrary, if we have S(B,A) and S(A,F), and need to estimate S(B,F) instead, chaining gives too small an estimated value: 0.4*0.5=0.2, instead of 0.8.

This basic problem is the motivation for investigating another algorithm called *topo-max*. It takes the largest computed similarity from all shortest paths between source and target, which gives u_{BDE} = 0.81 for the first example. We expect this algorithm to work best with graphs of high connectivity, as this increases the chance to find triangles with larger similarity values observed along the paths.

The third algorithm, called *modified topo-max*, includes paths of length $l+1$, $l+2$,... in addition to the shortest paths of length l. The estimated similarity value can be computed as linear combination or as maximum of similarity values from each path group of lengths l, $l+1$, *etc*. Extending the used paths to $l+1$ in the sample example includes paths of length 3, i.e., path B-A-D-E with u_{BADE}=0.216. Using a weight of 1 for averaging all path groups changes the average of S(B,E) to 0.5*(0.216+0.81)=0.513 when using *topo-max* for each group. However, it does not affect the maximum similarity, which remains S(B,E)=0.81.

Fourth, Dijkstra's algorithm (*dijkstra*) can be applied to find the path that yields the highest possible similarity between source and target based on multiplicative chaining. First, similarities need to be converted to cost c by inversion of the similarity values assigned to edges. In order to find the shortest path with the smallest cumulative multiplied cost, $log(c)$ is used for edge weights because $log(a*b) = log\ (a) + log\ (b)$. For example, a similarity value S=0.3 gives c=log(1/0.3)=0.5228, and S=1 gives c=log(1/1)=0. In the sample example, the *dijkstra* algorithm finds path B-D-E with $log(c)$ = 0.092, which equals to u_{BDE} = 0.81. The simulations have shown that for most source-target pairs, *dijkstra* and *topo-max* find the same path, as the shortest valued path (in terms of weighted edges) is most often also the path with the smallest graph distance.

The goal of this work is to identify how the combination of the algorithm, weighting factor p, and matrix scheme impact the quality of estimated values in the skeleton matrix. Two reference matrices are used for the simulation, namely Σ and A (see Section 2). Errors in the estimated values are computed through comparing estimated values to reference values in the reference matrices. An error is defined as the reference value minus the computed value. The smaller the absolute value of an error, the better the quality of the combination applied. We use three parameters to evaluate the quality of an algorithm: the mean (μ) and standard deviation (*SD*) of errors, and the correlation coefficient between reference and computed values.

6 Simulation and Results

Our simulations have shown that the *topo-max* algorithm and the *dijkstra* algorithm provide the best results for all matrix schemes. That is, when adapting the weighting factor p for each algorithm so that the mean error is 0, the standard deviations of errors are smallest for *topo-max* and *dijkstra*. This section will therefore show results that were found with the *topo-max* algorithm on the matrix schemes introduced in Section 4.2.

We used five variations for testing the *horizontal-vertical (HV)* scheme, each with a different central vertex, because results vary between different central vertices. Similar results were found for Σ and A, thus we show results only for Σ. As *HV* is a 1-connected graph scheme, each of the proposed algorithms would yield the same paths (and results) for this scheme. Four out of five variations provide significant correlations between reference and computed values (see Table 8, left block). We tried several weighting factors: Root weighting (p=1/2) provides a desirable μ of approximately zero, a large μ is found for linear weighting (p=1). The high $SD > 2$ for both weighting factors shows that estimated values are widely scattered around the reference values despite a significant correlation.

The *simple diagonal scheme* (see Table 8, right block) does not provide a significant correlation. Estimated matrix values are generally too small ($\mu > 0$) which is caused by the large λ of $n/2$. The *symmetric diagonal scheme* provides slightly better results with a significant correlation due to a smaller λ of $n/4$. The estimated similarity values are generally too small for both tested weighting factors.

The combination of *HV* and *diag* scheme, i.e., the *horizontal-vertical-diagonal* scheme, improves the quality of the results when compared to both schemes separately, providing significant correlations for all five variations with root weights (see Table 9). This improvement can be ascribed to a small λ of approximately 2.

Table 8. Statistical correlation (Pearson), *SD*, and μ for the horizontal-vertical scheme and the two diagonal schemes

	horizontal-vertical					**diagonal**	
root weight	**HV1**	**HV2**	**HV3**	**HV4**	**HV5**	**simple**	**symm**
Corr. coeff.	**.124***	**.249****	.112	**.169****	**.323****	.075	**.133***
σ (2-tailed)	*.041*	*.000*	*.064*	*.005*	*.000*	*.206*	*.028*
SD	2.53	2.45	2.67	2.58	2.47	2.56	2.70
μ	-0.07	0.70	-0.53	-0.31	-0.61	3.88	2.19
linear weight							
Corr. coeff.	**.102**	**.234****	.102	**.166****	**.337****	.083	**.127***
σ (2-tailed)	*.094*	*.000*	*.093*	*.006*	*.000*	*.161*	*.037*
SD	2.57	2.36	2.65	2.47	2.41	2.34	2.45
μ	2.14	2.82	1.69	2.42	1.48	3.35	3.38

** correlation is significant at the 0.01 level (2-tailed)
* correlation is significant at the 0.05 level (2-tailed)

Table 9. Statistical correlation (Pearson), *SD*, and μ for the horizontal-vertical-diagonal scheme

	horizontal-vertical-diagonal				
root weight	**HV1D**	**HV2D**	**HV3D**	**HV4D**	**HV5D**
Corr. coeff.	**.123***	**.267****	**.151***	**.218****	**.397****
σ (2-tailed)	*.049*	*.000*	*.015*	*.000*	*.000*
μ	-0.14	0.75	-0.60	0.20	-0.72
SD	2.59	2.41	2.61	2.51	1.46
linear weight					
Corr. coeff.	.103	**.255****	**.152***	**.217****	**.399****
σ (2-tailed)	*.100*	*.000*	*.014*	*.000*	*.000*
μ	2.08	2.82	1.60	2.34	1.38
SD	2.56	2.34	2.61	2.44	2.33

** correlation is significant at the 0.01 level (2-tailed)
* correlation is significant at the 0.05 level (2-tailed)

In the *multidiagonal schemes*, d is the only independent parameter. Therefore an optimal weighting factor p can be found for each d to yield $\mu(d)=0$. Table 10 shows the simulation results for both the Σ and the A-matrix, which reveals that best correlations are found for d=1/2 and d=1/3. Thus, at least every third similarity value needs to be observed to estimate gap values that correlate with the correct reference values.

Fig. 2a provides an exemplary correlation for d=1/3. The bar chart in Fig. 2b shows for Σ that, as expected, a smaller d (i.e., a larger λ) in matrix schemes causes an increasing *SD* of errors. It shows further that p decreases with longer paths in order to keep μ=0.

Table 10. Statistical correlation (Pearson) between estimated and reference values with two reference matrices (Σ, A) for the *multidiagonal scheme*. Weighting factors p provide a zero mean error (μ=0)

		d=1/2	**d=1/3**	**d=1/4**	**d=1/5**	**d=1/6**	**d=1/7**	**d=1/8**	**d=1/9**
Σ-matrix	***Corr. coeff.***	**.384****	**.381****	.086	**.204****	.074	.061	**.143***	.044
	σ (2-tailed)	*.000*	*.000*	*.218*	*.001*	*.242*	*.331*	*.019*	*.476*
	SD	2.22	2.18	2.50	2.46	2.61	2.68	2.56	2.77
	p	1.05	0.90	0.65	0.6	0.5	0.3	0.32	0.28
A matrix	***Corr. coeff.***	**.255****	**.186****	.083	**.142***	-.026	.018	**.146***	.013
	σ (2-tailed)	*.002*	*.009*	*.244*	*.028*	*.686*	*.774*	*.019*	*.831*
	SD	3.33	3.12	3.37	3.16	3.26	3.24	2.84	3.19
	p	1.8	1.35	1.15	0.6	0.45	0.4	0.6	0.35

** correlation is significant at the 0.01 level (2-tailed)
* correlation is significant at the 0.05 level (2-tailed)

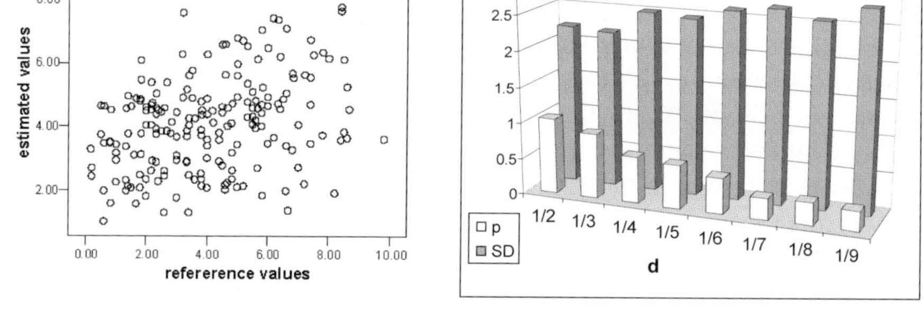

(a) (b)

Fig. 2. Visual correlation between reference and estimated values for the *multidiagonal scheme* for d = 1/3 (a); best weighting factors p, and standard deviations *SD* for each d (b)

7 Discussions and Outlook

When comparing the correlations between reference and estimated values for different matrix schemes, the predicted impacts of the topologic parameters on the errors of the estimated similarity values can be observed. Generally, with a smaller average connection density *d* and a larger characteristic path length λ, the correlation factor decreases. Matrix schemes provide a significant correlation between reference and estimated values especially with a connectedness > 1. However, correlation coefficients are relatively small, and all algorithms yield high standard deviations. This can be ascribed to several methodological simplifications, some of which are explained in the following paragraphs.

One potential problem mentioned in Section 5 is that product chaining tends to estimate similarity values that are too small for missing relations. In the simulations, chain selection was based on the *topo-max* algorithm to counteract this problem. With higher connectivity, its advantage becomes more effective, as seen by the smaller standard deviation for multidiagonal schemes (for chain lengths of 2 and 3 in Σ).

However, too high estimated similarity values, especially for the *topo-max* selection, can be ascribed to intransitivity of similarity. This may especially occur between concepts that are not hierarchically nested, i.e., not in superclass/subclass relations. One such example found in Σ (see Fig. 1) is between the terms "rivers" (*riv*), "hydrological data" (*hyd*) and "meteorological stations" (*met*). Deriving S(*riv,met*) through multiplicative chaining from S(*riv,hyd*)=0.84 and S(*hyd,met*)=0.55 gives 0.84*0.55=0.46, whereas the reference value in the matrix is found to be much smaller, namely S(*riv, met*)=0.02. This means that *riv* is similar to *hyd* in different aspects than *hyd* is similar to *met* (which is not captured by the algorithm). In dissimilarity terms, this result is an example of a non-metric measure, where pairwise dissimilarities do not follow the triangle inequality [d(a,b)+d(b,c)≥d(a,c)]. Setting dissimilarity=1-similarity, we get for this example: (1-0.84=0.16) + (1-0.55=0.45) < (1-0.02=0.98). However, the suggested indirect methods in this paper require metric measures in general, which, as shown in the example above, is not always true.

Small discrepancies may also be caused by the fact that scoring values in the matrix are ordinal, but are treated as interval. That is, the difference between scoring values of 0 and 1 for a pair of concepts may be perceived larger or smaller than the difference between scoring values 5 and 6.

Specifically terms that are found within a hierarchical classification seem to be problematic for indirectly estimating similarities. Whereas chaining along a hierarchy along subset/superset relations may work, it is problematic if chaining includes moving up one side and down another. The following example, which satisfies triangle inequality for dissimilarities, estimates two similarity measures for the sister terms "waterfall" and "river" through chaining. Assuming that all waterfalls are water bodies (S(*Wf*,*Wb*)=0.8), but not all water bodies are waterfalls (S(*Wb*,*Wf*)=0.4), and that rivers are water bodies (S(*Ri*,*Wb*)=0.8), but not all water bodies are rivers (S(*Wb*,*Ri*)=0.6), chaining gives S(*Wf*,*Ri*)=S(*Wf*,*Wb*)*S(*Wb*,*Ri*)= 0.48, whereas S(*Ri*,*Wf*)= S(*Ri*,*Wb*)*S(*Wb*,*Wf*)=0.32. Such relations which are not solely connected in a vertical direction require intransitive measures, which can for example be derived from common and different features between two classes based on a thesaurus (Rodriguez and Egenhofer 2004).

The effect of each potential modeling weakness on the dilution of the estimated values cannot be assessed from the results as a whole. We can, however, think of some general methodological modifications for future work, which should improve the accuracy of estimated values in sparse matrices. The problem that was already pointed in connection with the *topo-max* algorithm, namely the general problem of multiplicative chaining in triangles, suggests that only specific types of missing pairwise similarities can be estimated by specific types of methods. As dissimilarity measures along vertical direction in a superclass/subclass structure (e.g., thesaurus) satisfy triangulation inequality also in other approaches (Rodriguez and Egenhofer 2004), we assume that chaining will work for these directions once similarities between adjacent superclass/subclass hierarchies are given. Sister terms can be modeled to have symmetric similarities, thus asking one direction in the questionnaires will be sufficient and save time. Generally we need to explore whether a more advanced structure of an (incomplete) website query thesaurus that combines hierarchical taxonomies and similarity relations yields improved results for automated filling of gaps, and if this approach could lead towards metric dissimilarity measures. Further, we need to explore, how multiplicative chaining should be combined with other techniques for estimating gap values that do not presume similarity transitivity.

Acknowledgements

I want to thank Tom Philippi and Jennifer Fu for their valuable comments on an earlier version of this paper. Chuck Hostovsky helped me to improve the grammar and style of this work.

References

Binding C, Tudhope D (2004) KOS at your Service: Programmatic Access to Knowledge Organisation Systems. J of Digital Information 4(4)

Dagan I, Marcus S, Markovitch S (1993) Contextual Word Similarity and Estimation from Sparse Data. Proc of the 31st Annual Meeting on Association for Computational Linguistics, Association for Computational Linguistics, Morristown, NJ, USA, pp 164–171

Järvelin K, Kekläläinen J, Niemi T (2001) ExpansionTool: Concept-Based Query Expansion and Construction. Information Retrieval 4(3-4):231–255

Mandala R, Tokunaga T, Tanaka H (1999) Combining Multiple Evidence from Different Types of Thesaurus for Query Expansion. Proc of the 22th Int ACM-SIGIR Conf on Research and Development in Information Retrieval. ACM Press, New York, NY, pp 191–197

Rodriguez MA, Egenhofer MJ (2004) Comparing Geospatial Entity Classes: An Asymmetric and Context-Dependent Similarity Measure. Int J of Geographic Information Science 18(3):229–256

Shirky C (2005) Ontology is Overrated: Categories, Links, and Tags. Retrieved 02/20/2006 from http://shirky.com/writings/ontology_overrated.html

Sporns O (2002) Graph theory methods for the analysis of neural connectivity patterns. In: Kötter R (ed) Neuroscience databases. A practical guide. Kluwer Academic Press, Boston, pp 171–185

Tversky A (1977) Features of Similarity. Psychological Review 84(4):327–352

Voorhees EM (1994) Query Expansion Using Lexical-Semantic Relations. In: Croft WB, van Rijsbergen CJ (eds) Proc of the 17th Annual Int ACM-SIGIR Conf on Research and Development in Information Retrieval. ACM/Springer, pp 61–69

Widdows D (2003) Geometry and Meaning, University of Chicago Press

Using Metadata to Link Uncertainty and Data Quality Assessments

A.J. Comber, P.F. Fisher, F. Harvey, M. Gahegan, R. Wadsworth

Lead author Comber at: Dept. of Geography, University of Leicester, Leicester, LE1 7RH, UK

1 Introduction

In this paper we argue that the links between data quality reporting, metadata and subsequent assessments of data uncertainty need to be stronger. This builds on the ideas developed by Fisher (2003) who commented that data quality and uncertainty were like ships that pass in the night. Current data quality reporting is inadequate because it does not provide full descriptions of data uncertainty and allow assessments of data fitness (Comber et al. 2005a; Comber et al. 2005b). As Fisher (2003) noted "data quality as it has developed in the writing of data standards, and uncertainty as it has been researched in recent years, have followed two completely different tracks". The ideas presented in this paper are an attempt to set a research agenda to provide some glue to join together three related but distinct areas of scientific endeavor in spatial and geographical information sciences: uncertainty analysis, data quality / fitness for use assessments and metadata reporting.

For a variety of reasons data purporting to record the same real world features do so in a variety of different ways. Comber et al. (2003; 2005b) describe the impacts on the final data product of changing commissioning contexts, technical developments, advances in scientific understanding and on the final data products and the specific conceptualization embodied in the data of the objects under investigation. Institutional mandates and interests drive the specifications of data products. These issues strongly influence the meaning of the data in its widest sense. At present meaning is not provided in metadata and is not included in metadata standards, an absence that hinders data integration efforts as much as different source data and statistical processes.

A recent workshop on "Activating Metadata"[1] concluded that the case for current metadata standards (ISO, OGC) is not yet proven in relation to geographical information. This is in part because of the institutionally orientated and constructed nature of much geographical information (Harvey 2000; Comber et al. 2003), but also because of changing context within which such information is used: applications of geographical information are now ubiquitous; there are many more users; there are many more downloadable sources available to the user; geographical information (and systems) are used by many disciplines (Comber et al. 2005b).

In the dynamic and fluid context of more users accessing data, for a broad range of uses and with varying degrees of experience, the importance of appropriate metadata reporting increases. Metadata should provide information that helps users assess the *usefulness* of a dataset relative to their problem. They need to understand the meaning of the data relative to their uses. Importantly, it should facilitate assessments of the relationship between measures of data quality and uncertainty (for any specific application). Current metadata categories tend towards descriptions of *usability* and do not allow users to easily identify the suitability of data for their intended application. This paper pulls together various streams of literature to show that the data concepts and tools to allow users to assess data fitness for their application is a necessary requirement to link uncertainty and data quality.

2 Uncertainty, Data Quality and Metadata

2.1 Data Quality

Measures of data quality have traditionally been generated by the producers of spatial data. Data quality is described in terms of the 'big 5': Positional Accuracy, Attribute Accuracy, Logical Consistency, Completeness, and Lineage. Various standards for the components of metadata reporting have defined based on these measures of data quality (FGDC 1998; ISO 2003; OGC 2005). The interested reader is directed to Guptill and Morrison (1995) which provides an expansion and clarification of the concepts of quality used in those transfer standards. These measures originate from the historical cartographic legacy of geospatial data production

[1] http://www.niees.ac.uk/events/activating_metadata/index.html

(Fisher 1998) and the need to transfer information (Fisher 1999). In Guptill and Morrison, none of the authors discuss what the measures really mean or how they may be used in fitness-for-use assessments. Only Salgé (1995) discussed issues outside of the 'Big 5' and introduced the concept of semantic accuracy, noting that objects recorded in the data may not actually match the definition of those objects, which the database is based. This concept of quality has had little impact upon the standards process and data quality specifications (Fisher 2003). This is because the specification of data quality standards, previously dominated by the national mapping agencies and software companies, has now become the preserve of dedicated standards and industry specifications organizations such as ISO and the Open GIS Consortium.

The specification of quality standards continues to reflect data production interests, reporting the easily measurable and showing that the data producer can follow a recipe (Comber et al. 2005a) rather than more fully communicating the producer's knowledge of the data. A number of quality reporting paradigms have become established, principally the confusion matrix, user and producer accuracies and the kappa statistic. These describe how the map relates to an alternative, but hopefully compatible, source of information[2]. The net result of this legacy is that measures of data quality are difficult for users to interpret in relation to a specific application: users do not know how to apply data quality measures in their analyses (Hunter 2001) in order to assess the suitability of the data for their application. This assessment is done by expert analysts.

2.2 Spatial Data Uncertainty

Geographic objects are those structures created to impose order on the real world: objects are delineated, identified and placed into categories according to a set of criteria. Whilst many (non-geographic) objects have boundaries that correspond to physical discontinuities in the world, this is not the case for many geographic objects that may be less well defined[3]. Uncertainty in spatial data relates to doubt about the information that is recorded about a location (Fisher 1999). There are different types and directions of uncertainty relating to nature of the object under consideration. Fisher

---

[2] The 'accuracy' measures are commonly correspondences, e.g. when land cover from a field survey is compared to that derived from remote sensed data.

[3] Barry Smith and David Mark have written extensively on this subject exploring the concepts of *fiat* and *bone fide* boundaries, corresponding to *fiat* and *bone fide* geographic objects (Smith 1995, 2001; Smith and Mark 2001).

et al. (2005) have proposed taxonomy after Klir and Yuan (1995) where different types of uncertainty are related to how well the geographic objects are defined (see Fig. 1).

'Well defined objects' are those where the object classes (e.g. "building") are easily separable from other classes and where the individual instances are clearly distinct from other instances of the same class. For well-defined objects the main uncertainties are positional and attribute errors which can be analyzed using probabilities.

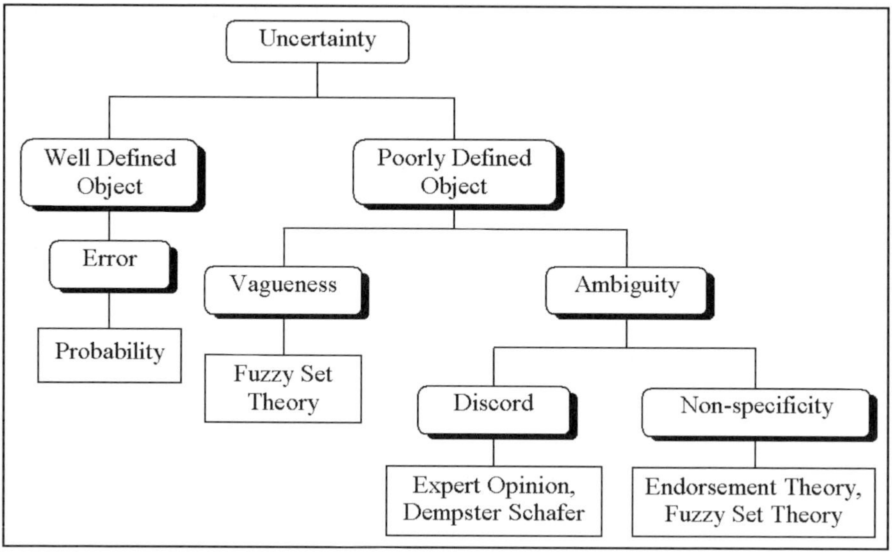

Fig. 1. A conceptual model of uncertainty in spatial data (from Fisher et al. 2005)

'Poorly defined objects' may be vague or ambiguous. Vagueness occurs where it is difficult to identify the spatial extent or to precisely define an object. That is, it is difficult to allocate unequivocally individual objects into a class. Ambiguity is composed of discord and non-specificity. Discord arises when one object is clearly defined but could be placed into two or more different classes under differing schemes or interpretation of the evidence. Non-specificity occurs when the assignment of an object to a class is open to interpretation.

The analysis of uncertainty for well-defined geographic objects is advanced. There are many examples in the literature where these aspects of data quality have been modeled using descriptions of error and accuracy, held by the metadata and combined with probabilistic approaches (e.g. Monte Carlo simulations, Bootstrapping). Most discussions of uncertainty in geographical information relate to these types of approaches to modeling error. This is not the case for poorly defined geographic objects, characterized by vagueness, ambiguity, discord and non-specificity, and yet we would argue that these are the most important and most frequent cases when working with geographical information.

2.3 Metadata

In contrast to the commonplace and simplified definition, 'metadata is data about data', we have defined metadata as 'information that helps the user assess the usefulness of a dataset relative to their problem'. That is, in the context of uncertainty assessments, it should allow the user to determine data quality and fitness for their analysis or data integration activity. Necessarily this involves relating one view of the world, as encapsulated by the specification used for a particular dataset, to another or to the objective of the analysis (Ahlqvist 2004). Assessing data fitness or suitability involves understanding data limitations (such as mismatches between user and data concepts) and quantifying the direction and magnitude of the uncertainties associated with integrating activities. In many cases environmental information supply is either a monopoly or an oligopoly and users do not have a choice about which dataset to use. Current metadata reporting does not give any information to the user about how to best exploit the data.

Whilst current metadata standards are adequate to guide assessment of technical constraints on data integration caused by Structure (raster to vector) or Scale (generalizations to lower level classes), they do not describe the concepts (ontologies) embedded in the data. Using the above definition they are inadequate: they do not facilitate user assessment of the applicability of the data to their problem; they convey nothing about the meaning of the data. Such information could include:

- The institutional (organizational) or epistemological context which gave rise to the data in the first place;
- The commissioning (often policy-related) context of the data;
- The experience of other users of the data.

Thus tools for user assessments that have been developed hitherto (e.g. Devillers et al. 2005) have to assume a perfect match between the concepts encapsulated in the data with those of the user application. They do not address one of the fundamental aspects of working with geographic information – it is relative, subjective and constructed.

3 Geographic Information – Origins and Use

The nature of geographic information and its increased use provide the context to expand metadata to include data concepts in order to make the link between uncertainty and data quality assessments.

The creation of geographic information is inherently relative. It involves the abstraction of the real world into some kind of data object and therefore results in information loss. There many choices to be made when constructing a geographic database – what to include, what not to include – for a number of data characteristics: granularity, scale, class numbers, class definitions etc. A complex specification usually guides the construction of geographic information in every case where institutions, e.g., Ordinance Survey, National Statistics, USGS, are involved. Every aspect of the construction contributes to the overall conceptualization of the real world that gets encoded into the database and each choice implicitly specifies some information loss. The degree of information loss depends on the application in hand.

The end result is variation, which requires far more than the simple class descriptions in order to understand. By way of example, consider how the concept of "Bog" changed between 1990 and 2000 land cover mappings of the UK. The 1990 Land Cover Map of Great Britain defined Bog in terms of standing water, permanent waterlogging, surface water and the presence of characteristic plant species (*Myrica gale* and *Eriophorum spp.*)[4]. In the Land Cover Map 2000 Bog was defined by peat depth greater than 0.5m[5]. The consequences of this change in conceptualisation are significant: for the 100km x 100km area (Ordnance Survey tile SK) in the English midlands there were:

- 12 pixels of 'bog' (<1 ha) in 1990;
- 120728 pixels of 'bog' (~75 km^2) in 2000.

It is worth noting that for both surveys the same class label was employed, the datasets were developed and constructed using similar remotely sensed data, by the same team from the same research institution.

The reasons for the change in construction of the class of 'Bog' are to be found in the different commissioning contexts of the 2 surveys (Comber et al. 2003). The 2000 dataset specifications were much influenced by changes in a number of behind the scenes factors from 1990:

[4] http://science.ceh.ac.uk/data/lcm/classM.htm

[5] http://science.ceh.ac.uk/data/lcm/lcmleaflet2000/leaflet3.pdf

- As a response to national legislation as a result of the Rio Earth summit;
- The shift of responsibility for the environmental away from the government to environmental agencies;
- The dynamic interaction of those agencies, the government and interpretations of policy with the processes on the ground that could be discerned using remote sensed data.

This is all information that would help the user understand more fully the data they incorporate into their analyses, but which currently is not included in metadata reporting paradigms.

The issues described above are involved in geographic data construction; its subjective nature and changes in ontology between surveys are established phenomena. They have always existed: different information collected by different agents for different purposes record the same real world features in different ways related to different conceptualizations and ontologies (although not under those labels). What have changed are the specifications and the institutional context within which this variation in data exists.

First, obtaining spatial data traditionally was a lengthy process. It meant entering into a dialogue with the data producer who would be concerned about whether their data would be used inappropriately, about their reputation, that of the data and the results of the original studies for which the data was commissioned. It was then an iterative exchange of information between user and producer. Potential users of the data would themselves be experienced and aware of the issues highlighted above: they would be spatial data-literate.

Second, the situation today is different. GI and GIS applications are ubiquitous in the public realm (e.g. GPS, in-car Sat-Nav, etc). The number of explicit users of GI and GIS has increased as reflected in the presence of GIS departments in local, regional and national government, health authorities, GIS is starting to be taught in high school and not just as a specialist post-graduate activity. Similarly, many new areas (such as insurance assessments) and academic disciplines (from archaeology to microbiology), now routinely use GIS where previously they did not. Such users may not be aware of variation in geographic information and may not understand specific uncertainties of the data they incorporate into their analyses. Users may not fully understand what the data represents – its meaning or semantics – and they will assume that it fits their conceptualizations because of familiar class names and labels that apparently match their prototypical categories with those names (Comber et al. 2005b).

Third, the number of applications and users are set to rise further with recent cyber infrastructure initiatives. The EU INSPIRE project seeks to make available "relevant, harmonized and quality geographic information to support formulation monitoring and evaluation of Community Policies"[6]. Similarly the development of the computing GRID is providing "pervasive, dependable, consistent and inexpensive access to advanced computational capabilities, databases, sensors and people"[7]. Broadly these cyber infrastructures and other eScience activities seek to connect users to spatial data without them having to go through the broker, the doorkeeper or the intermediary to the data and the process of dialogue that ensues.

Fourth, there has been a decline in the survey memoir as metadata. The book about the data describing the concepts and mapped features was always more interesting to researchers than the map it described, as noted by Fisher (2003). Coupled with the increased ease of access to digital data (ftp, web-portals, etc), these developments bring with them a risk of opening a Pandora's box of issues that have previously never had to be explicitly addressed because they were known.

However, users have to be pragmatic and use the data that is available, despite the fact that the existing data was (usually) collected for a different purpose (and those purposes change over time). Metadata does not communicate the data producer's model of the world embedded in data and, consequently, users are invited to treat information (an interpretation) as if it were an objective measurement. The consequences and uncertainties of unknown mismatches between the (remotely held) data objects and those of the user will be far more profound than those due to positional or attribute accuracy. In this context, assumptions may not generally be reported as a caveat to the "results" of a report or research project. This keeps the customer happy and allows the user to be seen as a "good" researcher.

4 Linking Uncertainty, Data Quality and GI

The preceding sections have commented on the implicit relationship between Uncertainty and assessments of Data Quality, and have indicated that current metadata reporting is insufficient to explicitly link them. This is in part due to the constructed and subjective nature of much geographic information and in part due to the increased analytical dangers of data misuse due to increased numbers of users and the ongoing development of

[6] http://inspire.jrc.it/ INfrastructure for SPatial InfoRmation in Europe

[7] http://www.escience-grid.org.uk

spatial data or cyber infrastructures. The questions that are thrown up buy this relationship are:

- How to provide appropriate information about the data such that users can either make informed decisions about which data is most suited to their analysis?
- How to enable users to understand the limitations of the results of any analysis using that data?
- How do we link uncertainty and data quality assessments to do this?

We believe that expanded metadata should provide information about the wider context of the data. It is instructive to review recent developments in thinking in the areas of uncertainty and data quality in spatial data.

4.1 Uncertainty

Early approaches to uncertainty in spatial data were concerned with 'error'. The focus was on the measures of data quality derived from the correspondence table (also referred to as the confusion matrix, data validation, correspondence table among others) and established a number of quality reporting paradigms: user and producer accuracies and the kappa statistic (e.g. Congalton 1991). These describe how the data relates to an alternative source of information allowing predictions to be compared with observations. There 2 main problems associated with global correspondence approaches: first, they are aspatial assuming error to be evenly distributed across the data; second, there is an implicit assumption that the predicted and observed are compatible with each other which is often not the case. For example, most remotely sensed land cover products are 'validated' by comparison with field or aerial photography survey: it is the major paradigm for reporting on 'accuracy', providing terms for 'error' – called variously user and producer accuracies, Group 1 and 2 uncertainties, Type I and Type II errors – which are then used in probabilistic approaches for modeling uncertainty: Monte Carlo simulations, epsilon error bands, bootstrapping.

More recent work has seen a movement away from probabilistic assessments of error toward uncertainty approaches that place greater emphasis on assessments of conceptual and semantic data aspects. Approaches based on formal ontologies (Frank 2001; Pundt and Bishr 2002) and concepts of interoperability (Bishr 1998; Harvey et al. 1999) have identified differences in semantic concepts as the major barrier to data integration. This has had as its focus a concern with the "what it is we are measuring" and necessarily draws on the work describing the indeterministic nature of many geographic features or objects (Burrough and Frank

1996; Smith 2001; Smith and Mark 2001; Gahegan and Brodaric 2002). The concept of 'fuzzy objects' has been explored by different authors for example by Cheng (2002) for the concept of a 'dune'. Dunes are well defined (that is, they have a strong "what") but are difficult to precisely locate (that is they have a weak "where"). Other geographic objects are more relative. For example defining and locating land cover classes (e.g. forest) is more difficult as they have a weak "what" and are difficult to locate because they have a weak "where".

Issues of indeterminacy and relativity, relating to different data conceptualizations, object definitions and semantics have been further discussed in relation to ontologies (Argawal 2005). For the uncertainties associated with specific integration problems (Gahegan and Ehlers 2000; Comber et al. 2004; Feng and Flewelling 2004; Comber et al. 2005c), applying an understanding of the definitional and conceptual aspects of the data has allowed more representative assessments uncertainty to be made.

4.2 Data Quality and Uncertainty

The objective in both reporting data quality (through metadata) and in much uncertainty research is to enable users to assess the limitations of using a particular dataset. As it is difficult to anticipate every use case, any producer oriented data quality description will at some point be found to be inadequate. As a result of this paradox Chrisman recommended that the need for user experience of the data be included in the specification of data quality in the original report from the standards committee (DCDSTF 1988). This was not included in the final specifications for the Spatial Data Transfer Standards that were included in the metadata content standards (FGDC 1998). This omission has been propagated through most standards specifications, despite the fact the need for user assessments were first identified in the draft standards in 1988. The FGDC metadata specification was an operationalized version of what came out of this committee (Chrisman, *pers* comm.)

The 'Big 5' have dominated data quality reporting, standards (e.g. STDS 1994; FGDC 1998) and reference texts (e.g. Guptill and Morrison 1995). Work describing the use of data quality parameters in actual applications is rare. DeBruin and Hunter (2003) describe an approach for assessing the value of different decisions about agricultural payment to farmers. A financial value was derived from the time stamp of remote sensed used to determine if a field was ploughed before a certain date, and the relative cost of inspecting it. DeBruin et al. (2001) describe an application that assesses the value of two DEMs with respect to the extent they control an error process for determining the volume of sand required to build a new port area. There is an "expected value of control" from being able to control how any uncertainty such as data positional accuracy / error resolves. In both of these examples, decision analytical techniques [a cost benefit analysis in DeBriun and Hunter (2003) and an expected value of control in DeBruin et al. (2001)] are used to make an informed trade-off between the improved decision quality and increased cost. That is, a real value is placed on the somewhat uncertain decision to use the data or not. In neither case is the quality measure of the information expressed in terms of any of the STDS / FGDC concepts. Rather an explicit value is placed on the assessment of uncertainty in using spatial data in the context of decision making.

5 Recommendations and Conclusions

In this paper we have identified a number of hitherto separate developments: increased use of spatial data, the ubiquity of geographic information applications, increased access to spatial data through cyber infrastructures and a decline in the tradition of metadata reporting (the survey memoir). In parallel we note a number of spatial data characteristics:

- Data are collected for all sorts of reasons, but we can't predict future use or value;
- The "real" metadata often resides with the individual scientist (and often only in their memory);
- There is increasing pressure on commercialization, IPR, "spin-offs" etc. which reduces desire to collaborate / cooperate;
- There is a naive belief that technology (e-science, grid) makes data integration trivial.

We have argued that in order to facilitate more robust data usage (i.e. re-use and appropriate use) and metadata needs to be expanded to link data quality and uncertainty assessments. A recent workshop on Activating Metadata for geographic information proposed that metadata be expanded to include the data *semantics and conceptualizations* and *user generated* metadata. The workshop outcomes identified the steps needed to do this:

1. Expansion of metadata slots to include free text descriptions of the data;
2. Development of text mining tools to populate slots;
3. Development of tools to mine metadata so created for matches between user application and data ontologies.

The provision of 'free text' based metadata slots would allow information such as transcripts of interviews or recordings given by the data producers on their understanding of the data *and* crucially data users on their experience of the data to be included in metadata. This acknowledges, firstly, that the provision and analysis of expanded metadata has a time and cost overhead for data producers. Secondly, that theirs is not the only valid experience of the data. Thirdly, that much of the information of interest to potential data users relates to the nuances and subtle verbal data descriptions. This is difficult to formalize and summarize in manner that will fit every potential user and to do so would be to impose a *de jure* limit to the data use.

The expansion of metadata ought to be complemented with the development and provision of tools to allow users to determine the usefulness of a dataset relative to their problem or application. Semantic web technologies offer platforms that would enable tools to match data provision (through INSPIRE, Grid and other cyber-infrastructures) with user applications and tasks through assessments of quality, semantic similarity and currency.

Such data services are needed to enable users to better identify the shortcomings and uncertainties of using 3rd party data in the context of expanding and increasingly pervasive cyber-infrastructures. Grid projects for instance have hitherto focused on large scale processing and data access rather than services. Data services are needed if wider use of cyber-infrastructures and spatial thinking is to be encouraged in fully operational environments (e.g. eGovernment or industry) and to facilitate a significant expansion of the transfer of Grid research. Automated tools for populating and mining metadata are essential to bridge the current gaps in the provision of joined-up, underpinning and spatially focused data infrastructures. This is important for a new tranche of spatial data users to ensure that they avoid conceptual mismatches between the user and data ontologies (Ahlqvist 2004) that they are able to assess the suitability of data for their use and that they are able to understand and assess uncertainties associated with any integrating activity.

If the benefits of the new cyber infrastructures for spatial data are to be realized then we need to expand metadata and to develop tools to allow users to assess data fitness for their application. Work to develop tools to populate and mine expanded metadata (user experience, deeper producer understanding) is ongoing.

Acknowledgements

The ideas in this paper were developed through a workshop sponsored by the National Institute for Environmental eScience (www.niees.ac.uk). The views expressed are those of the authors.

References

Agarwal P (2005) Ontological considerations in GIScience. Int J of Geographical Information Science 19(5):501–535

Ahlqvist O (2004) A parameterized representation of uncertain conceptual spaces. Transactions in GIS 8(4):493–514

Bishr Y (1998) Overcoming the semantic and other barriers to GIS interoperability. Int J of Geographical Information Science 12:299–314

Burrough PA, Frank AU (eds) (1996) Geographic Objects with Indeterminate Boundaries. Taylor & Francis, London

Cheng T (2002) Fuzzy objects: Their changes and uncertainties. Photogrammetric Engineering and Remote Sensing 68(1):41–49

Comber A, Fisher P, Wadsworth R (2003) Actor Network Theory: a suitable framework to understand how land cover mapping projects develop? Land Use Policy 20:299–309

Comber A, Fisher P, Wadsworth R (2004) Integrating land cover data with different ontologies: identifying change from inconsistency. Int J of Geographical Information Science 18(7):691–708

Comber AJ, Fisher PF, Wadsworth RA (2005a) You know what land cover is but does anyone else? ... an investigation into semantic and ontological confusion. Int J of Remote Sensing 26(1):223–228

Comber AJ, Fisher PF, Wadsworth RA (2005b) What is land cover? Environment and Planning B: Planning and Design 32:199–209

Comber A, Wadsworth R, Fisher P (2005c) Reasoning methods for handling uncertain information. In: Devillers R, Jeansoulin R (eds), Qualité de l'information géographique. Hermes, Paris, pp 153–168

Congalton RG (1991) A review of assessing the accuracy of classifications of remotely sensed data. Remote Sensing of Environment 37(1):35–46

DCDSTF (Digital Cartographic Data Standards Task Force) (1988) The proposed standard for digital cartographic data. American Cartographer 15(1):9–140

DeBruin S, Hunter GJ (2003) Making the trade-off between decision quality and information cost. Photogrammetric Engineering and Remote Sensing 69(1): 91–98

DeBruin S, Bregt A, Ven M. van de (2001) Assessing fitness for use: the expected value of spatial data sets. Int J of Geographical Information Science 15(5): 457–471

Devillers R, Bédard Y, Jeansoulin R (2005) Multidimensional Management of Geospatial Data Quality Information for its Dynamic Use Within GIS. Photogrammetric Engineering & Remote Sensing 71(2):205–215

Feng C-C, Flewelling DM (2004) Assessment of semantic similarity between land use and cover classification systems. Computers, Environment and Urban Systems 28:229–46

FGDC (Federal Geographic Data Committee) (1998) Content Standard for Digital Geospatial Metadata, FGDC-STD-001-1998, National Technical Information Service, Computer Products Office, Springfield, Virginia, USA

Fisher P, Comber A, Wadsworth R (2005) Approaches to Uncertainty in Spatial Data. In: Devillers R, Jeansoulin R (eds) Qualité de l'information géographique. IGAT, Hermes, France, pp 49–64

Fisher PF (1998) Improved Modeling of Elevation Error with Geostatistics. GeoInformatica 2(3):215–233

Fisher PF (1999) Models of Uncertainty in Spatial Data. In: Longley P, Goodchild M, Maguire D, Rhind D (eds.) Geographical Information Systems: Principles, Techniques, Management and Applications, vol 1. Wiley and Sons, New York, pp 191–205

Fisher PF (2003) Multimedia Reporting of the Results of Natural Resource Surveys. Transactions in GIS 7:309–324

Frank AU (2001) Tiers of ontology and consistency constraints in geographical information systems. Int J of Geographical Information Science 15(7):667–678

Gahegan M, Brodaric B (2002) Examining Uncertainty in the Definition and Meaning of Geographical Categories. In: Proc of the Fifth Int Symp on Spatial Accuracy Assessment in Natural Resources and Environmental Sciences, Melbourne

Gahegan M, Ehlers M (2000) A framework for the modeling of uncertainty between remote sensing and geographic information systems. ISPRS J of Photogrammetry & Remote Sensing 55:176–188

Harvey F (2000) The social construction of geographical information systems. Int J of Geographic Information Science 14:711–723

Harvey F, Kuhn W, Pundt H, Bishr Y, Riedemann C (1999) Semantic interoperability: A central issue for sharing geographic information. Annals of Regional Science 33(2):213–232

Hunter GJ (2001) Spatial Data Quality Revisited, Proceedings of GeoInfo 2001, 04–05 October, Rio de Janeiro, Brazil, pp 1–7

ISO (2003) 19115:2003 Geographic Information – Metadata (Int Organization for Standardization, Geneva)

Klir GJ, Yuan B (1995) Fuzzy Sets and Fuzzy Logic: Theory and Applications. Prentice Hall, Englewood Cliff

OGC (2005) OpenGIS Consortium. http://www.opengis.org/ (last date accessed: 10 June 2005)

Pundt H, Bishr Y (2002) Domain ontologies for data sharing-an example from environmental monitoring using field GIS. Computers and Geosciences 28(1): 95–102

Salgé F (1995) Semantic Accuracy. In: Guptill SC, Morrison JL (eds) Elements of Spatial Data. Elsevier, Oxford, pp 139–151

Smith B, Mark D (2001) Geographical categories: an ontological investigation. Int J of Geographical Information Science 15:591–612

Smith B (2001) Fiat Objects. Topoi 20:131–148

Spatial Data Transfer Standard (SDTS) (1998) New York: American National Standards Institute

An Evaluation Method for Determining Map-Quality

Markus Jobst, Florian A. Twaroch

Institute of Geoinformation and Cartography, Vienna University of Technology, Gusshausstrasse 27-29, A-1040 Vienna, Austria
email: markus@jobstmedia.at, twaroch@geoinfo.tuwien.ac.at

Abstract

The quality of maps, geo-visualization and usage of multimedia presentation techniques for spatial communication is an important issue for map creation, distribution and acceptance of these information systems (IS) by a public community. The purpose of this paper is to present an evaluation method based on stochastic reasoning for supporting map designers. We investigate the applicability of Bayesian Belief networks and present a prototypical implementation. We will give an outlook to future research questions.

1 Introduction

The quality of maps, geo-visualization and usage of multimedia presentation techniques for spatial communication is an important issue for map creation, distribution and acceptance of these information systems by a public community. In general the creation of an information system, its functionality and human-to-computer interface is supported by evaluations with aimed user groups. This time- and cost-consuming testing of users helps to specify efficient user interfaces and adapted functionality according to the provided knowledge basis. Every time an influencing parameter changes or becomes adapted, the investigation has to be done again to proof the results. Influencing parameters name isolated units, which take effect on the quality of the map, thus the understanding, perception and usefulness (in aspect of given situation) of presented information.

The present work addresses the possibility of modeling the influencing parameters in maps, by building an evaluation network and produce stochastic ratings for the quality of geospatial presentation forms. Different methods of uncertain reasoning have been considered. Bayesian Belief Networks got very popular in the recent years. They allow to model relations between causes and effects and can be used for causal as well as diagnostic reasoning. We exploited the Bayesian Network mechanism to build a conceptual model of a cartographic toolbox. It aims to support the evaluation process in user interface creation and functionality implementation.

Instead of a holistic access to the cartographic communication process, a segmentation of this process is intended, which should lead to small model able parts for further usage in the evaluation method. One main characteristic of this functional evaluation method is its ability of completion by expanding the Bayesian network. Thus a concentration on specific parts of the communication process in context with evaluations and preparations seem to be an appropriate way for a first implementation of a basic and straightforward functional model in order to identify and assess if the estimated use of this method would work.

The remainder of the paper is organized as following: Section 2 describes the cartographic communication process as found in the literature. In Section 3 we present the part of the process that is in the focus of the present paper. A conceptual model is worked out in Section 4. Section 5 introduces the Bayesian approach and how it can be used to create an evaluation method for map quality. A computational model and the results achieved are discussed. In the concluding sixth section we give an outlook to future research questions.

2 A cartographic Communication in Detail

In general the traditional, simple model of communication is presented by a sender to receiver relation using some medium for information transfer [1]. Information is coded to a signal and transmitted over a transmission line. For cartographic communication this transmission process seems to be expanded due to preparation of semiotic, semantic, pragmatic and cognitive requirements [2]. In its most comprehensive form the process spans from data acquisition to information dissemination with various available technologies. Thus it includes procedures and considerations concerning data quality due to acquisition and structuring techniques, topology and semantic model creation, generalization, structuring and preparation methods for the presentation model and cognitive aspects for the spatial information communication using various depth cues, sensual modes and metaphoric forms of semiotic. Kraak and Ormeling [3] offer a sequence of abstraction and transformation from "reality" (geographic real world objects) to the "mental-map" (the mental model of the user) as interpreted in the mind of the viewer. The steps of abstraction go from reality to digital landscape model and digital cartographic model and reach the visual map and finally the mental map. The transformations are processes leading from one state of abstraction to another. A similar approach is mentioned by Kelnhofer et al. [4] where the cartographic communication process is build up by primary, secondary and tertiary information models, which are connected with transformation procedures.

The primary model describes a state of data management and analysis, where problems of topology, various data qualities and similar aspects concerning structuring, combination and simplification of data coming from different kinds of measurements and acquisition techniques were mostly solved. This model forms the basis for analytical procedures used by the map-presentation afterwards. For instance a calculation of distance between two cities should use information of the primary data model in order to make use of "real" distances and not the simplified line distances of the cartographic model used within the presentation.

The secondary model names the cartographic presentation model, which takes account for the different cartographic presentation methods. According to a usable and chosen interface (paper, screen, real3d unit or similar) data of the primary model have to be transformed to fit a specific scale definition, resolution of the interface, effectiveness-, expressiveness-criteria and perceptional values. The aspect of perception seems to be one most important to enable information and knowledge extraction. Ignoring perceptional constraints may lead to massive information loss, although all requested data are put on the interface, as consequence of heavy information overload. The user is neither able to extract relevant information nor distinct various layers and identify any kind of information. In the same way the notions expressiveness and effectiveness explain a useful adaptation of information to the specific user interface.

Expressiveness refers to visualization capacity of the interface, which concerns the semiotic question of representing all important and necessary details of recorded objects in order to preserve semantic. Is it possible to present all the detailed information with the "low" resolution and "few" communication parameters the interface offers? For instance, if the resolution of the interface (e.g. screen) is lower than the number of desired detail values, the expressiveness criterion will not be met. Some detail values will then not be perceivable. Only if the number of resolution-pixels of the interface matches or is higher than the detail values, the desired univocal relationship becomes established [5] and all details of the object will be presentable on the interface. Mapping more detail values onto one single resolution-pixel makes determination impossible.

Effectiveness regards aesthetic concerns as well as perceptional information acquisition, immersive interface use, optimization processes for data simplification and visual rendering improvements. The quality of presentation and thus success of communication process is mainly depending on the understanding and acceptance of the user. The simulation or rebuilding of an interactive environment that is similar to the surrounding of everyday life by means of perception, multi-modality and interaction, seems to make the presentation more effective. By these means Egenhofer and Mark speak of "naive geography", where one main claim is that maps provide a very natural means to explore geographic space and that people perceive map space as more real then the experienced actual geographic environment itself [6].

The tertiary model of cartographic communication process names the user's mental model that forms the source for decision making and is build up by cognitive and psychological processes sequencing information perception and existing knowledge basis. Presented information becomes filtered by existing knowledge content [7]. Adaptable and understandable parts may be added to the existing knowledge base, whereas the rest may be rejected.

These three main states of cartographic communication and their transformations seem to clearly split the whole communication process into few parts. In fact this classification is more complicated. For instance a chosen interface needs a specific abstraction and scale of data in order to fulfill expressiveness criteria. This higher degree of abstraction calls for a given knowledge at the user-side to make information readable. Thus the transformation process from primary to secondary model is influenced by the user's mental model, if the map should be understandable, add new knowledge to the users knowledge and support decision making. The example may be the other way around, so that the used interface and presentation form supports spatial communication and there is no presumption for specific user knowledge, like surveys showed up for the communication of topographic data with the help of 3D presentations and interfaces [8].

The chain of the previously depicted communication model with its cross-connections makes some difficulties for the creation of an evaluation model obvious. This allows us to conclude that the cartographic communication process is simple utilizing the users aim and the according simplified data. Then, generally, discrepancies between the defined communication model and actual use of maps for knowledge acquisition may occur. When focusing on the main task of cartography, to efficiently transmit spatial correlated information [9], all influencing factors, from primary model data structure to creation of user's mental model, seem to be worthwhile considered for a determination of map quality.

3 Aspects of Map-Quality

The definition of map quality may be constituted on different parameters, which either focus on the consequence of cartographic communication and thus provide a holistic description of the communication process quality or concentrate on specific parts like semiotic (for a selected purpose of map-use), structuring of semantic map content, primary model data-quality (as result of consistency of a database) or similar.

A holistic approach mainly explores the effects of cartographic communication, thus how maps communicate spatial situations. The creation of formal processes of map production and use should help to judge map quality independent from map construction and map reading. Both imply intelligent human interpretation [10]. The factor of aesthetics based on the individual map-user is intended not to be useful within the judgment, be-cause it hardly may be modeled.

On the other hand the splitting of cartographic communication process into small segments, the evaluation of quality of these parts, identification of cross-connections and subsequent calculation of the "system" quality seems to be another potential strategy for quality evaluation. The idea of this method is particularly applied in the segment of user-interface design at present [11]. The access of functionality by dint of a drafted layout is tested with an arbitrary selection of users. The analysis of user activity results in redesign and revision of the interface design. This procedure is to be repeated several times. It is obvious that this time and cost consuming method is rarely applied for commercial products, in particular if one single segment of the whole communication process uses this kind of enormous expense.

The evaluation of map usability employs various quality definitions, whereas "usability" specifies among others understanding of map content by the user, possibility to utilize the map for desired purpose, compliance with effectiveness as well as expressiveness criteria and consistency of data (primary level) and content (secondary model). The definitions concern the quality of data, content, product or transmission.

Data quality describe consistency of data base on one hand and is structured to main indicators like completeness, legal consistency, positional, temporal, and thematic accuracy on the other [12].

Content quality concerns the group of map elements (secondary model) and their transported knowledge. A high content quality may be described by a good rate of cartographic completeness and geometric correctness. Following the rules of carto-semiotic for the most part ensures high content quality. For instance the cartographic principle of geometric correctness is violated by the enlarging of cartographic objects due to visual perception, because objects have to be displaced in order to be readable. In addition the cartographic principle of completeness becomes violated by simplifying, grouping and omitting objects in order to save space for incorporating additional information [13].

The depth of information, thus the complexity of integrated knowledge, in combination with a successful use of metaphoric description [14] and hierarchical structured functionality may characterize map application quality.

Quality of content transmission bases upon an unambiguous transfer of map information [2]. Instead of a straightforward transformation/ transmission from cartographic model (secondary model) to the mental model of the user, this process is under individual influence of existing knowledge conditions and issues of interface immersion. Existing knowledge and emotional response of the user to real-world objects may influence the transmission of map content and its understanding. As well as the design of the human-computer interface and its grade of immersion take influence on the quality and intuition of communication process.

The important role of carto-semiotic with its syntactic, semantic and pragmatic formulation for a successful map communication process provides a first structure for conceptualizing an automated evaluation tool for proofing map quality. In addition the complex structure of quality predefinition for maps seems to support the splitting of the cartographic communication model in very small parts, always concerning the question of communication quality for the specific part, identify cross-connections (e.g. from the scale independent data model to the mental model of the user) and incrementally implement these insights to a global geo-communication model.

4 Determining Simple Model Parameters

In order to keep our research question small a simple conceptual model for cartographic parameters has been setup. We investigate map quality by defining three parameters that influence the visual perception of symbols and text: element size, overlay and lightness. Figure 1 compares map presentations ignoring the chosen parameters and one considering cartographic guidelines.

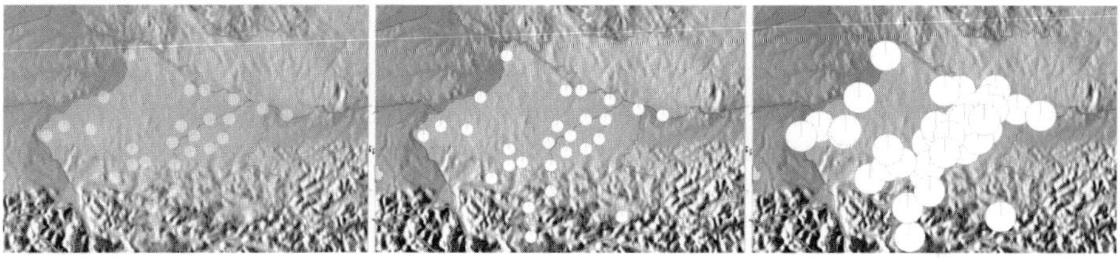

Fig. 1. A map representation that does not consider lightness (left) and element size (right) lowers the readability of a map. A sound cartographic design will raise map quality (middle)

The validity and functioning of an evaluation model in this context is presented by a very simple selection of parameters and their relations. The chosen segment within cartographic communication pertains to point symbols as graphic variables of visual map content. A selection of three parameters makes some reply on quality of perception.

The size of a point element in a map is an interplay of perceptibility and overlay. If the size of the element is to small it will neither be presentable on the interface nor ascertainable for the user. On the opposite an oversized element will overlay others and hide requested information.

Lightness is often used to visualize a rank order of information. Problems occur when point elements become too light or too dark, thus are troublesome visible and distinct able on the interface and result in information loss. In addition a high saturation of big elements attracts attention and lead to a distortion of map balance. Therefore bigger elements should be lighter.

The third parameter overlay simply assesses the rule that fewer information is presentable with increasing overlay. Figure 2 shows the parameters and their connections.

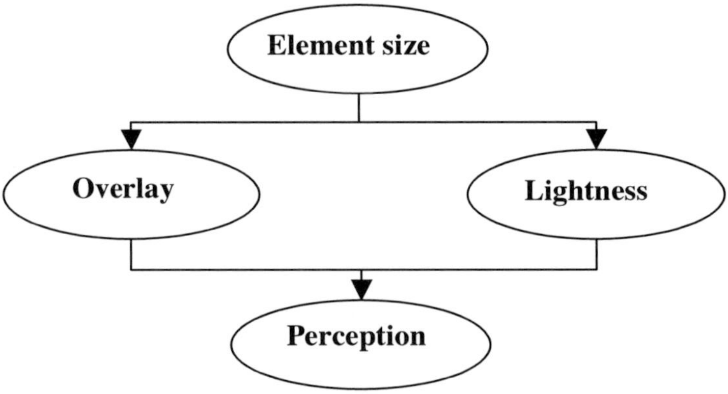

Fig. 2. A graph model of the investigated parameters and their influence on each other is shown in the figure above

5 Evaluation Tool

A simple evaluation tool is proposed that aims to support map designers. Not all users possess the expert knowledge to choose the appropriate parameters for designing maps. The intended tool shall support layman in the map creation process and contribute to an increase of high quality maps.

5.1 Bayesian Networks

Research in stochastic methods goes back to the 18th century. Nowadays stochastic reasoning can be found in various domains like diagnostic reasoning, natural language understanding, planning, scheduling, and learning [15]. Applications for searching minerals, filtering spam emails or provide help in troubleshooting a printer are just a few examples for the use of Bayesian Networks. In the field of geographic information systems Bayesian networks have been recently used to automatically provide the user with the appropriate data sets to a given question [16, 17]. We propose a model based on the Bayesian Network mechanism [18] to assist layman in cartographic design. The present article will not go into detail on the mathematics, but refers to the available literature [15, 18–20].

Bayesian networks are members of the family of graphical models. They can be represented in a directed acyclic graph structure. The nodes of the graph represent random variables or uncertain events in the world; the arcs are conditional probabilities between the variables. An arc that is directed from node A to node B can be translated to an event A causing event B.

The cause-effect relationships may be defined by an expert. Another method to obtain the relationships between the variables would be to learn them by statistical analysis from given data. The network helps to answer the question which factors influence a particular event.

The mathematical mechanism is Bayes formula that allows calculating a posteriori probabilities for the nodes given a priori probabilities and a new evidence from current data. The network reflects the change of beliefs in the light of new evidence. A Bayesian network can calculate the probabilities of the states of each node in the network after new evidence has arrived.

According to Heckerman Bayesian networks offer four benefits. They endow users to handle incomplete data sets. Users can learn about causal relationships, connect knowledge of experts with statistical analysis, and avoid over fitting of data [19].

Another benefit is that Bayesian Networks are close to human thinking. Experiments with children have shown that in tasks where conditional reasoning is required the predictive actions of the children can be simulated with a Bayesian Belief Network mechanism [21].

5.2 Computational Model

A simple computational model is proposed that aims to support a map designer in the creation of perceptible maps. The map designer is still assumed to be an expert on choosing the right parameters and the interactions between them. The computational model just helps him in his decision process.

In order to test our model we used a Bayesian Network library provided by Microsoft Research [20]. Bayesian Belief networks can be defined using a graphical editor and stored in a XML data structure. To extend the graph presented in Figure 2 towards a Bayesian Network, conditional probability distributions have to be defined for each of the nodes. The event visual perception has been defined by two causes the lightness and the overlay (see Fig. 3).

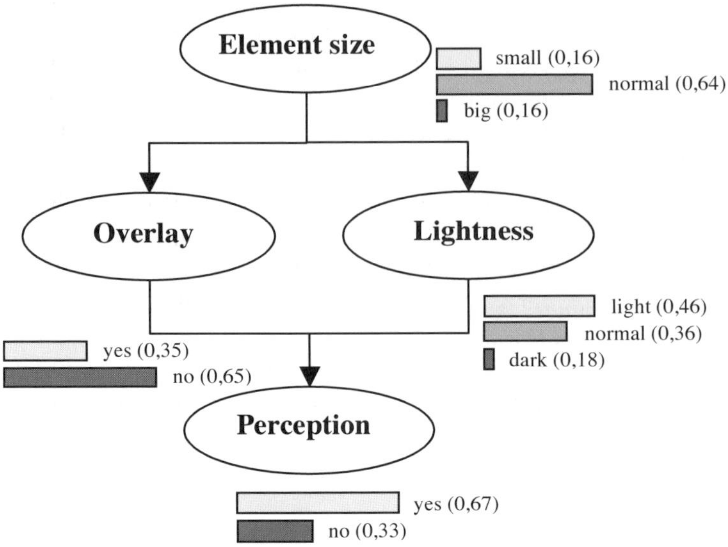

Fig. 3. The numbers, respectively bar charts at each node represent the conditional probabilities of the node states. The whole network represents a single joint probability distribution

In the case a user has the feeling that the map is not perceptible we can use the Bayesian network as a diagnostic tool and ask for the cause. For our simple model we can calculate if it is more likely that lightness or overlay influence the visual perception. This is bottom up reasoning. Given a certain element size we could also do a top down reasoning and ask to which degree the map is perceptible.

An advantage of the approach is that the network can start with an imperfect knowledge base, and incrementally, with new evidence the quality of the network will improve. The parameters of future models have to be chosen in interplay with empirical studies on map use. Once having enough data initial models could be also built by statistical analysis. At the time the approach relies on models built by experts, an extension to other model parameters is necessary.

6 Conclusions

A concept for an evaluation model for determining map-quality has been presented. A specific part of the cartographic modeling process has been chosen and its model parameters investigated. The present model is based on a mathematical approach that allows combining statistical analysis and expert knowledge. The intention is to save experimentation time when testing map design parameters. The model shall also support layman in the map creation process and increase to overall quality of available maps.

The simple model has to be extended in several directions. More model parameters have to be identified and included into the model. A tool for the layman has to be implemented with an easy to use interface. Other methods of uncertain reasoning have not been investigated and are of research interest for the development of such a tool. The definition of the interaction between the parameters to define the overall cartographic communication process has to be investigated.

The current findings motivate further research on an evaluation method for map quality and simplify the cartographic modeling process for layman towards automated mechanisms.

Acknowledgement

We thank all colleagues of the institute for geoinformation and cartography for their motivating discussions and valuable hints. This work has been partly funded by the E-TEN C517343 project GEOCOMPASS.

References

1. Weaver W, Shannon CE (1949) The mathematical theory of communication. University of Illinois Press, Urbana, Illinois
2. Brodersen L (2001) Maps as Communication, Theory and Methodology in Cartography. National Survey and Cadastre, Denmark
3. Kraak MJ, Ormeling FJ (1996) Cartography. Visualization of spatial data. Addison Wesley Longman, Harlow
4. Kelnhofer F, Lechthaler M, Brunner B (2002) Kartographie und Telekommunikation – Telekartographie and Location based Services. In: Geowissenschaftliche Mitteilungen. Vienna University of Technology, Vienna
5. Mackinlay J (1986) Automating the design of graphical presentations of relational information. ACM Transactions on Graphics 5(2):110–141
6. Egenhofer MJ, Mark DM (1995) Naive Geography. In: Frank AU, Kuhn W (eds) Spatial Information Theory – A Theoretical Basis for GIS. Springer-Verlag, Berlin, pp 1–15
7. Neisser U (1979) Kognition und Wirklichkeit. Klett-Cotta, Stuttgart
8. Buchroithner M (2002) Autostereoskopische kartografische 3D-Visualisierung. In: German Society for Cartography (ed) Kartografische Schriften, vol 6. Kirschbaum, Bonn

9. Gartner G (2002) Multimedia und Telekartographie. In: German Society for Cartography (ed) Kartografische Schriften, vol 6. Kirschbaum, Bonn
10. Frank AU (2000) Communication with maps: A formalized model. In: Freksa C et al. (eds) Spatial Cognition II (Int Workshop on Maps and Diagrammatical Representations of the Environment, Hamburg, August 1999). Springer-Verlag, Berlin Heidelberg, pp 80–99
11. Garret JJ (2002) The elements of user experience: user centered design for the web. American Institute of Graphic Arts – New Riders Press, Indiana
12. Devillers R, Bédard Y, Jeansoulin R (2005) Multidimensional management of geospatial data quality information for its dynamic use within GIS. Photogrammetric Engineering and Remote Sensing 71(2):205–215
13. Barkovsky T, Freksa C (1997) Cognitive requirements on making and interpreting maps. In: Hirtle S, Frank AU (eds) Spatial Information Theory: A theoretical basis for GIS. Springer, Berlin, pp 347–361
14. Cartwright W (2004), Using the web for focussed geographical storytelling via gameplay. In: First Int Joint Workshop on Ubiquitous, Pervasive and Internet Mapping – UBIMap 2004. Int Cartographic Association Commission on Ubiquitous Cartography, Tokyo, Japan
15. Luger GF (2005) Artificial Intelligence – Structures and strategies for Complex Problem Solving, 5th ed. Addision Wesley
16. Walker A, Pham B, Maeder A (2004) A Bayesian framework for automated dataset retrieval in Geographic Information Systems. In: IEEE Int Conf on Multimedia Modelling, MMM 2004, Brisbane, Australia
17. Stassopoulou A, Petrou M, Kittler J (1998) Application of a Bayesian network in a GIS based decision making system. Int J of Geographical Information Science 12(1):23–45
18. Pearl J (1988) Probabilistic Reasoning in Intelligent Systems: Networks of Plausible Inference. Morgan Kaufmann, Los Altos, CA
19. Heckerman D (1996) A Tutorial on Learning with Bayesian Networks. Microsoft Corporation, Redmond, WA
20. Kadie CM (2001) MSBNx: A Component-Centric Toolkit for Modeling and Inference with Bayesian Networks. Microsoft Research, Microsoft Corporation
21. Gopnik A, Schulz L (2004)) Mechanisms of theory-formation in young children. Trends in Cognitive Science 8(8):371–377

Efficient Evaluation Techniques for Topological Predicates on Complex Regions

Reasey Praing, Markus Schneider *

University of Florida, Department of Computer and Information Science & Engineering; email: {rpraing,mschneid}@cise.ufl.edu

Abstract

Topological predicates between spatial objects have for a long time been a focus of intensive research in a number of diverse disciplines. In the context of spatial databases and geographical information systems, they support the construction of suitable query languages for spatial data retrieval and analysis. Whereas to a large extent conceptual aspects of topological predicates have been emphasized, the development of efficient evaluation techniques for them has been rather neglected. Recently, the design of topological predicates for different combinations of *complex* spatial data types has led to a large increase of their numbers and accentuated the need for their efficient implementation. The goal of this paper is to develop efficient implementation techniques for them within the framework of the spatial algebra SPAL2D.

Key words: topological predicates, spatial relationships, efficient evaluation, complex spatial objects, complex regions, predicate implementation

1 Introduction

Topological predicates between spatial objects have always been a main area of research in spatial data handling, reasoning, and query languages in a number of disciplines like artificial intelligence, linguistics, robotics, and cognitive science. They characterize the relative position between two (or

* This work was partially supported by the National Science Foundation under grant number NSF-CAREER-IIS-0347574.

more) objects in space. The focus of this research has been on the conceptual design of and reasoning with these predicates. In contrast to this large amount of conceptual work, implementation issues for topological predicates have been widely neglected except for spatial index support as a pre-stage in query processing to identify candidate pairs of spatial objects that could possibly fulfill the predicate of interest. The main reason is probably the (simplifying) view that some plane-sweep algorithm is sufficient to implement topological predicates. Certainly a plane sweep plays an important role for the implementation of these predicates, but there are (at least) two aspects that make such an implementation much more challenging. The first aspect refers to the details of the plane sweep itself. Issues are how the plane sweep processes *complex* instead of *simple* spatial objects, whether spatial objects have been preprocessed in the sense that their intersections have been computed in advance (e.g., by employing a realm-based approach [7]), how intersections are handled, and what kind of information the plane sweep must output so that this information can be leveraged for predicate determination. The second aspect deals with the kind of query posed. Given two spatial objects A and B, we can pose (at least) two kinds of topological queries: (1) "Do A and B satisfy the topological predicate p?" and (2) "What is the topological predicate between A and B?". Only query 1 yields a Boolean value, and we call it hence a *verification query*. Query 2 returns a predicate (name) and we call it hence a *determination query*. We will see that these two kinds of query benefit from two different evaluation procedures.

The goal of this paper is to present efficient implementation strategies for topological predicates on simple and complex regions within the framework of the spatial algebra SPAL2D. Section 2 discusses related work on spatial data types and available design and implementation concepts for topological predicates. For *predicate execution* we distinguish two phases: In an *exploration phase*, described in Section 3, a plane sweep scans a given configuration of two spatial objects and collects metadata that can help us later derive the topological relationship between both objects. In the next phase, the *evaluation phase*, described in Section 4, these metadata are matched against characteristic properties of the topological predicates. This enables us to determine the Boolean result of a topological predicate (query 1) and the kind of topological predicate (query 2). Section 5 describes implementation results and performance analysis. Finally, Section 6 draws some conclusions.

2 Related Work

For the implementation of topological predicates we need two kinds of ingredients: a concept and implementation of spatial data types and a concept of topological predicates to be implemented. *Spatial data types* (see [7] for a survey) like *point*, *line*, and *region* have been accepted as fundamental abstractions for modeling the structure of geometric entities, their relationships, properties, and operations. Whereas in the past, spatial objects were only simple structures (single points, continuous lines, simple regions), the trend is now towards complex spatial objects allowing, e.g., multiple object components and holes in regions. The reason for this development lies in the need of applications and in the requirement of closure properties for spatial operations. Formal definitions of complex spatial data types can be found in [9]. Implementation descriptions of spatial data types are rare. Our implementation uses strategies found in [7].

The amount of conceptual work on *topological predicates* is large. The two most important approaches are the 9-*intersection model* [4] and the *Randell-Cui-Cohn model (RCC)* [3]. But since these approaches only deal with topological predicates for *simple* spatial objects, we have extended the approach in [4] to *complex* spatial objects in [9], which is the basis of our implementation. As we move to complex spatial objects, the number of topological predicates increases significantly. This requires more sophisticated and efficient evaluation techniques. For two complex regions, one obtains 33 topological predicates whereas only 8 exist between two simple regions. The details about the determination process can be found in [9]. Table 1 shows the matrix representations of the 33 predicates.

Only a few papers have dealt with the execution and implementation of topological predicates. The paper in [2] uses an optimization technique similar to our matrix thinning technique in Section 4.3 to minimize the number of needed computations. Ad hoc implementations of topological predicates have been proposed in [8].

3 The Exploration Phase: Collecting Topological Information

In this section, we give an overview of a process of exploring topological information between two spatial objects (here: regions). The detail of this process is discussed in [8]. In this process, we scan a given configuration of the two objects in order to collect data that help us later to derive the topological relationship between both objects in the evaluation phase. The objective is to discover the topological information of each object represented by overlap numbers using the plane sweep paradigm. The concepts resemble those described in [6, 7] but are different in the sense that they are not realm-based and that they allow intersecting segments of the argument objects. Thus, we describe the general case. Section 3.1 depicts the data structure implemented for the *region* data type. Section 3.2 sketches some needed geometric concepts like parallel object traversal, overlap numbers, and plane sweep with an emphasis on special features as they are relevant to our context. Section 3.3 explains the algorithm combining these concepts and the output information provided for further analysis in the evaluation phase.

3.1 Data Structure for the *region* Data Type

In the implementation described in this paper, we employ a new rational arithmetic called RATIO. RATIO provides a data type *rational* whose value representations can be of arbitrary, finite length and are only limited by main memory. This ensures numerical robustness and topological consistency in our implementation. Conceptually, complex regions can be considered from a "structured" and a "flat" view. The structured view defines an object of type *region* as a set of edge-disjoint faces. A face is a simple polygon possibly containing a set of edge-disjoint holes. A hole is a simple polygon. Two simple polygons are *edge-disjoint* if their interiors are disjoint and they possibly share single boundary points but not boundary segments. The flat view defines an object of type *region* as a collection of line segments which altogether preserve the constraints and properties of the structured view. For a formal definition see [5].

All coordinates are given as numbers of RATIO's data type *rational*. A value of type *point* is represented by a record $p = (x, y)$ where x and y are coordinates. We assume the usual lexicographical order on points. A *region* object is given as an *ordered* sequence (array) of *halfsegments*. Note that we omit all components of a *region* object representation that do not play a role in this context. The idea of halfsegments is to store each segment twice. A single segment corresponds to two halfsegments: a left halfsegment and a right halfsegment. A left (right) halfsegment is defined by the left/smaller (right/larger) point of the segment as its *dominating point*. The order relation on these halfsegments is defined first by the order of their dominating points, type, angle, and length. Detail on the formal definition of this order relation is discussed in [8].

An ordered sequence of halfsegments is represented as an array $\langle (h_1, a_1), \ldots, (h_m, a_m) \rangle$ of m halfsegments $h_i \in H$ with $h_i < h_j$ for all $1 \leq i < j \leq m$. Since inserting a halfsegment at an arbitrary position needs $O(m)$ time, in our implementation we use an AVL-tree embedded into an array whose elements are linked in sequence order. An insertion then requires $O(\log m)$ time. Each *left* halfsegment $h_i = (s_i, \mathit{left})$ has an attached set a_i of *attributes*. Attributes contain auxiliary information that is needed by geometric algorithms. It is usually unnecessary to attach attributes to *right* halfsegments since their existence in an ordered halfsegment sequence only indicates to plane sweep algorithms that the respective segment has to be removed from the *sweep line status*; in our implementation they are omitted. In the case for a region object r, the *left* halfsegments carry an associated attribute *InsideAbove* where the interior of r lies above or left of their respective segments.

3.2 Parallel Object Traversal, Overlap Numbers, Plane Sweep

Three main and well understood concepts are needed for the algorithm to be designed: parallel object traversal, the concept of overlap numbers, and the well-known plane sweep paradigm. *Parallel object traversal* allows us, during the plane sweep, to traverse the halfsegment sequences of both region operands in parallel since each sequence is already in halfsegment sequence order. This avoids expensive, initial sorting. Hence, by employing a cursor on both sequences, it is sufficient to check the halfsegments at the current cursor positions of both sequences and to take the lower one with respect to halfsegment sequence order for further computation. Below we will describe a slight extension of the parallel object traversal in the sense that it will traverse four instead of two halfsegment sequences and return the smallest halfsegment from the four current cursor positions.

Finding out the degree of overlapping of region parts is important for determining the topological relationship between two complex regions. For this purpose, we employ the concept of *overlap numbers* [6]. A point has overlap number k if k regions contain this point. For two regions F and G, a point p obtains the overlap number 2, iff $p \in F$ and $p \in G$. It obtains the overlap number 1, iff either $p \in F$ and $p \in G^-$, or $p \in F^-$ and $p \in G$. Otherwise, its overlap number is 0. Since a segment of a region separates space into two parts, an inner and an exterior one, during a plane sweep each (half)segment is associated with a *segment class* which is a pair (m/n) of overlap numbers, a lower (or right) one m and an upper (or left) one n. The lower (upper) overlap number indicates the number of overlapping *region* objects below (above) the segment. In this way, we obtain a *segment classification* of two *region* objects and speak about (m/n)-segments. Obviously, $m, n \leq 2$ holds.

Figure 1 shows the overlap numbers defined by two intersecting segments. The segment class of s_1 [s_2] left of the intersection point is $(0/1)$ [$(1/2)$]. The segment class of s_1 [s_2] right of the intersection point is $(1/2)$ [$(0/1)$]. That is, after the intersection point, seen from left to right, s_1 and s_2 exchange their segment classes. The reason is that the topology of both seg-

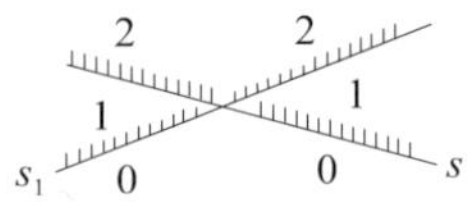

Fig. 1. Changing overlap numbers after an intersection

ments changes after the intersection point. To preserve these benefits and to enable the use of overlap numbers also for arbitrary segments, in the case that two segments from different *region* objects intersect, partially coincide, or touch each other within the interior of a segment, we pursue a splitting strategy that is executed during the plane sweep "on the fly". If segments intersect, they are split at their common intersection point so that each of them is replaced by two segments (i.e., four halfsegments) (see Fig. 2a). If two segments partially coincide, they are split each time the endpoint of one segment lies inside the interior of the other segment. Depending on the topological situations, which can be described by Allen's thirteen basic relations on intervals [1], each of the two segments either remains unchanged or is replaced by up to three segments (i.e., six halfsegments). From the thirteen possible relations, eight relations (four pairs of symmetric relations) are of interest here (see Fig. 2b). If an endpoint of one segment touches the interior of the other segment, the latter segment is split and replaced by two segments (i.e., four halfsegments) (see Fig. 2c).

This splitting strategy is feasible from an implementation standpoint since RATIO ensures numerical robustness, exactness, and topological consistency of intersection operations. Intersecting and touching points can be *exactly* computed, leading to representable points with rational coordinates provided by RATIO, and are thus precisely located on the intersecting or touching segments.

However, the splitting of segments entails some algorithmic effort. On the one hand, we want to keep the halfsegment sequences of the *region* objects unchanged since their update is expensive and is only temporarily needed for the plane sweep. On the other hand, the splitting of halfsegments has an effect on these sequences. As a compromise, for each *region* object, we maintain its "static" representation, and the halfsegments obtained by the splitting process are stored in an additional dynamic halfsegment sequence. The dynamic part is also organized as an AVL tree which is embedded in an array and whose elements are linked in sequence order. Assuming that k splitting points are detected during the plane sweep, we need additional $O(k)$ space, and to insert them requires $O(k \log k)$ time. After the plane sweep, this additional space is released.

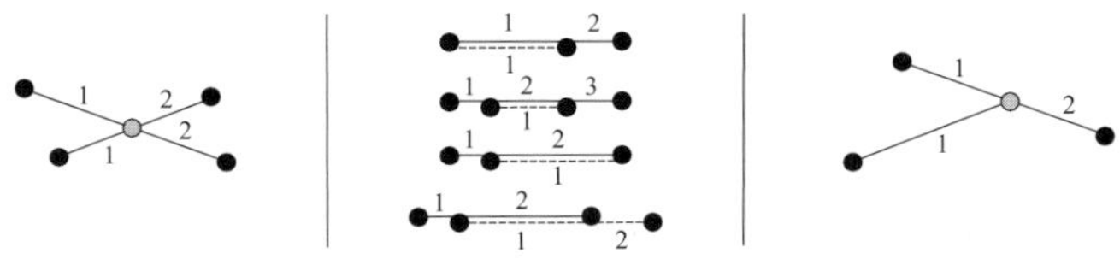

Fig. 2. Splitting of two intersecting segments **(a)**, two partially coinciding segments (without symmetric counterparts) **(b)**, and a segment whose interior is touched by another segment **(c)**. Digits indicate part numbers of segments after splitting

3.3 Topological Exploration Algorithm for the Region/Region Case

Using the three aforementioned concepts, the main goal of the exploration algorithm is to determine the segment classification of each *region* object. Each object is assigned a *Boolean segment classification vector*. This vector contains a field for the segment classes $(0/1)$ and $(1/0)$, a field for $(0/2)$ and $(2/0)$, a field for $(1/2)$ and $(2/1)$, and a field for $(1/1)$. That is, symmetric segment classes are merged since only either their non-existence or the existence of at least one of the two classes is later relevant. Each field is initialized with *false*. An additional flag *point_in_common* indicates whether any two segments of both objects meet in a common endpoint.

The segment classification is determined with the same sweep line status structure of the plane sweep that handles necessary segment splits. The pseudocode below presents the algorithm for its computation. When we encounter a right halfsegment in the event point schedule, its segment is looked up in the sweep line status and removed. For a left halfsegment, its segment is inserted into the sweep line status according to the y-coordinate of its left endpoint. Its lower overlap number is assigned the upper overlap number of its predecessor and its upper overlap number is assigned its incremented or decremented lower overlap number depending on whether the flag *Inside-Above* is true or false, respectively.

The algorithm uses some auxiliary operations which we briefly describe. The first two operations, *rr_select_first* and *rr_select_next*, support parallel object traversal. These operations check which of the two *region* objects F and G (i.e., the first, the second, or both) has the smaller halfsegment. If the status of the traversal is equal to *end_of_both*, both object representations have been traversed. The other needed operations refer to management of the sweep line status, which is initialized with *new_sweep*. If a left (right) half-segment of a *region* object is reached during a plane-sweep, the operation *add_left* (*del_right*) stores (removes) its segment component into (from) the segment sequence of the sweep line status. The operation *pred_exists* (*com-*

mon_point_exists) checks whether, for the segment currently considered in the sweep line status, a predecessor (a neighbored segment immediately below the current segment) exists. The operation *set_attr* (*get_pred_attr*) sets (gets) a set of attributes for (from the predecessor of) the segment currently considered in the sweep line status. Finally, the operation *get_attr* yields the attributes associated with a halfsegment.

```
algorithm SegmentClassification
input:   region objects F and G, initialized segment classification vectors v_F and v_G
output:  updated vectors v_F and v_G
begin
  S := new_sweep(); point_in_common := false; rr_select_first(F, G, object, status);
  while (status ≠ end_of_both) do
    if (object = first) or (object = both) then h := get_hs(F) /* Let h = (s, d). */
      else h := get_hs(G) endif; /* Let h = (s, d). */
    if d = right then del_right(S, s) else add_left(S, s);
      point_in_common := point_in_common or common_point_exists(S);
      if not pred_exists(S) then (m_p/n_p) := (*/0)
        else {(m_p/n_p)} := get_pred_attr(S) endif;
      m_s := n_p; n_s := n_p;
      if ((object = first) or (object = both)) and (InsideAbove ∈ get_attr(F))
        then n_s := n_s + 1 else n_s := n_s − 1 endif;
      if ((object = second) or (object = both)) and (InsideAbove ∈ get_attr(G))
        then n_s := n_s + 1 else n_s := n_s − 1 endif;
      S := set_attr(S, {(m_s/n_s)});
      if (object = first) then v_F[(m_s/n_s)] := true
      else if (object = second) then v_G[(m_s/n_s)] := true
      else if (object = both) then v_F[(m_s/n_s)] := true; v_G[(m_s/n_s)] := true; endif
    endif;
    rr_select_next(F, G, object, status);
  endwhile
end SegmentClassification.
```

If F has l and G has m halfsegments, the while-loop is executed at most $n = l + m$ times, because each time a new halfsegment is visited. The most expensive operations within the loop are the insertion and the removal of a segment into and from the sweep line status. We implement the status structure by an AVL tree which realizes each of the two update operations in time $O(\log n)$ and the other operations in constant time. Since at most n elements can be contained in the sweep line status, the worst time complexity of the algorithm is $O(n \log n)$.

4 Evaluation Phase: Matching Topological Relationships

So far, we are able to compute for two given complex regions F and G their segment classification vectors v_F and v_G. The values of both vectors depend on the relative positions of F and G to each other. The existence or non-existence of a segment class in v_F and v_G, respectively, conveys a topological information. For example, if v_F indicates the existence of $(0/1)$ segments in F, we can conclude that F contains segments that are located outside of G. But this does not enable us to derive the topological predicate between both objects, since it can be a disjoint-like, overlap-like, or covers-like relationship. To be able to decide this, we need to consider more information in both vectors. Before we get to the actual evaluation process, we give a definition for segment classification to better understand the semantic behind these classifications.

Definition 1. *The possible segment classes for a segment s of a complex region F with respect to another complex region G are given by a function* SC *as follows:*

$$SC(s, F; G) = \begin{cases} (0/1) \text{ iff } s \in \partial F \ \wedge \ IA(s, F) \ \wedge \ s \in G^- \\ (1/0) \text{ iff } s \in \partial F \ \wedge \ \neg IA(s, F) \ \wedge \ s \in G^- \\ (1/2) \text{ iff } s \in \partial F \ \wedge \ IA(s, F) \ \wedge \ s \in G^\circ \\ (2/1) \text{ iff } s \in \partial F \ \wedge \ \neg IA(s, F) \ \wedge \ s \in G^\circ \\ (0/2) \text{ iff } s \in \partial F \ \wedge \ IA(s, F) \ \wedge \ s \in \partial G \ \wedge \ IA(s, G) \\ (2/0) \text{ iff } s \in \partial F \ \wedge \ \neg IA(s, F) \ \wedge \ s \in \partial G \ \wedge \ \neg IA(s, G) \\ (1/1) \text{ iff } s \in \partial F \ \wedge \ s \in \partial G \ \wedge \ ((IA(s, F) \ \wedge \ \neg IA(s, G)) \ \vee \\ \qquad\qquad (\neg IA(s, F) \ \wedge \ IA(s, G))) \end{cases}$$

Of the nine possible combinations only seven describe valid segment classes. This is because a $(0/0)$-segment contradicts the definition of a complex *region* object, since then at least one of both regions would have two holes or an outer cycle and a hole with a common border. Similarly, $(2/2)$-segments cannot exist, since then at least one of the two regions would have a segment which is common to two outer cycles of the object.

With this definition, we can establish a connection to the topological predicates by constructing an evaluation technique called 9-*intersection matrix (9-IM) characterization* (Section 4.1). This technique is used to answer both verification and determination queries. Section 4.2 describes a fine-tuned approach called *minimum cost decision tree* for predicate determination. Section 4.3 delineates a sophisticated approach called *matrix thinning* for predicate verification.

4.1 9-Intersection Matrix Characterization

Instead of characterizing each topological predicate directly, the idea of this approach is to uniquely characterize each element of the 3×3-matrix of the 9-intersection model [4]. Such an element is a predicate that checks one of the nine intersections between the boundary ∂F, interior F°, or exterior

F^- of a spatial object F with the boundary ∂G, interior G°, or exterior G^- of another spatial object G for inequality to the empty set (see the left sides of the equivalence of Theorem 1). We call such an element *matrix predicate*. For each topological predicate, its specification is then given as the conjunction of the characterizations of the nine matrix predicates. In the region/region case, a *matrix predicate characterization* is again performed on the basis of a segment classification and finally on the regions' segment classification vectors. The goal of the following lemmas is to lead us to a unique characterization of all matrix predicates by means of segment classes. In all lemmas, let $H(F)$ and $H(G)$ be the set of all halfsegments (including any split halfsegment) in F and G respectively. The first lemma provides a translation of each segment class into a matrix predicate. Due to space limitations, detailed proofs are omitted.

Lemma 1. *Let F and G be two complex regions. Then we can infer the following implications and partial equivalences between segment classes and matrix predicates:*

$$\begin{array}{rll}
\text{(i)} & \exists\, h \in H(F) : SC(h, F; G) \in \{(0/1), (1/0)\} & \Leftrightarrow \partial F \cap G^- \neq \varnothing \\
\text{(ii)} & \exists\, g \in H(G) : SC(g, G; F) \in \{(0/1), (1/0)\} & \Leftrightarrow F^- \cap \partial G \neq \varnothing \\
\text{(iii)} & \exists\, h \in H(F) : SC(h, F; G) \in \{(1/2), (2/1)\} & \Leftrightarrow \partial F \cap G^\circ \neq \varnothing \\
\text{(iv)} & \exists\, g \in H(G) : SC(g, G; F) \in \{(1/2), (2/1)\} & \Leftrightarrow F^\circ \cap \partial G \neq \varnothing \\
\text{(v)} & \exists\, h \in H(F) : SC(h, F; G) \in \{(0/2), (2/0)\} & \Rightarrow \partial F \cap \partial G \neq \varnothing \ \wedge\ F^\circ \cap G^\circ \neq \varnothing \\
\text{(vi)} & \exists\, g \in H(G) : SC(g, G; F) \in \{(0/2), (2/0)\} & \Leftrightarrow \exists\, h \in H(F) : \\
 & & \quad SC(h, F; G) \in \{(0/2), (2/0)\} \\
\text{(vii)} & \exists\, h \in H(F) : SC(h, F; G) \in \{(1/1)\} & \Rightarrow \partial F \cap \partial G \neq \varnothing \ \wedge\ F^\circ \cap G^- \neq \varnothing \\
 & & \quad \wedge\, F^- \cap G^\circ \neq \varnothing \\
\text{(viii)} & \exists\, g \in H(G) : SC(g, G; F) \in \{(1/1)\} & \Leftrightarrow \exists\, h \in H(F) : \\
 & & \quad SC(h, F; G) \in \{(1/1)\}
\end{array}$$

The proof for this lemma follows directly from the segment classification definition (Definition 1) and the definition of complex region in [9]. The second lemma provides a translation of some matrix predicates into segment classes.

Lemma 2. *Let F and G be two complex regions. Then we can infer the following implications between matrix predicates and segment classes:*

$$\begin{array}{rll}
\text{(i)} & F^\circ \cap G^\circ \neq \varnothing \Rightarrow & \exists\, h \in H(F) : SC(h, F; G) \in \{(0/2), (2/0), (1/2), (2/1)\} \vee \\
 & & \exists\, g \in H(G) : SC(g, G; F) \in \{(0/2), (2/0), (1/2), (2/1)\} \\
\text{(ii)} & F^\circ \cap G^- \neq \varnothing \Rightarrow & \exists\, h \in H(F) : SC(h, F; G) \in \{(0/1), (1/0), (1/1)\} \vee \\
 & & \exists\, g \in H(G) : SC(g, G; F) \in \{(1/2), (2/1), (1/1)\} \\
\text{(iii)} & F^- \cap G^\circ \neq \varnothing \Rightarrow & \exists\, h \in H(F) : SC(h, F; G) \in \{(1/2), (2/1), (1/1)\} \vee \\
 & & \exists\, g \in H(G) : SC(g, G; F) \in \{(0/1), (1/0), (1/1)\}
\end{array}$$

This lemma is proved using the ovelap number concept in conjunction with complex region definition. The third lemma states some implications between matrix predicates.

Lemma 3. *Let F and G be two complex regions. Then we can infer the following implications between matrix predicates:*

(i) $point\_in\_common \Rightarrow \partial F \cap \partial G \neq \varnothing$
(ii) $\partial F \cap G^- \neq \varnothing \Rightarrow F^\circ \cap G^- \neq \varnothing \wedge F^- \cap G^- \neq \varnothing$
(iii) $F^- \cap \partial G \neq \varnothing \Rightarrow F^- \cap G^\circ \neq \varnothing \wedge F^- \cap G^- \neq \varnothing$
(iv) $\partial F \cap G^\circ \neq \varnothing \Rightarrow F^\circ \cap G^\circ \neq \varnothing \wedge F^- \cap G^\circ \neq \varnothing$
(v) $F^\circ \cap \partial G \neq \varnothing \Rightarrow F^\circ \cap G^\circ \neq \varnothing \wedge F^\circ \cap G^- \neq \varnothing$

The proof for this lemma is based on the definition of $point\_in\_common$ and point set topological concepts found in [9]. The following theorem collects the results we have obtained so far and proves the lacking parts of the nine matrix predicate characterizations.

Theorem 1. *Let F and G be two complex regions. Let H(F) and H(G) be the set of possibly split halfsegments of F and G. Then the matrix predicates of the 9-intersection matrix are equivalent to the following segment class characterizations:*

(i) $F^\circ \cap G^\circ \neq \varnothing \Leftrightarrow \exists h \in H(F) : SC(h, F; G) \in \{(0/2), (2/0), (1/2), (2/1)\} \vee$
$\exists g \in H(G) : SC(g, G; F) \in \{(0/2), (2/0), (1/2), (2/1)\}$
(ii) $F^\circ \cap \partial G \neq \varnothing \Leftrightarrow \exists g \in H(G) : SC(g, G; F) \in \{(1/2), (2/1)\}$
(iii) $F^\circ \cap G^- \neq \varnothing \Leftrightarrow \exists h \in H(F) : SC(h, F; G) \in \{(0/1), (1/0), (1/1)\} \vee$
$\exists g \in H(G) : SC(g, G; F) \in \{(1/2), (2/1), (1/1)\}$
(iv) $\partial F \cap G^\circ \neq \varnothing \Leftrightarrow \exists h \in H(F) : SC(h, F; G) \in \{(1/2), (2/1)\}$
(v) $\partial F \cap \partial G \neq \varnothing \Leftrightarrow \exists h \in H(F) : SC(h, F; G) \in \{(0/2), (2/0), (1/1)\} \vee$
$\exists g \in H(G) : SC(g, G; F) \in \{(0/2), (2/0), (1/1)\} \vee$
$point\_in\_common$
(vi) $\partial F \cap G^- \neq \varnothing \Leftrightarrow \exists h \in H(F) : SC(h, F; G) \in \{(0/1), (1/0)\}$
(vii) $F^- \cap G^\circ \neq \varnothing \Leftrightarrow \exists h \in H(F) : SC(h, F; G) \in \{(1/2), (2/1), (1/1)\} \vee$
$\exists g \in H(G) : SC(g, G; F) \in \{(0/1), (1/0), (1/1)\}$
(viii) $F^- \cap \partial G \neq \varnothing \Leftrightarrow \exists g \in H(G) : SC(g, G; F) \in \{(0/1), (1/0)\}$
(ix) $F^- \cap G^- \neq \varnothing \Leftrightarrow$ *true*

Theorem 1 provides us with a unique characterization of each individual matrix predicate of the 9-intersection matrix. This approach has several benefits. First, it is systematic and has a formal and sound foundation. Hence, we can be sure about the correctness of segment classes assigned to matrix predicates, and vice versa. Second, this evaluation method is independent of the number of topological predicates and only requires a constant number of evaluations for matrix predicate characterizations. Instead of nine, even only eight matrix predicates have to be checked since $F^- \cap G^- \neq \varnothing$ is always true (Theorem1(ix)). Third, we have proved the correctness of our corresponding implementation.

Based on this result, we can perform the predicate verification for a topological predicate p on the basis of p's 9-intersection matrix (see Table 1). In the case of a value 1 (*true*) for a matrix predicate, we take its equivalent, assigned segment classification on the right side in Theorem 1 and match it with the segment classification vectors v_F and v_G computed in the exploration phase. If there is a match, we proceed with the next value and matrix predicate in the 9-intersection matrix; otherwise p is *false*. In the case of a value 0 (*false*) for a matrix predicate, we pursue the same strategy but have to negate the assigned segment classification on the right side in Theorem 1 first.

For a predicate determination we take the following approach: For the first topological predicate p, we begin with the segment class characterization of the first matrix predicate on the right side in Theorem 1 and match it with v_F and v_G. For a value 1 for the matrix predicate, this means that v_F and v_G must satisfy the segment class characterization on the right side of the matrix predicate. For a value 0 of the matrix predicate, v_F and v_G must satisfy the negated segment class characterization. If they match, we know that this matrix predicate is satisfied and we continue with the segment class characterization of the next matrix predicate. Otherwise, p is *false* and we perform the whole procedure with the next topological predicate. In the worst case, this requires 33 tests of topological predicates.

4.2 The *MinCostDecisionTree* Algorithm

In this and the next section we fine-tune the approach of Section 4.1. A first observation is that for predicate determination we have to test all 33 topological predicates in the worst case. We propose a concept called *minimum cost decision tree* (MCDT) in this section which avoids this drawback and is similar to a technique introduced in [2] for topological predicates for simple regions. The idea is to construct a binary decision tree whose inner nodes are matrix predicates and whose leaf nodes are the 33 topological predicates. The tree partitions the search space at each node and progressively excludes more and more other topological predicates. In the best case, at each node of the decision tree the search space is partitioned into two halves. This requires $\log s$ computations where s is the number of topological predicates. For $s = 33$ the height of the tree is at least 6.

The pseudocode below shows our recursive algorithm for computing a minimum cost decision tree. Assuming that all topological relationships occur with equal probability, our cost model is to sum up all the path lengths from each topological predicate to the root. The algorithm takes as input the list of intersection matrices of the topological predicates (see Table 1). These matrices later become the leaves of the decision tree. In addition, a node list is required such that the algorithm may skip those decision branch elements that already appeared in the node path. This node list is empty at the start of the program and updated for every recursive call. The algorithm constructs the best tree by traversing through each valid decision branch using depth first search as it makes recursive calls. The recursion stack ends once a leaf is found, and at this point we have a sub-tree for which we can calculate the total cost. Then the recursion returns and recursively finds the next leaf. By comparing each alternative minimum cost sub-tree from each decision branch at a level, we obtain the minimum cost sub-trees at the parent level of the current level. This comparison takes place from the bottom up until the complete minimum cost tree is constructed and the root is chosen.

algorithm *MinCostDecisionTree*
input: A matrix list mat[] and a node list nodeList[]
output: The root node of a minimum cost decision tree.
begin
  $element := firstElement(); bestNode := newNode();$
  while $(element.isValid)$ **do** $node := newNode(element);$
    if $(node.isUnary)$ **then** $continue;$
    else if $(node.isLeaf)$ **then** $bestNode.id := mat[0].id; bestNode.cost := 0; break;$
    else $nodeList.add(node); node.lChild := MinCostTree(node.lChildren, nodeList);$
      $node.rChild := MinCostTree(node.rChildren, nodeList);$
      $node.cost := node.lChild.cost + node.rChild.cost + node.lChildren.length +$
        $node.rChildren.length;$
      if $(node.cost < bestNode.cost)$ **then** $bestNode := node;$ **endif**;
    endif; $element := nextElement();$
  endwhile;
  return $bestNode;$
end *MinCostDecisionTree*

Several trees exist with minimum total path length (minimum cost). For our case, we choose the first tree found. Due to space limitations, we cannot show the tree. It has height 6, and the total cost (path length) for all topological predicates is 170. Compared with the cost of $8 \cdot 33 = 264$ for the solution of Section 4.1, this reduces the cost for predicate determination to 64%.

4.3 Matrix Thinning

A second observation is that for predicate verification not all matrix predicates have to be evaluated. For example, for predicate 1 in Table 1 the combination that $F^\circ \cap G^\circ = \varnothing \ \wedge \ \partial F \cap \partial G = \varnothing$ holds (indicated by two 0's) is unique among the 33 predicates. Hence, only these two matrix predicates have to be tested in order to decide about *true* or *false* of the predicate. The question arises how the matrices can be "thinned out" and nevertheless remain unique among the 33 predicates. We have implemented a brute-force algorithm which for each intersection matrix checks all combinations of matrix predicate values for uniqueness among the 32 other intersection matrices. The algorithm begins with single matrix predicate values and

Table 1. Topological matrices | thinning matrices for the 33 topological predicates between two complex regions

$$1:\begin{pmatrix} 0|0 & 0|\ast & 1|\ast \\ 0|\ast & 0|0 & 1|\ast \\ 1|\ast & 1|\ast & 1|\ast \end{pmatrix}\quad 2:\begin{pmatrix} 0|0 & 0|\ast & 1|\ast \\ 0|\ast & 1|\ast & 0|0 \\ 1|\ast & 1|\ast & 1|\ast \end{pmatrix}\quad 3:\begin{pmatrix} 0|0 & 0|\ast & 1|\ast \\ 0|\ast & 1|\ast & 1|\ast \\ 1|\ast & 0|0 & 1|\ast \end{pmatrix}\quad 4:\begin{pmatrix} 0|0 & 0|\ast & 1|\ast \\ 0|\ast & 1|1 & 1|1 \\ 1|\ast & 1|1 & 1|1 \end{pmatrix}\quad 5:\begin{pmatrix} 1|\ast & 0|\ast & 0|0 \\ 0|\ast & 1|\ast & 0|\ast \\ 0|0 & 0|\ast & 1|\ast \end{pmatrix}$$

$$6:\begin{pmatrix} 1|\ast & 0|\ast & 0|0 \\ 0|0 & 1|\ast & 0|\ast \\ 1|1 & 1|\ast & 1|\ast \end{pmatrix}\quad 7:\begin{pmatrix} 1|\ast & 0|\ast & 0|0 \\ 1|\ast & 0|0 & 0|\ast \\ 1|\ast & 1|\ast & 1|\ast \end{pmatrix}\quad 8:\begin{pmatrix} 1|\ast & 0|\ast & 0|0 \\ 1|1 & 1|\ast & 0|\ast \\ 1|\ast & 0|0 & 1|\ast \end{pmatrix}\quad 9:\begin{pmatrix} 1|\ast & 0|\ast & 0|0 \\ 1|1 & 1|1 & 0|\ast \\ 1|\ast & 1|1 & 1|\ast \end{pmatrix}\quad 10:\begin{pmatrix} 1|1 & 0|0 & 1|1 \\ 0|0 & 1|\ast & 0|0 \\ 1|\ast & 1|\ast & 1|\ast \end{pmatrix}$$

$$11:\begin{pmatrix} 1|\ast & 0|0 & 1|1 \\ 0|\ast & 1|\ast & 1|\ast \\ 0|0 & 0|\ast & 1|\ast \end{pmatrix}\quad 12:\begin{pmatrix} 1|1 & 0|0 & 1|\ast \\ 0|0 & 1|\ast & 1|\ast \\ 1|1 & 0|0 & 1|\ast \end{pmatrix}\quad 13:\begin{pmatrix} 1|1 & 0|0 & 1|\ast \\ 0|0 & 1|\ast & 1|1 \\ 1|\ast & 1|1 & 1|\ast \end{pmatrix}\quad 14:\begin{pmatrix} 1|1 & 0|0 & 1|1 \\ 1|\ast & 0|0 & 1|\ast \\ 1|\ast & 1|\ast & 1|\ast \end{pmatrix}\quad 15:\begin{pmatrix} 1|\ast & 0|0 & 1|1 \\ 1|\ast & 1|\ast & 0|0 \\ 1|\ast & 0|0 & 1|\ast \end{pmatrix}$$

$$16:\begin{pmatrix} 1|\ast & 0|0 & 1|1 \\ 1|1 & 1|\ast & 0|0 \\ 1|\ast & 1|1 & 1|\ast \end{pmatrix}\quad 17:\begin{pmatrix} 1|\ast & 0|0 & 1|\ast \\ 1|1 & 1|\ast & 1|1 \\ 1|\ast & 0|0 & 1|\ast \end{pmatrix}\quad 18:\begin{pmatrix} 1|\ast & 0|0 & 1|\ast \\ 1|1 & 1|1 & 1|1 \\ 1|\ast & 1|1 & 1|\ast \end{pmatrix}\quad 19:\begin{pmatrix} 1|\ast & 1|\ast & 1|\ast \\ 0|\ast & 0|0 & 1|\ast \\ 0|0 & 0|\ast & 1|\ast \end{pmatrix}\quad 20:\begin{pmatrix} 1|1 & 1|\ast & 1|\ast \\ 0|0 & 0|0 & 1|\ast \\ 1|1 & 1|\ast & 1|\ast \end{pmatrix}$$

$$21:\begin{pmatrix} 1|\ast & 1|1 & 1|\ast \\ 0|\ast & 1|\ast & 0|0 \\ 0|0 & 0|\ast & 1|\ast \end{pmatrix}\quad 22:\begin{pmatrix} 1|\ast & 1|\ast & 1|\ast \\ 0|0 & 1|\ast & 0|0 \\ 1|1 & 0|0 & 1|\ast \end{pmatrix}\quad 23:\begin{pmatrix} 1|\ast & 1|1 & 1|\ast \\ 0|0 & 1|\ast & 0|0 \\ 1|\ast & 1|1 & 1|\ast \end{pmatrix}\quad 24:\begin{pmatrix} 1|\ast & 1|1 & 1|\ast \\ 0|\ast & 1|1 & 1|1 \\ 0|0 & 0|\ast & 1|\ast \end{pmatrix}\quad 25:\begin{pmatrix} 1|\ast & 1|1 & 1|\ast \\ 0|0 & 1|\ast & 1|1 \\ 1|1 & 0|0 & 1|\ast \end{pmatrix}$$

$$26:\begin{pmatrix} 1|\ast & 1|1 & 1|\ast \\ 0|0 & 1|1 & 1|1 \\ 1|\ast & 1|1 & 1|\ast \end{pmatrix}\quad 27:\begin{pmatrix} 1|\ast & 1|1 & 1|\ast \\ 1|\ast & 0|0 & 0|0 \\ 1|\ast & 1|\ast & 1|\ast \end{pmatrix}\quad 28:\begin{pmatrix} 1|\ast & 1|\ast & 1|\ast \\ 1|1 & 0|0 & 1|\ast \\ 1|\ast & 0|0 & 1|\ast \end{pmatrix}\quad 29:\begin{pmatrix} 1|\ast & 1|1 & 1|\ast \\ 1|1 & 0|0 & 1|1 \\ 1|\ast & 1|1 & 1|\ast \end{pmatrix}\quad 30:\begin{pmatrix} 1|\ast & 1|1 & 1|\ast \\ 1|1 & 1|\ast & 0|1 \\ 1|\ast & 0|0 & 1|\ast \end{pmatrix}$$

$$31:\begin{pmatrix} 1|\ast & 1|1 & 1|\ast \\ 1|1 & 1|1 & 0|0 \\ 1|\ast & 1|1 & 1|\ast \end{pmatrix}\quad 32:\begin{pmatrix} 1|\ast & 1|1 & 1|\ast \\ 1|1 & 1|1 & 1|1 \\ 1|\ast & 0|0 & 1|\ast \end{pmatrix}\quad 33:\begin{pmatrix} 1|\ast & 1|1 & 1|\ast \\ 1|1 & 1|1 & 1|1 \\ 1|\ast & 1|1 & 1|\ast \end{pmatrix}$$

then proceeds with pairs, triples, quadruples, quintuples, etc. Table 1 shows the result. The '*' elements correspond to "don't care" elements whereas other elements are the essential elements. We have found 6 matrices with 2 matrix predicates that have to be checked, 6 matrices with 3 matrix predicates to be checked, 10 matrices with 4 matrix predicates to be checked, and 11 matrices with 5 matrix predicates to be checked. The total cost is $6 \cdot 2 + 6 \cdot 3 + 10 \cdot 4 + 11 \cdot 5 = 125$. Compared with the cost of 8 per topological predicate for the solution of Section 4.1, this reduces the average cost for predicate verification to 3.8 ($= 47\%$) per topological predicate. [2] uses a similar approach which is based on a greedy heuristic. In contrast to this work, we also provide an implementation concept.

5 Implementation and Performance Analysis

The aforementioned techniques have been tested and verified through an implementation of the topological predicates as part of the SPAL2D package. The implementation makes use of a complex spatial data type system (SDT) which in turn is built on top of the rational number system (RATIO). This design framework ensures system-wide numerical robustness and topological

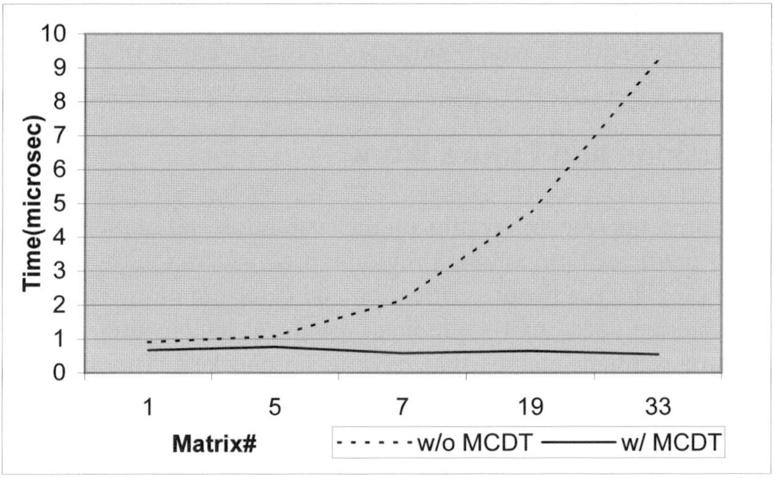

Fig. 3. Predicate Determination Comparison

consistency. Since performance is one of the goals for this implementation, we choose a popular compiled language *C++* for the development.

As far as topological predicate implementation is concerned, the 9-intersection matrix characterization technique makes it possible to process a single matrix predicate at a time. This gives rise to the possibility of using decision tree or thin matrices depending on the query type. An empirical study has been performed to verify both correctness and performance improvement in using these evaluation methods. Figure 3 illustrates the result for predicate determination.

Without using the decision tree, one would have to perform a linear search through all 33 predicates. Logarithmic search is not possible since there is no comparison relation between these predicates. Hence, the time function is a monotonely increasing function with the best case requiring 8 matrix predicate comparisons and worst case requiring upto $8 \cdot 33 = 264$ comparisons. In contrast, by using the decision tree, each predicate determination is limited to at most 6 matrix predicate comparisons. Our test cases are designed to cover all predicates for complex regions. Using a 1.8 GHz 64-bit processor, the average predicate evaluation time with decision tree is 0.6416 micro seconds as opposed to 3.6228 micro seconds without decision tree. This reduces the cost to 18% which indicates an improvement of 82%.

For predicate verification, evaluation without thin matrices requires 8 matrix predicate comparisons, whereas at most 5 comparisons are required with thin matrices. Furthermore, by using thin matrices, we can reject evaluation as soon as there is a mismatch of matrix predicates. For our test cases, the verification requires an average of 3.24 matrix predicate comparisons. This reduces the cost to 40.5% which indicates an improvement of 59.5%.

6 Conclusions and Future Work

This paper presents research results on the evaluation and implementation of topological predicates on complex regions. It considers the two main problems of topological predicate verification and determination. The main idea is to characterize each matrix predicate of the 9-intersection matrix by a unique set of segment classes that have to be checked. This characterization allows the use of minimum cost decision tree and matrix thinning, which significantly speed up the predicate determination and verification processes. The approach has been implemented in the SPAL2D software library which is currently under development and determined for an integration into extensible databases. In the future, we plan to consider the evaluation of topological predicates for all type combinations that also include the spatial data types *point* and *lines*.

References

1. Allen JF (1983) Maintaining Knowledge about Temporal Intervals. Communications of the ACM 26:832–843
2. Clementini E, Sharma J, Egenhofer MJ (1994) Modeling Topological Spatial Relations: Strategies for Query Processing. Computers and Graphics 18(6):815–822
3. Cui Z, Cohn AG, Randell DA (1993) Qualitative and Topological Relationships, 3rd Int. Symp. on Advances in Spatial Databases (= LNCS 692) pp 296–315
4. Egenhofer MJ, Franzosa RD (1991) Point-Set Topological Spatial Relations. Int. Journal of Geographical Information Systems 5(2):161–174
5. Güting RH, Schneider M (1995) Realm-Based Spatial Data Types: The ROSE Algebra. VLDB Journal 4:100–143
6. Güting RH, Ridder T de, Schneider M (1995) Implementation of the ROSE Algebra: Efficient Algorithms for Realm-Based Spatial Data Types. Int. Symp. on Advances in Spatial Databases pp 216–239
7. Schneider M (1997) Spatial Data Types for Database Systems – Finite Resolution Geometry for Geographic Information Systems (= LNCS 1288). Springer-Verlag, Berlin Heidelberg
8. Schneider M (2002) Implementing Topological Predicates for Complex Regions. Int. Symp. on Spatial Data Handling pp 313–328
9. Schneider M, Behr T (2006) Topological Relationships between Complex Spatial Objects. ACM Transactions on Database Systems (accepted for publication)

Implementation of a Prototype Toolbox for Communicating Spatial Data Quality and Uncertainty Using a Wildfire Risk Example

K.J. Reinke[1], S. Jones[1], G.J. Hunter[2]

[1] School of Mathematics and Geospatial Science, RMIT University GPO Box 2476V, Vic. 3001, Australia
[2] Department of Geomatics, The University of Melbourne Vic. 3010, Australia

Abstract

Current GIS are often described as rich in functionality but poor in knowledge content and transfer. This paper presents a prototype for communicating data quality in spatial databases using a hybrid design between data-driven and user-driven factors based upon traditional communication and cartographic concepts. The prototype aims to give data users a better understanding of the uncertainty that affects their information by utilizing a knowledge-based method where they can choose from multiple visualizations to represent the uncertainty in their data, as well as access information about why a particular visualization has been proposed. In doing so, decisions become more transparent to data users, which increases the capability of the prototype to act as a training aid. The example case study examines the data quality in a source dataset and illustrates how the concepts apply in an operational environment at different levels of communication.

1 Introduction

Frequently, data users do not possess the technical knowledge required to know how their data should be represented and what questions they should be asking of it. As such, visualization tools should be able to educate and guide non-experts, as well as performing the primary goal of communication. As a minimum such tools should provide easier ways to identify weaknesses in data and provide a choice of how to deal with the data presented. Effective communication of data quality will allow users to assess and better understand the usefulness of their data for a given application.

The key to successful communication of uncertainty lies in developing techniques by which spatial data users can interpret uncertainty correctly and apply this information to their decision tasks (Evans 1997). None of the current commercial GIS include mechanisms for visualizing or communicating uncertainty, nor do they provide visualization methods that ensure the correct application of graphic variables for particular types of datasets (Fisher 1998; DiBiase et al. 1992; Buttenfield and Beard 1991). Built-in visualization functionality generally exists with the intention of cartographic reproduction rather than for user understanding. There is a strong argument for developing appropriate tools to handle and communicate uncertainty. However, much of the current literature provides only hypothetical examples and isolated case studies, and neglects how this information may be delivered to the data user. Clearly, this deficiency needs to be addressed through the development of communication and visualization methods, evaluation of their usability and the implementation of these procedures as accessible tools (Hunter 1999).

2 Key Concepts of a Data Quality Communication Toolbox

The prototype toolbox was designed to treat the communication of uncertainty from the dual perspective of the user's needs as well as from traditional cartographic rules and followed similar design principles as described by Howard and MacEachren (1996). It is based upon the decomposition of the data into raw properties and then assigning suitable visualizations according to those properties in a similar way to that of Bertin (1983) and Mackinlay (1986). Many authors have also recommended the use of multiple but different presentation types (MacEachren 1995, 1994; Taylor 1994). Accordingly the prototype presents different visualizations as a set of multiple choices to the user who is then able to (repeatedly) select a suitable visualization for their individual preferences at the given level of communication. The conceptual framework that establishes the processes involved when communicating uncertainty is shown in Figure 1. It describes the transformation of data quality information into data quality displays by breaking down the data quality information according to the type of data quality parameter (as described in ANZLIC 2001), type of spatial distribution (based upon Fisher 1993), data model and level of measurement (based upon Chrisman 1995).

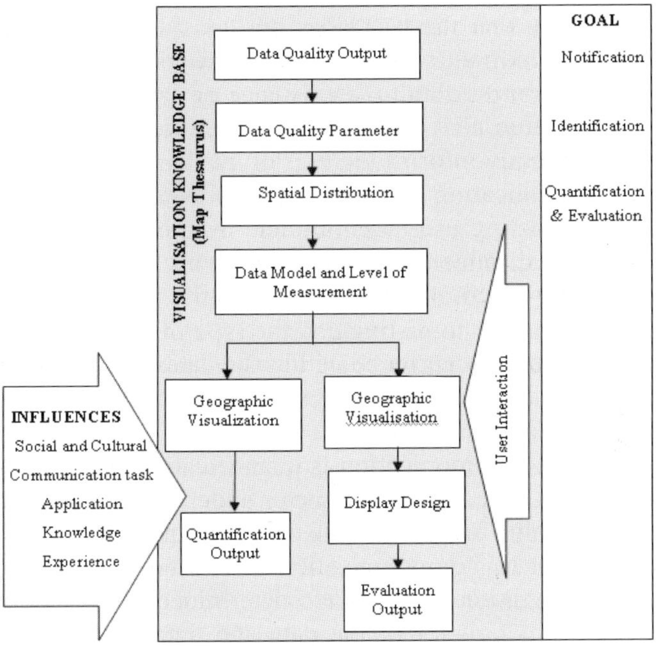

Fig. 1. A flowchart diagram summarizing the systematic process that occurs during the communication of a single data quality parameter, and potential user influences (after Reinke and Hunter 2002)

The concept behind communicating uncertainty follows four communication tasks. Beard and Mackaness (1993) originally proposed notification, identification and quantification as the steps required for communicating uncertainty. Notification indicates that data quality information is present in the data. Identification describes the type of data quality resident in the data and its distribution. Quantification describes the type, distribution and magnitude of the data quality. Evaluation has been added by Reinke and Hunter (2002) as a fourth communication task necessary to complete the conceptual model for communicating uncertainty in spatial data. Evaluation is the highest communication activity available and is used to show or estimate the significance uncertainty may have on the data being assessed. Each communication level has independent visualization goals that together form an overall approach in which to communicate data quality and uncertainty.

3 Four Levels of Communicating Data Quality Information

The first level of communication is notification. The aim of notification is to alert users to data quality information that exists in their datasets, and to do so in such a way that the user does not consider it an interference. It also performs the important roles of educating users about data quality and increasing user awareness about the existence of data quality. The primary message of notification alerts the user of a condition (that is, data quality exists) and subsequently informs the user of available functionality (that is, data quality communication tools exist). However, a fine balance is required between attracting user attention and minimizing user interruption. Indeed, these two requirements are at odds with each other.

The second level of communication is identification. The main objectives of identification are to distinguish the type of data quality parameter present and to locate its occurrence within the dataset. At a practical level, the goal of identification is to provide the user with a simple and rapid interpretation of data quality.

The third level of communication is quantification. Quantification aims to show the user the data quality parameter under investigation, the magnitude of the data quality measurements and how these measurements vary across space. It is at this communication level where decisions about the suitability of visualization methods are determined. This requires knowledge about the characteristics of the data and data quality being investigated, and requires users to make choices about the type of visualization that best meets their needs and expectations. Fundamental to the design of the prototype is the Map Thesaurus. This is essentially a cartographic visualization knowledge base that guides the user in the choice of graphic representation.

The fourth level of communication is evaluation. Evaluation differs from quantification in that the original data is incorporated into the output display to support the user task of assessment rather than observation. The purpose of evaluation is to estimate the significance the data quality has upon the given dataset for a given application. The question being addressed here is 'How do the data values vary according to the data quality?' This is achieved in one of two ways.

The first method estimates the significance the uncertainty has upon the given dataset (for a given application) by employing a technique whereby the original data is used to generate alternative datasets according to the data quality measurement. Actual underlying values are changed so magnitude of error no longer exists. It represents the amount of change from the original data value rather than the amount of data quality associated at that location. This new data provides a method for non-expert users to assess the significance different levels of uncertainty have on the data. This is an important distinction since a large amount of error may not always equate to a large amount of change in the data in the final decision layers. Uncertainty association physically combines the original data with the uncertainty data to produce a new dataset that can be visualized during evaluation.

The second method uses visual comparisons between the original data and the uncertainty measurements to assist in evaluating fitness for use. This task may use a variety of display compositions (that is, separated or merged display designs) and visualization techniques to visually combine the original data and uncertainty. Merged displays such as superimposed, composite and implicit designs, and the introduction of new visualization techniques such as simultaneously using different graphic variables to convey bivariate information are methods in which to visually combine data quality information with the original data. By including the actual data with the uncertainty information either visually or by uncertainty association, context is given to the uncertainty offering greater utility to the user compared to the preceding levels of communication.

4 The Map Thesaurus

The Map Thesaurus is the knowledge-based advisory component behind the prototype. The knowledge database is made transparent and accessible to users so they can become familiar with the reasoning behind the visualization options. The visualization outputs provided in the prototype are a few examples of potentially many chosen as suitable options. By presenting multiple methods of visualization and allowing interactive selection, user expectations and needs are better able to be met. Thus, the prototype guides users in selecting effective data visualizations and has the additional benefit of offering them the ability to discover and learn about visualizing data quality information.

The knowledge base works by classifying the data quality information to be visualized according to the components that define the uncertainty and user requirements (that is, data quality parameter, spatial variation, level of measurement and data model). Suitable visualizations are then assigned to the uncertainty data based on these characteristics. The type of information held in the knowledge base should be built upon uncertainty visualizations that have been tested and shown to support appropriate interpretation where possible, although most literature in uncertainty visualization is indicative rather than empirically tested.

The Map Thesaurus presents a series of different uncertainty displays for users to select from. This method allows some scope in selecting a preferred display without the potential for selecting an inappropriate visualization. The prototype is also able to educate the user by providing details about the choices made within the Map Thesaurus. The knowledge base applies a data-centric approach, but access to and utilization of the knowledge is driven by the user making the overall prototype a hybrid between the two design methods.

The Map Thesaurus takes advantage of the database facilities standard in GIS software by storing the rules for the different visualization methods along with the data characteristics to which the techniques have been successfully applied. In other words, the rules and properties for constructing visualization are stored within the database. The Map Thesaurus database is made up of several fields that describe the necessary data characteristics for assigning general visualizations. These fields are described in Table 1.

Table 1. The fields present in the Map Thesaurus database. The information stored in the Map Thesaurus is used to drive the quantification process

Field Name	Description
Method	The visualization method used. This references a look up table of scripts that generate the appropriate visualization
Display	Used only for evaluation
Data Quality Parameter (DQP)	The prototype uses spatial and thematic accuracy only. A full-scale implementation should use the generic characteristics of (1) data quality phenomena observed (space, theme, time) and (2) data quality type (accuracy, resolution, completeness, and lineage) to describe data quality parameters
Spatial Variation	Describes how the data quality measurements vary across space. This can be as multiple realizations, global, class, regional or feature level variations
Level of Measurement	The measurement scale of the data quality parameter
Data Model	The data model used to represent the data. It may be described as discrete (point, line), categorical (area) and continuous (TIN, grid)
Optional	Fields such as application, reference, weighting can be added

The advantage of the Map Thesaurus approach lies in the simplicity and flexibility in which the knowledge base can be modified. Users can easily add or delete methods, customizations and weightings as new knowledge is developed, and it would be easy to imagine a future with users logging on to the Internet to update their Map Thesaurus or to select an application-specific Map Thesaurus.

In order to interrogate the Map Thesaurus, the prototype must extract the information regarding the data quality parameter, level of measurement, spatial variation and data model used. The prototype does this by accessing the data itself or requesting the user for input. The data quality parameter is elected by the user although the available choices depend on available parameters. Level of measurement is currently not stored as a part of the data. Until this information is inherent in datasets, the level of measurement also requires user input. The remaining components of spatial variation and data model are determined by the prototype thereby minimizing user effort.

Spatial variation is used to decide the communication format and is calculated by the prototype upon selection of a data quality parameter. It is determined from a series of frequency calculations and subsequent queries, and distinguishes spatial variation into one of global, regional, class or feature variation. From this suitable visualizations are prescribed. For example, at the global level a simple text statement is sufficient, whereas class level is better-suited using graphs or multiple maps. It was considered unnecessary to differentiate between regional and feature level variation since the same visualizations could be applied to either level. The spatial variation at the regional level would be visually apparent in the display. It is at the feature level spatial variation that the opportunities for different visualizations are greatest.

As stated previously, the level of measurement should be able to be determined by the prototype but this information is not stored in most current GIS. Thus, the prototype requires this information to be entered by the user. For non-expert users this may pose difficulties when using the prototype, so it would be a valuable inclusion in future GIS to include the level of measurement (that is, nominal, ordinal, integer or ratio) as a field characteristic in the same way as the field type (that is, string, numeric, integer, ratio) is accessible.

Once the user has entered the required information about the data quality, and the data characteristics have been established by the prototype, the prototype selects suitable visualizations from the Map Thesaurus database. The prototype searches the Map Thesaurus database and iterates through a hierarchy-imposed set of Boolean queries to determine suitable visualizations. Using only the compulsory fields listed in Table 1 the prototype performs the following queries, working until at least one result is returned. If no results are returned from any of the queries then no visualizations are documented for that set of data quality characteristics. In these instances, a message box reporting no nominated visualizations would be generated. The first example query searches for all data characteristics to be met. The final query in the set requires only one condition to be met.

Select [Level of Measurement = <nominal>] AND [Data Quality Parameter = <Thematic Accuracy> AND [Data Model = <continuous>]
.
. {Combination of AND/OR operators}
.
Select [Level of Measurement = <nominal>] OR [Data Quality Parameter = <Thematic Accuracy> OR [Data Model = <continuous>]

Upon the first successful completion of a query, the set of selected representations is presented to the user. At this point the user may also elect to view the data properties that have been taken into consideration when determining the suitable visualizations based on the results of the query. Once a visualization selection has been made by the user, the prototype proceeds to generate the display using the relevant visualization script referenced by the Map Thesaurus database.

5 Visualization of a Wildfire Risk Example

5.1 Background

Fire intensity data is used by the Country Fire Authority (CFA) in Victoria, Australia, for wildfire management and planning in rural areas. The fire intensity data is a secondary product derived from wildfire threat models that use elevation, weather condition, vegetation, fuel loads and demographic data as inputs to the model. The output from the model indicates the potential fire danger (measured in kilowatts per metre) at a particular location. A CFA customized legend is used to classify and visualize the fire intensity data and ranks fire intensity into seven classes with an eighth class representing no data.

5.2 Method

The case study area is located around Ringwood (an outer eastern suburb of Melbourne, Australia) and covers an area measuring approximately 45 km by 55 km. It includes urban, semi-rural, rural, and forested landscapes. The areas surrounding Kinglake and the Dandenong Ranges are dominated by hilly terrain and natural vegetation up to 1000m above sea level. The communities located near these areas have historically been at some risk to wildfire. The consequence of wildfire in neighboring areas where the population density increases is of major concern.

The primary input to the wildfire model is elevation data. This data has known vertical errors in the elevation of ±5 m for 95% of the data. Error modeling using the Grid Cell Uncertainty Model (GCUM) (Hunter and Goodchild 1997) was employed to determine the amount of uncertainty that may be introduced into the wildfire risk maps as a result of these known elevation errors. The result of the error modeling is a number of equally probable realizations of fire intensity. Each individual realization or perturbation grid provides information about how the data could vary once known error is taken into account. An additional dataset was created based on the average of all the realizations. Each cell in this grid contained the mean value of all the realizations for that cell. This grid was subtracted from the original fire intensity grid to show the magnitude of variation in the original grid once errors were included in the analysis. The prototype was embedded within ArcView 3.2 GIS.

5.3 Visualizing Data Quality and Wildfire Uncertainty

Notification, Identification and Quantification of Data Quality

The notification message is activated whenever a dataset that contains data quality information is added or made active within the display. Notification has been implemented so the first instance of notification gains user awareness with the option to suspend ongoing notification. It is then up to the user to determine how important it is to be alerted to data quality content versus its interceptive nature. Permanent, passive notification is provided via the enabling/disabling of menu items. The notification of data quality sign was developed to be consistent with the software's interface and was accompanied with text to aid correct interpretation.

Next follows the identification phase requiring the user to nominate a data quality parameter. The identification output provides information about data quality in two ways. The first type is a map indicating the spatial distribution of the data quality parameter by color-coding features according to the presence/absence of data quality. The second type of reporting is non-spatial and describes the data in text form as a percentage of the affected data.

Upon initiation of the quantification process, the user is prompted to confirm a data quality parameter. The user is then requested to elect a visualization from the options presented to them by the Map Thesaurus (see Fig. 2). A description of the data characteristics used to generate the Map Thesaurus selections is also available through the Data Properties button (see Fig. 3) at both the quantification and evaluation levels.

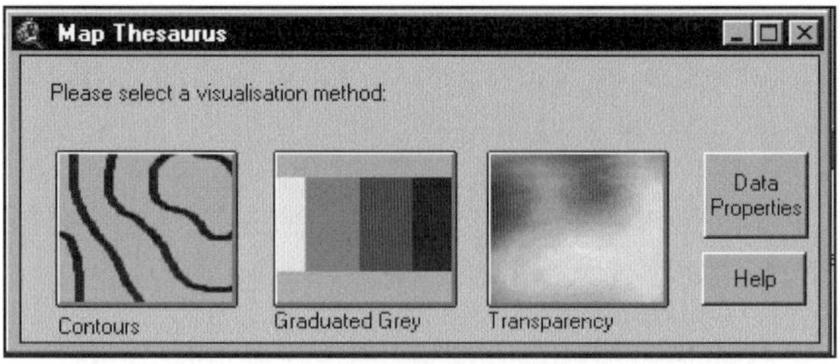

Fig. 2. The window presenting the visualization techniques available to the user for the given data quality characteristics of the sample dataset

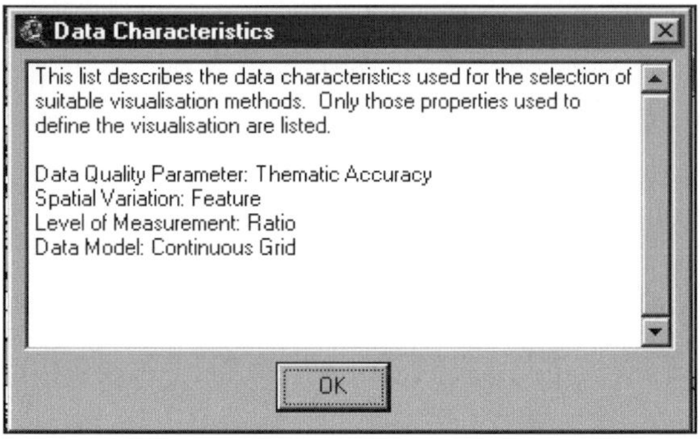

Fig. 3. The message box that provides a listing of the properties used to prescribe visualizations and what the characteristics were for that dataset

The resulting quantification display presents the data quality measurements in a separate window to the actual data. This is because the aim of quantification is to show only the magnitude and variation of a data quality parameter. Each visualization selection appears in a new window. The opportunity for the user to change the visualization method, proceed to the next level of communication (that is, evaluation) or obtain help is accessible from within the software display.

Evaluation of Data Quality and Uncertainty

Evaluation continues on from quantification and requires the user to select a display type (see Fig. 4) to enable both the data quality or uncertainty data, and the original data to be viewed. Diagrams and text are used to indicate the type of display design that will be presented, and help is available to explain each design option. The remaining display design options are merged, separated and uncertainty association displays. Separated displays use the same Map Thesaurus queries and results as for quantification. Because the actual data remains in its original user-defined representation, a visualization technique only needs to be assigned to the uncertainty data. Where possible the original dataset should be displayed in its original form. It is important to maintain the discipline specific legends and classifications used. Users should expect to be able to communicate with outputs using their own discipline specific graphics.

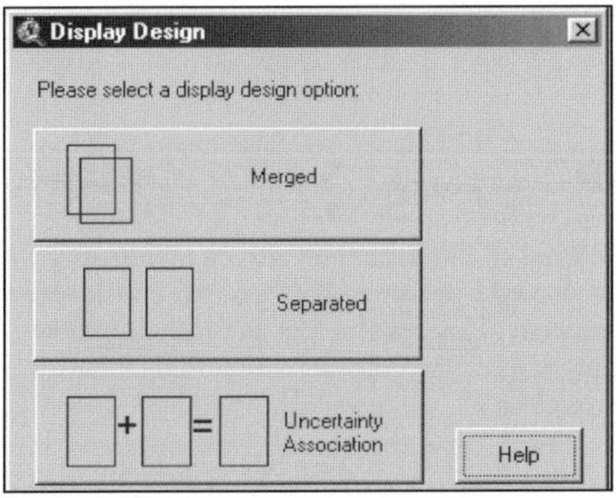

Fig. 4. The dialog used for the selection of display design during the evaluation process. A mixed modality approach is used to help users anticipate the display results

For this case study, uncertainty association was used to locate areas where a change in class has occurred, and to determine the magnitude and direction of class change. For instance, a change from a low fire danger rating to a high fire danger rating would be of extreme concern and allows the user to identify and prioritize areas that may be highly susceptible to wildfires otherwise overlooked in the original error modeling outputs.

The uncertainty association grid was generated using the process outlined in Figure 5. Both input grids were modified to use the same standard CFA classification system. The data of one grid was reclassified (to a multiple of 10) making it possible to add the two datasets together and view the magnitude and direction of class change. The uncertainty association grid can be described as a composition of the danger index for the mean of all perturbations, and the danger index for the original grid.

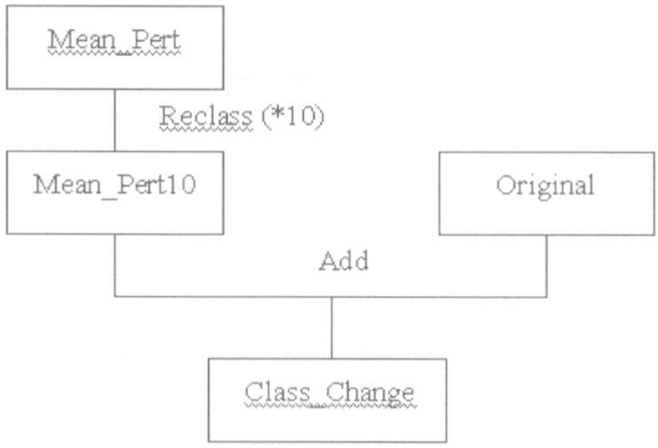

Fig. 5. The procedure used for calculating class changes into an uncertainty association grid

Table 2. The matrix used to describe the class change (using the Fire Danger Index classes 1–7) that occurs when error is taken into consideration. Italics indicate cells that have no change. The cells below the italics indicate a change from a lower class to a higher class. The cells above the italics indicate a change from a higher class to a lower class

		Original Fire Intensity Data						
		1 Very Low	2 Very Low	3 Low	4 Mod	5 High	6 Very High	7 Extreme
Mean Perturbations	10 V. Low	***11***	12	13	14	15	16	17
	20 V. Low	21	***22***	23	24	25	26	27
	30 Low	31	32	***33***	34	35	36	37
	40 Mod	41	42	43	***44***	45	46	47
	50 High	51	52	53	54	***55***	56	57
	60 V. High	61	62	63	64	65	***66***	67
	70 Extreme	71	72	73	74	75	76	***77***

The procedure resulted in the matrix described in Table 2. Table 2 presents the data in a way that is meaningful to the user, particularly one who is familiar with the application but not necessarily with uncertainty. The first digit in the uncertainty association matrix refers to the class after error modeling, and the second digit refers to the original class. For example, a cell with a value of 42 is interpreted as a change from fire index 2 to fire index 4. This is an increase of two fire danger classes.

A color scheme was developed to illustrate changes in fire risk categories. The display uses a simple color scheme of three colors show cells of no class change, a class downgrade of one or more classes and a class upgrade of one or more classes. The number of classes and interval breaks can be modified.

The prototype uses the display design selected and the data characteristics for that dataset to query the Map Thesaurus in order to identify and present appropriate visualization options (see Fig. 6). As with quantification, the user is able to examine the properties of the data that led to the visualizations being offered. The Map Thesaurus operates similarly to the quantification procedure but with slight changes depending upon the display design chosen.

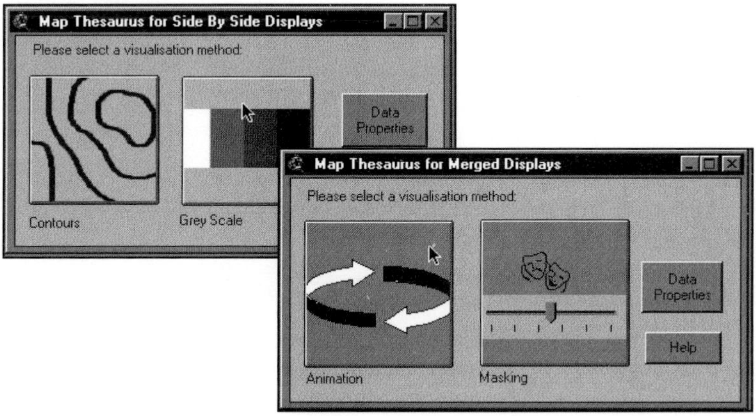

Fig. 6. The Map Thesaurus windows that appear after user selection of a display design. The examples shown include a merged and separated display design that have been generated for the sample CFA data quality information

The prototype presents animation and superimposed uncertainty masking as two examples of merged displays. For this case study, masking was selected (see Fig. 7) to visually portray the uncertainty. Uncertainty masking employs user-controlled thresholding to evaluate uncertainty in the data. The user is able to manipulate a slider widget to show all cells that are greater than the set threshold. The user can also choose the reverse to show all cells that fall below the specified threshold. This identifies non-uncertain cells. Finally, the user is able to toggle between the original data and the uncertainty display by checking on/off the uncertainty mask.

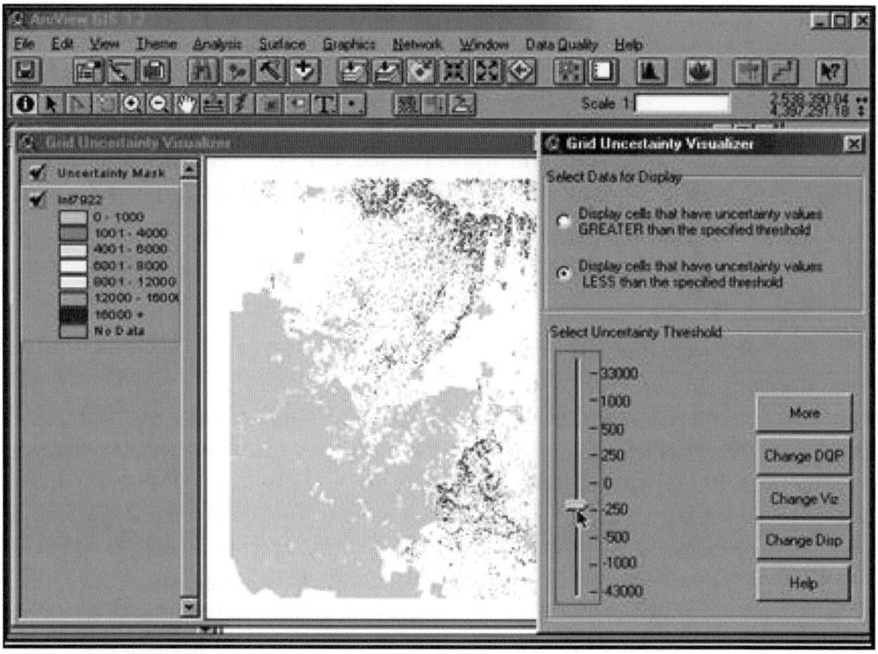

Fig. 7. This example of evaluation output uses a merged display design that employs the graphical variables of presence/absence to indicate levels of data quality. Unlike the other evaluation displays this example allows the user to interrogate the data real time by moving the slider widget to set uncertainty thresholds. The user is also able to control whether the uncertain or certain data are displayed at the given threshold

The interaction tasks enable interrogation via focusing (for example, show where features between specified values exist) and data re-expression (for example, modify class breaks). The data re-expression concentrates on re-expression in the thematic domain but can also be applied to the spatial domain by combining (upward fusion, for example, from feature to regional) or modifying (splitting features) attribute accuracy measures. This type of re-expression relies on cartographic generalization principles. Unlike the lower levels of uncertainty communication, evaluation must consider the context of the application, the data and the user.

Overall, the prototype is designed to permit users to circumvent selections simply by accepting prototype defaults. With the option for accepting defaults, it is possible for the user to proceed quickly and easily through the different levels of communication with minimal knowledge about spatial data quality.

Another benefit of providing default options is it removes the necessity for the user to pre-define ranges of quality measures that describe good, average and poor quality. This allows users who are unfamiliar with the dataset and its associated quality, or unfamiliar with the concept of uncertainty entirely, to generate a display. By providing a default type button and context-based help tools in parallel with the display allows users immediate insight into the quality of the data as quality measurements are manipulated.

5.4 Results and Discussion

The CFA example revealed an interesting behavior of data quality between primary and secondary data products. Initially the quality under investigation was a global positional error measurement. However, post error modeling revealed that the positional error translated into thematic accuracy that varied considerably at the individual feature level. So there is an observed change in the *type* of data quality and the *spatial distribution* of data quality.

This was perhaps one of the most important discoveries of this case study in that quantification could produce misleading yet truthful results. In the example of the CFA data, data in the raw DEMs showed high uncertainty in the hilly areas and low uncertainty in the flat areas. However, once changes in the original data resulting from the data quality (as determined by the Grid Cell Uncertainty Model) were translated into the CFA fire risk classification system it was revealed slopes are stable but flat areas show high change. So whilst slopes contain higher amounts of uncertainty, they are less prone to change compared to flat area.

Providing visual representations of the magnitude of error is limited for use by policy and decision-makers as a tool for assessing the confidence in their decisions (that is, locating and distributing resources to areas of higher risk), however, a visual representation that highlights areas where there is a change in risk class as a result of the uncertainty (for example, from a moderate to a high bushfire risk or vice versa) is expected to provide a more meaningful display to these users. The uncertainty association method was the most appealing to the researchers during model development as they felt they were able to obtain more information and a better understanding of how the uncertainty affected the prediction data. Users also preferred data inclusive of uncertainty compared to raw uncertainty displays. This example also highlights the influence application context has on communication methods.

6 Conclusions

Often uncertainty or data quality information is incomprehensible to non-expert users. Communication prototypes such as the one developed in this thesis will assist users in comprehending data quality and better understand the implications uncertainty may have on their data and potentially on their decisions. The ultimate goal of a communication system is for users to successfully incorporate data quality information and improve decisions. This paper has presented work that forms a satisfactory solid framework and design for meeting a semi-automated prototype that improves the way data quality information is currently used. It was shown that satisfaction plays an important role when developing a usable prototype. Usability characteristics such as efficiency and effectiveness are considered secondary to user satisfaction for an initial prototype for communicating thematic data quality. Two inhibitors to such data quality and uncertainty communication tools being developed and available commercially are (1) the lack of standards for how data quality data should be stored, and (2) the lack of data characteristics (such as measurement scale) able to be stored and accessed within the database. These problems occur throughout each level of communication, compounding as the higher levels of communication are reached.

References

ANZLIC Metadata Working Group (2001) ANZLIC Metadata Guidelines: Core Metadata Elements for Geographic Data in Australia and New Zealand, Version 2. Available from URL: http://www.anzlic.org.au/get/2358011755

Beard K, Mackaness W (1993) Visual access to data quality in geographic information systems. Cartographica 30(2&3):37–45

Bertin J (1983) Semiology of Graphics: Diagrams, Networks, Maps. The University of Wisconsin Press, Madison

Buttenfield BP, Beard KM (1991) Visualising the quality of spatial information. In: Proc of AUTO-CARTO 10 Baltimore 6, pp 423–427

Chrisman NR (1995) Beyond Stevens: A revised approach to measurement for geographic information. In: Proc of AUTO-CARTO 12 Charlotte 4, pp 271–280

DiBiase D, MacEachren AM, Krygier JB, Reeves C (1992) Animation and the role of map design in scientific visualisation. Cartography and Geographic Information Systems 19(4):201–214

Evans BJ (1997) Dynamic display of spatial data reliability: Does it benefit the map user? Computers and Geosciences 23(4):409–422

Fisher PF (1993) Conveying object-based meta-information. In: Proc of AUTO-CARTO 11 Minneapolis, pp 113–122

Fisher PF (1998) Is GIS Hidebound by the Legacy of Cartography? The Cartographic J 35(1):5–9

Howard DL, MacEachren AM (1996) Interface design for geographic visualisation: tools for representing reliability. Cartography and GIS 23(2):59–77

Hunter GJ (1999) New tools for handling spatial data quality: moving from academic concepts to practical reality. URISA J 11(2):25–34

Hunter GJ, Goodchild MF (1997) Modeling the uncertainty of slope gradient and aspect estimates in spatial databases. Geographical Analysis 29(1):35–49

MacEachren AM (1994) Visualisation in modern cartography: setting the agenda. In: MacEachren AM, Taylor DRF (eds) Visualisation in Modern Cartography, vol 2. Pergamon, Oxford, pp 1–12

MacEachren AM (1995) Approaches to truth in geographic visualisation. In: Proc of AUTO-CARTO 12 Charlotte 4, pp 110–118

Mackinlay J (1986) Automating the design of graphical presentations of relational information. ACM Transactions in Graphics 5(2):110–141

Reinke KJ, Hunter GJ (2002) A theory for communicating uncertainty in spatial databases. In: Goodchild MF, Fisher PF, Shi W (eds) Spatial Data Quality. Taylor and Francis, London, pp 71–102

Taylor DRF (1994) Cartographic visualisation and spatial data handling. Advances in GIS Research. In: Proc of the Sixth Int Symp on Spatial Data Handling, Edinburgh 1, pp 16–28

Changes in Topological Relations when Splitting and Merging Regions

Max J. Egenhofer, Dominik Wilmsen

National Center for Geographic Information and Analysis, Department of Spatial Information Science and Engineering, University of Maine, Orono, ME 04469-05711, USA; email: max@spatial.maine.edu

Abstract

This paper addresses changes in topological relations as they occur when splitting a region into two. It derives systematically what qualitative inferences can be made about binary topological relations when one region is cut into two pieces. The new insights about the possible topological relations obtained after splitting regions form a foundation for high-level spatio-temporal reasoning without explicit geometric information about each object's shapes, as well as for transactions in spatio-temporal databases that want to enforce consistency constraints.

1 Introduction

Efforts in spatio-temporal modeling have significantly enhanced the computational capabilities of otherwise static models of geographic space. In recent years the primary focus has been on moving objects (Wolfson et al. 1998), emphasizing point-like representations of objects and their trajectories. These investigations have led to a plethora of methods for querying and indexing of space-time samples as they are stored in and retrieved from spatio-temporal databases (Güting and Schneider 2005; Pfoser and Jensen 2003). Methods for making higher-level inferences about changes to spatial configurations, however, have been confined to objects that retain their identity over time, considering such changes as movement, rotation, expansion, and shrinking (Egenhofer and Al-Taha 1992).

More complex changes have been addressed at the level of the identity of objects (Hornsby and Egenhofer 1998), covering the splitting of objects into several autonomous pieces, the spawning off of parts from a continuing entity, the merging of several items into a unified object, or an item joining a collection. When such identity changes occur with respect to spatial objects these changes imply not only modifications at the level of the individuals' identities, but also involve spatial changes. Few considerations, however, have been given to the spatial ramifications of such spatio-temporal change operations, for instance topological changes when merging regions (Clementini et al. 1995; Tryfona and Egenhofer 1997) or by introducing holes into regions (Egenhofer et al. 1994).

This paper addresses changes in topological relations as they occur when splitting an object into two pieces. For example, when subdividing a land parcel with a building on it into two pieces, there are several possibilities for the building to be located with respect to the two newly created land parcels (see Fig. 1). Unless the exact location of the newly introduced boundary is known, the actual situation is one among several choices. Such inferences without graphical or detailed geometric information typically occur when analyzing and reasoning with verbal descriptions.

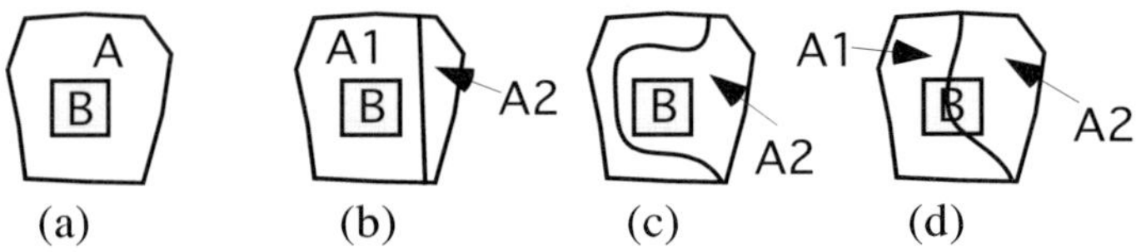

Fig. 1. Three scenarios of subdividing land parcel A into two, A1 and A2, such that building B has a different topological relation with respect to the two subdivisions, A1 and A2: **(a)** A1 *contains* B and A2 is *disjoint* from B; **(b)** A1 is *disjoint* from B and A2 *contains* B; and **(c)** A1 *overlaps* B and A2 *overlaps* B

A comprehensive understanding of all possible topological configurations would provide a basis for making temporal inferences about spatial relations, which may yield interesting, high-level information without the need of information about the actual geometric representations and, therefore, supports qualitative spatio-temporal reasoning. The inferences about the changes in topological relations are also critical in transactions so that one can assess whether a particular change was performed consistently with the operation's semantics.

The remainder of this paper is organized as follows: Section 2 summarizes the model used for describing binary topological relations as well as the inference mechanisms available for dealing with spatial objects that do not change their identities. Section 3 defines splitting and introduces the process used for deriving the set of topological relations that holds after splitting a region into two regions. Sections 4 and 5 determine potential and feasible relations, respectively, the results of which are integrated into achievable splitting configurations (Section 6). Section 7 draws conclusions and stimulates future work.

2 Binary Topological Relations between Regions

A *region* is a non-empty proper subset of a connected topological space such that the region's interior is connected and the region is identical to the closure of the region's interior. Each region is closed, bounded, homogeneously two-dimensional, and homeomorphic to a 2-disk. For pairs of such regions embedded in $\mathbb{R}^2$ a set of eight binary topological relations has been identified whose elements are mutually exclusive and provide a complete coverage between any two regions, that is, there holds exactly one of the eight topological relations (Egenhofer and Franzosa 1991). Their semantics are captured by the 4-intersections (Equations 1a-1i) among the two regions' interiors (A° and B°) and boundaries (∂A and ∂B). The regions' exteriors (denoted by A^- and B^-) capture their regions' complements (i.e., $\mathbb{R}^2 \setminus (A^\circ \cup \partial A)$ and $\mathbb{R}^2 \setminus (B^\circ \cup \partial B)$, respectively).

$$A\ disjoint\ B\text{:}\ A^\circ \cap B^\circ = \varnothing \ \wedge\ \partial A \cap \partial B = \varnothing \tag{1a}$$

$$A\ meet\ B\text{:}\ A^\circ \cap B^\circ = \varnothing \ \wedge\ \partial A \cap \partial B = \neg\varnothing \tag{1b}$$

$$A\ equal\ B\text{:}\ A^\circ \cap B^\circ = \neg\varnothing \ \wedge\ \partial A \cap \partial B = \neg\varnothing \ \wedge \\ A^\circ \cap \partial B = \varnothing \ \wedge\ \partial A \cap B^\circ = \varnothing \tag{1c}$$

$$A\ overlap\ B\text{:}\ A^\circ \cap B^\circ = \neg\varnothing \ \wedge\ \partial A \cap \partial B = \neg\varnothing \ \wedge \\ A^\circ \cap \partial B = \neg\varnothing \ \wedge\ \partial A \cap B^\circ = \neg\varnothing \tag{1d}$$

$$A\ inside\ B\text{:}\ A^\circ \cap B^\circ = \neg\varnothing \ \wedge\ \partial A \cap \partial B = \varnothing \ \wedge \\ A^\circ \cap \partial B = \varnothing \ \wedge\ \partial A \cap B^\circ = \neg\varnothing \tag{1e}$$

$$A\ contains\ B\text{:}\ A^\circ \cap B^\circ = \neg\varnothing \ \wedge\ \partial A \cap \partial B = \varnothing \ \wedge \\ A^\circ \cap \partial B = \neg\varnothing \ \wedge\ \partial A \cap B^\circ = \varnothing \tag{1f}$$

$$A\ covers\ B\text{:}\ A^\circ \cap B^\circ = \neg\varnothing \ \wedge\ \partial A \cap \partial B = \neg\varnothing \ \wedge \\ A^\circ \cap \partial B = \neg\varnothing \ \wedge\ \partial A \cap B^\circ = \varnothing \tag{1g}$$

$$A\ coveredBy\ B\text{:}\ A^\circ \cap B^\circ = \neg\varnothing \ \wedge\ \partial A \cap \partial B = \neg\varnothing \ \wedge \\ A^\circ \cap \partial B = \varnothing \ \wedge\ \partial A \cap B^\circ = \neg\varnothing \tag{1h}$$

U is the universal relation {*disjoint, meet, overlap, inside, covers, contains, coveredBy, equal*} and $topRel \in U$. If several topological relations are referred to, they are distinguished by indices $topRel_i$, $topRel_j$, etc. The set of eight topological region-region relations also enables qualitative spatial reasoning in the form of the *composition* of relations, that is, given a pair of topological relations *A* $topRel_i$ *B* and *B* $topRel_j$ *C* the composition derives candidates for the topological relation $topRel_k$ between *A* and *C* (Egenhofer 1994). With the composition table – the complete set of all possible compositions among the eight topological relations – one can make topological inferences among the set of regions of a spatial configuration using constraint propagation techniques (Egenhofer and Sharma 1993; Smith and Park 1992).

3 Splitting a Region into Two Regions

Splitting a region into two regions is defined in terms of the outcome of a geometric operation. A region *A* is *split* into two parts such that each part is a region as well and that the two parts *meet* (see Figs. 2a–d). Such splitting may be achieved by cutting *A* into two pieces with a non-self-intersecting simple line starting at a point in *A*'s boundary and extending through *A*'s interior back to a different point in *A*'s boundary than the starting point. This type of splitting excludes related operations, such as creating a hole in a region by cutting out an island (see Fig. 2e), or partitioning the region into more than two parts (see Fig. 2f). Regions with holes are known to fall into a different setting beyond simply connected spatial regions and their eight basic topological relations (Egenhofer et al. 1994). Likewise, splitting excludes a separation of the two parts by inserting a non-linear object, as it might be introduced when a flooded river ploughs through some terrain, carving out another extended spatial object. Subsequently, let A_1 and A_2 – the parts of *A* – be two regions such that A_1 *meets* A_2 and the union of A_1 and A_2 is *equal* to *A*.

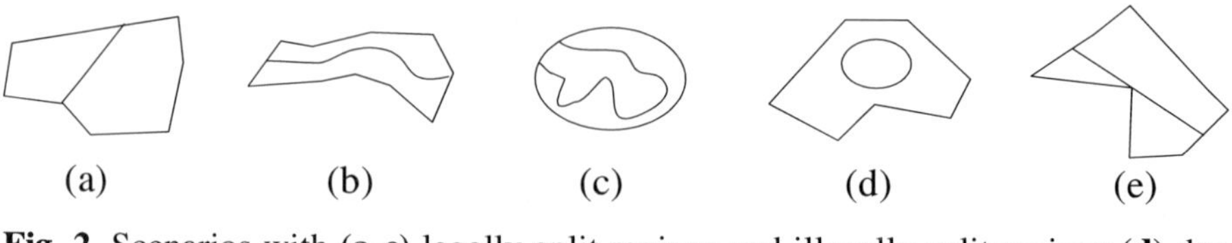

Fig. 2. Scenarios with **(a-c)** legally split regions and illegally split regions **(d)** due to the insertion of a hole and **(e)** due to splitting the region into more than two parts

The topological relations after splitting a region into two regions are derived through three successive steps:

- Identifying *potential splitting configurations* that are based on the constraint that the two split regions must *meet* (Section 4). This first step is performed as a consistency check using the composition property of binary topological relations.
- Deriving systematically the set of *feasible splitting configurations* based on the propagation of empty and non-empty intersections from the to-be-split region to its parts (Section 5). This second step requires a detailed elimination process based on constraints of the split parts' interior, boundary, and exterior intersections with the to-be-split object.
- Integrating the results of potential and feasible splitting configurations into *achievable splitting configurations* (Section 6).

4 Potential Splitting Configurations

When splitting region A into two parts, A_1 and A_2, the topology with respect to region B (i.e., A $topRel_i$ B) is captured by two binary topological relations, A_1 $topRel_j$ B and A_2 $topRel_k$ B. The domain of these topological relations is the set of eight topological region-region relations; therefore, $8^3 = 512$ different combinations would be possible, among the 64 combinations of $topRel_j$ x $topRel_k$. When considering all of these combinations, however, one does not take into account any constraints imposed by the splitting requirement that the two parts, A_1 and A_2, must *meet* and that both A_1 and A_2 must be *coveredBy* A; therefore, the set of possible post-splitting configurations is smaller. For instance, A_1 *contains* B and A_1 *contains* B cannot be realized, after splitting A into A_1 and A_2, because this conjunction is inconsistent with the constraint that A_1 *meets* A_2. On the other hand, A_1 *inside* B and A_2 *inside* B would be consistent with A_1 *meets* A_2.

We define *potential* relations as those that can be obtained by applying systematically a constraint satisfaction algorithm over the network of all binary topological relations among the regions A, A_1, A_2, and B (Egenhofer and Sharma 1993). Such constraint satisfaction enforces converse relations (through the arc consistency constraint) and, along paths in the network, ensures that inconsistencies based on the relations' compositions are eliminated. This approach implies that the set of binary topological relations that hold between each pair of each region is *equal* to itself; *A covers* A_1 and A_2; A_1 *meets* A_2; the unknown relations with B are the universal relation U, and converse relations are used consistently (see Fig. 3).

By replacing iteratively the universal relations U from A to B, from A_1 to B, and from A_2 to B with one concrete relation out of the set of eight topological relations, such that B $topRel_l$ A is converse to A $topRel_i$ B, B $topRel_m$ A_1 is converse to A_1 $topRel_j$ B, and A_2 $topRel_n$ B is converse to B $topRel_k$ A_2 (in order to satisfy the arc consistency constraint), one can perform a consistency check for all possible configurations, eliminating inconsistent and, therefore, impossible configurations.

	A	A_1	A_2	B
A	*equal*	*covers*	*covers*	U
A_1	*coveredBy*	*equal*	*meet*	U
A_2	*coveredBy*	*meet*	*equal*	U
B	U	U	U	*equal*

Fig. 3. The sixteen topological relations between region A, its split parts A_1 and A_2 and another region B

Whenever the path consistency constraint generates an empty relation, the configuration is *impossible*; however, the converse inference of possible configurations from a consistent network of binary topological relations does not always hold true (Papadimitriou et al. 1999); therefore a non-empty relation as the result of the path consistency constraint confirms that a particular configuration is a *potential* topological relation after splitting A into A_1 and A_2 (see Fig. 4).

A_1 $topRel_j$ B	potential topological relations for A_2 $topRel_k$ B
disjoint	*disjoint* ∨ *meet* ∨ *overlap* ∨ *covers* ∨ *contains*
meet	*disjoint* ∨ *meet* ∨ *overlap* ∨ *covers* ∨ *coveredBy* ∨ *equal*
overlap	*inside* ∨ *coveredBy* ∨ *overlap* ∨ *meet* ∨ *disjoint*
coveredBy	*inside* ∨ *coveredBy* ∨ *overlap* ∨ *meet*
inside	*disjoint* ∨ *meet*
covers	*disjoint* ∨ *meet*
contains	*disjoint*
equal	*meet*

Fig. 4. Potential topological relations for the parts A_1 and A_2 with respect to B

5 Feasible Splitting Configurations

Splitting a region into two parts requires the introduction of a new line, which extends from the boundary of the region, through its interior, to a point in the boundary. This line implies that some properties of the topological relations of the split regions can be derived from the topological properties before splitting. These properties rely primarily on the intersections of the interiors and boundaries of the to-be-split region and, therefore, trigger propagations of empty and non-empty interior, boundary, and exterior intersections from the to-be-split region to the parts (Sections 5.1–5.3). Since the newly introduced boundary runs through the to-be-split region's interior, corrective measures must be taken to account for the introduction of the corresponding boundary intersections (Section 5.4).

5.1 Interior Propagations

A's interior, A°, has three relations with respect to B and its parts:

R1: A° is a subset of B° ($A^\circ \subseteq B^\circ$).

R2: A° is a true subset of B's exterior ($A^\circ \subset B^-$).

R3: A° has non-empty intersections with all three parts of B ($A^\circ \cap B^\circ \neq \emptyset \wedge A^\circ \cap \partial B \neq \emptyset \wedge A^\circ \cap B^- \neq \emptyset$).

These three relations cover all possible cases and no other scenarios need to be considered. For instance, because a region's boundary has no extent it cannot contain the non-empty interior or non-empty exterior of another region ($A^\circ \not\subset \partial B$). Since the regions are embedded in $\mathbb{R}^2$, the interior of a region cannot coincide with the exterior of another region ($A^\circ \neq B^-$). Finally, if the interior of a region A contains another region's interior B, this implies that A's interior has non-empty intersections with all parts of B ($A^\circ \supset B^\circ \Rightarrow A^\circ \cap B^\circ \neq \emptyset \wedge A^\circ \cap \partial B \neq \emptyset \wedge A^\circ \cap B^- \neq \emptyset$), therefore, this last scenario is covered by R3. The three relations with respect to A's interior give rise to Theorems 1-3.

Theorem 1: $A^\circ \subseteq B^\circ \Rightarrow A_1^\circ \subset B^\circ \wedge A_2^\circ \subset B^\circ$

Proof: This follows from the definition of a subset (i.e., all parts of a contained connected set are also subsets of the containing set). Since $A_1^\circ \subseteq A^\circ$ and $A_2^\circ \subseteq A^\circ$, A_1° and A_2° are transitively contained in everything in which A° is contained. □

Theorem 2: $A^\circ \subset B^- \Rightarrow A_1^\circ \subset B^- \wedge A_2^\circ \subset B^-$

Proof: In analogy to the proof of Theorem 1, substituting B° with B^-. □

Theorem 3: $A^\circ \cap B^\circ \neq \varnothing \wedge A^\circ \cap \partial B \neq \varnothing \wedge A^\circ \cap B^- \neq \varnothing \Rightarrow$

$$(A_1{}^\circ \subset B^\circ \wedge A_2{}^\circ \subset B^-) \vee$$
$$(A_1{}^\circ \subset B^\circ \wedge A_2{}^\circ \cap B^\circ \neq \varnothing \wedge A_2{}^\circ \cap \partial B \neq \varnothing \wedge A_2{}^\circ \cap B^- \neq \varnothing) \vee$$
$$(A_1{}^\circ \subset B^- \wedge A_2{}^\circ \subset B^\circ) \vee$$
$$(A_1{}^\circ \subset B^- \wedge A_2{}^\circ \cap B^\circ \neq \varnothing \wedge A_2{}^\circ \cap \partial B \neq \varnothing \wedge A_2{}^\circ \cap B^- \neq \varnothing) \vee$$
$$(A_1{}^\circ \cap B^\circ \neq \varnothing \wedge A_1{}^\circ \cap \partial B \neq \varnothing \wedge A_1{}^\circ \cap B^- \neq \varnothing \wedge$$
$$A_2{}^\circ \cap B^\circ \neq \varnothing \wedge A_2{}^\circ \cap \partial B \neq \varnothing \wedge A_2{}^\circ \cap B^- \neq \varnothing)$$

Proof: When *A*'s interior has a non-empty intersection with all three parts of *B*, then five constellations for the split interiors ($A_1{}^\circ$ and $A_2{}^\circ$) are possible: (1) $A_1{}^\circ$ is completely contained in B° and $A_2{}^\circ$ is completely contained in the other extended part of *B* (i.e., B°) such that $A^\circ \setminus A_1{}^\circ \setminus A_2{}^\circ = \partial B \cap A^\circ$, which is non-empty; $A_1{}^\circ$ is completely contained in B° and $A_2{}^\circ$ has non-empty intersections with all three parts of *B*; (3) reversing in (1) $A_1{}^\circ$ and $A_2{}^\circ$; (4) reversing in (2) $A_1{}^\circ$ and $A_2{}^\circ$; and (5) $A_1{}^\circ$ and $A_2{}^\circ$ both extend through all three parts of *B*. □

5.2 Boundary Propagations

Similar to the propagation of non-empty interior intersections, non-empty boundary intersections between the to-be-split region and the related region are also propagated to the split regions' parts. Relevant for this propagation from *A*'s boundary to region *B* is that *A*'s boundary ∂A has six relations with respect to the parts of *B*:

R4: ∂A is a true subset of B° ($\partial A \subset B^\circ$).

R5: ∂A is a true subset of *B*'s exterior ($\partial A \subset B^-$).

R6: ∂A is a subset of *B*'s boundary ($\partial A \subseteq \partial B$).

R7: ∂A has non-empty intersections with *B*'s interior and *B*'s boundary ($\partial A \cap B^\circ \neq \varnothing \wedge \partial A \cap \partial B \neq \varnothing$), but no intersection with *B*'s exterior ($\partial A \cap B^- = \varnothing$).

R8: ∂A has non-empty intersections with *B*'s exterior and *B*'s boundary ($\partial A \cap B^- \neq \varnothing \wedge \partial A \cap \partial B \neq \varnothing$), but no intersection with *B*'s interior ($\partial A \cap B^\circ = \varnothing$).

R9: ∂A has non-empty intersections with all three parts of *B* ($\partial A \cap B^\circ \neq \varnothing \wedge \partial A \cap \partial B \neq \varnothing \wedge \partial A \cap B^- \neq \varnothing$).

Other set-theoretic combinations of ∂A and *B*'s parts are not meaningful or would not yield further insights when splitting *A*. For instance, considering only the non-empty intersections of ∂A with B's interior and B's exterior ($\partial A \cap B^{\circ} \neq \varnothing \;\wedge\; \partial A \cap B^{-} \neq \varnothing$), while assuming that $\partial A \cap \partial B = \varnothing$ is impossible, because of the role of a region's boundary as a Jordan curve, the non-empty intersections of $\partial A \cap B^{\circ} \neq \varnothing$ and $\partial A \cap B^{-} \neq \varnothing$ imply that $\partial A \cap \partial B \neq \varnothing$ as well. These six relations with respect to A's boundary give rise to Theorems 4–9.

Theorem 4: $\partial A \subset B^{\circ} \Rightarrow \partial A_1 \subset B^{\circ} \wedge \partial A_2 \subset B^{\circ}$
Proof: If the boundary of the to-be-split region A is fully contained in the interior of another region B, then the boundary of each split part (∂A_1 and ∂A_2) must be located in that region's interior (B°) as well. The newly introduced part of the boundary between A_1 and A_2 must be a subset of B°, because it falls into A°, which is a subset of B° at the same time as ∂A is a subset of B°. □

Theorem 5: $\partial A \subset B^{-} \Rightarrow \partial A_1 \subset B^{-} \wedge \partial A_2 \subset B^{-}$
Proof: In analogy to the proof of Theorem 4, substituting B° with B^{-}. □

Theorem 6: $\partial A \subseteq \partial B \Rightarrow$

$$\partial A_1 \cap \partial B \neq \varnothing \wedge \partial A_1 \cap B^{\circ} \neq \varnothing \wedge \partial A_2 \cap \partial B \neq \varnothing \wedge \partial A_2 \cap B^{\circ} \neq \varnothing$$

Proof: For region objects, $\partial A \subseteq \partial B$ implies $\partial A = \partial B$, that is, when splitting A into A_1 and A_2, the boundaries ∂A_1 and ∂A_2 will both have non-empty intersections with ∂B. In addition, a newly introduced boundary part, which belongs to both A_1 and A_2 such that it separates $A_1{}^{\circ}$ from $A_2{}^{\circ}$, will need to extend through B°, yielding non-empty intersections of B° with respect to ∂A_1 and ∂A_2. □

Theorem 7: $\partial A \cap B^{\circ} \neq \varnothing \wedge \partial A \cap \partial B \neq \varnothing \wedge \partial A \cap B^{-} = \varnothing \Rightarrow$

$$(\partial A_1 \subset B^{\circ} \wedge \partial A_2 \cap B^{\circ} \neq \varnothing \wedge \partial A_2 \cap \partial B \neq \varnothing \wedge \partial A_2 \cap B^{-} = \varnothing) \vee$$

$$(\partial A_1 \cap B^{\circ} \neq \varnothing \wedge \partial A_1 \cap \partial B \neq \varnothing \wedge \partial A_1 \cap B^{-} = \varnothing \wedge$$

$$\partial A_2 \cap B^{\circ} \neq \varnothing \wedge \partial A_2 \cap \partial B \neq \varnothing \wedge \partial A_2 \cap B^{-} = \varnothing)$$

Proof: If – after splitting A into A_1 and A_2 – ∂A_1 is completely contained in B°, then, since ($\partial A_2 \subseteq \partial A \backslash \partial A_1$) ∂A_2 must have non-empty intersections with B's interior and B's boundary (i.e., $\partial A_2 \cap B^{\circ} \neq \varnothing$ and $\partial A_2 \cap \partial B \neq \varnothing$) no intersection with B's exterior ($\partial A_2 \cap B^{-} = \varnothing$). Conversely, if ∂A_1 is not contained in B° then both ∂A_1 and ∂A_2 must extend through B's interior and boundary, but not through B's exterior. □

Theorem 8: $\partial A \cap B^- \neq \varnothing \wedge \partial A \cap \partial B \neq \varnothing \wedge \partial A \cap B^\circ = \varnothing \Rightarrow$

$$(\partial A_1 \subset B^- \wedge \partial A_2 \cap B^- \neq \varnothing \wedge \partial A_2 \cap \partial B \neq \varnothing \wedge \partial A_2 \cap B^\circ = \varnothing) \vee$$
$$(\partial A_1 \cap B^- \neq \varnothing \wedge \partial A_1 \cap \partial B \neq \varnothing \wedge \partial A_1 \cap B^\circ = \varnothing \wedge$$
$$\partial A_2 \cap B^- \neq \varnothing \wedge \partial A_2 \cap \partial B \neq \varnothing \wedge \partial A_2 \cap B^\circ = \varnothing)$$

Proof: In analogy to the proof of Theorem 7, exchanging B° and B^-. □

Theorem 9: $\partial A \cap B^\circ \neq \varnothing \wedge \partial A \cap \partial B \neq \varnothing \wedge \partial A \cap B^- \neq \varnothing \Rightarrow$

$$(\partial A_1 \cap B^\circ \neq \varnothing \vee \partial A_2 \cap B^\circ \neq \varnothing) \wedge$$
$$(\partial A_1 \cap \partial B \neq \varnothing \vee \partial A_2 \cap \partial B \neq \varnothing) \wedge$$
$$(\partial A_1 \cap B^- \neq \varnothing \vee \partial A_2 \cap B^- \neq \varnothing)$$

Proof: If $\partial A \cap B^\circ \neq \varnothing$ then it is impossible that $\partial A_1 \cap B^\circ = \varnothing$ and $\partial A_2 \cap B^\circ = \varnothing$, which is equivalent to $\partial A_1 \cap B^\circ \neq \varnothing \vee \partial A_2 \cap B^\circ \neq \varnothing$. The other three implications can be found accordingly by replacing B° with ∂B and B^-, respectively. □

5.3 Exterior Propagations

A's exterior has four relevant relations R10–R13 to A's and B's parts. R13 is a stronger statement than R11, but yields additional inferences. Likewise, R10 and R12 may coincide with R13, but there are configurations in which only R10 and R12 hold, but not R13. These four relations with A's exterior yield Theorems 10–13.

R10: A^- has a non-empty intersection with B°.
R11: A^- has a non-empty intersection with B^-.
R12: A^- has a non-empty intersection with ∂B.
R13: A^- is a superset of B^- ($A^- \supseteq B^-$).

Theorem 10: $A^- \cap B^\circ \neq \varnothing \Rightarrow A_1^- \cap B^\circ \neq \varnothing \wedge A_2^- \cap B^\circ \neq \varnothing$

Proof: Splitting A into A_1 and A_2 implies that $A_1^- \supset A_1$ and $A_2^- \supset A_2$. Also $A_1^- \supset A^-$ and $A_2^- \supset A^-$. Therefore, $B^\circ \cap A^- \neq \varnothing$ and $A^- \subset A_1^-$ implies $B^\circ \cap A_1^- \neq \varnothing$. Likewise, $A^- \subset A_2^-$ implies $B^\circ \cap A_2^- \neq \varnothing$. □

Theorem 11: $A^- \cap \partial B \neq \varnothing \Rightarrow A_1^- \cap \partial B \neq \varnothing \wedge A_2^- \cap \partial B \neq \varnothing$

Proof: In analogy to the proof of Theorem 10, substituting B° with ∂B. □

Theorem 12: $A^- \cap B^- \neq \varnothing \Rightarrow A_1^- \cap B^- \neq \varnothing \wedge A_2^- \cap B^- \neq \varnothing$

Proof: In analogy to the proof of Theorem 10, substituting B° with B^-. □

Theorem 13: $A^- \supseteq B^- \Rightarrow A_1^- \supset B^- \wedge A_2^- \supset B^-$

Proof: Splitting A into A_1 and A_2 implies that $A_1^- \supset A_1$. By transitivity $A_1^- \supset A_1 \supseteq B^- \Rightarrow A_1^- \supset B^-$. Substituting A_1^- with A_2^- it follows $A_2^- \supset B^-$. □

5.4 Boundary Overwrite

When splitting a region into two regions a new piece of boundary is introduced that must be connected to the to-be-split region's boundary and must run through its interior until it reaches the boundary again. Therefore, non-empty intersections of the to-be-split object's interior may overwrite empty boundary intersections of the copied boundary intersections.

Theorem 14: $A^\circ \subseteq B^\circ \Rightarrow \partial A_1 \cap B^\circ \neq \varnothing \wedge \partial A_2 \cap B^\circ \neq \varnothing$

Proof: The added boundary is part of A's interior. If A's interior is completely contained in some other component, then that component must intersect with the newly added boundary, which belongs to both A_1 and A_2; therefore, their boundaries must intersect with that component. □

Theorem 15: $A^\circ \subseteq B^- \Rightarrow \partial A_1 \cap B^- \neq \varnothing \wedge \partial A_2 \cap B^- \neq \varnothing$

Proof: In analogy to the proof of Theorem 14, substituting B° with B^-. □

6 Achievable Splitting Configurations

The feasible splitting configurations yield a set of pairs of candidate relations that might hold between the two split objects, depending on the relation that the to-be-split region held prior to splitting. Among these candidate sets, only those relations are *achievable* that lead to *potential* (Section 4) ***and*** *feasible* (Section 5) splitting configurations. We derived systematically those patterns of topological relations that fulfill Theorems 1–15. The stepwise elimination process leads to 21 achievable relations, each of which was confirmed by generating an example drawing (see Figs. 5–12). The stepwise elimination also enabled us to confirm that all sixteen theorems were necessary and no combination of a subset of these theorems would yield the same result as one of the sixteen theorems.

A_1 disjoint B	*A_2 disjoint B*	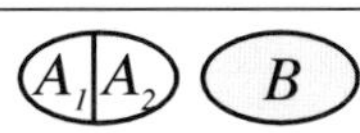

Fig. 5. Achievable splitting relations for *A disjoint B*

A_1 *disjoint* B	A_2 *meet* B	
A_1 *meet* B	A_2 *meet* B	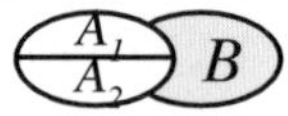

Fig. 6. Achievable splitting relations for *A meet B*

A_1 *covers* B	A_2 *disjoint* B	
A_1 *covers* B	A_2 *meet* B	
A_1 *equal* B	A_2 *meet* B	
A_1 *overlap* B	A_2 *overlap* B	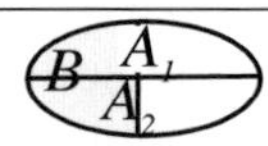
A_1 *coveredBy* B	A_2 *overlap* B	

Fig. 7. Achievable splitting relations for *A covers B*

A_1 *disjoint* B	A_2 *overlap* B	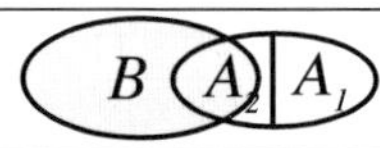
A_1 *meet* B	A_2 *overlap* B	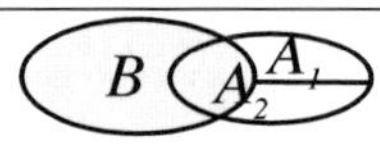
A_1 *meet* B	A_2 *coveredBy* B	
A_1 *overlap* B	A_2 *overlap* B	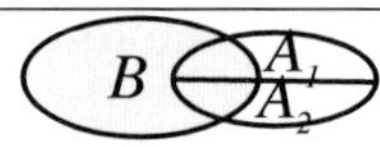
A_1 *overlap* B	A_2 *coveredBy* B	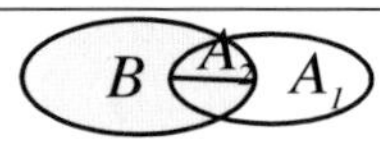
A_1 *overlap* B	A_2 *inside* B	

Fig. 8. Achievable splitting relations for *A overlap B*

A_1 *coveredBy* B	A_2 *coveredBy* B	
A_1 *coveredBy* B	A_2 *inside* B	

Fig. 9. Achievable splitting relations for *A coveredBy B*

A_1 *contains* B	A_2 *disjoint* B	
A_1 *covers* B	A_2 *meet* B	
A_1 *overlap* B	A_2 *overlap* B	

Fig. 10. Achievable splitting relations for *A contains B*

A_1 *coveredBy* B	A_2 *coveredBy* B	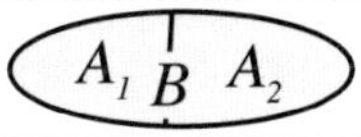

Fig. 11. Achievable splitting relations for *A equal B*

A_1 *inside* B	A_2 *inside* B	

Fig. 12. Achievable splitting relations for *A inside B*

This set of 21 splitting configurations enables a new sort of qualitative spatial reasoning about change from successive snapshots. For instance, with the knowledge that at some time *t1* three regions *X*, *Y*, and *Z* have the topological relations *X contains Z* and *Y disjoint Z*, then *X* and *Y* could have resulted from splitting region *W* into *X* and *Y* (see Fig. 10) and at an earlier time *t0*, prior to splitting, *W* would have *contained Z*. The cumulative inferences from Figures 5–12 show that such inferences about the pre-splitting relation of *X* to *Y* are typically unique, except for four ambiguous cases: (1) *X overlaps Z* and *Y* overlaps *Z* leads to *W overlaps*, *contains*, or *covers Z*; (2) *X covers Z* and *Y meets Z* leads to *W covers* or *contains Z*; (3) *X coveredBy Z* and *Y coveredBy Z* leads to *W equal* or *coveredBy Z*; and (4) *X coveredBy Z* and *Y overlaps Z* leads to *W overlaps* or *covers Z*.

7 Conclusions

We have derived the set of binary topological relations that may hold for each part if one splits a region into two region parts. Constraint satisfaction, establishing an arc-consistent and path-consistent network of topological relations, lead to a set of potential relations. An elimination process then propagated interior, exterior, and boundary properties from the to-be-split region to its parts, yielding feasible relations. The combination of potential and feasible relations led to 21 configurations that may occur for such a region-splitting process, which enables qualitative spatio-temporal reasoning from sequences of snapshots.

Acknowledgments

This work was partially supported by the National Geospatial-Intelligence Agency under grant numbers NMA201-01-1-2003 and NMA401-02-1-2009. Dominik Wilmsen was further supported by a Fulbright Fellowship.

References

Clementini E, di Felice P, Califano G (1995) Composite Regions in Topological Queries. Information Systems 20(7):579–594

Egenhofer M (1994) Deriving the Composition of Binary Topological Relations. J of Visual Languages and Computing 5(2):133–149

Egenhofer M, Al-Taha K (1992) Reasoning about Gradual Changes of Topological Relationships. In: Frank A, Campari I, Formentini U (eds), Theories and Methods of Spatio-Temporal Reasoning in Geographic Space (= LNCS 639), pp 196–219

Egenhofer M, Clementini E, di Felice P (1994) Topological Relations Between Regions with Holes. Int J of Geographical Information Systems 8(2):129–142

Egenhofer M, Franzosa R (1991) Point-Set Topological Relations. Int J of Geographical Information Systems 5(2):161–174

Egenhofer M, Sharma J (1993) Assessing the Consistency of Complete and Incomplete Topological Information. Geographical Systems 1(2):47–68

Güting R, Schneider M (2005) Moving Objects Databases. Morgan Kaufmann Publishers, Amsterdam

Hornsby K, Egenhofer M (1998) Identity-Based Change Operations for Composite Objects. In: Poiker T, Chrisman N (eds) Eighth Int Symp on Spatial Data Handling, Vancouver, Canada, pp 202–213

Papadimitriou C, Suciu D, Vianu V (1999) Topological Queries in Spatial Databases. J of Computer and System Sciences 58(1):29–53

Pfoser D, Jensen C (2003) Indexing of Network Constrained Moving Objects. ACM GIS 2003:25–32

Tryfona N, Egenhofer M (1997) Consistency among Parts and Aggregates: A Computational Model. Transactions in GIS 1(3):189–206

Smith T, Park K (1992) Algebraic Approach to Spatial Reasoning. Int J of Geographical Information Systems 6(3):177–192

Wolfson O, Xu B, Chamberlain S, Jiang L (1998) Moving Objects Databases: Issues and Solutions. In: Rafanelli M, Jarke M (eds) 10th Int Conf on Scientific and Statistical Database Management, pp 111–122

Integrating 2D Topographic Vector Data with a Digital Terrain Model – a Consistent and Semantically Correct Approach

Andreas Koch, Christian Heipke

Institut für Photogrammetrie und GeoInformation,Universität Hannover, Nienburger Str. 1, 30167 Hannover, Germany

Abstract

The most commonly used topographic vector data are currently two-dimensional. The topography is modeled by different objects; in contrast, a digital terrain model (DTM) is a continuous representation of the Earth surface. The integration of the two data sets leads to an augmentation of the dimension of the topographic objects, which is useful in many applications. However, the integration process may lead to inconsistent and semantically incorrect results.

In this paper we describe recent work on consistent and semantically correct integration of 2D GIS vector data and a DTM. In contrast to our prior work in this area, the presented algorithm takes into account geometric inaccuracies of both, planimetric and height data, and thus achieves more realistic results. Height information, implicitly contained in our understanding of certain topographic objects, is explicitly formulated and introduced into an optimization procedure together with the height data from the DTM. Results using real data demonstrate the applicability of the approach.

1 Introduction

Applications of geographic information systems (GIS) increasingly need consistent topographic data containing planimetric and height information. Examples include visualization in terms of true orthophotos and photorealistic perspective views, e.g. for navigation purposes, environmental simulations and traffic safety applications, in which a road must be adequately modeled in three dimensions in order to predict the forces acting on a car during turns. Checking the consistency between planimetric and height data is also useful to assess data quality.

Historically, planimetric and height data do not share many similarities: they have been modeled differently, they have been acquired using different techniques, at different times and resolutions, and for different purposes, and they are stored using different data structures. Therefore, based on existing topographic data bases the required consistency can in general not be guaranteed. Since it is neither desirable nor economically feasible to acquire a completely new, consistent data set, data integration techniques must be developed that meet the described requirements using existing data, i. e. two-dimensional topographic vector data and digital terrain models (DTMs). In many countries such data are being or will be provided by the respective National Mapping Agencies as part of the reference geoinformation. As a side note, we recall, that data integration techniques were also applied in topographic paper maps, of course in a manual fashion: for example, height contour lines cross roads perpendicular to the driving direction, and river beds are usually visible in the contour lines.

Besides consistency, also correctness with respect to gravity and construction principles and manuals must be ensured; the latter is relevant for man-made objects only. This correctness, which depends on the object class label, is termed *semantic correctness* in this paper. To give a few examples, (a) inland water bodies can be considered to be horizontal, if we neglect wind, water currents and local gravitational differences; (b) rivers have a monotonous slope, since water flows downhill only; and (c) roads have constant width, and limited curvature and slope, since otherwise they could not fulfill their function, namely to ensure safe traffic movement.

The integration of two-dimensional topographic GIS data and DTMs has been dealt with to some extent in the literature over the last decade or so. First suggestions go back to Fritsch (1991) and Weibel (1993). Pilouk (1996), Lenk (2001), and Stoter (2004) derive a TIN (triangular irregular network) data structure, in which the triangulation is constrained by using the existing vector data as edges, in addition Lenk (2001) makes sure that the surface shape of the original DTM is preserved. This geometric integration, however, does not pay attention to semantic aspects of the objects to be integrated. These are mentioned by Rousseaux, Bonin (2003), who focus on the integration of 2D linear data such as roads, dikes and embankments into a DTM. The linear objects are transformed into 2.5D surfaces by using attributes (e.g. road width) of the GIS data base and the height information of the DTM. Slopes and regularization constraints are used to check semantic correctness of the objects. However, in case of incorrect results the correctness is not established or re-established.

In this paper we propose an approach for integrating 2D topographic GIS vector data and a DTM in a consistent and semantically correct way. The approach captures the semantics in mathematical equations and inequations, the data integration problem is solved through an optimization approach based on least squares adjustment. We build upon earlier work (Koch, Heipke 2004), the extension presented in this paper consists in a formulation where not only DTM heights are subject to change to fulfill the formulated condition, but also the planimetric coordinates of the vector data are adjusted accordingly. In the next section we present the background and the mathematical description of the new algorithm, before presenting some results using real data sets from the State Surveying Authority of Lower Saxony.

2 An Algorithm for Consistent and Semantically Correct Integration

2.1 Overview

Inconsistency and semantic incorrectness between topographic GIS vector data and a DTM can in principle have two reasons: either the planimetric coordinates of the vector data or the DTM heights are incorrect. Of course, a combination of the two effects is also possible. In contrast to earlier work where we only dealt with incorrect DTM heights, we now present an approach, which can deal with both types of errors.

As in the earlier work we have chosen lakes, rivers, and roads as examples for topographic objects, because all of them contain implicit height information. The objects are modeled with the help of horizontal planes (lakes, road intersection areas) and tilted planes (roads, rivers). Details about object modeling are contained in Koch and Heipke (2004) and in Koch (2006).

The data structure we use for the integrated data set is a TIN. In a first step we convert linear objects to area objects through a buffering process, where the buffer width either comes from available attributes, or a default value is used. This conversion is necessary, since in the considered resolution the topographic objects we deal with all have a certain width in the landscape and thus are considered to be area objects.

The emphasis of our current work lies on the formulation of certain condition equations and inequalities for the vector data and the DTM in order to enforce consistency and semantic correctness. These constraints are taken into account in an optimization process based on least squares adjustment. The following assumptions have guided the selection of the constraints:

- The height information contained implicitly in the topographic objects must be captured explicitly in order to be introduced into the optimization process.
- The data sets to be integrated can contain random and systematic errors, but they do not contain any gross errors (gross errors can and should be eliminated in a pre-processing step).
- The topographic vector data is separated into man-made and natural vector data:
  - The shape of the man-made vector data (e. g. roads) is considered to be generally correct, because it follows construction principles. Therefore, their position can only be changed as a whole. We use a 2D similarity transformation for this task.
  - The border of natural vector data (e. g. lakes, rivers) can vary also locally, we therefore consider the individual border coordinates as unknowns in the adjustment.
- The shape of the terrain should be preserved as much as possible.
- Neighborhood consistency must be taken into account.

In most cases, an integration process involves a kind of compromise. We model the fact that some of the mentioned conditions can contradict each other by assigning weights to the individual equations. It is clear that a careful selection of the weights based on the quality of the input data is of major importance for obtaining meaningful results.

After the optimization we perform the actual integration using a triangulation based on Lenk's algorithm (Lenk 2001).

2.2 The Optimization Process

In the optimization process, the heights of the topographic objects as derived from the DTM, and the DTM heights in the neighborhood are considered as unknowns, together with the transformation parameters of the man-made topographic objects (4 per object) and all the planimetric coordinates of the natural topographic objects. These unknowns are estimated from a set of basic observation equations in a least squares adjustment, taking into account additional equation and inequality constraints.

The basic observation equations preserve the general position of the topographic objects, the shape of the terrain, and they ensure a smooth transition between changed and non-changed areas of the data set. The constraints capture consistency and semantic correctness. Equation constraints are formulated as observation equations with corresponding weights, thus the amount to which an equation constraint is actually fulfilled can be controlled by an adequate weight selection. The inequality constraints, on the other hand, are always fulfilled after the optimization process.

2.3 Observation Equations and Constraints for Planimetric Coordinates of Vector Data

Man-made objects: For man-made objects such as roads the coordinates X_i, Y_i of the border polygon are improved through a two-dimensional similarity transformation resulting in a set of new coordinates X_i^t, Y_i^t. The unknowns are the translation $\hat{X}_0, \hat{Y}_0$ and the rotation and scale parameters $\hat{a}$ und $\hat{b}$, X_S, Y_S represent the centre of gravity of the object:

$$
\begin{aligned}
X_i^t &= \hat{X}_0 + \hat{a}(X_i - X_S) + \hat{b}(Y_i - Y_S) + X_S \\
Y_i^t &= \hat{Y}_0 - \hat{b}(X_i - X_S) + \hat{a}(Y_i - Y_S) + Y_S
\end{aligned}
\tag{1}
$$

Points, which represent road intersections, are considered to be part of more than one road. Since for each road a separate set of equations of type (1) is used, this common point is lost without any further precautions. In order to preserve the topologic relationship between the roads, one constraint, formulated as an observation equation, is set up for each road ending in the intersection, where $\hat{X}_{int}, \hat{Y}_{int}$ denotes the unknown intersection point, and (as in all formulae throughout this paper) v stands for the residual of the observation equation:

$$
\begin{aligned}
0 + v &= \hat{X}_{int} - X_{int}^t \\
0 + v &= \hat{Y}_{int} - Y_{int}^t
\end{aligned}
\tag{2}
$$

For the remaining border polygon points X_i, Y_i of man-made objects basic observation equations to maintain the overall position are set up in the following way:

$$
\begin{aligned}
0 + v_i &= X_i^t - X_i \\
0 + v_i &= Y_i^t - Y_i
\end{aligned}
\tag{3}
$$

Natural objects: As mentioned above, for natural objects equations of type (1) are not used. Rather, individual border points can move separately, as shown in the basic observation equations (4). X_i, Y_i denote the original, $\hat{X}_i, \hat{Y}_i$ the unknown coordinates of the border polygon.

$$\begin{aligned} 0 + v_i &= \hat{X}_i - X_i \\ 0 + v_i &= \hat{Y}_i - Y_i \end{aligned} \tag{4}$$

It must be ensured that despite movements of individual points of the border polygon remains an area without loops. This constraint is formulated by allowing the polygon angle α_j, which is a function of the sequence of polygon points P_{j-1}, P_j, and P_{j+1}, to change only by a small predefined amount $\Delta\alpha$ to form the resulting polygon angle α_j^*. α_j^* is a function of the unknown coordinates of P_{j-1}, P_j, and P_{j+1}, which are estimated in the optimization procedure.

$$|\alpha_j^* - \alpha_j| \leq \Delta\alpha \tag{5}$$

This constraint can be formulated as a set of two inequalities:

$$\begin{aligned} \alpha_j^* - \alpha_j &\leq \Delta\alpha \\ \alpha_j^* - \alpha_j &\geq -\Delta\alpha \end{aligned} \tag{6}$$

Topological aspects valid for all vector objects: Point movements can lead to different objects overlapping each other in an undesired way. We require the topology of objects to remain unchanged during the optimization process. Figure 1 shows an example; two objects A and B change their outline after the optimization. Figure 1a depicts the original situation, Figure 1b and 1c show two results, which change the topology of the objects and must therefore be avoided. Figure 1d shows a possible point movement.

If the GIS vector data are triangulated without considering the DTM points, possible and impossible situations can be separated based on an inspection of the individual triangles. In order to preserve topology the sense of orientation of the triangles connecting different objects must be maintained. This sense of orientation can be expressed by the triangle determinant D and its change dD (O'Rourke 1998). Assuming the determinant of a triangle with points P_i, P_j, P_k to be negative, the following inequality captures the constraint:

$$-dD \geq \begin{vmatrix} (\hat{X}_j - \hat{X}_i) & (\hat{X}_k - \hat{X}_i) \\ (\hat{Y}_j - \hat{Y}_i) & (\hat{Y}_k - \hat{Y}_i) \end{vmatrix} \tag{7}$$

For man-made objects the transformed coordinates X^t, Y^t take the place of the unknown coordinates $\hat{X}, \hat{Y}$.

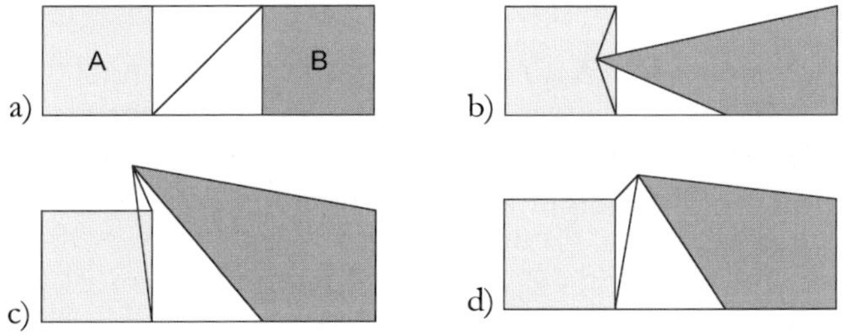

Fig. 1. Topologic relation between two objects A and B: **(a)** situation before optimisation, **(b)** and **(c)** invalid point movements, **(d)** valid point movement

2.4 Observation Equations and Constraints for Height Data

Observation equations, equation and inequality constraints for the heights of DTM points and the coordinates of the topographic objects were presented in detail in (Koch, Heipke 2004). Therefore, only, a short summary of these equations will be given here. The difference to our new formulation is that heights, which need to be interpolated from neighboring points, e. g. heights for the road centre axis, are now a function of the unknown planimetric position of the point under consideration. DTM heights are introduced as:

$$0 + v_i = \hat{Z}_i - Z_i \tag{8}$$

Z_i refers to the original height of the DTM, $\hat{Z}_i$ denotes the unknown height, v_i is again the residual. If the considered point is part of the border polygon of a topographic object, Z_i has to be interpolated using neighboring height information of the DTM.

In order to be able to preserve the slope of an edge connecting two neighboring points P_j and P_k of the DTM TIN where one is part of the polygon describing the object, and the other one is a neighboring point outside the object (and thus to control the general shape of the integrated DTM TIN) additional equations are formulated:

$$Z_j - Z_k + v_{jk} = \hat{Z}_j - \hat{Z}_k \tag{9}$$

The constraints used for horizontal and tilted planes are shortly described next. Heights Z_l of all points P_l lying in the area of a horizontal plane all have the same value $\hat{Z}_{HP}$. This fact is captured through the observation equation

$$0 + v_l = \hat{Z}_{HP} - Z_l \tag{10}$$

Heights Z_m for points P_m (X_m,Y_m) of the border polygon of the horizontal topographic objects are interpolated from the neighboring DTM TIN points P_u, P_v, P_w, and the height difference between the unknown object height and the interpolated height is used to formulate the constraint:

$$0 + v_m = \hat{Z}_{HP} - Z_m\left(\hat{X}_m, \hat{Y}_m, Z_u, Z_v, Z_w\right) \tag{11}$$

A further constraint expresses the fact that for lakes, surrounding terrain points must have a larger height Z_i than the lake:

$$0 < \hat{Z}_{HP} - \hat{Z}_i \tag{12}$$

As mentioned, roads and rivers are modeled with tilted planes. Points P_r on such planes must fulfill the following constraint, where $\hat{a}_0, \hat{a}_1, \hat{a}_2$ are the unknown plane parameters:

$$0 + v_r = \hat{a}_0 + \hat{a}_1 \hat{X}_r + \hat{a}_2 \hat{Y}_r - \hat{Z}_r \tag{13}$$

Roads and rivers are further constrained by requiring the slope along the object to be smaller than a certain predefined threshold. Also, roads are assumed to have horizontal cross sections, for further details see Koch and Heipke (2004).

The optimization problem including the inequality constraints is formulated as the linear complementary problem (LCP) and solved using the Lemke algorithm (Lemke 1968; Schaffrin 1981; Lawson & Hanson 1995). Since the unknowns appear in a nonlinear form, the solution can only be found iteratively. It should be noted that the number of equations may change from iteration to iteration, because due to the changes of the planimetric coordinates of the topographic objects, it may be necessary to consider different points of the neighborhood from iteration to iteration.

As mentioned above, adequate weights must be selected for all observation equations to obtain a meaningful result: the position and height coordinates have a certain geometric accuracy, and weights should be chosen accordingly. The weights of the equation constraints must be selected according to experience. Since the inequality constraints are automatically satisfied within the algorithm, the weights for the equality constraints together with the predefined thresholds (see above) determine the degree to which consistency and semantic correctness of the integrated data set is achieved.

3 Results

In this section we present results of a consistent and semantically correct integration of real topographic vector data and a DTM. We use the German *ATKIS Basis-DLM*[1] together with the DTM *ATKIS DGM5*. The geometric accuracy of the *Basis-DLM* is approximately ±3 m, the DTM heights have a standard deviation of about ±0,5 m.

The first data sets, called *3 lakes*, consists of three lake objects with 294 planimetric polygon points, covering a relatively flat area of 450 x 650 m^2. The corresponding DTM contains 1.961 grid points, and in addition additional 118 points representing geomorphologic information (break lines etc.). In a pre-processing step both groups were merged using a constrained Delaunay triangulation to form a TIN. Prior to the integration, at the border to the lakes inconsistencies were clearly visible.

The results for *3 lakes* data set are indeed consistent and semantically correct. They are shown in Table 1 and in Figure 2.

Table 1. Results for real data set *3 lakes*. For the main types of equations the table contains the standard deviations of the observations as well as the number and size of the residuals.

Type of equation		Standard deviation [m]	No.	Residuals Mean [m]	Min. [m]	Max. [m]
Planimetric position (4)	X	3.0	690	-0.47	-9.69	7.24
	Y	3.0	690	-0.28	-8.37	8.79
Heights of border polygon (11)		0.5	690	-0.24	-1.90	0.75
Heights outside the border (8)		0.5	531	-0.05	-0.35	0.95
Height differences (9)		2.0	3279	-0.19	-1.72	0.59

For the main types of equations the table contains the standard deviation of the observations as well as the number and size of the resulting residuals. It can be seen that major position changes occur in planimetry.

Although the shape of the objects remains more or less the same, the minimum and the maximum values of the residuals amount to three times the introduced standard deviation. From Figure 2 it is visible that these changes occur mainly at the border polygon points. Apparently the original

[1] ATKIS stands for Authoritative Topographic Cartographic Information system and represents the German national reference geoinformation database. The *Basis-DLM* (basic digital landscape model) contains the highest resolution and is approximately equivalent to a topographic map 1:25,000; the *DGM5* is a hybrid data set containing regularly distributed points with a grid size of 12.5 m and additional geomorphologic information.

border points of the lake polygons lie outside the actual lake and are now moved into the water, since the water height is mainly dictated by large number of points inside the lake, which were considered to be rather accurate. While this result is consistent and semantically correct, a somewhat smaller weight for the heights inside the lakes would have probably resulted in smaller and more realistic planimetric point movements.

The lake no. 1 to the upper right of Figure 2 shows a somewhat different behavior than the other two. Some heights in the middle of the lake become significantly lower. The reason was found to be a break line running through the lake, which constitutes a gross error in the data set.

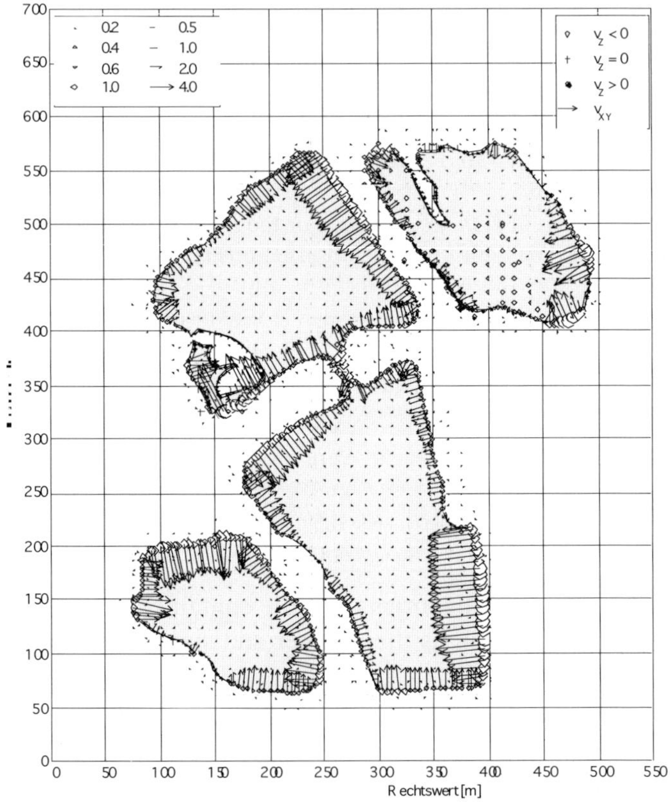

Fig. 2. Results of integration of *3 lakes* data set. White circles denote heights which became lower, black circles those, which became higher, arrows depict planimetric point movement

The second data set, called *roads*, consists of a network of 13 small roads. Most of them are connected at both ends, some are dead end roads. The data set consists of rolling terrain with height differences of about

50 m and covers an area of 575 x 400 m^2, it consists of 1,551 DTM grid points and 27 break lines. Before the integration, inconsistencies are clearly visible. The weights were again chosen according to the geometric accuracy of the input data, the constraints were introduced with high weights.

The results for the *roads* data set are similar to those for *3 lakes*. Again, a consistent and semantically correct result was achieved. The main changes could be observed in the planimetric position of the topographic objects, and in particular in the dead end roads in the rougher terrain. One of the points in steep terrain was moved by more than 10 m. In contrast to those roads ending in intersections, the position of the dead end roads is obviously not stabilized through equation (2).

4 Conclusions and Outlook

This paper presents an approach for the consistent and semantically correct integration of a DTM and 2D topographic GIS data. The algorithm is based on a Delaunay triangulation and a least squares adjustment including inequality constraints derived from the implicitly available height information of topographic objects, and is solved by converting the approach into a linear complementary problem (LCP). In contrast to our earlier work, we not only adjust DTM heights, but also the planimetric position of topographic objects. Thus, vector and height data can be introduced with their respective geometric accuracy.

The approach was tested using a number of real data sets, taken from the German ATKIS. The results of two of these data sets have been presented in this paper. In all cases, a consistent and semantically correct result was achieved, which is not self understood as such, because the equation constraints are introduced as observations equations and are controlled via weight selection.

While the results are very promising, the proper selection of weights remains a difficult problem which requires some experience. Another open question is whether our approach can be transferred from the aggregation level we currently work at (ATKIS Basis-DLM) to other scales, e.g. a more detailed scale, in which e.g. consistency plays a very important role for visualization. In addition, the geomorphologic information available in the DTM should be considered explicitly in an extended version of the algorithm. Finally, if a complete GIS data set is to be integrated with a DTM, aspects such as the propagation of planimetric changes from objects with implicit height information to neighboring objects also need to be dealt with. These are the issues we currently work on.

Acknowledgement

We express our deep gratitude to the State Surveying Authority of Lower Saxony (Landesvermessung und Geobasisinformation Niedersachsen; LGN) for significantly supporting this research.

References

Fritsch D (1991) Raumbezogene Informationssysteme und digitale Geländemodelle (= Reihe C, 369). DGK

Koch A (2006) Semantische Integration von zweidimensionalen GIS-Daten und Digitalen Geländemodellen. Dissertation, Universität Hannover

Koch A, Heipke C (2004) Semantically Correct 2.5D GIS Data – the Integration of a DTM and Topographic Vector Data. In: Fisher P (ed) Developments in Spatial Data Handling. Springer, Berlin, pp 509–526

Lawson CL, Hanson RJ (1995) Solving Least Squares Problems. Soc for Industr and Appl Mathematics. Philadelphia, 337 p

Lenk U (2001) 2.5D-GIS und Geobasisdaten – Integration von Höheninformation und Digitalen Situationsmodellen (= Reihe C, 546). DGK

O'Rourke J (1998) Computational Geometry in C, 2nd ed. Cambridge University Press, Cambridge, 376 p

Pilouk M (1996) Integrated Modelling for 3D GIS (= ITC Publication Series 40). PhD Thesis, Enschede

Rousseaux F, Bonin O. (2003) Towards a coherent integration of 2D linear data into a DTM. In: Proc of the 21st ICA Conf, pp 1936–1942

Schaffrin B (1981) Ausgleichung mit Bedingungs-Ungleichungen. AVN 6: 227–238

Stoter J (2004) 3D Cadastre. Netherlands Geodetic Commission, Publications on Geodesy 57, 327 p

Weibel R (1993) On the Integration of Digital Terrain and Surface Modeling into Geographic Information Systems. In: Proc AUTOCARTO 11. Minneapolis, Minnesota, pp 257–266

A Hierarchical Approach to the Line-Line Topological Relations

Zhilin Li, Min Deng

Department of Land Surveying and Geo-Informatics
The Hong Kong Polytechnic University, Hong Kong
e-mail: {lszlli; lsdmin}@polyu.edu.hk

Abstract

Topological relations have been recognized to be very useful for spatial query, analysis and reasoning. This paper concentrates on the topological relations between two lines in IR^2. The line of thought employed in this study is that the topological relation between two lines can be described by a combination of finite number of basic (or elementary) relations. Based on this idea, a hierarchical approach is proposed for the description and determination of basic relations between two lines. Seventeen (17) basic relations are identified and eleven (11) of them form the basis for combinational description of a complex relation, which can be determined by a compound relation model. A practical example of bus routes is provided for illustration of the approach proposed in this paper, which is an application of the line-line topological relations in traffic planning.

Key words: topological relations, topological invariant, formalism

1 Introduction

The representation of spatial relations, which essentially reflect the spatial configuration between spatial objects, is one of the key issues in GIS. Spatial relations are geometric constraints to spatial objects, and may be classified into topological, metric, and order three kinds (Egenhofer and Franzosa 1991). These relations have been found useful for spatial query, analysis and reasoning (Randell et al. 1992). For example, spatial inconsistency between rivers and roads, between contour lines and rivers, and between contour lines and roads, may have been created during the map updating process (Liu et al. 2004; Zhang et al. 2005). In order to automatically detect and remove such spatial inconsistency, a detailed model for the topological relations between two lines is required and indeed this paper concentrates on topological relations.

In the last decade, many researchers have paid much attention to the formal description and determination of the topological relations between spatial objects, as well as their applications in various areas such as GIS, spatial database, CAD/CAM systems, image databases, spatial analysis, computer vision, artificial intelligence, linguistics, cognitive science, psychology and robotics. A large body of literature on this topic has been available (Egenhofer and Franzosa 1991; Randell et al. 1992; Cui et al. 1993; Egenhofer 1993; Clementini et al. 1994; Egenhofer and Franzosa 1995; Bennett 1997; Renz and Nebel 1999; Chen et al. 2001; Li et al. 2002).

Through an analysis it can be found that the approaches used in existing work can be classified into two categories, i.e. decomposition-based and whole-based (Abdelmoty et al. 1994; Li et al. 2002). In the former, a spatial object is decomposed into two (i.e. boundary and interior) or three (i.e. boundary, interior and exterior) components and the topological relations between spatial objects are determined by the combinatorial relations of these components. In the latter, a spatial object is considered as a whole and the topological relations are determined by the interaction between spatial objects themselves. So far, most of topological relations models are built upon the decomposition-based approach (e.g. Güting 1988; Egenhofer and Franzosa 1991; Clementini et al. 1993), while there are only few researches based on the whole-based approach (Randell et al. 1992; Li et al. 2002).

Among the decomposition-based models, the 4-intersection model by Egenhofer and Franzosa (1991) has become the classic model, by making use of the interior and boundary of an object. This model is able to distinguish eight (8) relations between two lines in IR^1, and sixteen (16) relations between two lines in IR^2. Egenhofer and Herring (1991) expanded the 4-intersection to 9-intersection by introducing an additional component, i.e. the exterior of an object (which is represented by its complement). The 9-intersection model is able to distinguish thirty three (33) line-line relations. However, Chen et al. (2001) argued that such an extension from the 4- to 9-intersections is invalid in theory because there is a linear dependency between the interior, boundary and the complement. Chen et al. (2001) then proposed the use of Voronoi region to replace the complement and established the so-called Voronoi-based 9-intersection model. It has also been pointed out by Li et al. (2000) that the definition of two end points as the boundary is valid only in IR^1 (see Fig. 1a) and it will lead to topological paradox in IR^2 (see Fig. 1b), i.e. no need to pass across the boundary when traveling from exterior to interior. Li et al. (2002) then made use of the object as a whole and its Voronoi region to have developed a spatial algebraic model for spatial relations, which also made use of multiple operators (e.g. union, intersection, difference, etc) and multiple types of values for the computational results of set operations (e.g. content, dimension and number of connected components).

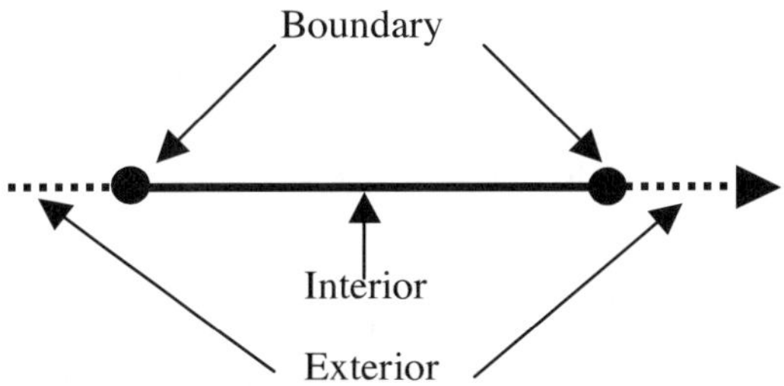

(a) Two end points form the boundary in IR^1

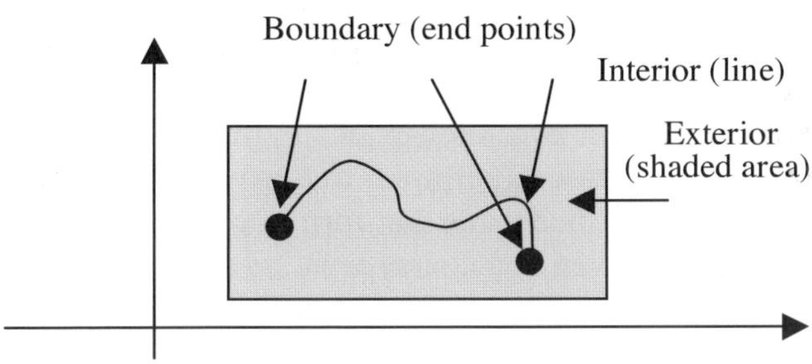

(b) Topological paradox in IR^2 after using the definition in IR^1 (i.e. exterior and interior connected

Fig. 1. Validity of defining two end points as boundary of a line (Li et al. 2000)

However, it should be pointed that the models mentioned above can only make a rough classification of line-line relations. This paper aims to present a model for more systematic discrimination of topological relations between two lines in IR^2.

This introduction is followed by a line of thought for the differentiation of line-line relations (Section 2), i.e. basic (or elementary) relations and compound (combined) relations. Section 3 describes the determination of the basic relations and Section 4 discusses compound relation model. Section 5 provides an example of bus routes to illustrate the implementation of the proposed approach in this study, which is also an application of the line-line topological relations. Some conclusions are made in Section 6.

2 A Line of Thought for Topological Relations between Two Lines

As reviewed in the Introduction, many models have been developed for topological relations but they are short in providing a comprehensive coverage of the relations between lines. To develop a new model, a new line of thought needs to be developed.

2.1 A Strategy

It has been claimed that the 9-intersection by Egenhofer and Herring (1991) is able to distinguish thirty three (33) types of topological relations between two lines. Two critical questions arising are "How many relations in total exist between two lines?" and "is the 9-intersection model sufficient to describe the topological relations between two lines?"

Through an analysis, one may find that there are an infinite number of potential relations between two lines in IR^2. Figure 2 is an example showing such a complex situation. Therefore, one may notice that the relations distinguished by existing models do not form a complete coverage of possible line-line relations, even at a coarse level. It can also be noted that different models differentiate line-line relations with different number and different types. That is to say, a model which is able to systematically distinguish line-line relations in hierarchy is very desirable.

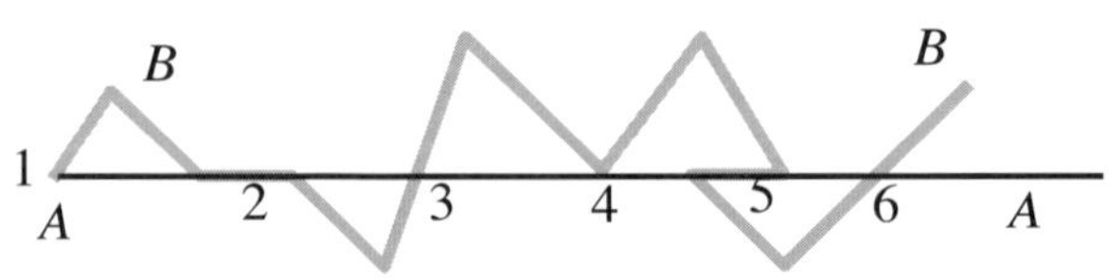

Fig. 2. An example of potential topological relations between two lines

The strategy employed in this study is to decompose the topological relations between two lines into a hierarchical structure. First, the relations are divided into two levels, i.e. basic (or elementary) and compound (combined) relations. For example, the compound topological relations between lines *A* and *B* consists of 6 basic relations, as marked in the figure. Second, the basic relations are hierarchically categorized.

2.2 Topological Invariants for Differentiation of Line-line Relations

Clementini and Di Felice (1998) have developed a set of topological invariants as follows:

- content,
- dimension,
- number of intersection components,
- intersection sequence,
- intersection type,
- collinearity sense, and
- link orientation.

Through an analysis, it can be found that there are two aspects of problems in their work. The first one is that some of the topological invariants defined in this study are closely related to the assumed orientation of the lines themselves, e.g. *intersection type* and *link orientation*. The second is that there is still a lack of investigation into prioritize these invariants and how to select an invariant or a sub-set of invariants for a given application.

In this study, separation number, dimension, component type and component-based order are used for the differentiation of topological relations between lines.

A compound relation can be decomposed into two or more elementary relations which can respectively belong to *partially overlap*, *cross* and/or *meet*. However, it is not possible to the elementary relations *wholly overlap* category into compound relations.

3 Hierarchical Descriptions for Basic (Elementary) Relations between Lines

In the previous section, it is argued topological relations between two lines can only be described by compound model, which may include an infinite number of basic (or elementary relations). A set of topological invariants is presented for the differentiation of topological relations between lines. In this section, a detailed discussion on the differentiation of basic topological relations between two lines is presented. The combination of basic relations into compound relations will be discussed in Section 4.

3.1 Separation Number and Dimension for Differentiation of Line-line Relations

For the two given lines *A* and *B*, *separation number* can be used to represent the number of components in the intersection between *A* and *B*, i.e. $\chi(A \cap B)$, as follows:

$$\chi(A \cap B)\begin{cases} = 0 & \Rightarrow \text{disjoint} \Leftrightarrow \text{order relations} \\ = 1 & \Rightarrow \text{connected} \Leftrightarrow \text{basic topological relations} \\ > 1 & \Rightarrow \text{multiple} \Leftrightarrow \text{compound topological relations} \end{cases} \tag{1}$$

When $\chi(A \cap B)$=0, then the topological relation between *A* and *B* is called as *disjoint* and further differentiation of *disjoint* relation can be made by the Voronoi-based K-order adjacency model (Chen et al. 2004). When $\chi(A \cap B) \geq 1$, *A* and *B* are *connected*. When $\chi(A \cap B) = 1$, the topological relations are termed as basic (or elementary) relations in this text. On the other hand, when $\chi(A \cap B) > 1$, then the topological relations are composed of more than one basic relation (as shown in Fig. 2) and thus termed as compound relations in this text.

If two lines *A* and *B* are connected, then one may ask whether or not they share a common piece. Then, dimension, i.e. $\dim(A \cap B)$, can be used to answer such a question, as follows:

$$\dim(A \cap B) = \begin{cases} 0 & \Rightarrow \text{Joint} \begin{cases} Cross \\ Meet \end{cases} \\ 1 & \Rightarrow \text{Overlap} \begin{cases} Partially\ Overlap \\ Whole\ Overlap \end{cases} \end{cases} \tag{2}$$

When dim($A \cap B$)=0, the two line are *joint* but intersection is a point. On the other hand, dim($A \cap B$)=1, the intersection is a line and thus the two lines *overlap* somehow.

As will be discussed in next subsection, the *joint* relation can be further divided into *cross* and *meet* and *overlap* relation can be further classified into *wholly overlap* and *partially overlap.* These relations can be further classified based on their component type in the line-line intersection.

3.2 Component Type and Local Order for Differentiation of Line-line Relations

Based on the component type, the *joint* relation can be further divided into five (5) types, denoted by p_a, p_b, p_c, p_d and p_e, as shown in Figure 3. In this figure, a solid circle is used to denote an endpoint of a line. These components are characterized by such names as *concatenation* point, T_A-*junction* point, T_B-*junction* point, *touching* point and *crossing* point. The corresponding relations are termed as *concatenate*, T_A-*intersect*, T_B-*intersect*, *touch* and *cross*.

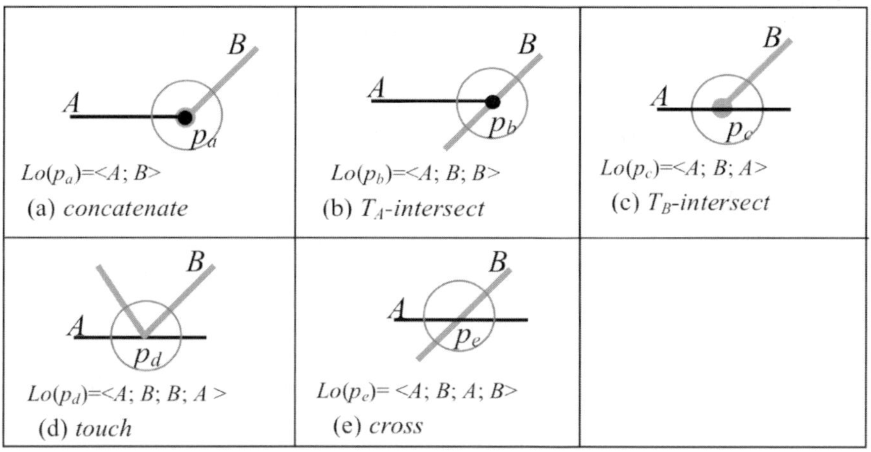

Fig. 3. Five types of 0-dimensional components in *joint* relation

These components can also be defined by another invariant -- local order. The local order, denoted by $Lo(p_i)$, of a 0-dimensional component p_i, denoted by $Lo(p_i)$, can be defined to be the order of the two intersecting lines within a very small circle centered at p_i. This small circle is equivalent to the concept of neighborhood. For instance, the local order of p_e in Figure 3e can be represented as $Lo(p_e) =< A; B; A; B >$.

This expression indicates that there are four intersection points between the two lines and the neighborhood, and the sequence of intersection is *A*, *B*, *A* and *B*. The number of intersection points is also called the connective degree. It should also be noted there, the starting direction for the local order is arbitrarily selected. As a result, the orders defined in Figure 3 can be equivalently represented as follows:

Type p_a: $<A; B> \equiv <B; A>$

Type p_b: $<A; B; B> \equiv <B; A; B> \equiv <B; B; A>$

Type p_c: $<A; B; A> \equiv <A; A; B> \equiv <B; A; A>$

Type p_d: $<A; B; B; A> \equiv <A; A; B; B> \equiv <B; B; A; A>$

Type p_e: $<A; B; A; B> \equiv <B; A; B; A>$

For 1-dimsional component, there are six (6) types which may occur in *partially overlap* relation, denoted as l_a, l_b, …, l_f, as shown in Figure 4. These components are characterized by such names as *concatenation-like*, *T_B-junction-like*, *branch-like*, *T_A-junction-like*, *crossing-like* and *touching-like overlap*. And there are five (5) types for *wholly overlapped* relation, denoted by l_g, l_h, …, l_k, as shown in Figure 5. These components are named as *equal*, *contain*, *containedby, cover*, and *coveredby.*

Similarly, local orders of its two endpoints of the overlapped line. For instance, the local orders of two endpoints of l_e in Figure 4e may be respectively represented as $Lo(l_{e1})=<A; B; A \cap B>$ and $Lo(l_{e2})=<A; A \cap B; B>$. Here, l_{e1} and l_{e2} denote left and right endpoints of l_e.

Indeed, each of the sixteen (16) component types in Figures 3, 4 and 5 represents a basic (or elementary) line-line relation. Therefore, for two lines in IR^2, there are the seventeen (17) fundamental relations (including *disjoint*) in total. Except the five (5) basic relations in the categories of *wholly overlapped* relation and *disjoint* relation, other eleven (11) basic relations form a basis of the description and determination of compound relations.

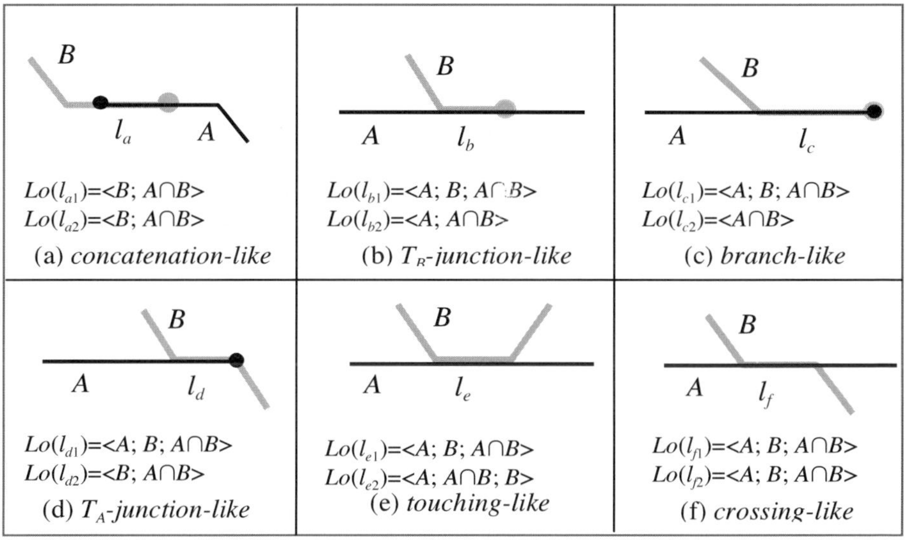

Fig. 4. Six types of 1-dimensional components in *partially overlap* relation

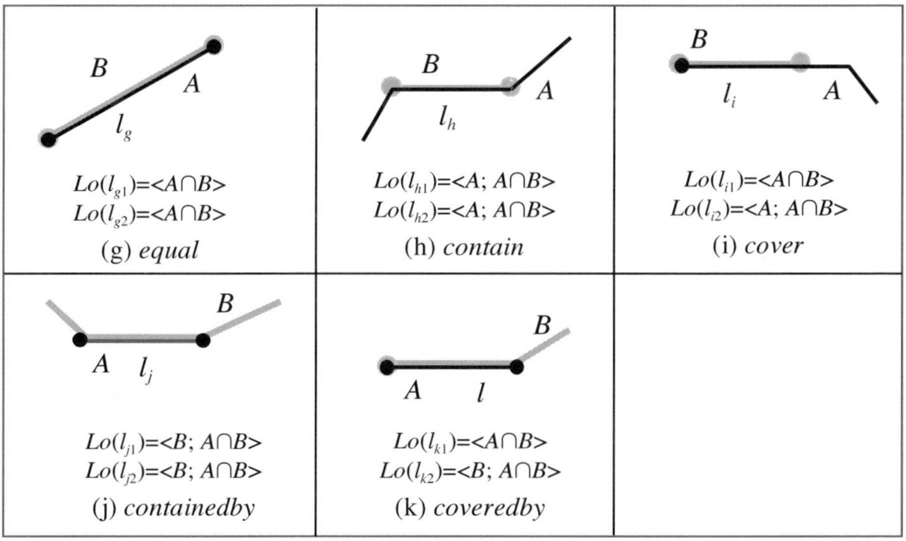

Fig. 5. Five types of 1-dimensional components in *wholly overlap* relation

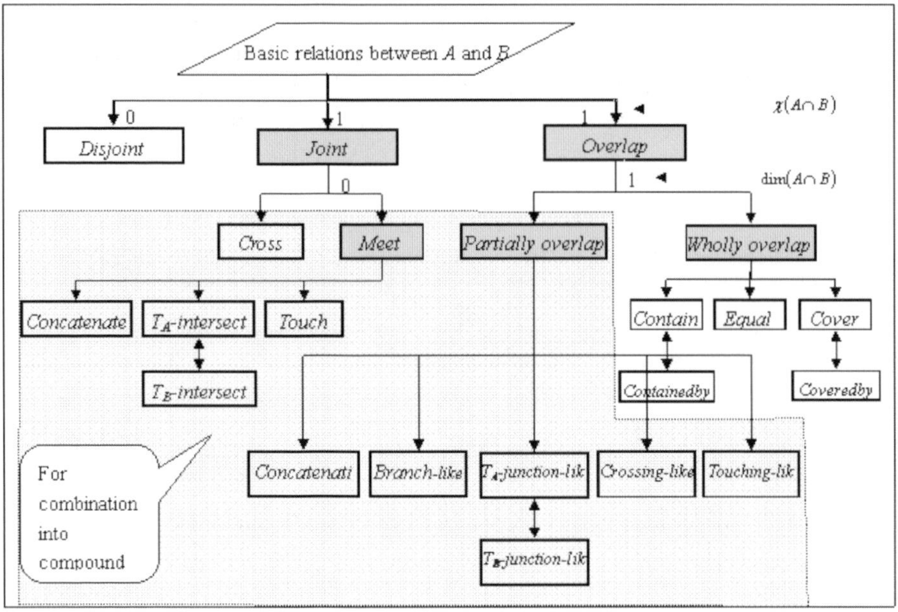

Fig. 6. Hierarchical descriptions of basic topological relations between two lines in IR^2

4 Compound Line-line Relation Model

As discussed previously, a compound relation is referred as such a relation with number of components in the line-line intersection being larger than one and may be decomposed into a set of basic relations. In this section, a discussion will be conducted on the sequential description of a compound relation.

4.1 Basic Order of Component Types for Compound Relations

First, the basic order (or sequence) of component types is introduced. The importance of such an adoption is illustrated in Figure 7. In this example, compound relations between *A* and *B* in both (a) and (b) can be decomposed into the same five basic relations, i.e. T_B*-intersect* (p_c), *cross* (p_e), *cross* (p_e), *cross* (p_e) and T_B*-junction-like overlap* (l_b). However, since these basic relations occur in different orders, thus topological relation between *A* and *B* in (a) should be different from that in (b).

Indeed, according to discrete mathematics, the order of its elements needs to be considered for a complete description of a set. In this study, four kinds of orders for the component types in the line-line intersection are introduced so that the compound relations can be completely described and determined.

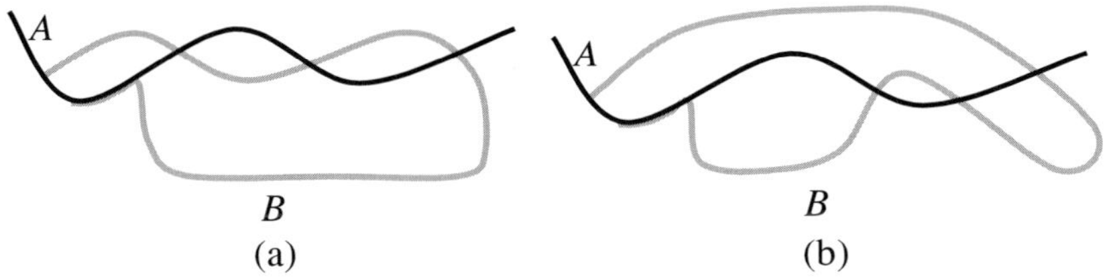

Fig. 7. Need of the order of component types in the line-line intersection

Basic order of component types (Oct) represents such an order that all the component types in the line-line intersection occur along the other line with one line used as a reference. It is a basic constraint for the combination of components, and also a topological invariant. The ordering starts from one endpoint of the reference line, e.g. *A*, and then traces all the components in sequence and records the components with numeric labels 0, 1, …, and *r*-1 (here $r = \chi(A \cap B)$). The next step is to record the numeric labels of components in order for the other line, i.e. *B*, also starting from one of its endpoints to the other. The order of component types (Oct) can then be defined as an alignment of numeric labels in occurrence order of the component types in the line *B*, which is represented as

$$Oct(B) = \left\langle c_0(t_0), c_1(t_1), \cdots, c_{r-1}(t_{r-1}) \right\rangle \quad (3)$$

In Equation (3), t_i ($0 \le i \le r-1$) denotes type of the ith component, and c_i ($0 \le i \le r-1$) denotes its numeric label. Due to the arbitrary selection of the starting endpoint of line *B*, Equation (3) can be equivalently represented as

$$Oct(B) = \left\langle c_{r-1}(t_{r-1}), c_{r-2}(t_{r-2}), \cdots, c_0(t_0) \right\rangle \quad (4)$$

Taking Figure 7a as example for illustration of the order of component types, where $\chi(A \cap B) = 5$. First, the components in line *A* are labeled with numeric numbers 0, 1, 2, 3 and 4 in the occurrence order. Next, the occurrence order of all the components in line *B* is 0, 2, 3, 4 and 1. As a result, the order of component types in the intersection between *A* and *B* is:

$$Oct(B) = \langle 0(p_c), 2(p_e), 3(p_e), 4(p_e), 1(l_b) \rangle$$

It can also be equivalently represented as

$$Oct(B) = \langle 1(l_b), 4(p_e), 3(p_e), 2(p_e), 0(p_c) \rangle$$

4.2 Incorporation of the Order of Characteristic Points

The characteristic points in $A \cap B$ also form a topological invariant and the order of them can be used to differentiate topological relations. The characteristics points can be computed as follows:

(i) Compute the difference set between set A and set B, i.e.

$$Diff_{AB} = A - B \tag{5}$$

(ii) Take the closure of $Diff_{AB}$, represented as $\overline{Diff_{AB}}$;

(iii) The set of characteristic points in $A \cap B$, denoted by S_{cp}, can be obtained by computing the difference between $Diff_{AB}$ and $\overline{Diff_{AB}}$, i.e.

$$S_{cp} = \overline{Diff_{AB}} - Diff_{AB} \tag{6}$$

The order of characteristic points can be obtained by using a similar approach to the order of components, and can be represented in such a form as

$$Ocp(B) = \langle cp_0, cp_1, \cdots, cp_n \rangle \tag{7}$$

Indeed, this kind of order is conducive to reduce the confusion of topological relations with different linkage. For example, the two spatial configurations in Figure 8 can be differentiated by order of characteristics points while they cannot be distinguished by the order of component types.

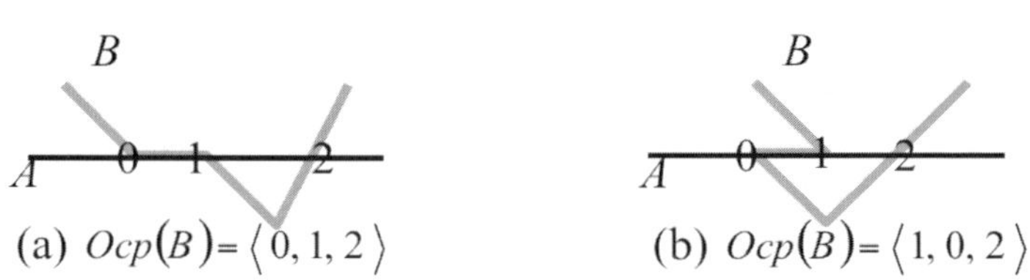

Fig. 8. Need for the order of characteristic points (where the orders of component types are same in (a) and (b), while the orders of their characteristic points are different)

4.3 Incorporation of the Orders of Loop Types and of their Linkage Relation

Only with the orders of component types and characteristic points as constraints, there is still a great freedom in the configuration of detailed topological relations. For example, Figure 9 shows two spatial configurations with different topological relations. However, they have completely identical orders of component types and characteristic points. To further differentiate topological relations, the order of linkage relations among all consecutive loops, which is based on the components in the line-line intersection, is utilized in this study.

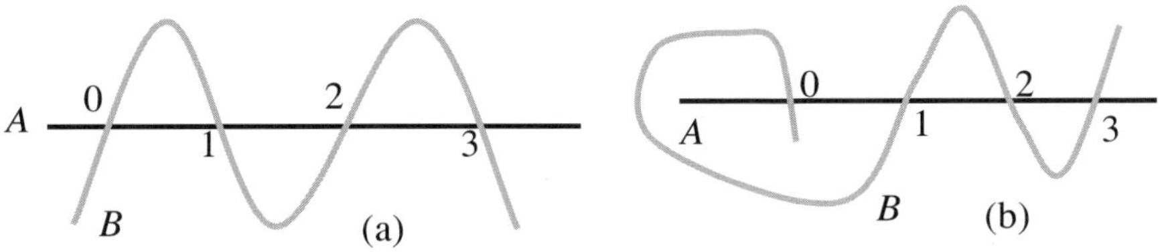

Fig. 9. Two spatial configurations with completely same orders of component types and characteristic points, i.e. $Oct(B)=\langle 0(p_e),1(p_e),2(p_e),3(p_e)\rangle$ and $Ocp(B)=\langle 0,1,2,3\rangle$, but in deed different in topology

Here, a loop is formed by adjacent components as well as the linking lines in *A* and *B*. Two types of loops can be distinguished: pure (or simple) loop (denoted as *pl*) and mixed loop (denoted as *ml*). The only difference between them is that the mix loop contains at least one of the endpoints. In addition, the number of the formed loops is one less than the number of components, i.e. equal to $r-1$. Therefore, the order of loop types (*Olt*) can de defined by

$$Olt(B)=\langle lt_0, lt_1,\cdots, lt_{r-2}\rangle \tag{8}$$

where lt_i ($1 \le i \le r-2$) denotes type of the ith loop.

Four kinds of linkage relations between consecutive loops are possible in terms of the region-region topological relations, i.e. *disjoint* (*d*), *meet* (*m*), *covers* (*c*), and *coveredby* (*cb*). Here each of the loops is regarded as its bounding region. Moreover, the number of linkage relations is one less than the number of the formed loops. Therefore, the order of the linkage relation (*Olr*) can be represented on the basis of the order of the loop types, i.e.

$$Olr(B)=\langle lr_0, lr_1,\cdots, lr_{r-3}\rangle \tag{9}$$

where lr_j $(1 \le j \le r-2)$ denotes linkage relation between the jth consecutive loops.

In summary, with the consideration of the four orders defined above, we can set up a compound relation model to describe and determine the topological relations between two lines in IR^2 as follows:

$$TR(A,B) = Oct(B) \cup Ocp(B) \cup Olt(B) \cup Olr(B) \tag{10}$$

With Equation (10), the topological relations of two spatial configurations in Figure 9 are

$TR(A,B) = \{ \langle 0(p_e), 1(p_e), 2(p_e), 3(p_e) \rangle \cup \langle 0,1,2,3 \rangle \cup \langle pl, pl, pl \rangle \cup \langle m, m \rangle \}$,

and

$TR(A,B) = \{ \langle 0(p_e), 1(p_e), 2(p_e), 3(p_e) \rangle \cup \langle 0,1,2,3 \rangle \cup \langle ml, pl, pl \rangle \cup \langle m, m \rangle \}$,

respectively.

5 An Example of Application

In GIS, a line object can be used to represent a river, a road, a pipeline, a contour, or a bus route. In this case study, we take bus routes as example. The effective design of bus routes has become an important traffic planning issue in urban traffic engineering. It is a premise to ensure balanced utilization of traffic resources and traffic service quality (e.g. convenience for transfer from one route to another). Indeed, this issue involves analysis of traffic information such as traffic carrying capacity, properties of roadway and spatial relations among the routes. Particularly, with speedy urbanization, the reconstruction and maintenance of bus routes would also have been necessary for the improvement of urban traffic, e.g. adding new routes, adjusting or removing some old routes. Figures 10(a) and (b) show Route 25 and Route 65 in 2002 and in 2005 in Nanjing city of China, respectively. Route 65 was adjusted in February 2003, and then Route 25 was adjusted in September 2004.

It can be found that the Route 25 has a clear change in route, and that a new configuration formed between Route 25 and Route 65 in 2005. As a result, the topological relation between the two routes in 2005 also differs from that in 2002. It is understood that the spatial relations between routes have been considered in the process of redesign of the routes. That is to say, it needs to query and analyze the topological relations between the existing routes, and between the existing routes and the redesigned routes.

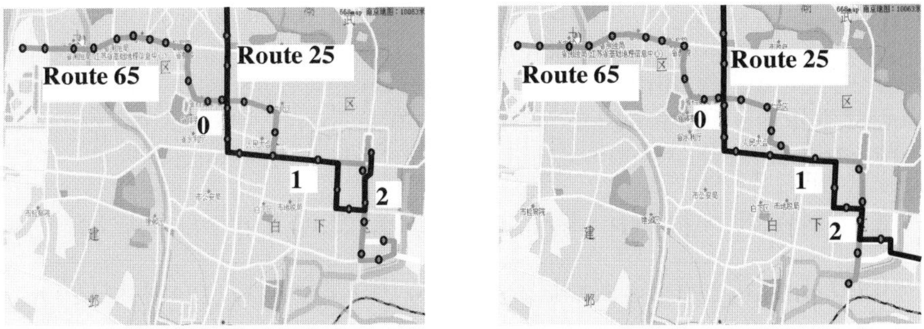

(a) Routes 25 and 65 in 2002 (b) Routes 25 and 65 in 2005

Fig. 10. A practical example about reconstruction of bus routes with the consideration of their spatial relations (modified from http://www.668map.com)

The following paragraphs are designed to illustrate how to hierarchically describe and determine the topological relations between two involved routes, i.e. route 25 and route 65. For convenience, A_1 and B_1 are used to represent the routes 25 and 65 in 2002, A_2 and B_2 to represent the routes 25 and 65 in 2005.

At first, line algorithms (e.g. intersection of lines) is used to perform intersection operation between two routes (Worboys and Duckham 2004), so that one can determine whether route A_1 (or A_2) intersects with route B_1 (or B_2). And then, all the components in the intersection will be computed and recorded. In Figure 10, three components occur between A_1 and B_1, and A_2 and B_2, i.e. $\chi(A_1 \cap B_1) = 3$ and $\chi(A_2 \cap B_2) = 3$, which are labeled by numbers 0, 1 and 2, respectively. From these three components, one may further know the topological information at two different levels. At first level, one knows $\dim(A_i \cap B_i) = 1$, and at a second level one knows that the relations between A_i and B_i (i=1, 2) can be decomposed into three basic relations.

The type of each component can be determined if needed, so that one can know that route A is *crossing* or *meeting* or *overlapping* route B in some intersection location. For a 0-dimensional component, its type can be determined by using basic segment intersection algorithms (Lee and Preparata 1984), in which the coordinates of its adjacent points in A_1 (or A_2) and B_1 (or B_2) are involved, while type of a 1-dimensional component depends on the combinatorial type of its two endpoints. According to definitions given in Figures 4 and 5, one can obtain types of the components between routes A_i and B_i (i=1, 2) in Figure 10 as follows:

$$\text{(a): } 0 \rightarrow p_e,\ 1 \rightarrow l_e,\ 2 \rightarrow l_f;\quad \text{(b): } 0 \rightarrow p_e,\ 1 \rightarrow l_e,\ 2 \rightarrow l_f$$

A complete order of all the components in the intersection between routes A_i and B_i (i=1, 2) can then be set if it is needed. They include order of component types, order of characteristic points, order of loop types, and order of linkage relations among consecutive loops. The results obtained by the compound relation model are respectively represented as:

$$\text{(a): } TR(A_1,B_1)=\left\{\left\langle 0(p_e),1(l_e),2(l_f)\right\rangle \cup \langle 0,1,2,4,3\rangle \cup \langle pl,pl\rangle \cup \langle d\rangle\right\}$$

$$\text{(b): } TR(A_2,B_2)=\left\{\left\langle 0(p_e),1(l_e),2(l_f)\right\rangle \cup \langle 0,1,2,3,4\rangle \cup \langle pl,pl\rangle \cup \langle d\rangle\right\}$$

These results can also be expressed in a table format as shown in Figure 11. From these tables one can clearly see the difference in the topological relations (i.e. between A_1 and B_1, and A_2 and B_2).

Ocp(B_1	Oct(B_1)	*Olt*(B_1)	*Olr*(B_1)
0	0(p_e)	*pl*	*d*
1	1(l_e)		
2		*pl*	
4	2(l_f)		
3			

(a)

Ocp(B_2)	*Oct*(B_2)	*Olt*(B_2)	*Olr*(B_2)
0	0(p_e)	*pl*	*d*
1	1(l_e)		
2		*pl*	
3	2(l_f)		
4			

(b)

Fig. 11. The description results of the topological relations between A_1 and B_1 in (a), and between A_2 and B_2 in (b)

6 Conclusions

In this paper, a hierarchical approach is proposed for the topological relations between two lines in IR^2. A set of topological invariants is developed based on the line-line intersection set, including separation number, dimension, component type and complete order of components. Based upon this approach, line-line topological relations can be described and determined hierarchically to meet need of topological information at various levels. A practical example is given for the illustration of the approach presented.

Indeed, in this paper,

- the hierarchical classification for the description of basic topological relations between two lines in IR^2 is comprehensive and complete. A total seventeen basic relations have been identified;

- the set of topological invariants are unrelated to the geometric measure (orientation, size, etc). Particularly, they can be used to classify topological relations in hierarchy;
- the approach used in this study is directly based on the line objects themselves, instead of the topological components. Therefore, there is no such problem as the inadequacy of adopting the boundary definition in IR^1 into IR^2.

It has been mentioned in the Introduction that spatial conflicts may be created during the updating, e.g. intersection, touching and/or overlap between rivers and roads. The model developed in this study can be used for automated detection of spatial conflicts. Indeed, the next step of development is to develop methodology for and automated resolution based on this model.

Acknowledgements

This work is supported by the Hong Kong Polytechnic University Research Grant under the grant number G-T873.

References

Abdelmoty AI, Williams HM (1994) Approaches to the representation of qualitative spatial relationships for geographic databases. In: Molenaar M, de Hoop S (eds), Advanced Geographic Data Modelling: Spatial data modelling and query languages for 2D and 3D applications. Delft, The Netherlands, pp 204–216

Bennett B (1997) Logical representations for automated reasoning about spatial relations. PhD Thesis, School of Computer Studies, University of Leeds

Chen J, Li C, Li Z, Gold C (2001) A Voronoi-based 9-intersection model for spatial relations. Int J of Geographical Information Science 15(3):201–220

Chen J, Zhao R, Li Z (2004) Voronoi-based k-order neighbour relations for spatial analysis. ISPRS J of Photogrammetry and Remote Sensing 59(1-2): 60–72

Clementini E, di Felice P, Oosterom Peter van (1993) A small set for formal topological relationships suitable for end-user interaction. In: Abel D, Chin Ooi B (eds), Advances in Spatial Databases. Springer-Verlag, New York, pp 277–295

Clementini E, Sharma J, Egenhofer MJ (1994) Modeling topological spatial relations: Strategies the query processing. Computer & Graphics 18(6):815–822

Clementini E, Di Felice P (1998) Topological invariants for lines. IEEE Transactions on Knowledge and Data Engineering 10:38–54

Cohn AG, Bennett B, Gooday J, Gotts NM (1997) Representation and reasoning with qualitative spatial relations about regions. In: Stock O (ed), Spatial and Temporal Reasoning. Kluwer Publishing Company, pp 97–134

Cui Z, Cohn AG, Randell DA (1993) Qualitative and topological relationships in spatial databases. In: Proc of SSD-93, pp 296–315

Egenhofer MJ (1993) A model for detailed binary topological relationships. Geomatica 47(3&4):261–273

Egenhofer MJ, Franzosa R (1991) Point-set topological spatial relationships. Int J of Geographical Information Systems 5(2):161–174

Egenhofer MJ, Herring J (1991) Categorizing binary topological relationships between regions, lines, and points in geographic databases. Technical report, Department of Surveying Engineering, University of Maine, Oronoi, ME

Egenhofer MJ, Franzosa R (1995) On the equivalence of topological relations. Int J of Geographic Information Systems 9(2):133–152

Güting R (1988) Geo-Relational algebra: A model and query language for geometric database systems. In: Schmidt J, Ceri S, Missikoff M (eds), Advances in Database Technology-EDBT' 88 (Proc of Int Conf on Extending Database Technology), Venice, Italy

Lee DT, Preparata FP (1984) Computational geometry: a survey. IEEE Transactions on Computers C-33,12:1072–1101

Li ZL, Li YL, Chen YQ (2000) Basic topological models for spatial entities in 3-dimensional space. GeoInformatica 4(4):419–433

Li ZL, Zhao RL, Chen J (2002) A Voronoi-based spatial algebra for spatial relations. Progress in Natural Science 12(7):528–536

Liu WZ, Chen J, Li ZL, Zhao RL, Cheng T (2005) Detection of spatial conflicts between rivers and contours in topographic database updating. In: Gold C (ed), The 4th Workshop on Dynamic and Multi-dimensional GIS, Pontypridd, Wales, UK, 5-8 September, pp 99–105

Randell DA, Cui Z, Cohn AG (1992) A spatial logical based on regions and connection. In: Kaufmann M, Mateo S (eds), Proc of 3rd Int Conf on Knowledge Representation and Reasoning, pp 165–176

Renz J, Nebel B (1999) On the complexity of qualitative spatial reasoning: A maximal tractable fragment of the region connection calculus. Artificial Intelligence 108(1–2):69–123

Worboys M, Duckham M (2004) GIS. A computing perspective, 2nd ed. *CRC* Press, London

Zhang M, Shi W, Meng LQ (2005) A generic matching algorithm for line networks of different resolutions. The 8th ICA Workshop on Generalization and Multiple Representation, A Coruna, 7-8 July, 2005

Coastline Matching Process Based on the Discrete Fréchet Distance

Ariane Mascret[1], Thomas Devogele[1], Iwan Le Berre[2], Alain Hénaff[2]

[1] Naval academy Research Institute (IRENav), Lanvéoc, BP 600, F-29240 Brest Naval, France; e-mail: {mascret, devogele}@ecole-navale.fr

[2] GEOMER Laboratory, LETG UMR 6554 CNRS, Institut Universitaire Européen de la Mer (UBO), Technopôle Brest-Iroise, F-29280 Plouzané, France; e-mail: {iwan.leberre, alain.henaff}@univ-brest.fr

Abstract

Spatial distances are the main tools used for data matching and control quality. This paper describes new measures adapted to sinuous lines to compute the maximal and average discrepancy: Discrete Fréchet distance and Discrete Average Fréchet distance. Afterwards, a global process is defined to automatically handle two sets of lines. The usefulness of these distances is tested, with a comparison of coastlines. The validation is done with the computation of three sets of coastlines, obtained respectively from SPOT 5 orthophotographs and GPS points. Finally, an extension to Digital Elevation Model is presented.

Key words: data fusion, quality control, data matching, Fréchet distance, coastline monitoring

1 Introduction

Computing the distance between two objects is a basic tool of geographic information systems. The most commonly used distance is the Euclidean distance (d_E) between two points. Others geometries (line and area in a two dimensional Cartesian system) need additional measures. In daily life, the notion of distance stand for the minimal effort required to reach one place from another. For example, the minimal distance between a pipeline and a river is the Euclidean distance between the two closest points from the river and the pipeline. Mathematically, a distance verifies three properties: non-negative, symmetry, triangle inequality. Thus, minimal distances that measure the distance between the closest points of geometries, can be completed by other distances like average distances or maximal distances. These lasts measure the average or maximal Euclidean distance between points of both geometries.

Maximal and average distances are useful to control or match data from different datasets. For example, in a quality control, they give the discrepancy between the encoded location and the location as defined in the specification (Veregin 1999). Likewise during the matching process, those measures permit to identify sets of data representing the same real world phenomenon in different data sets (Devogele et al. 1996).

Two different maximal distances are employed to calculate the maximal gap between lines: the Hausdorff distance and the Fréchet distance. The Hausdorff distance is the most popular maximal distance between two lines (L_1, L_2) (Deng et al. 2005) (Alt and Godau 1995). The Hausdorff distance (d_H) is defined as follows:

$$d_H(L_1, L_2) = \max\left(\sup_{p_1 \in L_1} \inf_{p_2 \in L_2} (d_E(p_1, p_2)), \sup_{p_2 \in L_2} \inf_{p_1 \in L_1} (d_E(p_1, p_2)) \right) \tag{1}$$

A line is an ordered set of points. Unfortunately, the Hausdorff distance does not take into account this property. Two lines can have a small d_H, without being similar each other at all. The inconvenient of the Hausdorff distance is the computation of Euclidean distance between closer points and not between homologous points (points, which can be visually matched). Hence, Hausdorff distance cannot be used for sinuous lines. For this kind of lines, the Fréchet distance is more appropriated (Alt and Gadau 1995).

In the maritime context, the majority of the lines, like coastlines or isolines, used for making studies, are sinuous. The aim of this paper is to detail how calculate the Fréchet distance, measure the average discrepancy for those kind of lines, and illustrate the result with a comparison between coastlines. This last part is realized thanks to the implementation of new methods based on Average Fréchet distance.

The remainder of this paper is as follows. Section 2 describes the discrete Fréchet distance, which is a good approximation of the Fréchet distance. In Section 3, this discrete distance is extended to introduce an average linear distance: the average Fréchet distance. Moreover it also explains how to compute this measure. A global process defined to match homologous objects from two datasets is proposed in Section 4. Section 5 illustrates this matching process and these two distances by a real example of quality control on coastlines datasets. Related works on digital elevation model are described and discussed in Section 6 and finding are summarized in Section 7.

2 Discrete Fréchet Distance

The Fréchet distance is the maximal distance between two oriented lines. Each oriented line is equivalent to a continuous function f: [a, a']→V (g: [b, b']→V) where a, a', b, b' ∈ $\Re$, a < a' (b<b') and (V, d) is a metric space. d_F denotes their Fréchet distance defined as follows:

$$d_F(f,g) = \inf_{\substack{\alpha:[0,1]\to[a,a'] \\ \beta:[0,1]\to[b,b']}} \max_{t\in[0,1]} d(f(\alpha(t)), g(\beta(t))) \quad (2)$$

Let us give an illustration of the Fréchet distance: a man is walking with a dog on a leash. This man is walking on the one curve, the dog on the other one. Both may vary their speed, but backtracking is not allowed. Then the Fréchet distance of the curves is the minimal length of a leash that is necessary. The Fréchet method has the advantage of computing distances only on homologous points and not between closest points as for the Hausdorff distance.

Eiter and Mannila (1994) gave an approximation: the discrete Fréchet distance (d_{dF}) that computes in time O(n m). L_1 and L_2 are interpreted as two oriented finite sets of points: $<L_{1.1}...L_{1.n}>$ and $<L_{2.1}...L_{2.m}>$. d_{dF} is the minimal length of leash such as a way from the pair of beginning points $(L_{1.1},L_{2.1})$ to the pair of ending points $(L_{1.n},L_{2.m})$ is possible. The path gives an ordered set of $(L_{1.i},L_{2.j})$ such as the following pair of $(L_{1.i},L_{2.j})$ is one of these three pairs:

- $(L_{1.i+1},L_{2.j+1})$ man and dog are walking,
- $(L_{1.i+1},L_{2.j})$ only the man is.
- $(L_{1.i},L_{2.j+1})$ only the dog is.

These sets of points include end points of line segments (vertices). Some points on a line segment can also be integrated by resampling these sets.

d_{dF} is a good estimation of d_F because the approximation is limited by the maximal distance between two consecutive points (LengthMaxSeg) (Eiter and Mannila 1994):

$$d_F(L_1, L_2) \leq d_{dF}(L_1, L_2) \leq d_F(L_1, L_2) + \text{LengthMaxSeg} \quad (3)$$

In order to limit this approximation to ε, a resampling can be applied to both lines.

The discrete Fréchet between L_1 and L_2 can be computed recursively as follows:

$$d_{Fd}(L_1, L_2) = \max\left(\begin{array}{l} d_E(L_{1.n}, L_{2.m}) \\ \min\left(\begin{array}{l} d_{Fd}(< L_{1.1}...L_{1.n-1} >, < L_{2.1}...L_{2.m} >) \forall n \neq 1 \\ d_{Fd}(< L_{1.1}...L_{1.n} >, < L_{2.1}...L_{2.m-1} >) \forall m \neq 1 \\ d_{Fd}(< L_{1.1}...L_{1.n-1} >, < L_{2.1}...L_{2.m-1} >) \forall n \neq 1, m \neq 1 \end{array} \right) \end{array} \right) \quad (4)$$

$<L_{1.1}\ldots L_{1.n-1}>$ and $<L_{2.1}\ldots L_{2.m-1}>$ represent lines. Hence, it is possible to recursively apply this d_{dF} process with parameters: $<L_{1.1}\ldots L_{1.n-1}>$, $<L_{2.1}\ldots L_{2.m-1}>$. This process is terminated when the two lines are reduced to two single points ($<L_{1.1}>$, $<L_{2.2}>$).

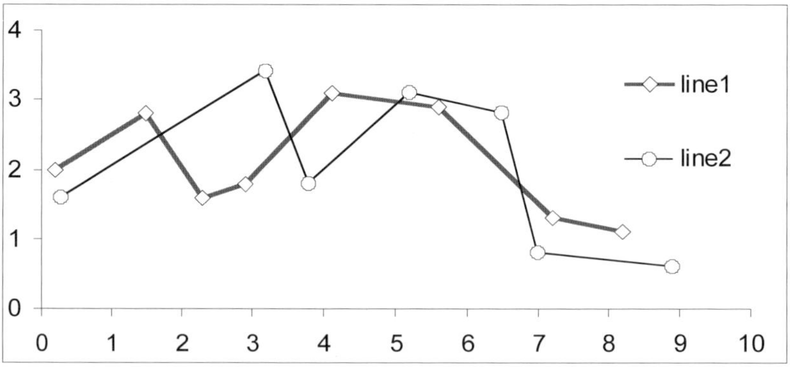

Fig. 1. Example of a couple of lines

The example presented in Figure 1, illustrates the computing of the discrete Fréchet distance. The line 1 (L_1) is composed by 8 vertices ($L_{1.1}$ to $L_{1.8}$) and the line 2 (L_2) is composed by 7 vertices ($L_{2.1}$ to $L_{2.7}$).

The computing of the two matrices replaced the recursive process. The dimension of these two matrices is n × m (see Table 1), where n and m are the number of vertices of L_1 and L_2. These matrices are:

- The matrix of Euclidean Distance (MD). The value of the cell $MD_{i,j}$ is the distance between $L_{1.i}$,$L_{2.j}$.
- The Fréchet matrix (FM) which allows to calculate iteratively the Fréchet distance (Eiter et Mannila 94). The formula to compute $MF_{i,j}$ is:

$$MF_{i,j} = \max\left(d_E(L_{1.i},L_{2.j}),\ \min(MF_{i-1,j}\ ,\ MF_{i,j-1}\ ,\ MF_{i-1,j-1})\right) \tag{5}$$

The discrete Fréchet distance is the value of $MF_{n,m}$. Table 1 gives the d_{dF} between the two lines of Figure 1: 1.8. d_{dF} is equal to the Euclidean distance between $L_{1.2}$ and $L_{2.2}$. These two points are homologous. For these two lines, the Hausdorff distance is smaller and is not significant in term of matching.

Table1 present a simple sample. The partial discrete Fréchet distance (d_{pdF}), with two smaller matrices (7×8) is computed in this table. If a re-sampling is processed with LengthMaxSeg equal to 0.1, two matrices (107×129) are computed and d_{pdF} is equal to 1.2.

Table 1. Matrix of Euclidean Distance and Fréchet Matrix for lines of Figure 1

L1.i.x	0.2	1.5	2.3	2.9	4.1	5.6	7.2	8.2
L1.i.y	2	2.8	1.6	1.8	3.1	2.9	1.3	1.1

L2.j.x	L2.j.y
0.3	1.6
3.2	3.4
3.8	1.8
5.2	3.1
6.5	2.8
7	0.8
8.9	0.6

Matrix of Euclidian distance between (L1.i, L2.j)

	1	2	3	4	5	6	7	8
1	0.41	1.70	2.00	2.61	4.09	5.46	6.91	7.92
2	3.31	**1.80**	2.01	1.63	0.95	2.45	4.52	5.50
3	3.61	2.51	1.51	0.90	1.33	2.11	3.44	4.46
4	5.12	3.71	3.26	2.64	1.10	0.45	2.69	3.61
5	6.35	5.00	4.37	3.74	2.42	0.91	1.66	2.40
6	6.91	5.85	4.77	4.22	3.70	2.52	0.54	1.24
7	8.81	7.72	6.68	6.12	5.41	4.02	1.84	0.86

Fréchet Matrix

	1	2	3	4	5	6	7	8
1	0.41	1.70	2.00	2.61	4.09	5.46	6.91	7.92
2	3.31	1.80	2.01	2.00	2.00	2.45	4.52	5.50
3	3.61	2.51	1.80	1.80	1.80	2.11	3.44	4.46
4	5.12	3.71	3.26	2.64	1.80	1.80	2.69	3.61
5	6.35	5.00	4.37	3.74	2.42	1.80	1.80	2.40
6	6.91	5.85	4.77	4.22	3.70	2.52	1.80	1.80
7	8.81	7.72	6.68	6.12	5.41	4.02	1.84	**1.80**

New Measures derived from discrete Fréchet distance are introduced in (Devogele 2002):

- The partial discrete Fréchet distance (d_{pdF}). This measure is useful to match a line L_1 with a part of another line L_2. d_{pdF} detects the partial homologous line $< L_{2.begin}\ldots L_{2.end} >$ and computes d_{pdF}. d_{pdF} is equal to $d_{dF}(L_1, < L_{2.begin}\ldots L_{2.end} >)$. Figure 2a shows a case where the computing of d_{pdF} is necessary.

- The discrete Fréchet distance between 2 polygon borderlines. The process defines a function T to translate polygon borderlines P_1 and P_2 into lines L_1 and L_2 such as the d_{Fd} between L_1 and L_2 is minimal. This process can also inverse the ordering of points. For example, d_{dF} between the two polygon borderlines of Figure 2b can be measured. No partial discrete Fréchet distance between 2 polygons borderlines are defined.
- The partial discrete Fréchet distance (d_{pdF}) between a line L_1 and a part of polygon borderlines P_2. This measure is a mix of the two first measures. Figure 2c shows an example where the d_{pdF} can be computed between a line and a polygon borderline.

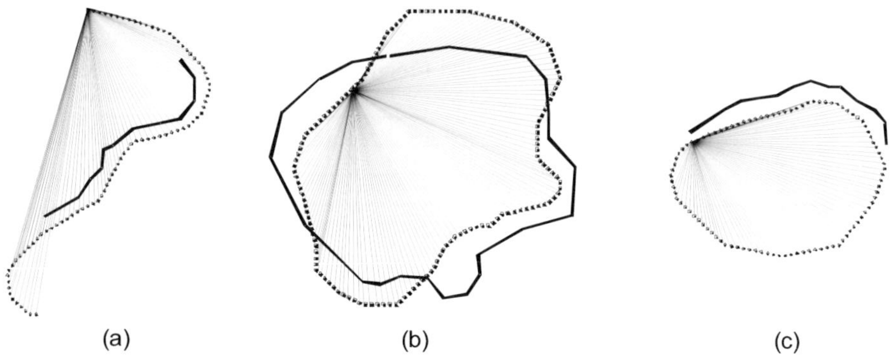

Fig. 2. Examples of interesting pairs of geometries for the computation of measure derived from the Fréchet distance

3 Average Fréchet Distance

A new distance is defined from discrete Fréchet distance: the average Fréchet distance (d_{aF}). d_{aF} is the average Euclidean distance between points of pairs, which is based on the minimum path (MP).

As pre-processing, the path between pairs of points ($L_{1.1}$, $L_{2.1}$) and ($L_{1.n}$, $L_{2.m}$) is computed. This one is MP, compatible with discrete Fréchet distance. Several paths of the man and the dog with a length of leash equal to Fréchet distance are possible. MP is the one where the man and the dog walk one their curves but choose to be as closer as possible. The two matrices MD and FM and an inferior or equal (<=) operation between two pair of real numbers are used to compute MP.

This operation <= is defined as follow:

$$\begin{gathered}(a,b) \text{ and } (c,d) \in \Re^2 \\ (a,b) <= (c,d) \text{ if } a < c \text{ or if } a == c \text{ and } b <= d\end{gathered} \tag{6}$$

The minimal path is constructed by backtracking through the matrix. So the last pair ($L_{1.n}$, $L_{2.m}$) is added, while the previous one look for pair help to <= operation. For ($L_{1.i}$, $L_{2.j}$), three previous candidate pairs are possible: ($L_{1.i-1}$, $L_{2.j-1}$), ($L_{1.i-1}$, $L_{2.j}$), ($L_{1.i}$, $L_{2.j-1}$). In order to chose the previous pairs of points, an associated pairs of real $C_{i,j}$ is defined, where $C_{i,j}$ is equal to (FM($L_{1.i}$,$L_{2.j}$), MD($L_{1.i}$,$L_{2.j}$)). The candidate pair, where the associated pair of real is inferior or equal to the two other real pair, is chosen. This construction is finished when i and j equal to 1.The algorithm is given by Figure 3.

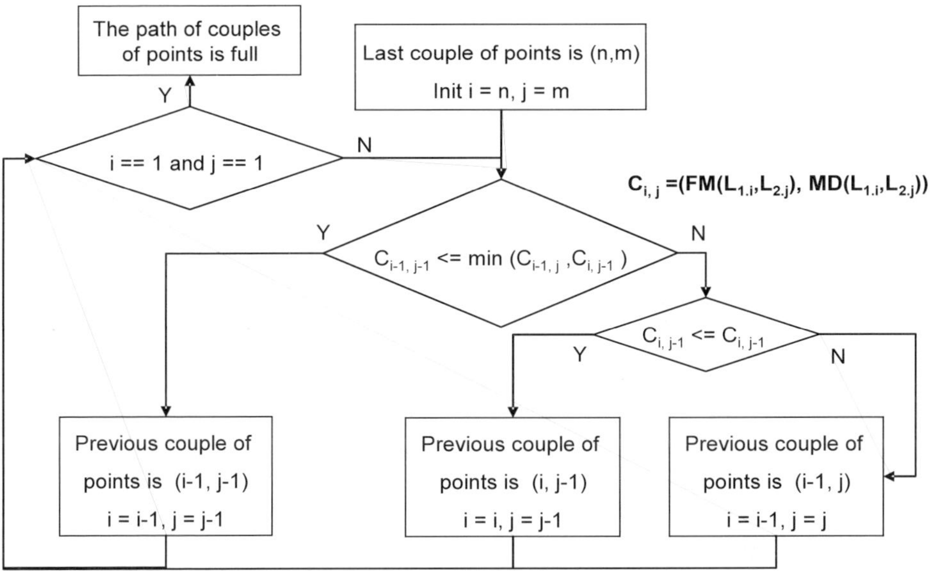

Fig. 3. Algorithm for computing the minimal path (MP)

In the example of Figure 1, the minimal path is represented by the pairs of real in the grey cells of Table 2. After added the last couple ($L_{1.7}$, $L_{2.8}$), the previous couple ($L_{1.6}$, $L_{2.7}$) is chosen because FM($L_{1.6}$,$L_{2.7}$) is inferior or equal to both FM($L_{1.7}$, $L_{2.7}$) and FM($L_{1.6}$, $L_{2.8}$). Indeed (1.80, 0.54) is inferior or equal to (1.84, 1.84) and (1.80, 1.24). This process is reused until i and j equal 1. So the minimal path is the order set of nine couples: ($L_{1.1}$, $L_{2.1}$), ($L_{1.2}$, $L_{2.2}$), ($L_{1.3}$, $L_{2.3}$), ($L_{1.3}$, $L_{2.4}$), ($L_{1.4}$, $L_{2.5}$), ($L_{1.4}$, $L_{2.6}$), ($L_{1.5}$, $L_{2.6}$), ($L_{1.6}$,$L_{2.7}$) , ($L_{1.7}$,$L_{2.8}$). Figure 4 shows couples of ($L_{1.i}$, $L_{2.j}$) associated with minimal path. Points from pairs are homologous.

Table 2. Minimal path from ($L_{1.1}$, $L_{2.1}$) to ($L_{1.n}$, $L_{2.m}$) is defined by selecting the couple of real in grey cells. In cell i, j, the first number is FM($L_{1.i}$, $L_{2.j}$) and the second one is MD($L_{1.i}$, $L_{2.j}$)

	1	2	3	4	5	6	7	8
1	0.41 0.41	1.70 1.70	2.00 2.00	2.61 2.61	4.09 4.09	5.46 5.46	6.91 6.91	7.92 7.92
2	3.31 3.31	1.80 1.80	2.01 2.01	2.00 1.63	2.00 0.95	2.45 2.45	4.52 4.52	5.50 5.50
3	3.61 3.61	2.51 2.51	1.80 1.51	1.80 0.90	1.80 1.33	2.11 2.11	3.44 3.44	4.46 4.46
4	5.12 5.12	3.71 3.71	3.26 3.26	2.64 2.64	1.80 1.10	1.80 0.45	2.69 2.69	3.61 3.61
5	6.35 6.35	5.00 5.00	4.37 4.37	3.74 3.74	2.42 2.42	1.80 0.91	1.80 1.66	2.40 2.40
6	6.91 6.91	5.85 5.85	4.77 4.77	4.22 4.22	3.70 3.70	2.52 2.52	1.80 0.54	1.80 1.24
7	8.81 8.81	7.72 7.72	6.68 6.68	6.12 6.12	5.41 5.41	4.02 4.02	1.84 1.84	1.80 0.86

Other processes have been described to define others minimal paths. In order to reconstruct three-dimensional solid from serial sections, Fuchs et al. (1977) propose to find the minimum cost cycles in a directed toroidal graph. To compute the minimum cost cycles, the matrix of Euclidean distance is transformed into a graph and a Dijkstra's Algorithm (Dijkstra 59) is employed to find the shortest path from the vertex ($L_{1.i}$, $L_{2.j}$) to the vertex ($L_{1.n}$, $L_{2.m}$).

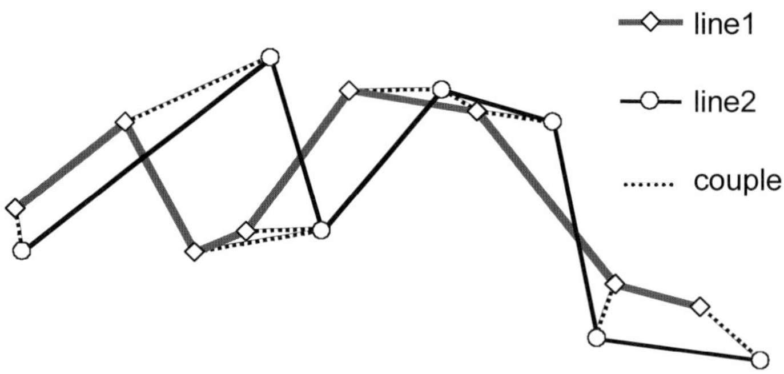

Fig. 4. Couple of ($L_{1.i}$,$L_{2.j}$) of the minimal path represented by dot lines

This method was translated in other domains such as 2D objects morphing (Sederberg and Greewood 1992). These methods minimize the sum of Euclidean distances between the points of pairs. Thus, a large distance can be chosen if the other distances are small. For computing average distance, MP has the advantage over these last paths to not select pair of points with large distance.

A few softwares (TCI 2005) (Geomod 2005) also employ list of pairs of homologous points as pre-processing for rubber sheeting. Nevertheless, these pre-processing are semi-automatic and each point can only appear in one couple.

The average Fréchet distance (d_{aF}) used the MP. In the example of Figure 1, the d_{aF} is equal to 0.95 and with a resampling pre-process with 0.1 as LengthMaxSeg, d_{aF} is equal to 0.51. Table 3 complete the example, showing that resampling is necessary to obtain a value for d_{Fd} closed to a value of d_F. Moreover the approximation is inferior to the one of LengthMaxSeg.

Table 3. d_{dF} and d_{aF} computed with different values for LengthMaxSeg

	without	**1**	**0.1**	**0.01**	**0.001**
$\mathbf{d_{dF}}$	1.8028	1.2260	1.2015	1.2012	1.2012
$\mathbf{d_{aF}}$	0.9524	0.5843	0.5116	0.5030	0.4997

The average Fréchet distance is an accurate measure to match data, to control quality and to merge data. For quality control, the d_{aF} with resampling, is an appropriate measure of the average discrepancy of lines whatever sinuosity.

4 Global Process

The previous methods allow computing measures (d_{dF} and d_{aF}) between homologous lines but a global process is required to work with sets of lines. Among all couple of lines, the program selects pair of lines: L_1 from the first set and L_2 from the second set as the distance between them could be inferior to a maximal distance (MaxDist). This process is divided in three steps (see Fig. 6).

For the first step, the process makes a query on the line's "open" attribute to distinguish three different cases for matching:

- total or partial matching between open lines,
- partial matching between open line and a close one,
- total matching between close lines.

For the second step, the process determines if lines can be homologous. This step used two bounding-boxes for each line (shows Figure 5):

- BB_i: The bounding-box of L_i: BB_i.
- BBE_i: The enlarged bounding-box of L_i. This rectangle is the BB_i enlarged to contain all points that distance is inferior to MaxDist to BB_i.

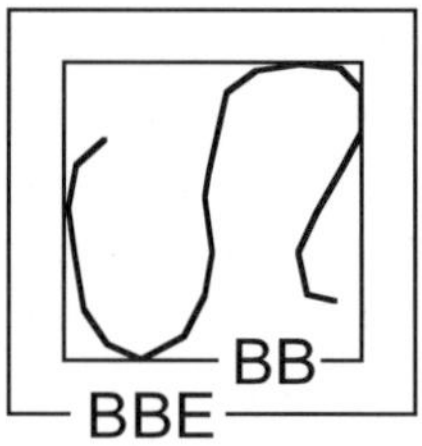

Fig. 5. Bounding-box (BB) and enlarged bounding box (BBE) of one line

If the bounding-box of L_1 is included in the Enlarge bounding-box of L_2, consequently the Fréchet distance between those two lines can be inferior to the MaxDist previously defined. Hence there is a real probability for these lines to be homologous.

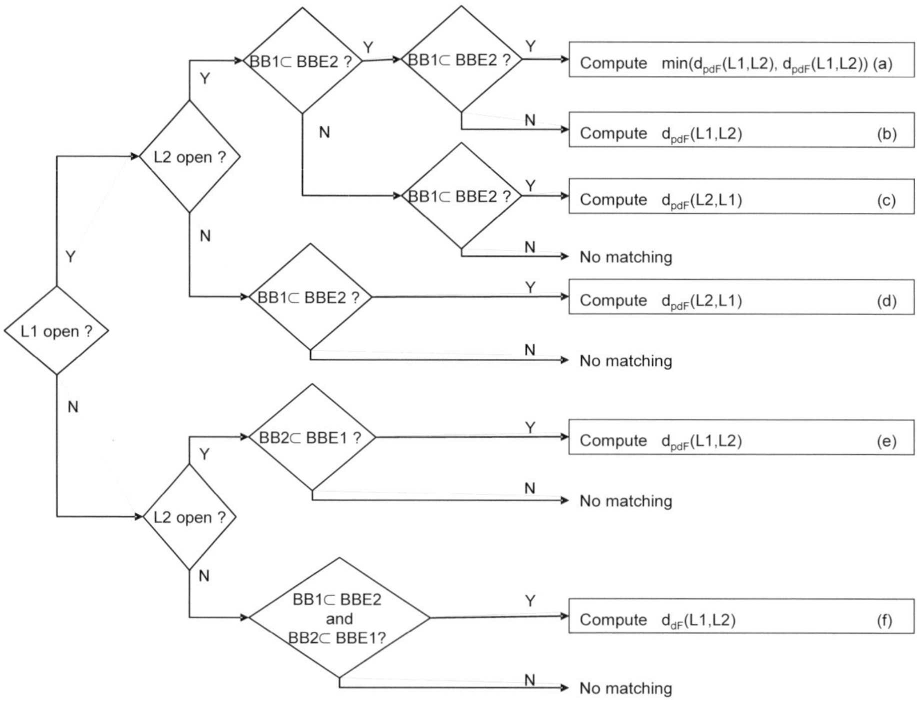

Fig. 6. Global process algorithm for each pair of (L_1, L_2)

The most common and tricky case is when a line is included in the enlarge bounding-box of another (see Fig. 6, case (b) or (c)) but the reverse is false. This means that lines are homologous but one is shorter, a partial matching process between them is computed. If the reverse is true (see Fig. 6, case (a)), the two partial Fréchet distances are calculated

($d_{pdF}(L_1,L_2)$, $d_{pdF}(L_2,L_1)$) and the minimal distance is kept. The total matching is considered as a sub-set of a partial matching.

In the case of a partial matching between one open line and one close line (see Fig. 6, case (d) or (e)), the close line is cut according to the open line, then the process calculates the Fréchet distance.

For two close lines (see Fig. 6, case (f)), if one BB is include in other BBE, the discrete Fréchet distance between two polygons borderlines is computed.

Finally the global process selects only the homologous lines when their Fréchet distances are inferior to MaxDist. The result is a set of pairs of matching lines with their Fréchet distances (average and discrete) and a list of homologous points.

5 Example of Coastline Matching Process

Within the framework of CNES-IFEN (French space agency and French environmental Institute) littoral monitoring, Le Berre et al. (2004) has tested the capabilities of SPOT 5 data as a relevant tool for coastal zone mapping and coastline updating. The aim of the project was to assess the potentialities of a high resolution sensor (satellite SPOT 5) to delineate a reference coastline used in many coastal applications like offshore dynamic monitoring, protection works against sea erosion or coastal landcover mapping.

In addition to their visual interpretation, the measure derives from Fréchet distance complete the study by giving a quantitative evaluation of the distance between different digitized coastlines.

The shoreline is still an ambiguous concept despite a common use as a reference boundary between sea and land. Indeed, a gradual change on both sides and a permanent evolution during time don't allow defining an accurate and permanent boundary. The coastline is defined by IHO (International Hydrographic Organization 2005) as the line where shore and water meet. Although the terminology of coasts and shores is rather confused, shoreline and coastline are generally used as synonymous.

For marine application, the coastline is defined as the coast limit reached by the highest level of water (high seasonal tide). For the SHOM (French Marine Hydrographic and Oceanographic service) or Ifremer (French Institute for marine research) it is the conventional limit of the coastal domain at the neighborhood of the High water line (Coastchart project 2004). A theoretical definition could be either the highest astronomical tide or the extreme level high water limit on a period from 10 years.

From another point of view, one of geomorphological definitions is "morphological discontinuity area where the sea reaches the coast". Thus using sedimentary, morphological and botanical features, it's possible to avoid the problem of tide and therefore to use remote sensing data to map the coastline.

This last definition is chosen to digitize the shoreline from SPOT and orthophotographs data using together vegetation limits on cliffs, sand dunes and schorres, the foot sand dune, the erosion slope, the beach vegetation boundary, or the high water spring tide mark.

5.1 Datasets and Methodology

The site of the experimentation is located in the North-West of Brittany (France). It has been chosen according to the availability of reference datasets along with its coastline diversity: cliffs of various heights and rocks (soft or hard), beaches and sand dunes, tidal flats, estuaries, and artificial coast.

The digitization of the shoreline is both based on SPOT 5 image and orthophotographs data from BD ORTHO® (IGN 2003):

- The satellite image dated 2003/04/17 has 2.5 m resolution with a multispectral band (THR + XS) during low neap tide (tidal range: 114);
- The BD ORTHO® is produced by the French National Mapping Agency (IGN) with aerial photography shot at a 1:25,000 scale in June 2000. They are geometrically and orthogonally corrected with a Digital Elevation Model of the natural ground only (and not the superficial relief). The final product is a real colour picture, with a 50 cm spatial resolution and can be use at a scale of above 1:1,000.

In addition, several detailed topographical surveys of shoreline section were made in order to compare the digitized shoreline to reference lines.

These surveys were made at the same period than the Spot image acquisition with a laser tacheometer or with a differential Global Positioning System (GPS), with a precision close to centimeter, on various parts of the coast (namely sand dunes, low height soft cliff, artificial coast, cobble ridge and shore). The station position was determined with georeferenced positioning points and the topographical survey fit with the plot of the shoreline inflexion point layout in a plan.

As described previously, the Hausdorff distance is not accurate for sinuous lines like coastline (see Fig. 7), so Fréchet distance is used for this numerical comparison.

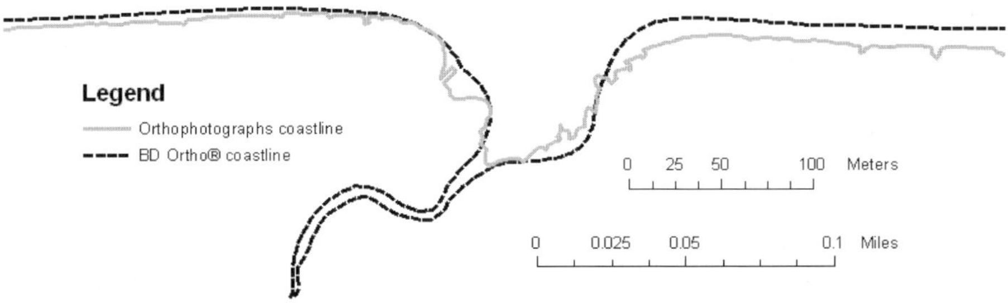

Fig. 7. Illustration of two different coastlines in estuary area

5.2 Results

All the lines are resampled with a range equal to 0.5 meter. Empirically the distance accuracy is about 10 cm. The next tables summarize the results: in each cell, the first number is the discrete Fréchet distance and the second one between brackets is the average Fréchet distance, in meter.

For the digitalization two scales were employed: 1:1,500 (SPOT 5 1500) and 1:6,000 (SPOT 5 6000).

Table 4. Results for Artificial area

	GPS	Ortho
SPOT 5 1500	14.01 (3.58)	15.75 (3.70)
SPOT 5 6000	6.26 (2.37)	10.49 (2.67)
Ortho	7.37 (1.77)	×

Table 5. Results for Cobble ridge

	GPS	Ortho
SPOT 5 1500	3.39 (1.45)	5.13 (2.78)
SPOT 5 6000	4.18 (2.01)	6.10 (3.28)
Ortho	5.08 (1.82)	×

Table 6. Results for Vegetated cliffs

	GPS	Ortho
SPOT 5 1500	10.84(2.77)	12.20 (4.25)
SPOT 5 6000	11.98 (4.54)	10.32 (6.07)
Ortho	7.25 (2.8)	×

Table 7. Results for Shore area

	GPS	Ortho
SPOT 5 1500	5.20 (1.22)	4.93 (1.67)
SPOT 5 6000	3.90 (0.96)	4.97 (1.47)
Ortho	5.38 (1.41)	×

Table 8. Results for Cliffs top

	tacheometer	Ortho
SPOT 5 1500	3.78 (1.62)	3.84(1.10)
SPOT 5 6000	6.07 (3.34)	6.43 (2.62)
Ortho	3.10 (1.03)	×

Table 9. Results for Beach shoreline

	tacheometer	Ortho
SPOT 5 1500	3.79 (1.17)	10.22(2.37)
SPOT 5 6000	6.08 (2.25)	12.33 (4.14)
Ortho	8.64 (2.1)	×

First of all, for all tables two kinds of discrete Fréchet distance discrepancies are present: one upper to 10 meters (see Table 4 and Table 6), and a second in average of 6 meters.

The discrepancy of ten meters could be explained either by a difference of interpretation between images and reality or changes in morphology (circle in Fig. 8). The second is mostly due to a problem of data resolution (see Fig. 8). Indeed, the digitization of points itself is an uncertain process. Even though we choose a precise scale to get the points, it's impossible to be sure that the coordinates are correct (Harvey and Vauglin 1996).

In artificial area (see Table 4) the surprising best digitalization at 1:6000 than SPOT 5 1500 is only a consequence of a difference of interpreted features. Despite that, the best support remains orthophotographs (25% better than SPOT 5) due to its resolution, which allow a better identification of build up areas.

The good results of average distance in "shore area" with satellite image (see Table 7) may be due to an evolution of the shore between 2000 and 2003, the dates the orthophotographs and the Spot image have been acquired. The same comment can be done for the cobble ridge (see Table 5) and especially for the sand dune results (see Table 9).

The measures obtained for cliffs are not homogeneous: Table 6 gives a bigger Fréchet distance than Table 8. For "vegetated cliffs" the interpretations is disturbed, both with SPOT 5 or the orthophoto, by the vegetation that can mask the location of the *de facto* coastline. In comparison, "cliffs top" results (see Table 8) show that orthophotographs are most convenient for the shoreline digitization.

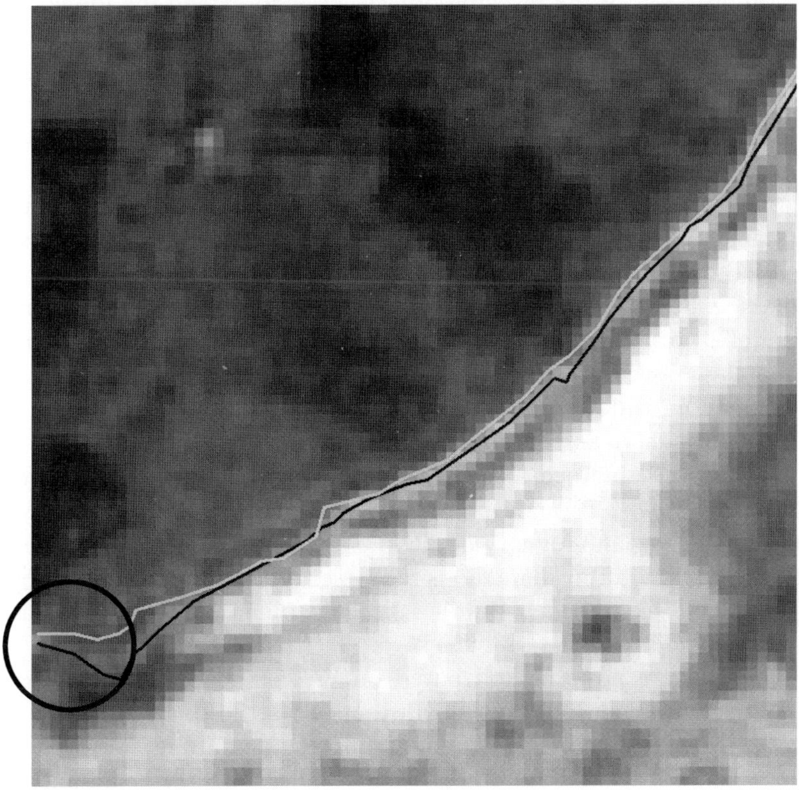

Fig. 8. Illustration of two kinds of errors: maximum Fréchet distance and average Fréchet distance. In the circle, the Fréchet distance between lines is bigger than between the reminder

To summarize, the global process application had demonstrated that SPOT 5 and the orthophotographs may be the support of the digitizing of coastlines with comparable planimetric accuracy. Actually, these quantified results were really surprising as we felt during the whole experimentation process that the identification of coastlines was more difficult on Spot 5 according to its lower spatial resolution. Indeed the results given by the program of discrete Fréchet distance show that SPOT 5 data are a relevant support for short term shoreline monitoring. This method can also be useful for other types of data matching. For example, the processes concerning the matching of different networks (road, hydrological, electrical) presented in (Mustière 2006) could be improved by integrating this method.

6 Discussions

An important point of the research objectives is to extend the global process to the 3D. Thus future works tends to develop an integration matching method whatever the type of the Digital Elevation Model (DEM) is. The main issue is to solve the question of matching two DEMs in coastal area: by using matching surfaces between them, or forced lines obtained with a DEM enhancement.

In the context of seamless elevation model integration previous studies give a start point. In coastal domain, Gesch and Wilson (2001) propose an ad-hoc method, which first converts each DEM in the same common vertical reference. Then, after remove false old bathymetric points, a raster surface model is produced from topographic and bathymetric points in the zero area elevation. Finally they merge the two first DEMs together using the third one to avoid interpolation edge effects.

In the continental area, Podobnikar (2005) proposes to average each DEM cell. As the obtained DEM is smoother than the input DEM, he enhances the result with geomorphological feature such as landmarks, hydrological network, or land registry points.

The problems with these methods are first a partial or total overlapping is required and second they didn't take care of an eventual planimetric shift between cells.

To solve that, a new method based on Fréchet matching must be defined. One of the most interesting contributions could be to process a geomorphological enhancement after landscape segmentation in order to obtain morphological forced lines. The landscape segmentation uses taxonomy of different land types: rock or reef formation, beach, estuaries, build up areas, and so on. For each type, main features of these relief elements (equivalent to "forced line") are determined as ridge, thalweg, line of slope break, roughness.

Then the homologous force lines are identified and matched on each DEM. Finally, the distorted DEM are merged.

This method gives the advantage of merging different coordinate cells with the consideration of the type of landscape. So it's equivalent as making an integration with adapted enhancement and merging method with a planar control, taking care planimetric shifts and not only according the altimetric shift.

7 Conclusions

In the context of data matching and quality control, the Fréchet distance, where a good approximation is the discrete Fréchet distance (d_{Fd}), is a relevant tool for measure differences. A new distance is also defined from it: The Average Fréchet Distance (d_{aF}). While the d_{Fd} represents the maximal gap somewhere between two homologous points of lines, d_{aF} gives the average difference whatever sinuosity. These two measures are complementary to control lines quality. Furthermore, a global process is implemented in order to automatizes the matching process between two datasets.

The coastline matching process was done with the discrete Fréchet distance, as it's the most appropriate tool to find sinuous homologous line, instead of Hausdorff distance.

In order to determine if SPOT 5 is a relevant tool for coastline mapping, we have implemented a program based on this method. The program gives a quantitative methodology to compare two chosen coastlines. The measures demonstrate that there is a good accuracy between SPOT data and reference data, despite a lower spatial resolution and temporal mismatch.

As theses processes are appropriate to 2D data, further developments will extend the matching to 3D data, according to the type of relief in the coastal area.

References

Alt H, Godau M (1995) Computing the Fréchet distance between two polygonal curves. Int J of Computational Geometry & Applications 5(1-2):75–91

Coastchart project (2004) Demonstration of an EO based service for the update of marine charts, http://www.logicacmg.com/COASTCHART/index.aspx.html

Deng M, Chen X, Li Z (2005) A Generalized Hausdorff Distance for Spatial Objects in GIS. In: Proc of the 4th ISPRS Workshop on Dynamic and Multidimensional GIS, University of Glamorgan, Archives of ISPRS, University of Glamorgan, pp 10–15

Devogele T (2002) A new Merging process for data integration based on the discrete Fréchet distance. In: Richardson D, van Oosterom P (eds) Proc of the 10th Int Symp on Spatial Data Handling (SDH). Springer, pp 167–181

Devogele T, Trevisan J, Raynal L (1996) Building a multi-scale database with scale-transaction relationships. In: Kraak, Molenaar (eds) Proc of the 7th Int Symp on Spatial Data Handling (SDH). Taylor & Francis, pp 337–351

Dijkstra E (1959) A note on Two Problems in Connection with Graphs. Numerische Mathematik 1:269–271

Eiter T, Mannila H (1994) Computing Discrete Fréchet Distance. Technical report of Christian Doppler Labor für Expertensensyteme. Vienna University of technology, num. CD-TR 94/64

Fuchs H, Kedem ZM, Uselton SP (1977) Optimal Surface reconstruction from planar contours. Graphics and image processing. ACM 20(10):693–702

Geomod (2005) CadSIS, http://www.geomod.fr/logiciels/dvpt/dvpt.htm

Gesch D, Wilson R (2001), http://nauticalcharts.noaa.gov/bathytopo/ESRI_UC2001/ESRI_UC_paper.html

Harvey F, Vauglin F (1996) Geometric Match-processing: Applying Multiple Tolerances. In: Kraak, Molenaar (eds) Proc of the 7th Int Symp on Spatial Data Handling (SDH). Taylor & Francis, pp 155–171

IGN (2003) BD ORTHO® Version 1, Descriptif de contenu http://www.ign.fr/telechargement/MPro/produit/BD_ORTHO/DC_BDORTHO.pdf

International Hydrographic Organization (2005) HYDROGRAPHIC DICTIONARY 5th ed, Special Publications N°32 of IHO

Le Berre I, Henaff A, Wenzel F, Giraudet J (2004) Cartographie synthétique de l'environnement littoral du Finistère, exploitation de SPOT pour la cartographie de l'estran, du trait de côte et de l'occupation du littoral. In: Rapport final, Appel à proposition CNES/IFEN "Suivi du littoral par SPOT 5", GEOMER laboratory/Cetmef/DDE29

Mustière S (2006) Results of experiments on automated matching of networks at different scales. In: Joint ISPRS Workshop on Multiple Representation and Interoperability of spatial Data, Hannover, Germany, February 2006

Podobnikar T (2004) Production of integrated digital terrain model from multiple datasets of different quality. Int J of Geographical Information Sciences 19(1), January 2005:69–89

Sederberg TW, Greenwood E (1992) A Physically based Approach to 2-D shape blending. SIGGRAPH'92, 26(2):25–34

TCI software (2005) Adjust – True Rubbersheeting inside of AutoCAD, http://tcicorp.com/html/adjust.html

Veregin H (1999) Data quality parameters. In: Longley, Goodchild, Maguire, Rhind (eds), Geographical Information systems. John Willey & Sons Inc., pp 177–189

Whitfield M, Pepper J (2003) Integrated coastal zone-Data research project (ICZMap®). http://www.thsoa.org/hy03/4b_2.pdf

Characterizing Land Cover Structure with Semantic Variograms

Ola Ahlqvist[1], Ashton Shortridge[2]

[1] The Ohio State University, Deptartment of Geography, 1049 Derby Hall, 154 N Oval Mall, Columbus, OH 43210, USA
email: ahlqvist.1@osu.edu
[2] Michigan State University, Deptartment of Geography, 235 Geography Building, Michigan State University, East Lansing, MI 48824-1115, USA; email: ashton@msu.edu

Abstract

This paper introduces the semantic variogram, which is a measure of spatial variation based upon semantic similarity metrics calculated for nominal land cover class definitions. Traditional approaches for measuring spatial autocorrelation for nominal geographical data compare classes between pairs of observations to determine a simple binary measure of similarity (identical/different). These binary values are summarized over many sample pairs separated by various distances to characterize some spatial metric of correlation, or variation. The use of binary similarity measures ignores potentially substantial ranges in similarity between different classes. Through the development of category representations capable of producing quantifiable measures of pair wise class similarity, descriptive spatial statistics that operate upon ratio data may be employed. These measures, including the semantic variogram proposed in this work, may characterize spatial variability of categorical maps more sensitively than traditional measures. We apply the semantic variogram to National Land Cover Data (NLCD) for three different study sites, and compare results to those from a multiple class indicator semivariogram. We demonstrate that substantial differences exist in observed short-range variability for the two metrics in all sites. The semantic variograms detect much lower short-range variability due to the tendency of semantically similar classes to be closer together.

1 Introduction

Land use and land cover data sets are of prime utility for investigating many current geographic issues such as site-specific effects of global environmental change, climate models, urban sprawl, and habitat loss (DeFries et al. 1995; Dale et al. 1998; Wu and Webster 1998; Nunes and Augé 1999; Lambin et al. 2001). These data are also employed as input to models of climate and ecological dynamics, and are employed as framework layers for traditional GIS-based land administration activities (DeFries et al. 1995; Wu and DeFries et al. 1995). Spatial variability is a fundamental quality of land cover data with important implications for environmental modeling and ecological analysis. A number of methods have been developed to quantitatively describe spatial variability, including measures of autocorrelation such as Joins Count statistics, Moran's I, and semivariogram techniques (O'Sullivan & Unwin 2003). For example, variograms are frequently employed to characterize the spatial variability of continuous soil properties (e.g. Wang et al. 2002; Wu et al. 2002). Although most of these techniques have been developed for both categorical and measurement variables, the categorical versions are typically restricted because of the lower information granularity inherent in categorical data. Many authors have argued for ways to define nominal categories with the help of parameterized descriptions (c.f. Ahlqvist 2004) of defining characteristics. In this way, typical values for several attributes can essentially be inferred using the class as a proxy for ranges of plausible values. These category representations also enable various quantitative measures of similarity between any two categories. Consequently we seek to investigate the possibility of engaging spatial analytical statistics that require quantitative measurement variables through such similarity measures of nominal categories.

This paper introduces the idea of a semantic variogram based on semantic similarity metrics calculated on parameterized class definitions. We also evaluate the semantic variogram and its utility as a descriptive statistic for land cover information. To do so we contrast semantic variograms with indicator variograms (Goovaerts 1997), on land cover data for three regions with differing spatial characteristics. The particular study regions themselves hold little significance for us; they serve simply as useful proving grounds for the proposed metric. The next section reviews the standard indicator variogram and extends this to handle multiple classes. It then formally describes the semantic variogram. Section Three is a brief description of the case studies, including some detail about the land cover data, and indicates the methodology employed to define the classes as parameterized concept definitions and to calculate the metrics. Section rameterized concept definitions and to calculate the metrics. Section Four discusses the results derived variograms for the three study sites along with comments on the similarities and differences between them. The paper then concludes with some thoughts about the utility of the metric for distinguishing different land cover spatial structures.

2 Variograms and Semantic Distances

A standard approach for characterizing spatial variability of categorical data is the indicator semivariogram (Goovaerts 1997), which is normally defined on a class-by-class basis. An indicator variable for class s_k is defined:

$$i(\mathbf{u}_?;s_k)=\begin{cases}1 & \text{if } s(\mathbf{u}_\alpha)=s_k\\ 0 & \text{otherwise}\end{cases} \tag{1}$$

Then the indicator semivariance for class s_k for N location pairs separated by spatial lag $\mathbf{h}$ is:

$$\gamma_I(\mathbf{h};s_k)=\frac{1}{2N(\mathbf{h})}\sum_{\alpha=1}^{N(\mathbf{h})}\left[i(\mathbf{u}_\alpha;s_k)-i(\mathbf{u}_\alpha+\mathbf{h};s_k)\right]^2 \tag{2}$$

A semivariogram is composed of a sequence of these lags, and semivariances can be plotted against distance to determine the degree of spatial autocorrelation in the data. Large values (closer to 1) correspond to low correlations, while small values (closer to 0) correspond to high correlations.

In the present case we wish to characterize overall spatial variability for a multiclass map. We define the following indicator variable for coordinate pairs:

$$i\left[s(\mathbf{u}_\alpha);s(\mathbf{u}_\alpha+\mathbf{h})\right]=\begin{cases}1 & \text{if } s(\mathbf{u}_\alpha)\neq s(\mathbf{u}_\alpha+\mathbf{h})\\ 0 & \text{otherwise}\end{cases} \tag{3}$$

where $s(\mathbf{u}_\alpha)$ is the category at location $(\mathbf{u}_\alpha)$ and $s(\mathbf{u}_\alpha+\mathbf{h})$ is the category at a location separated by vector $\mathbf{h}$ from $\mathbf{u}_\alpha$. Then the multiclass indicator semivariogram is:

$$\gamma_{\mathrm{Im}c}(\mathbf{h}) = \frac{1}{2N(\mathbf{h})} \sum_{\alpha=1}^{N(\mathbf{h})} i\left[s(\mathbf{u}_\alpha); s(\mathbf{u}_\alpha + \mathbf{h})\right]^2 \tag{4}$$

This multiclass indicator semivariogram is a special formulation of the multivariable variogram developed by Bourgault & Marcotte (1991). A large semivariance (close to 1) is evidence for lack of category correspondence for lag **h**, while a number close to 0 indicates high correspondence at that lag. Indicator semivariograms treat the relationship between unlike classes very simply. The relationship between a class 41 cell (deciduous forest) and a class 43 cell (mixed forest) is identical (e.g. 1) to that between a class 41 cell and a class 11 cell (open water).

In contrast, to the indicator approach, the semantic distance methodology provides us with a more sensitive way to characterize the pairwise relationship between unlike classes. Semantic distance is based on the notion that knowledge about the deeper meaning of a class definition enables a user to think of some classes as more similar (semantically less distant) than other classes. This knowledge may be formalized in various ways; in this work we follow the procedure in Ahlqvist (2004) and use characteristic attributes and their values elicited from the documentation (Anderson et al. 1976) to formally define each class in the Anderson land cover classification system. Figure 1 illustrates how a textual class description has been translated into a semi-formal parameterized list of defining characteristics. This is subsequently decoded into a formal representation that uses a rough fuzzy set approach as detailed in Ahlqvist (2004). This representation makes it possible both to acknowledge graded boundaries for attribute values (indicated by graded shades in Fig. 1) as well as overcome the limited granularity of nominal attribute values that pertain to a certain class definition.

Based on a complete set of parameterized class descriptions, the semantic distance is determined based on a weighted summary of pair wise comparisons of fuzzy attribute values. A cross product of semantic distances for all combinations of land cover classes produces a semantic distance matrix. The semantic distances for the Anderson class codes are shown in Figure 2. Continuing the example from a previous paragraph, a cell with land cover class 41 (deciduous forest) is semantically quite close (0.23) to a cell with land cover class 43 (mixed forest), and quite distant (0.77) from a cell classed as 11 (open water).

We introduce the semantic variogram as a measure to more sensitively characterize the spatial structure of categorical data. This measure characterizes the average relationship of the (squared) semantic distance of two

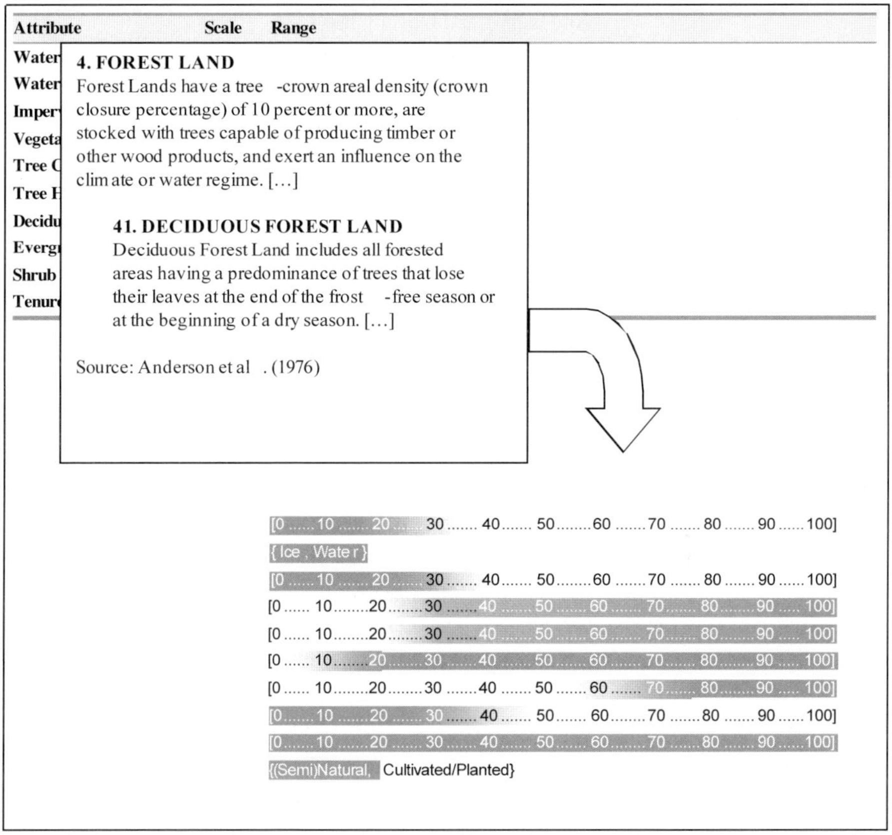

Fig. 1. Parameterizing category relationships from class definitions

point pairs and the ground distance between those pairs. Formally the semantic semivariance for a lag **h** is defined:

$$\gamma_{SD}(\mathbf{h}) = \frac{1}{2N(\mathbf{h})} \sum_{\alpha=1}^{N(\mathbf{h})} sd\left[s(\mathbf{u}_\alpha); s(\mathbf{u}_\alpha + \mathbf{h})\right]^2 \tag{5}$$

where $sd[s(\mathbf{u}_\alpha); s(\mathbf{u}_\alpha + \mathbf{h})]$ is the semantic distance between the land cover class of cell $\mathbf{u}_\alpha$ and the land cover class of cell $\mathbf{u}_\alpha + \mathbf{h}$. Interpretation is identical to that for any semivariogram: lower values for any particular lag **h** indicate higher spatial autocorrelation, while higher values indicate greater variability.

	Anderson Land Cover Class Code																			
	11	**21**	**22**	**23**	**31**	**32**	**33**	**41**	**42**	**43**	**51**	**61**	**71**	**81**	**82**	**83**	**84**	**85**	**91**	**92**
11	0	0.75	0.79	0.75	0.82	0.82	0.82	0.55	0.55	0.51	0.46	0.66	0.62	0.7	0.7	0.7	0.7	0.76	0.64	0.57
21	0.77	0	0.23	0.5	0.84	0.84	0.84	0.51	0.51	0.47	0.45	0.62	0.59	0.68	0.68	0.68	0.68	0.74	0.73	0.68
22	0.77	0.23	0	0.53	0.82	0.82	0.82	0.55	0.55	0.51	0.46	0.66	0.62	0.7	0.7	0.7	0.7	0.76	0.76	0.68
23	0.77	0.5	0.53	0	0.83	0.83	0.83	0.51	0.51	0.47	0.45	0.62	0.59	0.68	0.68	0.68	0.68	0.74	0.73	0.68
31	0.77	0.73	0.73	0.72	0	0.58	0.82	0.55	0.55	0.51	0.46	0.66	0.62	0.7	0.7	0.7	0.7	0.64	0.76	0.68
32	0.77	0.75	0.79	0.75	0.58	0	0.82	0.55	0.55	0.51	0.46	0.66	0.62	0.7	0.7	0.7	0.7	0.76	0.76	0.68
33	0.77	0.73	0.71	0.72	0.82	0.82	0	0.55	0.55	0.51	0.43	0.33	0.64	0.57	0.57	0.57	0.57	0.66	0.74	0.7
41	0.77	0.73	0.81	0.75	0.86	0.86	0.86	0	0.44	0.23	0.34	0.64	0.36	0.7	0.7	0.7	0.7	0.76	0.59	0.59
42	0.77	0.73	0.81	0.75	0.86	0.86	0.86	0.44	0	0.23	0.34	0.64	0.36	0.7	0.7	0.7	0.7	0.76	0.59	0.59
43	0.77	0.73	0.81	0.75	0.86	0.86	0.86	0.23	0.23	0	0.29	0.64	0.36	0.7	0.7	0.7	0.7	0.76	0.59	0.59
51	0.77	0.73	0.81	0.74	0.85	0.85	0.85	0.38	0.38	0.32	0	0.64	0.29	0.68	0.68	0.68	0.68	0.74	0.62	0.57
61	0.77	0.73	0.81	0.75	0.86	0.86	0.27	0.54	0.54	0.5	0.46	0	0.62	0.54	0.54	0.54	0.54	0.64	0.72	0.7
71	0.77	0.73	0.81	0.75	0.86	0.86	0.86	0.32	0.32	0.24	0.22	0.67	0	0.63	0.63	0.63	0.63	0.71	0.65	0.55
81	0.77	0.73	0.81	0.75	0.86	0.86	0.64	0.55	0.55	0.51	0.46	0.66	0.5	0	0.45	0.45	0.45	0.58	0.76	0.67
82	0.77	0.73	0.81	0.75	0.86	0.86	0.64	0.55	0.55	0.51	0.46	0.66	0.5	0.45	0	0.45	0.45	0.58	0.76	0.67
83	0.77	0.73	0.81	0.75	0.86	0.86	0.64	0.55	0.55	0.51	0.46	0.66	0.5	0.45	0.45	0	0.45	0.58	0.76	0.67
84	0.77	0.73	0.81	0.75	0.86	0.86	0.64	0.55	0.55	0.51	0.46	0.66	0.5	0.45	0.45	0.45	0	0.58	0.76	0.67
85	0.77	0.73	0.81	0.75	0.64	0.86	0.64	0.55	0.55	0.51	0.46	0.66	0.5	0.45	0.45	0.45	0.45	0	0.76	0.67
91	0.16	0.73	0.81	0.75	0.86	0.86	0.86	0.23	0.23	0.1	0.24	0.61	0.4	0.72	0.72	0.72	0.72	0.77	0	0.26
92	0.16	0.73	0.81	0.75	0.86	0.86	0.86	0.31	0.31	0.23	0.22	0.66	0.13	0.64	0.64	0.64	0.64	0.58	0.27	0

Fig. 2. Semantic dissemblance matrix

Multi-class indicator semivariograms were calculated on the data points from each of the three study regions. Semantic distances between NLCD land cover classes (see Fig. 2) were used to construct semantic semivariograms for each of the three study regions. Semivariance (gamma) values for these two different measures cannot be compared directly. Instead, the relative trajectory for each plot can be contrasted. Valuable perspective may be gained either by contrasting both metrics for a single study region, or by identifying differences between regions. To facilitate graphical comparison, semantic distance variances are standardized:

$$\gamma_{SD}(\mathbf{h}) = \frac{\gamma_{SD}(\mathbf{h})}{\sigma(sd[\text{all pairs}])} \tag{6}$$

where the denominator is the standard deviation of the semantic distance between all pairs of points in the study region. This standardization has no impact on the shape of the curve, but simply rescales values so that they are roughly similar in magnitude to the indicator semivariances, enabling simpler plotting.

3 Experimental Data and Methods

National Land Cover Data (NLCD) from 1992 (USGSa 2005) were obtained for three study regions in the eastern United States using the USGS Seamless Data server (USGSb 2005). NLCD is based on satellite remote sensing imagery classified into 21 different land cover classes similar to the Anderson land use/cover classification system (Anderson et al. 1976). It was produced as a cooperative effort between the U.S. Geological Survey (USGS) and the U.S. Environmental Protection Agency (US EPA) to produce a consistent, land cover data layer for the conterminous U.S. using early 1990s Landsat Thematic Mapper (TM) data. The end product has a pixel resolution of 30x30m and has an estimated overall classification accuracy of about 60% (Yang et al. 2001) The first study region occupies a portion of the Connecticut River valley in central Massachusetts (see Fig. 3). It is 38 by 43 kilometers in extent. The town of Amherst is in the lower right; Greenfield is in the upper center. The second region includes the middle reaches of the Muskegon River valley in the western lower peninsula of Michigan (see Fig. 4). This region is 31 by 34 km in extent. The town of Newago is in the central portion of the map. The third region covers a portion of south central Ohio (see Fig. 5). The region is 40 by 45 kilometers in extent. The city in the upper right is Lancaster. Random samples of 500 points (pixels) were extracted from each land cover dataset. The sampled locations (black dots) are displayed in Figures 3, 4, and 5. Some summary characteristics of these samples are provided in Table 1. While land cover proportions in the Ohio and Michigan samples are roughly similar, the Massachusetts sample has substantially more forest and less agriculture.

Data from these three study sites were analyzed to explore the utility of the measures presented in section Two. We used the textual class definitions provided with the NLCD data and Anderson et al. (1976) to elicit formalized representations of the land cover classes. Characterizing parameters to describe the classes were selected from the LCCS system (Di gregorio and Jansen 1998). After encoding the characterizing parameters as fuzzy numbers we employed an existing distance measure for fuzzy numbers, the dissemblance index, δ (Kaufman and Gupta 1985).

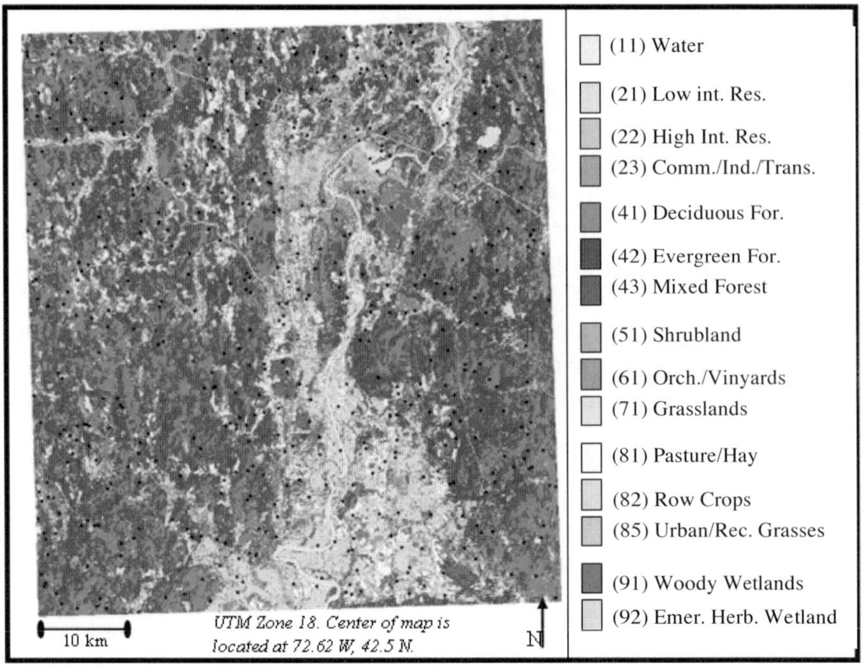

Fig. 3. Landcover for the Massachusetts Study Area

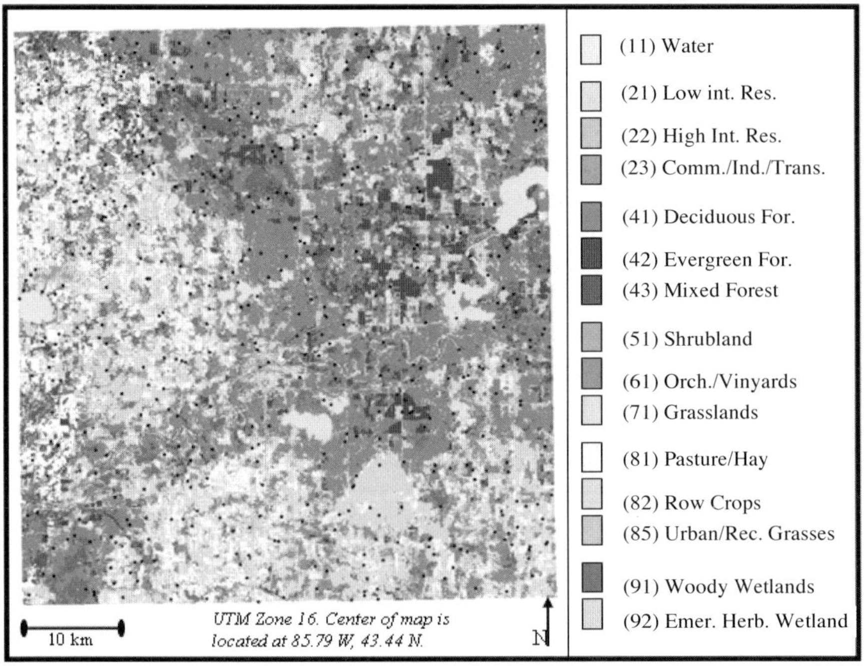

Fig. 4. Landcover for the Michigan Study Area

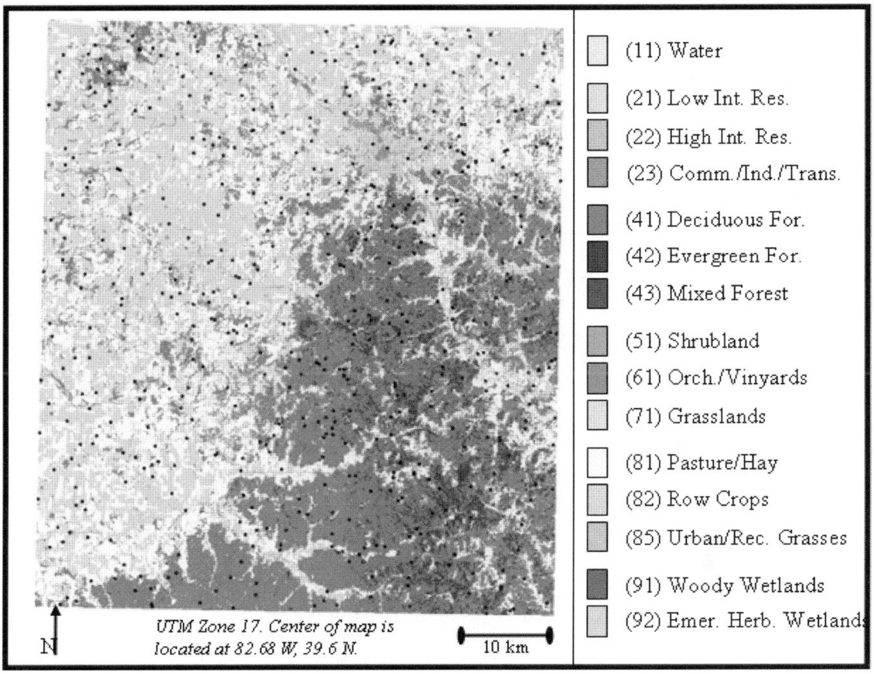

Fig. 5. Landcover for the Ohio Study Area

This metric is relatively straightforward to compute and generates a real number in the range [0,1] where 0 corresponds to identical descriptions and 1 corresponds to maximum difference between descriptions. The overall semantic distance between two classes is then calculated as a weighted average of all parameter-wise dissemblance values, again calibrated to give a [0,1] value range similar to the parameter wise dissemblance metric. The full cross-tabulation of all pair wise land cover class distances is summarized in table. Variography was performed separately on each set of 500 sampled points from the study areas using code developed by the authors in the R statistical programming language (R Development Core Team 2005). Both semantic and multiple class indicator variograms were calculated for the three sites. These variograms are graphically portrayed in Figures 6–8.

Table 1. Sample land cover data summary

NLCD Codes	Description	Mass. cells (%)	Mich. cells (%)	Ohio cells (%)
11	Water	9 (1.8)	16 (3.2)	3 (.6)
21	Low Int. Resid.	23 (4.6)	5 (1.0)	4 (.8)
22	High Int. Resid.	1 (0.2)	1 (0.2)	2 (.4)
23	Comm./Ind./Trans.	7 (1.4)	1 (0.2)	2 (.4)
41	Deciduous Forest	169 (33.8)	198 (39.6)	178 (35.6)
42	Evergreen Forest	66 (13.2)	34 (6.8)	16 (3.2)
43	Mixed Forest	113 (22.6)	4 (0.8)	8 (1.6)
51	Shrubland	2 (0.4)	0 (0)	0 (0)
61	Orchards/Vineyards	1 (0.2)	0 (0)	0 (0)
71	Grasslands	0 (0)	3 (0.6)	0 (0)
81	Pasture/Hay	25 (5.0)	64 (12.8)	155 (31.0)
82	Row Crops	44 (8.8)	135 (27.0)	125 (25.0)
85	Urban/Rec. Grasses	13 (2.6)	0 (0)	6 (1.2)
91	Woody Wetlands	22 (4.4)	35 (7.0)	1 (.2)
92	Emerg. Herb. Wetlands	5 (1.0)	4 (.8)	0 (0)

4 Results

The analysis results are first summarized with the help of site specific variograms followed by a short cross site comparison in Section 4.2.

4.1 Site Specific Results

4.1.1 Massachusetts Data

Figure 6 plots the standardized semantic variogram and the multiclass indicator variogram for the Massachusetts data. Distances are in meters. The indicator semivariogram may be interpreted as having very little spatial dependence, since gamma is high for all distances and rises only slightly from the shortest values. Any spatial autocorrelation appears to be in the first 8,000 meters. The semantic variogram suggests much lower variation at the shortest lag (1,000 meters) relative to all longer lags, and perhaps a somewhat greater range of spatial autocorrelation, to a distance of about 10 km. In general, the two variograms correspond quite closely to one another.

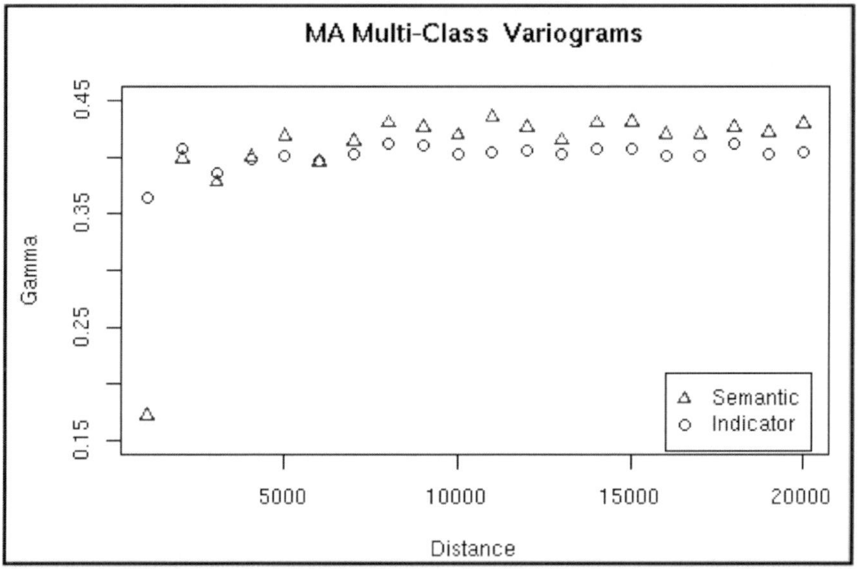

Fig. 6. Multiple Class Variograms for Massachusetts data

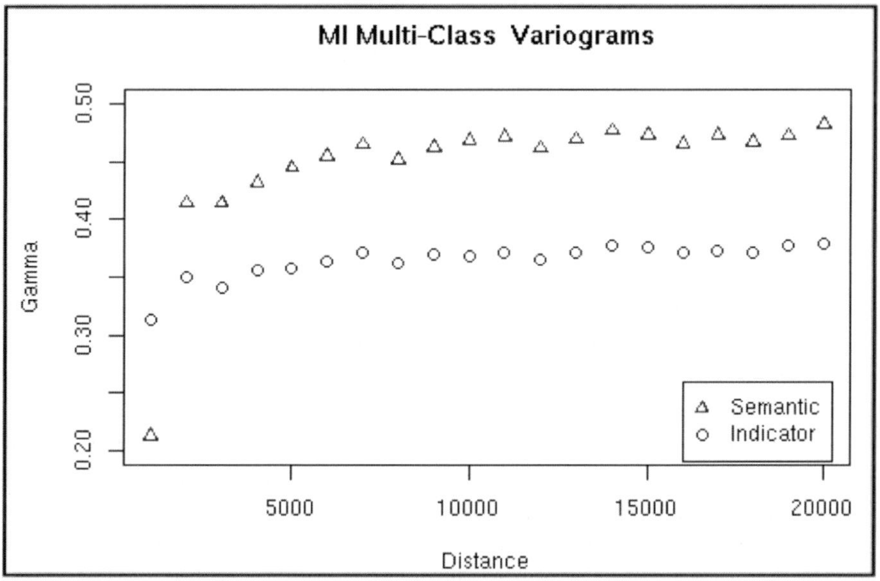

Fig. 7. Multiple Class Variograms for Michigan data

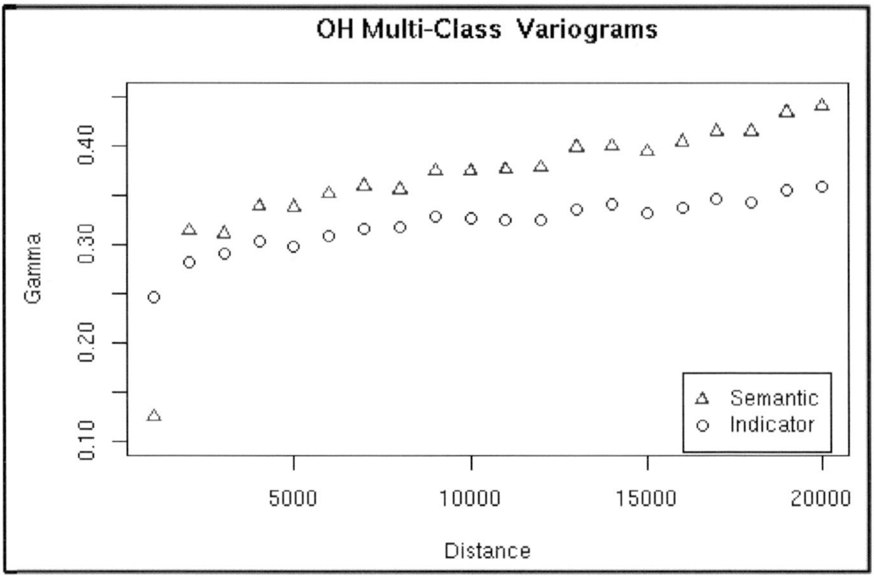

Fig. 8. Multiple Class Variograms for Ohio data

4.1.2 Michigan Data

Results for the Michigan study data are illustrated in Figure 7. The indicator semivariogram shows evidence of spatial autocorrelation. Semivariances rise to a sill of about 0.36 at a range of approximately 7,000 meters. The semantic variogram again demonstrates substantially lower variation in the first lag, followed by an abrupt jump at 2,000 meters and gradual increase in semivariance up to 13 km. The forms of the two variograms are similar.

4.1.3 Ohio Data

The multiclass indicator semivariogram for the Ohio data in Figure 8 suggests substantial persistence of autocorrelation to beyond 20,000 meters, since semivariances rise gradually from shorter to longer distances with no clear sill. Variance is substantially lower at very short lags (< 4,000 m). The semantic variogram again shows very low variances at the shortest lag, with an abrupt jump at 2,000 m. Semivariance rises consistently with longer distances from this point. The rate of increase with distance is more rapid for the semantic variogram than for the indicator semivariogram. No sill is evident in this plot.

4.2 Cross-Site Comparison

Variogram forms are substantially affected by the spatial patterns evident in the maps (see Figs. 3–5). Variograms for the Massachusetts study region suggest that little spatial autocorrelation is present, while variograms for the Ohio region exhibit no apparent leveling off at longer distances. The Michigan study area evinces a spatial pattern intermediate between the others.

5 Discussion / Conclusions

Indicator multiclass variograms and semantic variograms provided some very similar information about the spatial variability of land cover in the studied areas. However, they also varied in important ways. In all three regions, semantic semivariance at the shortest lag was relatively much lower than that of the indicator semivariance at the shortest lag. The incorporation of inter-class similarity results in substantially greater spatial autocorrelation at short distances. Differences are much less pronounced for larger lags. The Ohio dataset shows some evidence of a more rapid rise in semivariance for the semantic variogram than for the indicator variogram, while the other study areas showed no sign of different behavior at longer distances.

The observed differences in variograms for the three study areas may be due to substantially different spatial configurations of land cover. In general, the Ohio map is divided evenly between agriculture to the west and north, and forest to the south and east. This results in rather homogenous land cover over large areas, and substantial persistence of spatial autocorrelation at long distances. Massachusetts land cover, in contrast, appears to be more heterogeneous, resulting in little evidence of spatial autocorrelation at any but the shortest lags, even when employing semantic distances.

One caveat of using semantic distance measures is that we do not know exactly how these are scaled. Shepard (1987) does provide empirical evidence that perceived similarity decays exponentially with distance measured in a psychological space. We still do not have sufficient knowledge about how well the formal representation used in this work approximates a psychological space for the land cover categories. Extensive user testing is needed to verify this and this is beyond the scope of this work. A different route would be to assume an ordinal scaling for the similarity values and use the standard indicator variogram formula to derive the experimental variogram. In any case we would still have gained the additional insight of graded similarity between the nominal categories. Further research may shed some light on these issues.

This paper has extended the notion of semantic distances to the problem of measuring spatial variability of categorical data. We introduced the semantic variogram as a metric to characterize variability more sensitively than standard measures based on binary measures of category equivalence. The case studies demonstrated that the semantic method may offer substantially more information, especially for short distances, about spatial variability of categorical data.

Acknowledgements

The authors thank Phaedon Kyriakidis, who provided assistance with the notation employed to communicate the multiclass indicator variograms used in the study.

References

Ahlqvist O (2004) A parameterized representation of uncertain conceptual spaces. Transactions in GIS 8(4):493–514

Anderson JR, Hardy EE, Roach JT, Witmer RE (1976) A land use and land cover classification system for use with remote sensor data. U.S. Geological Survey Professional Paper 964

Bourgault G, Marcotte D (1991) Multivariable variogram and its application to the linear model of coregionalization. Mathematical Geology 23(7):899–928

Dale VH, King AW, Mann LK, Washington-Allen RA, McCord RA (1998) Assessing Land-Use Impacts on Natural Resources. Environmental Management 22(2):203–211

DeFrie RS, Field CB, Fung I, Justice CO, Los S, Matson PA, Matthews E, Mooney HA, Potter CS, Prentice K, Sellers PJ, Townshend JRG, Tucker CJ, Ustin SL, Vitousek PM (1995) Mapping the land surface for global atmosphere-biosphere models: toward continuous distributions of vegetation's functional properties. J of Geophysical Research: Atmosphere 100:20867–20882

Di Gregorio A, Jansen LJM (1998) Land Cover Classification System: Classification Concepts And User Manual. FAO, Rome, 179 p

Goovaerts P (1997) Geostatistics for Natural Resources Evaluation. Oxford University Press, New York

Kaufman A, Gupta MM (1985) Introduction to fuzzy arithmetic. Van Nostrand Reinhold Company, New York, 351 p

Lambin E, Turner B, Geist H, Agbola S, Angelsen A, Bruce J, Coomes O, Dirzo R, Fischer G, Folke C, George P, Homewood K, Imbernon J, Leemans R, Lin

X, Moran E, Mortimorep M, Ramakrishnan P, Richards J, Skanes H, Steffen W, Stone G, Svedin U, Veldkamp T, Vogel C, Xu J (2001) The causes of land-use and land-cover change: moving beyond the myths. Global Environmental Change 11:261–269

Nunes C, Augé JI (eds) (1999) Land-Use and Land-Cover Change (LUCC): implementation strategy. IGBP report(48), IHDP report (10). International Geosphere-Biosphere Programme, Stockholm, 125 p

O'Sullivan DO, Unwin DJ (2004) Geographic Information Analysis. Hoboken, Wiley, NJ

R Development Core Team (2005) R: A Language and Environment for Statistical Computing. URL: http://www.R-project.org/. R Foundation, Vienna, Austria. Accessed 12/1/2005

USGSa (2005) National Land Cover Dataset 1992. http://landcover.usgs.gov/natllandcover.asp. United States Geological Survey, Reston, Virginia. Accessed 12/1/2005

USGSb (2005) Seamless Data Distribution System. http://seamless.usgs.gov/ United States Geological Survey, Reston, Virginia. Accessed 12/1/2005

Wang H, Hall CAS, Cornell JD (2002) Spatial dependence and the relationship of soil organic carbon and soil moisture in the Luquillo Experimental Forest, Puerto Rico. Landscape Ecology 17(8):671–684

Wu F, Webster CJ (1998) Simulation of land development through the integration of cellular automata and multicriteria evaluation. Environment and Planning B 25:103–126

Wu J, Norvell WA, Hopkins DG, Welch RM (2002) Spatial variability of grain cadmium and soil characteristics in a durum wheat field. Soil Science Society of America J 66:268–275

Yang L, Stehman SV, Smith JH, Wickham JD (2001) Thematic sccuracy of MRLC land cover for the eastern United States. Remote Sensing of Environment 76:418–422

Semantic Similarity Measures within the Semantic Framework of the Universal Ontology of Geographical Space

Marjan Čeh[1], Tomaž Podobnikar[2], Domen Smole[3]

[1] Faculty of Civil and Geodetic Engineering, University of Ljubljana, Ljubljana, Slovenia
[2] Institute of Anthropological and Spatial Studies at the Scientific Research Centre of the Slovenian Academy of Sciences and Arts, Ljubljana, Slovenia
[3] DFG CONSULTING, d.o.o., Ljubljana, Slovenia

Abstract

The objective of this paper is to discuss our methodology for comparing, searching and integrating geographic concepts. Searching for spatially oriented datasets could be illustrated by the complexity of the communication between the producer and user. The common vocabulary consists of a set of concepts describing the geographic space called universal ontology of geographical space (UOGS). We have defined the semantic parameters for measuring semantic similarities within the UOGS semantic framework and described our applicative approach to the similarity analyses of spatial databases. In order to test our results we have implemented the entire vocabulary as a set prolog fact. Following this we also implemented functionality such as the querying mechanism and the simple semantic similarity model, again as a set of prolog clauses. In addition to this, we applied prolog rules for the purpose of extracting semantic information describing geographic concepts and extracting it from natural language texts.

Key words: spatial data, geographical concept, semantic similarity measures, ontology, prolog, machine learning

1 Introduction

Providing an efficient and user-friendly approach for comparing the meaning and searching for spatial oriented data within the distributed environments has been a research motivation in the geo-information community for some time. Numerous issues have already been solved, however many still remain. The toughest issue refers to resolving the meaning or semantics of the words exchanged in the communication between the user and the system. However, the geo-information community is far from being alone in this endeavor. For example, in 1960s the artificial intelligence research community started with their efforts to develop a system that would interact with the human user in a natural language (Bilotti 2004). As a result, a number of so called question answering systems that can be characterized by its convenient, natural interface for accessing information have emerged since then, some of which have proven to be surprisingly successful.

In this paper we will outline our efforts oriented towards developing an approach that will hopefully provide users with an accurate and convenient spatial data searching system. This includes the option of comparing the meaning of geographic concepts. In the following chapters, we will present some of the preliminary results on two interconnected objectives our research group has been focusing on. The first direction of our research has been the estimation of the semantic similarity of two different spatial databases by means of certain predefined measures.

Could the data broker in this kind of an interaction be an artificial system, and not a human being, or is this merely an idea of a science fiction novel? Implementing such a data broker that would satisfy the use-case scenario is surely not an easy task. A number complex and still in the research-phase issues have yet to be addressed, some of which are beyond the scope of this paper. However, we will demonstrate the feasibility of certain aspects that are an essential part of such a system.

2 Universal Ontology of Geographical Space (UOGS) and Semantic Measures

For the purpose of the semantic interoperability assessment of spatial databases we have developed the following semantic measures (Čeh et al. 2004):

- semantic distance between concepts,
- absolute and relative semantic depth of a concept,
- semantic depth of a spatial database,
- semantic dispersion of a spatial database,
- semantic weight of an unavoidable level.

The core methodology that enabled the creation of the *universal ontology of the geographical space* (UOGS) is derived from the notion of *human activities* (Laurel 1993, Fellbaum 1999, Kaptelinin et al. 1999, Camara et al. 2000, Bowker and Star 2000, Kuhn 2001). The simultaneous treatment of human activities and objects is a key element in the modeling of geographical space. The hierarchy of activities spans from high-level activities such as intention and purpose to goal oriented operations. It is necessary to recognize how certain agents act in the universe of a discourse. The activities within the geographical space have to be realized independently from the methods for modeling human knowledge within an information system. The geographical space has to be treated simultaneously to the system of objects and activities.

Our ambition was to build a semantic reference system of the geographical space (Čeh 2003) as later designated by Kuhn and Raubal (2003) in the form of semantic data and a semantic reference system. In the development of the formal ontology the extent of the top level has to be limited. This presumption preserves the clarity of the understanding and keeps it easy to use. In the geographical space numerous spatially materialized objects with human origin exist, i.e. artificial objects. Besides them there are also spatial schemes within the geographical space such as abstract spatial formations with consensually defined boundaries and observed spatial formations of natural phenomena and social phenomena Figure 1.

		SPATIAL OBJECTS	
		Physical	**Abstract**
HUMAN ACTIVITIES	**Basic**	Physical Basic	Abstract Basic
	Advanced	Physical Advanced	Abstract Advanced

Fig. 1. Spatial objects of human activities

The conceptual framework of the geographical space is designed as a multilayer composition of domains consisting of heterogeneous secondary spatial concepts. Symbols and words that are commonly used for naming various spatial objects are attributes of (spatial) activities within the described ontology (for instance housing – house, apartment; trading – store, warehouse).

The criterion for defining a new basic concept in the ontology of geographical space can be found in the difference of the purpose of its activity versus the existing one, which produces an object such as a physical spatial thing or an abstract scheme of space. Complex objects and human activities schemes within geographical space can be presented by combining simple categories (objects or schemes) defined in UOGS. The purposes of human activities and their appearances overlap in space and time.

The designed ontology is independent from the data of the objects in existing databases as well as real spatial objects. While it treats the complete domain of the space of geographical extent we call it the universal ontology of geographical space (UOGS). Due to the requirement of the hierarchic structure, the method for building UOGS is divided into three phases: defining the origin of the ontology, developing the induced level of the ontology and designing the reality levels of the ontology.

Since the aim is to treat objects of the geographical space, we have used the parameter of extensiveness for limiting the general concept of space. The geographical extent has been defined by Egenhofer and Mark (1995) as an extension of objects which are larger than the human body and cannot be completely perceived in a single perception act. At the induced level of UOGS we have used the methodology of constructive induction, which is the method of classification also used in inductive logic programming. It has been used for treating problems at which the existing attributes and relations cannot reasonably explain the given concept. Such problems are solved by the automatic or manual addition of new, intermediate concepts. We have linked three concepts from the general ontology of knowledge representation by Sowa (2000), which can be formally expressed, namely the concepts of purpose, object and scheme. The first and second level of the UOGS hierarchy represents semantic data, which formally defines the meaning of the basic concepts outside the system. With this method we tackle the problem of the invisibility of the classification (Bowker and Star 2000) in the semantic part of the spatial information infrastructure. Furthermore, we have also developed concepts on the second and lower levels of the ontology by using the approach of natural learning with understanding and insight (Bratko et al. 1998), specifically with the strategy of broad searching and knowledge research as a strategy of in depth search.

Currently, UOGS is a semantic reference system that consists of 588 concepts expressed in terms familiar to the users. The concepts are logically organized into five hierarchic levels. UOGS omits geometrical relations and is an evolving ontology, which means that a set of concepts is not exhaustible, but upgradeable. The root concept or level 0 of the UOGS is the geographical space. Before enumerating the list of concepts on the next ontological level it must be noted that human activities in the geographical space are defined by the category *purpose*. According to this, we grouped human activities into two categories: *basic* and *advanced*. The purpose of the *basic human activities* is in satisfying the needs of an individual, while the purpose of the *advanced human activities* is in satisfying the needs of a community or a nation as a whole. Further on, the phenomenon of geographical space is introduced either as a physical object or as an abstract scheme. According to this, level 1 of the UOGS consists of four induced categories of human activities, which are:

- basic physical human activities,
- advanced physical human activities,
- advanced abstract human activities,
- basic abstract human activities.

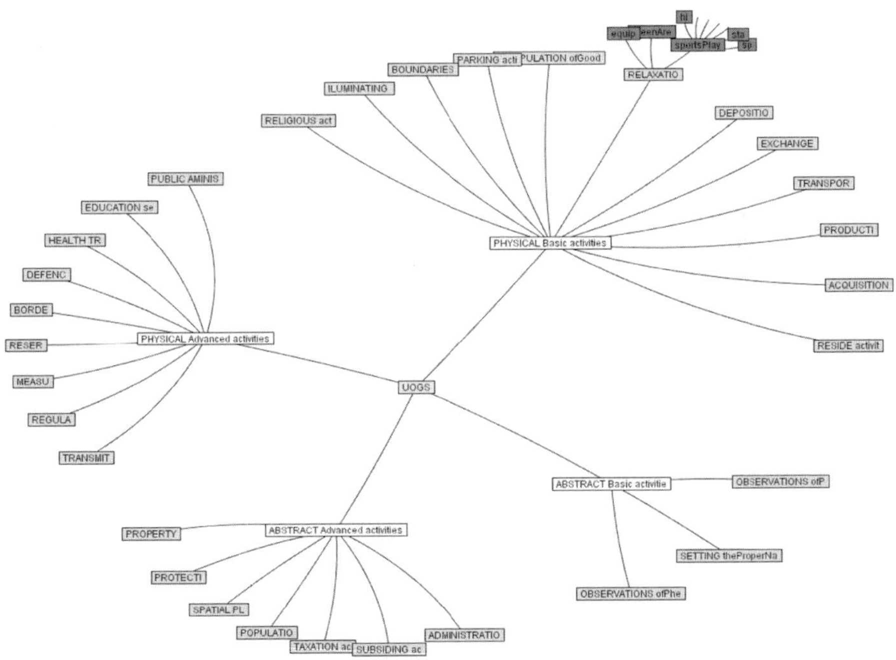

Fig. 2 Fish-eye view of the UOGS hierarchy

In the UOGS the activities are presented as concepts and resemble functions (*verbs*), which are features of entity classes (see Fig. 2). In this manner, human activities may define the user context or application domain. In contrast, the lower layers of UOGS, i.e. level 3 and the final level 4, consist of concepts that resemble geospatial entity classes (*nouns*) and should be read as a cause or consequence of the higher *human activities* level concepts.

3 Estimation of Semantic Similarity among the Concepts in Spatial Databases

For the purpose of spatial database integration it is necessary to extract information from the database as regards the conceptual (thematic) detail, i.e. semantic resolution and semantic concentration. Information on the semantic resolution and concentration of the database provides the mechanism for comparing the semantic metadata of the databases. The representation quality of the semantics for spatial data is one of the measures for estimating the development level of certain geographical information systems.

For the purpose of estimating semantic similarity of spatial databases we have developed various measures at which we consider them as an entity or partially as separate concepts held within the data. The semantic harmony of the spatial database is defined as a level of similarity of spatial databases dealing with two main parameters: *semantic resolution* and *semantic concentration/dispersion* of the spatial database. The greater the similarity of the semantic parameters of the compared databases, the greater is the semantic harmony among them. For this purpose we established all possible permutations of the semantic relations for all concepts within the database, calculating their semantic distances and other developed semantic parameters as below.

3.1 Semantic Parameters for Measuring Semantic Similarities in the UOGS Semantic Framework

Semantic Distance of concepts in relation (SDist) is the sum of segments within the semantic reference system, needed to establish the relationship among two concepts. The distance among neighbouring semantic levels has the norm value of 1. The semantic distances of relations between concepts (see Eq. 1) are presented in Figure 3.

$$\mathrm{SDist(A, B) = 3, SDist(C, D) = 5, SDist(E, F) = 6} \quad (1)$$

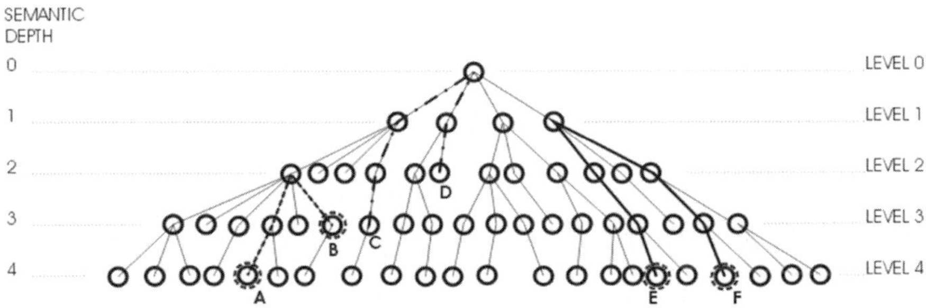

Fig. 3. Semantic graph (tree) with mapped concepts of the spatial database SD1{A, B, C, D, E, F} and some relations expressed with semantic distances

Absolute Semantic Depth of the Spatial Concept (ASDpthC) is calculated as the semantic distance from the UOGS origin at Level 0 (Fig. 3) to the concept (see Eq. 2).

$$\mathrm{ASDpthC(A) = 4} \quad (2)$$

We also determined the *unavoidable level of relation* as the depth at the position of the first common concept within the semantic reference system. The unavoidable level of relation is expressed with the *Semantic Depth of the unavoidable level of relation* (SDpthUnavLvl). In Figure 3, the unavoidable level of relationship between the concepts A and B is 2 (see Eq. 3).

$$\begin{aligned} &\mathrm{SDpthUnavLvl(A, B) = 2} \\ &\mathrm{SDpthUnavLvl(C, D) = 0} \\ &\mathrm{SDpthUnavLvl(E, F) = 1} \end{aligned} \quad (3)$$

Absolute Semantic Difference in Depth (ASDiffDpth) of the concepts in relation is the difference of relevant absolute depths as seen in Eq. 4:

$$\begin{aligned} &\mathrm{ASDiffDpth(B, A) =} \\ &\mathrm{ASDpthC(B) - ASDpthC(A)| = |3 - 4| = 1} \end{aligned} \quad (4)$$

Relative Semantic Depth of relation (RelSDpth) is calculated for every semantic relationship as a coefficient of the absolute semantic difference in depth of the concepts in relation and the *Semantic Depth of Unavoidable Level of relation* (SDpthUnavLvl) raised by 1, in order to avoid dividing by 0 at SdpthUnavLvl(x, y) = 0 (Eq. 5):

$$\begin{aligned} &\mathrm{RelSDpth(x, y) =} \\ &\mathrm{ASDiffDpth(x, y) / (SdpthUnavLvl(x, y) + 1)} \end{aligned} \quad (5)$$

The relative semantic depth of relation expresses the semantic concentration/dispersion of the two concepts in relation. The smaller the relative depth among the concepts, the smaller is the dispersion, and the higher semantic concentration.

3.2 Definition of the Semantic Resolution of the Database

The semantic resolution of the spatial database explains generality or the opposite speciality of the meaning of the data in the spatial database. We express the semantic resolution of the database with the average level of semantic depth of the concepts in the considered database. *Absolute Semantic Depth of Spatial Database* (ASDpthSD) is defined as an average depth of all concepts within the database (see Eq. 6):

$$\text{ASDpthSD} = \Sigma\,(\text{ASDpthC})/n \tag{6}$$

In the case of the spatial database SD1{A, B, C, D, E, F}, the absolute semantic depth of the spatial database is: ASDpthSD = (4 + 3 + 3 + 2 + 4 + 4)/6 = 3.33.

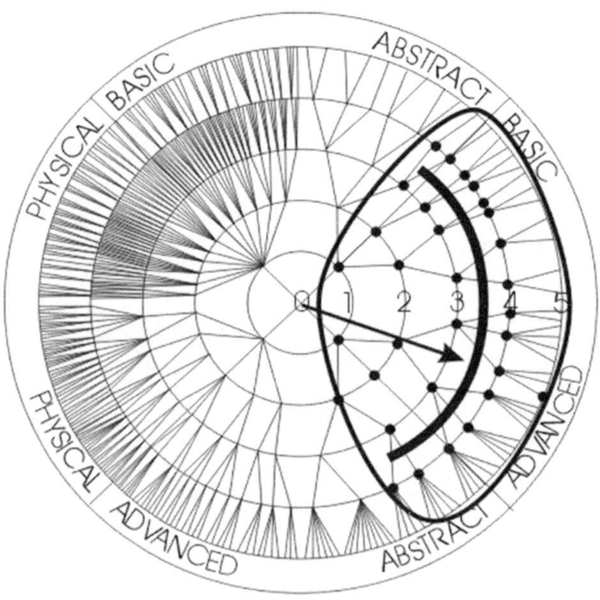

Fig. 4 Graphic representation of the absolute semantic depth of the spatial database (ASDpthSD)

Figure 4 presents the semantic reference system (UOGS) in which the spatial database concepts are mapped as black dots. The arc and the arrow represent the absolute semantic depth of a spatial database within UOGS.

3.3 Determining the Semantic Concentration/Dispersion of the Database

We express the semantic concentration of the database with the *Dispersion of Spatial Database* (SDispSD) parameter (see Fig. 5). This is calculated from the semantic distances between all concepts and semantic weights of their semantic relationships. For this purpose we introduced the *Weight of Unavoidable Level of relation* (WUnavLvl).

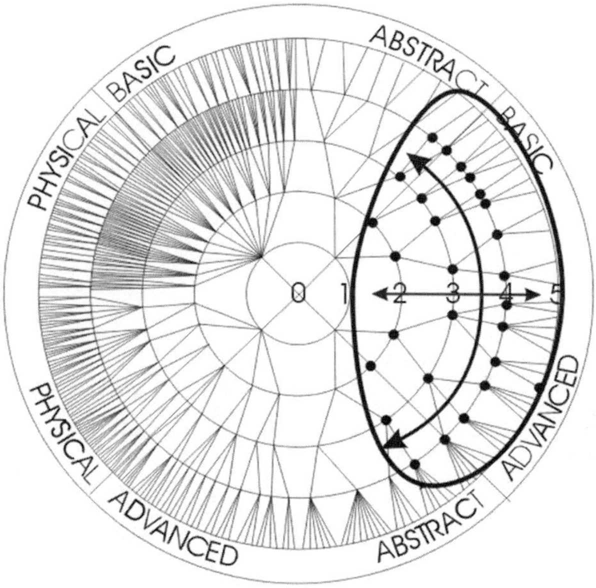

Fig. 5 Semantic concentration/dispersion of the database

The deeper the unavoidable level of relation, the smaller is the semantic dispersion of the concepts in relation and greater the semantic concentration of the related concepts (see Eq. 7).

$$\text{WUnavLvl}(x, y) = 1 / (\text{SDpthUnavLvl}(x, y) + 1) \quad (7)$$

Therefore the *Absolute Dispersion of Spatial Database* (ASDispSD) is the sum of products between the semantic distances and the corresponding semantic weights of the relations (Eq. 8):

$$ASDispSD = \Sigma\,(Sdist(i, j) \bullet WUnavLvl(i, j)) \quad (8)$$

The *Relative Dispersion of Spatial Database* (RSDispSD) can be calculated as the coefficient of the *Absolute Dispersion of Spatial Database* (ASDispSD) and the *Absolute ssssss Dispersion of the Semantic Reference System* (ADispSRefSys) (see Eq. 9).

$$RSDispSD = ASDispSD\,/\,ADispSRefSys \quad (9)$$

3.4 Validating the UOGS Suitability as a Semantic Mapping and Similarity Identification Tool in Spatial Databases

The semantic suitability of the universal ontology of geographical space (UOGS) as an independent (of applicative ontologies) semantic reference system was estimated against the data catalogues of six databases. Mapping the relations between various symbols and concepts of applicative ontologies into a semantic framework is held in tables. Table 1 presents the aggregate table with cumulative results.

Table 1 Cumulative results of the UOGS semantic suitability analyses: e – equivalent, a – aggregate, u – unmatchable, t – total

	Semantic accordance				Semantic accordance of UOGS		
DB Catalogue	**Equivalent**	**Aggregate**	**Non-matchable**	**Total**	**e + a**	**e + a/t (%)**	**u (%)**
Land use – cadastre	31	5	6	42	36	86	14
Land use – agriculture	34	15	9	58	49	84	16
Topographic map 1:500	155	56	3	214	211	99	1
Topographic map 1:5000	10	1	5	16	11	69	31
CORINE Land Cover	27	27	5	59	54	92	8
Topographic scheme	95	60	7	162	155	96	4
Total	352	164	35	551	516	94	6
	64%	30%	6%	100%			

According to the applicative ontologies the result of 94% semantic expressiveness of UOGS confirms the semantic suitability of the developed ontology for the purpose of searching for semantic similarity among various spatial databases.

4 Implementing UOGS and its Searching Capabilities

Currently, the UOGS ontology and the search engine functionality are implemented as a prolog program. Prolog is a programming language for symbolic, non-numeric computation. It is especially suited for solving problems that involve objects and the relations between them (Bratko 2001). We started developing a custom Windows application that is capable of operations such as adding, deleting, editing and inserting concepts into the UOGS hierarchy in a relatively user-friendly fashion. In addition, the tool enables any hierarchy to be exported in the form of a prolog database in a pre-defined way.

We have defined the entire hierarchy of UOGS concepts with one predicate, namely *implies/2*. For example, *implies (physicalAdvancedActivities, uogs)* denotes that the concept *physicalAdvancedActivity* is a child of the *uogs* concept, which happens to be the root concept. The entire hierarchy consists of a large set of prolog facts similar to the one given above. However, merely making explicit concepts is in itself of no use without an appropriate mechanism supporting the query. In its effort searching through a list of appropriate databases the system might use the following functionalities, of which the last one provides a simple semantic similarity model based on the edge counting approach:

- find all parent/child concepts for a given concept,
- find the most specific parent for two or more given concepts,
- acquire the depth level for a given concept,
- acquire the distance between two or more given concepts,
- acquire the average semantic depth of all concepts found in a database,
- acquire a set of concepts found in a certain semantic similarity neighborhood for a given concept.

Predicate *getPredecessor/2* is able to find a child or a parent of a given concept. For example, *getPredecessor(C,uogs)* tries to find all child concepts of the *uogs* concept. Similarly, the same predicate written as *getPredecessor(department,P)* finds a parent for a given concept department. Following is a definition of *getPredecessor/2* predicate in prolog form:

```
predecessor(X,Z):-
        implies(Z,X).
predecessor(X,Z):-
        implies(Z,Y),
        predecessor(X,Y).
```

Finding the most specific parent for the two concepts is another useful functionality. Predicate *getMSP/3* is defined as follows:

```
getMSP (A,A,A):-!.
getMSP (A,B,A) :-
        predecessor(A,B),!.
getMSP (A,B,B) :-
        predecessor(B,A),!.
getMSP (A,B,C):-
        predecessor(C,A),
        predecessor(C,B),!.
```

Given the two concepts as the first and second parameter, the third parameter, the result, is the most specific parent of both concepts according to the UOGS hierarchy. The same predicate can be used to find all combinations of the concepts, which have the given concept as the most specific concept. This predicate will be particularly useful for calculating the semantic similarity described in the next few lines.

4.1 Semantic Similarity Model

The next step was to choose and implement *the most appropriate* semantic similarity model. A semantic similarity model is a computational model for the semantic similarity assessment, which is an essential part of any modern information retrieval method. The new trends in the research of information retrieval stress the advantages of using domain knowledge and semantic similarity functions in order to compare words or documents. Psychologists and cognitive scientists have analyzed the way people evaluate similarity and have defined models based on features or descriptors of concepts (Rodriguez 2000). However, we followed the approach familiar to computer scientists. This is an approach that defines semantic similarity as the distance between concepts within a hierarchical structure according to the definition of semantic distance given in Rada et al. (1989).

First of all, we must be able to find the level of a given concept within the hierarchy. The question *getConceptLevel(house,L)* returns the (semantic) depth of a given concept. Similarly, *getConceptLevel(C,3)* returns all concepts found on level 3. The predicate *getConceptLevel/2* calls the *getConceptDistance/3* relation which counts the number of edges among the two different concepts found in the hierarchy using getMSP/3. The number of edges represents the distance and consequently the semantic similarity/dissimilarity among the given geographical concepts.

So far, only a distance parameter has been used in our application to calculate the semantic similarity in the process of finding similar concepts to the one selected by the user. However, not merely the distance, but also the direction should be taken into account when defining a useful semantic similarity measure. The search direction of a similar concept may be guided by means of weights linked to distances (like the above defined relative semantic depth of relation (RelSDpth) or the weight of unavoidable level of relation (WUnavLvl)).

In the example, if the user sets the value of the semantic similarity measure (let us assume that it equals the sum of distances between concepts) to 3, the resulting semantically similar concepts may only be children of sibling (same-level) concepts and not, for example, a grand grandparent concept.

The predicate, which estimates the similarity between two concepts *findSimilar/3* was defined using the state-of-the-art Inductive Logic Programming (ILP) system, called CProgol (Muggleton 1995). As with all ILP algorithms, CProgol constructs logic programs from extensional or intentional background knowledge and from positive and negative examples of the target relation, which is in our case *findSimilar/3*. Given the background knowledge the goal of the ILP system is to learn a hypothesis consisting of sentences that cover all positive and none of the negative examples.

The input file consisted of intentional background knowledge *getMSP/2*, *getConceptLevel/2* and of positive and negative examples of the target relation. The result of the learning process follows the hypothesis consisting of five clauses:

findSimilar(A,A,high) :- getConceptLevel(A,4).
findSimilar(A,B,high) :- msp(B,A,C), getConceptLevel(C,3).
findSimilar(A,B,high) :- msp(B,A,C), getConceptLevel(C,2).
findSimilar(A,B,medium) :- msp(A,B,C), getConceptLevel(C,1).
findSimilar(A,B,low) :- msp(B,A,C), getConceptLevel(C,0).

As we can see, we have defined three levels of semantic similarity: low, medium and high. For example, the question *findSimilar(A,B,high)* returns all combinations of concepts, for which the most specific parent is found on level 3 and all appropriate concepts that are found on the lowest level, i.e. level 4. We can conclude that the upper hypothesis is quite straightforward and could be acquired without using the ILP approach. However, our aim was to demonstrate the advantage of using the logic programming approach in the light of solving much more complicated problems that are to be solved in the field of Natural Language Processing (NLP) by means of machine learning algorithms.

For the purpose of assessing semantic similarity of two databases, we can use the predicate *getASD/2*, where the first argument represents a list of concepts described within a given spatial database, and the second argument represents the average semantic depth of the listed concepts. We believe this is only a rough estimation from which we can draw conclusions for the overall semantic (di)similarity of two or more spatial databases. Similarly, we believe further research is needed in this direction, which would focus on the additional measures that could give more accurate and reliable information on the overall semantic (di)similarity of two or more spatial databases.

Up to now, our work was based on the assumption that spatial databases are already described with UOGS concepts. Some of them are indeed described in Čeh (2003). What we would like to enhance further is the process of describing spatial datasets with UOGS concepts. Currently this is performed manually, however, this should be replaced with a (semi)automatic process since immense amounts of spatial datasets and their descriptions in the form of metadata emerge on a daily basis.

4.2 Natural Language Processing of Object Catalogues and Metadata Descriptions

The question answering system is often viewed as a combination of two related, established information access tasks known as information retrieval (IR) and information extraction (IE), but unlike them, the goal of the question answering system is to provide exact, precise answers to the users' questions posed in a natural language (Bilotti 2004). As found in Kavouras et al. (2005) geographical concept definitions are a rich source of knowledge with special structure and content. These definitions are usually found in object catalogues and/or metadata documents. Our research group's aim has been to outline the characteristics and implement a system that would be able to recognize spatially related concepts and its meaning expressed in a natural language. This would be the core functionality of a system with capabilities similar to the capabilities of a typical question answering system. Usually, the question answering system is divided into five main components (Abney et al. 2000): passage retrieval, entity extraction, entity classification, query classification and entity ranking. In the following section we will describe merely the first two components.

Our work started with finding sentences that define geographical concepts and are located in a given catalogue or metadata description document. We decided to focus on analyzing documents in the Slovenian language. This is important, for almost any NLP analysis requires a certain amount of pre-processing, e.g. part-of-speech tagging, which depends on the particular language. There are many POS taggers available. We have been provided with a POS tagger developed for the Slovenian language.

As a result of the POS tagging process, each word of the document is tagged with one or more of the ten possible tags, e.g. noun, adjective, verb, adverb conjunction, etc. Each tag refers to further information, for example nouns and adjectives are described with number, case, sex, etc. In the first research phase, the tagged text was the input for learning rules, which extract a list of geographical concepts. Finding the list of geographical concepts from definitions might seem a trivial task, but providing high accuracy requires dealing with exceptions. For example, there might be more than one geographical concept defined in the definition, or the concept being defined might be found at the end and not at the beginning of the definition. The predicate *getHeadwords/2* returns a list of geographical concepts for a given definition.

According to the preliminary results, we think genus or hypernim extraction as well as other semantic relations such as purpose, adjacency, material and so on, could be extracted by means of prolog programming or the means of machine learning, for example inductive logic programming. The latter would require the preparation of a substantial amount of POS tagged sentences, along with a number of positive and negative examples as well as background knowledge.

Further on, we intend to prepare a list of the most frequent questions that the users are interested in regarding spatial datasets. This will not be an infinite list of possible questions. For example, it might be a list of all geographical concepts that can be found in a given spatial database, or it might describe these concepts (e.g. the genus, the differences including the purpose, size, adjacency and other semantic relations). Similarly, the user might pose typical questions regarding the dataset, such as the date of data capturing, contact person information, etc. Answers to these questions could be extracted from documents by means of the first order predicate rules defined in the prolog program.

5 Conclusions

Unsurprisingly, there is a gap in the communication between the spatial data producer and the spatial data user. If they communicate at all, they speak their own language, consisting of different vocabularies. The question is whether an efficient approach that could be applied to bridge this gap exists? Currently a great deal of research effort in the GIS community is oriented towards finding methods that would enable automatic interpretation of the meaning of different vocabularies. The ontology-based approach for resolving semantic heterogeneity problems is amongst the most commonly referred ones. We are of the opinion that the ontological approach should be combined with natural language processing (NLP) techniques.

We mapped the concept of geographical space in the centre of the UOGS semantic framework, thus obtaining a radial organization of the mental model, as did Lakoff (1987) in his Idealized Cognitive Model (ICM). However, contrary to his approach, the features close to the centre of the UOGS characterize the category of the geographical space most generally. The features on the deeper levels enrich the meaning of the geographical space category. We represented such a model with a conical frustum that enables the general public to understand the model, which is a necessary condition if we wish to achieve the desired acceptance of the ontologic obligation of such a mental model giving a value to the implementation attempts in the GIS domain. Mapped on the conical frustum and the hyperbolic (fish eye) browser, conceptual hierarchy enhances spatial representation and communication of the conceptual model of geographical space as well as indirectly enhances the spatial representation of geometric models within the spatial domain, offering a possibility to improve the decision process at the automatic map generalization. This fact and the developed semantic measures raise the possibility of valuable appreciation of the approach among spatial data handling scientists and users within the web-GIS community. The hierarchical approach offers the possibility for the instrumentation of the concept of the semantic distance introduced as a distinguishing parameter on which we can develop other semantic measures explained in this paper, thus raising the value of the proposed model and methodology.

In the paper we have theoretically defined the semantic parameters for measuring semantic similarities within the semantic framework of the universal ontology of geographical space (UOGS) as well as shown our applicative approach to the similarity analyses of spatial databases. We have tried to demonstrate our experience and advantages of using the predicate logic approach in overcoming problems connected with the spatial data search. As a first order predicate logic programming environment prolog offers capabilities that suit the needs of, for example, defining a hierarchy of ontological concepts and the necessary querying mechanism. Not only this, logic programs could be learned with the inductive logic programming (ILP) approach what could be of particular interest when figuring out the rules for extracting semantic relations from the natural language text.

The challenges that should be tackled in the future are of a versatile nature. Firstly, the fine-tuning of the existent and the definition of additional measures for assessing the overall semantic similarity of two spatial databases by means of UOGS should be carried out. Next on the waiting list is an implementation of the first order predicate rules for extracting additional semantic information found in the object catalogue and metadata documents, along with assessing the accuracy of the learned rules on several different spatial data catalogue documents. Beside the implementation of the other components of the question answering system (e.g. question classification, answer ranking) and recalling the use-case scenario, the toughest of all will probably be the explication of the question answering system's knowledge in a way in which the system will be: "cognizant of the context in which the user is asking the question, and of his or her purpose in asking it" (Bilotti 2004). In this manner the system will be able to deliver an answer tailored to the information needed by the user by posing additional questions and/or useful tips.

Acknowledgements

We would like to thank the company Amebis for their unselfish application of their POS tagger to a set of sentences provided for the purpose of our research. Domen Smole would like to thank the Ministry of Higher Education, Science and Technology of the Republic of Slovenia for financing his research program.

References

Abney S, Collins M, Singhal A (2000) Answer Extraction. In: ANLP-2000

Bilotti MW (2004) Query Expansion Techniques for Question Answering. Thesis, Massachusetts Institute of Technology

Bowker G, Star S (2000) Sorting Things Out: Classification and its Consequences. MIT Press, Cambridge, MA

Bratko I (2001) Prolog Programming for Artificial Intelligence, 3rd ed, Addison-Wesley

Bratko I, Džeroski S, Kompare B, Walley W (1998) Analyses of Environmental Data with Machine Learning Methods. Center for Knowledge Transfer in Information Technologies, Jožef Stefan Institute, Ljubljana

Camara G, Monteiro V, Paiva J, Souza R (2000) Action-driven Ontologies of the Geographic Space. GIScience 2000, AAG, Savannah, GA

Čeh M (2003) Semantic Integration of Spatial Databases. Založba ZRC, ZRC SAZU, Ljubljana

Čeh M, Smole D, Podobnikar T (2004) Geodata – are they Accessible and Useful? In: Toppen F (ed) AGILE, Heraklion, Crete, pp 789–794 (winning poster)

Egenhofer M, Mark D (1995) Naive Geography. In: Frank A, Kuhn W (eds) COSIT '95 (= LNCS 988). Springer Verlag, Berlin, pp 1–15

Fellbaum C (ed) (1999) WordNet: An Electronic Lexical Database. MIT Press, Cambridge, MA

Kaptelinin V, Nardi B, Macaulay C (1999) The Activity Checklist: a Tool for Representing the Space of Context. Human-Computer Interaction, vol 1(4). Lawrence Erlbaum Associates, Inc., Mahwah, New Jersey

Kavouras M, Kokla M, Tomai E (2005) Comparing Categories among Geographic Ontologies. In: Gould M (ed) Computers & Geosciences, vol 31(2). Special Issue, pp 145–154

Kuhn W (2001) Ontologies in Support of Activities in Geographical Space. Int J of Geographic Information Science 15(7):613–631

Kuhn W, Raubal M (2003) Implementing Semantic Reference Systems. In: Gould M, Laurini R, Coulondre S (eds) AGILE, Lyon

Lakoff G (1987) Women, Fire, and Dangerous Things: What Categories Reveal about the Mind. University of Chicago Press, Chicago

Laurel B (1993) Computers as Theatre. Addison-Wesley, Reading, Massachusetts, USA

Muggleton S (1995) Inverse Entailment and Progol. New Generation Computing J 13:245–286

Rada R, Mili H, Bicknell E, Blettner M (1989) Development and Application of a Metric on Semantic Nets. IEEE Transactions on System, Man, and Cybernetics, vol 19(1):17–30

Rodriguez MA (2000) Assessing Semantic Similarity among Spatial Entity Classes. Thesis, University of Maine

Sowa J (2000) Knowledge Representation, Logical, Philosophical, and Computational Foundations. Brooks Cole Publishing Co, Pacific Grove, CA

A Quantitative Similarity Measure for Maps

Richard Frank, Martin Ester

School of Computing Science, Simon Fraser University
Burnaby B.C., Canada V5A 1S6
email: rfrank@cs.sfu.ca, ester@cs.sfu.ca

Abstract

In on-demand map generation, a base-map is modified to meet user requirements on scale, resolution, and other parameters. Since there are many ways of satisfying the requirement, we need a method of measuring the quality of the alternative maps. In this paper, we introduce a uniform framework for measuring the quality of generalized maps. The proposed Map Quality measure takes into account changes in all local objects (Shape Similarity), their neighborhoods (Location Similarity) and lastly across the entire map (Semantic Content Similarity). These three quality aspects measure the major generalization operators of simplification, relocation and selection, exaggeration and aggregation, collapse and typification. The three different aspects are combined using user-specified weights. Thus, the proposed framework supports the automatic choice of best alternative map according to preferences of the user or application.

1 Introduction

With on-demand maps attracting increasingly more attention from large organizations such as Google (maps.google.com), MapQuest (www.mapquest.com) and Microsoft (mappoint.msn.com), the quality of maps generated by these companies becomes a competitive factor. Some of these systems allow for user customization through selection of different object-types while others allow for the visualization of buildings in 3D (Google Earth). Within a single system maps can be generated in different ways and from different resources, each of which must satisfy a different set of requirements from the client. Previously, paper maps were evaluated by the geographer during the design and creation process to match the requirements as best as possible. Now, with large quantities of on-demand maps being generated almost instantly without human involvement, the quality of the final output must also be determined automatically by a computer [16].

In on-demand map generation, a given base-map is modified to the user requirements on scale, resolution, and other parameters such as themes or symbolic representation [11]. Due also to constraints on the output, compromises have to be made during this process: shapes must be modified, combined or not displayed [12]. Since there are many ways of doing generalization and satisfying the user requirement, we need a method of determining the best alternative map generated so that user preferences can be met. Current map creation processes do not include a way of measuring the quality of the result, nor do they allow customization in the way of preferences [3, 5, 13]. Currently no consolidated measure exists, one that incorporates the shapes of the individual objects, the relationship between them, and their distribution on a map. The methods that have been suggested in the literature [5, 13, 14 and 18] measure some or all of these aspects, but lack a uniform framework and cannot combine the individual aspects into a single metric; they display the results, sometimes as many as 7 [13], to the user for interpretation. These methods are not applicable in the increasingly important scenario where an algorithm automatically needs to select the best-generalized map. The goal of this paper is to find the best map from alternatives by establishing a uniform framework, which allows the calculation of a single metric quantifying the quality of a map generated regardless of source. Quality measures that attempt to do this are proposed in [13], [14] and [18], none of which are based on a uniform conceptual framework.

Our approach categorizes measurements into three aspects, which are similar to those in [13]. When analyzing shapes in isolation, we measure change through differences in the outline of the object. Then, instead of measuring absolute changes in location [13, 14], we propose a method to quantify the change of location relative to its immediate neighbors. Neighborhood relationship is determined through the use of adjacent Voronoi regions. Finally, we've modified the standard entropy metric so that it's more suitable for information displayed on a map when compared to the entropy metric developed in [18]. Taking into account user preferences, these three aspects will combine into one metric representing the quality of a map.

This approach allows the users to input their preference to which aspect should be given priority during generalization since different applications will assign different importance to the above three aspects. Given preferences, it's possible to tailor the map-generation process by generating several alternative maps meeting the user requirements of scale, area and resolution. After comparing them to the base map, the map with the highest quality is selected; the user is presented the best map that was tailored to their desires rather than a generic map that only meets user specifications of scale and resolution.

The rest of the paper is organized as follows: Section 2 surveys related work while Section 3 details our proposed approach along with detailed discussion of each aspect that it measures. Section 4 experimentally evaluates our approach and highlights the benefits. Section 5 concludes the paper with a summary and ideas for future work.

2 Related Work

[13] proposes an approach for quality assessment based on comparing data characterization before and after generalization. It describes 'micro', 'meso' and 'macro' level analysis, which is similar in breakdown to 'shape', 'location' and 'semantic' similarity measure in our method, respectively. 'Micro' contains individual objects of all classes, which are described by their properties, such as area, orientation, position, concavity and elongation. Meso level considers groups of objects, a city block for example, and measures such properties as density and proximity relations within these groups. Their grouping of objects limits the usefulness of, for example, a density function, since density is now not calculated across all objects or classes but only in a group, whatever a group happens to be. Hence map quality partially depends on the quality of, or meaning behind, the grouping. 'Macro' properties consider all objects and the distribution of the properties of the individual objects, with the possibility of a restriction to only a single class of object. Unlike our proposed Semantic Content measure, they have no 'global' measure that takes into account, for example, the spatial distribution of the objects (not their properties). Their algorithm provides them with 5 measures from the micro level and 1 from the meso level. The authors themselves state that the measures need to be aggregated since a human cannot understand the relevance of so many numeric measures at once.

In [18] the authors introduce 'Region of Influence' by using Voronoi diagrams to create 'Entropy of Voronoi Regions'. Instead of using the probability of an object of class i existing, the area of the Voronoi regions is used. This measure does not make a distinction between a class that is made of one object with Voronoi area of 10, and n objects with a total Voronoi area of 10.

[14] discusses how agents can co-ordinate and co-operate during the generalization process. Their quality measures are simple statistics regarding the objects (like size, minimum width, orientation, position, angle deviation and separation from other objects). This paper presents only the framework; it however does state their need for a map quality measure.

A method comparing and matching objects by using a shape description function based on the curvature, not structure, of the object is presented in [8]. For shape similarity they discuss the 'turning function', a plot of the length vs. slope of the curvature of the original object. The turning function allows the comparison of two shapes by numerically quantifying the difference between two objects by finding the area between two turning-functions.

[5] presents a method to generalize by attempting to calculate all possible maps that can be generalized. It tries to move each object into k possible places, for n objects, there are k^n possible maps. The evaluation function they adopt is based on minimizing the total number of conflicts (pairwise overlaps) within a particular region. As stated by the authors, they "require the development of more advanced evaluation functions that take account of a wider range of constraints, including those of form and structure".

3 Our Approach

During a typical generalization process multiple maps can be considered acceptable even when generated from the same source, but a single map has to be selected to be returned to the user. We propose a framework for comparing two maps generated from the same source, resulting in one metric. The final results of each comparison can be evaluated with respect to the user preferences and the optimal map then given to the user in response to their original request. This process is illustrated in Figure 1. Since we are given multiple maps generated from the same source, we assume that correspondence between objects from different maps is established through Object-ID's retrieved from the database when the map is created. If the method is applied to maps generated from different sources then extra metadata is required to establish correspondence between objects.

Fig. 1. Comparison of two maps

Since all computer-based output medium is made of pixels, be it a monitor or printer, the map that is presented to the user must be in raster format, regardless of whether sourced from bitmap or vector data. The generalization procedures vary for bitmaps or vectors, but the result must be bitmap based. This paper presents a method of measuring not the quality of the individual generalization operators but the quality of the map that is presented to the user. When a map is put through a generalization process, there are several operations that can be used to solve conflicts between objects on the final map. The choices are outlined in [1, 5, 12, 14, 16, 17], these are:

- not displaying some objects (selection/elimination)
- enlarging objects (exaggeration)
- combining objects (aggregation)
- moving objects to avoid overlap (relocation/displacement)
- removing features of the original object (simplification/reduction)
- reducing the dimensionality of the object, from spatial to a line, for example (collapse)
- representing using distribution patterns (typification)

Which operation deals with the conflict depends on the generalization, but regardless of the algorithm, the above types of changes can be classified into 3 different aspects.

The first is shape similarity and would measure the changes from exaggeration, collapse and simplification operations defined above. This measure takes place on the object level because only the single object is changed.

If an object is moved or the relative distances change due to a change of shapes, then the location of it, and hence relatively its neighbors', is changed, but this change is restricted to the local neighborhood. A change on this level impacts the location similarity and measures location changes due to collapse, displacement, exaggeration and simplification.

The last type measures changes due to aggregation, omission and typification. These changes all impact the entire map and hence have a global effect. Also, each object belongs to a class and the removal of an object will influence the entire class distribution across the map impacting the importance of the objects.

To combine the different aspects of map quality, it is desirable that the underlying conceptual framework for all aspects is uniform. When working with maps, Voronoi diagrams and Delaunay triangulations are prevalent since they clearly represent the spatial relationships between the objects [4, 7, 9]. These Voronoi diagrams split an area of space into regions, called Voronoi cells, which contain all the points that are closest to the object contained in the Voronoi cell. The size of these cells gives an indication of how dense an area a certain object is in or how large an object is and hence is a good aid when required to calculate the amount of information contained in an area. The Voronoi cell structure also yields the Delaunay triangulation, which easily allows calculating an object's immediate neighbourhood, a prerequisite to our calculations on the local level.

Skeletonizations, [2, 10 and 15], also called shock graphs, aim at calculating the similarity of two shapes; the skeleton is very similar to a Voronoi diagram (but within an object). However, for certain objects, such as those with indentations, the shock graph is very different although the original shapes can be considered the acceptable result of shape simplification, as in Figure 2.

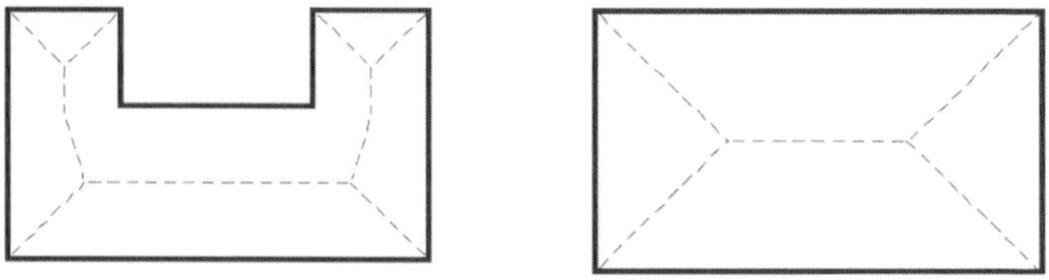

Fig. 2. Object, its generalized shape, their shock graphs

3.1 Shape Similarity

In order to compare the individual objects, some kind of descriptor is needed which describes the amount of change between two shapes. Typically shapes are described by certain properties, which yield a number, and either the absolute or relative change between the two values results in the *Shape Similarity* (SS) measure. The *Turning Function* is a step function, which describes a shape by its perimeter vs. slope; the x-coordinate denotes the distance along the perimeter and the y-coordinate denotes the value of the slope [8].

This function uniquely identifies the shape and is rotation independent since rotation of the original shape is equivalent to a translation in the turning-function [8]. However, in order to be able to compare two shapes, their turning-function must have the same length, i.e. the perimeter has to be normalized. A small disturbance in the perimeter of the object could cause the normalized turning-function to not align properly during matching because sides that should be the same will not be due to one object now having a larger perimeter even if the two shapes are very similar otherwise.

To isolate disturbances, a matching procedure must be performed [8]. The goal is for the shapes to be compared and the largest identical segment removed from normalization and distance calculations. Our matching algorithm compares the lengths and turns of all edges in both shapes and removes the largest identical subset.

After matching, Area(O_i), the non-overlapping area between the two turning-functions, is calculated, where O_i represents object i on both maps. For an example see solid areas in Figure 3. This yields the absolute difference between two shapes with a numerical value that can theoretically be fairly large. In order to restrict the result to be between 0 and 1, the relative SS for object i on Map A and Map B is defined using (see Eq. 1):

$$SS_i(O_i) = 1 - \frac{Area(O_i)}{Max[Area(TF_{i,A}), Area(TF_{i,B})]} \quad (1)$$

- where $Area(TF_{i,A})$ is the area under the turning-function for object i on map A.

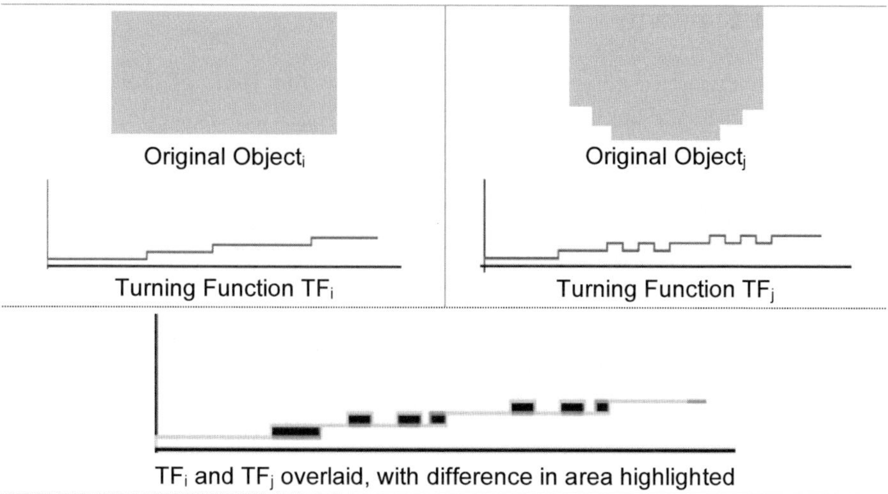

Fig. 3. Turning Function Comparison

If two shapes being compared are identical, then their turning-functions will be identical, resulting in an SS of 1. Globally, each shape on Map *A* is compared to its corresponding shape on Map *B* and each shape will yield one SS number. The Shape Similarity for Maps *A* compared to *B*, is defined as the average of the individual SS values using (see Eq. 2).

$$SS(Map_A, Map_B) = \sum_i SS_i(O_i)/n \tag{2}$$

3.2 Location Similarity

Objects can be displaced during generalization and this displacement becomes greater as scale decreases [6]. This change can be reflected and measured in the change in distance relative to other objects. A local neighborhood for an object is composed of the object itself and its immediate neighbors, where the neighbors are defined based on adjacency of Voronoi cells. Calculating pair-wise distances for all objects is computationally expensive and not necessary since changes in the Voronoi structure do not propagate past neighboring cells, hence, *Location Similarity* (LS) can be restricted to the immediate neighborhood, as illustrated in Figure 4.

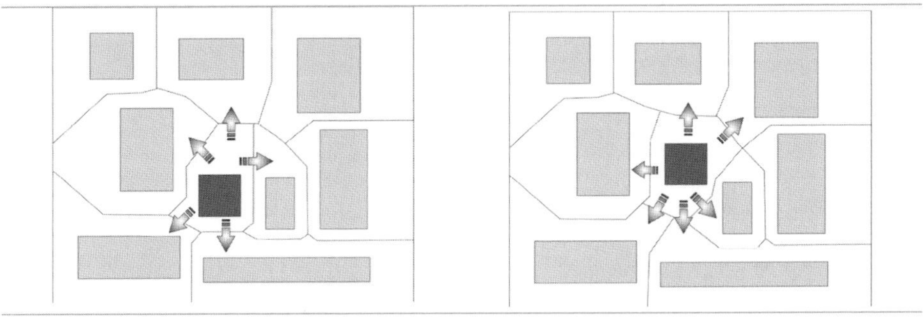

Before Generalization After Generalization

Fig. 4. Relocation

Distance between two spatially extended objects is typically based on distance between representative points on these objects. Defining representative points to be *center-of-gravities* can lead to points being inside other objects. Representative points of *A* can also be ones *closest* to *B*; this measure seems intuitive because this is the way humans also define distance, from the closest edge between two locations. Hence this type of measurement is used.

Pair-wise distances between two objects can be stored in a matrix $dist_A(i,m)$, where *i* and *m* are two objects on Map *A*. The *LS* for object *i* can then defined as (see Eq. 3)

$$LS_i(A,B) = 1 - \frac{\left[\sum_{m=1}^{p} \left| \frac{dist_A(i,m) - dist_B(i,m)}{\max(dist(i))} \right| \right]}{p} \tag{3}$$

where *max(dist(i))* is the maximum distance between object *i* and all of its Voronoi neighbors and *p* is the number of neighbors.

$LS_i(A,B)$ can be interpreted as the average change in distances between a central object *i* and objects around the central object. Since LS_i is a local neighborhood measure, it generates a measure for every single object, hence those measures must be combined into a single global measure. Thus the global *Location Similarity* for a map is defined by (see Eq. 4)

$$LS(Map_A, Map_B) = \left(\sum_i^n LS_i \right) / n \tag{4}$$

where LS_i is the Location Similarity metric for object *i* and *n* the number of objects on the map.

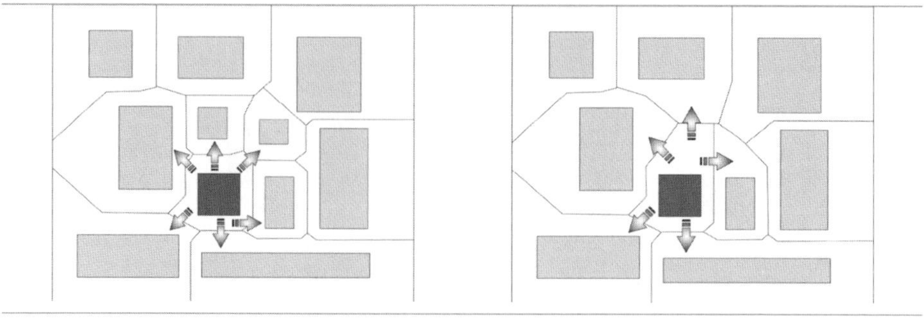

Fig. 5. Deletion (shaded object used for comparison)

This definition is assuming that all objects are of equal importance, and if this assumption does not hold then weights can be introduced for each class or object.

Since the location-similarity metric is normalized with respect to the largest distance between the object and its neighboring objects, the largest possible value that can occur is 1, hence the average of all $LS_i(A,B)$-s (i.e.: LS index for the map) will also be bounded between 0 and 1.

Due to operations done during generalization, such as deletion or move, it is possible that the set of neighbors before and after generalization will not be the same, as illustrated by the central object in Figure 5, in which case $dist_A(i,m) - dist_B(i,m)$ for the shaded object will not be defined since pair-wise distance cannot be determined before or after generalization. Neighborhood definition must be restricted to objects that existed before and after generalization and thus LS of objects is only defined on the common set of neighbors between two objects.

Another possible outcome of generalization where the neighborhood set is not the same before and after generalization is aggregation. In this case the aggregate object is not the same as the individual objects (before generalization) and hence would not be placed into the intersection of the two sets. An example is shown with the shaded objects in Figure 6 where the central four objects are aggregated. These differences are not measured by Location Similarity, but by Semantic Content Similarity (SCS).

3.3 Semantic Content Similarity

During the generalization process objects can be aggregated into one, or be deleted. These operations imply a loss of information and there have been statistics developed in [18] for measuring the amount of information in a

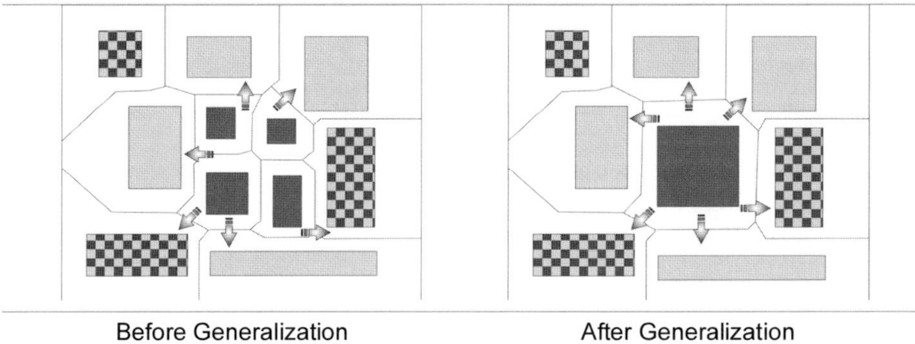

Fig. 6. Aggregation (different shading represent different classes)

set of objects. The existing entropy calculation measures the information within a set of non-spatial objects and is based solely on the number of objects within each class on the map. In reality, on a map the objects are spatially distributed, two maps with identical number of objects from the same classes could have drastically different spatial distributions as discussed in [18]. The method we propose takes into account spatial information by weighing each entry in the Entropy sum by their respective Voronoi areas, as a percent of the total map area.

Before attempting to define a modified entropy measure, it is important to define what the entropy measure is based on, i.e.: what an object is. Objects that intersect or are networks that span entire maps (ex: roads) should be treated as segments of individual objects: an object with a general shape of '+' can be split into 4 non-intersecting segments. This allows any shape-similarity measure to work since there are no objects that are not closed or contain holes. This also is required in order to measure disappearance or displacement of road segments of a road-network since a disappearance of a segment would be detected by SCS while a displacement would be captured using Location Similarity.

As discussed in [18], the size of the Voronoi cell is a good indicator of how large the object is and also indicates to a certain degree the distribution on the map. It would be beneficial to modify the existing Entropy measure by incorporating the relevance of objects according to their Voronoi regions. True, information is lost when something disappears, but objects remaining become more important; diversity is lost, but fewer objects are more influential. We define the *Voronoi Entropy* of Map *A* using (see Eq. 5).

$$VE(Map_A) = \sum\left[P_i \times \ln(P_i) \times \%V_i\right] \tag{5}$$

where

- V_i – total area of the Voronoi regions of objects in class i,
- K_i – number of objects of class i,
- N – total objects on the map,
- and $P_i = K_i/N$.

A class is a collection of objects that have the same semantic attributes but can vary in position, size and other attributes, such as the set of objects 'banks'. For example, using Figure 6, information in Table 1 be collected as:

Table 1. Semantic Content Similarity

		Map 1		Map 2	
Object	**Class**	**# objects**	**% Voronoi area**	**# objects**	**% Voronoi area**
Clear	Stores	4	50%	4	50%
Dark	Residence	4	12.5%	1	12.5%
Pattern	Parking-lot	34	37.5%	3	37.5%

$$\begin{aligned} VE(Map_A) &= [4\times\log(4)\times.5]+[4\times\log(4)\times.125]+[3\times\log(3)\times.375] \\ &= 2.0419 \\ VE(Map_B) &= [4\times\log(4)\times.5]+[1\times\log(1)\times.125]+[3\times\log(3)\times.375] \\ &= 1.7408 \end{aligned}$$

The largest value that the entropy measure of a single map can take can be arbitrarily large, but the change between two maps, when expressed relative to the larger entropy, is between 0 and 1. SCS can hence be defined as (see Eq. 6), the amount the two entropy measures have changed.

$$SCS(Map_A, Map_B) = \frac{MIN\left[VE(Map_A), VE(Map_B)\right]}{MAX\left[VE(Map_A), VE(Map_B)\right]} \tag{6}$$

The relative similarity can be *1* when Maps *A* and *B* are identical in composition, although they don't have to be identical, just the number and class of objects along with their Voronoi Areas have to be the same. This type of change, however unlikely, will give a result of *1*, but changes on these maps would be captured by the other measures.

Since the importance of the different aspects of map quality depends on the user or application, the weights for each can be user defined.

3.4 Combining the Three Quality Aspects

Although results from the three components are meaningful individually, in order to calculate one metric for each map, the three results have to be consolidated into one. Since all three numbers purposefully have the same range of values (0 to 1) and all behave in a similar fashion (i.e.: small change is indicated by a small value) it is possible to combine them. Simple average can be a good indicator; it however gives equal weight to all components when that assumption might not be appropriate for the application or user, hence weights are used as in (see Eq. 7).

$$Quality(Map_A, Map_B) = w_1 * SS(Map_A, Map_B) + \\ + w_2 * LC(Map_A, Map_B) + w_3 * SCS(Map_A, Map_B) \qquad (7)$$

Where w_i is the weight for the metric and $\sum w_1 = 1$.

The resulting quality value would be associated with the map that it was calculated from. A lower quality value would indicate that the map had to undergo a larger amount of change and hence quality is worse than a map with a higher measure.

In some instances the generalization operators can be considered to be dependent, for example the change of shape of one object could cause the distance between two objects to change. In these instances, two (or all three) measures would simultaneously change to reflect the change in the object itself. This not only is intuitive but desired.

4 Experimental Results

To illustrate the meaningfulness of the proposed map similarity measure, we present experimental results on some maps generated using generalization methods developed in the GEMURE project [16]. One of the goals of the GEMURE project was to improve on-demand cartographic information delivery through generalization and multiple-representation, and they required a method to evaluate the results, which our method was designed to do.

The first map is a large-scale map (see Fig. 7) that acts as the 'base-map' while two others (see Figs. 8 and 9) alternative generalizations of the same area. These alternative maps are then compared to the 'base-map' in order to find out their accuracy. The maps depict a small section of Quebec City with three different classes of objects: roads, residential buildings and commercial buildings.

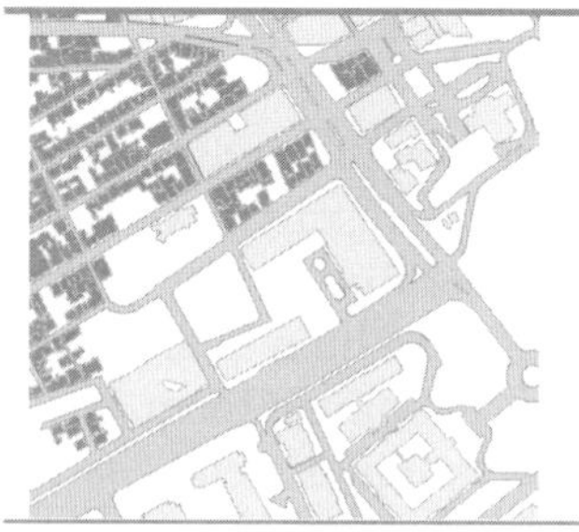

Fig. 7. Base-Map

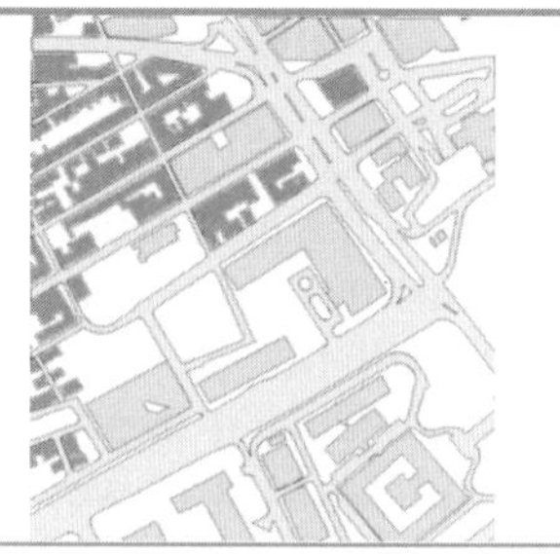

Fig. 8. Map 1

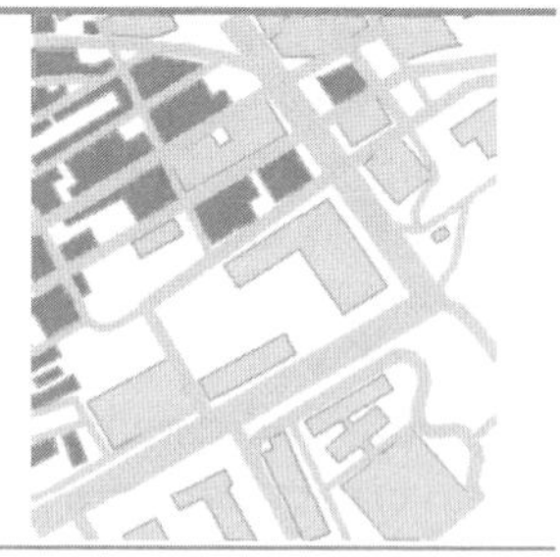

Fig. 9. Map 2

The first generalized map, *Map1* (Fig. 8) includes generalization effects of shape simplification along with merges of nearby smaller objects. The second map, *Map2* (see Fig. 9), is at a smaller-scale than *Map1* and also uses selection with a large number of objects disappearing or merging with neighboring objects. The results of the comparison are shown in Table 2.

The SS values are also somewhat intuitive: in *Map1* all objects that are both in *Base-Map* and *Map1* are fairly similar (note that this does not mean that all the objects are the same, some small objects were aggregated into bigger objects, but these are not counted for SS) whereas the outline of a lot of the objects were drastically changed going from the *Base-Map* to *Map2*. Since the objects on *Map1* are relatively close to their original counterparts, while *Map2* underwent a lot of change, hence the SS quality of *Map1* is higher.

Table 2. Results using a different set of weights

	Comparing	
*weight indicated in parenthesis	**Base-Map to *Map1***	**Base-Map to *Map2***
Shape Similarity (15%)	0.996	0.285
Location Similarity (70%)	0.821	0.828
Semantic Similarity (15%)	0.437	0.270
Final Similarity	**0.790**	**0.663**

The location-similarity is relatively high for both maps. *Map1* is able to preserve the location well since it is not generalized very much, *Map2* uses selection in abundance and hence is also able to preserve location relatively well.

Semantic Content Similarity was low for both maps because they underwent significant aggregation during generalization. *Map1* however had the best SCS since it contained the most objects while *Map2* contained fewer.

By defining a set of weights, the user is able to place more emphasis on different aspects of map quality. It is possible that a different map is selected depending on what the weights are. In this case, *Map1* is determined to be of better quality.

5 Conclusions

In this paper we presented an approach that can calculate the quality of alternative generalized maps using a uniform framework and present a single number that quantifies the quality. The approach takes into account changes in individual objects in the form of Shape Similarity, groups of objects using the Location Similarity and changes across the entire map using Semantic Content Similarity. The framework also allows for the user to specify a set of preferences which then influences the final metric and hence the choice of "best" map. Our experimental evaluation on real maps demonstrates that the proposed Map Quality measure produces intuitive results and, thus, supports the automatic map selection according to the preferences of the user.

Our method could also lead to a novel way of performing spatio-temporal data mining by allowing the calculation of the changes that areas of a map have gone through over time. By comparing two maps of the same area, but different time-periods, and applying a data-mining algorithm to sub-areas of the two maps, it would be possible to use the map as the search-space to search out the sub-area with the most change. The sub-area with the smallest Map Quality metric would indicate the region that has undergone the largest amount of. Using this method, it would be possible to answer such queries as "Where did most of the development occur between $Time_1$ and $Time_2$?" with nothing more than the temporal database available.

References

1. Jones CB, Abdelmoty AI, Lonergan ME, van der Poorten P, Zhou S, Multiscale Spatial Database Design For Online Generalisation. 9th Int Symp on Spatial Data Handling
2. Gold C (1999) Crust and Anti-Crust: a one-step boundary and skeleton extraction algorithm. Annual Symp on Computational Geometry, pp 189–196
3. Dijk SF van, Kreveld MJ van, Strijk T, Wolff A (2002) Towards an evaluation of quality for names placement methods. Int J of Geographical Information Science 16(7):641–661

4. Aurenhammer F (1991) Voronoi diagrams – a survey of a fundamental geometric data structure. ACM Computing Surveys 23(3):345–405
5. Ware JM, Jones CB (1998) Conflict Reduction in map generalization using iterative improvement. GeoInformatica 2(4):383–407
6. João EM (1998) Causes and consequences of map generalization. Taylor & Francis, London
7. Hoff III KE, Culver T, Keyser J, Lin M, Manocha D (1999) Fast computation of generalized Voronoi Diagrams using graphics hardware. SIGGRAPH 1999.
8. Longin JL, Lakamper R (2000) Shape similarity measure based on correspondence of visual parts. IEEE Transactions on Pattern Analysis and Machine Intelligence 22(10)
9. Mayya N, Rajan VT (1996) Voronoi Diagrams of Polygons: A framework for shape representation. J of Mathematical Imaging and Vision 6(4): 355–378
10. Klein PN, Sebastian TB, Kimia BB (2001) Shape matching using edit-distance: an implementation. Symp on Discrete Algorithms 2001. Twelfth annual ACM-SIAM Symp on Discrete Algorithms, pp 781–790
11. Rivest S, Bedard Y, Marchand P (2001) Toward better support for spatial decision making: defining the characteristics of spatial on-line analytical processing (SOLAP). GEOMATICA 55:539–555
12. Spaccapietra S, Parent C, Vangenot C (2000) GIS Databases: From multiscale to multi-representation. In: Proc of the Int Workshop on Emerging Technologies for Geo-Based Applications, LNAI 1864. Springer
13. Bard S (2002) Quality Assessment of Generalized Geographical Data. Accuracy 2002 Symp
14. Lamy S, Ruas A, Demazeu Y, Jackson M, Mackaness W, Weibel R (1999) The Application of Agents in Automated Map Generalisation. 19th Int Cartographic Conf
15. Sebastian TB., Klein PN, Kimia BB (2001) Recognition of Shapes by Editing Shock Graphs. ICCV:755–762
16. Bedard Y, Bernier E (2002) Supporting multiple representations with spatial databases views management and the concept of VUEL. Joint Workshop on Multi-Scale Representations of Spatial Data, ISPRS WG IV/3, ICA Commission on Map Generalization
17. Li Z, Yan H, Ai T, Chen J (2004), Automated building generalization based on urban morphology and Gestalt theory. Int J of Geographical Information Science 18(5):513–534
18. Li Z, Huang P (2002) Quantitative measures for spatial information of maps. Int J of Geographical Information Science 16(7):699–709

A Semantic-based Approach to the Representation of Network-Constrained Trajectory Data

Xiang Li[1,2], Christophe Claramunt[1], Cyril Ray[1], Hui Lin[2]

[1] Naval Academy Research Institute, Lanvéoc-Poulmic, BP 600, 29240 Brest Naval, France
[2] Joint Laboratory for GeoInformation Science, The Chinese University of Hong Kong, Shatin, N.T., Hong Kong

Abstract

Recent technological advances in urban traffic systems engender the availability of large trajectory data sets. However, the potential of these large urban databases are often neglected. This is due to a twofold problem. First, the volumes generated represent gigabytes of information per day, thus making data processing and analysis a computationally costly operation. Secondly, there is a lack of analysis of the semantics revealed by urban trajectories, at both the representation and data manipulation levels. The research presented in this paper addresses these two issues. We introduce an optimized representation approach that can efficiently reduce trajectory data volumes and facilitate data access and query languages. Our approach is a semantic-based representation model that characterizes significant trajectory points within a network. Key points are selected according to a combination of network, velocity, and direction criteria. This semantic approach facilitates trajectory data queries, the implicit modeling of trajectory processes. The proposed model is illustrated by a prototype implemented in a district of Hong Kong.

1 Introduction

With recent advances of positioning and telecommunication technologies, location-based databases are becoming widely available, particularly in the field of urban traffic systems (Jagoe 2003). Nowadays, one of the emerging forms of data within urban systems relies on the generation of trajectory-based data that represent the location of moving objects over time. In the context of urban traffic systems, vehicle trajectory data are integrated as an input of macro- and micro-modeling simulation systems that rely on the local properties and behaviors of individual vehicles (Peytchev and Claramunt 2001). For example, traffic data collection methods such as loop detector and closed circuit television collect information of individual vehicles in an indirect, infrastructure-based manner at the macro-modeling level, whereas trajectory data, at a micro-modeling level of granularity, can record such information in a direct and vehicle-based manner (Barron et al. 2004).

Despite the fact that trajectory data are fully integrated within urban traffic systems, there is still a need to make a better use of the large traffic databases generated to assist urban engineers and researchers. This will be of interest for monitoring and planning tasks in order to adjust time-dependant traffic regulations, discover the factors, which cause frequent traffic accidents, and design new transportation schemas. The research challenge to address is a twofold problem. First, the data volumes generated represent gigabytes or even terabytes of information per day, thus making data processing and analysis a computationally costly operation. Secondly, there is a lack of analysis of the semantics revealed by urban trajectories at both the representation and data manipulation levels. For example, studying the average travel times on a given path (e.g. from train station to city hall via roads *A*, *B*, and *C*, sequentially) during different time periods (e.g. every two hours from 10 a.m. to 8 p.m.) may require a trajectory data query language that includes spatial, temporal, and thematic dimensions.

The research presented in this paper introduces a semantic-based trajectory data model constrained by the underlying urban network that also presents the advantage of facilitating storage efficiency. The semantic model is based on characteristic *key points* that aggregate trajectory data without a loss of information. The reminder of the paper is organized as follows. Section 2 introduces basic network modeling concepts and a literature review of trajectory data representation approaches. Section 3 introduces the semantic-based approach. Section 4 introduces the interpolation algorithm. Section 5 illustrates the potential of the approach by a prototype development. Finally, Section 6 summarizes the paper and discusses further work.

2 Network-Based modeling

Let us first introduce some basic principles of network modeling (Sheffi 1985; Miller and Shaw 2001). Let G=<A, N> denote a network, where $A=\{a_1, a_2, \ldots a_n\}$ denotes the set of arcs and $N=\{n_1, n_2, \ldots n_m\}$ the set of nodes. For $\forall p \in [1..n]$, a_p=<n_f, n_t, *Geoline*>, where n_f and n_t denote the start and the end nodes, respectively, $f,t \in [1..m]$, and *Geoline* = <(x_1, y_1), (x_2, y_2), ...> is an ordered list of points representing the spatial extent of a_p. A *Trajectory* T is a time-ordered list of sampled points, so-called *location points*, that represent an object's trajectory with $T=\{(l_1, t_1), (l_2, t_2), \ldots (l_q, t_q) \ldots\}$; A tuple (l_q, t_q) denotes a location point, that is, the object's location l_q at time t_q.

Trajectory data can also be represented using a *network-based* model that retains the offset from the precedent node to l_q along an arc of the network to locate l_q. The *network-based* model also supports a logical representation that favors graph-based operations. The second modeling difference relies on the filtering of network data using a representation that retains the main points of a given trajectory or not, denoted hereafter as semantic-based or non semantic-based models. Therefore, we classify trajectory data representation approaches according to these criteria (see Table 1).

Table 1. Trajectory data representation approaches

Type	Modeling approach		Spatial reference	
	Location point / non-semantic-based	Key point / semantic-based	network-based	coordinate-based
A	√			√
B	√		√	
C		√		√
D		√	√	

The most straightforward trajectory data representation approach (type A), is often used in existing industrial applications such as fleet management and automatic vehicle location (Theodoridis et al. 1996; Nascimento et al. 1999; Saltenis et al. 1999; Tao and Papadias 2001; Hadjieleftheriou et al. 2002). Despite its simplicity, the critical drawback relies on the large data volumes generated. Network-based models (type B) explicitly represent data trajectories using a network coordinate system (e.g. Guting et al., 2004). However, trajectories are represented by either points or segments, without explicit consideration of the temporal dimension. Furthermore, as these models do not explicitly make the difference between basic trajectory points and trajectory points of importance in the network, the data volumes generated is still very large. Semantic representations have been associated to coordinate-based approaches (type C). (Sistla et al. 1997; Vazirgiannis and Wolfson 2001; Wolfson 2002) introduce a model where a trajectory is modeled as a dynamic attribute, using a temporal function to dynamically calculate moving object's location. Points of importance in the network are selected using velocity and direction criteria. However, trajectory data are referenced by a coordinate-based spatial reference. Also no developments have been made at the query language level. Recently, a semantic network-based representation (type D) has been introduced (Frentzos 2003). The model is developed at the logical level, but does not integrate the geometrical characteristics of the network. Criteria to filter points of importance in the network do not consider some important parameters such as changes of velocity.

3 Semantic-Based Approach

3.1 Key Points Modeling Principles

Our objective is to design an appropriate and efficient semantic modeling approach for the selection and characterization of network key points. A suitable selection criterion should model significant changes in all the dimensions of trajectory data. This involves for example a difference of velocity or change of direction between two location points (note that other changes such as acceleration can still be derived from velocity). In order to derive information on changes of velocity and direction, the geometrical and topological properties of the underlying network should be considered. The geometry of the network constrains these variables, particularly in the case of curvilinear networks. When an object mostly moves on straight arcs, less key points are likely to be selected when applying direction criterion than an object moving on curvilinear arcs. In fact, a network-constrained trajectory can be considered and modeled as a logical path sequentially consisting of an ordered series of arcs and nodes. Thus, the identification of key points should consider both network primitives, as most network models do. Representing key points in the network should allow identifying the nature of network changes and the underlying processes. The analysis of the semantic relationships and patterns between moving objects and the network should be also possible.

In order to select key points of interest in the network, we introduce a selection criterion as a change of *Object-Network Relationship* (ONR), where an ONR refers to the semantic relationships between a moving object and the underlying network. An ONR can be a relationship of two types, an 'object on an arc' or an 'object on a node', denoted respectively as OA and ON relationships. When considering key points, changes in the network are of four types: OA->ON, ON->OA, OA->OA, ON->ON. The last two cases represent discontinuous changes of ONR in the network when the sampling rate is low. Figure 1 gives some examples of trajectories projected onto a one-dimensional network path consisting of arcs and nodes, and where stars represent a time-ordered list of location points. In Figure 1b, the sampling rate is lower than that of Figure 1a, and thus the ONR value varies between any two adjacent location points. Accordingly, all location points in Figure 1b are key points.

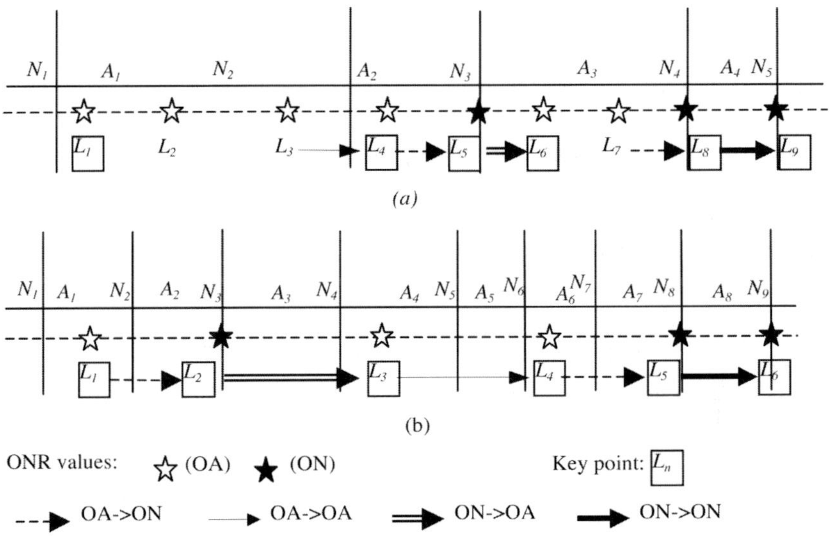

Fig. 1. Changes of ONR values with different sampling rates

The basic principle for selecting key points is as follows: a key point is generated when there is a change of ONR between two successive trajectory points in the network. Figure 2 exemplifies the four types of ONR changes, where K is the immediate key point before location point L_m.

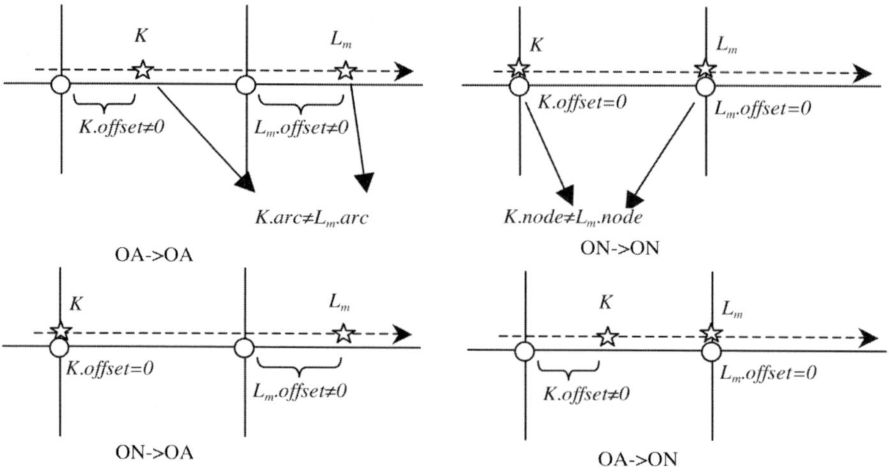

Fig. 2. Four types of ONR changes

3.2 Key Points Logical Representation and Processing

This section introduces the logical specification of the key point selection. Trajectory location points and key points are defined as a tuple (MLP): *<arc, node, offset, velocity, time>*. ONR changes are modeled as follows.

- OA->OA: $K.offset \neq 0 \wedge Lm.offset \neq 0 \wedge K.arc \neq L_m.arc$
- OA->ON: $K.offset \neq 0 \wedge Lm.offset = 0$
- ON->OA: $K.offset = 0 \wedge Lm.offset \neq 0$
- ON->ON: $K.offset = 0 \wedge Lm.offset = 0 \wedge K.node \neq L_m.node$

The value *offset* is used to detect the changes of ONR value. When the *offset* is the value *null*, the object is on a node (i.e. ONR value is ON), and otherwise the object is on an arc (ONR value is OA). Velocity criterion (C_v), direction criterion (C_d), and network criterion (C_n) are defined as follows (see Fig. 3):

- C_d: $K.direction \neq L_m.direction$
- C_v: $K.velocity \neq L_m.velocity$
- C_n: (OA->OA) ∨ (OA->ON) ∨ (ON->OA) ∨ (ON->ON)

Changes of velocity and direction have been already used to select key points (Sistla et al. 1997; Vazirgiannis and Wolfson 2001; Wolfson 2002). We also consider a change of network component as a significant semantic change to be considered in the selection of a key point, but combined to a directional criterion. Overall, a key point is selected if either the velocity of the moving point changes (C_n), or if both the direction and the ONR change (i.e. $C_v \vee C_d$).

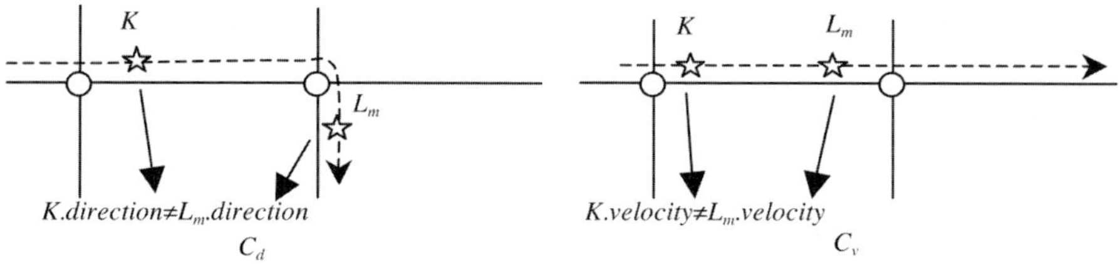

Fig. 3. Criteria C_d and C_v

3.3 Discussion

As illustrated in Figure 4, an advantage of this approach is that the number of key points selected with criteria based on velocity, network and direction together is smaller than that of a solution based on velocity and direction, or velocity and network criteria. In particular, this approach is most likely to not select redundant and not significant key point within the network. For instance, Figure 4 illustrates a situation where a velocity criterion is not retained as the object considered is moving at the same velocity on every location point.

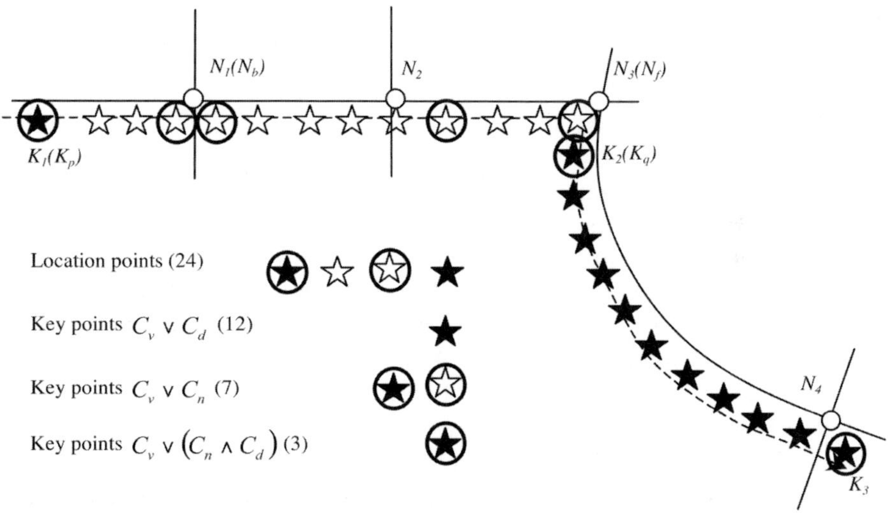

Fig. 4. Comparison of key point selection approaches

The second advantage of the approach is that key points can be selected from any location in the network, at a node or within an arc in the network. In comparison, the approach suggested by Frentzos (2003) requires that a key point must be coincident with a node, which is not always the case, and also not always straightforward to guaranty from a data integration point of view.

The third advantage of our approach is that the network-based spatial reference can efficiently reduce the spatial dimensionality of trajectory data from two dimensions to one dimension, while maintaining the relationships between moving objects and the underlying network. This advantage simplifies the physical index structures of network-constrained trajectory data access methods, and facilitates the evaluation of network-based trajectory data queries.

In order to be implemented, our approach requires some pre-processing that guaranties the map-matching of converting coordinate-based trajectory data to network-based trajectory data and MLP tuples. However, one can consider that a map-matching process is always inevitable regardless of representation choices. As previously mentioned, queries of network-constrained trajectory data usually include network-based query conditions such as shortest paths.

4 Prototyping

A prototype that implements the proposed trajectory data representation approach has been developed on top of the ESRI's ArcGIS software. The prototype is a trajectory data management system based on a part of the Hong Kong road network. Trajectory data has been received and recorded using vehicles equipped with GPS receivers. The prototype integrates and compiles trajectory data according to our modeling approach and with respect to the underlying network.

The road network is represented using the node-arc model including 1099 arcs and 1025 nodes. The sampling rate of GPS receiver is one location point per second. An algorithm introduced by Li et al. (2005) is used to conduct the map-matching process. A key point selection program is implemented with ESRI's ArcObjects. Figure 5 illustrates a trajectory represented either by location points or key points, and selected using different criteria. Each screen snapshot presented in Figure 5 is labeled and related to the corresponding logical expression, and key point or location point numbers.

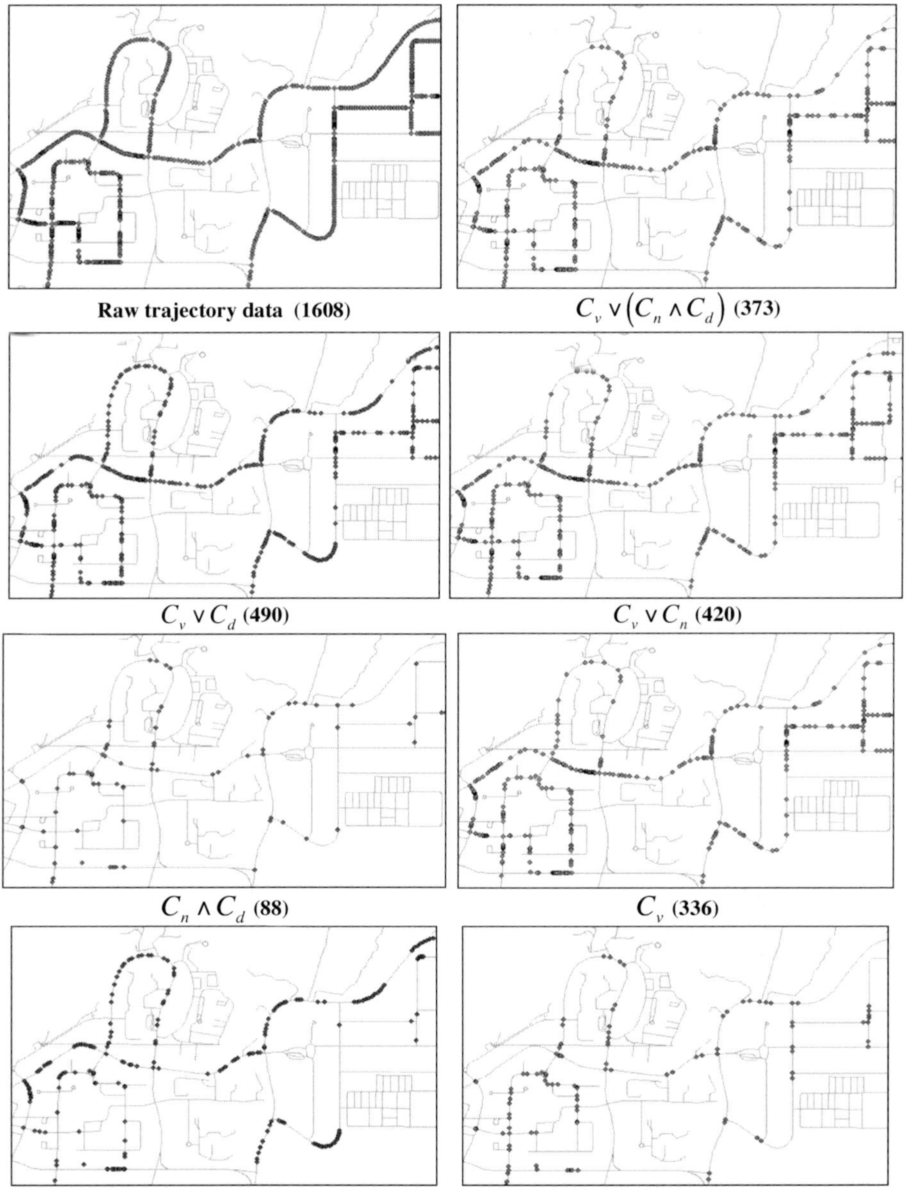

Raw trajectory data (1608) | $C_v \vee (C_n \wedge C_d)$ **(373)**

$C_v \vee C_d$ **(490)** | $C_v \vee C_n$ **(420)**

$C_n \wedge C_d$ **(88)** | C_v **(336)**

C_d **(217)** | C_n **(145)**

Fig. 5. Trajectories represented by different logical expressions

Figure 5 presents the different selection criteria combinations in order to demonstrate their relative capabilities. These criteria combinations can be classified in different groups:

1. $C_v \vee (C_n \wedge C_d)$, $C_v \vee C_n$, $C_v \vee C_d$ to preserve the maximum semantics of trajectory data.
2. $C_n \wedge C_d$, C_n, C_d to keep all information of trajectory data except velocity.
3. C_v to keep velocity information of trajectory data

Each combination of selection criteria can be applied to different sorts of applications. For example, group 1 may be well suited for real-time traffic applications which need concise spatial and temporal information of trajectory data, whereas group 2 for long-term traffic applications, such as traffic planning and traffic pattern mining, where vehicle route choice information is much more important than concise velocity information. The selected key points retained for group 1 and 2, and the effects of the selection criteria in the resulting numbers of selected key points are shown in Figure 6. Two trends can be inferred from these two figures. First, the results of the case study coincide with our earlier modeling analysis. For instance, the number of key points selected with the criteria based on either the velocity of the direction and network criteria, that is, $C_v \vee (C_n \wedge C_d)$ or $(C_n \wedge C_d)$, gives the lower number of key point whatever the group of criteria. Secondly, the velocity criterion plays a more dominating role than the others. In the first group, all three algorithms select much more key points than the other scenario. One possible explanation to this trend stems from the nature of the moving objects, i.e. vehicles, which usually change their velocities before and after changing directions and passing through intersections (nodes).

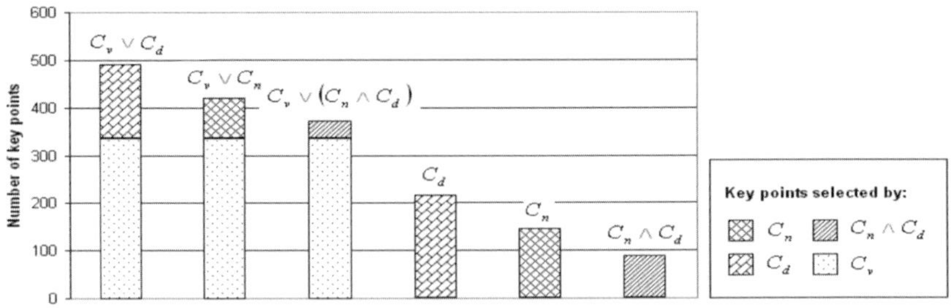

Fig. 6. Selected key point numbers comparison

The prototype also helps to evaluate the impacts of the ONR changes (i.e., ON->ON, ON->OA, OA->OA, and OA->ON) on the network with respect to the different selection algorithms such as $C_v \vee (C_n \wedge C_d)$, $C_n \wedge C_d$, $C_v \vee C_n$, and C_n. Results, as illustrated in Figure 7.

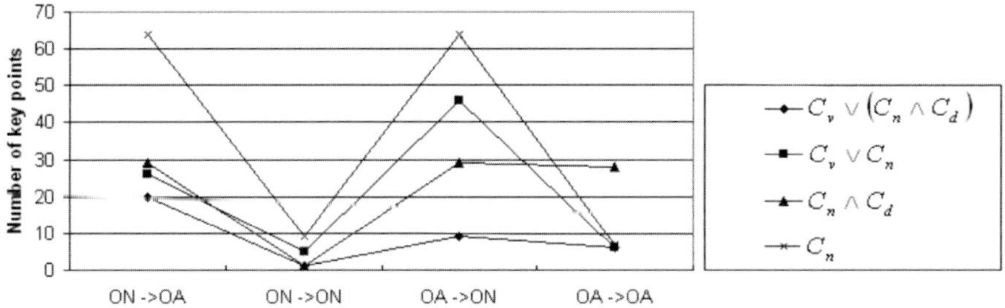

Fig. 7. C_n scenarios comparison

The output shows that ON->OA and OA->ON happen more frequently than the two others. One can remark that the numbers of key point associated with different scenarios are not the same due to the respective velocity and direction criteria. With respect to urban traffic management, the above analysis can be used to analyze the traffic performance of a road network according to the physical meanings of the different changes. For example, when considering trajectory data collected on a given network, larger numbers of ON->ON or OA->OA changes are likely to reflect free traffic flows, while larger numbers of OA->ON and ON->OA changes are likely to correspond to relatively obstructed traffic. It is noted that the above statement is also related to GPS sample rate. If the sampling rate is high, the changes of ONR values of vehicles under different traffic conditions might be similar. Therefore, a pre-defined GPS sampling rate is needed according to the geometry of the underlying road network, if one wants to apply the analysis of ONR changes (e.g. Fig. 7) to the evaluation of traffic performance.

Respective performances of selections based on network and distance criteria are illustrated in Figure 8. The experiment is made on two different sorts of network structures; curvilinear arcs (see Figs. 8a and 8b) and straight arcs (see Figs. 8c and 8d). It is worth observing that network selection criterion C_n performs better than direction selection criterion C_d for curvilinear arcs, and the result is reversed for straight arcs. This shows that application of a composite logical expression (i.e. $C_n \wedge C_d$) makes the proposed approach adaptable to different geometrical features of the underlying network.

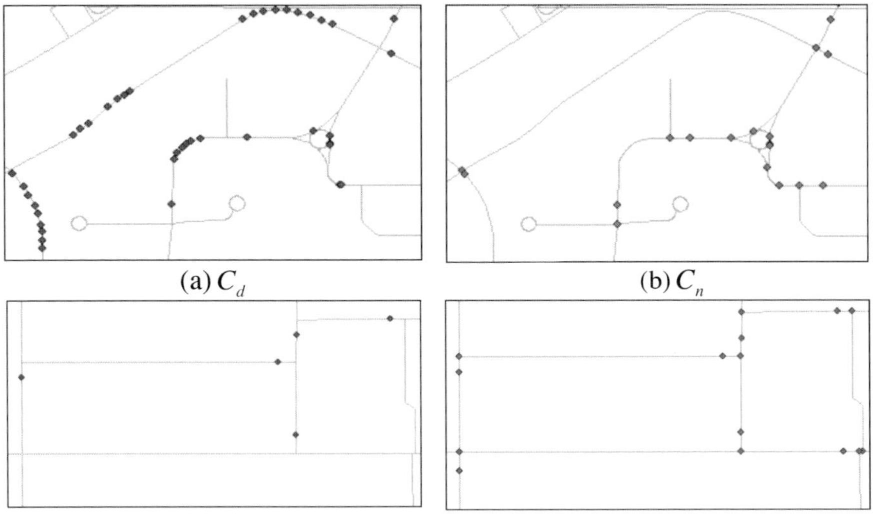

(a) C_d (b) C_n

(c) C_d (d) C_n

Fig. 8. Performance comparison between C_n and C_d

5 Conclusions

The research presented in this paper introduced a semantic-based representation approach oriented to network-constrained trajectory data. The approach is based on a key point selection method that integrates different network-based selection criteria. These criteria include network, velocity, and direction parameters. This semantic-based model is adapted to network structures and trajectory properties on different network environments. The proposed approach has been implemented in a demonstrative prototype in a district of the city of Hong Kong. The semantic relationships between moving objects and the network, and their different changes have been analyzed and formally defined. The experiments show that the proposed model represents trajectory data with less data volume than existing approaches, and facilitates the analysis of the semantic characteristics of trajectory data.

Further work concerns the development of a semantic-based query language and appropriate index structures that integrate the temporal dimension as a core component. We also plan to extend the prototyping development towards a comprehensive traffic analysis system applied to the city of Hong Kong in association with existing urban traffic systems.

References

Barron K, Collins J, Derr R, Jacobson L, Lomax T, Mudge D, Robinson JR, Row S, Smith B, Smith C, Tarnoff P (2004) The future of travel time data – a paradigm shift. Retrieved 14 October 2005 from http://depts.washington.edu/ahb20/reports/FinalTrvTime2-3-2004.pdf

Frentzos R (2003) Indexing moving objects on fixed networks. Proc of the 8th Int Symp on Spatial and Temporal Databases (SSTD), Santorini Island, Greece, pp 289–305

Guting RH, Almeida VTd, Ding Z (2004) Modeling and querying moving objects in networks. Research Report. Fern Universitat Hagen.

Hadjieleftheriou M, Kollios G, Tsotras VJ, Gunopulos D (2002) Efficient indexing of spatiotemporal objects. Proc of the 8th Int Conf on Extending Database Technology, Prague, pp 251–268

Jagoe A (2003) Mobile location services: the definitive guide. Prentice Hall, Upper Saddle River, NJ

Li X, Lin H, Zhao YB (2005) A connectivity-based map matching algorithm. Asian J of Geoinformatics 5:69–76

Miller HJ, Shaw SL (2001) Geographic information systems for transportation: principles and applications. Oxford University Press, New York

Nascimento MA, Silva JRO, Theodoridis Y (1999) Evaluation for access structures for discretely moving points. Proc of the Int Workshop on Spatio-Temporal Database Management (STDBM'99), Edinburgh, Scotland, pp 171–188

Peytchev E, Claramunt C (2001) Experiences in building decision support systems for traffic and transportation GIS. Proc of the 9th Int ACM GIS Conf, Atlanta, pp 154–159

Saltenis S, Jensen CS, Leutenegger ST, Lopez MA (1999) Indexing the positions of continuously moving objects. Research Report. TIMECENTER

Sheffi Y (1985) Urban transportation networks: equilibrium analysis with mathematical programming methods. Prentice Hall, Englewood Cliffs, NJ

Sistla AP, Wolfson O, Chamberlain S, Dao S (1997) Modeling and querying moving objects. Proc of the Int Conf on Data Engineering, Birmingham, UK, pp 422–432

Tao Y, Papadias D (2001) MV3R-tree: a spatiotemporal access method for timestamp and interval queries. Proc of the 27th Int Conf on Very Large Databases, Roma, Italy, pp 431–440

Theodoridis Y, Vazirgiannis M, Sellis T (1996) Spatio-temporal indexing for large multimedia applications. Proc of the 3rd IEEE Conf on Multimedia Computing and Systems, Hiroshima, Japan, pp 441–448

Vazirgiannis M, Wolfson O (2001) A spatiotemporal model and language for moving objects on road networks. In: Jensen CS, Schneider M, Seeger B, Tsotras VJ (eds) Advances in spatial and temporal databases. Springer, Berlin, pp 20–35

Wolfson O (2002) Moving objects information management: The database challenge. Proc of the 5th Workshop on Next Generation Information Technologies and Systems (NGITS'2002), Caesarea, Israel, pp 75–89

Towards an Ontologically-driven GIS to Characterize Spatial Data Uncertainty

Ashton Shortridge, Joseph Messina, Sarah Hession, Yasuyo Makido

Department of Geography and Center for Global Change and Earth Observations, Geography Building, Michigan State University, East Lansing, MI 48824-1115, USA
email: ashton@msu.edu, jpm@msu.edu, makidoya@msu.edu

Abstract

Current data models for representing geospatial data are decades old and well developed, but suffer from two major flaws. First, they employ a one-size-fits-all approach, in which no connection is made between the characteristics of data and the specific applications that employ the data. Second, they fail to convey adequate information about the gap between the data and the phenomena they represent. All spatial data are approximations of reality, and the errors they contain may have serious implications for geoprocessing activities that employ them. As a consequence of this lack of information, users of spatial data generally have a limited understanding of how errors in data affect their particular applications. This paper reviews extensive work on spatial data uncertainty propagation. It then proposes development of a data producer focused ontologically-driven GIS to implement the Monte Carlo based uncertainty propagation paradigm. We contend that this model offers tremendous advantages to the developers and users of spatial information by encapsulating with data appropriate uncertainty models for specific users and applications.

1 Introduction

The propagation of spatial data uncertainty to application results is not a new idea. Examples from the research literature include a host of implementations, a few examples of which follow to illustrate the point. Canters et al. (2002) employ a Monte Carlo approach to assess the impact of input uncertainty on landscape classification using land cover and terrain data. Aerts et al. (2003) demonstrate its utility for spatial decision support. Ehlschlaeger et al. (1997) investigate the spatial and cost variability of cross-country routing due to elevation uncertainty, while Fisher (1991) considers viewshed uncertainty. Variability in soil properties has been examined in numerous studies via this technique, including Heuvelink (2002).

In spite of substantial work by the research community, the expectations of early adventurers into spatial data uncertainty propagation (e.g. Openshaw 1989) for widespread adaptation have not been realized. A variety of reasons for this failure exist: complicated models, lack of necessary information, and difficulties with existing GIS, including both the propagation itself, as well as the preprocessing and fusion of data from different sources and spatial resolutions. The following paragraph treats each of these factors in turn.

First, uncertainty models are highly sophisticated, typically requiring specialized knowledge of geostatistics to comprehend (Heuvelink 2002) and employ. While software capable of implementing such models is becoming more accessible (e.g. Pebesma and Wesseling 1998; R Development Core Team 2005), the concepts behind these models as well as their parameterization remain forbidding. Second, the information required to parameterize such models, particularly the spatial structure of error, is often unavailable to users (Heuvelink 2002). In general, current metadata standards are inadequate because they fail to convey meaningful information about the fitness for use of data (Fisher 1998; Frank 1998). Approaches to modify the metadata paradigm to address data quality more robustly have been identified. Devillers et al. (2005) propose a framework to structure and communicate data quality about a hierarchy of attributes. Duckham et al. (2001) introduce ontology of imperfection encompassing error, imprecision, and vagueness, and demonstrate how rough sets might be used to incorporate an assessment of vagueness in retail sitting. Third, GIS must be extended to easily handle Monte Carlo simulation. (Karssenberg and de Jong 2005). Methods for incorporating model-based uncertainty have also been proposed (Hwang et al. 1998). Methods for integrating model uncertainty and data quality have also been developed (Crosetto and Tarantola 2001). Fourth, the interpretation of results is not straightforward, especially if the desired output scale is not the same as the input (Heuvelink 2002; Heuvelink 1998).

We suggest that the overarching problem is the general manner in which spatial data are distributed, and employed, and that its solution requires fundamental changes in this system. The challenge of handling spatial data uncertainty is in fact linked to the challenge of spatial data handling. The following section describes the current state of practice. Section three is concerned with the gap between this current state and what is evidently desirable, and a superior method. Section four integrates this methodology with an ontologically-driven geographic information system, and illustrates what such a system might look like. Section five concludes with a discussion of key implementation challenges and expectations.

2 Metadata, Fitness, and Propagation

The days when individual researchers digitized data from paper maps are largely behind us. Most digital spatial data – many gigabytes per day – are developed from primary sources and distributed by providers. Global networking and the continuous development of new application domains have been responsible for these important changes to information dissemination and application processes (Fonseca et al. 2002). Contemporary information systems are becoming increasingly distributed and heterogeneous, while digital libraries are now a significant component of this emerging trend towards knowledge-based distributed environments. A significant challenge is how to integrate geographic information of different kinds at different levels of detail (Longley et al. 1999), and to provide spatial data users with appropriate data for their particular applications.

Because spatial data producers and users are generally not the same, carefully designed metadata specifications are critical for enabling the fullest possible use of spatial data (Hill et al. 2000). Indeed, the rationale for these specifications is to enhance the sharing of spatial information, to encourage consistency in data generation and use, and to reduce redundancy in data compilation (SDTS 1996). To achieve these admirable goals, metadata reports generally consist of information about the spatial location and extent of the data, the methods by which the data were collected, the storage format, the producer and distributor, and, most relevant for this paper, the quality of the data (Guptill 1999).

While the Federal Geospatial Data Committee (FGDC) standard was released in 1994, FGDC development efforts began in 1992 (FGDC 1998) and the standard owes much of its content to work dating from early in the previous decade by the National Committee for Digital Cartographic Data

Standards (Chrisman 1984). In fact, the five primary elements of spatial data quality are present in the Chrisman paper, published a decade before the release of the FGDC standard. So, the notion of characterizing data quality in terms of these five components has been around for a long time. The fitness for use objective has been around for at least as long (Chrisman 1984). The highly practical intent of the data quality specifications is clearly to enable data users to determine whether particular data sets are appropriate for their applications. This language is repeated throughout the various map and accuracy standards documentation (FGDC 1998; SDTS 1996; SDTS 1998).

Typical data users are not interested in the data set itself, but rather in the phenomenon that the data set imperfectly represents. They need to know how imperfect this representation is, as it relates to their applications. Consider a forester who wishes to use a USGS DEM to help identify promising sites for a new fire tower. The forester has calculated the size of the viewshed for a set of locations, and is interested in determining how closely the calculated viewshed matches the actual viewshed at these sites. That the RMSE for the quadrangle does not exceed 7 meters is not a detail that can easily be used to determine the quality of the viewshed calculations (Fisher 1991).

This situation is not unique to the USGS DEM case, but is a general failing of data quality standards (Goodchild 1995). Current standards fail to satisfy their government mandated objective to, "report what data quality information is known", so that users can make informed decisions about the applicability of the data for their applications (USGS 1996). Specifically, currently reported global measures of data quality are inadequate for developing the models that drive the propagation method, since they provide no information about spatial structure (Fisher 1994; Wong and Wu 1996). Indeed, current map and data accuracy standards in general are not sufficient to characterize the spatial structure of uncertainty (Goodchild 1995; Unwin 1995). Without a model characterizing local spatial variation, data users are unable to complete a fitness for use assessment.

We suggest that the only general-purpose method for achieving the fitness for use criterion is direct propagation of uncertainty from data to the specific application result. This requires a model capable of generating multiple realizations of the phenomenon of interest and a Monte Carlo simulation implementation to apply these realizations to a specific spatial application (see Fig. 1). The simulation method is completely general, in the sense that any spatial application may be handled. This appears to be the most adequate way of satisfying the fitness for use requirement (examples and arguments in favor of the approach appear in Heuvelink et al.

1989; Fisher 1991; Lee et al. 1992; Englund 1993; Ehlschlaeger et al. 1997). The implications for modeling uncertainty stochastically are profound:

- 'fitness for use' mandate for Federal spatial data producers is satisfied
- multiple sources of data may be combined to improve information content
- data developers have more control over distribution of information
- users understand the effects of uncertainty on a case-by-case basis
- analysis results are more realistic, more defensible, and more useable

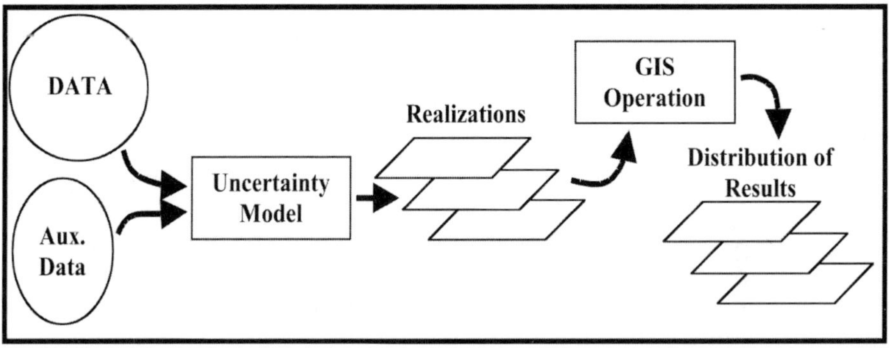

Fig. 1. Propagation of uncertainty from data to results

3 Ontology-driven GIS

In an ontology-driven geographic information system (ODGIS), an ontology is a component, such as the database, designed and functioning to fulfill the systems objectives (Guarino 1997). The first step to build an ODGIS is to specify the ontologies that are stored and translated into a formal computing language. The stored ontologies also serve as a library of information about the knowledge embedded within the system. The ontologies are reformed as objects or classes that contain the operations and attributes that constitute the system's functionality (Fonseca and Davis 1999). With this research, special emphasis is given to using the ontological structures for semantic information integration between geographic information systems and geospatial data as both product and producer of multidimensional data.

The use of ontologies in GIS has been discussed by Frank (1998, 1997) and Smith (1998). Ontology plays a central role in the construction of next generation GIS through the establishment of correspondences and interrelations among different domains of spatial entities and relations (Smith 1998). Frank (1997) believed that the use of ontologies would contribute to better information systems by avoiding the problems associated with multiple data model representations in existing GIS. Kemp and Vckovski (1998) note that although certain types of geographic phenomena, like discrete objects, have been the subject of ontological study, spatially continuous phenomena have received little attention. The proposed approach provides dynamic and flexible information exchange and allows partial integration of information when security or reliability constraints make absolute confidence impossible. The ODGIS approach is dramatically different than the simple import or data transfer options already in existence. The ODGIS model allows for a common ground in which multiple data models, algorithms, error specifications, and data sets with very different spatial and temporal characteristics can interact (Fonseca et al. 2002).

The ODGIS structure has two main components: knowledge generation and knowledge use. Knowledge generation involves the specification of the ontologies, the generation of new ontologies from existing ones, and the translation of the ontologies into software components (Matsuyama and Hwang 1990). The ontologies or definitions are available to the end user and contain metadata. ODGIS are infused with two basic premises: ontologies are explicit before the GIS is developed, and a hierarchical structure. Explicit ontologies are not database schemas, but rather an agreement on the nature of the entities to be generated. The data, e.g., terrain data, must be adapted to fit the different ontological frameworks. User communities create diverse ontologies, but it is our belief that these diverse ontologies can be explicitly specified and reconciled given the finite nature of the data. In an ODGIS implementation it is necessary to assemble and specify ontologies at different scales. An example of this procedure can be found in the document produced by the FAO of the United Nations where high-level ontologies were developed for the classification of different soil types (Gregorio and Jansen 1998). The components of the hierarchy of ontologies are classes modeled or defined by their specific characteristics (e.g. parts, functions, attributes, and roles). These components, if completely specified, allow for effective representation with targeted audiences.

The ODGIS framework offers a way to explicitly develop geospatial data models that are suitable for specific and different applications. From a use perspective it is a top-down approach to conceptualize data. Data uncertainty is one data-specific component of the conceptual model, albeit a crucially important one. It is also a bottom-up perspective, beginning with data and error specification and rising to match specific applications. Both approaches embrace an object-oriented implementation, in which the data model encapsulates appropriate methods to describe, depict, and employ geospatial information. The framework encourages multiple representations, enabling the implementation of truly stochastic spatial data models.

4 Tying Uncertainty Models to ODGIS

We propose a prototype model to implement the uncertainty propagation model presented in Figure 1 for geospatial data with a specific goal of incorporating digital terrain data. We envision a system resembling that depicted in Figure 2; this design separates users from the spatial data maintained by the data producer and provides the data producer with substantial security and flexibility in representation. The left side of Figure 2 represents that part controlled by the spatial data producer or maintainer. This includes both a database with all terrain data and information necessary for each specific error model, along with multiple error models. Some of these models are 'better' than others, in that they use higher quality information to produce more accurate and/or higher spatial resolution realizations.

The right side of Figure 2 represents different spatial data users – or classes of users. Requests for spatial data by different users result in the provision of spatial data realizations produced by different error models. Each user would then subject those realizations to the GIS application, thereby creating a distribution of results, as shown in Figure 1 and discussed earlier. The ODGIS framework described in the previous section provides the basis for linking particular applications or users to appropriate data models.

The system would enable the producer to determine an appropriate error model for a particular class of data users. Data users would not have direct access to the original spatial data; in effect, the server delivers appropriate realizations on-demand to the user's computer (Shortridge 2000). This model offers security and flexibility to the data maintainer, satisfies the fitness-for-use mandate, and integrates two decades of GIScience research on uncertainty into the GIS knowledge production process.

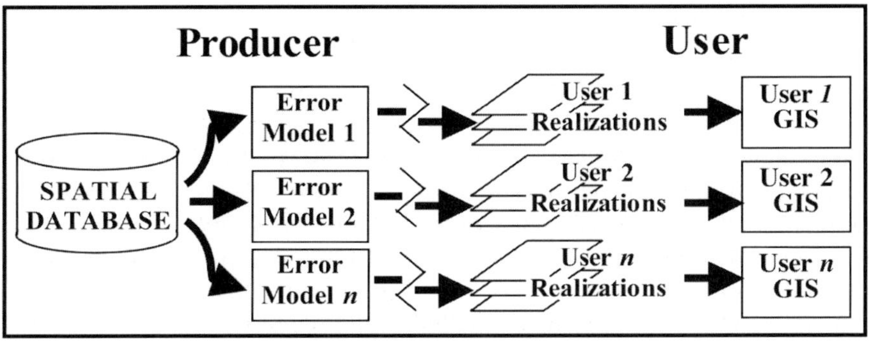

Fig. 2. A framework for uncertain data distribution

5 Implementation and Evaluation

The initial case study for the ODGIS model is set in East Africa. The Climate Land Interaction Project (CLIP), an NSF-funded project, supports an interdisciplinary team that is developing regional climate and land use models for this area (CLIP 2006). The goal is to study interactions between climate change and land use/cover. As climate conditions change, it is anticipated that people will make decisions that affect the cover and use of the landscape. These changes can in turn alter critical surface conditions, with consequent feedback impacts on regional climate.

Important input data for these models include land cover and digital elevation models; currently several continental and global data sets (e.g. Africover land cover data and SRTM elevation data) are available for the region. In addition to these readily available datasets, substantial reference data have been gathered in portions of the study region by CLIP researchers and others. An ODGIS will be developed for these inputs with representative models, enabling CLIP scientists to assess the sensitivity of their models to data uncertainty. Preliminary results for the study region illustrate the talk given in Vienna at SDH 2006.

Designing a modeling framework is, in many ways, simpler than parameterizing, calibrating, and validating the model. We define calibration as the process of refining the parameter settings to more closely mimic reality. However, parameterization and validation of simulated systems are frequently ignored. Parameterization in this context refers to the process of assigning temporal constants to the structures, procedures, or functions within the model itself, and the process of justifying the existence of those same structures (Messina and Walsh 2001). In process-based models, parameterization occurs via direct measurement or statistical prediction. Both statistical and non-statistical inferences are made with the goal of fitting the appearance of the model to reality or a spatial reference, defined by initial conditions, spatial or temporal constraints, or predefined stochastic limits. Parameterization in simulation system models is often confused with calibration through the use of commercial software packages or architectures. In existing software packages, parameters or at least their structures are often hard-coded, leaving the user to fit of a list of options. Even in the most liberal of packages the list of options is distinct and predetermined; the user is left with calibration as the primary science tool. Validation is commonly performed with process-based models. With system simulation models, validation is often confused with calibration. Historically, models were designed to fit Markovian reality, measured against control or reference images, as the measure of successful validation. A Markov process is a

Markov process is a sequence of experiments with a set of properties including a fixed number of states, dependence upon the present, and an outcome defined by movement to another state or maintenance of the same state. However, the most significant limit is that the neighborhood has no impact on the Markov process and successfully meeting a Markovian threshold in no way guarantees a reasonable approximation of reality.

However, justifying a simulation result through theoretical interpretations also does not justify the use of the model or allow an expression of confidence in the result. One of the great challenges of simulation modeling is finding a suitable validation technique. In the ODGIS case, validation is possible through the comparison of realizations and the evaluation of (limited) reference data. While it may be argued that this test only quantifies stochastic uncertainty, it serves as a basic framework for the future development of pattern-based validations. By evaluating the fit of multiple model runs, both the endpoints of the simulation and the mean terrain or land use and cover surface can be derived, and predictive evaluations accomplished by adding all of the realizations together. A test of validation for these simulations may ultimately involve a test of context. The context here is the nature of the landscapes and terrain represented though multiple realizations for some future time period.

6 Conclusions

Large spatial data producers have employed the fitness for use criterion as the objective for metadata quality reporting. However, this paper argues that traditional metadata accuracy reports must change. Those who use spatial data increasingly demand to know how reliable their GIS results are, and standard accuracy statistics are not able to supply answers. This is especially unfortunate since U.S. government spatial data producers (and many others) are mandated to provide adequate measures of spatial data quality to users.

Simulation-based uncertainty propagation models have been developed for spatial data. These propagation models are uniquely suited to providing general fitness for use assessments, but they remain difficult to understand and utilize for most end users. Because the data producer is in the best position to develop appropriate models, it seems appropriate that the producer should be responsible for model development and documentation.

Substantial challenges remain. Currently employed statistical uncertainty models are poor at reproducing key geomorphological neighborhood and global characteristics. Over the coming months we intend to develop a working ODGIS prototype to propagate spatial data uncertainty to specific applications. We will encapsulate and model error for a variety of spatial data types (DEM, satellite image, and land cover) for multiple study sites around the Earth. We will implement error models employing different degrees of information and capable of different spatial resolutions to demonstrate the capacity of the framework to handle different user and application types. The ODGIS framework will be employed on standard and state-of-the-art geoprocessing objectives, such as change detection, spatial environmental modeling, and spatial decision support. We believe the ODGIS approach combined with basic research on innovative stochastic data objects shows great promise for linking geospatial information, models of data uncertainty, and specific applications.

References

Aerts JCJH, Goodchild MF, Heuvelink GBM (2003) Accounting for Spatial Uncertainty in Optimization with Spatial Decision Support Systems. Transactions in GIS 7(2):211–230

Canters F, Genst WD, Dufourmont, H (2002) Assessing effects of input uncertainty in structural landscape classification. Int J of Geographic Information Science 16(2):129–149

Chrisman NR (1984) The role of quality information in the long-term functioning of a geographic information system. Cartographica 21(2-3):79–87

CLIP (2006) Climate Land Interaction Project. http://clip.msu.edu

Crosetto M, Tarantola S (2001) Uncertainty and sensitivity analysis: tools for GIS-based model implementation. Int J of Geographical Information Science 15(5):415–437

Devillers R, Bédard Y, Jeansoulin R (2005) Multidimensional Management of Geospatial Data Quality Information for its Dynamic Use Within GIS. Photogrammetric Engineering & Remote Sensing 71(2):205–215

Duckham M, Mason K, Stell J, Worboys M (2001) A formal approach to imperfection in geographic information. Computers, Environment, and Urban Systems 25:89–103

Ehlschlaeger CR, Shortridge AM, Goodchild MF (1997) Visualizing spatial data uncertainty using animation. Computers & Geosciences 23(4):387–395

Englund EJ (1993) Spatial simulation: environmental applications. In: Goodchild MF, Parks BO, Steyaert LT (eds) Environmental modeling with GIS. Oxford Press, New York, pp 432–437

FGDC (1998) Content Standards for Digital Geospatial Metadata (2.0) FGDC-STD-001-1998. Federal Geographic Data Committee: Washington, DC. http://www.fgdc.gov/standards/documents/standards/metadata/v2_0698.pdf

Fisher PF (1991) First experiments in viewshed uncertainty: the accuracy of the viewshed area. Photogrammetric Engineering and Remote Sensing 57(10): 1321–1327

Fisher PF (1994) Sources and consequences of error in spatial data. In: Int Symp on the Spatial Accuracy of Natural Resource Data Bases: Unlocking the Puzzle. ASPRS, Williamsburg, VA, pp 8–17

Fisher PF (1998) Improved modeling of elevation error with geostatistics. Geoinformatica 2(3):215–233

Fonseca F, Davis C (1999) Using the internet to access geographic information: An OpenGIS prototype. In Goodchild M, Egenhofer M, Fegeas R, Kottman C (eds) Interoperating Geographic Information Systems. Kluwer, Norwell, MA, pp 313–324

Fonseca F, Egenhofer M, Agouris P, Camara G (2002) Using Ontologies for Integrated Geographic Information Systems. Transactions in GIS 6(3):231–257

Frank A (1997) Spatial ontology. In: Stock O (ed) Spatial and Temporal Reasoning. Academic Publishers, Dordrecht, pp 135–153

Frank A (1998) Metamodels for data quality description. In: Goodchild M, Jeansoulin R (eds) Data Quality in Geographic Information – from Error to Uncertainty, pp 15–29

Goodchild MF (1995) Sharing imperfect data. In Onsrud HJ, Rushton G (eds) Sharing Geographic Information. Rutgers University Press, New Brunswick, NJ, pp 413–425

Gregorio A, Jansen L (1998) Land Cover Classification System: Classification and User Manual. WWW document, http://www.fao.org/WAICENT/FAOINFO/SUSTDEV/Eidirect/EIre0062.htm.

Guarino N (1997) Semantic matching: Formal ontological distinctions for information organization, extraction, and integration. In: Pazienza MT (ed) Information Extractions: A Multidisciplinary Approach to an Emerging Information Technology. Springer, Berlin, pp 139–170

Guptill SC (1999) Metadata and data catalogues. In: Longley PA, Goodchild MF, Maguire DJ, Rhind DW (eds) Geographical Information Systems, 2nd ed. Wiley, New York, pp 677–692

Heuvelink GB, Burrough PA, Stein A (1989) Propagation of errors in spatial modelling with GIS. Int J of Geographical Information Systems 3(4):303–322

Heuvelink GBM (1998) Uncertainty analysis in environmental modelling under a change of spatial scale. Nutrient Cycling in Agroecosystems 50(1-3):255–264

Heuvelink GBM (2002) Analysing Uncertainty Propagation in GIS: Why is it not that Simple? In: Foody GM, Atkinson PM (eds) Uncertainty in Remote Sensing and GIS. John Wiley & Sons, Hoboken, NJ, pp 155–165

Hill LL, Carver L, Larsgaard M, Dolin R, Smith TR, Frew J, Rae M-A (2000) Alexandria digital library: user evaluation studies and system design. J of the American Society for Information Science 51(3):246–259

Hwang D, Karimi HA, Byun DW (1998) Uncertainty analysis of environmental models within GIS environments. Computers and Geosciences 24(2):119–130

Karssenberg D, de Jong K (2005) Dynamic environmental modelling in GIS: 2. Modelling error propagation. Int J of Geographical Information Science 19(6):623–637

Kemp K, Vckovski A (1998) Towards an ontology of fields. In: Proc of the Tenth Int Conf on GeoComputation, Bristol, United Kingdom

Lee J, Snyder PK, Fisher PF (1992) Modeling the effect of data errors on feature extraction from digital elevation models. Photogrammetric Engineering & Remote Sensing 58(10):1461–1467

Longley PA, Goodchild MF, Maguire DJ, Rhind DW (1999) Geographical Information Systems, 2nd ed. Wiley, New York

Matsuyama T, Hwang VS (1990) A Knowledge-based Aerial Image Understanding System. Plenum, New York

Messina JP, Walsh SJ (2001) 2.5D morphogenesis: modeling landuse and land cover dynamics in the Ecuadorian Amazon. Plant Ecology 156(1):75–88

Openshaw S (1989) Learning to live with errors in spatial databases. In: Goodchild MF, Gopal S (eds) Accuracy in Spatial Databases. Taylor & Francis, London, pp 263–276

Pebesma EJ, Wesseling CG (1998) Gstat: a program for geostatistical modelling, prediction and simulation. Computers and Geosciences 24(1):17–31

R Development Core Team (2005) R: A Language and Environment for Statistical Computing. URL:http://www.R-project.org/ R Foundation, Vienna, Austria. Accessed 12/20/2005

SDTS (1996) The Spatial Data Transfer Standard: Guide for Technical Managers. U.S. Dept. Interior

SDTS (1998) Spatial Data Transfer Standard – ANSI NCITS 320-1998. American National Standards Institute, Washington, DC

Shortridge AM (2000) Compact data models for spatially continuous phenomena. First Int Conf on Geographic Information Science, Savannah, GA, Oct 28–31

Smith B (1998) An introduction to ontology. In: Peuquet D, Smith B, Brogaard B (eds) The Ontology of Fields. National Center for Geographic Information and Analysis, Santa Barbara, CA, 10-4

Unwin DJ (1995) Geographical information systems and the problem of 'error and uncertainty'. Progress in Human Geography 19(4):549–558

USGS (1996) DEM/SDTS transfers. In: The SDTS Mapping of DEM Elements. U.S. Dept. Interior, U.S. Geological Survey

Wong DWS, Wu CV (1996) Spatial metadata and GIS for decision support. 29th Annual Hawaii Int Conf on System Sciences, IEEE, pp 557–566

Structuring Kinetic Maps

Maciej Dakowicz, Chris Gold

School of Computing, University of Glamorgan
Pontypridd, Wales CF37 1DL UK
email: mdakowic@glam.ac.uk; cmgold@glam.ac.uk

Abstract

We attempt to show that a tessellated spatial model has definite advantages for cartographic applications, and facilitates a kinetic structure for map updating and simulation. We develop the moving-point Delaunay/Voronoi model that manages collision detection snapping and intersection at the data input stage by maintaining a topology based on a complete tessellation. We show that the Constrained Delaunay triangulation allows the simulation of edges, and not just points, with only minor changes to the moving-point model. We then develop an improved kinetic Line-segment Voronoi diagram, which is a better-specified model of the spatial relationships for compound map objects than is the Constrained Triangulation. However, until now it has been more difficult to implement. We believe that this method is now viable for 2D cartography, and in many cases it should replace the Constrained approach. Whichever method is used, the concept of using the moving point as a pen, with the ability to delete and add line segments as desired in the construction and updating process, appears to be a valuable development.

1 Introduction

Constructing digital maps with line data has various difficulties, especially with the connectivity ("topology") between various polylines ("arcs" or, in computer graphics, a continuous line composed of one or more line se-ments). This is implicit in the spatial model used – one-dimensional features floating loose in two-dimensional space. A more hopeful approach is based on tessellating the whole map space, so that all cells are bounded by other cells. Connectivity is thus automatic.

Tessellations may be regular or irregular. A regular grid ("raster") has simple spatial relationships but, because it is based on the coordinate system and not on the features being represented, has difficulties in managing complete features or polylines. Irregular tessellations may be triangulations or other cell structures, e.g. the Voronoi diagram (VD). Triangulations connecting data points may follow any chosen properties, but the Delaunay triangulation (DT), the dual of the VD, has a geometric (coordinate-based) specification that gives a unique solution, except in degenerate cases, and most importantly it can be updated locally while perturbing only the immediately neighboring triangles. However, the triangle edges may not exactly follow the desired polyline. Voronoi cells are usually constructed for sets of data points, giving the proximal region to each point, but they are not readily aggregated into polylines.

A topological model is required for two purposes: to perform analysis on the final map (e.g. network flow) or to construct the map in the first place (e.g. snapping one polyline to another). Saving individual features (polygons, polylines) to a database and reconstructing the connectivity as required may suffice for some types of analysis (but not network flow), but this still leaves the construction problem. Node construction from the intersections of individual polylines is a classical GIS problem, caused because the intersections of individual one-dimensional entities, and their subsequent merging to form nodes (in a polygon map for example) do not always give a well-defined ordering of edges around the nodes – a requirement for an elementary topology on a 2-manifold. Here a tessellation is an attractive option.

In addition, traditional topology-building algorithms are batch-oriented: all polylines are inserted at once, and any local changes require a complete rebuild (although in some cases a "patch" may be recalculated and reinserted into the larger map). In practice, many construction and editing operations are incremental, mimicking the manual use of a pen (and an eraser) to make local changes. This suggests the use of a kinetic algorithm to simulate pen movement.

We here propose an integrated approach to tessellated map construction, which uses locally-updated tiles as a consistent and kinetic definition of adjacency. The first component is the moving-point VD/DT, which manages the topology of the moving pen. The second component is the Constrained DT, which permits the specification of certain triangle edges, even if they fail the Delaunay conditions, thus allowing the construction and connection of line segments. The third component is the Line-Segment VD, where the VD defines proximal regions around line segments as well as points. This integrated approach uses the Moving-Point VD (or DT) to "draw" the desired line segment in both modes, thus applying a common underlying algorithm to both processes. This approach to preserving topology assists in the construction process, and also in analysis of the final map, for example for map generalization or network flow. Its main drawback, as with all large-scale graph structures, is the lack of an efficient mapping to a database system – although some object-oriented databases may assist in this. This algorithmic approach has the following stages:

1: The dynamic point VD and its dual DT.
Here the simple point VD/DT is constructed incrementally, and the inverse point deletion algorithm is included.

2: The Moving-Point VD or DT
This requires the insertion and deletion of data points. In addition, individual points may be moved from their previous location to some subsequent location – the "trajectory". In order to maintain the VD/DT geometric properties there must be a predictive tool to specify at what location the neighboring VD/DT edges must be modified – a "topological event" (TE).

3: The Kinetic Constrained DT
The Moving Point (MP) in the MPVD is split from a previous "old" point and moved towards its "new" destination. The initial triangle edge between the old and new points is flagged as constrained (CE), and any TE generated by the moving point is ignored if it involves switching the CE.

4: The Kinetic Line-Segment VD
Instead of flagging the CE in the initial position of the CDT, a pair of "half-lines" is generated instead. These are two new generators in the VD – one for each side of the line, in addition to the two end points. Each of these is the potential generator of a Voronoi proximal region. As the MP moves, TEs are identified as before, and the topology updated, thus giving an expanding region associated with each half-line.

In this model the topological events are the same as before, but the circumcircle (CC) calculation must be expanded in order to work with distances from line segments as well as points. In earlier work (Gold 1990; Yang and Gold 1995) a direct calculation of Voronoi boundary intersections was used to find the circumcentre. This failed on occasion as arithmetic precision limitations could place the centre on the wrong side of a line segment, thus destroying the node-ordering necessary for topology maintenance. A new iterative algorithm was developed (Anton and Gold 1997; Anton et al. 1998) that converged on the correct solution from an initial condition while preserving the necessary order of the generator locations around the circumcircle.

All the operations used have their inverses, as MP movement may expand or contract the trailing line. Preserving the topological relationships during construction means that potential collisions may be detected in advance, and the appropriate join operations implemented. This is simplified as the lines and their proximal regions are embedded in the two-dimensional space, guaranteeing that, for example, one VD line segment may detect an imminent collision and form the appropriate junction that preserves the correct node and region ordering around the junction point.

2 The Dynamic Point VD and its Dual DT

The simple VD can be constructed in many ways, (Aurenhammer 1991; Okabe et al. 1992) but the incremental algorithm has often been found to be both stable and simple (Guibas and Stolfi 1985). In simple terms, each new point is inserted into the existing DT by first finding the enclosing triangle, using the CCW test of (Guibas and Stolfi 1985), splitting it into three triangles using the new point, and then testing each edge recursively to see if it conforms to the Delaunay criterion: that neither of the adjacent triangles' CCs have an interior point. If they do, the common diagonal is switched and the new edges are added to the stack of edges to be tested (Guibas and Stolfi 1985). This CC test (INCIRCLE, Guibas and Stolfi 1985) can be shown to be equivalent to calculating the VD vertices and testing if the VD edges cross. Point deletion can be performed by approximately following the inverse process: switch DT edges if the result gives an exterior triangle whose CC is empty except for the point being deleted. When only three triangles remain the central point is deleted. There are two similar approaches: (Devillers 1999; Mostafavi et al. 2003). Thus the VD is updated at the same time as the DT. Both insertion and deletion may be considered as partitioning the DT into two parts: the valid DT exterior area and the valid DT interior area. Boundary edges are then switched until the two parts merge.

3 The Moving-Point VD or DT

When a point MP moves as part of a DT/VD it may either travel a short distance without requiring a topology update, or else triangle edges must be switched to maintain the Delaunay criterion. These TEs (Roos 1990; Guibas et al. 1991) occur when MP moves into or out of a CC. "Real" CCs are those formed from triangles immediately exterior to the "star" or set of triangles connected to MP. "Imaginary" CCs are formed by triangles that would be created if MP was moved out of its CC, and are formed by triples of adjacent points around MP's star. Thus if MP moved into a constellation of points in a DT it would first enter the CC of a triangle, causing a triangle edge switch and adding the furthest point of the triangle to the star of MP. The original real CC is now preserved as an imaginary one. As MP continued to move, at some later time it would move out of this imaginary CC, the original triangle would be recreated and the CC would become real again.

As MP moves in any direction, it may enter or leave many CCs, and the triangle edges must be updated accordingly. Two lists are therefore preserved, one of the real CCs surrounding MP's star, and one of the imaginary CCs formed by triples around MP's star. Real CCs are dropped if they are re-formed from the imaginary ones, so the real list consists only of CCs likely to be entered during MP's current trajectory. (Initially, before a point is first moved, all surrounding real CCs are found and tested against the proposed trajectory.) To maintain the DT during MP's movement it is therefore necessary only to find the intersection of its trajectory with the first CC in either the real or imaginary list, switch the affected edge, update the two lists, move MP to the intersection point and then repeat the process (see Fig. 1). To avoid problems due to degenerate cases, e.g. when several CCs are superimposed due to a regular square grid of data points for example, the "first" intersection must be clearly defined. In practice it is critical not to "forget" an intersection because it is behind MP's current position, usually due to computer arithmetic limitations. This is achieved by always using the earliest intersection, even if it is behind MP, subject to a test that the intersection point is associated with an arc of the triangle edge to be switched.

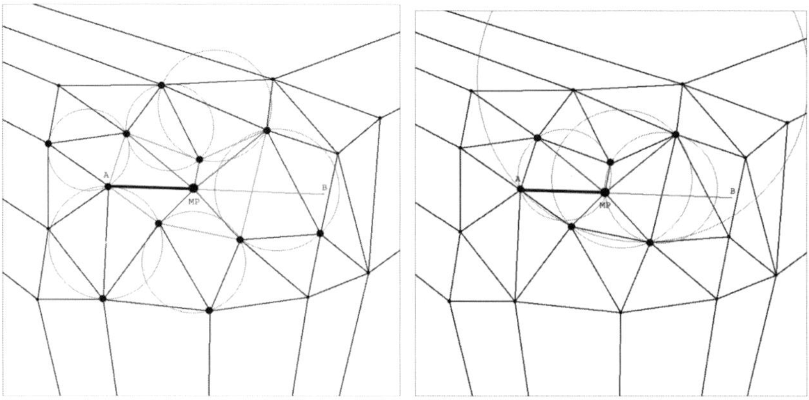

Fig. 1. a) "Real" circumcircles to MP **b)** "Imaginary" circumcircles to MP

This loop: find the next intersection with a CC; switch the DT edge; update the lists; move MP – is continued until MP's destination is reached. It is, however, possible that MP collides with an existing point, destroying the DT structure. Thus at each iteration of the loop the distance from MP to each new neighboring point is tested, and if it is below some tolerance the destination coordinates of MP are adjusted to those of the collision point.

4 The Kinetic Constrained DT

Where MP collides exactly with a neighboring point, the point is removed along with the two triangles adjacent to the edge between these points. The reverse process may be used to create a new MP from a previous point. This creates a zero-length edge between them, which expands as MP moves away. In the normal course of events this edge will be switched once MP moves outside the imaginary CC formed by the previous point and its two adjacent points in the star. However, for many applications it would be desirable if the edge was preserved, and MP used to "draw" a triangle edge between two locations. In this case the trailing edge of MP is flagged "do not switch", and all tests to switch it are ignored. This generates the Constrained DT, where specific triangle edges are fixed (constrained, CE) and do not follow the DT/VD condition (Chew 1998; Shewchuk 1996) – see Figure 2.

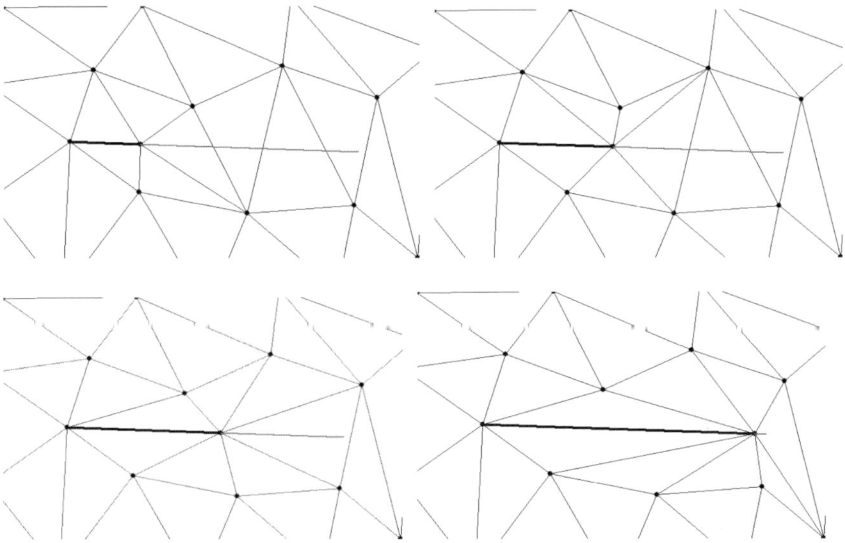

Fig. 2. Four stages in the construction of a constrained edge

The interesting thing about this approach, as opposes to those in the literature, is that it is incremental, allowing the addition of further points or constraints as required. A further property is that it is reversible – MP may be moved back along its (constrained) trajectory, "rolling it up" as it goes, until it reaches its starting point, with which it may be merged. Thus each operation has its inverse, giving a kinetic data structure, which allows the construction and incremental or interactive modification of the desired map. In addition, the construction commands may be preserved as a "log file" for later reconstruction or modification. If timestamps are associated with each command the map may be rolled forwards or rolled backwards to any desired time state.

Because of this reversibility, intersections of pairs of CEs may be managed. If MP finds a CE as part of its star, and attempts to switch it, then a potential collision exists. Then, if necessary, MP's movement is stopped; an intersection point between the two CEs is calculated, the previous CE is rolled back to the new intersection point; a new CE is constructed to close the gap, leaving the intersection as a new point on the CE; and MP is moved again until it merges with the intersection point. A new portion of the CE is then drawn from the intersection point to the original destination of MP. This may be repeated for as many intersections as necessary (see Fig. 3).

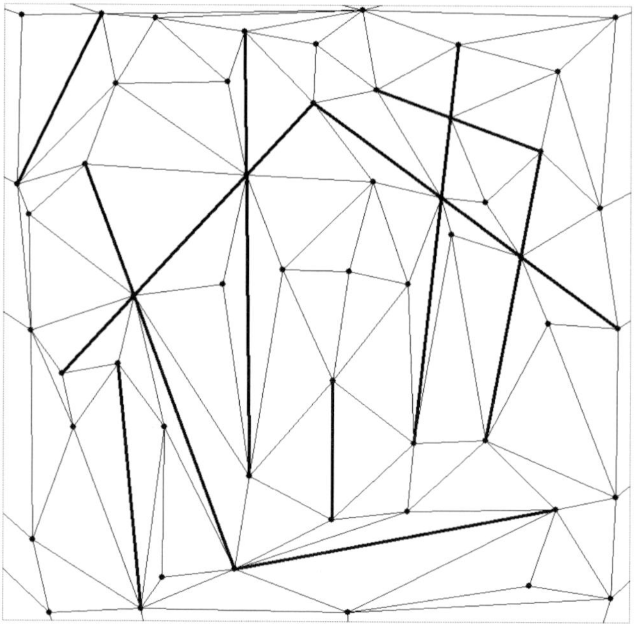

Fig. 3. Constrained DT with intersections

Figure 4 shows the construction of the Constrained DT for a UK urban data set. This is of particular interest because of the work of (Jones et al. 1995; Jones and Ware 1998; Ware and Jones 1998) who constructed the CDT of roads and building outlines and then used the adjacency information to modify and move the buildings as part of the process of map generalization. Figure 5 shows the same approach for several buildings and roads.

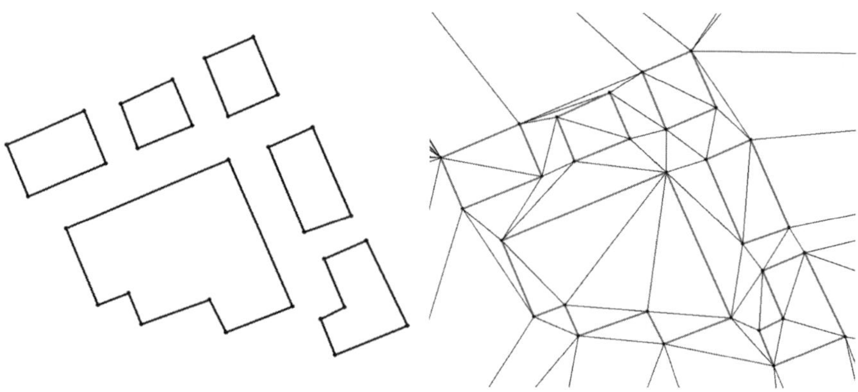

Fig. 4. Building boundaries and Constrained DT

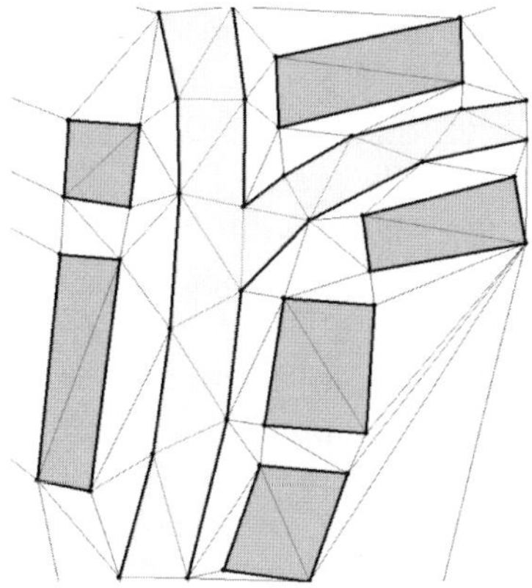

Fig. 5. Constrained DT for buildings and roads

5 The Kinetic Line-Segment VD

The primary problem with the Constrained DT is the confusion of entities. For a simple point DT/VD the primary objects are data points, which are the generators of the proximal Voronoi cells. The DT merely describes the dual relationships of the Voronoi edges: Delaunay edges are merely pointers expressing which pairs of data points are separated by Voronoi boundaries. For a simple TIN model it is convenient to imagine that these are geometrically defined as "straight", as the triangle is a 2D simplex and hence forms a basis for linear interpolation within, but their real function is to support the set of equidistant boundaries that form the VD of a set of generators. (These generators may, if required, be any set of non-overlapping objects: the dual DT remains a triangulation.)

However, with the Constrained DT there is confusion between triangle edges that express duals of VD edges and those that have been manually added as objects – in the sense that a building outline is formed of point and line-segment objects. Thus the VD of a Constrained DT is broken at each constrained edge, and the VD edges that are correct on one side of a constrained edge are invalid when they penetrate to the other side. It is more correct to define the mapped objects separately (perhaps composed of points and line segments) and then to construct the DT/VD expressing the spatial relationships between them.

Unfortunately the construction of the VD of points and line segments has proved to be a difficult task, primarily due to the limited precision of computer arithmetic. This causes no great difficulty for well-separated individual line segments (e.g. Gold et al. 1995) but map objects constructed from connected points and line segments need to have tight guarantees that, for example, the circumcircle for a line segment, its end-point, and a line segment connected to that point, falls on the correct side of the polyline. (Geometrically it falls precisely on the common end-point, but topologically it must be associated with the correct side.) This has proved difficult to achieve, and workers have spent a great deal of time attempting to construct robust algorithms (e.g., Held 2001; Imai 1996).

In addition, we have wanted to allow incremental, rather than batch, construction, so we followed the approach of Gold et al. (1995), which was based on the concept of the moving point VD described above. Instead of preserving a trailing triangle edge, as described above for the CDT, the "old point" OP and the "moving point" MP are connected with additional map objects: half-lines connecting OP and MP that stretch as MP moves away. As both end points and both half-lines are map objects they are therefore generators of the VD, and thus they are vertices of the dual DT, as shown in Figure 6. Karavelas (2004) produced an incremental algorithm for exact arithmetic, which allowed intersections by splitting line segments in advance, using exact arithmetic, but not by using our "moving point" approach, which permits deletion.

Thus calculation of the circumcircles of these triangles is more complex than for point data sets. In Yang et al.'s work this was calculated from the intersections of the curves forming the Voronoi boundaries, but this suffered from the arithmetic precision problems mentioned above.

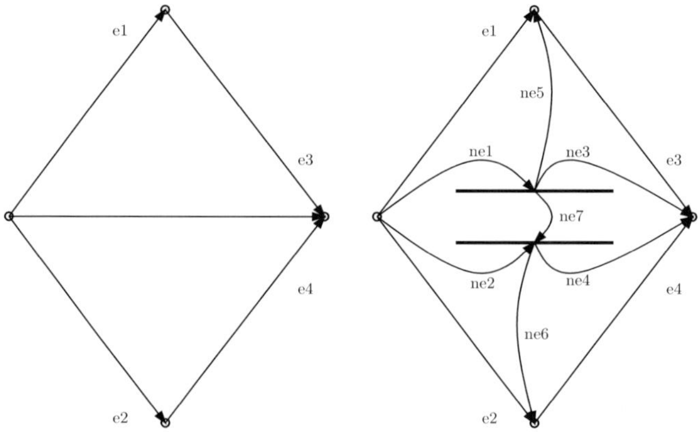

Fig. 6. Half-lines between two data points

6 Circumcircle

In our new work we use the approach of Anton et al. (1998) where the simple point circumcircle (CC) calculation (INCIRCLE of Guibas and Stolfi 1985) was given an initial estimate based on the configuration of the points/line-segments used. Initially the mid-points of valid portions of the line segments were used for the INCIRCLE test. The centre was then projected onto each line segment, and a new CC calculated based on INCIRCLE. This was iterated to a suitable level of precision, and the method was guaranteed to preserve the order of the generating points around the CC, thus keeping the initial Voronoi edge order around the circumcentre (and thus the correct DT order as well). Figure 7 shows the rapid convergence of the method for two initial configurations. The key improvement over the work of Anton et al. was the trimming of the potential space for the initial estimate: the centre had to be on the correct side of the line segments, given the original anticlockwise order of the vertices obtained from the data structure; line segments had to be trimmed to be on the correct side of other line segments; data points had to be on the correct sides of line segments and projected within the trimmed segment, etc. A tolerance was used to "snap" the projections onto an endpoint of a line-segment if necessary. While all "real" CCs, as defined previously, had to have valid solutions, as they were part of a valid VD, "imaginary" CCs might not have valid solutions, and needed to be rejected in order to avoid false topological changes. After extensive testing our current algorithm appears robust.

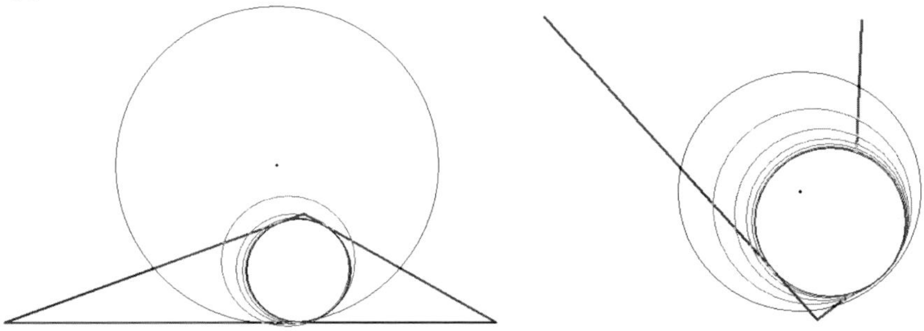

Fig. 7. Iterative circumcircle calculations

As with the Constrained DT, the MP is used to draw the line segment using the half-lines described above. As MP moves it acquires and loses Voronoi neighbors, as with the simple moving point, but when it loses them they are transferred to the trailing line segment: since this is the locus of MP, it retains all the neighborhood relationships previously held by MP (see Fig. 8).

Fig. 8. Two stages in drawing a line segment VD

For a simple line segment with two end points there are four Voronoi regions: one for each end point and one for each half-line. This permits the querying of each side of a line, e.g. to find if a point is inside or outside a polygon. As shown in Gold (1990) the partitioning of the map space into proximal regions also makes buffer-zone generation an elementary operation on each region. Figure 9 shows the line segment VD for the same urban dataset as shown previously. When the DT is also displayed, note that each line segment is also a DT vertex. This clearly distinguishes the DT function of expressing the adjacency relationships, and not being part of the map object. Each of the map objects may be edited by the insertion or deletion of line segments (half-edge pairs) and free vertices. The method is dynamic, in the sense of being locally updatable, and kinetic, in the sense that MP may move within the map space. However, line segments may only expand or shrink, and not sweep sideways, as collision detection and topology maintenance are based on MP alone.

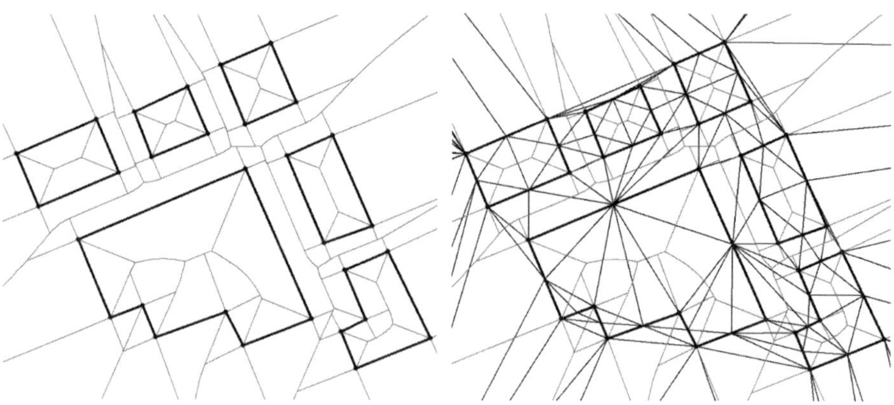

Fig. 9. a) Line segment VD; **b)** VD plus DT for the simple buildings of Figure 4

7 Robustness

Space does not permit the description of all the details of the suite of algorithms described in this paper. The key question in practice is the robustness of the method for all types of data input, given the problems of arithmetic precision. The underlying method described in this paper consists of two parts: a geometric test and a topological update. Any arithmetic operation not resulting in a topological change cause no robustness problems – for example calculating the CC centre (Voronoi node) for display purposes, or the projection of a point onto the interior of a line segment. Only geometric tests used to trigger topological changes can cause robustness problems, and there are only two – calculation of CCs and a sidedness test ("walk" in Gold 1977, "CCW" in Guibas and Stolfi 1985). CCW and INCIRCLE (for three points) are geometric predicates that have been studied extensively, and arbitrary-precision solutions are readily available (Shewchuk 1997). Anton's iterative circle calculation for line-segments uses INCIRCLE, and although it is iterative (and therefore approximate) this causes no problem where the centre projects onto the interior of the line-segment. In practice, a tolerance value is needed (to allow for arithmetic imprecision) only in the specific cases below.

For moving points, when a point is selected for movement or splitting, the CCs for "exterior" triangles must be put on a list if their projections onto the trajectory are in front of MP. A tolerance is used here. When finding the next topological event for MP, the intersection of the trajectory with the CC is imprecise. A tolerance is used to check if the intersection of a "real" or "imaginary" CC is too close to the trajectory start or end – in which case it is ignored. There is also a tolerance check for collisions with objects in MP's path. In the specific cases of the Constrained DT or the Line-segment VD there are no additional tolerance tests, except those used for the moving point or the iterative CC.

8 Applications

Our previous examples have been urban applications, showing building and street boundaries, for potential application in map generalization (Jones et al. 1995; Jones and Ware 1998). We will briefly show two others. Figure 10 shows the Line-segment VD for a portion of a contour map. Both the points and line-segments forming the contours are map objects, as would have been the intention of the compilers. In addition, the medial axis, or skeleton, between or within the contours is clearly seen (see Gold and Snoeyink 2001, for further discussion of the skeleton). This map is directly editable if required.

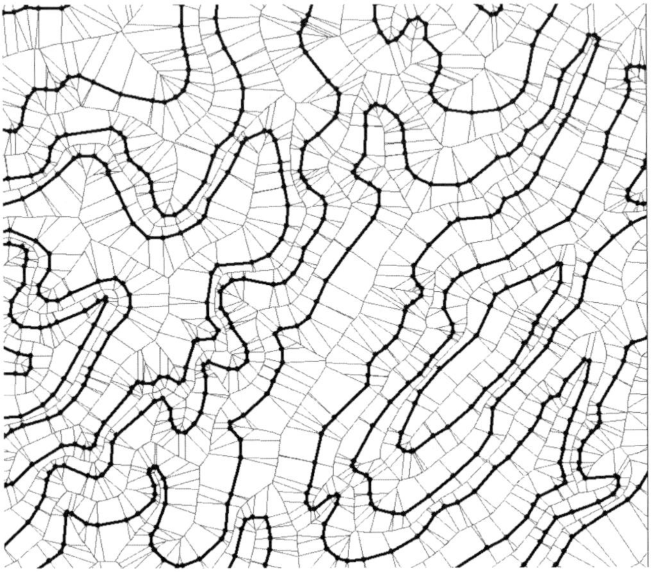

Fig. 10. Line-segment VD for contours

Figure 11 shows the Line-segment VD and the Constrained DT for a road network. Again, the relationships between map objects and relationships are clearer in the VD, and both maps are directly editable.

9 Conclusions

We have attempted to show that a tessellated spatial model has definite advantages for cartographic applications, and facilitates a kinetic structure for map updating and simulation. Firstly, the moving-point DT/VD model approximates human thinking, and manages collision detection, snapping and intersection at the data input stage by maintaining a topology based on a complete tessellation. Secondly, the Constrained DT allows the simulation of edges, and not just points, with only minor changes to the moving-point model, but at the cost of confusing map objects and topological entities. Thirdly, the Line-segment VD is a better-specified model of the spatial relationships for compound map objects built from points and line segments than is the Constrained DT. However, until now it has been more difficult to develop.

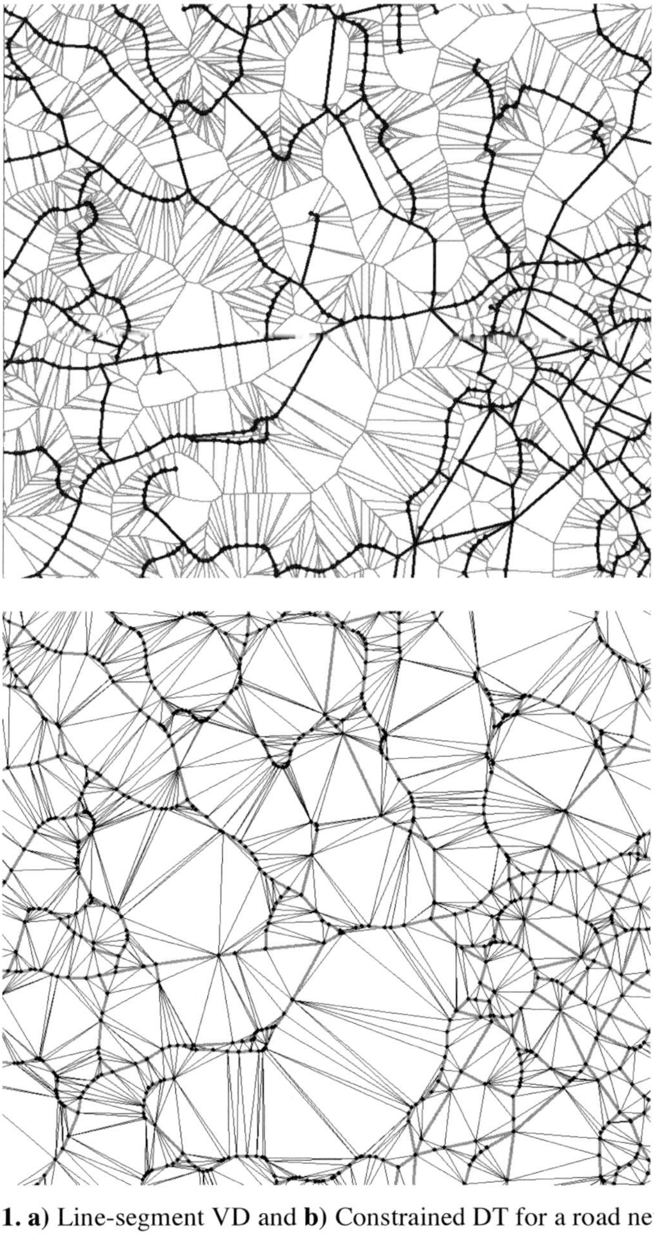

Fig. 11. a) Line-segment VD and b) Constrained DT for a road network

We believe that this method is now viable for 2D cartography, and in many cases it should replace the Constrained DT. However, whichever method is used, the concept of using the moving point as a pen, permitting interactive navigation within the map under construction, together with the ability to delete and add line segments as desired in the construction and updating process, appears to be a very useful approach.

Acknowledgements

We would like to acknowledge the financial support of the EU Marie-Curie Chair in GIS at the University of Glamorgan.

References

Anton F, Gold CM (1997) An iterative algorithm for the determination of Voronoi vertices in polygonal and non-polygonal domains. In: Proc Ninth Canadian Conf on Computational Geometry, Kingston, ON, Canada, pp 257–262

Anton F, Snoeyink J, Gold CM (1998) An iterative algorithm for the determination of Voronoi vertices in polygonal and non-polygonal domains on the plane and the sphere. In: Proc 14th European Workshop on Computational Geometry (CG'98), Barcelona, Spain, pp 33–35

Aurenhammer F (1991) Voronoi Diagrams – A Survey of a Fundamental Geometric Data Structure. ACM Computing Surveys 23(3):345–405

Chew P (1988) Constrained Delaunay Triangulations. Algorithmica 4:97–108

Devillers O (1999) On deletion in Delaunay triangulations. 15th Annual ACM Symp on Computational Geometry, pp 181–188

Gold CM (1990) Spatial Data Structures – The Extension from One to Two Dimensions. In: Pau LF (ed) Mapping and Spatial Modelling for Navigation (= NATO ASI Series F No 65). Springer-Verlag, Berlin, pp 11–39

Gold CM, Remmele PR, Roos T (1995) Voronoi diagrams of line segments made easy. In: Gold CM, Robert JM (eds) Proc 7th Canadian Conf on Computational Geometry, Quebec, QC, Canada, pp 223–228

Gold CM, Snoeyink J (2001) A one-step crust and skeleton extraction algorithm. Algorithmica 30:144–163

Guibas L, Mitchell JSB, Roos T (1991) Voronoi diagrams of moving points in the plane. In: Proc 17th Int Workshop on Graph Theoretic Concepts in Computer Science, Fischbachau, Germany (= Lecture Notes in Computer Science 70). Berlin, Springer-Verlag, pp 113–1 25

Guibas L, Stolfi J (1985) Primitives for the manipulation of general subdivisions and the computation of Voronoi diagrams. Transactions on Graphics 4: 74–123

Held M (2001) VRONI: an engineering approach to the reliable and efficient computation of Voronoi diagrams of points and line segments. Computational Geometry, Theory and Application 18(2):95–123

Imai T (1996) A topology oriented algorithm for the Voronoi diagram of polygons. In: Proc 8th Canadian Conf on Computational Geometry, Carleton University Press, Ottawa, Canada, pp 107–112

Jones CB, Bundy GL, Ware JM (1995) Map generalization with a triangulated data structure. Cartography and Geographic Information Systems 22(4): 317–331

Jones CB, Ware JM (1998) Proximity Search with a Triangulated Spatial Model. Computer J 41(2):71–83

Karavelas MI (2004) A robust and efficient implementation for the segment Voronoi diagram. In: Int Symp on Voronoi Diagrams in Science and Engineering (VD2004), pp 51–62

Mostafavi M, Gold CM, Dakowicz M (2003) Dynamic Voronoi / Delaunay Methods and Applications. Computers and Geosciences 29(4):523–530

Okabe A, Boots B, Sugihara K (1992) Spatial Tessellations – Concepts and Applications of Voronoi Diagrams. John Wiley and Sons, Chichester, 521 p

Roos T (1990) Voronoi Diagrams over Dynamic Scenes. In: Proc Second Canadian Conf on Computational Geometry, Ottawa, pp 209–213

Shewchuk JR (1996) Triangle: Engineering a 2D Quality Mesh Generator and Delaunay Triangulator. First Workshop on Applied Computational Geometry (Philadelphia, Pennsylvania), Ass for Computing Machinery, pp 124–133

Shewchuk JR (1997) Adaptive Precision Floating-Point Arithmetic and Fast Robust Geometric Predicates. Discrete and Computational Geometry 18(3): 305– 363

Ware JM, Jones CB (1998) Conflict Reduction in Map Generalization Using Iterative Improvement. GeoInformatica 2(4):383–407

Yang W, Gold CM (1995) Dynamic spatial object condensation based on the Voronoi diagram. In: Chen J, Shi X, Gao W (eds) Proc Fourth Int Symp of LIESMARS'95 – Towards three-dimensional, temporal and dynamic spatial data modelling and analysis, Wuhan, China, pp 134–145

Advanced Operations for Maps in Spatial Databases

Mark McKenney, Markus Schneider *

University of Florida, Department of Computer & Information Science & Engineering, Gainesville, FL 32611, USA
email: {mm7, mschneid}@cise.ufl.edu

Abstract

Maps are a fundamental spatial concept capable of representing and storing large amounts of information in a visual form. Map operations have been studied and rigorously defined in the literature; however, we identify a new class of *map join* operations which cannot be completed using existing operations. We then consider existing operations involving connectivity concepts, and extend this class of operations by defining new, more complex operations that take advantage of the connectivity properties of maps.

1 Introduction

Spatially-oriented disciplines such as cartography and geography, as well as computer assisted systems like spatial database systems (SDBMS), geographic information systems (GIS) and image database systems rely heavily on the idea of *maps* or *spatial partitions*. A map is a fundamental spatial concept that is capable of representing and storing large amounts of information in a visual form.

Although maps are an intuitive basis for geometric applications such as SDBMS and GIS, these applications do not incorporate maps as a fundamental data type. The fundamental data types are instead geometric data

* This work was partially supported by the National Science Foundation under grant number NSF-CAREER-IIS-0347574.

types such as points, lines, and regions, which are compiled together to form map representations. Constructing maps as an amalgam of many more simple types presents the user with a map representation; however, the underlying operations on the map must then be defined based on these more simple types. Thus, maps are not considered "first class citizens" in spatial software systems, but are merely used for visualization purposes. Operations over maps represented as collections of points, lines, and regions are difficult to implement and cannot use the structural information inherent in maps because each individual simple data type has no knowledge of that structure. Furthermore, the regions in maps satisfy certain topological relationships, namely all regions are either disjoint, or share a common boundary. By using regions to represent maps, this constraint cannot be enforced by the data type itself, but must be managed by the system using the maps, such as the SDBMS. By representing maps as a fundamental data type, SDBMS resources do not need to be used to enforce such constraints because the data type (including the operations defined over that data type) will enforce them implicitly.

Part of the reason that points, lines, and regions are used in spatial systems instead of maps is that manipulating these data types in a database setting is a relatively well understood concept. The basic operations over these objects have been well defined and implemented in various main stream systems. The complex structure of maps, however, causes them to be much more difficult to understand and formalize, as opposed to points, lines, and regions. While much research has been done on map operations, we show scenarios that require new operations that have not previously been considered.

The shortcomings of the currently defined set of map operations rest on three problems. First, the complex structure of maps and the need for operations that take advantage of that structure indicate that a small set of basic operations will not provide enough general functionality to a user. The current set of map operations was defined without considering certain aspects of map structure that a user may wish to take advantage of. For example, there are only few operations that manipulate maps at the level of their component regions. Secondly, the operations currently defined over maps have been formalized in an ad-hoc fashion. Therefore, in addition to merely introducing new operations, we classify the new operations into general categories that provide some structure to the set of map operations, and will help in further studying map operations. Finally, map operations have largely been defined without considering the embedding of maps into databases. Therefore, some fundamental database concepts have not yet been applied to maps. Such concepts, such as the database join operation, lead to new, powerful map operations.

In this paper, we formalize new operations over maps based on an abstract, point set topological model of maps. While such a model is not suited to discrete representation, it provides an exact mathematical model which can be leveraged to precisely define map operations. Section 2 discusses work related to our research. Section 3 provides motivating query scenarios that highlight the requirements of new map operations. The formal model of spatial partitions and basic operations over them are given in Section 4. Based on this model, new map operations are formalized in Section 5. Finally, in Section 6, we draw some conclusions.

2 Related Work

Attempts in the field of spatial databases have been made to handle maps. Unfortunately, these attempts have been unsatisfactory. In [1], a spatial type *area* is suggested to model constraints on maps. However, the maintenance of these constraints is not supported by the model, but must be enforced by the user. [2] informally proposes a generic data type of partitions called *tessellation* which can be parametrized with an attribute of a yet unspecified type. The concept of *restriction types*, proposed in [3], allows the general type for regions to be restricted to subtypes whose values all satisfy a topological predicate. However, it is unclear how these constraints are controlled by the database.

In this paper, we present operations over maps that are based on the formal model of *spatial partitions* presented in [4]. These operations add to the set of known map operations that have been defined in the literature. For a review of existing map operations, the reader is directed to [5, 6, 7, 8]. While these papers result in a large number of map operations, operations that deal specifically with the structure of maps are generally overlooked.

The study of the traditional data types of point, line, and region and their use in databases [9] is closely related to the concept of maps in databases. For example, the component regions of maps are similar to complex spatial region models and the border structure is similar to complex spatial lines. Furthermore, the regions in maps must satisfy certain topological predicates [10, 11] with each other, namely they can only meet or be disjoint.

3 Motivating Query Scenarios

In this section, we consider map operations from the user's perspective. We assume that a user will have access to a system that provides map access and the ability to perform operations over maps (i.e., a map based GIS system

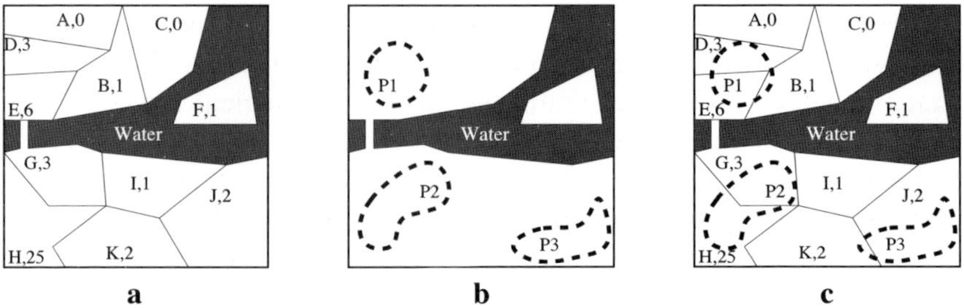

Fig. 1. Figure **a** shows a map of counties. Figure **b** shows areas of high population density. Figure **c** shows **b** superimposed over **a**

or a map database). A series of scenarios is presented in which the user requires certain information from the maps in the system. We then specify, in English sentences, the specific results required by the user in such scenarios. By examining these specifications in later sections, we discover the need for new map operations.

Throughout this paper, we study map operations as they pertain to specific examples; however, the operations presented are general operations that are relevant to situations other than the ones we show here. In fact, we find that we can define only a few powerful operations that can be generalized to apply to many cases through the use of different operands. We merely present example scenarios to motivate the utility of the operations and to help explain their semantics. For the sake of clarity, we will only introduce three sample maps, shown in Figure 1, against which queries requiring map operations will be posed. The first map in Figure 1a depicts counties around a body of water. Each county is labeled with an identifying letter, and the number of flu vaccine units currently in that county. County G contains a bridge over the water which is represented by an extension of G that touches county E. Figure 1b contains regions enclosed by dashed lines which represent areas of high population density. The third map, shown in Figure 1c shows the areas of high population density superimposed over the map of counties. Operations will be performed over the first two maps, the third is shown for reference. Note that the water shown in the figures is not part of the actual map stored in the database, but is shown merely to provide additional context.

3.1 Map Joining Scenarios

For the remainder of this paper, we will assume that the flu virus has been detected in all of the areas of high population density represented in Figure 1b.

Frequently, supplies such as flu vaccines are distributed through local governments, such as county governments in the United States. Therefore, state officials need to know which counties contain the infected areas of high population density so that additional vaccinations may be sent there. The officials can obtain this information by combining information from the population density map and the county map to calculate a map containing only counties that overlap a region of high population density (the term 'overlap' indicates the topological predicate *overlap* between two complex regions). This type of query cannot be calculated using existing map operations. Therefore, a new operation is required that allows a user to create a map consisting of complete faces of regions from two source maps that satisfy a given predicate. We call this type of operation a *map join*. The result of such a scenario is shown in Figure 2a. Only regions from the county map that overlap a region in the population density map remain in the result map. Note is that this particular example represents a one-sided join, i.e., only regions from one of the argument maps are being returned. Two sided joins are also possible, in which regions from both argument maps that satisfy the predicate are shown in the result map.

Query 1. *Return the map consisting of all counties from the county map that overlap a region in the population density map.*

The map join operation is parametrized by a user supplied predicate; thus, different queries can be posed over data by simply changing the predicate used in the join. For example, instead of finding counties that overlap a region of high population density, the official might want to know which counties meet a region of high population density. Such a query would identify counties in which the flu is likely to spread. To express this query, we simple replace the word 'overlap' in Query 1 with the word 'meet' . The resulting map is shown in Figure 2b.

A second map join scenario considers the connectivity information which is inherently represented in a map's structure. Suppose that the particular strain of the flu that the authorities are tracking is highly contagious. In this case, the authorities need to know all counties that are connected through one or more counties to a county that overlaps an infected region of high population density. Such a map (depicted in Fig. 2c) represents all possible counties in danger of becoming infected.

Query 2. *Return the map consisting of all counties that overlap a region of high population density, and all counties that are connected through one or more counties to county that overlaps a high population density region.*

3.2 Complex Connectivity Scenarios

One of the advantages of using maps as a data type is that topological and connectivity information is inherently stored in the map in a visual form. While some work has been done on identifying and defining operations that utilize connectivity information, the extent to which such information can be used has not been explored. Here we consider scenarios in which connectivity calculations are required beyond those offered by existing map operations.

A more complex extension of the basic notion of connectivity is the idea of connectivity through a specific construct. For example, county H in Figure 1a contains far more units of flu vaccine than the other counties. Assume that a large number of transport vehicles are stationed in county J. A useful query in this situation is to find out if county C is connected to county J through county H. If the connection exists, then the transport vehicles can drive to county H, pick up vaccines, and deliver them to county C which has no vaccines. There are two possible versions of such a query, one that returns the map of counties that connect C and J through H, and one that simply returns a true or a false value. A query that calculates the latter would be stated:

Query 3. *Are counties C and J connected through county H?*

Instead of transporting vaccines around the map, the authorities determine that if the sum of vaccine units in a county and all of its neighboring counties is 6 or greater, that county has access to enough vaccines within a short distance to effectively combat the spread of the virus. A query to calculate which regions have enough units of vaccine close by uses the notion of a region's neighborhood, i.e., the region and the regions immediately surrounding it. The authorities need to know which regions are not part of a neighborhood containing at least 6 vaccine units. The result of this scenario is shown in Figure 2d.

Query 4. *Return the map consisting of all regions that are not part of a neighborhood containing at least 6 vaccine units.*

4 The Spatial Partition Model

In this section we briefly review the definitions for the formal model of spatial partitions. Furthermore, we review the basic spatial partition operations that form the basis for nearly all existing spatial partition operations. The reader is directed to the literature for a complete list of known partition operations.

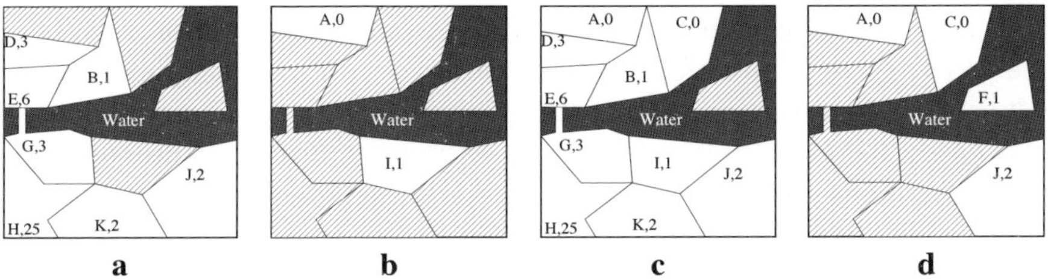

Fig. 2. The resulting partitions from the sample queries. Shaded regions and the water areas are not included in the actual map, but are shown for reference

A spatial partition, in two dimensions, is a subdivision of the plane into pairwise disjoint *regions* such that each region is associated with an attribute having simple or complex structure, and these regions are separated from each other by *boundaries*. All points within the spatial partition that have an identical attribute as a particular region are part of that region. Topological relationships are implicitly modeled among the regions in a spatial partition, and can be considered integrity constraints. For instance, neighborhood relationships where different regions share common boundaries are visible in a map. An implication of this property is that, neglecting common boundaries, the regions of a partition are always disjoint; this property causes maps to have a rather simple structure. From this point forward, we use the term "partition" to refer to a spatial partition.

We stated above that each region in a spatial partition is associated with a single attribute. A spatial partition is modeled by mapping Euclidean space to such *attributes* or *labels*. The regions of the partition are then defined as consisting of all points which contain an identical label. Adjacent regions each have unique labels in their interior, but their common boundary is assigned the labels of both regions.

4.1 Formal Spatial Partition Model

We begin by briefly summarizing the mathematical notation used throughout the following sections. The application of a function $f : A \to B$ to a set of values $S \subseteq A$ is defined as $f(S) := \{f(x)|x \in S\} \subseteq B$. In some cases we know that $f(S)$ returns a singleton set, in which case we write $f[S]$ to to denote the single element, i.e.$f(S) = \{y\} \Longrightarrow f[S] = y$. For doubly nested singleton sets, we use $f[[\cdot]]$ similarly.

The inverse function $f^{-1} : B \to 2^A$ of f is defined $f^{-1}(y) := \{x \in S|f(x) = y\}$. It is important to note that f^{-1} is a total function and that f^{-1} applied to a set yields a set of sets. We define the range function of a function $f : A \to B$ as $rng(f) := f(A)$. It is useful to denote the range of a particular partition.

Thus, for a set-valued function $f : A \to 2^B$, we define the notation $s\text{-}rng[f] := \{b \in B | \{b\} \in rng(f)\}$ which gives the values occurring in singleton sets.

It is useful to sometimes denote functions as parameters for operations. In these cases, the *Lambda notation* $\lambda x : S.E(x)$ where E is an expression using x is used. This is simply an abbreviation for the set expression $\{(x, E(x)) | x \in S\}$.

Let (X, T) be a topological space with topology $T \subseteq 2^x$, and let $S \subseteq X$[1]. The *interior* of S, denoted by Int$S$, is defined as the union of all open sets that are contained in S. The *closure* of S, denoted by $\overline{S}$ is defined as the intersection of all closed sets that contain S. The *exterior* of S is given by Ext$S$:= Int$(X - S)$, and the *boundary* or *frontier* of S is defined as FrS := $\overline{S} \cap \overline{X - S}$. An open set is *regular* if $A = \text{Int}\,\overline{A}$. The type of regular open sets is closed under intersection. In this paper, we deal with $\mathbb{R}^2$ topological space.

A *partition* of a set S, in naive set theory, is a complete decomposition of the set S int non-empty, disjoint subsets $\{S_i | i \in I\}$, called blocks:

(i) $\forall i \in I : S_i \neq \emptyset$,
(ii) $\bigcup_{i \in I} S_i = S$, and
(iii) $\forall i, j \in I, i \neq j : S_i \cap S_j \neq \emptyset$.

where I is an index set used to name different blocks. A partition can equivalently be regarded as a total and surjective function $f : S \to I$. However, a partition cannot be defined simply as a set-theoretic partition of the plane, that is, as a partition of $\mathbb{R}^2$ or as a function $f : \mathbb{R}^2 \to I$, for two reasons: first, f cannot be assumed to be total in general, and second, f cannot be uniquely defined on the borders betwen adjacent subsets of $\mathbb{R}^2$. Furthermore, from an application point of view, it is desirable to to require blocks (which model regions of a common label) to be regular open sets [13].

In [4], spatial partitions have been defined in several steps. First a *spatial mapping* of type A is a total function $\pi : \mathbb{R}^2 \to 2^A$. The existence of an undefined element $\perp_A$ is required to represent undefined labels (i.e., the exterior of a partition is a block $b \in \mathbb{R}^2$ with $\pi[p] = \perp_A$ for all $p \in b$). The labels on the borders of regions are modeled using the power set 2^A; a *border* of π is a block that is mapped to a subset of A containing two or more elements, as opposed to a *region* of π which is a block mapped to a singleton set. The *interior* of π is defined as the union of π's regions. The *boundary* of π is defined as the union of π's borders.

---

[1] In topological space, the following three axioms hold [12]: (i) $U, V \in T \Longrightarrow U \cap V \in T$, (ii) $S \subseteq T \Longrightarrow \bigcup_{U \in s} U \in T$, and (iii) $X \in T, \emptyset \in T$. The elements of T are called *open sets*, their complements in X are called *closed sets*, and the elements of X are called *points*.

Definition 1. Let π be a spatial mapping of type A

(i) $\rho(\pi) := \pi^{-1}(rng(\pi) \cap 2^A)$ (*regions*)
(ii) $\omega(\pi) := \pi^{-1}(rng(\pi) \cap 2^A)$ (*borders*)
(iii) $\iota(\pi) := \bigcup_{r \in \rho(\pi)} r$ (*interior*)
(iv) $\beta(\pi) := \bigcup_{b \in \omega(\pi)} b$ (*boundary*)

A *spatial partition* of type A is then defined as a spatial mapping of type A whose regions are regular open sets and whose borders are labeled with the union of labels of all adjacent regions:

Definition 2. A *spatial partition* of type A is a spatial mapping π of type A with:

(i) $\forall r \in \rho(\pi) : r = \text{Int}\overline{r}$
(ii) $\forall b \in \omega(\pi) : \pi[b] = \{\pi[r] | r \in \rho(\pi) \wedge b \subseteq \overline{r}\}$

4.2 Basic Partition Operations

Three basic spatial partition operations have been defined that can be used to form the formal definitions of almost all other known partition operations: *intersection*, *relabel*, and *refine*. Each of these operations is closed over the set of valid spatial partitions, meaning that if valid partitions are supplied as arguments to these operations, a valid partition will be returned. The intersection of two partitions π and σ, of types A and B respectively, returns a spatial partition of type $A \times B$ such that each interior point p of the resulting partition is mapped to the pair of labels $(\pi[p], \sigma[p])$, and all border points are mapped to the set of labels of all adjacent regions. Formally, the definition of intersection of two partitions π and σ of types A and B can be described in several steps. First, the regions of the resulting partition must be known. This can be calculated by a simple set intersection of all regions in both partitions, since $\cap$ is closed on regular open sets.

$$\rho_\cap(\pi, \sigma) := \{r \cap s | r \in \rho(\pi) \wedge s \in \rho(\sigma)\}$$

The union of all these regions gives the interior of the resulting partition: $\iota_\cap(\pi, \sigma) := \bigcup_{r \in \rho_\cap(\pi,\sigma)} r$. Next, the spatial mapping restricted just to the interior is calculated by mapping each interior point $p \in I := \iota_\cap(\pi, \sigma)$ to the pair of labels given by π and σ:

$$\pi_I := \lambda p : I.\{(\pi[p], \sigma[p])\}$$

Finally, the boundary labels are derived from the labels of all adjacent regions. Let $R := \rho_\cap(\pi, \sigma)$, $I := \iota_\cap(\pi, \sigma)$, and $F := \mathbb{R}^2 - I$. Then we have:

$$intersection : [A] \times [B] \rightarrow [A \times B]$$
$$intersection(\pi, \sigma) := \pi_I \cup \lambda p : F.\{\pi_I[[r]] | r \in R \wedge p \in \bar{r}\}$$

Relabeling a partition π of type A by a function $f : A \rightarrow B$ is defined as $f \circ \pi$, i.e., in the resulting partition of type B each point p is mapped to $f(\pi(p))$ (recall that $\pi(p)$ yields a singleton set, e.g. $\{a\}$, and that f applied to this yields the singleton set $\{f(a)\}$).

$$relabel : [A] \times (A \rightarrow B) \rightarrow [B]$$
$$relabel(\pi, f) := \lambda p : \mathbb{R}^2.f(\pi(p))$$

The refinement of a partition identifies the connected components of the partition. This is achieved by relabeling the connected components of a partition with consecutive numbers. A connected component of an open set S is a maximum subset $S \subseteq T$ such that any two points of T can be connected by a curve lying completely inside T [12]. Let $\gamma(r) = \{c_1, ...c_{n_r}\}$ denote the set of connected components in a region r. Then, *refine* can be defined in several steps. The regions of the resulting partition are the connected components of all regions of the original partition:

$$p_\gamma(\pi) := \bigcup_{r \in \rho(\pi)} \gamma(r)$$

The union of all these regions results in the interior of the resulting partition: $\iota_\gamma(\pi) := \cup_{r \in p_\gamma(\pi)} r$. This means that the set of interior and boundary points are not changed by refine.

We can now define the resulting partition on the interior:

$$\pi_I := \{(p, \{\pi[p], i)\}) | r \in \rho(\pi) \wedge \gamma(r) = \{c_1, ..., c_{n_r}\} \wedge i \in \{1, ..., n_r\} \wedge p \in c_i\}$$

Finally, we derive the labels for the boundary from the interior, much like the definition for intersection. Let $R := \rho_\gamma(\pi)$, $I := \iota_\gamma(\pi)$, and $F := \mathbb{R}^2 - I$. Then:

$$refine : [A] \rightarrow [A \times \mathbb{N}]$$
$$refine(\pi) := \pi_I \cup \lambda p : F.\{\pi_I[[r]] | r \in R \wedge p \in \bar{r}\}$$

5 Formalization of Novel Operations

The scenarios presented in Section 3 provide specific instances in which new types of operations are required. In this section, we define new, operations which provide the functionality to calculate those queries. Note that while we used the example queries to motivate the need for these operations, we define operations to satisfy the general types of queries that the examples represent.

5.1 Formalization of Map Join Operations

To calculate Query 1, the query must collect entire regions from one map that satisfy a particular topological predicate with a region from the opposite map. In effect, an operation must calculate a new map of entire regions from argument maps by joining two maps according to a specific constraint. We denote the set of operations that perform this function *map join* operations.

We define the map join in two parts. First, we must define a method by which regions in one of the argument partitions that satisfy the join constraint can be collected. We achieve this in the *collect* operation which takes two partitions, π and σ, and a predicate P (which takes two complex regions), and relabels any region from π that does not satisfy the predicate with at least one region from σ to $\perp$.

$$\begin{aligned}&\mathit{collect} : [A] \times [B] \times (2^{\mathbb{R}^2} \times 2^{\mathbb{R}^2} \to \mathbb{B}) \to [A]\\&\mathit{collect}(\pi,\sigma,P) := \mathit{relabel}(\pi, \lambda x : A.\ \text{if } \exists y \in B : P(\overline{\pi^{-1}(x)}, \overline{\sigma^{-1}(y)}) \text{ then } x \text{ else } \perp_A)\end{aligned}$$

The join operation can then be defined using the operations *intersection* and *collect*. Since a single region may consist of multiple faces within the partition, we use the refine operation to break such regions so that each face is considered a single region. Because refine extends each region label by appending an integer, we will also use a relabel operation after each collect operation that simply removes the appended integer from each region. We denote this operation *truncate* and omit a formal definition due to space considerations. Given two partitions π and σ and a predicate P, we define the *left-join* and *right-join* operations which create a result partition consisting of regions from the first and second argument partitions, respectively (i.e., they are one-sided joins):

$$\begin{aligned}&\mathit{left\text{-}join} : [A] \times [B] \times (2^{\mathbb{R}^2} \times 2^{\mathbb{R}^2} \to \mathbb{B}) \longrightarrow [A]\\&\mathit{left\text{-}join}(\pi,\sigma,P) := \mathit{truncate}(\mathit{collect}(\mathit{refine}(\pi), \mathit{refine}(\sigma), P))\\&\mathit{right\text{-}join} : [A] \times [B] \times (2^{\mathbb{R}^2} \times 2^{\mathbb{R}^2} \to \mathbb{B}) \longrightarrow [B]\\&\mathit{right\text{-}join}(\pi,\sigma,P) := \mathit{truncate}(\mathit{collect}(\mathit{refine}(\sigma), \mathit{refine}(\pi), P))\end{aligned}$$

The two-sided join, which we denote simply as *join*, returns the partition consisting of all regions from both argument partitions π and σ that satisfy a given predicate P.

$$\begin{aligned}&\mathit{join} : [A] \times [B] \times (2^{\mathbb{R}^2} \times 2^{\mathbb{R}^2} \to \mathbb{B}) \longrightarrow [A \times B]\\&\mathit{join}(\pi,\sigma,P) := \mathit{intersection}(\mathit{left\text{-}join}(\pi,\sigma,P), \mathit{right\text{-}join}(\pi,\sigma,P))\end{aligned}$$

These three map join operations are general enough to calculate any join that can be expressed based on a predicate that is satisfied between regions from opposing partitions. However, Query 2 introduces the idea of incorporating connectivity with a map join operation. The result partition for this query must contain the regions that satisfy a left-join operation using the overlap predicate, and regions from the left partition that are connected (through one or more partitions) to the result of the left-join. For example, to compute Figure 2c, Query 2 must first compute the partition in Figure 2a using a left-join with the overlap predicate, then add more regions from the county map to the result if they satisfy a second predicate with another region in the county map that is part of the left-join, namely if they are connected to a region in the result of the join. We denote this new operation the *extended-join* between two partitions. To define this new operation, we must use the *difference* [4] operation, which takes two partitions and returns the first partition minus any area that is overlapped by the second partition. We begin by specifying the *left-extended-join* (*right-extended-join*). Given two partitions, π and σ, a predicate P for the join, and a predicate Q for the extended portion of the join, we first calculate the left-join (right-join) based on predicate P. We then include, using the operations intersection and collect, any regions in the difference of π (σ) and the result of the left-join (right-join) that satisfy the predicate Q with a region in the left-join (right-join).

$$\begin{array}{l}
\textit{extended-left-join} : [A] \times [B] \times (2^{\mathbb{R}^2} \times 2^{\mathbb{R}^2} \to \mathbb{B}) \times (2^{\mathbb{R}^2} \times 2^{\mathbb{R}^2} \to \mathbb{B}) \longrightarrow [A] \\
\textit{extended-left-join}(\pi,\sigma,P,Q) := \textit{intersection}(\textit{collect}(\textit{difference}(\pi, \\
\quad \textit{left-join}(\pi,\sigma,P)), \textit{left-join}(\pi,\sigma,P), Q), \textit{left-join}(\pi,\sigma,P)) \\
\textit{extended-right-join} : [A] \times [B] \times (2^{\mathbb{R}^2} \times 2^{\mathbb{R}^2} \to \mathbb{B}) \times (2^{\mathbb{R}^2} \times 2^{\mathbb{R}^2} \to \mathbb{B}) \longrightarrow [B] \\
\textit{extended-right-join}(\pi,\sigma,P,Q) := \textit{intersection}(\textit{collect}(\textit{difference}(\sigma, \\
\quad \textit{right-join}(\pi,\sigma,P)), \textit{right-join}(\pi,\sigma,P), Q), \textit{right-join}(\pi,\sigma,P))
\end{array}$$

The two-sided extended join can then be defined as the intersection of the left and right extended joins. Because the second predicate used by the left and right extended joins is evaluated between regions of the first and second argument partition, respectively, we pass two additional predicates to the two-sided extended join, one which is used by the first argument partition, and the other which is used by the second argument partition.

$$\begin{array}{l}
\textit{extended-join} : [A] \times [B] \times (2^{\mathbb{R}^2} \times 2^{\mathbb{R}^2} \to \mathbb{B}) \times (2^{\mathbb{R}^2} \times 2^{\mathbb{R}^2} \to \mathbb{B}) \\
\quad \times (2^{\mathbb{R}^2} \times 2^{\mathbb{R}^2} \to \mathbb{B}) \longrightarrow [A \times B] \\
\textit{extended-join}(\pi,\sigma,P,Q,R) := \textit{intersection}(\textit{extended-left-join}(\pi,\sigma,P,Q), \\
\quad \textit{extended-right-join}(\pi,\sigma,P,R))
\end{array}$$

Query 2 can be computed using an extended left join operation. However, to complete the query, a predicate is required that takes two regions in a partition and returns a value of true if those regions are connected, and returns a value of false otherwise. This notion of connectivity has been studied in previous work; however, we will provide a slightly more general definition of connectivity. We define the *connected* predicate that takes two regions r and s each belonging to a separate argument partition. This predicate returns a value of true if there exists a chain of regions from r to s such that each region in the chain is not disjoint with its neighbors in the chain (i.e., the regions share at least one common point). Furthermore, each region in the chain can be from either argument partition. For example, if the two argument partitions were the county map and the population density map, then the chain J, P3, K, H, P2 connects regions J and P2. Given two regions r and s from partitions π and σ respectively, we define the connected predicate as follows:

$$
\begin{aligned}
&\mathit{connected}: [A] \times A \times [B] \times B \longrightarrow \mathbb{B}\\
&\mathit{connected}(\pi, r, \sigma, s) :=\\
&\begin{cases}
\text{true} & \text{if } \exists t_1 \ldots t_n | t_1 = r \wedge t_n = s \wedge \forall 1 \leq i < n : \\
& ((t_i, t_{i+1} \in A \wedge \pi^{-1}(t_i), \pi^{-1}(t_{i+1}) \in \rho(\pi) \wedge \neg \mathit{disjoint}(\overline{\pi^{-1}(t_i)}, \overline{\pi^{-1}(t_{i+1})})) \\
& \vee (t_i, t_{i+1} \in B \wedge \sigma^{-1}(t_i), \sigma^{-1}(t_{i+1}) \in \rho(\sigma) \wedge \neg \mathit{disjoint}(\overline{\sigma^{-1}(t_i)}, \overline{\sigma^{-1}(t_{i+1})})) \\
& \vee (t_i \in A \wedge t_{i+1} \in B \wedge \pi^{-1}(t_i) \in \rho(\pi) \wedge \sigma^{-1}(t_{i+1}) \in \rho(\sigma) \\
& \wedge \neg \mathit{disjoint}(\overline{\pi^{-1}(t_i)}, \overline{\sigma^{-1}(t_{i+1})})) \vee (t_i \in B \wedge t_{i+1} \in A \wedge \sigma^{-1}(t_i) \in \rho(\sigma) \wedge \\
& \pi^{-1}(t_{i+1}) \in \rho(\pi) \wedge \neg \mathit{disjoint}(\overline{\sigma^{-1}(t_i)}, \overline{\pi^{-1}(t_{i+1})}))) \\
\text{false} & \text{otherwise}
\end{cases}
\end{aligned}
$$

The purpose of Query 2 is to find all counties in the county map that the flu is likely to spread to in the future based on the connectivity of a county to a county which already has flu infections present in it. However, no limit is set on the number of counties in a connected chain. Instead, the user may wish to calculate an extended join containing counties that are connected to an infected county by only one or two counties. This query allows the user to see all infected counties, and all counties which are likely to experience flu infections in the near future. To define such a predicate, we introduce the concept of degree of connectivity between two regions.

The *degree of connectivity* between two regions indicates the minimum number of regions, not counting the source and destination regions, that are needed in a particular partition to connect a source region and a destination region. If two regions with labels r and s in a partition π are not connected, then their connectivity degree is -1. If they are connected directly (meaning they are not disjoint, or they share at least one common point), then their connectivity degree is 0. If a single region connects them, then their connectivity degree is 1, etc. Note that some regions may be connected by more than one path of regions, in which case the connectivity degree is the degree of the minimal path. For example, counties H and K in the county map have a connectivity degree of zero even though they are connected by paths *H,K* and *H,G,I,J,K*.

$$
\begin{aligned}
&\mathit{degree\text{-}connected} : [A] \times A \times A \longrightarrow \mathbb{N}\\
&\mathit{degree\text{-}connected}(\pi, r, s) :=\\
&\quad\begin{cases}
-1 \text{ if } \neg \mathit{connected}(\pi, r, \pi, s)\\
0 \quad \text{if } \mathit{connected}(\pi, r, \pi, s) \wedge \neg \mathit{disjoint}(\overline{\pi^{-1}(r)}, \overline{\pi^{-1}(s)})\\
n \quad \text{if } \exists t_1, \ldots, t_n \in A | \pi^{-1}(t_1), \ldots, \pi^{-1}(t_n) \in \rho(\pi) \wedge \mathit{connected}(\pi, r, \pi, s)\\
\quad\quad \wedge \neg \mathit{disjoint}(\overline{\pi^{-1}(r)}, \overline{\pi^{-1}(t_1)}) \wedge \ldots \wedge \neg \mathit{disjoint}(\overline{\pi^{-1}(t_n)}, \overline{\pi^{-1}(s)})\\
\quad\quad \text{where } n \text{ is minimal}
\end{cases}
\end{aligned}
$$

5.2 Formalization of Complex Connectivity Operations

The notions of connectivity defined in the previous section build upon the idea of determining if two regions in a map are connected. In this section, define new, more advanced connectivity operations.

Query 3 asks if two regions are connected through a specific region in the map. In order to determine whether two regions are connected through a specified region, we define the operation *degree-connected-through*, which takes a partition, a source and destination region label, and a region label which identifies a region that the connection must pass through. The degree of connectivity returned is the minimum number of regions, including the region required to be part of the path, that are needed to connect the source and destination regions, or -1 if such a connection does not exist:

$$
\begin{aligned}
&\mathit{degree\text{-}connected\text{-}through} : [A] \times A \times A \times A \longrightarrow \mathbb{N}\\
&\mathit{degree\text{-}connected\text{-}through}(\pi, r, s, q) :=\\
&\quad\begin{cases}
-1 & \text{if } \neg \mathit{connected}(\pi, r, \pi, q)\\
& \quad \vee \neg \mathit{connected}(\pi, q, \pi, s)\\
0 & \text{if } \mathit{degree\text{-}connected}(\pi, r, s) = 0\\
& \quad \wedge (r = q \vee s = q)\\
\mathit{degree\text{-}connected}(\pi, r, q) & \\
\quad + \mathit{degree\text{-}connected}(\pi, s, q) & \text{if } (r = q \vee s = q)\\
\mathit{degree\text{-}connected}(\pi, r, q) & \\
\quad + \mathit{degree\text{-}connected}(\pi, s, q) + 1 & \text{otherwise}
\end{cases}
\end{aligned}
$$

It is sometimes useful to calculate the partition containing the connection of two regions. We use the *connect* operation to provide a partition containing a source and destination region, and the minimum number of regions that connect them. This operation has been previously defined in [5], so we omit a detailed description. The *connect-through* operation calculates the partition consisting of a source and destination region, and the minimum number of regions that connect them, including a specified region. This operation can be reduced to a relabeling problem if the set of region labels contained in the resulting partition is known (i.e., the set of labels that correspond to the minimal chain of regions connecting a source and destination region). Using the degree of connectivity, we first define the operation *connected-through-set* which calculate this set of region labels. The connect-through operation then relabels any region in the argument partition whose label is not in that set to the empty label:

$$\begin{aligned}
&\textit{connected-through-set} : [A] \times A \times A \times A \longrightarrow A \\
&\textit{connected-through-set}(\pi, r, s, q) := \{t_1, \ldots, t_n | \\
&\quad n = \textit{degree-connected-through}(\pi, r, s, q) + 2 \wedge t_1 = r \wedge t_n = s \wedge \\
&\quad (\exists 1 \leq i \leq n : t_i = q) \wedge \forall 1 \leq k \leq n : \overline{\pi^{-1}(t_k)} \in \rho(\pi) \\
&\quad \wedge \forall 1 \leq j < n : \neg \textit{disjoint}(\overline{\pi^{-1}(t_j)}, \overline{\pi^{-1}(t_{j+1})})\} \\
&\textit{connect-through} : [A] \times A \times A \times A \longrightarrow [A] \\
&\textit{connect-through}(\pi, w, x, y) := \\
&\quad \textit{relabel}(\pi, \lambda z : A.\ \text{if } z \in \textit{connected-through-set}(\pi, w, x, y) \text{ then } z \text{ else } \bot_A)
\end{aligned}$$

A final connectivity operation suggested in Query 4 is the neighborhood operation. In this query, the user is interested in the number of vaccine units available in the immediate neighborhood of a region (which includes that region and all regions adjacent to that region). In general, a neighborhood can be specified to include any regions within a specified degree of connectivity to a source region. For instance, the 1-neighborhood of a region consists of the source region and all regions adjacent to it. The 2-neighborhood consists of the source region and all regions with a connectivity degree of 0 or 1 to the source region. We define the neighborhood operation to take a partition π, a region label r, and an integer indicating the size of the neighborhood desired. The neighborhood is then computed by assigning the labels of all regions that are not in the neighborhood of r to $\bot$:

$$\begin{aligned}
&\textit{neighborhood} : [A] \times A \times \mathbb{N} \longrightarrow [A] \\
&\textit{neighborhood}(\pi, r, n) := \\
&\quad \textit{relabel}(\pi, \lambda x : A.\ \text{if } \textit{degree-connected}(\pi, r, x) < n \text{ then } x \text{ else } \bot_A)
\end{aligned}$$

6 Conclusions

In this paper we have defined several new operations on maps. By examining maps from the user's perspective in a spatial database context, we have identified the new class of map join operations, and demonstrated the power of those operations with example scenarios. We have further defined new operations that deal with complex connectivity concepts. These operations, coupled with existing map operations, provide a powerful operational foundation to further explore the use of maps in spatial databases. This research was conducted as part of an effort to integrate maps as fundamental data type into database systems. Future work includes implementing a model of spatial partitions in a database and implementing the operations over them. Another aspect of this research is the integration of map operations into a query language such as SQL. In addition to operations, we plan to explore predicates over maps in a future paper.

References

1. Güting RH (1988) Geo-relational algebra: A model and query language for geometric database systems. In: Int Conf on Extending Database Technology (EDBT), pp 506–527
2. Huang Z, Svensson P, Hauska H (1992) Solving spatial analysis problems with geosal, a spatial query language. In: Proc of the 6th Int Working Conf on Scientific and Statistical Database Management. Institut fuer Wissenschaftliches Rechnen, Eidgenoessische Technische Hochschule Zürich, pp 1–17
3. Güting RH, Schneider M (1995) Realm-Based Spatial Data Types: The ROSE Algebra. VLDB J 4:100–143
4. Erwig M, Schneider M (1997) Partition and conquer. In: 3rd Int Conf on Spatial Information Theory (COSIT) (= LNCS 1329). Springer-Verlag, pp 389–408
5. Erwig M, Schneider M (2000) Formalization of advanced map operations. In: 9th Int Symp on Spatial Data Handling, pp 8a.3–17
6. Scholl M, Voisard A (1990) Thematic map modeling. In: SSD 1990, Proc of the 1st Symp on Design and Implementation of Large Spatial Databases, New York, NY, USA. Springer-Verlag, New York, pp 167–190
7. Chan EPF, Zhu R (1996) Ql/g: A query language for geometric databases. In: 1st Int Conf on GIS in Urban and Environmental Planning, pp 271–286
8. Frank AU (1987) Overlay processing in spatial information systems. In: 8th Int Symp on Computer-Assisted Cartography, AUTOCARTO, pp 16–31
9. Schneider M (1997) Spatial Data Types for Database Systems – Finite Resolution Geometry for Geographic Information Systems (= LNCS 1288). Springer-Verlag
10. Egenhofer MJ, Herring J (1990) Categorizing binary topological relations between regions, lines, and points in geographic databases. Technical report, National Center for Geographic Information and Analysis, University of California, Santa Barbara
11. Schneider M, Behr T (2006) Topological relationships between complex spatial objects. ACM Transactions on Database Systems (accepted for publication)
12. Dugundi J (1966) Topology. Allyn and Bacon
13. Tilove RB (1980) Set membership classification: A unified approach to geometric intersection problems. IEEE Trans on Computers 29:874–883

A Tangible Augmented Reality Interface to Tiled Street Maps and its Usability Testing

Antoni Moore

Spatial Information Research Centre, Department of Information Science, University of Otago, PO Box 56, Dunedin, New Zealand
email: amoore@infoscience.otago.ac.nz

Abstract

The Tangible Augmented Street Map (TASM) is a novel interface to geographic objects, such as tiled maps of labeled city streets. TASM uses tangible Augmented Reality, the superimposition of digital graphics on top of real world objects in order to enhance the user's experience. The tangible object (i.e. a cube) replicates the role of an input device. Hence the cube can be rotated to display maps that are adjacent to the current tile in geographic space. The cube is capable of theoretically infinite movement, embedded in a coordinate system with topology enabled. TASM has been tested for usability using heuristic evaluation, where selected experts use the cube, establishing non-correspondence with recognized usability principles. While general and vague, the heuristics helped prioritize immediate geographic and system-based tasks needed to improve the usability of TASM, also pointing the way towards a group of geographically oriented heuristics. This addresses a key geovisualization challenge – the creation of domain-specific and technology-related theory.

Key words: geographic, navigation, heuristic evaluation

1 Introduction

1.1 A Tangible Augmented Reality Interface to Street Maps

A Tangible Augmented Street Map (TASM) has been developed to address certain issues associated with conventional analogue and digital street maps (Moore and Regenbrecht, 2005). These are:

- maps in book form do not strictly follow principles of autocorrelation. In following a route, closeness in reality does not necessarily mean likewise in the book, as non-contiguous pages may be accessed.
- street maps in fold-up form have contiguity; are geographically intuitive, but can be hard to use in outdoor environments
- digital map sources in discrete (e.g. city street maps, such as Dunedin's – 2006) and continuous form (Google Earth – Google 2006) are prevalent on the Internet (and have contiguity), but the absence of a tangible object weakens the mental model that a user has to navigate around an environment (Crampton 1992; Ratti et al. 2004).

TASM uses Augmented Reality (AR), "composite systems that use a combination of a real scene viewed by users and a virtual scene generated by a computer, [thus] augment[ing] the real scene with additional information" (Ratti et al. 2004, p 409). More specifically, tangible AR (Hedley et al. 2002) uses physical objects to seamlessly interact with digital content. The Magic Book application (Billinghurst et al. 2001) is an example of tangible AR, in which the recognition of patterns from the pages of a children's book (imaged through a web cam) generates virtual 3D images that appear to rest on the book page (as seen on a monitor window or Head Mounted Display – HMD). Ratti et al. (2004) describe Illuminating Clay and Sandscape, which uses conventional modeling media (clay and sand) to manipulate a digital landscape in DEM format. Hedley et al. (2002) use tangible user interfaces (e.g. paddles) for collaborative geovisualization of virtual terrain model data, enabling immersive fly through and zooming.

For TASM, a cube assumes the role of the tangible object (see Fig. 1a), with the six faces corresponding to book pages (to borrow the Magic Book terminology). The cube forms the interface to a tiled digital street map, and can be rotated across an edge to alter what can be virtually displayed on a cube face. Moreover, the face itself can be aligned so that the street map orientation agrees with reality, reducing the amount of cognitive processing needed to navigate. The test set-up comprises the cube, a web cam and a laptop providing the processing power and display means. TASM represents an indirect delivery of visual AR (Höllerer and Feiner 2004), since digital output is not related to real geographic objects *in situ*.

1.2 Research Challenges in Geovisualization

The TASM project addresses a current research challenge in cartographic visualization, the creation of new interface paradigms "to take advantage of recent (and anticipated) technological advances in both hardware and data formats" (MacEachren and Kraak 2001). The use of tangible AR addresses a call for "more natural" interfaces (Cartwright et al. 2001).

Whilst the potential of the new technology is great, there is a need for cognitive and usability evaluation of the novel interface paradigms, as such studies are rare for highly interactive visual environments (MacEachren and Kraak 2001; Slocum et al. 2001; Bowman et al. 2005). This is to establish the worth of the interface in its chosen operating situation. From a geovisualization point-of-view, the shift from technology-driven to user-driven development of interfaces (Fuhrmann et al. 2005; Tobón 2005) is seen as a way of providing theory to this research area (Slocum et al. 2001; Fuhrmann et al. 2005; Wood et al. 2005), or in other words, a phenomenological approach to augment the predominant positivist paradigm (Bodum 2005). Paper maps have a history of theory-based development, unlike 3D cartography, which follows technology, the "zeitgeist" (Wood et al. 2005). These produces novel applications at a rapid rate (noted by Bowman et al. 2005), but are left devoid of theory.

The creation of theory is not simply a matter of transplanting HCI (Human Computer Interaction) theory to geovisualization (Fuhrmann et al. 2005) – with pure HCI, the focus is on the system itself, whilst HCI for geovisualization aims to support the modeling of the application domain. Therefore there is a need to address what a given geographic application actually needs: are they being satisfied by the emergent interfaces? (echoed by Hedley et al. 2002). Researchers have been examining what sets the geovisualization domain apart, and the main facet is the propensity for unstructured tasks, particularly with exploration and knowledge discovery (Slocum et al. 2001; Fuhrmann et al. 2005; Tobón 2005). Walsh et al. (2002) point to the geographic emphasis on the graphical as well as the lexical, whereas conventional interfaces tend to be purely lexical in nature.

1.3 Using TASM to Generate Theory

TASM has been tested using heuristic evaluation (HE), a method of interface usability assessment. A small group of experts individually explore the interface in HE, establishing non-correspondence with recognized usability principles, or heuristics (Nielsen 1994a). As the heuristics pertain to conventional interfaces, it will be interesting to observe how well they address both the geographic application and the tangible AR technology presented here. In this way, the heuristics will form a loose structure, against which any paradigmatic mismatch would point towards a theory specific to this type of application. Bowman et al. (2005) state that no overall heuristics for 3D visualization currently exist, due to the sheer variety of interfaces that exist in this domain. However, TASM will be working in a narrow domain and its function of map presentation indicates that structured geographic tasks will be possible. This "custom tailored" usability test is in keeping with Tobón (2005).

The design of TASM is based on geographic principles and the use of the cube is intended to be geographically intuitive. The requisite contiguity is achieved by having adjacent map tiles accessed via neighboring faces on the cube. A further requirement is that of an unbounded space – the cube could be rotated endlessly in any of the four major cardinal directions to access an infinite amount of map tiles (though practically the user will be limited by the physical extent of the data). The global analogy is the ceaseless ability to navigate along the equator. The four directions used (North, South, East and West) are also cognitively intuitive; such qualitative operators are ubiquitously used for spatial reasoning in small-scale space (Frank 1996). Finally, the use of a coordinate system is conventional in mapmaking, and will be applied discretely to the map tiles.

The next section expands on the short introduction to TASM in Moore and Regenbrecht (2005), providing more detail on implementation and operation, and an overview of heuristic evaluation. The paper describes the evaluation results, and then presents the discussion and conclusion.

2 Tangible Augmented Reality Street Map (TASM)

TASM has been built using the open source AR development environment ARToolKit (ARToolKit 2006). An application written for ARToolKit takes as input a video feed from a web cam. The input is processed for recognition of markers, each of which consists of a square surrounding a pattern (see Fig. 1). The dimension of the square is precisely known, and acts in the same way as fiducial marks on an aerial photograph to spatially register the pattern within. Then, the pattern is compared with a stored bank of patterns, each of which is linked to a VRML (Virtual Reality Modeling Language) object. If there is a match, the 3D object is displayed in the coordinate space of the marker (see Fig. 2). Hedley et al. (2002) enhanced ARToolKit to develop their geovisualization user interfaces.

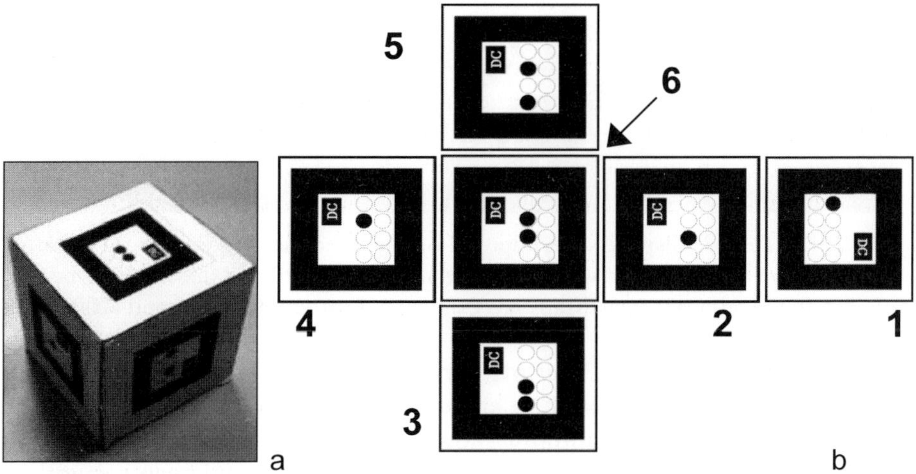

Fig. 1. **(a)** The TASM cube, and **(b)** the cube laid out flat – marker patterns and orientations are correct. The faces are numbered 1 to 6 for future reference

For TASM, there are six markers, one per cube face. Upon recognition of a marker, a planar street map tile (a VRML object or texture) is displayed on the face on which the marker is mounted. For the test scenario, 25 map tiles covering the centre of Dunedin, New Zealand can be accessed. The initial map tile displayed is right at the centre of the 5x5 array, and is given a discrete coordinate of (0, 0). The coordinate system has a lower-left origin, which means the tile to the north will be (0, 1).

To display the north map tile, the cube is rotated towards the user (across the top edge). The new marker will elicit display of the northern tile (see Fig. 2). Similarly the cube can be rotated away from the user to move south on the map; a rotation to the left will pan eastwards and a rotation to the right will pan westwards. This satisfies the requisites of spatial association and use of cognitively intuitive cardinal directions. Furthermore, Figure 2b demonstrates that two faces can be displayed simultaneously, ensuring the visual association of neighboring tiles, further strengthening the induced mental model.

Four 90-degree rotations in a specific direction (apart from in the plane of the current cube face itself) will cause the original marker to be displayed again. Conventionally, a marker is explicitly linked with a specific 3D object. Clearly (and with data allowing), this would result in the original map tile being displayed again, which would not fulfill the requirement for a geographically realistic extent to space. TASM solves this by explicitly linking each marker to its four neighboring markers, ensuring each marker's topology is coded. Each marker is numbered 1 to 6.

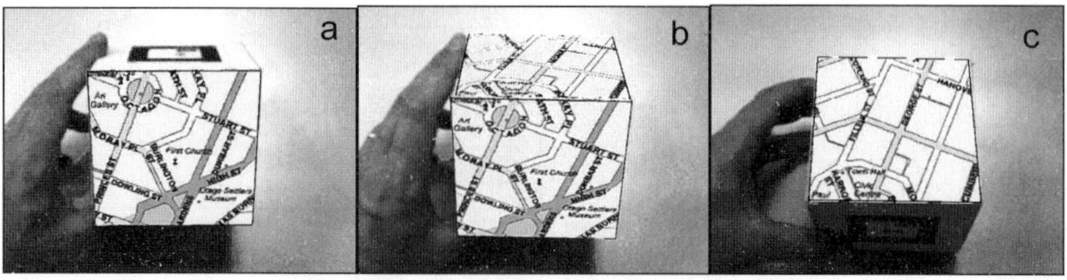

Fig. 2. (a) Texture display superimposed on the cube face marker. **(b, c)** Rotating the cube towards the user will cause the neighboring tile to the north to be displayed on the adjacent face. Maps are Copyright Dunedin City Council (2006)

For example (see Fig. 3), if marker 6 is initially recognized by the camera, the central map tile will be shown (the same would be true if any of the other markers were initially shown). If the map were aligned so that north is upwards, a rotation to the left (moving east) would cause marker 5 to be shown (see Table 1 for a full list of topologies). A coordinate shift of +1 in x would be recorded, and the new coordinate used to calculate the identity of the new map tile to be displayed (the tiles are explicitly identified by their row order numbers to facilitate this). This system enables repeated rotations in a sustained direction, fulfilling the unbounded space criterion.

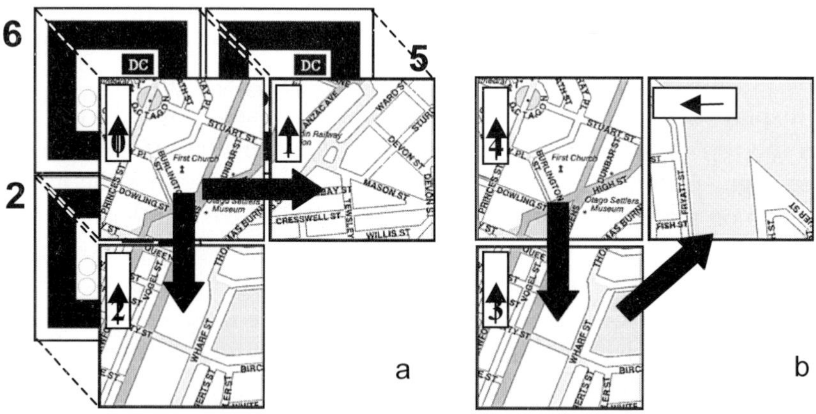

Fig. 3. (a) Moving eastwards or southwards from the centre tile. Part of the cube layout with marker numbers is revealed beneath. **(b)** Moving southwards *then* eastwards. See text for explanation. Maps Copyright Dunedin CC (2006)

Display of a map tile is aligned to a specific orientation of the marker. In moving to an adjacent cube face, a different marker orientation is likely (see Table 1 for a list of orientation shifts). Therefore a local angular correction needs to be made for the map to be displayed at the correct alignment. Furthermore, an overall alignment value (relative to the first map tile displayed) needs to be maintained and propagated, since each correction is relative only to the previous face displayed.

For example (see Fig. 3), moving from marker 6 to marker 5 or to marker 2 would require no shift in orientation for the map tile to be displayed at the correct alignment. However, moving from marker 2 to marker 5 would necessitate a correcting 90-degree anti-clockwise rotation of the map for it to be correctly aligned. Table 2 shows the list of coordinate changes that can be made with different face orientation changes. Extending the example, a move eastwards (from marker 2 to marker 5 – look up Table 1) would naturally be a positive increment in x, which is verified in the –90 corrective rotation row in Table 1. Furthermore, the slightly different routes taken to the same marker (6 to 5 as opposed to 6 to 2 to 5) critically alter both the orientation and the identity of the map tile displayed.

Table 1. Cube face topologies and orientation changes relative to adjacent faces

Cube Face	North Neighbor		East Neighbor		South Neighbor		West Neighbor	
	Face	Change in orientation (degrees)	Face	Change in orientation (degrees)	Face	Change in orientation (degrees)	Face	Change in orientation (degrees)
1	4	180	3	0	2	180	5	0
2	6	0	5	90	1	180	3	-90
3	4	-90	6	0	2	90	1	0
4	1	180	5	-90	6	0	3	90
5	4	90	1	0	2	-90	6	0
6	4	0	5	0	2	0	3	0

Table 2. Discrete coordinate changes to be applied (for various corrective rotations) when moving to a neighboring cube face

	Move North		Move East		Move South		Move West	
	X change	Y change	X change	Y change	X change	Y change	X change	Y change
0	0	1	1	0	0	-1	-1	0
90	1	0	0	-1	-1	0	0	1
180	0	-1	-1	0	0	1	1	0
-90	-1	0	0	1	1	0	0	-1

3 Usability Testing

3.1 Usability and Types of Test

Interface usability "...encompasses everything about an artifact and a person that affects the person's use of that artifact." (Bowman et al. 2005, p 351). Usability testing is essential, as even the most fully realized designs can go wrong in practice. There are a number of tests for usability, mostly user-based, including cognitive walkthrough, formative evaluation, summative evaluation, questionnaires, interviews and video analysis [Davies and Medyckyj-Scott (1994, 1996) used the latter three techniques to collect user opinion on established GIS implementations.] Gabbard et al. (1999) combined these techniques and (expert-based) heuristic evaluation (see 3.2) for their sequential evaluation approach, stressing iteration and participatory testing. Other authors use (or plan to use) linear combinations of testing techniques (Walsh et al. 2002; Fuhrmann et al. 2005).

3.2 Heuristic Evaluation

Heuristic evaluation was devised to discover problems associated with a user interface design, usually in the initial stages (Nielsen 1994a). A small number of expert usability evaluators are used to judge the interface's degree of compliance with widely accepted usability principles (heuristics – listed in Table 3). Walsh et al. (2002) use heuristics as part of their testing process of web-based data clearinghouses (two later stages involved a user expectations survey and user testing).

Table 3. Nielsen's heuristics in outline (derived from Nielsen 1994a)

Heuristic Number	Description of heuristic
1	Visibility of system status
2	Match between system and the real world
3	User control and freedom
4	Consistency and standards
5	Error prevention
6	Recognition rather than recall
7	Flexibility and efficiency of use
8	Aesthetic and minimalist design
9	Help users recognize, diagnose and recover from errors
10	Help and documentation

For the testing of TASM, seven evaluators were used, chosen for their familiarity with usability testing methods and / or general experience in the human-computer interaction field. Normally, 3–5 evaluators are recommended for HE, as together they can normally be relied upon to identify the majority of problems associated with using an interface (Nielsen 1994a) without being too costly.

An evaluation session was conducted with one evaluator at a time exploring the TASM interface, verbalizing usability problems to the author as they were encountered. The exploration of the interface occurred in two phases, firstly to get an overall feel of the system and secondly for a deeper focus on specific elements. During the latter phase, a specific circular route across Dunedin was dictated to the evaluator, who was expected to "walk" the route whilst navigating through the map tiles. After this, the evaluator was encouraged to explore TASM further, until satisfied that they had a good understanding of the issues associated with using the interface.

During the session, the evaluators were encouraged to elaborate where necessary on the problems they encountered. If they were unsure of interface elements or having general trouble with operating the interface, then feedback was actively sought since the difficulty is likely to be associated with usability problems. Steps were taken to ensure that there was no communication between evaluators on the subject of TASM and its evaluation until all testing was completed. This was to ensure independence of results and reduction of bias. After the session with TASM, the usability problems were revisited and associated with one or more of the heuristics in Table 3. If no heuristic was readily appropriate to a specific problem, then feedback on this was sought, to fully document it.

Lastly, the problems identified were collated and sent back to the evaluators in the form of a post hoc questionnaire. For each of the problems (some of which may be unfamiliar to the individual evaluator) the evaluators were invited to assign a severity rating. The five-point scale suggested by Nielsen (1994a) was used, ranging from a zero rating (not a usability problem at all) to a rating of 4 (usability catastrophe – it is imperative to fix this). Factors to take into account when assigning severity ratings include the frequency with which the problem occurs, the impact of the problem if it occurs and the persistence of the problem once identified.

4 Results

Table 4 lists the 21 usability problems encountered by the evaluators during the testing sessions. For the purposes of reporting, these problems were split into two groups: system-related and geographic. System-related issues were sub-divided further, into ergonomics (e.g. the cube induces user fatigue), AR display-related problems (e.g. when display is interrupted due to the marker no longer being tracked by the camera) and peripheral elements (e.g. lack of instructions). Geographic issues were sub-divided into navigational aids (e.g. North Arrow), cartographic (e.g. inconsistent street labeling) and spatial cognition problems (e.g. cube does not agree with the tester's spatial mental model). These groupings are by no means independent of each other; for instance, the navigation and AR display issues heavily influence, and are influenced by, cognition; such an overlap is a feature of the heuristics themselves.

Table 5 shows which of the heuristics were broken by each problem, as identified only by the evaluators that found the problem during testing. In most cases, more than one heuristic was broken by any specific problem. A measure of how much the evaluators agreed on what heuristic(s) had been broken for a specific problem (such as the kappa coefficient[1]) was not used in this case. This is due to the subjective conditions under which the heuristics were originally labeled and the degree to which the heuristic categories can overlap (Nielsen 1994b). However, the most common heuristics associated with each problem were identified and counted.

The rankings on usability problem severity ratings, arranged by evaluator, are given in Table 6. To establish whether agreement existed on the assignment of ratings, Kendall's coefficient of concordance (W) for ordinal data was applied. Ties in ranking were inevitable, and a correction for this was applied (Siegel and Castellan 1988). There was a significant amount of concordance between the evaluators ($W = 0.28$; $\chi^2 = 39.28$; $df = 20$; $p<0.01$). Therefore, we can conclude that there was significant agreement (i.e. there is confidence that the agreement is higher than it would be had the rankings been random or independent) among the evaluators ($W=1$ for complete agreement; $W=0$ for no agreement).

[1] This is to ascertain the amount of agreement on the nominal data categories within the pool of evaluators on heuristics chosen. A modified version of kappa developed in Mezzich et al. (1981) for inconsistent numbers of judges and multiple diagnoses would fit the evaluator-heuristic scenario in this study.

Table 4. The 21 usability problems identified by the seven evaluators, subjectively arranged by theme

PROBLEM ID	DESCRIPTION OF PROBLEM
SYSTEM – ERGONOMIC	
A	Ergonomic problem – is the cube too big? Affordance depends on the user (i.e. child as opposed to adult)
B	Controlling the camera for optimum view of cube can be hard
C	The cube may be used to access one tile out of very many, requiring much rotation to navigate; user fatigue may ensue
SYSTEM – ANCILLARY	
D	There is no "history" of map tiles visited by the user, which would be useful to direct the user to a previous map ("undo")
E	There are no initial instructions on use of the cube
SYSTEM – AR DISPLAY	
F	If cube is turned very fast, the map shown doesn't change
G	As the viewer gets closer to the map image, text gets harder to read and graphics become harder to interpret
H	"Third tile problem": Tracking error results in unintended tile being displayed with wrong orientation, leading to confusion
I	Vision tracking can cut out (e.g. occlusion by hand, marker at least partially out of frame), which is disconcerting
J	Cube can lose identity and orientation of current tile to display if off camera and randomly rotated
GEOGRAPHIC – NAVIGATIONAL	
K	There is no indication in the display of where the user has come from (a "history") or going to (result of gazetteer query)
L	Lack of orientation cues, such as a North Arrow
M	Lack of explicit coordinate display with the map – (0,1) or j12
N	Lack of zoom functionality; output limited to only one scale
O	Make the displayed map bigger than the cube face, less tiled and more continuous. Alternatively display neighboring tiles
GEOGRAPHIC – CARTOGRAPHIC	
P	Geographic features are "ignored" by the rigid tiles and the NSEW movement between them (e.g. to explicitly follow a road)
Q	Street names can occur partially shown on the boundary of tiles – messy and navigation is difficult
R	Street names not shown on every tile, making navigation hard
GEOGRAPHIC – COGNITIVE	
S	The cube gives access to a small tile at one time, which does not induce as comprehensive or useful a mental picture / map as a conventional large flat map would
T	Confusion exists with the relationship of direction of cube rotation and geographically moving in a corresponding direction. i.e. upon rotating the cube left, a move to the east is expected. But the opposite effect has been felt, which is counter-intuitive
U	Marker content is unrelated to the application – more pleasing to have marker content that relates to maps and navigation

Table 5. Heuristics broken, as identified by evaluator (f = frequency). Refer to Table 4 to look up the usability problem and Table 3 to look up heuristic numbers (Heuristic 11 is new and concerns the ergonomics of the hardware and cube)

ID	f	HEURISTICS BROKEN AS IDENTIFIED BY EVALUATOR							TOP HEUR.	f
		1	2	3	4	5	6	7		
SYSTEM – ERGONOMIC										
A	2					7, 8		11	11, 7, 8	1
B	4			5	2	11	11		11	2
C	1	1, 7							1, 7	1
SYSTEM – ANCILLARY										
D	1		1, 2, 5, 9						1, 2, 5, 9	1
E	1	10							10	1
SYSTEM – AR DISPLAY										
F	2		2			1, 5			1, 2, 5	1
G	2	2, 6, 7				1, 2			2	2
H	6	2, 4, 5	1, 2, 4, 9	2, 9	2, 3, 4, 5, 9	1, 2, 6		2, 4, 5, 9	2	6
I	3	1, 4			4	1, 2, 3, 5			4,1	2
J	3		1, 2, 3, 5, 6, 9	7	2				2	2
GEOGRAPHIC – NAVIGATIONAL										
K	1				1, 2, 3, 5, 6, 7				1, 2, 3, 5, 6, 7	1
L	6	1, 2, 6	1, 4, 5, 6	2, 5		1, 2, 6	1, 2	1, 2, 6	1, 2	5
M	2		1, 2, 4, 6		1, 2, 5, 6, 7				1, 2, 6	2
N	3	2, 4, 6	1, 2, 3, 6, 7		1, 2, 5, 6, 7				2, 6	3
O	3			2	1, 2, 5, 6, 7			2, 6	2	3
GEOGRAPHIC – CARTOGRAPHIC										
P	1						1, 2, 6		1, 2, 6	1
Q	1							8	8	1
R	7	2, 4, 6, 10	1, 4, 6, 10	4, 6	1, 2, 5, 6	2, 3, 6	2, 4	1, 2, 6	6	6
GEOGRAPHIC – COGNITIVE										
S	1		2						2	1
T	2			4			2, 4		4	2
U	2	2		2, 6					2	2

Table 6. Severity rankings, grouped by evaluator. Problems (see Table 4; geographic problems italicized) are arranged in order of ascending overall rank

PROBLEM ID	EVALUATORS' RANKING OF SEVERITY 1	2	3	4	5	6	7	SUM of RANKS
H	4	3.5	2	7	3	7.5	5	32
J	17	3.5	2	7	3	7.5	5	45
L	*1.5*	*11.5*	*2*	*7*	*8*	*7.5*	*12*	*49.5*
R	*9.5*	*3.5*	*5*	*16.5*	*8*	*7.5*	*5*	*55*
T	*1.5*	*11.5*	*9.5*	*7*	*20*	*7.5*	*5*	*62*
B	9.5	18.5	5	7	3	7.5	17.5	68
I	17	11.5	9.5	16.5	1	7.5	5	68
N	*9.5*	*3.5*	*16*	*7*	*8*	*20*	*5*	*69*
Q	*17*	*11.5*	*5*	*16.5*	*8*	*7.5*	*5*	*70.5*
S	*4*	*11.5*	*9.5*	*16.5*	*14*	*7.5*	*17.5*	*80.5*
F	9.5	18.5	9.5	16.5	8	16.5	5	83.5
E	9.5	21	13	1	20	7.5	12	84
G	4	11.5	20	16.5	8	7.5	17.5	85
P	*9.5*	*11.5*	*9.5*	*16.5*	*14*	*7.5*	*17.5*	*86*
O	*9.5*	*11.5*	*9.5*	*7*	*20*	*16.5*	*12*	*86*
K	*9.5*	*3.5*	*20*	*7*	*14*	*16.5*	*17.5*	*88*
A	17	18.5	16	16.5	8	7.5	5	88.5
C	17	11.5	16	7	14	16.5	12	94
D	17	3.5	20	7	14	20	17.5	99
M	*17*	*11.5*	*16*	*7*	*20*	*20*	*12*	*103.5*
U	*21*	*18.5*	*16*	*21*	*20*	*7.5*	*21*	*125*

Finally, a Spearman's rank correlation calculation (corrected for ties – Siegel and Castellan 1988) was performed on the "sum of ranks" column in Table 6 and the frequency (with which a problem was identified by the evaluators) from Table 5. The results ($r_S = 0.64$; one-tailed test: $p<0.0025$) show that the problems with the highest severity ratings were also in the main those that were identified with most frequency.

5 Discussion

The evaluators' significant agreement on the severity ratings means that some inferences can be drawn for prioritization in the subsequent development of TASM (in preparation for the user test, the next substantive stage in the usability testing process – Gabbard et al. 1999).

The top two problems in Table 6 were related to the Augmented Reality display – the "third tile problem" and inconsistent tile display after the cube has returned in view having been off-camera. The latter can be simply rectified; the system can be made to record the last tile displayed before the cube went off camera, ready to display it again at the next opportunity. Outlining the first problem, the "third tile" is the cube face that has the lowest angle of incidence with the camera (assuming that no more than three tiles are visible at one time). The system has been set to display two or less tiles simultaneously, which are normally the cube faces that the user is meant to view. However, occlusion or inferior lighting conditions may mean that the third face displays a tile. Referring to Figure 3b, it can be seen how the third tile can facilitate an abrupt tile and orientation change when either of the premier faces are displayed to the user again. Any solution would rely on identifying the third face (through incidence angle comparison) and controlling what is displayed thereon (e.g. displaying a projected combination of two tiles – the neighboring tiles to the two main tiles that the user is viewing – on the third face).

The major geographic usability problems (3rd, 4th and 5th in Table 6) can be solved simply, and encompass the navigational (orientation cue), cartographic (inconsistent street naming) and cognitive (rotation is counter-intuitive) categories. Given an initial awareness of map orientation and that all subsequent rotations are relative to that initial state, display of a North arrow is a straightforward task. The addition of street names to tiles is also conceptually simple but possibly cartographically complex (though a solution could lie in the display of neighboring tiles – problem O). The coupling of a GIS and having geographic data (rather than bitmap tiles) feed into the TASM scenario (e.g. Ratti et al. 2004 and Hedley et al. 2002 use DEM data) would be desirable for helping solve these and the majority of the geographic problems identified in the navigational and cartographic categories. A GIS would be less useful in addressing the spatial cognition issues; the solution to counter-intuitive rotation would lie with the AR software, and may involve gesture-based input (a common form of AR interaction) to simply reverse the rotation scheme.

There was significant agreement of these severity rankings with the frequency with which a problem was identified. The top three identified problems were inconsistent street naming (by all seven evaluators), the third tile issue and need for an orientation cue. Although the heuristics assigned were not specific enough for any global pattern to emerge, the "2: match between system and real world" (all) and "6: recognition rather than recall" (for the geographic problems only, which occurred for want of aids such as a North Arrow and consistent labeling, facilitating recognition) heuristics had a particularly high frequency for these three issues.

In developing a theory of geographic (and AR) heuristics (see 1.2, 1.3), it is pertinent that the most application-based of the original heuristics (no. 2) figures highly in the evaluators' results. Given that geovisualization interfaces are application-driven (Fuhrmann et al. 2005) there is scope here to subdivide heuristic 2 into three geographically-based heuristics. These would be based on the navigational, cartographic and cognitive groupings in Table 4 (this is subject to a more rigorous approach, such as factor analysis – Nielsen 1994b). These are:

2a: *Location- and orientation-aware through geovisualization system functionality*
2b: *Representation (and data) fits purpose*
2c: *Match between system and spatial mental model*

In addition, a few of the evaluators identified an extra ergonomic heuristic, applicable to the AR system set-up, including the camera and cube:

11: *Ergonomically appropriate*

On a general note, the number of problems found per successive evaluator was found to be consistent with Nielsen (1994a), with diminishing returns beyond the 3–5 evaluator interval. Having seven evaluators was beneficial though, as extra problems were yielded, for minimal cost. However, like Habbard et al. (1999), Nielsen's heuristics were too general and vague for in-depth analysis (but useful as a rough structure); those authors have since developed their own usability guidelines for virtual environments.

6 Conclusions

The Tangible Augmented Street Map (TASM), a tangible AR interface to street maps, has been detailed and it's testing with heuristic evaluation described. 21 usability problems were found, with erroneous "third tile" display, off-camera effects on AR display (system-based), lack of a North Arrow, street labeling errors and counter-intuitive rotation (geographic) being ranked as the most severe (backed up by heuristic votes).

These will now become the immediate tasks in the development of TASM, prior to a formative user-centered evaluation (following the sequential approach of Habbard et al. 1999; Walsh et al. 2002; Fuhrmann et al. 2005). A series of navigation tasks will be set, and the users will be assessed both qualitatively and quantitatively. The qualitative techniques used will be cognitive walkthrough, think-aloud protocol (to gather data –

Hackos and Redish 1988) and emergent themes analysis (Wong and Blandford 2002). Quantitative techniques (e.g. Koua and Kraak 2005; Tobón 2005) will include the measurement of user performance attributes such as task completion times using the cube and while using conventional paper maps. This will be an indication of navigation efficiency; correctness of navigation will also be assessed.

Although general and vague, Nielsen's heuristics were useful to stimulate development of application specific heuristics pertaining to navigation, cartography and cognition, as well as ergonomics. This addresses a key challenge in geovisualization, the creation of a geospatial theory relating to novel technologies such as Augmented Reality. In the future, a "wearable computing" set-up (Clarke 2004) for TASM will be devised, coupled with access to data from GIS, GPS and web sources.

Acknowledgements

I would like to thank Holger Regenbrecht and William Wong for discussions and contribution to the TASM implementation. I would also like to thank the heuristic evaluators Mark Billinghurst, Raphael Grasset, Scott Green, Julian Looser (of HitLabNZ), Holger, Sam Moyle and Rochelle Villanueva (University of Otago) for their revelations, insights and advice. I am also grateful to Kirsten Donald for proofreading the manuscript. This research has been funded by a University of Otago School of Business Research Grant.

References

ARToolKit (2006) Home Page http://www.hitl.washington.edu/artoolkit [Accessed 6th January, 2006]

Billinghurst M, Kato H, Poupyrev I (2001) The Magic Book: A Transitional AR Interface. Computers and Graphics, Nov:745–753

Bodum L (2005) Modelling Virtual Environments for Geovisualization: A Focus on Representation. In: Dykes J, MacEachren AM, Kraak M-J (eds) Exploring Geovisualisation. Elsevier, London, pp 389–402

Bowman DA, Kruijff E, LaViola JJ, Poupyrev I (2005) 3D User Interfaces: Theory and Practice. Morgan Kaufmann. Chapter 11

Cartwright W, Crampton J, Gartner G, Miller S, Mitchell K, Siekierska E, Wood, J (2001) Geospatial Information Visualisation User Interface Issues. Cartography and Geographic Information Science 28(1):45–60

City of Dunedin (2006) Street Map http://www.cityofdunedin.com/city/?page=searchtools_street
Clarke KC (2004) Mobile Mapping and Geographic Information Systems. Cartography and Geographic Information Science 31(3):131–136
Crampton J (1992) A cognitive analysis of wayfinding expertise. Cartographica 29(3/4):46–65
Davies C, Medyckyj-Scott D (1994) GIS usability: recommendations based on the user's view. Int J of Geographical Information Systems 8(2):175–189
Davies C, Medyckyj-Scott D (1996) GIS users observed. Int J of Geographical Information Systems 10(4):363–384
Frank AU (1996) Qualitative Spatial Reasoning: Cardinal Directions as an Example. Int J of Geographical Information Systems 10(3):269–290
Fuhrmann S, Ahonen-Rainio P, Edsall RM, Fabrikant SI, Koua EL, Tobón C, Ware C, Wilson S (2005) Making Useful and Useable Geovisualization: Design and Evaluation Issues. In: Dykes J, MacEachren AM, Kraak M-J (eds) Exploring Geovisualisation. Elsevier, London, pp 553–566
Gabbard JL, Hix D, Swan JE (1999) User-centered design and evaluation of virtual environments. IEEE Computer Graphics and Applications 19(6):51–59
Google (2006) Google Earth. http://earth.google.com [Accessed 6th January, 2006]
Hackos JT, Redish JC (1998) User and task analysis for interface design. John Wiley & Sons Inc., New York
Hedley NR, Billinghurst M, Postner L, May R, Kato H (2002) Explorations in the Use of Augmented Reality for Geographic Visualization. Presence 11(2): 119–133
Höllerer TH, Feiner SK (2004) Mobile Augmented Reality. In: Karimi HA, Hammad A (eds) Telegeoinformatics – Location-Based Computing and Services. CRC Press, Baton Rouge, FLA, pp 221–260
Koua EL, Kraak M-J (2005) Evaluating Self-Organizing Maps for Geovisualization. In: Dykes J, MacEachren AM, Kraak M-J (eds) Exploring Geovisualisation. Elsevier, London, pp 627–644
MacEachren A, Kraak M-J (2001) Research Challenges in Geovisualization. Cartography and Geographic Information Science 28(1):3–12
Mezzich JE, Kraemer HC, Worthington DRL, Coffman GA (1981) Assessment of Agreement Among Several Raters Formulating Multiple Diagnoses. J of Psychiatric Research 16:29–39
Moore AB, Regenbrecht H (2005) The Tangible Augmented Street Map. In: ICAT 2005: Proc of the 15th Int Conf on Artificial Reality and Telexistence, pp 249–250
Nielsen J (1994a) Heuristic Evaluation. In: Nielsen J, Mack RL, Usability Inspection Methods. John Wiley & Sons, United States, pp 25–62
Nielsen J (1994b) Enhancing the explanatory power of usability heuristics. In: Proc ACM CHI'94 Conf, Boston, MA, pp 152–158
Ratti C, Wang Y, Ishii H, Piper B, Frenchman D (2004) Tangible User Interfaces (TUIs): A Novel Paradigm for GIS. Transactions in GIS 8(4):407–421
Siegel S, Castellan Jr NJ (1988) Nonparametric Statistics for the Behavioral Sciences, 2nd ed. McGraw-Hill, New York, NY

Slocum TA, Blok C, Jiang B, Koussoulakou A, Montello DR, Fuhrmann S, Hedley NR (2001) Cognitive and Usability Issues in Geovisualisation. Cartography and Geographic Information Science 28(1):61–76

Tobón C (2005) Evaluating Geographic Visualization Tools and Methods: An Approach and Experiment Based upon User Tasks. In: Dykes J, MacEachren AM, Kraak M-J (eds) Exploring Geovisualisation. Elsevier, London, pp 645–666

Walsh KA, Pancake CM, Wright DJ, Haerer S, Hanus FJ (2002) "Humane" Interfaces to Improve the Usability of Data Clearinghouses. In: Egenhofer MJ, Mark DM (eds) Geographic Information Science: 2nd Int Conf, GIScience 2002. Springer, Berlin, pp 333–345

Wong BLW, Blandford A (2002) Analysing ambulance dispatcher decision making: Trialing emergent themes analysis. HF2002, Human Factors Conference "Design for the whole person – integrating physical, cognitive and social aspects", Melbourne, Australia

Wood J, Kirschenbauer S, Döllner J, Lopes A, Bodum L (2005) Using 3D in Visualization In: Dykes J, MacEachren AM, Kraak M-J (eds) Exploring Geovisualisation. Elsevier, London, pp 295–312

A Linear Programming Approach to Rectangular Cartograms

Bettina Speckmann[1], Marc van Kreveld[2], Sander Florisson[3]

[1] Department of Mathematics and Computer Science, TU Eindhoven
email: speckman@win.tue.nl
[2] Department of Information and Computing Sciences, Utrecht University
email: marc@cs.uu.nl
[3] email:sanderq@xs4all.nl

Abstract

In [26], the first two authors of this paper presented the first algorithms to construct rectangular cartograms. The first step is to determine a representation of all regions by rectangles and the second – most important – step is to get the areas of all rectangles correct. This paper presents a new approach to the second step. It is based on alternatingly solving linear programs on the x-coordinates and the y-coordinates of the sides of the rectangles. Our algorithm gives cartograms with considerably lower error and better visual qualities than previous approaches. It also handles countries that cannot be present in any purely rectangular cartogram and it introduces a new way of controlling incorrect adjacencies of countries. Our implementation computes aesthetically pleasing rectangular and nearly rectangular cartograms, for instance depicting the 152 countries of the World that have population over one million.

Key words: rectangular cartogram, algorithm, linear programming

1 Introduction

Cartograms are a useful and intuitive tool to visualize statistical data about a set of regions like countries, states, or counties. The size of a region in a cartogram corresponds to a particular geographic variable. The most common variable is population: in a population cartogram, the sizes (measured in area) of the regions are proportional to their population. Cartograms are also called *value-by-area maps* [8, 20]. In a cartogram the sizes of the regions are not the true sizes and hence the regions generally cannot keep both their shape and their adjacencies. A good cartogram, however, preserves the recognizability in some way.

Globally speaking, there are four types of cartogram. The standard type – also referred to as contiguous area cartogram – has deformed regions so that the desired sizes can be obtained and the adjacencies kept. Algorithms for such cartograms were given, among others, by Tobler [24], Dougenik et al. [10], Kocmoud and House [17], Edelsbrunner and Waupotitsch [11], Keim et al. [16], and Gastner and Newman [13]. The second type of cartogram is the non-contiguous area cartogram [12, 21]. The regions have the true shape, but are scaled down and generally do not touch anymore. Sometimes the scaled-down regions are shown on top of the original regions for recognizability. A third type of cartogram is based on circles and was introduced by Dorling [9]. The fourth type of cartogram is the rectangular cartogram introduced by Raisz in 1934 [23]. Each region is represented by a single rectangle, which has the great advantage that the sizes (area) of the regions can be estimated much better than with the first two types. However, the rectangular shape is less recognizable and it imposes limitations on the possible layout. Hybrid cartograms of the first and fourth type exist as well. The regions are rectilinear polygons with a small number of vertices instead of rectangles. Figure 1 shows a population cartogram where all but two countries are rectangular; these two countries are L-shaped.

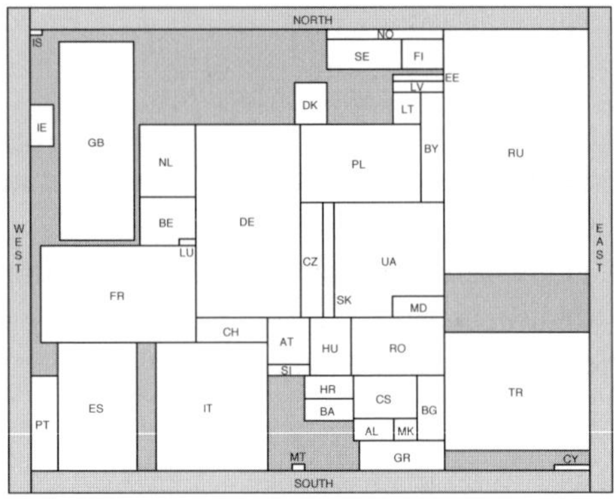

Fig. 1. The population of Europe (country codes according to the ISO 3611 standard)

1.1 Quality Criteria

Whether a rectangular cartogram is good is determined by several factors. One of these is the *cartographic error* [10, 11], which is defined for each region as $|A_c - A_s| / A_s$, where A_c is the area of the region in the cartogram and A_s is the specified area of that region, given by the geographic variable to be shown. The following list summarizes all quality criteria:

- Average and maximum cartographic error.
- Correct adjacencies of the rectangles (e.g., the rectangles for Germany and France should be adjacent and the rectangles for Germany and Spain should not be adjacent).
- Maximum aspect ratio.
- Suitable relative positions (e.g., the rectangle for The Netherlands should be West of the one for Germany).

For a purely rectangular cartogram we cannot expect to simultaneously satisfy all criteria well. Figure 2 shows an example where correct adjacencies must be sacrificed in order to get a small cartographic error. In larger examples, bounding the aspect ratio of the rectangles also results in larger cartographic error.

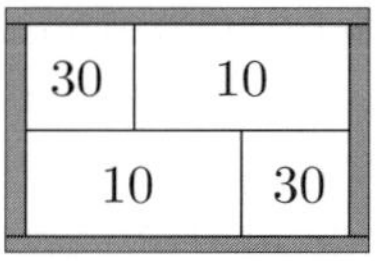

Fig. 2. The values inside the rectangles indicate the preferred areas

1.2 Related Work

Rectangular cartograms are closely related to *floor plans* for electronic chips and architectural designs. Floor planning aims to represent a planar graph by its *rectangular dual*, defined as follows. A *rectangular partition* of a rectangle R is a partition of R into a set $\mathcal{R}$ of non-overlapping rectangles such that no four rectangles in $\mathcal{R}$ meet at the same point. A *rectangular dual* of a planar graph G is a rectangular partition $\mathcal{R}$, such that (i) there is a one-to-one correspondence between the rectangles in $\mathcal{R}$ and the nodes in G, and (ii) two rectangles in $\mathcal{R}$ share a common boundary if and only if the corresponding nodes in G are connected. A *triangle* is a cycle of G consisting of three arcs (a 3-cycle). A cycle C of G divides the plane into an interior and an exterior region. If C contains at least one vertex in its interior and in its exterior, then C is called a *separating cycle*. The following theorem was proven in [2, 18]:

Theorem 1. *A planar graph G has a rectangular dual R with four rectangles on the boundary of R if and only if*

1. *every interior face is a triangle and the exterior face is a quadrangle*
2. *G has no separating triangles.*

Planar graphs that do not have a rectangular dual can still be represented by using other shapes than rectangles. It was shown in [19] that L- and T-shapes in addition to rectangles are always sufficient to represent a planar graph. A rectangular dual is not necessarily unique. When converting a rectangular dual into a cartogram, the rectangles must obtain the specified areas while respecting all quality criteria.

The only algorithm for standard cartograms that can be adapted to handle rectangular cartograms is Tobler's pseudo-cartogram algorithm [24] combined with a rectangular dual algorithm. However, Tobler's method is known to produce a large cartographic error and is mostly used as a preprocessing step for cartogram construction [20]. Tobler states in a recent survey [25] that none of the existing cartogram algorithms are capable of generating rectangular cartograms. Shortly thereafter, the first two authors of this paper presented the first methods for the automated construction of rectangular cartograms [26]. One problem of those methods was the handling of sea regions, and an experimental study of possible solutions was performed in [27].

A few other papers exist that discuss constructions related to rectangular cartograms. Biedl and Genc [3] show NP-hardness of certain orthogonal layout problems with prescribed areas, while Heilmann et al. [14] and Rahman et al. [22] provide algorithms to construct rectangular or orthogonal layouts with prescribed areas. Recently, de Berg et al. [7] show how to construct orthogonal layouts with prescribed areas and correct adjacencies, while using at most a constant number of edges per region. Their result holds for any input graph.

1.3 Results

We present a new algorithm for the computation of rectangular cartograms which improves upon the previous ones in all aspects. In particular, our algorithm produces cartograms with a lower cartographic error, a smaller aspect ratio, and with better global shape. Several preprocessing steps are the same as described in [26], but the main computation of obtaining correct areas for all rectangles is significantly different. We also present new variations of the method by allowing L-shaped regions in the cartogram. Furthermore, we show how to control the degree up to which incorrect adjacencies between countries are allowed. Our new method gives satisfactory rectangular cartograms of all countries of the World, whereas the previous method often does not.

We give the outline of the algorithm from [26] in Section 2, and present the new, linear programming based algorithm – including variations – extensively in Section 3. We implemented the linear programming approach, evaluated its output for different settings, and compared it with the approach from [26], which is reported in Section 4.

2 Algorithmic Outline

Assume that we are given an administrative subdivision into a set of regions, like the countries of the World or the States of the USA. The regions and adjacencies can be represented by nodes and arcs of a graph F, which is the face graph of the subdivision.

2.1 Preprocessing

To satisfy the conditions of Theorem 1 we have to ensure that the face graph F is triangulated and contains no internal nodes of degree less than four. F is in most cases already triangulated, except for its outer face and for inner seas that are adjacent to several regions. We triangulate any remaining non-triangular faces, like the face formed by the nodes for Nevada, Utah, New Mexico, and Arizona, and the four-country point in Africa.

Regions with only three neighbors and no adjacency to the seas, like Luxembourg and Burundi, cannot be represented in a rectangular cartogram where all adjacencies are correct. The same is true for regions with only two neighbors, like Mongolia. We discuss these issues further in Section 3.

2.2 Directed Edge Labels

Any two nodes in the face graph have at least one direction of adjacency which follows naturally from their geographic location (for example, The Netherlands lies clearly West of Germany). While in theory there are four different directions of adjacency that any two nodes can have, in practice only one or two directions are reasonable (for example, Germany can be considered to lie West or North of Austria). We employ a simple heuristic to extract the possible directions of adjacency from the administrative subdivision: we consider the line through the centers of mass of the two regions and let its orientation determine the directions. If two directions are reasonable, we have a so-called *layout option*. While in theory many pairs of adjacent regions can have a layout option, in practice there is often only one natural choice for the direction of adjacency between two regions.

Our algorithm will go through all possible combinations of layout options and determines which one gives a correct or the best result. If the input subdivision has m layout options, we will test 2^m different combinations of adjacency directions. A particular choice of adjacency directions gives rise to a *directed edge labeling*.

Observation 1. *A face graph F with a directed edge labeling can be represented by a rectangular dual if and only if*

1. *every internal region has at least one North, one South, one East, and one West neighbor.*
2. *when traversing the neighbors of any node in clockwise order starting at the western most North neighbor we first encounter all North neighbors, then all East neighbors, then all South neighbors and finally all West neighbors.*

For example, a realizable directed edge labeling for the US cannot let Nevada have California to the West, Oregon and Idaho to the North, and Utah and Arizona to the East, because then Nevada would miss a South neighbor. A realizable directed edge labeling constitutes a *regular edge labeling* for F as defined in [15] which immediately implies our observation.

2.3 Rectangular Layout

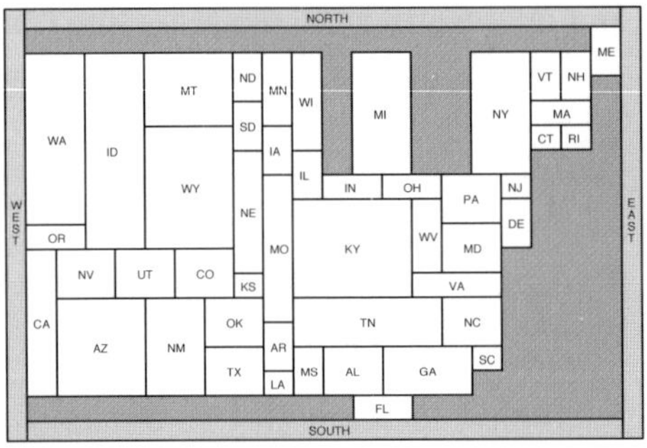

Fig. 3. One of 4608 possible rectangular layouts of the US

To actually represent a face graph together with a realizable directed edge labeling as a rectangular dual we have to pay special attention to the nodes on the outer face since they may miss neighbors in up to three directions. To compensate for this we add four special nodes NORTH, EAST, SOUTH, and WEST. Furthermore, we add additional nodes (*sea nodes*) to F that represent bodies of water and so help to preserve the original outline. For example, our face graph of the World contains one node each for the Black and the Caspian Sea. Larger bodies of water (like the oceans) and bodies of water with a more complex shape (like the Mediterranean Sea) are represented by many sea nodes.

Now we can employ the algorithm by Kant and He [15] to construct a *rectangular layout*, i.e., the unique rectangular dual of a realizable directed edge labeling. Every sea node is represented by a *sea rectangle*. The land and the sea rectangles together form a partitioning of the whole map (a rectangle) into smaller rectangles. The output of our implementation of the algorithm by Kant and He is shown in Figure 3.

2.4 Area Assignment

For a given set of area values and a given rectangular layout we would like to compute a rectangular cartogram that has a small cartographic error, while maintaining reasonable aspect ratios and relative positions. In this step the main information to be conveyed by the cartogram – the rectangle sizes – is determined.

In a previous paper [26] we introduced three methods for computing cartograms from rectangular layouts. One is the simple segment moving heuristic, which incrementally moves segments by fixed steps if this reduces the error. This method was implemented. Secondly, if the rectangular layout is L-shape destructible (see [26]) we can compute a zero-error cartogram if one exists. The third method is based on bilinear programming and can produce a cartogram with minimum maximal error, provided a good bilinear program solver is available. Unfortunately, bilinear programs are notoriously difficult and none of the available solvers is guaranteed to work [1].

In this paper we formulate area assignment as a linear program which takes only the vertical or only the horizontal segments into account. We then alternatingly solve a linear program for the vertical and the horizontal segments. We implemented this approach – using the well-known CPLEX program [6] – and it clearly improves upon the segment moving heuristic. The next section shows how to formulate area assignment as a linear program if either the horizontal or the vertical segments are considered to be at fixed positions.

3 Linear Programming

Assume that we are given a rectangular layout $\mathcal{L}$. $\mathcal{L}$ is uniquely determined by the x-coordinates of its maximal vertical segments and the y-coordinates of its maximal horizontal segments. (See for example Fig. 3: the horizontal segment which has Illinois, Indiana, and Ohio above and Kentucky and Wyoming below is a maximal horizontal segment.) The area of a rectangle R of $\mathcal{L}$ is determined by the coordinates of exactly four maximal segments: the y-coordinates of the two horizontal segments bounding it from above and below and the x-coordinates of the two vertical segments bounding it from the left and the right. Similarly, the aspect ratio of a R is also

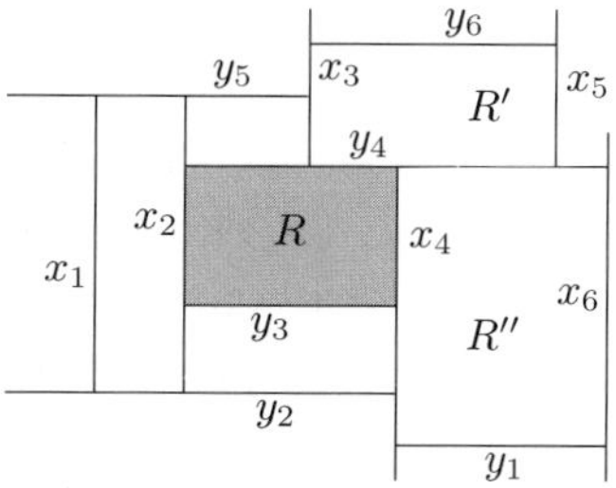

Fig. 4. Linear constraints for rectangle R

determined by these four segments. Minimizing area error, bounding the aspect ratio, and preserving adjacencies of rectangles can all be formulated using bilinear constraints in the x- and y-coordinates of the maximal segments. These constraints are linear in x if the y's are treated as constants and vice versa.

We create constraints for every rectangle R of a given rectangular layout $\mathcal{L}$. The specified area of the region represented by R is denoted by $A_s(R)$. Please refer to Figure 4 for the notation used in the following paragraphs. Note that the horizontal segment with y-coordinate y_4 is simultaneously the top of R and R'', and the bottom of R' and two other rectangles. Hence the constraints associated with several different rectangles will impose restrictions on the value of y_4.

There are four different types of constraints associated with every rectangle R: *error constraints*, *aspect ratio constraints*, *planarity preserving constraints*, and *adjacency preserving constraints*. We explain these constraints in detail below, including the modifications we make for sea rectangles.

Area Constraints. The area of R in the layout is $(x_4 - x_2) \cdot (y_4 - y_3)$. This implies that the error of R is

$$\frac{|(x_4 - x_2) \cdot (y_4 - y_3) - A_s(R)|}{A_s(R)} .$$

We introduce a variable $\mathrm{ERR}(R)$ that bounds the error from above, that is

$$\mathrm{ERR}(R) \geq \frac{|(x_4 - x_2) \cdot (y_4 - y_3) - A_s(R)|}{A_s(R)} .$$

Absolute values cannot be used in linear programs, so we replace this constraint by two others:

$$(x_4 - x_2) \cdot (y_4 - y_3) \geq (1 - \mathrm{ERR}(R)) \cdot A_s(R) ,$$
$$(x_4 - x_2) \cdot (y_4 - y_3) \leq (1 + \mathrm{ERR}(R)) \cdot A_s(R) .$$

Note that these constraints are indeed linear if either the x's or the y's are treated as constants.

Aspect Ratio Constraints. The aspect ratio of R is the maximum of its *height:width* ratio and its *width:height* ratio, so it is always at least 1. Given a maximum aspect ratio D for all rectangles, we obtain the following two constraints for R:

$$(x_4 - x_2) \leq (y_4 - y_3) \cdot D ,$$
$$(y_4 - y_3) \leq (x_4 - x_2) \cdot D .$$

Planarity Constraints. These constraints preserve the planarity of the layout. For example, no solution with $x_2 > x_4$ is acceptable, because R would be "inverted", and the layout would no longer be planar. Hence, for every rectangle we require its left segment to be left of its right segment, and its bottom segment to be below its upper segment:

$$x_2 < x_4 \quad \text{and} \quad y_3 < y_4 .$$

Only one of these constraints is in effect in any linear program, because only the x's or only the y's are the variables.

Adjacency Constraints. To make sure that rectangles R and R' remain adjacent when vertical segments (x-coordinates) are changed, we require that $x_3 < x_4$. Note that violating this constraint is not disastrous for the final cartogram, because planarity is not affected. Only the adjacency of rectangles is influenced.

3.1 Sea Rectangles

There are no error constraints or aspect ratio constraints associated with sea rectangles. However, planarity and adjacency constraints must be defined. Adjacency constraints can become less strict. If in Figure 4, both R and R' are sea rectangles, then the constraint $x_3 < x_4$ can be dropped: the segment that separates R and R'' is not shown in the final cartogram anyway. Finally, we add two new constraints for every sea rectangle R to make sure that it does not get zero height or width (otherwise a sea may become invisible and opposite regions would visually be adjacent):

$$x_2 < x_4 - d \quad \text{and} \quad y_3 < y_4 - d .$$

These constraints guarantee that any sea rectangle has at least some width and height d, where d is some well-chosen constant.

To generate a cartogram that has minimum error, we must minimize $\mathrm{ERR}(R)$ for all rectangles R. The objective function of the linear program therefore is

$$\text{minimize} \sum_{\text{rectangles } R} \mathrm{ERR}(R) .$$

This is a linear objective function, and hence we have derived a linear program for minimizing the total or, equivalently, the average error of the rectangles.

Our whole algorithm is as follows. Starting with the rectangular layout produced by the algorithm of Kant and He [15], generate all constraints for the vertical segments and solve a linear program where all x's are variables. Then fix the new x-coordinates and generate all constraints for the horizontal segments and solve a linear program where all y's are variables. Repeat this for a fixed number of iterations, or until no more reduction of error occurs.

3.2 Nearly Rectangular Cartograms

In Section 2 we mentioned the issue of regions with three or fewer neighbors, and the fact that they cannot be represented in a rectangular cartogram with correct adjacencies. One option is to use more general shapes than rectangles, like L-shaped and C-shaped regions. For example, we can place Luxembourg in the lower right corner of Belgium, which then becomes L-shaped. It is straightforward to set up error, aspect ratio, planarity, and adjacency constraints for Belgium and Luxembourg in this case. Hence, we can still use the linear programming approach. Similar examples on the world map are Paraguay, Malawi, and Burundi.

There are also countries that are adjacent to only two other countries, like Nepal and Moldova. Incorporating these can be done by using C-shaped regions. A country like Lesotho inside South Africa can be incorporated by making South Africa O-shaped.

In cases where L-shaped, C-shaped and O-shaped regions appear instead of rectangles, we must adapt the constraints and optimization function for the linear program. For example, in Figure 5, the area of an L-shaped region R is given by $(x_2 - x_1) \cdot (y_5 - y_1) + (x_4 - x_2) \cdot (y_5 - y_2)$, and this can be converted into a bilinear constraint including an error term $\mathrm{ERR}(R)$ as before. The changes required are straightforward and we omit them from this paper.

The other option is to maintain a purely rectangular layout and change the correct adjacencies locally. For example, Luxembourg could be made adjacent to the North Sea between Belgium and France (which would no longer be adjacent), or Luxembourg could be made adjacent to the Netherlands, in which case Belgium and Germany are no longer adjacent. If the initial layout is adapted in this manner, a purely rectangular layout can be made. Regions with only two adjacent regions like Mongolia can be handled in a similar way. An advantage of using only rectangles over using L-shaped, C-shaped, and O-shaped regions besides rectangles is that areas can be estimated better, which important for cartograms.

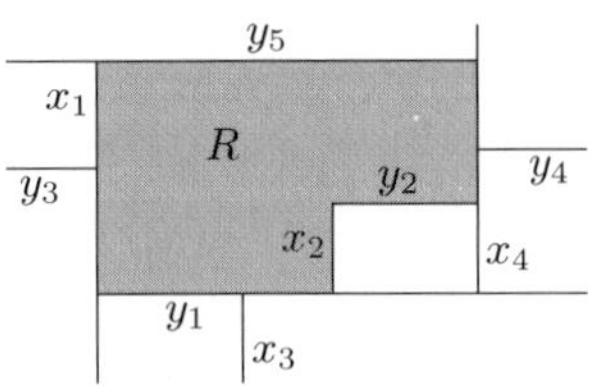

Fig. 5. Linear constraints for L-shaped region R

3.3 Limiting Incorrect Adjacencies

The example in Figure 2 shows that requiring correct adjacencies can make rectangular cartograms with low error impossible. Therefore, it is often necessary to sacrifice correct adjacencies. As noted before, not including the adjacency constraints in the linear program leads to rectangular cartograms that are still planar, rectangular layouts, but where two countries may no longer be adjacent, or are adjacent although they are not in reality. Consider the cartogram of Figure 1. Without adjacency constraints and a theme where Norway and Turkey have very high values, it can happen that these two countries become adjacent. Similarly, in Figure 3, North Dakota (ND) and Louisiana (LA) can potentially become adjacent.

So clearly, if we omit all adjacency constraints, the relative positions of regions can be disturbed too much. Therefore, we introduce *levels* of adjacency error and the corresponding linear constraints. For consistency, we call the correct adjacency constraints *level-0 adjacency constraints*, like $y_1 < y_2$ in Figure 4. A *level-1 adjacency constraint* allows each vertical or horizontal segment to pass at most 1 other vertical or horizontal segment, but no more. An example is $y_1 < y_3$ in Figure 4. There are no other level-1 adjacency constraints in the figure, and also no constraints of even higher level. Figure 6 shows the situation for a level-k adjacency constraint. If k is set to a high value, corresponding to the situation without any adjacency constraints, we speak of *false adjacencies*.

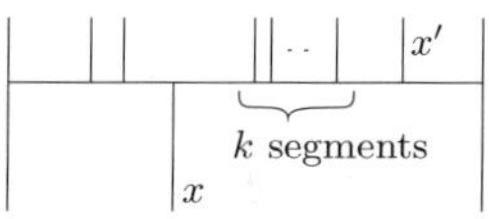

Fig. 6. Illustration of the level-k adjacency constraint $x < x'$

4 Implementation and Test Results

The linear programming approach to construct rectangular cartograms has been implemented and tested. The main objective was to test if it produces cartograms that are both visually pleasing and have low error. Also, we were interested in the difference between purely rectangular cartograms and cartograms with L-shaped regions, and we wanted to know how the linear programming approach compares with the segment moving heuristic. Finally, we wanted to investigate if level-1 adjacency constraints provide a good compromise between low error and only mildly incorrect adjacencies.

We used three different subdivisions: the contiguous states of the USA, the countries of Europe with population over 100,000, and the countries of the World with population over 1,000,000. Themes that were tested are population, area, gross domestic product, and total highway length (data was taken from the CIA World Factbook 2005 [5]). The USA has 13 layout options, Europe has 10, and the World 22. For Europe and the World, we considered both purely rectangular layouts and layouts with L-shaped countries.

The rectangular cartograms in the figures have regions that are colored based on their error[1]. Shades of red show that a region is too small and shades of blue show that a region is too large. If the error is below 0.05, the region is white. The lightest shades of red and blue are used for errors between 0.05 and 0.1, and darker shades are used for errors between 0.1 and 0.2, between 0.2 and 0.3, and above 0.3.

During implementation, we made two adaptations to the method as described so far. Test runs of the linear programming approach often gave output where one or two countries had a high error, while all others had no or negligible error. The reason is that the optimization function minimizes the sum of errors, which means that it is equally good to have one country with error 0.4 and another country with error 0.0, as to have those two countries with error 0.2 each. However, we clearly prefer to have two countries with error 0.2, and, more generally, to have low maximum error. This is easily done by minimizing the sum of *squared* errors instead of the sum of errors. CPLEX [6] allows for minimizing the sum of squared errors, even though technically speaking, we no longer have a linear program. All results given in this section have been computed by minimizing the sum of squared errors.

The other adaptation is an efficiency issue. Recall that our algorithm allows for layout options, that is, a region may for instance be either North or East of an adjacent region in a rectangular layout. For the USA subdivision we have 13 of these layout options, implying that we try to generate $2^{13} = 8,192$ different rectangular layouts on which the algorithm is run. Not all rectangular layouts are realizable, so fewer layouts are actually generated. For the World data set, there are 22 layout options, potentially giving over four million layouts. This would make the cartogram computation prohibitively time-consuming.

To deal with this problem we partition the layout options into groups. The idea is that layout options in Europe do not have to be tested in combination with layout options in South America, for instance. So the best layout option for Europe can be determined independently from the best layout option for Europe. We partitioned the 22 layout options into 7 groups of sizes up to 4, which leads to testing only 74 rectangular layouts. This adaptation may result in slightly larger errors in the cartogram, but the time savings are drastic. Figure 7 shows two cartograms of the World with the theme population. Only countries with population over one million are included. The first map

[1] The electronic version of this paper contains color figures and can be found at http://www.win.tue.nl/∼speckman/pub.html

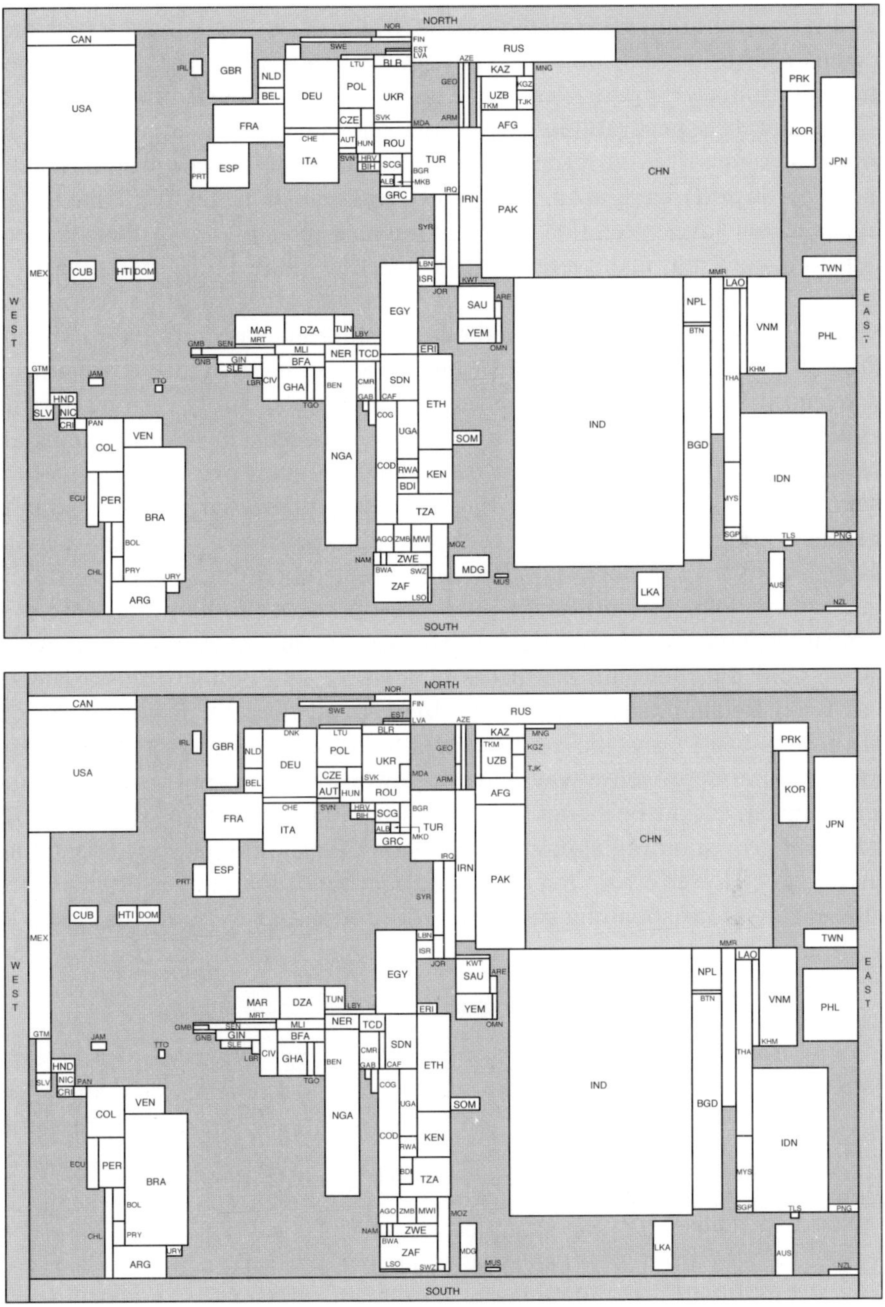

Fig. 7. A purely rectangular cartogram of the World with the theme population, and a corresponding cartogram that includes L-shaped regions

is based on a purely rectangular layout whereas the second uses L-shaped regions for Moldova, Mongolia, Gambia, Lesotho, Swaziland, Malawi, and Burundi. Both cartograms were made with the linear programming approach. The maximum aspect ratio is 12, the adjacency constraints are of level 1, the overall sea area is 40% of the map, and 50 iterations (linear programs on x and y per layout) were used. The purely rectangular cartogram has slightly higher errors (average 0.003 and maximum 0.099 in China) than the one with L-shaped regions (average 0.002 and maximum 0.063 in China). We observed the same small differences on tests with other data (gross domestic product, total highway length) and other maximum aspect ratios (8, 16, and 20). Usually the cartograms with L-shaped regions have slightly lower errors.

We used basically the same settings to produce a purely rectangular cartogram with the segment moving heuristic. However, false adjacencies are allowed, and 400 iterations were used. The average error is 0.103 and the maximum is 0.534, which is significantly worse than what we obtained with the linear programming approach, despite the fact that adjacencies may be disturbed more. The corresponding cartogram (not shown) is also visually much worse than the cartogram produced by linear programming. We also compared the segment moving heuristic and linear programming approach on the USA data set with the theme population, and on the Europe data set with the theme highway length, see Table 1. In all cases the maximum aspect ratio was set to 12 and the total sea area to 20%. We observe that the linear programming approach always gives a lower average error. The maximum occurring error can be better for either method. We also observe that for the linear programming approach, inclusion of level-1 adjacency constraints gives a much lower error than correct adjacencies, albeit not as good as false adjacencies. Corresponding cartograms for four cases are given in Figure 8.

Table 1. Linear programming versus segment moving

Method	Data	Adjacencies	Av. error	Max. error
LP	US pop.	correct	0.086	0.873
LP	US pop.	level 1	0.032	0.189
LP	US pop.	false	0.013	0.107
Segment moving	US pop.	correct	0.204	0.443
Segment moving	US pop.	false	0.018	0.058
LP	EU highway	correct	0.022	0.166
LP	EU highway	level 1	0.000	0.001
LP	EU highway	false	0.000	0.000
Segment moving	EU highway	correct	0.099	0.375
Segment moving	EU highway	false	0.052	0.151

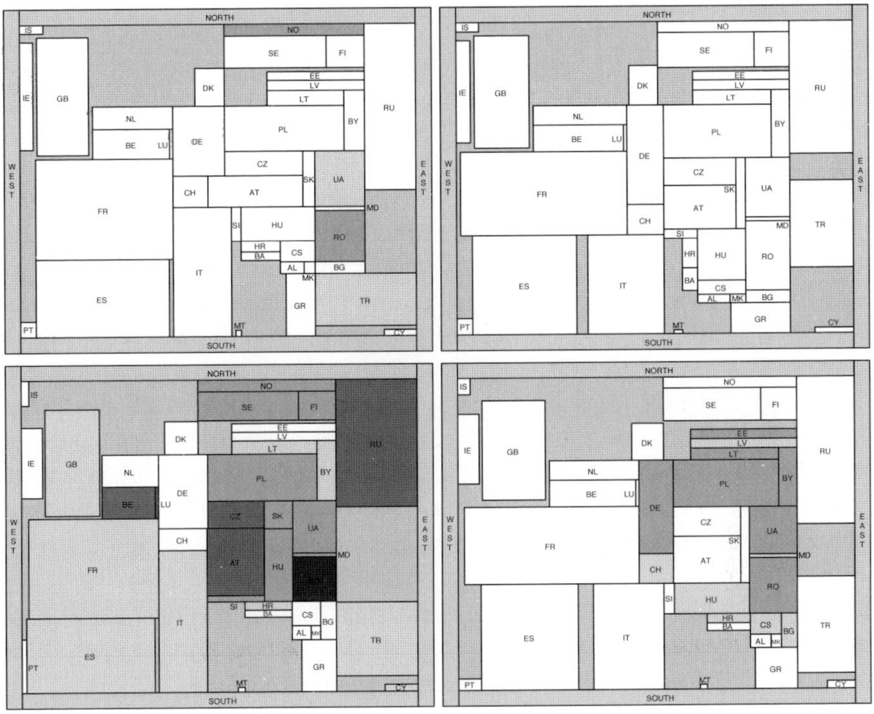

Fig. 8. Purely rectangular cartograms with correct adjacencies (left) and false adjacencies (right) produced by linear programming (top) and segment moving (bottom)

We present one more rectangular cartogram related to the FIFA World Cup in 2006. Figure 9 shows all countries or regions with population over one million who participated in the qualification rounds. The qualified 32 countries are shown by their FIFA country code and flag with a fixed size, whereas all other countries are shown in grey by their actual area. For qualified countries we added minimum width and height constraints to make sure that the flag and name fit inside the rectangle. During the World Cup we will grow countries that do well to show their increased chance of winning the cup.

5 Conclusions

We presented a new solution for the automated construction of rectangular cartograms. It is based on iterative linear programming, and produces better results than the previous, segment moving method. We also presented and

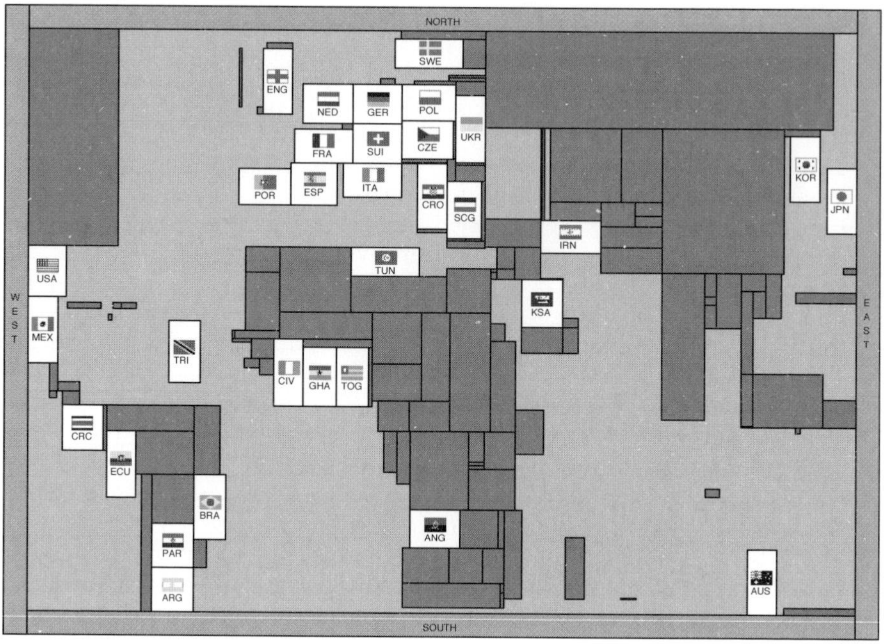

Fig. 9. FIFA World Cup 2006 qualified countries

implemented two extensions. The first allows L-shaped countries in the cartogram, and the second limits the adjacency errors of rectangles. The new algorithm makes it possible to generate rectangular cartograms of the World that both have low error and are aesthetically pleasing.

The difference in error and visual quality of purely rectangular cartograms, and cartograms with L-shaped regions is not significant, and which type is preferred is a matter of taste. Concerning adjacency constraints, it appears that level-1 provides a significant improvement in error over correct adjacencies, but a higher level does not seem worthwhile and may give cartograms of lower visual quality. We note that the rectangular cartograms of the Bosatlas [4] have incorrect adjacencies of various types, and our level-1 constraints give at least as good results in terms of relative positions.

References

1. Bazaraa M, Sherali H, Shetty C (1993) Nonlinear Programming – Theory and Algorithms, 2nd ed. John Wiley & Sons, Hoboken, NJ
2. Bhasker J, Sahni S (1987) A linear algorithm to check for the existence of a rectangular dual of a planar triangulated graph. Networks 7:307–317

3. Biedl T, Genc B (2005) Complexity of orthogonal and rectangular cartograms. In: Proc 17th Canad Conf on Computational Geometry, pp 117–120
4. De Grote Bosatlas (2001) Wolters-Noordhoff, 52nd ed. Groningen
5. CIA World Factbook (2005) http://www.cia.gov/cia/publications/factbook
6. CPLEX: High-performance software for mathematical programming and optimization. http://www.ilog.com/products/cplex/
7. Berg M de, Mumford E, Speckmann B (2005) On rectilinear duals for vertex-weighted plane graphs. In: Proc 13th Int Symp on Graph Drawing (= LNCS 3843), pp 61–72
8. Dent B (1999) Cartography – thematic map design, 5th ed. McGraw-Hill
9. Dorling D (1996) Area Cartograms: their Use and Creation (= Concepts and Techniques in Modern Geography 59). University of East Anglia, Environmental Publications, Norwich
10. Dougenik J, Chrisman N, Niemeyer D (1985) An algorithm to construct continuous area cartograms. Professional Geographer 37:75–81
11. Edelsbrunner H, Waupotisch E (1997) A combinatorial approach to cartograms Comp Geom Theory Appl 7:343–360
12. Fabrikant S (2000) Cartographic variations on the presidential election 2000 theme. http://www.geog.ucsb.edu/~sara/html/mapping/election/map.html
13. Gastner M, Newman M (2004) Diffusion-based method for producing density-equalizing maps. Proc of the National Academy of Sciences of the United States of America (PNAS) 101(20):7499–7504
14. Heilmann R, Keim D, Panse C, Sips M (2004) Recmap: Rectangular map approximations. In: Proc IEEE Symp on Information Vis, pp 33–40
15. Kant G, He X (1997) Regular edge labeling of 4-connected plane graphs and its applications in graph drawing problems. Theor Comp 172:175–193
16. Keim D, North S, Panse C (2004) Cartodraw: A fast algorithm for generating contiguous cartograms. IEEE Trans Vis and Comp Graphics 10:95–110
17. Kocmoud C, House D (1998) A constraint-based approach to constructing continuous cartograms. In: Proc Symp Spatial Data Handling, pp 236–246
18. Koźmiński K, Kinnen E (1985) Rectangular dual of planar graphs. Networks 5:145–157
19. Liao C-C, Lu H-I, Yen H-C (2003) Floor-planning using orderly spanning trees. J Algorithms 48:441–451
20. NCGIA / USGS. Cartogram Central (2002) http://www.ncgia.ucsb.edu/projects/Cartogram_Central/index
21. Olson J (1976) Noncontiguous area cartograms. Prof Geographer 28:371–380
22. Rahman M, Miura K, Nishizeki T (2004) Octagonal drawings of plane graphs with prescribed face areas. In: Proc 30th Graph-Theoretic Concepts in Computer Science (= LNCS 3353), pp 320–331
23. Raisz E (1934) The rectangular statistical cartogram. Geogr Review 24:292–296
24. Tobler W (1986) Pseudo-cartograms. The American Cartographer 13:43–50
25. Tobler W (2004) Thirty-five years of computer cartograms. Annals of the Assoc American Cartographers 94(1):58–71
26. Kreveld M van, Speckmann B (2004) On rectangular cartograms. In: Proc 12th Europ Symp Algorithms (= LNCS 3221), pp 724–735

27. Kreveld M van, Speckmann B (2005) Rectangular cartogram computation with sea regions. In: Proc 22nd Int Cartographic Conf

Automated Construction of Urban Terrain Models

Henrik Buchholz[1], Jürgen Döllner[1], Lutz Ross[2], Birgit Kleinschmit[2]

[1]University of Potsdam and [2]Technical University of Berlin

Abstract

Elements of urban terrain models such as streets, pavements, lawns, walls, and fences are fundamental for effective recognition and convincing appearance of virtual 3D cities and virtual 3D landscapes. These elements complement important other components such as 3D building models and 3D vegetation models. This paper introduces an object-oriented, rule-based and heuristic-based approach for modeling detailed virtual 3D terrains in an automated way. Terrain models are derived from 2D vector-based plans based on generation rules, which can be controlled by attributes assigned to 2D vector elements. The individual parts of the resulting urban terrain models are represented as "first-class" objects. These objects remain linked to the underlying 2D vector-based plan elements and, therefore, preserve data semantics and associated thematic information.

With urban terrain models, we can achieve high-quality photorealistic 3D geovirtual environments and support interactive creation and manipulation. The automated construction represents a systematic solution for the bi-directional linkage of 2D plans and 3D geovirtual environments and overcomes cost-intensive CAD-based construction processes. The approach both simplifies the geometric construction of detailed urban terrain models and provides a seamless integration into traditional GIS-based workflows.

The resulting 3D geovirtual environments are well suited for a variety of applications including urban and open-space planning, information systems for tourism and marketing, and navigation systems. As a case study, we demonstrate our approach applied to an urban development area of downtown Potsdam, Germany.

1 Introduction

Photorealistic virtual 3D city models and virtual 3D landscape models form a basis for an increasing number of applications and systems. They can be used, for example, in landscape and open-space planning to present planning scenarios to the public (e.g., Danahy 2005; Lange and Hehl-Lange 2005; Stock and Bishop 2005; Warren-Kretzschmar and Tiedtke 2005; Werner et al. 2005) or in tourism to allow visitors exploring a city virtually. Many geovirtual environments such as the example in Figure 1 (left) are based on a set of 3D building models and 2.5D terrain model draped by aerial images to represent areas that are not covered by buildings. These areas consist of manmade surface structures (e.g., roads, pavement, walls, stairs, or squares), natural terrain surfaces, which are often covered by vegetation (e.g., woodland, agricultural land, grassland, or rocks) and water surfaces (e.g., rivers, canals, or lakes). Aerial images are well suited for representing such surface cover information from a bird's eye view but do not provide sufficient detail for visualization from a pedestrian's point-of-view (see Fig. 1 right). In addition, the represented terrain elements cannot be analytically identified and modified because there is no concise object-oriented representation of these elements and the linkage between GIS planning data and virtual 3D model has been lost. This makes aerial images unsuitable for land use related planning tasks. Therefore, many applications require more detailed virtual 3D terrain models (see Fig. 2).

There are three common approaches for creating detailed terrain surface representations: *manual modeling triangulated irregular network* (TIN) *models*, and *image draping*.

Manual Modeling

Using standard 3D modeling tools or CAD tools for manual modeling of terrain surface structures provides maximum flexibility. It involves, however, high manual efforts and requires a high degree of expertise. In addition, the result consists of computer graphics 3D models that are detached from any underlying 2D geo-data that have been originally used as a basis for modeling. The following limitations result:

- No object-specific modeling techniques: For most elements of urban terrain models, editing techniques could take advantage of object-specific construction rules and constraints (e.g., distribution, form, or height of stairs).

Fig. 1. Aerial images are well suited for bird's eye views (left) but fail for ground-level views as illustrated by the flattened trees (right)

- No automated updating: The computer graphics 3D models cannot be reused if the 2D geo-data changes.
- No geo-data context: The computer graphics 3D models have to be created separately from related geo-data that could be helpful for editing. For instance, showing aerial images or topographic maps can be useful to validate a 3D model visually while editing. Since generic 3D editing tools are not aware of the geo-spatial context of the data, they do not provide such functionality.
- No geo-data linkage: The computer graphics 3D models do not preserve data semantics and thematic information contained in or associated with 2D plans.

TIN-based Modeling

TIN models can also be used to represent detailed terrain models. They allow for integrating land use information and built surface structures by using line or polygon features as break lines and by assigning different colors or textures to certain triangles. This way, detailed terrain models can be created from GIS input data. The TIN-based approach, however, has two major disadvantages:

- No linkage to 2D plan elements: No direct link is established between a TIN and the original data from which it has been created. Therefore, thematic and semantic information is lost and changing the original data requires rebuilding the TIN.
- Restricted geometry: As an inherent limitation, vertical faces cannot be modeled by TINs.

Fig. 2. Snapshot of the automatically created urban terrain model of our case study, a redevelopment of a downtown area in Potsdam

Modeling Based on Image Draping

The most common approach to create detailed urban terrain models in landscape visualization is based on the principles of image draping. Specialized modeling tools for virtual landscapes such as World Construction Set, ArcScene, or VirtualGIS allow for integrating GIS data into the 3D scene by draping vector-based information onto the terrain and assigning colors or textures to the polygons. Such tools reduce the manual modeling effort and are able to create satisfying results at landscape scale (Appleton et al. 2002). In landscape planning, they are primarily used to visualize terrain surfaces carrying vegetation by textures (Muhar 2001) and prototype 3D plant models. In addition, the tools mentioned above are suitable to visualize roads or other manmade terrain surfaces. However, in order to integrate manmade surface structures that include textured vertical faces or that have to be created from polylines by buffering a line feature, manual modeling effort and extensive data preparation is required.

Smart Terrain Models

In this paper, we propose *smart terrain models*, an approach for generating high-quality urban terrain models automatically from 2D vector-based plans. Instead of converting available 2D data to new, detached 3D models, our approach is based on the idea to keep the original 2D data and to add complementary information to enhance them to a complete 3D scene specification. The approach is similar to the image draping method but extends it by defining specialized representations for typical terrain elements. This increases the modeling flexibility while providing significant advantages compared to generic 3D models:

- The underlying 2D vector data that form the core of the smart terrain model specification can still be accessed and edited by external 2D GIS tools and can be obtained and updated from 2D data sources.
- Thematic information and data semantics can be directly queried within the 3D visualization.
- Due to the permanent link between 2D data and their 3D representation, interactive visualization of the generated 3D models can be directly combined with editing tools for the underlying 2D data. This way, the effect of modifications of the 2D data can be directly shown in the 3D model. Particularly, it allows for immediate corrections of any errors in the underlying 2D data when they become visible in the 3D visualization.
- Since the data semantics is preserved in the 3D representation, smart terrain models form a basis for the development of visualization tools that are aware of the data semantics and use it to apply specialized real-time rendering techniques (e.g., Finch 2004; Shah et al. 2005) for certain surface elements to optimize the rendering performance or to improve the visual quality.

We demonstrate our approach in a case study, in which we modeled an urban area of downtown Potsdam, Germany. The combination of smart terrain models with 3D building models, e.g., provided by Smart Buildings (Döllner and Buchholz 2005) or CityGML (Kolbe et al. 2005), and 3D vegetation models (Deussen 2003) leads to photorealistic 3D city models that are suitable for real-time visualization at a pedestrian's point-of-view. Our approach has been implemented on the basis of the graphics library VRS (Döllner and Hinrichs 2002) and the LandXplorer geovisualization system (Döllner et al. 2003).

2 Smart Terrain Models

In this section, we describe our system for modeling and editing smart terrain models. Section 2.1 gives an overview of the system and defines the data components that specify the model. Section 2.2 gives a classification of the terrain elements that constitute a smart terrain model. In section 2.3, we describe the appearance specification for terrain elements.

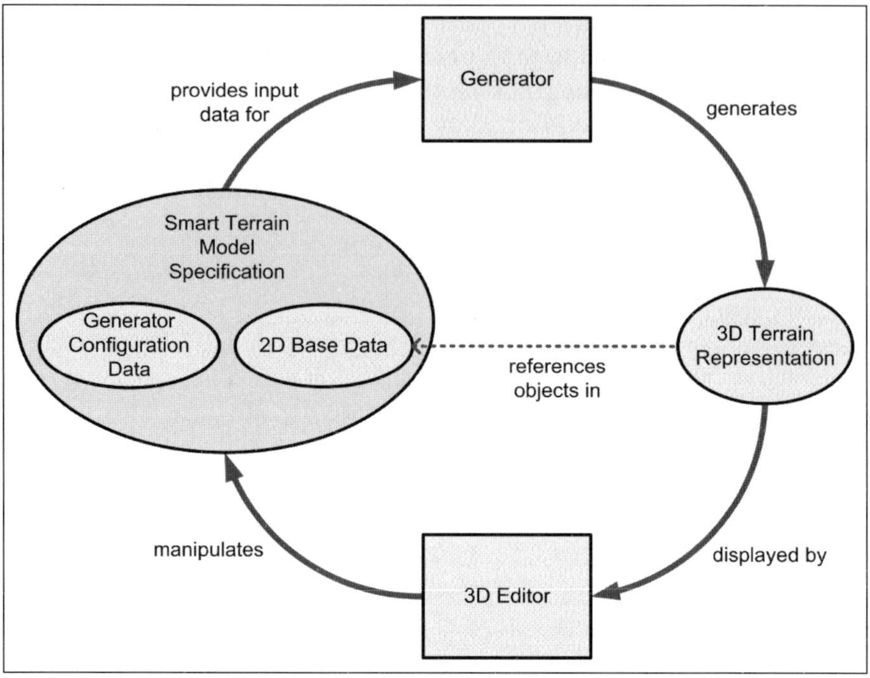

Fig. 1. System overview for the automated construction of urban terrain models

2.1 System Overview

The principal workflow of our approach is illustrated in Figure 3. The specification of a smart terrain model provides the complete information for automatic generation of a 3D model representation and consists of two parts, the *base data* and the *generator configuration*.

The *base data* consist of any 2D vector-based land use data containing polylines and polygons, in the following referred to as *base vector-objects*, with attached attribute tables. Each attribute table defines a set of numerical, textual, or Boolean values that can be accessed via certain attribute-table key strings. The base data can be obtained from existing geo-data bases but can also be edited directly in the 3D-editor, e.g., based on a given aerial image. The base data can be specified in any format that allows for describing 2D vector-data with associated attribute tables, e.g., ESRI shapefiles or GML.

The *generator configuration* specifies the way in which base data and related attributes are interpreted to generate the 3D terrain model. For instance, in a given base-data set polygons representing water areas may be indicated by defining the value "Water" for the attribute-table key "Area Type". The generator configuration also defines one or more material catalogues. A *material catalogue* consists of a set of material descriptions that are referenced by 2D base vector-objects. Materials are discussed in Section 2.3. The generator configuration and the material catalogues are stored as separate XML files.

The *generator* takes the smart terrain model specification as input and creates a 3D representation of the terrain that is used for real-time rendering. The generator also maintains for each generated 3D object a reference to the underlying base vector-object. This information is used by the editor to support selection and editing of surface-model elements directly in the 3D scene.

The *interactive 3D editor* allows for creation and management of 2D base data, generator configuration, and material catalogues. It provides real-time visualization of both the underlying base data as well as the resulting 3D representation (see Fig. 4). The 2D base data can be displayed on the surface of a digital elevation model using the technique described by Kersting and Döllner (2002). If a terrain element is selected and edited by the user, the modification is applied to the underlying base data, and the generator immediately updates the corresponding parts of the 3D representation, so that changes of the base data are directly visible in the 3D environment. The 3D representation itself, however, is never changed by the user but always fully determined by the smart terrain model specification. Hence, editing effort is never lost when the 3D model has to be re-built by the generator.

2.2 Smart Terrain Model Elements

The elements of a smart terrain model are organized in the classes `GroundArea`, `WaterArea`, `Stair`, `Wall`, `Kerb`, and `Barrier` (see Fig. 5). Instances of these classes are not explicitly part of the specification but are defined implicitly by the base vector-objects and their related attributes. Most terrain elements correspond to exactly one base vector-object. The only exception holds for irregular stairs, which require multiple polygons for specification. The attributes of a base vector-object determine the class of the corresponding terrain element. Depending on the respective class, the required attributes for 3D shape and appearance of a terrain element are also taken from the attribute table of the base vector-object.

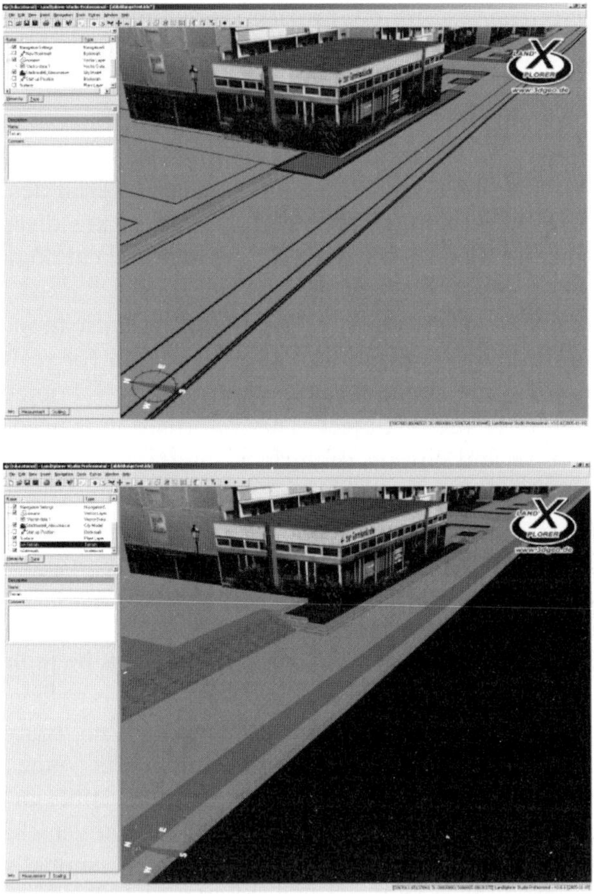

Fig. 4. Snapshot of the editor system: The user can switch between the 2D view of the base data (top) and the resulting 3D representation (bottom)

Our primary design goal was to make the refinement of pure 2D base vector-data to a full smart surface-model specification as simple as possible for typical cases. If the class model would provide the unrestricted geometric flexibility of a generic 3D tool, it would not be possible anymore to specify the terrain elements completely via 2D base data, and the editing process would become very complicated. Thus, the main advantages of the system would be lost. Therefore, we restricted the class model to cases that can be intuitively described via 2D polygons or polylines and support optional refinement of individual objects by external 3D tools. For this, for each terrain element, the automatically generated 3D model can be exported, externally refined, and finally referenced by the terrain element. This way, the effort for fitting an externally created 3D model correctly in the scene is avoided and the 3D model is still related to the underlying base data.

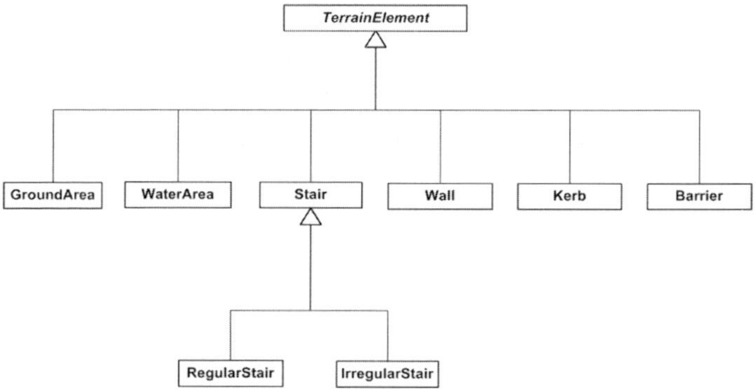

Fig. 5. Overview of the terrain element classes

Ground Areas

A `GroundArea` represents a polygonal surface on the ground that is displayed with a certain appearance. For instance, a ground area may represent a part of a street, a sidewalk, a lawn, or a wasteland area. Since the `GroundArea` class covers several kinds of terrain elements, it could theoretically be split up into several subclasses but there are some reasons to use a single class instead:

- To enforce a generic sub-classification of ground areas can be impossible or ambiguous. For example, a single asphalted area might be interpreted as a part of a parking area, a part of a street, or a part of a square.
- From a technical point-of-view, a finer classification is not necessary because the created 3D representations differ only by material.
- From a user's point-of-view, a finer classification is not necessary because thematic classification can be obtained from the base vector-objects' attribute tables, and visually thematic sub-classification can be achieved by assigning different materials.

Each `GroundArea` is defined by a single base-data polygon or a polyline that is buffered to a certain width. If the data format used for the base data supports 2,5D vector data, i.e., if it allows for specifying per-vertex height values for each vector object, as in the case of shapefiles or GML, the surface geometry can be completely defined by the geometry of the underlying base vector-object. If height values are not provided by the initial base data, they can either be automatically derived from a digital elevation model or edited manually. To assign height values by projecting the dataset onto a digital elevation model corresponds to the principle of image draping and is a good solution for continuous surfaces without vertical breaks.

In the general case, however, `GroundAreas` do not necessarily define identical height values at their borders. Therefore, some `GroundAreas` must be rendered with vertical border faces to avoid holes in the terrain model. For this, an optional extrusion depth can be specified, by which the surface is extruded downwards. If desired, a separate material can be specified for the vertical border faces.

Water Areas

A `WaterArea` represents a polygonal surface that appears as a water surface in the 3D visualization. The geometry of a `WaterArea` is specified in the same way as a `GroundArea` but it must always be fully horizontally and allows for specifying a desired flow direction of the water. We integrated water areas as an own class to allow future viewing applications for rendering the water different to solid surfaces, e.g., by animated water textures.

Stairs

In the general case, a `Stair` is described by multiple base-data polygons, whereby each polygon represents a single step and is extruded downwards (`IrregularStair`) to obtain the 3D shape of the step. Many stairs can be described more simply via a single rectangular footprint by moving certain edges of the rectangle inwards to describe the footprints of the upper steps (`RegularStair`, see Fig. 6). A `RegularStair` can be specified by a sin-

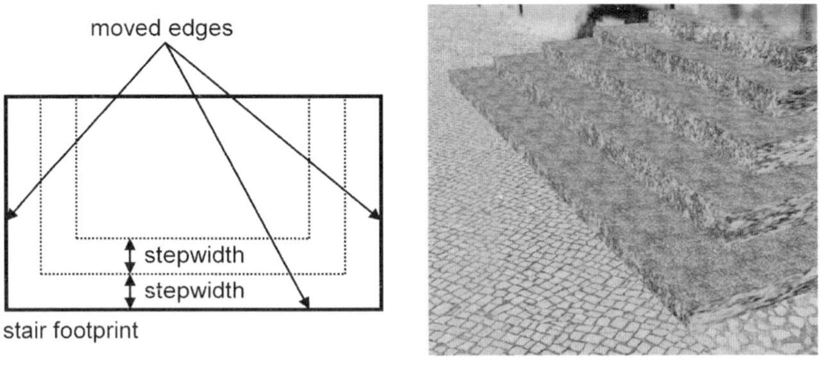

Fig. 6. Construction of a simple stair: The footprints of the upper steps are defined by moving inwards certain edges of the full stair footprint (left). The 3D shape is then defined by extruding each step polygon downwards (right)

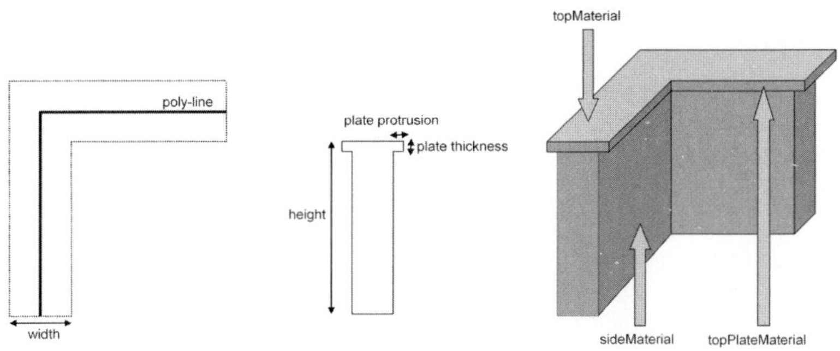

Fig. 7. Construction of walls: **(a)** Specification of the footprint (left); **(b)** 3D extrusion parameters (middle); **(c)** materials for each surface (right)

gle base-data polygon, whose attribute table defines number of steps, step height, step width, and the information which edges to move inwards. Each stair defines a material for its surface and optionally an additional material for its vertical faces.

Walls

The class `Wall` represents freestanding walls and retaining walls. Many walls contain a separate top plate, which might differ from the rest of the wall by another material and by slightly exceeding the wall footprint. Therefore, the `Wall` class provides optional parameters to specify a separate top plate. Figure 7 shows an example. Walls are represented by polygons or polylines of the base data and their attributes. By buffering line features to a width that is specified by the attribute table, we obtain the footprint of the wall (see Fig. 7a). The 3D shape is now determined by the extrusion parameters shown in Figure 7b, and finally, materials are specified for each side (see Fig. 7c).

Kerbs

The class `Kerb` represents boundary objects between two `GroundAreas` with a separately specified appearance, e.g., by an own kind of stone or by a different height (see Fig. 8). This corresponds to typical construction methods for, e.g., pavements and roads, where a kerbstone is necessary to stabilize the construction. A `Kerb` is defined by a polyline of the base-data set.

For the frequent case of two `Kerbs` along a single boundary line, both `Kerbs` should appear side by side and not overlapping. For this, each `Kerb` must specify one of the two adjacent `GroundAreas` as the *parent area* to

Fig. 8. Examples of kerbs along the border lines of streets and sidewalks

which the `Kerb` belongs. This defines the order in which the `Kerbs` appear between the two adjacent `GroundAreas`. The parent area, i.e., its underlying base vector-object, is referenced in the attribute table of the `Kerb`'s underlying polyline. To allow for referencing, the base-data vector-objects must provide unique ID values in their attribute tables. In the rare case of three or more parallel boundary objects along a single boundary line, the middle objects must be represented by additional `GroundAreas`.

The footprint of the `Kerb`'s 3D representation is obtained by buffering the `Kerb`'s underlying poly-line to a certain width. Since the Kerb shall appear completely as a part of its parent area, the buffering is performed only in the direction pointing inside the parent area. Finally, the 3D shape is obtained by extruding the footprint to a certain height. Buffering width and extrusion height are specified via the attribute table.

Barriers

The class `Barrier` represents boundary objects such as fences or balustrades. In contrast to walls, barriers do not cover any area on the ground but their footprint is only line-shaped. Each `Barrier` object is defined by a polyline in the base-data set which specifies the height of the `Barrier` in its attribute table. The generator provides a set of standard barrier types. In our current implementation, `Barriers` are simply represented by extruded lines covered with partially transparent texture-images. More advanced future `Barrier` types could include actual 3D detail geometry. If the standard types are not sufficient, the `Barrier` can be refined by an external 3D modeling tool.

2.3 Appearance Properties

The material catalogue provides a set of materials that can be referenced by the base vector-objects via unique names. A material can be of one of two types: Color and texture. A *color material* defines an RGB color value.

A *texture material* defines a reference to a texture image file and scaling parameters that define to which width and height the texture is stretched when it is applied to a surface. Each terrain element that references a texture material must define an anchor point onto which the texture origin is mapped in the 3D model and an orientation angle of the texture. Since these values are usually different even for objects of the same material, they are not stored as a part of the material itself.

3 Case Study

We used our approach in an open-space planning task concerning a part of the German city Potsdam. The aim of our modeling project was to provide a detailed photorealistic 3D geovirtual environment that could be explored interactively. The result can be seen in Figures 2 and 4.

The initial data of the modeling project were provided to us by the city of Potsdam and were part of the digital municipal town map. These data originate from ground survey, are geometrically very accurate, and hold detailed information about buildings, surface cover, installations and vegetation. Information about the terrain surface cover is maintained in the data by mapping borderlines between different surface types and placement of cartographic symbols for different surface materials and vegetation areas. Borderlines can either represent a change in surface materials or can represent kerbstones. Buildings are represented by polygons with attached information about the number of floors and additional information about balconies and car passages. Walls and stairs are represented through polylines or polygons depending on their size. Installations and trees are represented through point symbols.

To create a smart terrain model from the data, all features representing a change in surface type or material were used to create an area wide polygon dataset. Thematic information, stored in point features, was then transferred to the polygons through a point in polygon selection. Using the original thematic information all objects were classified into the smart terrain element classes. The resulting dataset contained all polygonal terrain elements that were represented by the town map. In order to integrate walls and kerbstones that were represented by polylines, these were classified and specified as well. In the next step a material table was created and assigned to the features. It holds material names, the names of the texture files and texture scaling parameters. Finally, height information was assigned to the terrain elements. Height information for `GroundAreas` and `Kerbs` were derived by projecting the respective features onto a digital elevation model. For features representing `WaterAreas` and `Stairs` constant values were assigned. Objects representing walls were assigned constant height values or the height was calculated as an offset of the terrain.

The resulting terrain model was combined with models for buildings and plants as can be seen in Figure 2. For the plants, we used plant models and the plant rendering engine of the project Lenné3D (Paar & Rekittke 2005). For modeling and representation of buildings we used the approach of Döllner and Buchholz (2005). The buildings were automatically generated from 2D GIS input data and refined afterwards, e.g., by textures for roofs and facades. As the footprints of the buildings were used in creating the terrain model, no inconsistencies could appear between buildings and terrain. Furthermore any thematic information that was stored within the attribute table of the input data was preserved.

4 Conclusions and Future Work

The presented approach simplifies and enhances the construction, manipulation, and usage of complex urban terrain models. Its major advantages include the persistent linkage to 2D vector-based plans, the rule-based and heuristic-based automated model generation, and the inherent functionality and smartness of urban terrain objects. Base 2D geo-data can be taken from GIS and integrated seamlessly into the 3D modeling process. The ability of smart terrain models to maintain semantic and thematic information provides a technical basis for smart 3D geo-visualization tools. In addition, the approach represents a step towards 3D geovisualization from a pedestrian's point-of-view in contrast to "fly-through" based systems.

As future work, we are investigating the integration of algorithms for procedural generation of textures and geometric details such as asphalted streets or stone mosaics. In addition, we are working on related real-time

3D rendering techniques to improve photorealism, including complex 3D vegetation models and shadows.

Acknowledgements

We would like to thank the urban surveying office Potsdam for the data set used in our case study, the Remote Sensing Solutions GmbH (www.rssgmbh.de) for the city model of Berlin used in Figure 1, and the 3D Geo GmbH (www.3dgeo.de) for providing us support to the landXplorer technology. We are also grateful to the German environmental foundation Deutsche Bundesstiftung Umwelt for supporting our research within the Lenne3D project (www.lenne3d.de) and to the Lenné3D GmbH for providing the tree models used in Figure 2. Finally, we would like to thank our students Oleg Dedkow, Anselm Kegel, Manuel Wellmann, Wolfgang Braunisch, Christian Schubert, Andrea Hentschke, Sebastian Kuhn, Jan Rehders, and Jan Zimmermann, who contributed to the SmartTerrain editor implementation.

References

Appleton K, Lovett A, Sünnenberg G, Dockerty D (2002) Rural landscape visualisation from GIS: a comparison of approaches, options and problems. Computer, Environment and Urban Systems 26:141–162

Danahy JW (2005) Negotiating public view protection and high density in urban design. In: Bishop I, Lange E (eds) Visualization in landscape and environmental planning. Spon Press, London, pp 203–211

Deussen O. (2003) A framework for geometry generation and rendering of plants with applications in landscape architecture. Landscape and urban planning 64 (1-2):105–113

Döllner J, Hinrichs K (2002) A generic rendering system. IEEE transactions on visualization and computer graphics 8(2):99–118

Döllner J, Baumann K, Kersting O (2003) LandExplorer – ein System für interaktive 3D-Karten (= Kartographische Schriften 7). DGK, pp 67–76

Döllner J, Buchholz H (2005) Continuous level-of-detail modelling of buildings in Virtual 3D City Models. In: Proc of the 13th ACM Int Symp of Geographical Information Systems, ACM GIS 2005, pp 173–181

Döllner J, Hagedorn B, Schmidt S (2005) An approach towards semantics-based navigation in 3D city models on mobile devices. In: Proc of the 3rd Symp on LBS & TeleCartography, Vienna (*to appear*)

Finch M (2004) Effective water simulation from physical models. GPU Gems, Addison Wesley, pp 5–29

Kersting O, Döllner J (2002) Interactive visualization of 3D vector data in GIS. In: Proc of the ACM GIS 2002. ACM Press, pp 107–112

Lange E, Hehl-Lange S (2005) Future scenarios of peri-urban green space. In: Bishop I, Lange E (eds), Visualization in landscape and environmental planning. Spon Press, London, pp 195–202

Muhar A (2001) Three-dimensional modelling and visualisation of vegetation for landscape simulation. Landscape and urban planning 54(1-4):5–17

Paar P, Rekittke J (2005) Lenné3D – Walk-through visualization of planned landscapes. In: Bishop I, Lange E (eds) Visualization in landscape and environmental planning. Spon Press, London, pp 152–162

Shah MA, Kontinnen J, Pattanaik S (2005) Real-time rendering of realistic-looking grass. In: Proc of the 3rd Conf on Computer Graphics and Interactive Techniques in Australasia and South East Asia, pp 77–82

Stock C, Bishop I (2005) Helping rural communities envision their future. In: Bishop I, Lange E (eds) Visualization in landscape and environmental planning – technology and applications. Taylor & Francis, Oxon, UK

Warren-Kretzschmar B, Tiedtke S (2005) What role does visualization play in communication with citizens. In: Buhmann E, Paar P, Bishop I, Lange E (eds) Trends in real-time landscape visualization and participation. Proc at Anhalt University of Applied Science 2005. Wichmann Verlag, Heidelberg

Werner A, Deussen O, Döllner J, Hege HC, Paar P, Rekittke J (2005) Lenné3D – Walking through landscape plans. In: Buhmann E, Paar P, Bishop I, Lange E (eds) Trends in real-time landscape visualization and participation. Proc at Anhalt University of Applied Science 2005. Wichmann Verlag, Heidelberg

A Flexible, Extensible Object Oriented Real-time Near Photorealistic Visualization System: The System Framework Design

Anthony Jones, Dan Cornford

Knowledge Engineering Group, School of Engineering and
Applied Science, Aston University, Birmingham, B4 7ET, UK
email: d.cornford@aston.ac.uk

Abstract

In this paper we describe a novel, extensible visualization system currently under development at Aston University. We introduce modern programming methods, such as the use of data driven programming, design patterns, and the careful definition of interfaces to allow easy extension using plug-ins, to 3D landscape visualization software. We combine this with modern developments in computer graphics, such as vertex and fragment shaders, to create an extremely flexible, extensible real-time near photorealistic visualization system. In this paper we show the design of the system and the main sub-components. We stress the role of modern programming practices and illustrate the benefits these bring to 3D visualization.

Key words: real-time, visualization, object-oriented, data driven, plug-in, extensible

1 Background

The visual output of most current GIS visualization software often exhibits either a low level of visual realism with high levels of user interaction[1], or

[1] For example, the viewpoint may be altered in position and focus in an interactive way, images are often animated (albeit with low quality textures), there is a low

has a high degree of visual realism with low levels of user interaction[2].

Commercial applications producing computer-based renderings based on geographical information, for example, WorldPerfect[3] and LandXplorer[4] are relatively easy to use, but very difficult to customize to specific requirements, or to add desired features and behaviors. Recently alternatives [1, 2] have been proposed based on modern computer game engines that place a heavy emphasis on geospatial representation such as Microsoft Flight Simulator2004[5]). Game engines appear attractive because of the high levels of immersion they are required to produce, and the increasingly realistic quality of the graphics this entails. The development of game engines occurs within a problem domain whose focus greatly influences the constraints and capabilities of the software. For example, a game engine can be written to optimize enclosed environments with a relatively small number of dynamic objects, focussing on the use of immersive lighting and special effects in order to increase user emersion. While it has been demonstrated that existing game engines can be applied to visualization in a wide range of applications [3], the applicability of game engines will ultimately be restricted by the context of their original problem domains.

The application framework we describe in this paper is based upon a novel design that extends and improves upon an existing model intended for computer games developers [4]. In contrast to the fixed information processing pipeline common to current GIS visualization software [5], the geospatial rendering system incorporates an information processing pipeline that is highly flexible and extensible. This will be achieved through the combination of plug-ins and data-driven content and behavior which will allow the application framework to support a diverse range of applications. Introducing data-driven design to the field of GIS visualization provides an exciting opportunity to create a uniquely flexible visualization system. We go on to describe the application framework's use of modern methodologies in order to produce near photo-realistic renderings at interactive frame rates. While a small number of GIS visualization software developers are currently using modern rendering technology to produce non-realtime images[6], the appli-

number of polygons in the scene, and a small number of objects are depicted in total.

[2] For example, the viewpoint position, focus and path are predetermined; the scene contains high fidelity textures, accurate lighting and shadows, and detailed objects.

[3] http://www.metavr.com/products/worldperfect/worldperfect.html

[4] http://www.landex.de/

[5] http://www.microsoft.com/games/flightsimulator/

[6] For example, see http://www.3dnature.com/index.html

cation of emerging techniques to produce highly detailed, interactive near photo-realistic renderings of spatiotemporal scenes has yet to be realized in the field of GIS visualization.

1.1 A Running Application Example

In order to provide a more concrete illustration of the application framework's capabilities, we present an example problem domain that will be used and extended throughout the paper. The example focusses on the development of a traffic simulation system with the following functional requirements:

- The system will import both Integrated Transport Network (ITN)[7] and topography data in order to inform the modeling of a traffic simulation.
- The system will maintain a real-time traffic simulation occupying the given ITN. The simulation will include dynamic traffic elements (such as traffic lights), a range of vehicle types, and pedestrians. While passive items such as roads and buildings can simply be represented, active items such as traffic lights, vehicles and pedestrians must exhibit appropriate runtime behavior.
- The system is also required to produce a real-time visual output that will provide the user with an illustrative summary of the simulation's changing state over time.

1.2 Additional Applications

It is important to stress here that these example applications are supposed to illustrate the potential of the system, and have not been implemented at this early stage. Our focus is visualization, but our longer term goal is an integrated modeling environment that is fast, efficient, extensible and flexible. The framework supports user interaction, and so users will be able to orient themselves in a depicted scene by manipulating the camera's position and orientation; similarly, queries and modifications of the modeled environment could be made via an extensible range of input devices. The types of applications we envisage include:

- *Visual assessments* The framework has been designed to support the simulation and visualization of a wide variety of dynamic spatiotemporal environments, and this fundamental functionality could easily form the ba-

[7] Further information: http://www.ordnancesurvey.co.uk/oswebsite/products/osmastermap/itn/

sis of a visual assessment application. Due to the use of data-driven programming (as described in Section 2.3), designated study areas can be described incrementally through an XML based (and likely tool-oriented) definition of object properties, types and instances. Visualizations can contain animation, such as ind farm blades that turn, trees that appear to sway in the wind, and crops that grow over time. Visualizations can also include Artificial Intelligence (AI) so that cyclists make use of a proposed bicycle route and pedestrians explore a new shopping center.
Due to the text-based nature of the framework's scene descriptions, users will be able to quickly influence the detail, realism and overall visual appearance of the study area. Changes can be made to emphasize specific details and modifications (such as a user's home), or to highlight objects with similar properties (all proposed elements, for example).
- *Soil erosion* The framework is designed to be flexible, and thus can also incorporate non-trivial spatiotemporal models, for example it would be possible to produce a simplified erosion model to act on a ground model or Triangulated Irregular Network to model (in a naive manner it must be admitted) erosion and render the output realistically, in accelerated time. The application maintains both a real time clock and a model time clock, so this is very easy to undertake.
- *Process-based models, polling input via the task, input, and binding system* If a more detailed model were required the framework could be readily coupled to a more complex process based model, via the task, input, and binding systems (see Section 2.2). A good example of this might be the coupling of the simulation world with a numerical weather prediction model to provide realistic weather conditions with the correct timing and location with respect to the model forecast.

2 The Application Framework

We are developing an application framework that will form the basis of a modeling and visualization environment, where the client is able to tailor the application according to their own requirements through the use of customizations of, and extensions to the framework's runtime behavior. Figure 1 illustrates the core components that make up the application framework. The framework's overall design is loosely based on the design of an object composition framework presented in [4]. The framework separates overall application functionality into a number of coherent, loosely coupled responsibilities, each of which is represented in the framework by an abstract interface illustrated via the inner octagon in Figure 1. The concrete implementation,

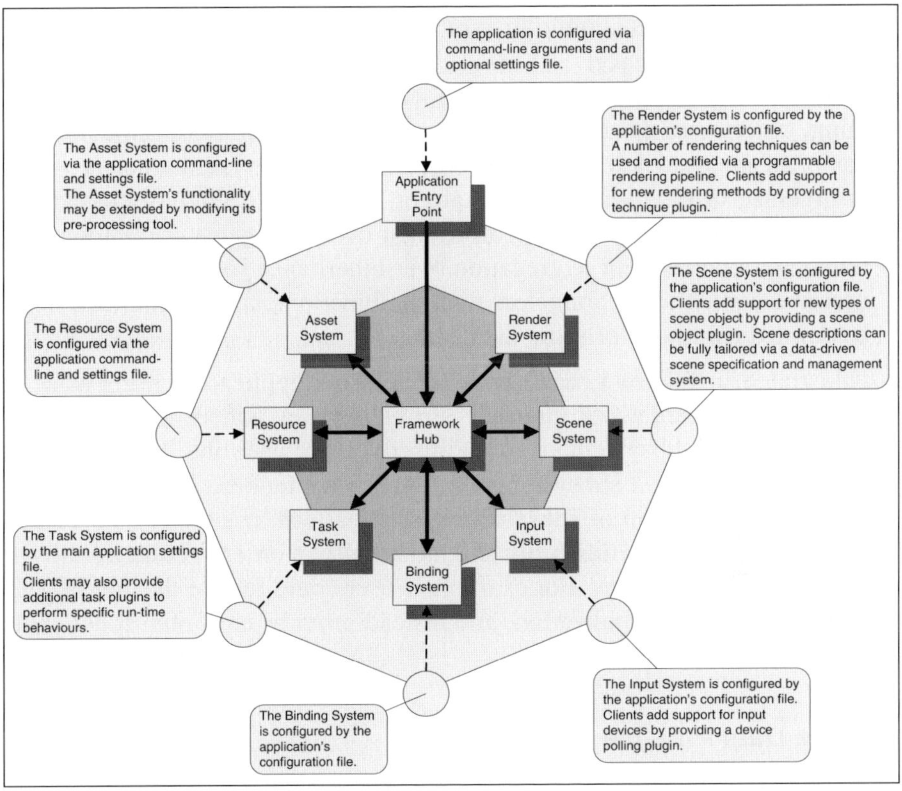

Fig. 1. An overview of the application framework showing the core systems

and thus the run-time behavior, of each subsystem may be provided by the user in the form of a dynamically linked library or through the use of our pre-supplied default concrete implementations as shown by the outer octagon in Figure 1. During application execution, the central framework hub performs dynamic allocation and binding of sub-system implementations to their respective interfaces; the hub also acts as an intermediating interface between the various subsystems.

A number of concepts are used throughout the application framework in order to increase its extensibility and flexibility, and these form a basis upon which further functionality can be built.

- *Plug-ins* A plug-in is a portion of code that is compiled into a dynamically linked library file, commonly extending a predefined interface that is exposed by the application code. At runtime, each plug-in is bound to the

application, and is then able to exhibit its contained runtime behavior via the predefined interface.
- *Data driven programming (DDP)* In traditional object oriented programming, objects are described using classes, which define the state and behavior of the modeled real-world object. Inheritance hierarchies are used to organize objects with shared state or behavior. In DDP, state and behavior are described separately from their owning objects as components, reflecting a favoring of aggregation over inheritance [6]. By using data to describe component parameters and combinations, a class hierarchy can be defined using one or more data files.

As illustrated in Figure 1, each subsystem in the application framework is accessed via its framework interface. Providing they adhere to the contract described by the subsystem interface, users can replace the default behavior of most application subsystems with their own tailored implementation. Through a combination of subsystem specializations, users can take advantage of the framework's flexibility and modularity in order to build a series of very different applications. For example, a user could reduce the complexity of an applications's visualization, and instead provide additional functionality for data input and analysis.

2.1 The Data Pipeline

Figure 2 shows the data processing pipeline represented by the framework's data preprocessing tool, asset system and resource system. The framework's data pipeline makes use of a fixed enumeration of asset types, each of which corresponds to an intended mode of data usage, as shown by Table 1.

Preprocessing Tool

The preprocessing tool embodies a conversion process, where one or more source files may be compressed and encrypted, and are ultimately written as one or more binary files whose format correspond to the framework's asset types. The behavior of the preprocessing tool is driven by a combination of command-line arguments plus an optional configuration file, and can be extended through the use of format plug-ins. Each format plug-in represents the conversion process from source data to asset data for a single source data format. The use of format plug-ins results in a preprocessing tool that supports a diverse, extensible range of input formats, providing their data corresponds to one or more framework asset types.

To continue the running example presented in Section 1, the preprocessing tool will be responsible for converting the traffic simulation's ITN, topography and other data into their corresponding framework asset file types.

Table 1. Asset Types

Asset Type	Description
Text Asset	Text assets represent a contiguous block of immutable textual characters; as such, they are the most fundamental of application asset types. Anticipated uses of the text asset include documentation and fixed-length string storage, for example the user may wish to store interpreted-language AI scripts or user instructions as text assets.
Tree Asset	Tree assets correspond to a tree of named nodes, each of which may contain zero or more named attribute values of a supported type. Example uses of the configuration asset type include the definition of tree-like run-time structures such as scene graphs, and the storage of hierarchical data such as object and property inheritance trees.
Array Asset	Array assets are intended to store multi-dimensional arrays of values of a range of data types. Example uses of the array asset data format include the storage of n-dimensional tables, and n-dimensional textures.
Mesh Asset	Mesh assets correspond to a collection of 3D vertices alongside a system of specifying interconnection based on sequences of vertex indices. The mesh asset format will support a variety of 3D concepts, including animated 3D models with texture coordinates and volumes, such as bounding volumes.

We will assume that the ITN data is described using Geography Markup Language (GML); the user must either obtain or develop a format plug-in that can validate and convert the ITN data to the application framework's

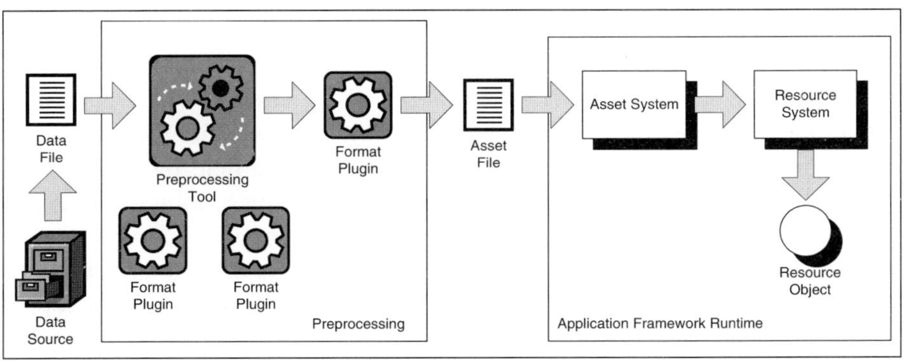

Fig. 2. An illustration of the data processing pipeline. Note that the preprocessing tool maintains a collection of format plug-ins, from which it selects an appropriate plug-in to process any given data file

array asset format. In this case, we choose to represent the ITN as a two-dimensional table that maintains the original's topology information. During the preprocessing tool's execution, the specified ITN files will be read and their data processed by the assigned plug-in(s), which will in turn output one or more array asset files to be read by the framework's asset system. The traffic simulation system's other data files will be similarly processed according to the user's configuration.

Asset System

The asset system represents a repository of asset files, and is responsible for maintaining this collection and providing efficient access to its data. The asset system is therefore synonymous to a file system, albeit one with a fixed range of file types. For example, a given implementation may represent a locally stored directory tree of asset files, a networked or ftp-based cache of asset files, or a web (service) based catalogue of compressed and encrypted asset archives.

In the context of our running example, the traffic simulation's data may be distributed as a number of compressed archive files. The ITN array asset produced above, plus a number of other representations of the same road network (for example, the roads' geometry in the form of mesh assets), are supplied as a single archive file. Further archive files contain data-driven descriptions for the various vehicles that will populate the simulation. A final archive file contains the scene descriptions and application configuration files in the form of one or more tree assets. The asset system implementation will be responsible for locating a given asset file within these archives, and providing access to asset data when required.

Resource System

The framework's resource system is responsible for maintaining a collection of run-time objects, each of which represents the data held by a single asset file. While the asset system provides low-level access to asset data, the resource system's framework interface requires that any given implementation is capable of mapping an asset identification string (such as a file path or URI) to a run-time object that provides the corresponding asset type's modus operandi. For example, resources can be constructed from asset data in a background thread in order to hide load-time delays from the user or resources can be incrementally or partially constructed according to the application's data requirements, e.g. in level of detail implementations.

The traffic simulation's resource system implementation will build run-time objects that allow the data to be used in a meaningful way: tree assets will be represented as hierarchical data structures, array assets will be represented as N-dimensional arrays of data items of a described type, and so on. The default implementation constructs such resources in a background thread.

Summary of the Data Pipeline

The data pipeline described here represents an optimized route for static file-based data. A data pipeline for more dynamic data, such as streaming input, and data that is not file-based, such as web (service) content, is realized by a combination of the input and binding systems (see Section 2.2).

The definition of a fixed range of asset types results in a predetermined format for data manipulation, which in turn allows data processing (that is, parsing and validation) to be reassigned to an offline stage. A fixed range of asset types also aids the design of a concrete asset system interface, which allows the details of asset collection and access to be decoupled from the application framework's other responsibilities. The resulting data pipeline, shown in Figure 2, can be optimized for efficient throughput of a known range of data formats.

The traffic simulation example demonstrates how the data processing pipeline can be tailored in order to support a given use of data. The preprocessing tool has been extended to support a variety of input files, and the asset system has been specialized to support a chosen asset distribution scheme.

2.2 Application Kernel

The application kernel represents the processing heart of the application framework. The task system encapsulates application behavior and functionality, while the binding system allows subsystem implementations to communicate effectively and store arbitrary data in a type-safe, centrally controlled manner. Together, these two core subsystems provide a backbone of functionality that forms the basis of further application behavior.

Task System

At a high level of abstraction, the task system's overall behavior takes the form of iteration over a number of distinct time slices, each consisting of a number of subsystem operations or events occurring in a given order; this is illustrated by Figure 3. The task system thus represents what is traditionally termed an *application loop*. Tasks submitted to the task system are ordered and subsequently triggered according to their priority value, which is provided by the submitter. When triggered, a task is supplied with a summary of the task system's status, along with access to the framework hub and hence the state of the application framework as a whole. The task objects themselves are both defined and supplied by subsystem implementations or as task plug-ins.

Each task plug-in provides a single task object to be submitted to, and thus processed by, the task system. Task plug-ins represent one way in which users can extend existing framework functionality, by providing additional behavior to be exhibited at run-time. For example, a user could write and submit a task plug-in that regularly monitors congestion levels along a number of inner city roads. Similarly, another task plug-in could randomly dispatch emergency response vehicles in order to test alternate routes through the traffic network.

Tailored implementations of the task system can take advantage of dual core processors or distribute available tasks over a number of clustered machines, and can thus represent a customized task scheduling and distribution policy. A user in need of greater flexibility could extend the task plug-in concept in order to expose application framework functionality to scripting languages such as Python[8] and LUA[9].

Binding System

The framework's binding system is a repository for run-time data of any type. Data is associated with an identifier, and is stored as part of a hierarchical collection of *namespaces*. Subsequent to storage, data can be accessed through the use of a type-safe binding object. While subsystem interfaces present a fixed channel for inter-system communication, the binding system can be used for more implementation dependent storage and interaction.

To illustrate binding system use, a summary of the traffic system's current state can be stored as a dedicated compound type bound to an appropriate location in the binding system. Specializations of one or more application subsystems, or alternatively application plug-ins, could then bind to and use this information in order to affect task scheduling, select resource construction policies, or inform the user as part of a graphical user interface (GUI). The traffic system's summary information could also be fed back into the simulation itself, so that emergency vehicles avoid congested areas and vehicles choose alternate routes to avoid icy roads.

---

[8] http://www.python.org/

[9] http://www.lua.org/

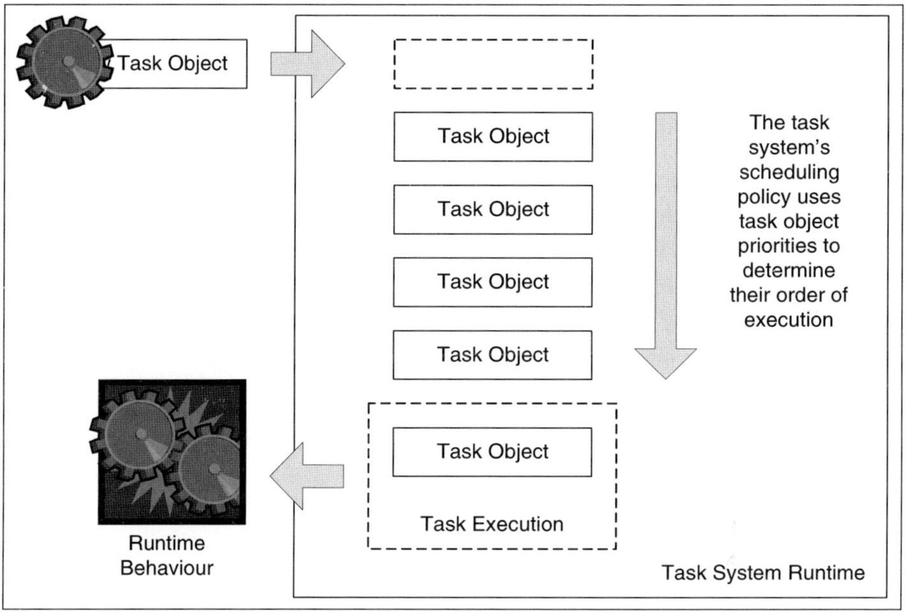

Fig. 3. An overview of the task system's functionality

Input System

The input system is a logical extension of the binding system; it represents a collection of objects which periodically update data bindings. Input plug-ins, each representing a single data source, are registered with the input system, which polls each data source during the main application loop. Typically, data sources will be input devices such as the mouse or keyboard, but these could also include other sources such as database or internet connections, value generators, procedural models, and so on. This is in contrast to the data processing pipeline presented above, which is intended to provide access to static data sources, and does not support dynamic streaming content by default.

An input plug-in can easily represent a device such as a mouse or keyboard. An input plug-in, along with its associated data bindings, could also represent a more complicated model, such as a numerical weather prediction system. In this case, the input plug-in may be connected to an online database providing current or forecast weather data. In the context of the traffic simulation example, the weather data stored by the binding system could be accessed by other parts of the application framework and adjust vehicle spacing due to altered stopping distances and visibility or increase the probability of an accident occurring in icy conditions.

Summary of the Application Kernel

The application kernel forms a collection of low-level functionality upon which further developments can be made. The task system provides an abstraction of the application loop, represents a systematic processing of runtime behavior, and allows users to inject additional behavior where required. At runtime, certain tasks may be polling a diverse range of input devices and data sources, and writing values to bound variables. Other tasks may be querying the value of variables that have been identified by name via the binding system's interface.

2.3 Simulation and Visualization Components

The subsystems described here build upon the functionality provided by the lower level framework subsystems in order to support the simulation and visualization of many different spatiotemporal environments. While the scene system maintains the topological and spatial representations of a given environment, the render system makes use of modern developments in rendering technologies in order to present a powerful yet flexible visualization pipeline.

Scene System

The scene system is responsible for maintaining the runtime state and content of a given spatiotemporal simulation. The scene system maintains two representations of the simulation environment and its constituent objects: a spatial partitioning system to maintain the spatial relationships between objects, and a scene graph to embody the high level topological aspect of the environment, which are shown in Figure 4. While the former representation will allow for efficient spatial queries such as proximity and collision detection, the latter representation makes heavy use of data driven programming (DDP), which introduces a further aspect of extensibility and flexibility to the framework's overall design. The application framework's scene system will use data driven objects to populate its simulated environments, which means that users will not only be able to stipulate scene composition using data, but will also be able to describe new types, and extend the definition of existing ones, via data manipulation.

The scene system describes objects as a composition of object components, or *facets*, as described in Table 2. Additional facet types may be supplied as *facet plug-ins*, which define subtypes of those presented in Table 2,

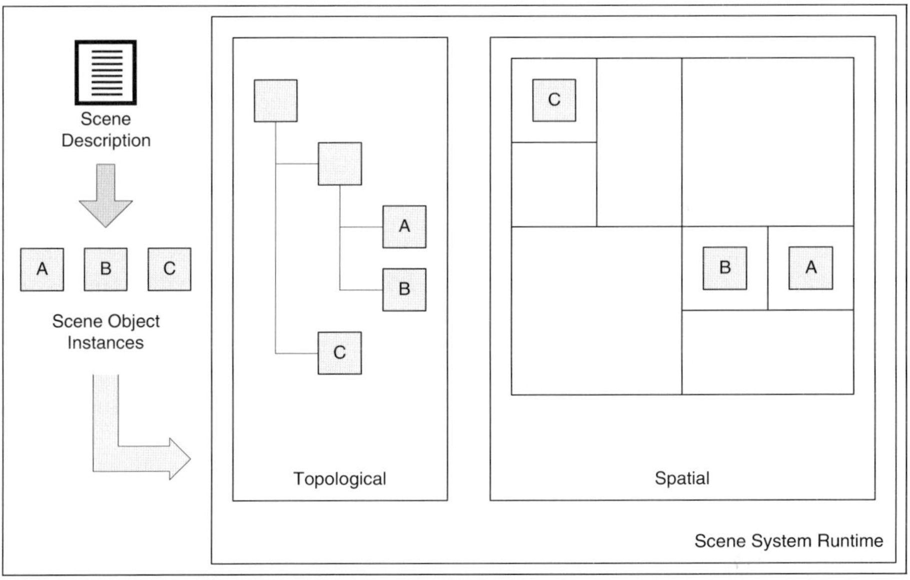

Fig. 4. An overview of the scene system's organization

allowing users to identify new ways in which to describe scene objects without having to develop or modify the scene system implementation. For example, a behavior facet plug-in could allow Python scripts to stipulate the runtime behavior of vehicles in the traffic simulation. In practice, scene object *types* will be defined by the user using XML. The object type definition will include an element for each facet that contributes towards the object's functionality, and facet plug-in developers will typically provide XML schema that can be used to validate descriptions for their facet types.

Object types can also form part of a data-driven object hierarchy, through the use of object type *inheritance*. When defining a new object type, users can also specify that the new object type is a subtype of an existing parent object type. Conforming to traditional object oriented software concepts, child object types *inherit* or *override* properties of their parent types. The inheritance scheme described here is applied at the facet level, so a child type description is free to override some behavioral parameters while inheriting others.

The benefits of DDP are increased extensibility and flexibility; the definition of new object types and behaviors when linked to a scripting language, as well as the modification and instantiation of existing ones, can all be achieved via data manipulation *without access to application code*. In the context of visualizing GIS information, the use of DPP enables the user to

Table 2. Scene Object Facet Types

Facet Type	Description
Data Facet	A data facet represents a collection of named variables whose initial values may be specified as part of a scene object description. For example, a car object may have an engine size, fuel level and registration associated with it.
Behavior Facet	A behavior facet's functionality is similar to that of a task plug-in, although a behavior facet also has access to the instance to which it belongs, along with that instance's constituent facets.
Bounding Volume Facet	A bounding volume facet simply specifies the spatial boundary of its owning scene object. For example, the bounding volume of a vehicle may be defined as an axis-aligned bounding box.
Scene Graph Facet	A scene graph facet represents its owning object's node in the scene system's hierarchical representation of the simulated environment.
Geometry Facet	A geometry facet stores the geometry associated with a given scene object, although this will typically take the form of a reference to a mesh asset file.
Appearance Facet	An appearance facet is used to determine visual appearance of its owning scene object. A scene object's appearance is described using nVidia's CgFx format (see later). For example, the appearance facet of a car object type may provide a CgFx fragment alongside default color parameters that together give all cars a glossy gray appearance.

assemble information rich virtual environments through the combination and extension of existing scene object type descriptions.

Render System

The render system is responsible for the visualization of the spatiotemporal simulation. While geometry and texture properties are supported by the framework's various asset and resource types, appearance properties are described using nVidia's Cg language and the CgFx effect framework. Recent developments in graphics hardware are now able to bring the rendering capabilities of even basic machines close to that of dedicated systems. While past incarnations of both hardware and software APIs have utilized a fixed functionality pipeline for transform and rasterization, today's hardware and software interfaces support a flexible programmable pipeline that exposes key functionality to the client. Programs written in a dedicated language, known as *shaders*, stipulate the appearance of objects in a given virtual scene by

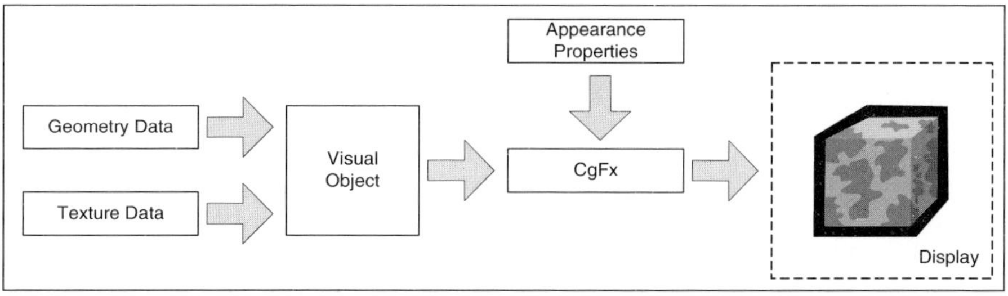

Fig. 5. An overview of the render system's functionality

specifying light, material, and surface characteristics alongside scene-wide effects such as shadows.

The rendering subsystem is based around the use of nVidia's CgFx effect framework [7], that aims to maximize flexibility without sacrificing run-time efficiency or ease of use. Improving image quality in the context of 3D visualization traditionally requires additional detail and accuracy, which in turn translates to increased geometrical and computational overheads, thus reducing application response and user interactivity [8]. A better method of improving the level of visual realism would be to focus on image-based techniques like bump mapping and shadow mapping [9, 10]; such techniques have demonstrated that additional detail can be produced via attribute maps and vertex and fragment manipulation [11].

Summary of the Simulation and Visualization Components

The scene and render systems together form a modeling and visualization environment that is capable of supporting a wide range of applications. The data driven implementation means the user is thus able to influence the modeling of a given simulation, and its visual output, via text-based modifications. In the traffic simulation example, users can describe a basic car type though a combination of facet parameterizations using XML. Within the render system users can provide a catalogue of vehicle geometry files and material properties, which are used in various combinations to illustrate an assortment of different vehicle types and color schemes.

2.4 Summary of the Running Example Application

Examples throughout this paper have demonstrated how elements of a real-time traffic simulation system could be implemented using specializations of, and extensions to, the application framework.

Section 2.1 explains how the application framework's data pipeline can be modified in order to support a given input format and distribution method. The modifications allow the traffic simulation to use a variety of data formats, including a GML based data. A specialization of the asset system allows the application framework to locate and access the resulting asset data, which provide application content and are used to drive runtime behavior.

Section 2.2 describes a number of alterations that allow users to define application behavior. Further examples show how the binding and input systems can be used to obtain and store data from real-time sources, such as a numerical weather prediction database, and use this data to influence the traffic simulation.

Section 2.3 gives examples of how a data-driven object and scene description system can be used to provide users with a flexible, extensible tool for defining the state, behavior and appearance of runtime objects.

3 Conclusions

In this paper we have shown the framework for a novel visualization system we are developing. Central to the design of the framework is careful attention to the ease with which the application can be extended or modified to suit particular visualization tasks. Through the design shown above we have been able to ensure that almost all parts of the system can be modified or extended, some using plug-ins, others through a data driven approach, including the use of fast scripting languages. Our aim is to create an open source base platform that can be extended by us, other members of the visualization programming community, or users of the system to address a range of requirements. More fundamentally we expect that a range of plug-ins for import of a range of data formats will be created, and possibly a range of plug-ins for driving specialist visualization hardware. The careful design of the system means that this can be achieved easily without any need to recompile or, for the data driven aspects, even code.

In future work we are looking at extending the application framework to add GIS functionality to create an integrated modeling and visualization package. We are also exploring the links that we can usefully make between GML3.1 and the data driven components in the scene system.

References

1. Herwig A, Paar P (2002) Game Engines: Tools for Landscape Visualization and Planning? Wichmann, pp 161–171
2. Fritsch D, Kada M (2004) Visualisation using game engines. In: Geo-Informations-Systeme, June, pp 32–36
3. Kot B, Wuensche B, Grundy J, Hosking J (2005) Information visualization utilising 3d computer game engines case study: a source code comprehension tool. In: CHINZ '05: Proc of the 6th ACM SIGCHI New Zealand Chapter's Int Conf on Computer-Human Interaction. ACM Press, New York, NY, USA, pp 53–60
4. Patterson S (2002) An object–composition game framework. In: Treglia D (ed) Game Programming Gems 3, ch 1.2. Charles River Media, pp 15–25
5. Appleton K, Lovett A, Sünnenberg G, Dockerty T (2002) Rural landscape visualisation from gis databases: a comparison of approaches, options and problems. Computers, Environment and Urban Systems 26:141–162
6. Shalloway A, Trott JR (2005) Design Patterns Explained: A New Perspective on Object-Oriented Design, 2nd ed. Addison-Wesley, London, p 429
7. Fernando R, Kilgard MJ (2003) The Cg Tutorial: The Definitive Guide to Programmable Real-Time Graphics. Addison-Wesley
8. Peercy MS, Olano M, Airey J, Ungar PJ (2000) Interactive multi-pass programmable shading. In: SIGGRAPH '00: Proc of the 27th Annual Conf on Computer Graphics and Interactive Techniques. ACM Press/Addison-Wesley Publishing Co, New York, NY, USA, pp 425–432
9. Wang J, Sun J (2004) Real-time bump mapped texture shading based-on hardware acceleration. In: VRCAI '04: Proc of the 2004 ACM SIGGRAPH Int Conf on Virtual Reality Continuum and its Applications in Industry. ACM Press, New York, NY, USA, pp 206–209
10. Stamminger M, Drettakis G (2002) Perspective shadow maps. In: SIGGRAPH'02: Proc of the 29th Annual Conf on Computer Graphics and Interactive Techniques. ACM Press, New York, NY, USA, pp 557–562
11. Claude AJ, Stevens M (2004) Leveraging high-quality software rendering effects in real-time applications. In: Fernando R (ed) GPU Gems, ch 35, 1st ed. Addison Wesley, Boston, MA, pp 581–599

A Tetrahedronized Irregular Network Based DBMS Approach for 3D Topographic Data Modeling

Friso Penninga[1)], Peter van Oosterom[1)], Baris M. Kazar[2)]

[1] Delft University of Technology, OTB, section GIS technology,
Jaffalaan 9, 2628 BX Delft, The Netherlands
email: F.Penninga@otb.tudelft.nl, oosterom@geo.tudelft.nl

[2] Oracle USA, One Oracle Drive, Nashua, NH 03062, USA
email: baris.kazar@oracle.com

Abstract

Topographic features such as physical objects become more complex due to increasing multiple land use. Increasing awareness of the importance of sustainable (urban) development leads to the need for 3D planning and analysis. As a result, topographic products need to be extended into the third dimension. In this paper, we developed a new topological 3D data model that relies on Poincaré algebra. The internal structure is based on a network of simplexes, which are well defined, and very suitable for keeping the 3D data set consistent. More complex 3D features are based on this simple structure and computed when needed. We describe an implementation of this 3D model on a commercial DBMS. We also show how a 2D visualizer can be extended to visualize these 3D objects.

1 Introduction

The 3D data models should enable 3D analysis, whereas early 3D GIS developments often focused on visualization, often in Virtual Reality-like environments. Another important characteristic of topographic data sets is the wide variety of applications, thus disabling optimization of the data model for a specific task. Due to current developments in sensor techniques (Vosselman 2005) more and more 3D data becomes available. Furthermore, the point density and thus data volume is increasing. An example of the capabilities of terrestrial laser scanning is illustrated in Figure1.

Fig. 1. Terrestrial laser scanning provides insight in complex 3D objects

There are a vast number of studies (amongst others Arens et al. 2005; Guibas et al. 1985; van der Most 2004; van Oosterom 1994, 1997, 2002, Penninga 2005; Verbree et al. 2005; Zlatanova 2000a, 2000b, 2002a) on 3D data modeling. Most of these studies are summarized and neatly compared in (Zlatanova 2002b). Extending topographic data models into 3D is most relevant at large scale topography. However, this will lead to a substantial increase of data volume. With this increase, ensuring data integrity and maintaining performance become important requirements. As a result, implementing the 3D data structure in a spatial database is a sensible thing to do. In this research, the Tetrahedronized Irregular Network (TEN) is selected as an internal data structure. The selection of this structure, motivated by computational advantages, the well-definedness of the triangles (always flat), the presence of well-known topological relationships (Guibas and Stolfi 1985), easy maintenance, visualization of triangles (Zlatanova 2002a, 2002b), and flexibility of forming more complex objects, is described by (Penninga 2005). Our model is based on the Poincaré algebra (and has therefore a solid foundation) and does pay attention to DBMS issues, such as indexing, updating and locking mechanisms.

In Section 2 an introduction to the theory behind our conceptual TEN-model is given with ingredients such as n-dimensional simplexes, boundary and coboundary, Poincaré algebra, network of simplexes (in 3D called the TEN). DBMS issues related to the 3D TEN-based modeling are introduced in Section 3, while Section 4 describes an (partial) implementation of the in the DBMS with a 'toy' example. The paper concludes with summarizing the most important results and indication on future research in Section 5.

2 3D Topographic Data Modeling in a TEN Data Structure

As we see topography as the collection of physical objects, two observations can be made regarding 3D topographic data modeling:

1. Physical world objects have by definition a volumetric shape. There are no such things as point, line or area features; only point, line and area representations at a certain level of generalization. Which representation to use should be stated in the DCM (Digital Cartographic Model) but not in the DLM (Digital Landscape Model), which contains our 3D topography?
2. The real world can be considered as a volume partition: a set of non-overlapping volumes that form a closed modeled space. As a consequence, objects like 'earth' or 'air' are explicitly part of the real world and thus have to be modeled.

As a result, the topographic data set consists of volume features. However, in some cases area features might be useful. Area features can be modeled in our approach to mark important boundaries between two volume features (and can have their own properties, such as surface material and color). Therefore, they cannot exist without the presence of these volume features; an area feature is the first derivative of a volume feature (and this is repeated for line features and point features). In the UML class diagrams in Figure 3 and Appendix A, these area features are modeled as association classes.

The decision to explicitly include 'air' and 'earth' features – thus modeling 'empty' space in between physical objects – is influenced by the fact that this empty space is subject of many analyses. In case of modeling air pollution or flooding, the user is interested in what happens in this empty space. The remainder of this section will discuss the Poincaré algebra (2.1) and the resulting conceptual TEN model in UML class diagram (2.2).

2.1 Poincaré Algebra

The Tetrahedronized Irregular Network (TEN) is the three-dimensional variant of the well-known Triangulated Irregular Network (TIN). Besides nodes, edges and triangles, a TEN also consists of tetrahedrons for representing volumetric shapes. Nodes, edges, triangles and tetrahedrons are all simplexes, i.e., the simplest possible geometry in every dimension. Modeling 3D features by the use of simplexes is described by Carlson (1987). Using simplexes has three advantages:

1. Simplexes are well-defined: a kD simplex is bounded by k+1 (k-1)D simplexes (Egenhofer et al. 1989a). For instance: a 2D simplex (triangle) is bounded by 3 1D simplexes (edges).
2. Flatness of the faces: every face can be described by three points.
3. Every simplex is convex, regardless of its dimension.

A direct result of the well-defined character of simplexes and thus of a TEN is the availability of 3D topological relationships. Whereas in the two-dimensional case, (the TIN) the important relationships are on edge level (i.e. an edge has a face on the left and one on the right, thus defining adjacency of faces), in three dimensions the important relationships are on face level. Each face (triangle) bounds two tetrahedrons. Left and right are meaningless in 3D, but due to the ordering of the edges in the triangle one can determine the direction of the normal vector and thus relate to tetrahedrons in the positive and negative direction. The n-dimensional simplex is defined by n+1 nodes and has the following notation $S_n = <x_0,\ldots,x_n>$.

So, the first four simplexes are $S_0 = <x_0>$, $S_1 = <x_0, x_1>$, $S_2 = <x_0, x_1, x_2>$, and $S_3 = <x_0, x_1, x_2, x_3>$. With (n+1) nodes, there are (n+1)! combinations of these nodes, that is for the four simplexes, there are respectively 1, 2, 6 and 24 options. For S_1 the two combinations are $<x_0, x_1>$ and $<x_1, x_0>$, of which the first one (from start to end) is called positive (+) and the other one negative (-), indicated as: $<x_0, x_1> = - <x_1, x_0>$. The two-dimensional simplex has six combinations S_2: $<x_0,x_1,x_2>$, $<x_1,x_2,x_0>$, $<x_2,x_0,x_1>$, $<x_2,x_1,x_0>$, $<x_0,x_2,x_1>$, and $<x_1,x_0,x_2>$. The first three have the opposite orientation from the last the three combinations, so one can state $<x_0,x_1,x_2> = - <x_2,x_1,x_0>$. The positive orientation is counter clockwise (+) and the negative orientation is clockwise (-). For the three-dimensional simplex $S_3 = <x_0, x_1, x_2, x_3>$ there are 24 different combinations of which 12 are related to positive oriented tetrahedrons (+, all normal vectors outside) and the other 12 are negative oriented tetrahedrons -, all normal vectors inside). As there are several equivalent notations (combinations), it is possible to agree on a preferred notation; e.g., the combination related to a positive orientation with the nodes with lowest id's (indices) first. According to the Poincaré algebra (Geoghegan 2005), the boundary of a simplex is defined by the following sum of (n-1) dimensional simplexes (omitting the ith node and with alternating + or – sign):

$$\partial S_n = \sum_{i=0}^{n} (-1)^i < x_0, \ldots, \hat{x}_i, \ldots, x_n > \qquad (1)$$

So, the boundary of $\partial S_1 = <x_0,x_1>$ is $<x_1>$ - $<x_0>$ and the boundary of $\partial S_1^{neg} = <x_1,x_0>$ would be $<x_0>$ - $<x_1>$. The boundary of $\partial S_2 = <x_0,x_1,x_2>$ is $<x_1,x_2>$ - $<x_0,x_2>$ + $<x_0,x_1>$. In a similar way, the boundaries related to the 5 other combinations of S_2 can be given. Finally, the boundary of $\partial S_3 = <x_0,x_1,x_2,x_3>$ is $<x_1,x_2,x_3>$ - $<x_0,x_2,x_3>$ + $<x_0,x_1,x_3>$ - $<x_0,x_1,x_2>$ (and the same for the other 23 combinations). Going to the boundaries of the boundaries of a tetrahedron, that is the boundary of the triangles (edges), it can be observed that every edge is exactly used once in the positive direction and once in the negative direction (within the tetrahedron). Another interesting result from the Poincaré algebra is the number of lower dimensional simplexes used as (in)direct boundary (face) of a given simplex:

$$S_n \text{ has } \binom{n+1}{p+1} \text{ faces of dimension } p \text{ with } (0 \le p < n) \qquad (2)$$

So, S_2 (triangle) has 3 0D simplexes (nodes) and 3 1D simplexes (edges) as boundary 'faces'. The simplex S_3 has respectively 4, 6 and 4 0D, 1D and 2D simplexes as boundary 'faces'. When neighbor simplexes of the same dimension are joined or merged, then their shared boundary is removed as shown in Figure 2. For example, take neighbor triangles $<x_0,x_1,x_2>$ and $<x_0,x_2,x_3>$ then adding the boundaries results in: $(<x_1,x_2> - <x_0,x_2> + <x_0,x_1>) + (<x_2,x_3> - <x_0,x_3> + <x_0,x_2>) = <x_1,x_2> + <x_0,x_1> + <x_2,x_3> - <x_0,x_3> = <x_1,x_2> + <x_0,x_1> + <x_2,x_3> + <x_3,x_0>$. Note that the shared boundary $<x_0,x_2>$ is removed. Similarly, when merging the two neighbor tetrahedrons $<x_0,x_1,x_2,x_3>$ and $<x_0,x_2,x_4,x_3>$, then adding the boundaries (triangles) results in $<x_1,x_2,x_3>$ + $<x_0,x_1,x_3>$ + $<x_2,x_1,x_0>$ + $<x_2,x_4,x_3>$ + $<x_3,x_4,x_0>$ + $<x_4,x_2,x_0>$. When looking at the edges again, then it can be observed that every edge is used once in the positive direction and once in the negative direction. A set of merged (joined) neighbor n-simplexes is called a simplicial complex (or n-cell). It is also possible to create a topological structure consisting of connected n-simplexes (with all their lower level boundaries: 0,...,n-1 simplexes) partitioning the whole n-dimensional domain. In 3D, this is then called the tetrahedronized network (TEN). In such a network, it is not only interesting to give the boundary of a simplex, but also to give the coboundary. For example, the boundary of a triangle is formed by the 3 edges and the coboundary is formed by the 2 tetrahedrons. Similarly, the boundary of an edge is formed by 2 nodes and the coboundary is formed by 2 or more triangles.

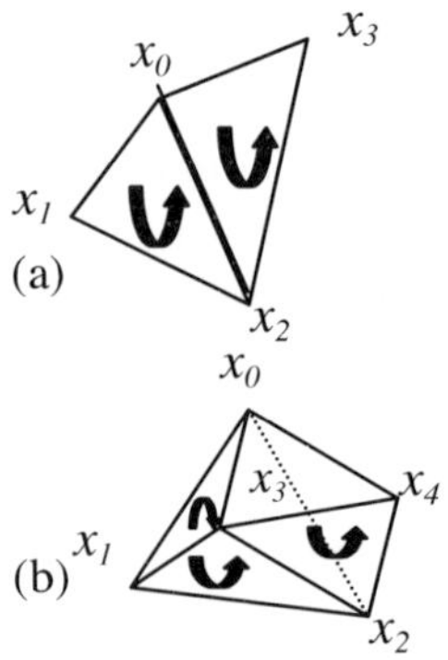

$$S_{21} =< x_0, x_1, x_2 > \text{ and } S_{22} =< x_0, x_2, x_3 >$$
$$C_2 =< x_1, x_2 > - < x_0, x_2 > + < x_0, x_1 >$$
$$+ < x_2, x_3 > + < x_0, x_2 > + < x_3, x_0 >$$
$$=< x_1, x_2 > + < x_0, x_1 > + < x_2, x_3 > + < x_3, x_0 >$$

$$S_{31} =< x_0, x_1, x_2, x_3 > \text{ and } S_{32} =< x_0, x_2, x_4, x_3 >$$
$$C_3 =< x_1, x_2, x_3 > - < x_0, x_2, x_3 > + < x_0, x_1, x_3 >$$
$$- < x_0, x_1, x_2 > + < x_2, x_4, x_3 > - < x_0, x_4, x_3 >$$
$$+ < x_0, x_2, x_3 > + < x_4, x_2, x_0 >$$
$$=< x_1, x_2, x_3 > + < x_0, x_1, x_3 > - < x_0, x_1, x_2 >$$
$$+ < x_2, x_4, x_3 > - < x_0, x_4, x_3 > + < x_4, x_2, x_0 >$$

Fig. 2. Adding simplex with a neighbor to form complex of simplex in **(a)** 2D and **(b)** 3D

2.2 3D Topography: TEN Based Conceptual Model

The closest data models to our model are implemented in the Panda system (Egenhofer et al. 1989b) and Oracle Spatial (Kothuri et al. 2004; Oracle 2005), both of which consider up to two-dimensional spaces. Panda is based on complexes, which are unions of neighbor simplexes also called n-cells. There is no attention called for the feature modeling as opposed to our system.

Although the topographic model is very strict in its way of handling only volume (and implied area, line and point) features, the actual TEN implementation will be more generic and also support true point, line and area features (which might be good representations at smaller scales for features that are in reality also volumes). For other cases, line and point features might be useful, too. Three different conceptual models (UML class diagrams) have been created. These are all more or less identical as they all capture the tetrahedronized network structure and the features embedded. However, there are already remarkable differences in these conceptual models. As these conceptual models form the start for the logical and technical models (implementation), this is an important issue. We will therefore present three different conceptual models (in UML class diagrams). The first one is given in Figure 3 as model 1.

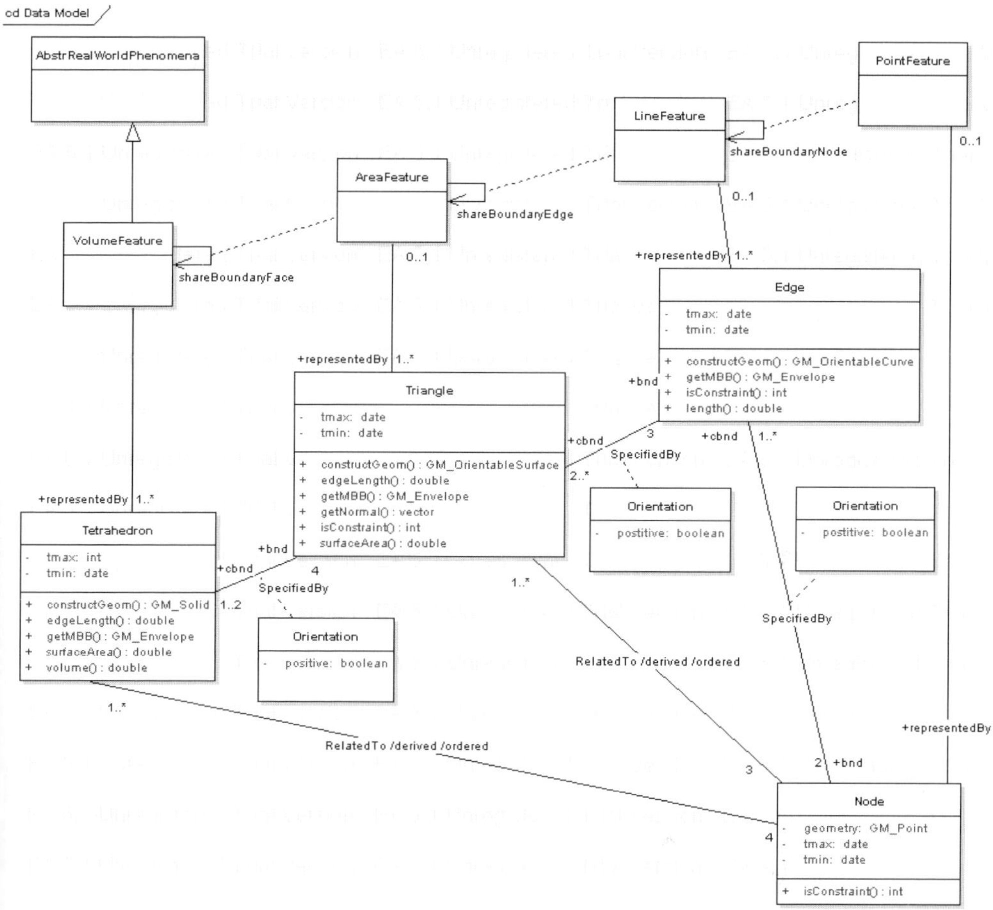

Fig. 3. UML class diagram of our first TEN model (i.e., model 1)

The primitives are directed (positive) and the boundary/coboundary associations between node and edge, edge and triangle, and triangle and tetrahedron are signed (indicated by the association class Orientation). An efficient implementation of this model will not explicitly include the association class Orientation, but will use signed (+/-) references to encode the orientation. The association between tetrahedron (or triangle) and node can be derived (and gives a correct ordering of the nodes within the primitive).

The second model (presented in Appendix A) is based on the first model, but instead of the association classes with explicit Orientation indicating the sign/direction (+/-), new undirected classes are created, which are the counterparts of their directed origins. The model may suggest that both positive and negative versions of the directed primitives are stored (as the undirected counterpart is a composition of a positive and negative directed primitive), but this is not the case: only the positive oriented primitives are stored. Similar to the first model, the association between tetrahedron (or triangle) and node can be derived and is again ordered.

Finally, the third model (again depicted in Appendix A): this is the conceptual model based on Poincaré formulas: direct associations from node to all three other primitive classes (edge, triangle, tetrahedron). The ordering of the nodes to define the primitives is important as it implies the orientation. Based on these associations, now the other boundary/coboundary associations can be derived (tetrahedron-triangle and triangle-edge), and it should be noted that these are again signed. The main differences between all these models are: (1) which associations are 'explicit' and which are derived and (2) in case of signed association (references), is this modeled with an association class or with an additional undirected primitive? Somehow, the third model seems to be the least redundant with respect to references. However, it is common use to explicitly model (and store) the references between a primitive and its boundary. Therefore we will continue in this paper with the first (or second) model, but the orientation is implemented via +/- sign in the references (and not via explicit classes).

3 Incorporating the TEN Structure in a Spatial DBMS

In this section the following issues will be further discussed: incremental update within the TEN feature model (3.1), primitive update functions (3.2), feature level updates (3.3), and storage requirements (3.4).

3.1 Implementing an Incremental Algorithm

All data are stored in a spatial database. Initially the database is empty, and via incremental updates is should be brought from one consistent state into the next consistent state. The most straightforward implementation of the TEN structure consists of four tables with nodes, edges, triangles and tetrahedrons and a table with volume features. If one wants to add a feature (for instance, a building), one needs to ensure correct representation in the TEN model by enforcing the boundary faces of this building to be present. As tetrahedronization algorithms can only handle constrained edges, the building's surface first needs to be triangulated. The resulting edges are the input for the building tetrahedronization, which is performed separately from the TEN network. The complete set of edges is then inserted as constraints into the TEN model by an incremental tetrahedronization algorithm. Note that this is one specific procedure and more efficient/direct procedures and associated algorithms could be imagined (though not so easy to realize). As a last step, the volume feature table needs to be updated. A new record is created which links the building to the representing tetrahedrons and the previous 'air' tetrahedrons on the specific location are removed.

If one wants to remove this building from the data set, for instance because it is demolished, the record from the volume feature table can be deleted. At the same time, the TEN needs to be updated. The constraints on the edges of the surface triangulation can be removed only if this building is the only feature that is bounded by this constrained edge. In the case of the demolished building, constrained edges on the building's floor also bound the earth surface and therefore needs to remain present in the TEN model. The tetrahedrons that were previously representing the building now need to be re-classified, in this case, most likely just as 'air'. This reclassification is necessary to maintain the volume partition. At this moment, the building is entirely removed from the model, both on TEN and on feature level, but the deletion process is not finished. As a last step, it is necessary to check whether the TEN can be simplified by creating larger tetrahedrons or can be optimized by creating better-shaped tetrahedrons (by flipping; 3.2). As an alternative, one might delete directly all edges that were part of the building, except for (constrained) edges that also contribute to the shape of other features. The resulting hole in the TEN needs to be retriangulated and the created tetrahedrons will be linked to the 'air' feature.

New (volume) features that are inserted take over the space of the existing features. This will be not a problem in case of the air and ground tetrahedrons. However, in case of tetrahedrons belonging to other types of features, the correctness of this occupation has to be checked (by the user) before committing. Further, it should be noted that most features are also (indirectly) connected to the earth surface, and also this has to be translated into constraints, which can be used to validate changes. Now that the update process is described, the algorithm requirements can be extracted. For creating and maintaining the TEN, an incremental algorithm is required. Due to the potential enormous amount of data, this incremental algorithm has to work in the database and should preferably impact the TEN structure as locally as possible. In the TEN, all simplexes should be available. As the tetrahedrons represent volume features, the triangles contain most topological relationships, the edges contain the constraints and the nodes contain the geometry. Another requirement is the need for numerical stability through detection and repair of ill-shaped triangles and tetrahedrons. Shewchuk has performed a lot of research (Shewchuk 1997; Shewchuk 2004) in the field of Delaunay mesh refinement in both 2D and 3D.

3.2 Basic Updating of the Topological Elements

The basic update procedures to modify an existing topology complex are described here, which are low level editing operations and are usually done in three steps: First, the user decides a window to be updated in a user session. A lock operator is executed after specifying the area of interest. The database will lock all of the features and topological elements overlapping the specified area of interest to prevent other users from updating the same area (Oracle 2005). Second, the user selects all of the required pieces of topology and features into memory and operates only on them. At this stage, new topological elements can be added or old ones can be removed as well. Third, all of the changes done in the session are committed back to the database. Next, we describe some of the basic functions available in the interface. The operations on the primitives consist of the following:

1. Move node (only allowed without destroying topology structure)
2. Insert node and incident edges/triangles/tetrahedrons (or the reverse operation 'remove node'), where 3 cases can be distinguished depending on where the node is inserted (see Fig. 4):
   - Middle of tetrahedron (one tetrahedron involved) and added are +1 node, +4 edges, +6 triangles, and +3 tetrahedrons (respectively the 0/1/2/3-simplexes).
   - Middle of triangle (2 tetrahedrons involved) and added are +1 node, +5 edges, +7 triangles, and +4 tetrahedrons.
   - Middle of edge (n tetrahedrons involved) and added are +1 node, $+(n+1)$ edges, $+2n$ triangles, $+n$ tetrahedrons.
3. Flipping of tetrahedrons, two cases, depending on configuration (see Figure 5):
   - 2-3 bistellar flip
   - 4-4 bistellar flip

The feature mapping when topology is updated can be described as follows: Since spatial features are defined on topological elements, it is very important to keep the integrity of spatial features even when updates are allowed on the underlying topological elements. The features define the constraint (nodes,) edges and triangles and care must be taken that these constraint simplexes are included within the TEN, in order to be able to represent the features. The other type of edges and triangles are introduced to make the TEN structure a 'complete' tetrahedronization (and during manipulation there is more freedom for these simplexes: flip, remove, etc.).

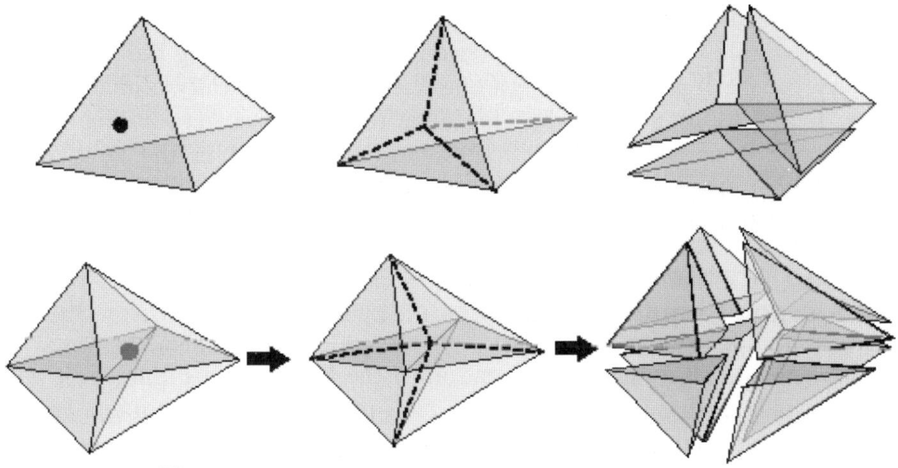

Fig. 4. Inserting a node: in triangle, neighbor tetrahedron not displayed) (above) and in edge with four incident tetrahedrons (below), both taken form (van der Most 2004)

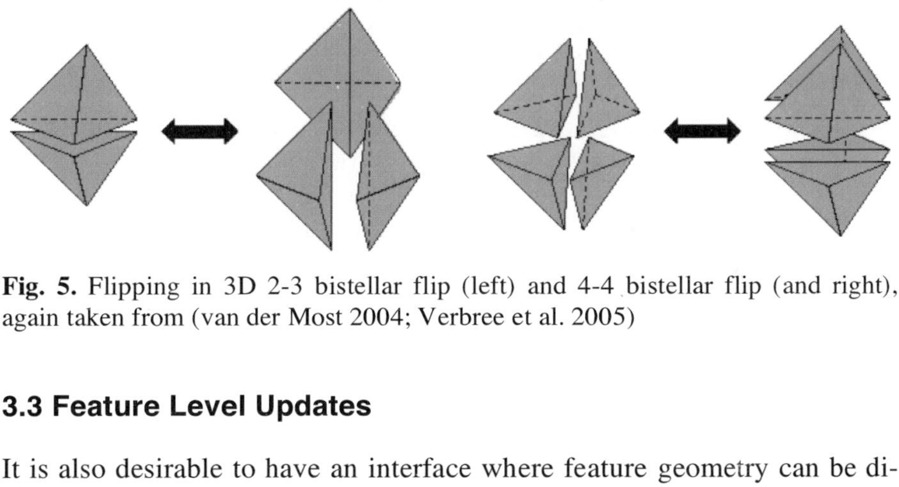

Fig. 5. Flipping in 3D 2-3 bistellar flip (left) and 4-4 bistellar flip (and right), again taken from (van der Most 2004; Verbree et al. 2005)

3.3 Feature Level Updates

It is also desirable to have an interface where feature geometry can be directly inserted into the topology complex returning a list of corresponding topological elements. This is more natural to users only dealing with the feature geometry and not necessarily caring about the actual storage model used for the geometry. In such cases, the interface can take the geometry as the input and insert into the topology complex. This operation will translate into a series of lower level topology update operations. At the end of this step, a list of primitives mapping to the input geometry are returned.

Feature creation from the primitives: In some cases, an existing feature needs a minor modification for some reason. For example, a segment of the road needs a small adjustment, which results in a new shape for the road. In simple feature model, this would result in updating the whole geometry for the road, even though one needs to change only a piece of the road. In the topological model, the edge (or edges) corresponding to the new shape is updated and the road automatically derives the new shape from its corresponding edges. In case of a 3D building, this could be a small extension to the back of a building.

3.4 Storage Requirements

If one considers the tetrahedronization of the building in Figure 6, it will be clear that storing the building in a TEN requires a lot of storage. In Table 1 the required number of tetrahedrons, triangles, edges and nodes is compared to the number of volumes, faces, edges and points in a polyhedron approach.

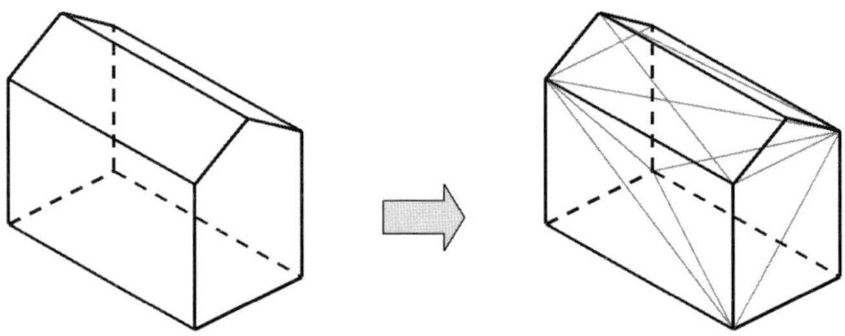

Fig. 6. Tetrahedronized building

Table 1. Comparison between polyhedron and TEN model of the building

Building as polyhedron	Building as TEN
(1 volume)	8 tetrahedrons
7 faces	24 triangles
(15 edges)	25 edges
(10 points)	10 nodes

In order to reach acceptable performance, it has to be decided which relationships (as modeled in the class diagrams in Figure 3 and Appendix A) will be stored explicitly. The performance requirements do not tolerate full storage of all possible relationships. Several approaches exist in 2D to reduce storage requirements of TINs by either working with an edge or a triangle based approach, in which not both triangles, edges and nodes are stored explicitly. However, in the 3D situation and in the case of constraints in the TEN this is very difficult.

4 Implementation: First Experiences

A small 'toy' data set is created by hand. It consists of an earth surface with a road on top and a single building with a saddle roof. This dataset was tetrahedronized by hand. In order to get 'air' and 'earth' tetrahedrons two extreme points were added, one on top and one at the fat bottom. Figure 7 shows the small data set, with the building and the road in front of it. This small data set, consisting of three volume features (building, air, earth), is composed by 56 tetrahedrons, 120 triangles, 83 edges and 20 nodes in Oracle Spatial (Kothuri et al. 2004).

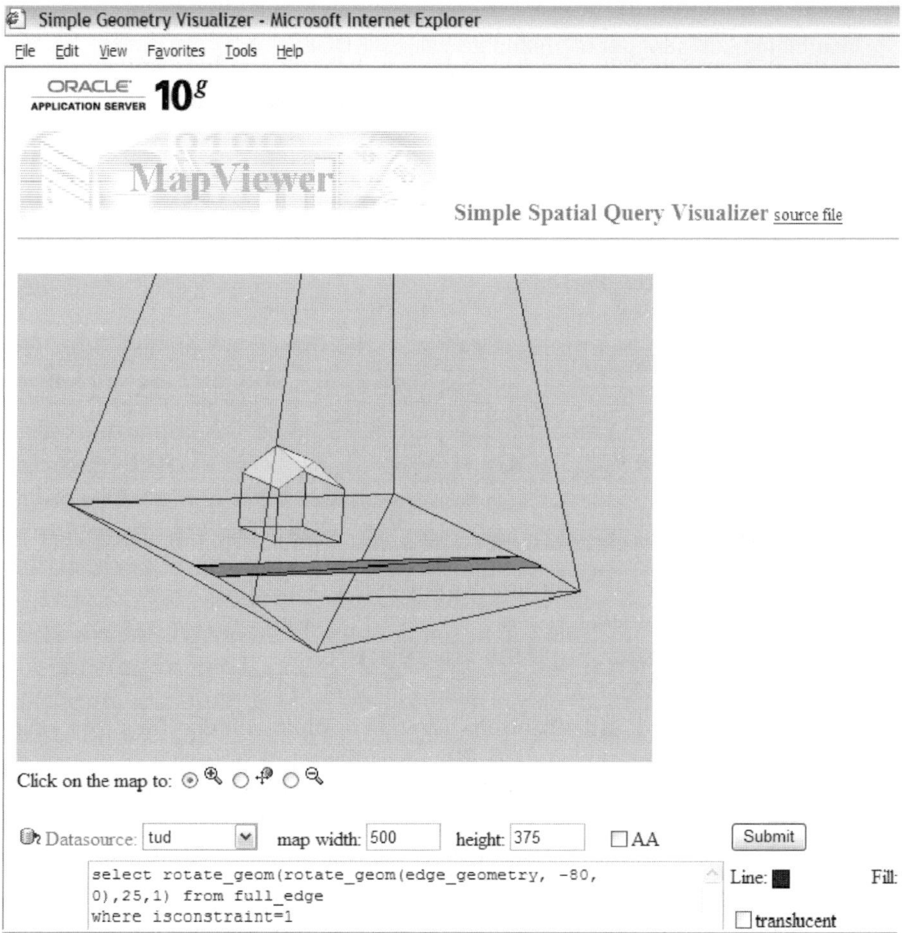

Fig. 7. Small test data set rotated '3D view'

The database tables (node, edge, triangle, tetrahedron) contain mainly references, geometry is only stored in the node table (note this is the physical model corresponding to conceptual model 1 in Fig. 3).

Functions are defined to obtain the geometry associated with the edges (get_edge_geometry) and the triangles (and test). These are then the counterparts of the 'constructGeom' methods in the conceptual UML models. One example will be given to obtain the geometry for the edge (including the definition of a view with geometry 'full_edge' and a functional spatial index on the 3D geometry of the edge):

```
create view full_edge as
select a.*, get_edge_geometry(eid) edge_geometry from edge a;

insert into user_sdo_geom_metadata values('EDGE',
 'TUD.GET_EDGE_GEOMETRY(EID)',
 sdo_dim_array(sdo_dim_element('X', -100000, 100000, 0.05),
 sdo_dim_element('Y', -100000, 100000, 0.05),
 sdo_dim_element('Z', -100000, 100000, 0.05)), null);

drop index edge_Sidx;
create index edge_sidx on edge(get_edge_geometry(eid))
indextype is mdsys.spatial_index
parameters('sdo_indx_dims=3')
```

The realization of a simple 3D viewer was based on an available 2D viewer i.e., Oracle AS MapViewer (Kothuri et al. 2004) and the implementation of one 3D rotation function 'rotate_geom' (suitable for any type of Oracle spatial geometry). Further by depth sorting (after rotation) also hidden line hidden surface may be obtained (painter algorithm). Further nice features are semi-transparency. Further simple improvements could be the development of a GUI that defines the rotation angles for a set of views.

5 Conclusions and Further Research

Extending topographic data models into 3D is most relevant at large scale topography. However, this will lead to a substantial increase of data volume. With this increase, ensuring data integrity and maintaining performance become important requirements. As a result, implementing the 3D data structure in a spatial database is a sensible thing to do. In this paper, we developed a new topological 3D data model that relies on Poincaré algebra. We described an implementation of this 3D model on a commercial DBMS. We also showed how a 2D visualizer can be extended to visualize these 3D objects. In this research, the Tetrahedronized Irregular Network (TEN) is selected as an internal data structure. The selection of this structure, motivated by computational advantages, the well-definedness of the triangles (always flat), the presence of well-known topological relationships, easy maintenance, visualization of triangles, and flexibility of forming more complex objects. As future work, we will work on temporal topology where there are no holes and overlap in time.

Alternative physical models than the one presented in Section 4 (based on conceptual model 1) are possible; for example the direct specification of the nodes of a tetrahedron and views for edges and triangles. Issues to be kept in mind are: 1. What to do with the 'isconstraint' attributes? (these attributes might also be part of the view and could be computed). 2. Would these views be efficient enough for manipulation (updating and querying)? This is very difficult to estimate upfront. Future work will consist of experiments needed to discover these aspects. The well-definedness of the TEN model comes with a prize of a large number of (conceptual) simplexes, therefore it is important to investigate in detail the actual size of tables and indices and evaluate what to store explicitly and what to derive (and present as a view).

Acknowledgements

This publication is the result of the Bsik Space for Geo-information '3D Topography' research project. Thanks to John Herring, Siva Ravada, Ravi Kothuri and Han Wammes for their constructive discussions on 3D TEN modeling, which formed the basis of this paper. Their ideas and comments were of great help, but all (remaining) errors are the sole fault of the authors. Further, Friso Penninga and Peter van Oosterom participate in the research program 'Sustainable Urban Areas' (SUA) carried out by Delft University of Technology.

References

Arens C, Stoter J, van Oosterom P (2005) Modeling 3D spatial objects in a geo-DBMS using a 3D primitive. Computers & Geosciences 31(2):165–177

Carlson E (1987) Three-dimensional (3D) modeling in a geographical database. In: Auto-Carto 8, pp 336–345

Egenhofer MJ, Frank AU, Jackson JP (1989a) A topological data model for spatial databases. In: Design and Implementation of Large Spatial (= Lecture Notes in Computer Science – LNCS 409), pp 271–286

Egenhofer M, Frank AU (1989b) PANDA: An Extensible DBMS Supporting Object-Oriented Software Techniques Database Systems in Office, Engineering, and Science, Zurich, Switzerland. In: Harder T (ed) Informatik Fachberichte vol 204. Springer-Verlag, pp 74–79

Geoghegan R (2005) Topological Methods in Group Theory; to appear in Springer

Guibas L, Stolfi J (1985) Primitives for the manipulation of general subdivisions and the computation of voronoi diagrams. ACM Transactions on Graphics 4(2):74–123

Kothuri R, Godfrind A, Beinat E (2004) Pro Oracle Spatial: The essential guide to developing spatially enabled business applications, Apress

van der Most A (2004) An algorithm for overlaying 3D features using a tetrahedral network. Master's Thesis, TU Delft, 96 p

van Oosterom P, Vertegaal W, van Hekken M, Vijlbrief T (1994) Integrated 3D Modeling within a GIS. Int Workshop on Advanced Geographic Data Modeling:80–95

van Oosterom P (1997) Maintaining Consistent Topology including Historical Data in a Large Spatial Database. In: Proc Auto-Carto 13, Seattle WA, 8-10 April 1997, pp 327–336

van Oosterom P, Stoter J, Quak W, Zlatanova S (2002) The balance between topology and geometry, In: Richardson D, van Oosterom P (eds) Advances in Spatial Data Handling, 10th Int Symp on Spatial Data Handling, pp 209–224

Oracle Spatial Topology and Network Data Models (2005) http://www.oracle.com/technology/documentation/spatial.html

Penninga F (2005) 3D Topographic data modeling: Why rigidity is preferable to pragmatism. In: Spatial Information Theory, Cosit 2005 (= Lecture Notes in Computer Science – LNCS 3693), pp 409–425

Shewchuk JR (1997) Delaunay refinement mesh generation. PhD Thesis, Carnegie Mellon University

Shewchuk J (2004) General-Dimensional Constrained Delaunay and Constrained Regular Triangulations I: Combinatorial Properties. To appear in: Discrete & Computational Geometry. Available at: http://www-2.cs.cmu.edu/jrs

Verbree E, van der Most A, Quak W, van Oosterom P (2005) Towards a 3D Feature Overlay through a Tetrahedral Mesh Data Structure. Cartography and Geographic Information Science 32(4):303–314(12)

Vosselman G (2005) Sensing Geo-information, Inaugural address, ITC Enschede

Zlatanova S (2000a) On 3D Topological Relationships. In: Int Workshop on Database and Expert System Applications, pp 913–919

Zlatanova S (2000b) 3D GIS for urban development. PhD Thesis, Graz University of Technology

Zlatanova S, Rahman AA, Pilouk M (2002a) 3D GIS: current status and perspectives. In: Proc of the Joint Conf on Geo-Spatial Theory, Processing and Applications, 8-12 July, Ottawa, Canada, 6 p

Zlatanova S, Rahman AA, Shi W (2002b) Topology for 3D spatial objects. In: IntSymp and Exhibition on Geoinformation 2002, 22-24 October, Kuala Lumpur, Malaysia, 7 p

Appendix A

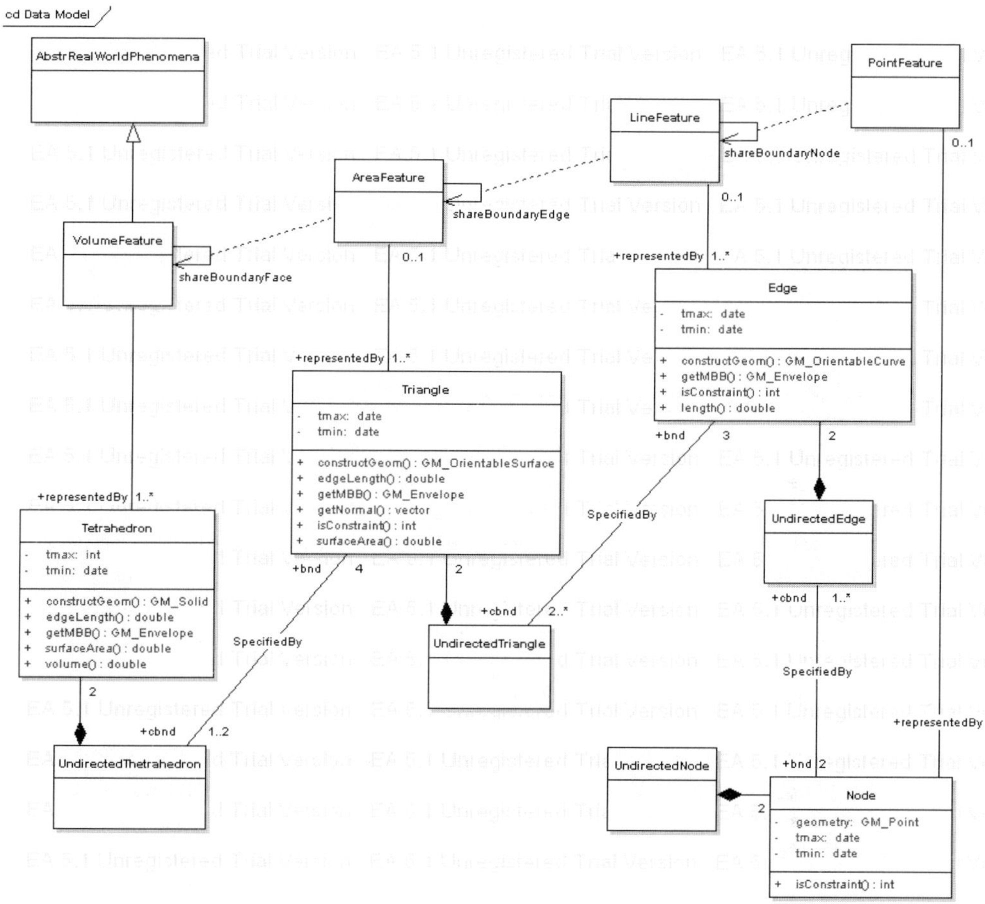

UML class diagram of our second model

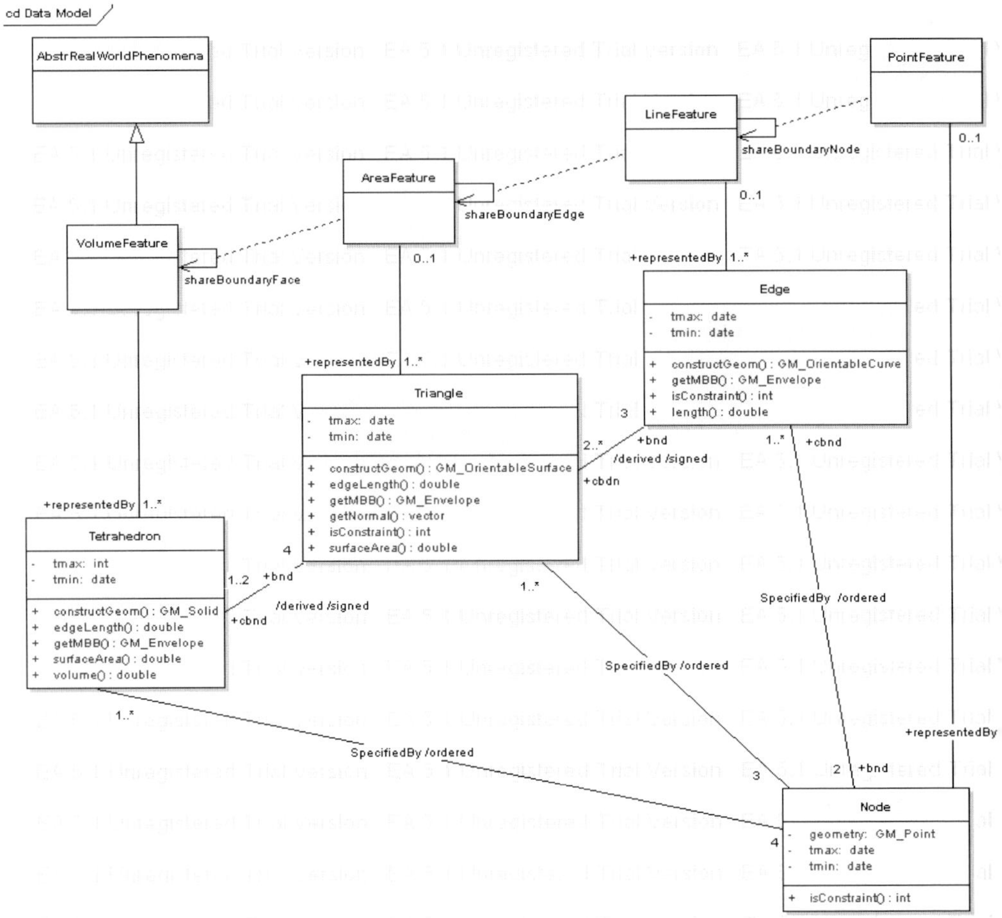

UML class diagram of our third TEN model (based on Poincaré)

3D Analysis with High-Level Primitives: A Crystallographic Approach

Benoit Poupeau, Olivier Bonin

IGN / COGIT, 2-4 avenue Pasteur, F-94165 Saint-Mandé CEDEX France
email: {benoit.poupeau,olivier.bonin}@ign.fr

Abstract

This paper introduces a new approach to 3D handling of geographical information in the context of risk analysis. We propose to combine several geometrical and topological models for 3D data to take advantage from their respective capabilities. Besides, we adapt from crystallography a high-level description of geographical features that enables to compute several metric and cardinal relations, such as the "lay on" relation, which plays a key-role for geographical information.

1 Introduction

3D GIS are relevant for many applications such as geosciences, urban planning, estate-market or telecommunication (Stoter and Ploeger 2002). However, their usage is still impeded by severe drawbacks (Zlatanova et al. 2002). For instance, natural hazard analysis requires advanced functionalities in terms of spatial and temporal modeling, visualization and spatial analysis because the phenomena of interest (e.g. Katrina storm in August 2005, Kashmir earthquake in October 2005) are peculiarly complex. 3D analysis of these phenomena shall be helpful to understand, manage and sometimes forecast hazardous events.

3D visualization is the first by-product of 3D GIS. It plays an essential role to deliver information, and is the basis for most negotiations, policies and decision-makings concerning risk. Part of the power of 3D visual analysis is due to the fact that 3D data is closer to our perception of reality than the usual 2D cartographic data.

Moreover, the third dimension makes room for a semantic enrichment of geographical data. For example, 2D buildings are simple polygons, whereas 3D buildings can be subdivided into walls, levels, roofs and even details such as gutters, rooftops and chimneys. These objects may play different roles when exposed to some hazard.

As geometrical modeling is concerned, several models can be found in the literature (see de la Losa 2000; de Cambray 1994; Foley 1990; Requicha 1980 for reviews of these models). Recent works focus either on a simplified Boundary Representation (BRep) or on spatial enumeration by tetrahedra (Penninga 2005). Let use emphasize on the fact that visualization is always performed by simplicial complexes (i.e. a BRep composed of points, segments and triangles). In the BRep approach, geometric primitives are the point (0-dimension), the line string (1-dimension) and the polygon (2-dimension). Bodies (3-dimensional objects) are defined by their boundaries composed of 2-dimensional objects. Topological structures are classically associated to these geometric models (Ramos 2003; Zlatanova 2000; de la Losa 2000; Pilouk 1996; Molennar 1990; Carlson 1987). The usual model of simplicial complexes limits the possible shapes of 2D and 3D primitives, and associates these primitives to an n-GMap (a graph with several edges between vertices and with faces defined by cycles of edges). As an alternative to this model, the cell-tuple model (Mesgari 2000; Pigot 1995; Brisson 1990) allows a more generic topological description at the expense of increased complexity and data volume. Globally, these approaches stay close to the usual 2D models, with the exception of the spatial enumeration model that suffers from the same drawbacks as the raster model in 2D (numerous elements and weak spatial structure).

While efficient for data storage and visualization, classical 3D models are not very adequate for complex analyses. To illustrate this point, let us examine the case of an underground collapse (for example, catacombs or old quarries of Lutetian limestone in the surrounds of Paris). The analysis of such an event requires to model the underground geological bends, the cavity itself and its pillars, and the soil and buildings atop the risk area. It can range from simple volume calculations to actual event simulation. Thus, a 3D GIS should enable to compute volumes, to store superimposition relationships and to export data towards geosciences simulation software. It is noteworthy that the superimposition relationship ("lay on" after de Cambray, 1994) plays a key role in landslide phenomena. Making this relation explicit enables to propagate a collapse to the objects on the surface of the ground. In this way, territorial diagnosis (evaluation of geographical objects at stake) becomes achievable.

Other relations than superimposition are useful for natural hazards analysis. In case of lava flow, the exposition of building walls to the lava can be described by adjacency relationships.

Both examples illustrate the interest of a description level above the geometrical level. The very simple description level introduced by our examples (*above*, *below*, *next to*, *close to*, *aligned with*) bears some resemblances with topology. It could be qualified to be a high-level description, by opposition to the geometric level.

This paper introduces a new model to perform this high-level description of 3D data in the context of risk analysis. This model is adapted from crystallography.

Geographical science is not the only field to tackle with 3D data. Crystallography is the science of solid characterization. In crystallography, solids are crystals. Each crystal is characterized by its symmetry properties. This characterization enables to determine which crystallographic system the crystal belongs to. A crystal is analyzed as a network of meshes. Each mesh is characterized by its symmetry properties, and its faces are indexed in the *Miller* system.

This approach can be transposed to the geographical field, as many anthropic 3D objects are regular enough to be though of as crystals. It can also be extended to irregular objects at the expense of shape simplification. As an example, a house in a L-shape can be thought of as the association of four elementary meshes of two different geographical natures. This requires subdividing objects into strictly convex bodies. Then, the *Miller* indexation of faces enables to extract superior, inferior and lateral parts of objects, and thus simplifies the computation of high-level description.

The use of crystallographic primitives to analyze 3D geographical objects relies on geometrical and topological structures. Thus, this approach is complementary to existing works on 3D modeling, and provides a high-level description of the shape and the relative position of objects.

2 A Plea for the Third Dimension in GIS

This first section argues for 3D geographical information in the context of risk studies. Compared to 2D data, 3D allows among others:

- Realistic visualization of objects. This topic is largely studied, as visualization is a privileged medium for analysis, communication (ArcGIS) and decision-making. It is widely spread among people responsible for town and country planning (see Stoter and Ploeger 2002; Zlatanova 2000 among others). It benefits from the development of games and vir-

tual reality systems. 3D visualization is less abstract than the usual cartographic representation. Let us note however that it lacks part of the objectivity of maps;
- Volume calculation;
- View-shed determination demanded by telecommunication firms.

Beyond these characteristics, 3D offers several interesting capabilities. The first one is a semantic enrichment of objects. For instance, a building in 2D is represented by a polygon that is the projection of all the constituting elements of this building. It is impossible to make a distinction between the roof and the floors as those features have the same 2D spatial extent (see Fig. 1). A full 3D representation allows the modeling of as many details as required by the analysis. Moreover, vertical faces can act the role of obstacles for wave propagation (e.g. noise, water, lava). The second one is the explicit notion of superimposition (or tiling) of objects, according to verticality. This allows relating objects with differences in thematic, nature or geometric dimension. Figure 2 illustrates a building (3D) onto a terrain model (2D) which itself relies onto geological bends. This tiling relationship (named "lay on" by (de Cambray 1994)) enables to determine which objects are concerned by an underground phenomenon.

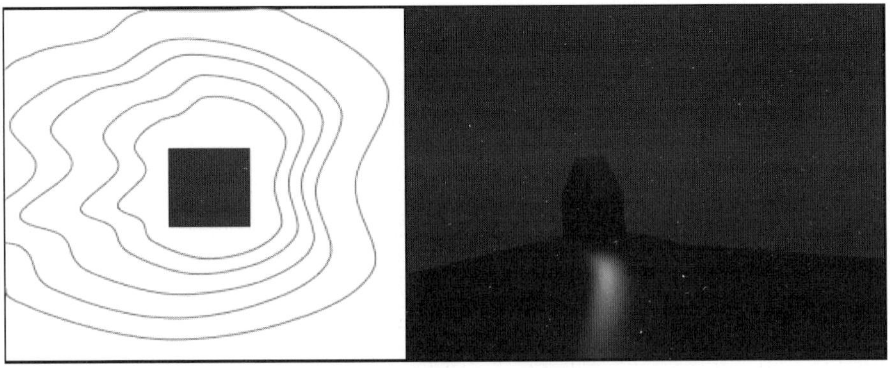

Fig. 1. 2D and 3D representation of the same scene

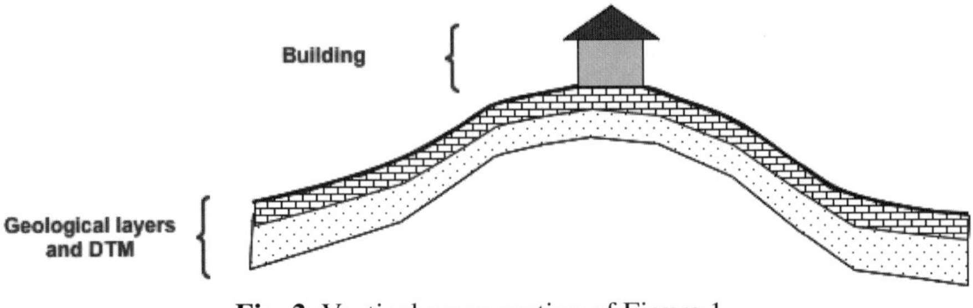

Fig. 2. Vertical cross-section of Figure 1

A third capability of 3D geographical information is the refinement of lateral relationships between objects according to altitude. Figure 3 illustrates in 3D and 2D two buildings connected by a footbridge. In 2D, a line connecting polygons represents this footbridge. With this representation, it is impossible to determine which levels are connected. In a 3D representation, this piece of information can be recorded.

Fig. 3. 3D modeling enables the refinement of lateral relationships

The capabilities of 3D geographical information are relevant for many natural hazard analyses. For example, when studying a cave collapse, vertical relationships help to focus on the geological formations and the urban area at stake. Thus, it is possible to model the propagation of the phenomenon and to calculate the volume of the collapsed area.

In case of a flow (e.g. lava, water, mud), the hazardous phenomenon is generally modeled in 2D, because the flow itself is on the surface of the ground. However, the determination of the impact of obstacles on the path of this flow (building walls, rocks) requires a 3D model for geographical objects and phenomena. This enables to refine the flow simulation, and to assess the vulnerability of the area.

This short and partial enumeration of the advantages of 3D data over 2D for risk assessment enlightens the need for 3D models allowing visualization, analysis and simulation. Moreover, synthetic relations such as "lay on" seem useful to perform many analyses. The remaining of this paper reviews existing 3D GIS models and proposes, on the basis of classical models, a high-level analysis of geographical objects deriving from crystallography.

3 Striving for 3D Models Suited to Geographical Information Analysis

Several authors have addressed the problem of 3D geographical information modeling (Ramos 2003; Billen 2002; de La Losa 2000; Zlatanova 2000; Pilouk 1996; Molennar 1990). Most models are both geometrical and topological models. In the sequel, this distinction will be omitted.

Unfortunately, there is more than one candidate model for geometry, and each candidate has forces and weaknesses. Moreover, no model is the obvious extension of the 2D GIS model (points, lines and surfaces). Topology in 3D can be peculiarly difficult to describe (de la Losa 2000; Pigot 1995; Brisson 1990; Mesgari 2000), and data volume must be taken into account for practical applications (Zlatanova 2000). Topological and metric queries are sometimes difficult to answer.

Basically, 3D information can be modeled in three ways (see Fig. 4):

1. Spatial enumeration (e.g. voxels, octrees and tetraedrons);
2. Constructive Solid Geometry (CSG) where Boolean operations are applied to predefined solids (e.g. sphere, cube, and cylinder);
3. Boundary Representation (BRep) where solids are defined by their bounding surfaces.

Some criteria can be used to assess the relevance of a model for a specific data set (Requicha 1980; Breunig 1996; Bernard 2000). Comparative studies of these models (Billen 2002; de la Losa 2000; de Cambray 1994) conclude that the BRep model is the most suitable candidate for 3D GIS. Spatial enumeration suffers from high data volume and weak data structure, whereas CSG is best used for engineering.

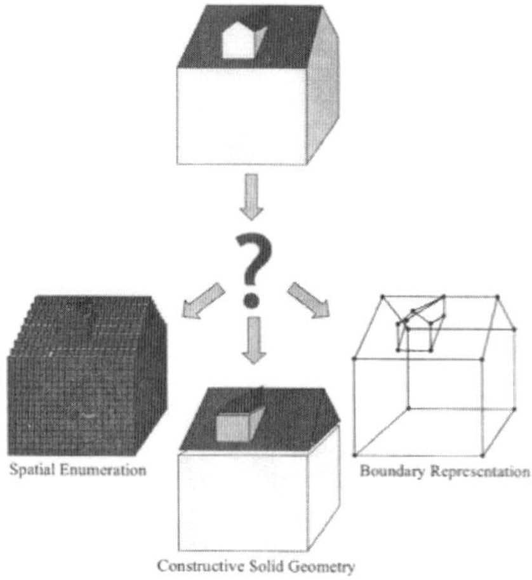

Fig. 4. Model candidates for 3D GIS (after Pfund 01)

The BRep model handles surfaces, often restricted to polygons, and thus data are easy to acquire. Moreover, these surfaces are required for visualization: graphics cards display at best surfaces composed of triangles, because they use the model of simplicial complexes.

Very shortly, a simplicial complex is a composition of simplexes. In case of 3D data, only three simplexes are used: the point (0-simplex), the segment (1-simplex) and the triangle (2-simplex). To keep the models simple and efficient, BReps for geographical information are often restricted to handle only planar surfaces, to be close to simplicial complexes (Coors 2002; de la Losa 2000; Zlatanova 2000; Pilouk 1996; Molenaar 1990; Carlson 1987).

To preserve coherence in 3D data sets, topological models are associated to the geometrical models Figure 5 illustrates an example of such a model. With a BRep relying on simplicial complexes, the natural topological model is the model of n-GMaps. A n-GMap is a graph where more than one edge can link two nodes, and where cycles of edges define faces. Some authors also consider that cycles of faces define bodies. However, things are much more complicated in 3D than the 2D topological map, as bodies can have holes or degenerate topology (e.g. the Klein bottle). Authors tackle these problems by restricting hypothesis or at the expense of additional primitives (de la Losa 2000).

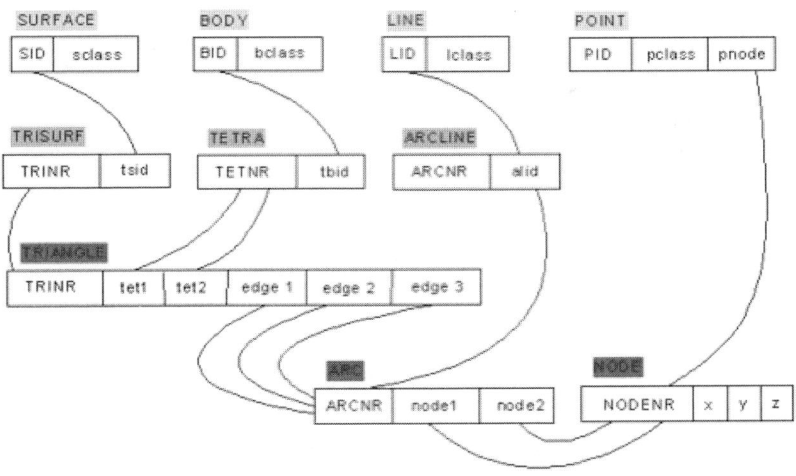

Fig. 5. Example of geometrical and topological model involving simplexes (after Pilouk 1996)

When designing the geometrical model the other way, i.e. deriving the geometrical primitives from the topological model, the model of cell-tuples has been successfully adapted to the case of geographical data

(Mesgari 2000; Pigot 1995; Brisson 1990). This ensures that no ambiguity can be introduced in the data set and that a full topological description of data is available. Unfortunately, this model suffers from data volume, and also resorts to simplexes for implementation.

This short review of literature reveals that the perfect model is yet to be discovered. A weakly structured BRep model with simplification hypothesis seems to most sensible way to go. Such a model does not allow for complex analysis or extensive data coherence assessment, but suits well to visualization and data storage. Further analysis can be performed with the help of a change in representation (a Delaunay tetraedrization of objects for instance), and a higher-level description of objects such as the one from crystallography we introduce in this paper.

In our 3D GIS system, objects are acquired as a weakly structured BRep. These objects are stored in a BRep model derived from (Zlatanova 2000). This model is used for visualization, data storage and data exchange (e.g. web applications). Tetrahedrization of objects is performed on demand to allow basic queries (e.g. volume calculations, topological queries), and the crystallographic description of objects provides a high-level description of geographical information (tiling, alignments, shape analysis and semantic compositions). The next section presents the crystallographic description of 3D geographical objects. The last section reports the progress of our implementation, and gives an outlook of this work.

4 Object Description with Crystallography

The review of 3D models shows that whatever the completeness of a model, complex analyses are impeded by the limits of the geometrical and topological modeling. We argue that rather than improving these models, a complementary description of geographical data can gain access to high-level properties of the objects. This complementary description is taken from crystallography.

The science of crystallography studies the nature of crystals according to their geometrical properties, especially symmetry. This discipline analyses crystals at a higher level than the geometric level. According to its symmetry properties, any crystal can be classified into on of the seven crystallographic systems (see Table 1), leading up to 230 different shapes.

A crystal is defined as a "natural homogeneous solid bounded by surfaces (usually planar) forming well-defined angles" (Pomerol et al. 2003). This definition could be freely applied to most geographical objects (see Fig. 6).

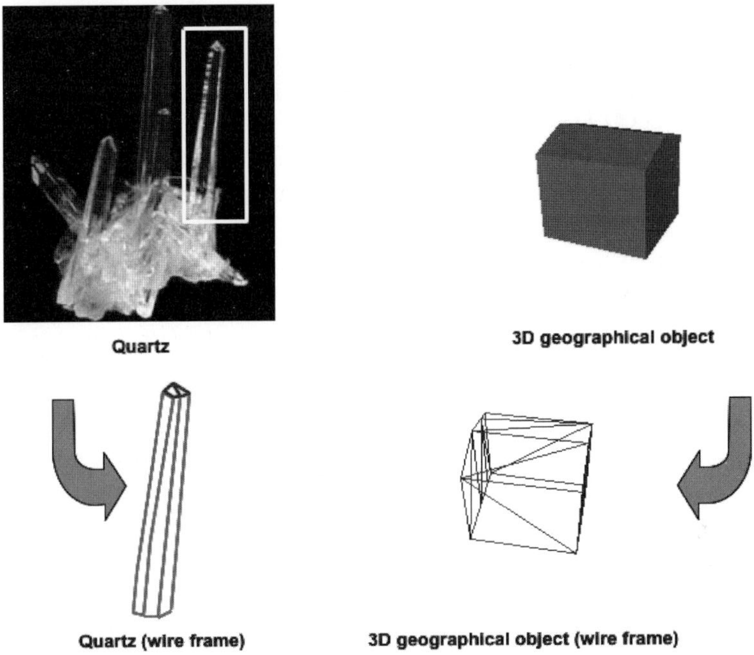

Fig. 6. Comparison of a crystal and a 3D geographical object

4.1 Some Notions of Crystallography

As a result from its internal structure, a crystal is characterized by the geometrical and topological arrangement of its constituting elements. It is composed of elementary meshes, defined as the smallest bounding boxes of the elements for which the geometrical, physical and chemical properties are preserved. As crystals are networks of identical meshes, the characterization of one mesh is sufficient to classify each crystal.

The geometry of a mesh is defined by three vectors (a local coordinate system Ox, Oy and Oz), three length (a, b and c) and three angles (α, β, γ). In that local coordinate system, each plane of the mesh can be characterized by a simple system of indices introduced by Miller (see Fig. 7). The Miller indexes of the face in Figure 7 are (h, k, l) if the equation of plane ABC is $hx + ky + lz = N$, where N is some constant.

The analysis of the symmetry of a mesh is performed in terms of punctual, axial and planar symmetries (see Fig. 8). The results of this analysis enable the classification of crystals into one of the seven crystallographic systems (see Table 1).

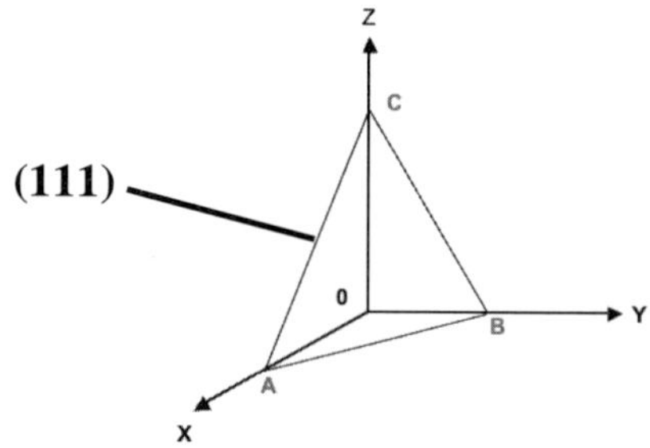

Fig. 7. Miller indexation of plane ABC

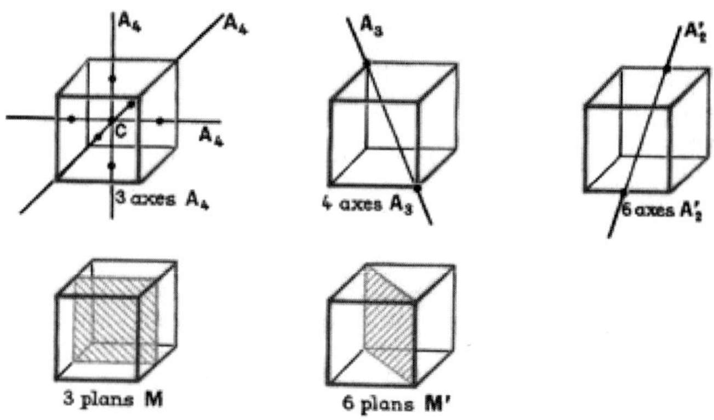

Fig. 8. Symmetry elements of a cube (after Pomerol 2003)

Table 1. The seven crystallographic systems (after Pomerol 2003)

	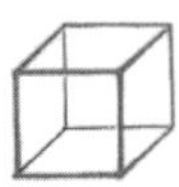	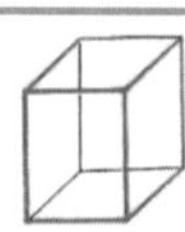	
Name	Isometric	Tetragonal	Hexagonal
Length	$a = b = c$	$a = b \neq c$	$a = b \neq c$
Angles	$\alpha = \beta = \gamma = 90°$	$\alpha = \beta = \gamma = 90°$	$\alpha = \beta = 90$ $\gamma = 120°$
Symmetry elements	C, 3A4, 4A3, 6A2, 9M	C, 1A4, 4A2, 5M	C, 1A6, 6A2, 7M

	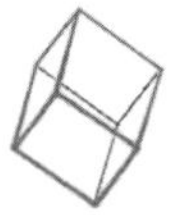	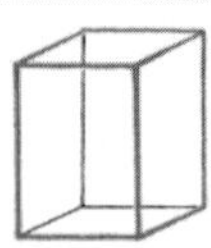	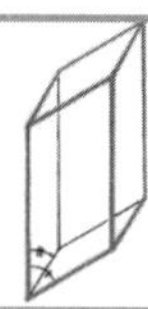	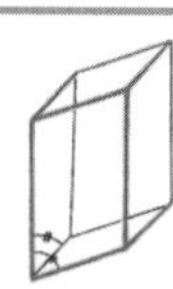
Name	Trigonal	Orthorhombic	Monoclinic	Triclinic
Length	$a = b \neq c$	$a \neq b \neq c$	$a \neq b \neq c$	$a \neq b \neq c$
Angles	$\alpha \neq \beta \neq \gamma$	$\alpha = \beta = \gamma = 90°$	$\alpha = \gamma = 90° \neq \beta$	$\alpha \neq \beta \neq \gamma$
Symmetry elements	C, 1A3, 3A2, 3M	C, 3A2, 3M	1A2, C, 1M	C

While the similarity between geographical objects and crystals is noteworthy, some of the crystallographic systems will likely not be relevant. Let us recall that those seven elementary forms generate 32 simple crystallographic forms (for a total of 230 possible forms), by parameter variation and loss of symmetry (see Fig. 9). The composition of different crystallographic forms leads to macles (when crystals are of the same nature) or epitaxies.

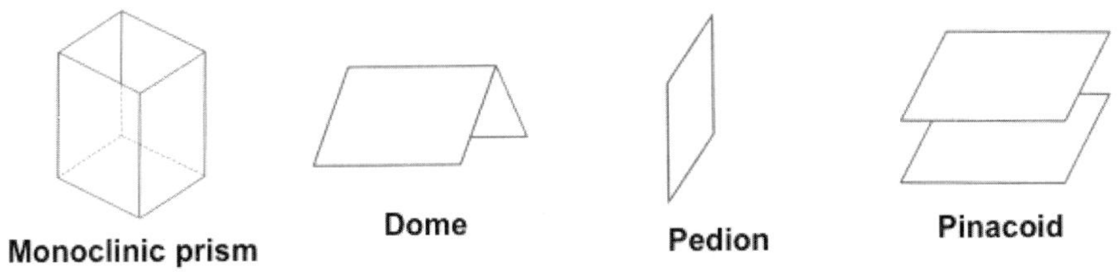

Fig. 9. A few crystallographic forms belonging to the monoclinic system

4.2 Interest of the Crystallographic Model for 3D Geographical Objects

The benefits of the crystallographic approach are twofold. First, the crystal associated to a geographical object will be the most similar shape in the crystal catalog. Thus, it constitutes a simplified representation of the geometry of this object. Then, the relative position of faces constituting the object is described in the synthetic and homogeneous Miller system. Comparison of local coordinate systems of two objects allows extending this approach to the analysis of spatial relations between objects. Crystals associated to objects provide a high-level description of these objects.

To obtain the crystallographic characterization of a geographical object, our system performs the following steps:

1. Tetrahedrization of the object, to subdivide it into strictly convex parts;
2. Matching of each part of the object to one or several crystallographic forms;
3. Description of the composition of geographical forms (macle, epitaxy);
4. Indexation of faces of the crystallographic forms in the Miller system;
5. Association of faces of the object to faces of the crystallographic forms.

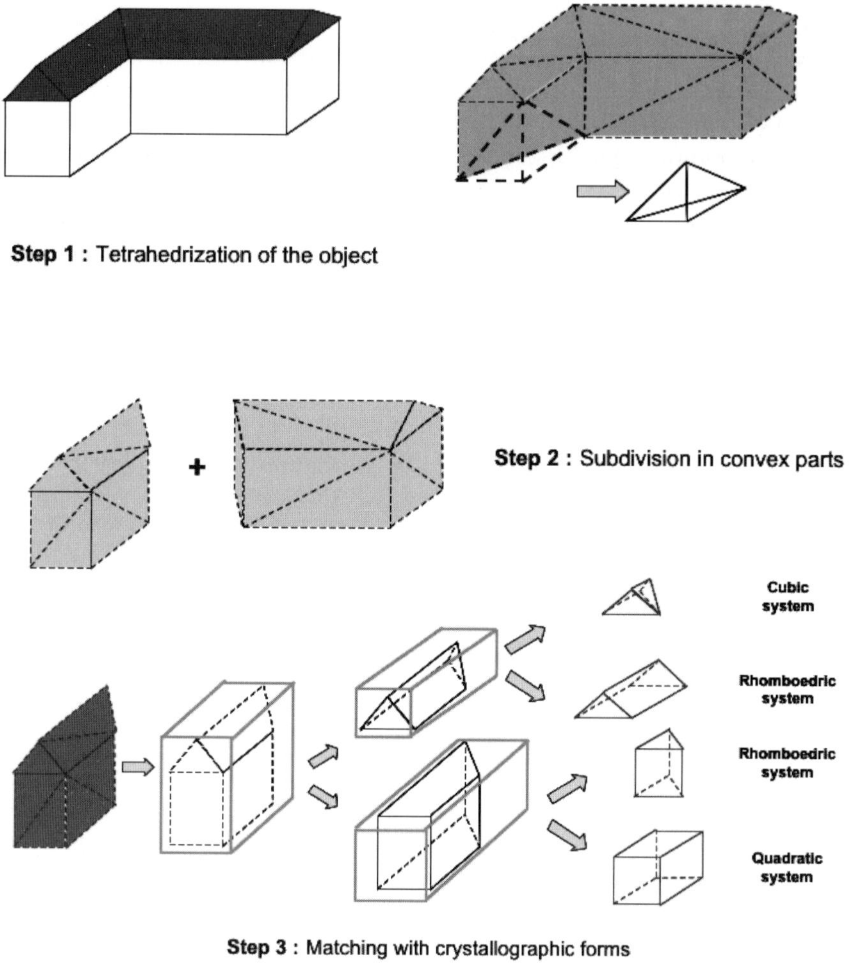

Fig. 10. Steps necessary to perform the crystallographic characterization of a geographical object

Figure 10 illustrates the operations performed on a L-shaped building. First, tetrahedrization of the object constrained by its faces enables to obtain a solid whose volume is known. Then, the solid (composition of tetrahedra) is subdivided into strictly convex bodies (two bodies in the example of Fig. 10). The algorithm is inspired from (Grumbach et al. 1997). Both bodies are matched to a crystallographic form. The algorithm tries first a quadratic form, which corresponds in 3D to the 2D minimum bounding rectangle, and then refines the form until it fits well enough. During this process, the body can be subdivided again if it appears to be the composition of several crystallographic forms (macles and epitaxies). The nature of the composition depends on the semantics of the bodies: it is a macle if the crystallographic forms belong to the same system, and an epitaxy in the opposite case. Then, each face of the crystallographic forms is linked to faces of the geographical object, and is indexed by Miller.

When extending this approach to less regular objects, such as a geological bend, the process of matching the body of the object to crystallographic forms may not lead to a satisfying solution. In that case, we favor the hexagonal and the quadratic systems, and forbid any subdivision of the object. The aim is to compute the relative position of the boundaries of the object rather than to obtain an approximate form. For instance, the crystallographic description of a geological layer enables to characterize the upper and lower parts of the formation.

By its crystallographic characterization, each object is associated to very simple volumic primitives (the crystals) whose volumes, orientations and symmetries are known. Any query on geographical objects can be quickly handled at the crystallographic level to deliver an approximate answer. This enables to focus on areas of interest, and to analyze properties that do not directly derive from geometry. Finer analyses can then be performed at the geometric and topological levels of objects (BRep or tetrahedra). Let us emphasize on the fact that the crystallographic description can be used to compute relations of different natures:

1. Metric relations: distances, volumes, some alignments (with Miller indexes);
2. Topological relations (Egenhofer and Herring 1992; Egenhofer 1990)
3. Cardinal relations (with local coordinate system orientation and Miller indexes)
4. Ordinal relations

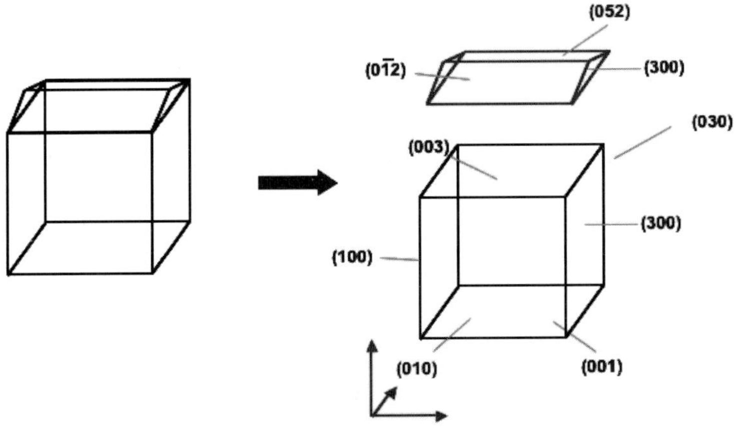

Fig. 11. Miller indexes on a 3D geographical object

Figure 11 illustrates the indexation of the faces of a simple building. This building is matched to a quadratic form (for its body in black) and to a rhombohedral form (for its roof in red). As both crystallographic forms are used to describe the same geographical object, the Miller system of indexes is extended to express the relative position of all faces in only one local coordinate system. The local coordinate system is such as the lower left corner of the front face of the cube has coordinates (111) and the upper right corner of the back face of the cube has coordinates (333). In this system, the coordinates of the intersection of planes with the axis give the Miller indices. Thus, the walls are indexed by (010), (100), (300), and (030), and the roofs by (0-12) and (052). All vertical faces have indexes of the form (ab0), all horizontal faces have indexes of the form (00c). If this house is exposed to a rock fall, the determination of faces exposed to this phenomenon is simplified by the Miller indices: let us assume that the rock trajectory is orthogonal to the plane of Figure 11, i.e. along the y axis. Then, the faces at stake have indices of the form (0bc). Let us note that face (030) will only be concerned if the rock goes through the building!

As regards implementation, our approach introduces two layers to handle 3D geographical data. The first one is the high-level layer where analyses are performed: objects are described by crystallographic forms, by their boundaries (possibly non plane) and by tetrahedra. The second one is dedicated to visualization: boundaries are mapped to a BRep model, which is decomposed into simplicial complexes to stay close to graphic card models. Queries are performed onto the high-level layer: the crystallographic forms enable to derive cardinal relations and some metric relations; tetrahedra enable to compute volumes and intersection relations (as tetrahedra are very simple solids to handle); and the boundaries, stored as equations, enable exact representation of objects. Visualization and consistency assessment are performed on the visualization layer, where the BRrep offers the advantages of an n-GMap structure.

5 Outlook of the 3D GIS with Crystallographic Primitives

Our 3D GIS prototype implements the model of Figure 12 on top of GeOxygene (http://oxygene-project.sourceforge.net/). This model is based on three main ideas. The first one is the distinction between a geometric level dedicated to visualization (in red on the diagram), and a geometric and semantic level dedicated to analysis (in blue on the diagram). The visualization level is the model of simplicial complexes and thus ensures a good integration with the Java 3D API and provides a topological description of objects helpful for consistency. The analysis level is objects helpful for consistency. The analysis level is composed of a BRep dedicated to data acquisition and classical algorithms (connected for compatibility purpose to the 2D topological map), of a spatial enumeration by tetrahedra for volumic operations, and of the high-level crystallographic description.

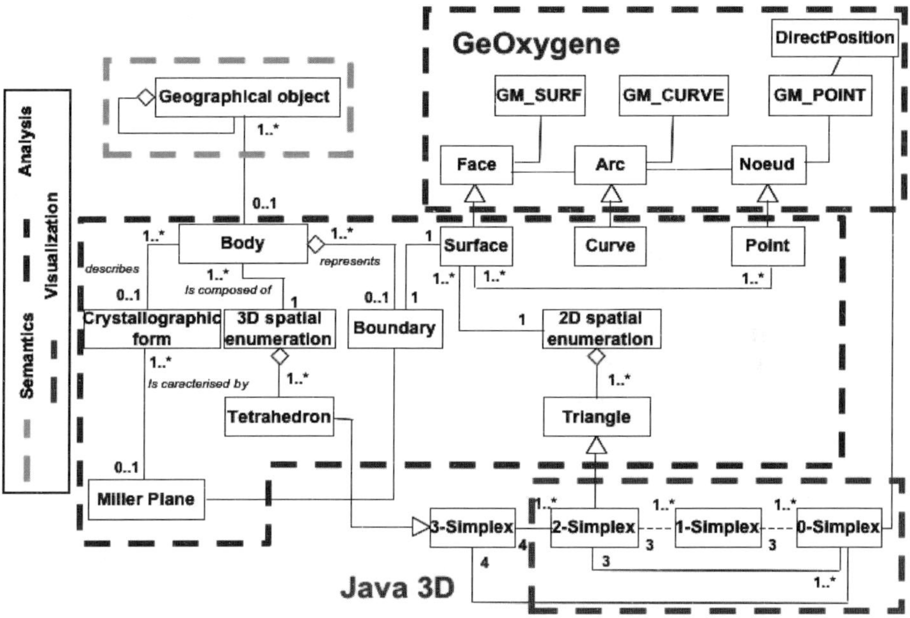

Fig. 12. UML diagram of the 3D GIS

The second idea is to exploit the redundancy of information in the analysis level to optimize queries. The BRep is used to operations that require handling the boundary of the object. As no topological information is computed for relations involving several objects, this BRep is restricted to analyses involving only one object at the same time. The tetrahedra enable to compute volumes, and intersection of several objects. They are used to compute relations involving several objects. Note that tetrahedrizations are always performed on–the-fly. The crystallographic primitives are used to help computing spatial analysis queries, especially some cardinal and metric relation determination.

The third idea is to help extracting areas of interest in geographical objects with the help of Miller indices. This is peculiarly useful to analyze the propagation of phenomena. The relationships between the ground, the underground geological formations and objects on the ground are handled more easily thanks to the crystallographic description.

To sum up, the spatial extent of a geographical object is modeled as a body in our system (see Fig. 13). This body has a boundary, some composing tetrahedra and crystallographic forms. The boundary is represented in a BRep model, whose faces are decomposed into simplicial complexes. The tetrahedra composing the object can also be visualized into the simplicial complex model. Last, the crystallographic forms have faces indexed by Miller and linked to the boundaries of the object.

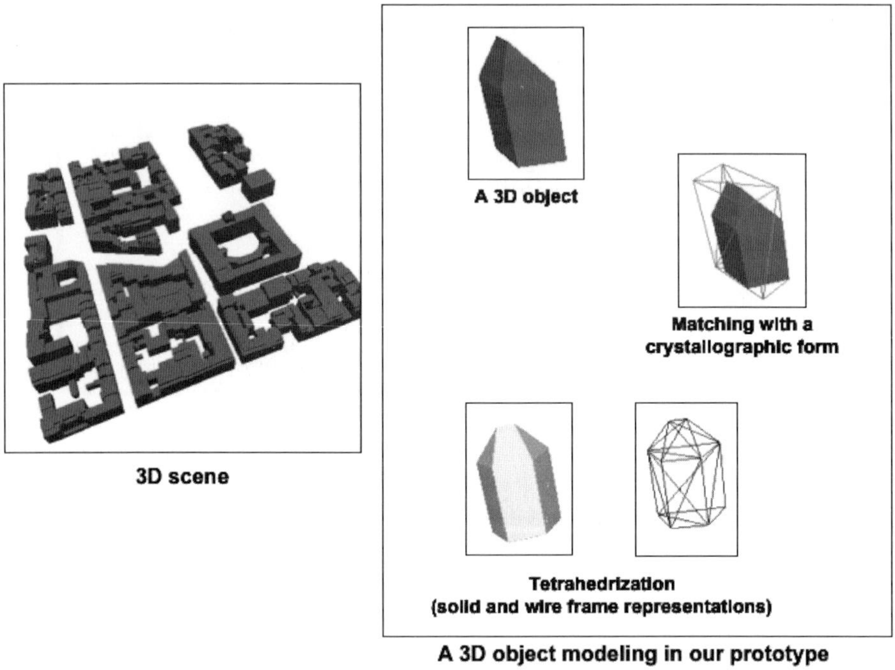

Fig. 13. Screenshots of 3D object analysis and representation in our GIS prototype

6 Conclusions

The 3D GIS model and prototype presented in this paper is a trial to overcome the limitations of geometric and topologic modeling of geographical objects. It makes profit from existing works in the field of 3D modeling, and introduces a high-level description of objects dedicated to analysis.

The model emphasizes on the distinction between a visualization model and a model dedicated to analysis. Moreover, the analysis level makes complementary use of three kinds of geometric description: the high-level crystallographic forms, a BRep and a spatial enumeration. Note that data is duplicated only during analysis, as data storage is performed at the BRep level.

Besides, the crystallographic description of the objects can be thought of as a kind of approximate CSG model. Each of the three models in use in our system (BRep, spatial enumeration and CSG) is used according to its capabilities. Data coherence is maintained at the BRep level, and topology between objects is computed on demand.

So far, the entire model has been implemented. Algorithms to determine the matching crystallographic forms are being developed. The next step of the implementation will be to test the system on real field data to simulate a complex event, such as an underground collapse.

The adaptation of the crystallographic approach will be investigated further in a near future, especially the identification of forms relevant to geographical features, and extensions of the Miller indexing system.

References

Bernard G, Ramos F, Lebard A (2000) Rapport de synthèse sur l'analyse des différents modèles 3D, EADS, Geomatics

Billen R (2002) Nouvelle perception de la spatialité des objets et de leurs relations. Éveloppement d'une modélisation tridimensionnelle de l'information spatiale. PhD Thesis, Université de Liège

Breunig M (1996) Integration of Spatial Information for Geo-Information Systems (= Lecture Notes in EarthSciences 61). Springer-Verlag, Berlin Heidelberg

Brisson E (1990) Representation of d-Dimensional Geometric Objects. PhD Thesis, University of Washington

Carlson E (1987) Three dimensional conceptual modelling of subsurface structures. In: Technical Papers of ASPRS/ACSM Annual Convention 4 (Cartography), pp 188–200

Coors V (2002), 3D GIS in Networking environments. CEUS 17

de Cambray B (1994) Etude de la modélisation, de la manipulation et de la représentation de l'information spatiale 3D dans les bases de données géographiques. PhD Thesis, Université Paris VI

De La Losa A (2000) Modélisation de la troisième dimension dans les bases de données géographiques. PhD Thesis, Université de Marne-La-Vallée

Egenhofer MJ, Herring JR (1992) Categorizing topological relations between regions, lines and points in Geographic databases. Technical report, University of California 94-1

Egenhofer MJ, Herring JR (1990) A mathematical framework for the definition of topological relationships. In: Proc of Fourth Int Symp on SDH, pp 803–813

Foley JD (1990) Computer graphics, Principles and Practice. Addison – Wesley Systems Programming Series

Grumbach S, Rigaux P, Scholl M, Segoufin L (1997) DEDALE, A Spatial Constraint Database, Workshop on Database Programming Languages (DBPL'97)
Mesgari SM (2000) Topological cell-tuple structures for three – dimensional spatial data. PhD Thesis, ITC dissertation number 74
Molenaar M (1990) A formal data structure for 3D vector maps. In: Proc of EGIS'90, 2, pp 770–781
Penninga F (2005) 3D Topographic Data Modelling: Why Rigidity is Preferable to Pragmatism, COSIT 2005, Ellicottville, pp 409–425
Pfund M (2001) Topologic data structure for a 3D GIS. In: Proc of Int Workshop on Dynamic and Multi-dimensional GIS, Beijing
Pigot S (1995) A topological model for a 3D Spatial Information System. PhD Thesis, University of Tasmania
Pilouk M (1996) Integrated modelling for 3D GIS. PhD Thesis, ITC, The Netherlands
Pomerol C, Lagabrielle Y, Renard M (2003) Éléments de géologie. Dunod
Ramos F (2003) Modélisation et validation d'un système d'information géographique 3D opérationnel. PhD Thesis, Université de Marne-La-Vallée
Requicha A (1980) Representation of Rigid Solids: Theory, Methods and Systems. Computing Surveys 12:87–93
Stoter JE, Ploeger HD (2002) Multiple use of space: current practice of registration and developpement of a 3D cadastre. Proc of UDMS 2002
Zlatanova S (2000) 3D GIS for urban development. PhD Thesis, ITC, The Netherlands
Zlatanova S, Rahman AA, Pilouk M (2002) 3D GIS: current status and perspectives. In: Proc of the Joint Conf on Geo-spatial theory, Processing and Applications

The Hierarchical Watershed Partitioning and Data Simplification of River Network

Tinghua Ai[1], Yaolin Liu[1], Jun Chen[2]

[1] School of Resource and Environment Sciences, Wuhan University, 129 LuoYu Road, Wuhan, 430072, P.R. China
email: tinghuaai@gmail.com, Yaolin610@163.com
[2] National Geometrics Center of China, Zizhuyuan, Bejing, 100044, P.R. China; email: chenjun@nsdi.gov.cn

Abstract

For the generalization of river network, the importance decision of river channels in a catchment has to consider three aspects at different levels: the spatial distribution pattern at macro level, the distribution density at meso level and the individual geometric properties at micro level. To extract such structured information, this study builds the model of watershed hierarchical partitioning based on Delaunay triangulation. The watershed area is determined by the spatial competition process applying the partitioning similar to Voronoi diagram to obtain the basin polygon of each river channel. The hierarchical relation is constructed to represent the inclusion between different level watersheds. This model supports to compute the parameters such as distribution density, distance between neighbor channels and the hierarchical watershed area. The study presents a method to select the river network by the watershed area threshold. The experiment on real river data shows this method has good generalization effect.

Key words: delaunay triangulation, map generalization, river network, spatial analysis

1 Introduction

Map generalization is the process of "information abstraction" rather than a "data compression" although two processes have associations with each other. The generalization operation should first make decision of object importance at geographic level, which relates not only to the geometric properties of independent object but also the other context objects. Some researches think the true generalization should investigate geographical nature and so called geo-oriented generalization should move from critical points to sub-features (Poorten and Jones 1999). In this sense, the generalization is the task of structured analysis requiring to extract spatial knowledge (Wu 1997). In Brassel and Weibel's (1988) generalization model, structure recognition is regarded as the first step among five procedures. Plazanet (1998) presents a learning method of feature simplification and from the expert system point of view stresses the importance of line structure knowledge. In this field the constraint-based generalization gets emphasis and this concept is also under the control of structure characteristics.

For the generalization of hydrographic features, such as river network, we try to get the simple representation but remaining the main properties of hydrographic meaning. We need to consider the river channels as a whole at geographic level to pre-determine the important information contained in river network. If we just see the independent river without considering its context, the abstraction will dramatically destroy the original structure. So the assessment of geographic and hydrographic meaning plays an important role in river network generalization.

Indeed, the generalization of river network has to answer three questions: (1) How many branches to be selected? (2) Which channel is unimportant? (3) How to simplify the selected channel? The Töpfer law (Töpfer and Pillewizer 1966) has answered the question one during the catchment data transformation from large scale to small scale by the computation of scale rate. For question three, there are lots of algorithms to conduct the line simplification under the consideration of special properties of river feature. But for question two, it is not easy to answer requiring intelligent decision based on the analysis of river geographic properties and the context. At least three aspects has to be considered at different levels: the spatial distribution pattern at macro level, the distribution density and proximity relationship at meso level and the individual geometric properties at micro level. The study in this domain tries to find parameters and models to represent such interesting information. The Horton order and Strahler order which describe the topological organization of catchment channels have something to do with the importance role of one river channel playing in the catchment. But we cannot simply judge one channel of importance if it has high order number. Also the channel length can't serve as the only judgment condition. Finding an integrated parameter to decide the importance of channel in catchment becomes an interesting topic.

As the main feature in GIS and traditional map representation, the river network generalization has attracted interests over years. Richardson (1993) presents a method to select river based on Horton order and river length. Thomson and Brooks (2000) applies the Gestalt recognition principles in river network generalization judging the main channel and removing unimportant channels. Since the distribution of river network associates with the terrain surface, Wolf (1988) builds a weighted network data structure integrating the drainage, ridge, peak and pit point. This data structure supports to determine the significance of river channel. The river tree has various patterns leading to the generalization strategies different. Wu (1997) investigates the characteristics of river tree and develops a method based on buffer spatial analysis to establish the river tree structure.

The previous studies aim at the data abstraction concerning the geographic meaning of hydrographic features from different perspectives. But the operational model and algorithm is not available. Some researches build parameters to express the river structure just at conceptual level without providing computation approaches. This study considers the watershed area an important parameter in river network and presents a partitioning method to represent it based on the network analysis by Delaunay triangulation. The model contributes to extract the distance between neighbor rivers, the river watershed area and the inclusion relation between hierarchical watersheds. The rest paper is structured as follows. After the discussion of hierarchical structure of river network, section 2 presents the partitioning model and the establishment algorithm. Some parameter computations are offered in Section 3. Based on the model in Section 2 Section 4 investigates the selection of river in catchment. Section 5 gives the conclusion with the future improvement works.

2 Watershed Partitioning

The rainfall down on the earth collects into basin or river. The watershed area of one river channel reflects the ability of its corresponding catchment to compete for the rainfall. The watershed line is the competition result between neighbor rivers. It implies the larger area one river has the more significant role of the river plays in catchment. The watershed area and separation line has something to do with the river length, the distance between neighbor rivers and the hierarchical structure among parent-child rivers. So the watershed can be regarded as the main aspect in river decision-making during the river network generalization. Based on this idea, this section presents a watershed-partitioning model. First we examine the important characteristics of the watershed, namely the hierarchy.

2.1 The Hierarchical Properties of River Network

An outstanding property of river network is the hierarchy. A catchment can be divided into main channel, 2nd order, 3rd order channels and so on. In hydrographic research domain, three well-known ordering, namely the Strahler order, Horton order and Shreve order describes the hierarchical organization from different perspectives. In Horton order, the highest order N corresponds to the main channel, the order N-1 the next important channel, and so on, as shown in Figure 1.

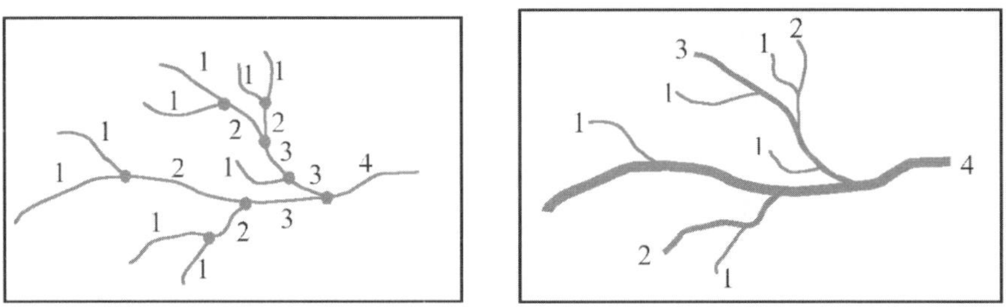

Fig. 1. An illustration of Strahler order(left) nad Horton oreder(right)

It is the hierarchical structure that makes the river network the self-similarity and fractal nature (Rosso 1991). There exists self-similarity among the channels in several geometrical parameters. Horton (1945) discovers the ratios of the number of streams and the mean length of streams between successive orders area approximately constant. Ros (1997) presents the similar relationships expressed as the following formula, called Horton law:

$$\frac{N_{W-1}}{N_W} = R_B \quad \frac{L_{W-1}}{L_W} = R_L \quad \frac{A_{W-1}}{A_W} = R_A \quad \frac{S_{W-1}}{S_W} = R_S \tag{1}$$

Where for each order w, N_w denotes the number of streams, L_w, A_w, S_w are the mean length, mean area, mean slope. R_B, R_L, R_A, R_S area termed the bifurcation, length, area, and slope ratio respectively.

The hierarchy of channel network at one dimension correspondingly leads to the hierarchical structure in the watershed partitioning at two dimensions, acting as the inclusion relation between different level watershed polygons. All watershed areas of child level rivers are within that of parent-river. The watershed generation and the inclusion relation construction is an interesting topic in GIS research and hydrographic domain. A lot of methods based on DEM data have been built over years. See the review by Tribe (1992). Here we focus on the data generalization target presenting a method of watershed construction from the river network under the suppose that the terrain distribution is uniform. So the watershed is just determined by the geometrical characteristics at 2 dimensions.

2.2 Data Structure

We first define the data structure of river network. From the point of view of geographic elements, the river network is defined as four levels: catchment, channel, segment and nodes as shown in Figure 2. The channel is the element with complete geographic meaning under the high level catchment. It corresponds to the element in Horton order, which is composed of serial of segments from outlet to joint node. The component of channel is the segment element which corresponds to the element in Strahler order with two terminal node, namely start node and end node along water flow direction.

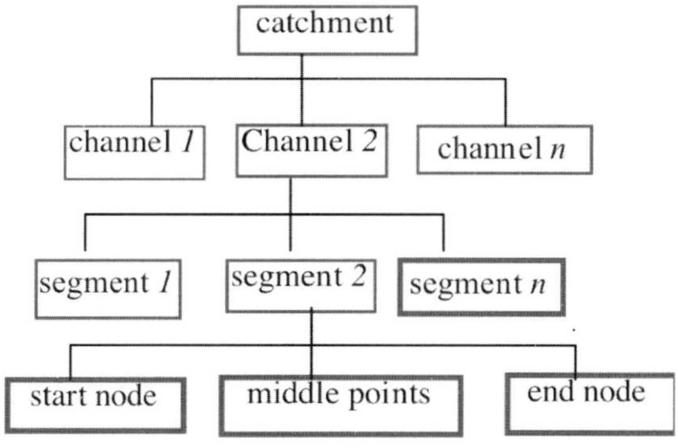

Fig. 2. The hierarchical tree of river network

The joint relationship between different level channels is recorded with Horton order. To organize such hierarchical relation, the key method is the judgment of flow direction and the extraction of mainstream. This work is out of the paper. We just apply the existed method. Pavia and Egenhofer (2000) investigates this question in detail and develops an approach to automatically build the flow direction.

2.3 Triangulation Construction and Watershed Extraction

Each channel ant its child channels make a sub-catchment, as the part of the whole catchment. It receives the water flow by a competition extension with its proximity sub-catchments. In geographic analysis, such spatial competition question is usually resolved by Voronoi diagram. Here we use the idea similar to Voronoi diagram partitioning to represent the watershed competition. The applied model is Delaunay triangulation with skeleton supporting to represent the spatial competition result. This method has been used to settle several similar questions in proximity object detection, spatial conflict judgment, neighbor object aggregation and so on in map generalization (Ai and Oosterom 2002; Jones 1992; Sester 2000, Ai et al. 2000). Next we first generate the watershed line between proximity channels and then link the sequent lines to build the inclusion of watershed polygons. The whole process contains four steps.

2.3.1 Triangulation Construction

Get all vertex points of catchment to make a point set *S*. For those segments containing too long direct line between two points, we interpolate serial middle points between terminal points to generate additional points to avoid the intersection between triangle edge and river segment in later triangulation construction. Add three outside points making a triangle to envelope the point set *S* and construct the Delaunay triangulation of *S*. Remove those triangles which are related to outside three points and the remained triangles compose the coverage of the river catchment as shown in Figure 3. If the distance between two neighbor points is short enough, the triangulation does not result in the intersection between triangle edges and the channel segments. Otherwise replace the normal DT with the constrained DT.

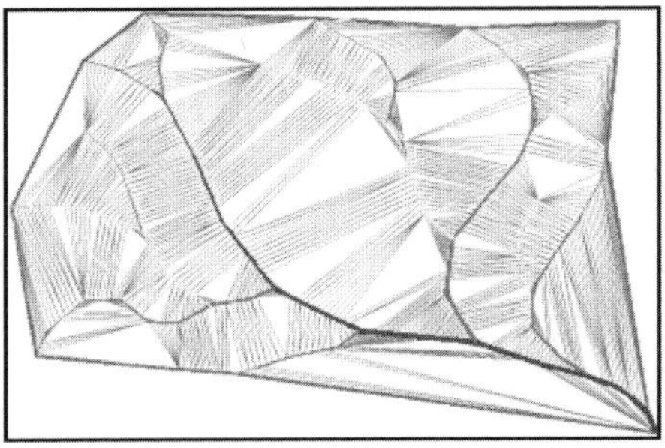

Fig. 3. Construct the Delaunay triangulation in the coverage of river catchment

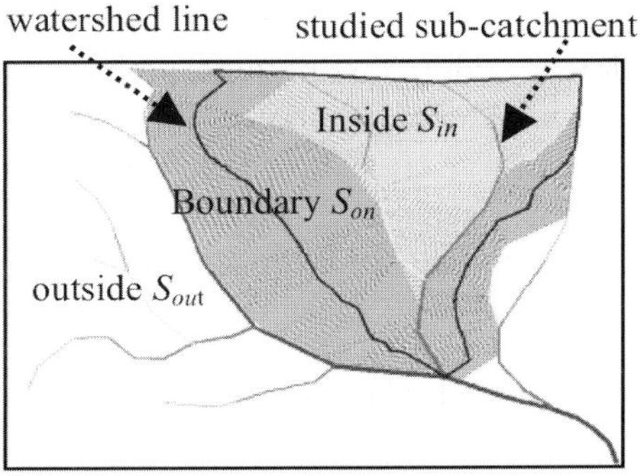

Fig. 4. For one sub-catchment, identify three sorts of triangles: inside, outside and boundary triangle regions

2.3.2 Triangle Classification

One channel *a* together with its descent channels makes a sub-catchment as part of its parent sub-catchment. According to the relationship between a triangle and the sub-catchment *a*, we can divide the triangles into three classes.

Take one segment *b* of sub-catchment *a* into account. The triangle with at least one vertex locating on the segment *b* is assigned to be segment-related triangle. All segment-related triangles of sub-catchment *a* (including current channel *a* and its descent channels) are assigned to be sub-catchment-related triangles which makes the coverage region of sub-catchment *a* . These triangles are further able to be classified as two types. One is the completely related triangle with all three vertexes locating on the segments of sub-catchment *a*, and partially related triangle with one or two vertexes on the segments of sub-catchment *a*. The other vertex of partially related triangle locates either on the segment of parent channel of sub-catchment *a* or the segments of brother sub-catchments of sub-catchment *a*. Finally for one sub-catchment *a*, the triangles are classified as three types, namely (1) the outside triangles S_{out} without relation to the sub-catchment *a*; (2) the inside triangles S_{in} being completely related to *a*, and (3) the boundary triangles S_{on} partially related to *a*. Three types of triangle are illustrated in Figure 4. The core red-colored channel is currently studied channel. The light blue shaded triangle region is the inside S_{in} and the deep blue shaded triangle region the boundary S_{on}. The other white region belongs to the outside S_{out} .

Suppose the terrain slope, the soil filter and the vegetation abstraction distributes in a uniform way. The watershed area will be determined by the spatial relation of river channels. There exists the close principle that the rainfall finds the nearest path to collect into basin or river. Take the sub-catchment *a* into account, the rainfall on the S_{in} region will completely flow into *a* no matter the flow directly down to channel *a* or through its descent channels. The rainfall on S_{out} region will completely down flow to other sub-catchments having nothing to do with sub-catchment *a*. But for S_{on}, the rainfall faces the competition since it locates as a bridge between sub-catchment *a* and context neighbors. The buffer area as shown in deep shaded in Figure 4 needs to be divided into two parts by some way.

2.3.3 Watershed Line Extraction

Based on the analysis above, the watershed extraction should be conducted in the area of boundary triangles S_{on}. We use the skeleton of DT method (Ai and Oosterom 2002) to extract the watershed line. Just consider the triangles in S_{on} and distinguish them three types according to the number of neighbor triangles, namely type I with only one neighbor, type II with two and type III with three. The skeleton connection way for three types of triangle is described in Figure 5, where P_1, P_2, P_3 is the midpoint of corresponding triangle edge, and O is the triangle center. The skeleton segments are created by means of the next paths:

Type I : $A \rightarrow P_1$;
Type II: $P_1 \rightarrow P_2$;
Type III: $O \rightarrow P_i$, i=1,2,3

The skeletonization result of S_{on} is illustrated as black line, namely the watershed line, in Figure 4. As the skeleton line is closed, the watershed area automatically generates.

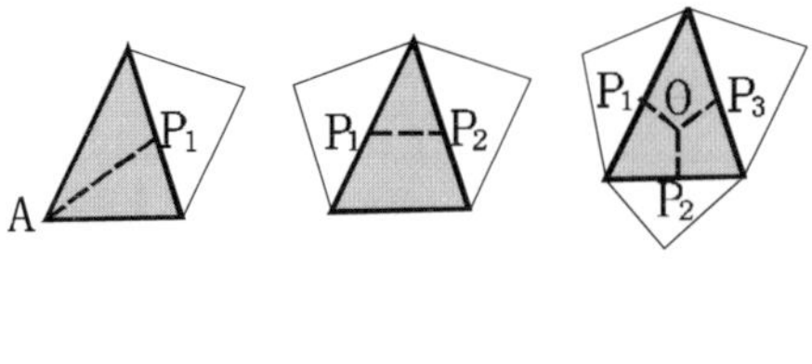

Fig. 5. Center-line connection ways for three types of triangle

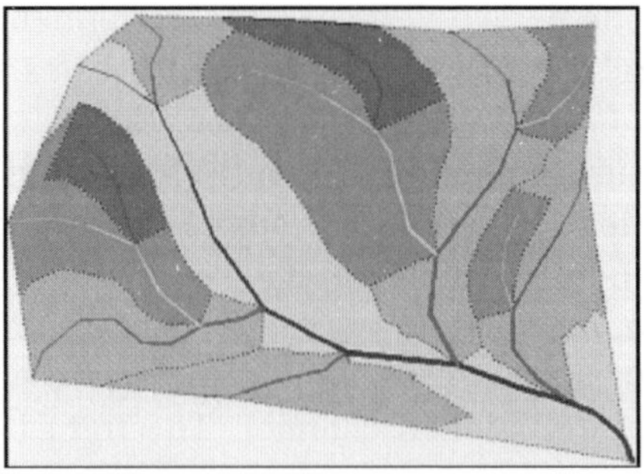

Fig. 6. The result of hierarchical partitioning of the river catchment

Trace the channels of river tree one by one to repeat the method above. After all sub-catchments have been processed, we finally get the watershed line distribution result as shown in Figure 6.

2.4 Inclusion Relationship Establishment

The hierarchical structure of river channels acts as the minor channel joins into the main channel. This hierarchical relationship is mapped as the inclusion between watershed polygons. It means the sub-catchment a with child sub-catchments b1, b2,...bn corresponds to the watershed polygon of a enveloping that of b1, b2,...bn. Based on this association, it is easy to build the inclusion relationship of watershed polygons. The child watershed polygon must be included within the parent watershed polygon. But the integration of all child watershed does not equal to the parent watershed. Some regions belong to the parent channel rather than any child channel. At same level, the watershed of brother channels does not overlap to each other. Note that the boundary line between neighbor watershed be exactly the same, since it is extracted from the same sub-triangulation by the same skeletonization method. Figure 6 illustrates the inclusion among the hierarchical watershed by the different level of shaded color.

3 Parameter Computation

The model based on the triangulation and skeleton is able to express the local distribution and geometric computation. Besides the area of sub-catchment, some other useful parameters can also be computed.

3.1 Distribution Density

For sub-catchment *a*, the area of the watershed polygon is defined as *area(a)*. All segments contained in sub-catchment *a* is defined as set $\{d_i\}$. The integration length of $\{d_i\}$ is $\sum \mathrm{length}(d_i)$. Then in local region of sub-catchment *a* the river distribution density is

$$\mathrm{density} = \sum \mathrm{length}(d_i)/\mathrm{area}(a). \tag{2}$$

3.2 Distance Between Proximity Channels

As an important parameter to describe the local distribution of river network, the distance between proximity channels is useful in river network generalization. It is usually computed in the same side of main channel. As two proximity channels are not parallel to each other in nature, the distance between different locations among the channel is different. Here we present a weighted distance computation based on triangulation.

We take the calculus idea. The channel curve is divided into serial of small fragments. The gap distance between the fragments from different channel is determinately computed. The triangle can serve as such small fragments as shown in Figure 7. The skeleton goes across serial of triangles. For the terminal triangle, the local gap distance is computed as the length of triangle edge that is across by the skeleton. For the middle triangle, the local gap distance is computed as the length of the height edge perpendicular to the skeleton. The distance in two situations is illustrated as W_1W_2 in Figure 7.

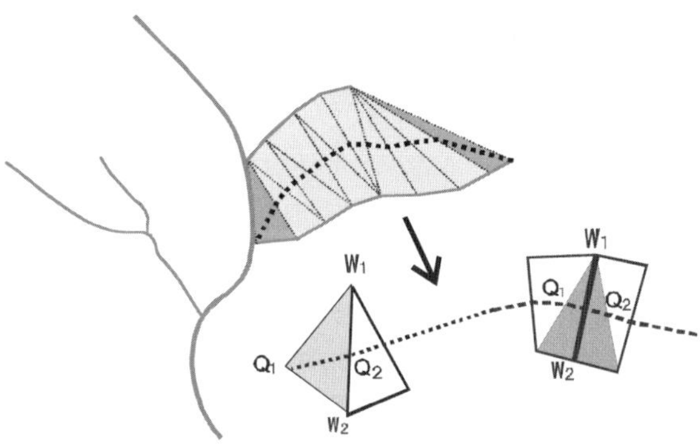

Fig. 7. The illustration to compute the weighted distance between neighbor rivers

The length of the part skeleton locating across one triangle is $\| Q_1Q_2 \|$. Then the weight of the distance in current local triangle

$$\overline{w} = \sum_{i=0}^{k} \frac{\|Q_i Q_{i+1}\|}{1} \|W_{i1} W_{i2}\| \tag{3}$$

can be represented as the rate of $\| Q_1Q_2 \|$ to the length of the whole skeleton. The weighted distance is computed as
where l is the length of the whole skeleton and k the number of triangles being across by the skeleton.

4 River Network Generalization

The channel selection during the river network generalization has to consider the order, the length, the distribution pattern and other parameters. The selection only based on one parameter condition cannot get ideal result. In Figure 8, the original river network is represented as *A* with river 1 the main channel, river *2,3,4* the second order and river *5,6* the third order. If we just consider the length tolerance, the selection result is illustrated as *B* in which short channel *2* is removed but the child channel *5, 6* remained as the dangled branches. If we just consider the Horton order, the selection result is as *C* in which the channels of third order namely *5,6* are removed. But channel *5* is very long although the order is low. The correct selection should be *C* in which the integration of length and order is taken into consideration. How to find a simple parameter which integrates the length, the order and the distance between proximity rivers as the importance decision condition in the selection of river network? In this study we try to let the watershed area playing this role. In some degree the watershed area describes the integration of three aspects.

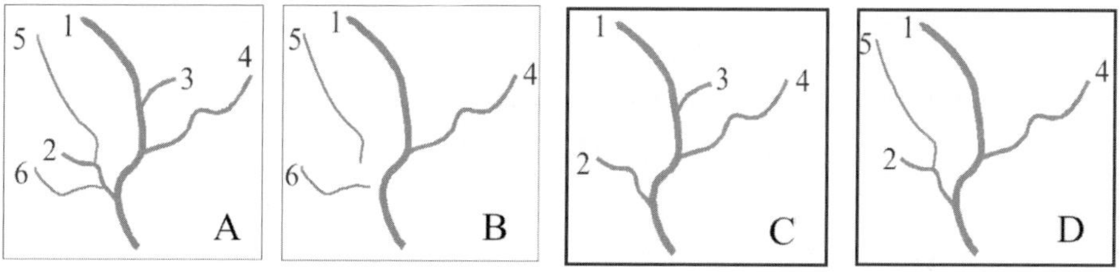

Fig. 8. The river selection from catchment based on different parameter tolerances respectively. (**A** – Original river network; **B** – Selection by length; **C** – Selection by Horton order; **D** – Correct selection)

The watershed area strongly depends on the river distribution density. If the river distributes in a dense way, the channel can just compete to obtain a small watershed area. The watershed area considers the context impact that the same river in a high-density area is less important than that distributes in a sparse region. (2) The watershed area has considered the order impact. The hierarchical structure shows the watershed area of high order channel is not smaller than the integration of that of all its child channels. This principle guarantees the parent channel has preference to its child channels in selection, not generating the case such as in Figure 8 *B* that the short parent channel removed but long child channel remained. (3) The watershed area has considered the channel length. Obviously the long channel extends in a large range obtaining a large watershed area. For the watershed-area-based selection conducting on the river network in Figure 8 A, we may see channel 3 has smaller watershed than that of channel 5 with higher order than channel 3. So channel 3 is removed but channel 5 remained. The generalization result is the same as manual operation.

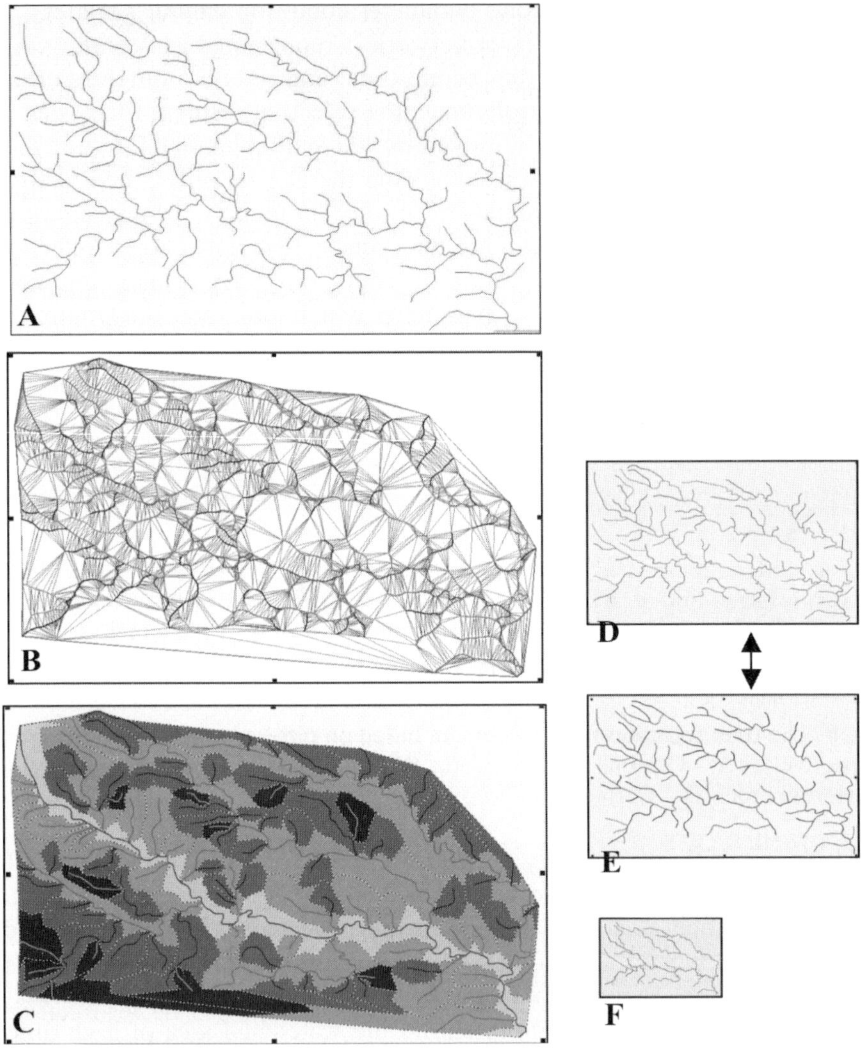

Fig. 9. The experiment of river network selection based on the watershed area and neighbor distance, and the comparison with manual generalization result. **A** – Original network (1:100,000); **B** – Build Delaunay triangulation; **C** – Hierarchically partition the watershed; **D** – Selection result (1:200,000); **E** – Manual generalization result (1:200,000); **F** – Selection result (1:500,000)

We carry out the generalization experiment with the watershed-area-based selection. The data exported from real map database with 1:100,000 scale is a typical tree-structured catchment distributing in the middle of China. The river network contains four orders with 99 channels. For the purpose scale 1:200,000, the generalized result should remain 70 channels according to Töpfer principle. For the purpose scale 1:500,000, it should remain 19 channels. First through the Delaunay triangulation partition the hierarchical watershed and then sort the channels on the size of watershed area from large to small. Based on the channel sequence, progressively select the channel until the number respects Töpfer law. The original data, the process of watershed partitioning and the generalized result is illustrated in Figure 9. For the comparison, we present the selection result by this method in *D* and that of manual operation in *E* for 1:200,000 generalization. Note the presented data is just conducted selection without line simplification. Through the comparison between the generalized data and the original data and the manual selected data respectively, we can see main structure is remained by the watershed-based selection. The short channel and the channel locating in the crowd context have more chances to be deleted. The result is similar to manual generalization. Compared with that just based on single variable such as channel length or channel order, the generalization gets better effect.

Generally the watershed area of different channel in a catchment is obviously varied and so it can make the channel different in importance decision. But for some special cases such as feather shaped catchment, the channels locating on the same side of mainstream are parallel to each other with approximately the same length. The watershed area are almost the same and without supporting to select the important channels.

5 Conclusions

The decision in map generalization on one hand has to consider all related impacts, on the other hand usually is difficult to find a multi-parameter model to simulate the intelligent reasoning as man does. If an integrated parameter is able to be found, the decision question can be resolved in a simple way. Usually such geometric parameter belongs to the geographic level, which integrates several basic parameters at geometric level. The geometric parameters such as length, distance, angle, area and others are easy to compute. But the geographic parameter is usually difficult to compute requiring complex model to derive interesting information.

This study focuses on the decision of channel importance during the river network generalization applying the integrated hydrographic concept, namely watershed area to replace several geometric parameters of river feature. The model construction uses the idea similar to Voronoi diagram on spatial competition by Delaunay triangulation and skeletonization. The partitioning watershed model inherits the hierarchical properties from the channel joint in river network. Through the hydrographic analysis, the watershed area can serve as a significant parameter to represent the importance role, and the experiment on real river data indicates the case.

But the watershed area is not a complete condition to generalize river network. First it cannot describe the river network pattern, which grows in a special geological and hydrographic condition. In manual generalization, the pattern recognition is the first high-level decision and then different generalization strategies conducting on corresponding river structure, such as grid, spider network, feather and other shape. Some channel has small watershed area but plays an important role in pattern component, it should be selected. So to find an operational model representing the river network pattern is the first improvement work in the future.

Secondly, even in local decision, some special case like the feather-structured river makes the watershed of proximity same level channel approximately has the same size. Then the watershed area will not serve as the decision parameter. For this situation, the watershed-based-selection should be improved.

Thirdly, the watershed partitioning in this study just considers the topological and geometric distribution in plan against the real natural situation that the 3rd dimension terrain strongly impacts the watershed distribution. The method integrating with the DEM based watershed extraction will improve the decision condition.

Acknowledgements

This research is supported by the Program for New Century Excellent Talents in University (NCET), the Ministry of Education, China.

References

Ai T, Oosterom P van (2002) GAP-tree Extensions Based on Skeletons. In: Richardson D, Oosterom P van (eds) Advances in Spatial Data Handling. Springer-Verlag, Berlin, pp 501–514

Ai T, Guo R, Liu Y (2000) A Binary Tree Representation of Bend Hierarchical Structure Based on Gestalt Principles. In: Proc of the 9th Int Symp on Spatial Data Handling, Beijing, pp 2a30–43

Band LE (1986) Topographic Partition of Watersheds with Digital Elevation Models. Water Resource Research 22(1):15–24

Brassel KE, Weibel R (1988) A Review and Framework of Automated Map Generalization. Int J of Geographical Information Systems 2(3):229–244

Horton RE (1945) Erosion Development of Streams and Drainage Basins: Hydrophysical Approach to Quantitative Morphology. Bulletin of the Geological Society of America 56(3):275–370

Jones CB, Bundy GL, Ware JM (1995) Map Generalization With A Triangulated Data Structure. Cartography and Geographic Information System 22(4): 317–331

Paiva J, Egenhofer MJ (2000) Robust Inference of the Flow Direction in River Networks. Algorithmica 26(2)

Plazanet C, Bigolin NM, Ruas A (1998) Experiment with Learning Techniques for Spatial Model Enrichment and Line Generalization. Geoinformatica 2(4): 315–333

Poorten P, Jones CB (1999) Customisable Line Generalization Using Delaunay Triangulation. In: CD-Rom Proc of the 19th ICC, Ottawa, Canada, Section 8

Richardson DE (1993) Automatic Spatial and Thematic Generalization using a Context Trans-formation Model. PhD Thesis, Wageningen Agricultural University

Ros DD, Borga M (1997) Use of Digital Elevation Model Data for the Derivation of the Geomorphological Instantaneous Unit Hydrograph. Hydrological Process 11:13–33

Rosso R, Bacchi B, La Barbera P (1991) Fractal Relation of Mainstream Length to Catchment Area in River Networks. Water Resource Research 27:381–387

Sester M (2000) Generalization Based on Least Squares Adjustment. In: IAPRS Vol XXXIII, Part B4/3, Comm IV, pp 931–938, ISPRS Congress, Amsterdam

Thomson RC, Brooks R (2000) Efficient Generalization and Abstraction of Network Data Using Perceptual Grouping. In: Proc of the 5th Int Conf on GeoComputation

Töpfer F, Pillewizer W (1966) The principles of selection: a means of cartographic generalization. The Cartographic J 3(1):10–16

Tribe A (1992) Automated Recognition of Valley Lines and Drainage Networks from Grid Digital Elevation Models: A Review and a New Method. J of Hydrology 139:263–293

Wolf GW (1998) Weighted Surface Networks and Their Application to Cartographic Generalization. In: Barth W (ed) Visualization Technology and Algorithm. Springer-Verlag, Berlin, pp 199–212

Wu H (1997) Structured Approach to Implementing Automatic Cartographic Generalization. In: Proc of the 18th *ICC*, Stockholm, Sweden, vol 1, pp 349-356

Grid Typification

Karl-Heinrich Anders

Institute of Cartography and Geoinformatics, University of Hannover, Appelstrasse 9a, 30167 Hannover, Germany
email: karl-heinrich.anders@ikg.uni-hannover.de

Abstract

In this paper the detection and typification of grid structures in building groups is described. Typification is a generalization operation that replaces a large number of similar objects by a smaller number of objects, while preserving the global structure of the object distribution. The typification approach is based on three processes. First the grid structures are detected based on the so-called *relative neighborhood graph*. Second the detected grid structures are regularized by a least square adjustment of an affine or Helmert transformation. The third process is the reduction or simplification of the grid structure, which can be done using the same affine or Helmert transformation approach.

Key words: map generalization, typification, pattern recognition, relative neighborhood graph, affine transformation, Helmert transformation

1 Introduction

Map generalization is needed in order to limit the amount of information in a map by enhancing the important information and dropping the unimportant one. Triggers for generalization are on the one hand limited space to present all the information and on the other hand the fact that different scales of an object are needed in order to reveal its internal structure. Typification is a generalization operation that replaces a large number of similar objects by a smaller number of representative objects, while ensuring that the typical spatial structure of the object arrangement is preserved. As an example consider a set of buildings in a city: when looking at this spatial situation at a small scale or resolution, the typical distribution and structure of the buildings should still be preserved, while the number of buildings and details of the building geometries can be reduced. In general there are two classes of typification approaches: typification with structural knowledge and typification without structural knowledge. Approaches without explicit structural knowledge try to preserve the overall distribution and structure. Müller and Wang (1992) use mathematical morphology to typify natural areal objects. Their principle is to enhance big objects and reduce small ones – unless they are important. Sester and Brenner (2000) describe an approach based on Kohonen Feature Maps. Kohonen Feature Maps are self organizing maps which try to preserve the original structure by moving the remaining objects in the direction of the removed one to minimize a certain error measure.

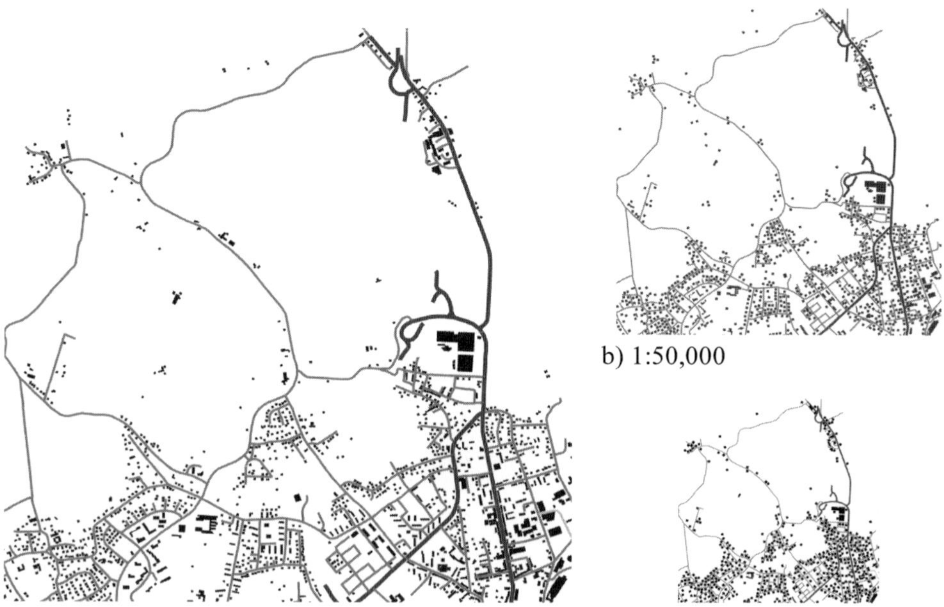

b) 1:50,000

a) 1:25,000

c) 1:75,000

Fig. 1. Automatic generation of different target scales from cadastral building data

This approach works very well for irregular object distributions (see Fig. 1), but fails on dominant regular structures, like linear or grid structures (see Fig. 2).

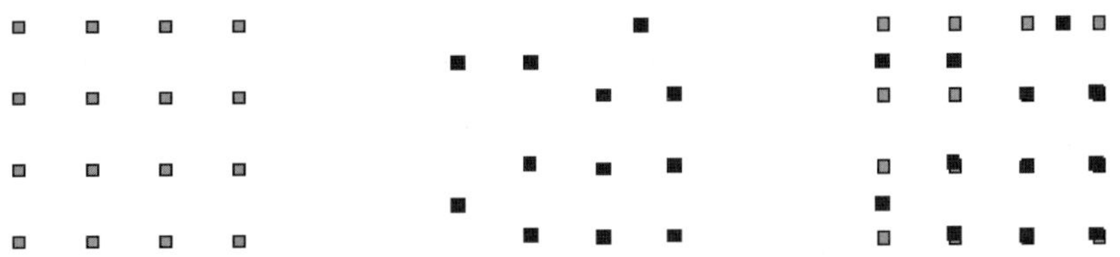

a) b) c)

Fig. 2. Regular grid structure of objects, which cannot be preserved; initial situation (**a**), result (**b**), overlay of initial situation and result (**c**)

Structural knowledge is used in approaches, which try to detect geometrical structures in the object groups which should be preserved by the generalization process. The first step in structural typification processes is always a segmentation process of the given objects. This segmentation process can be seen as a classification of a point set into point groups of certain geometric shape. In general one can distinguish between linear, circular, grid, star and irregular shapes with homogenous density (see Fig. 3). Typification for linear structures is proposed by (Regnauld 1996). Based on a minimum spanning tree clustering groups are detected; then the relevant objects within these groups are replaced by typical exemplars. This approach for building typification is motivated by the phenomenological property of buildings being aligned along streets – thus a one-dimensional approach is feasible. Another approach, which tries to find linear building structures, is described in Christophe and Ruas (2002). In Anders and Sester (2000) an approach to detect two-dimensional irregular structures with homogenous density is described. After clustering, the number of objects within the clusters has to be reduced. The reduction factor can be derived using e.g. the black-and-white–ratio, which is to be preserved before and after generalization, or Töpfer's radical law. The problem now is to decide which object has to be removed. This question is decisive, since the removal of one-object results in gaps. In the following sections an approach will be introduced which detects and preserves grid like structures without gabs. First the grid detection based on the so-called relative neighbor graph is described. Then the regularization and the reduction (simplification) of the grid structure by a least square adjustment approach are explained.

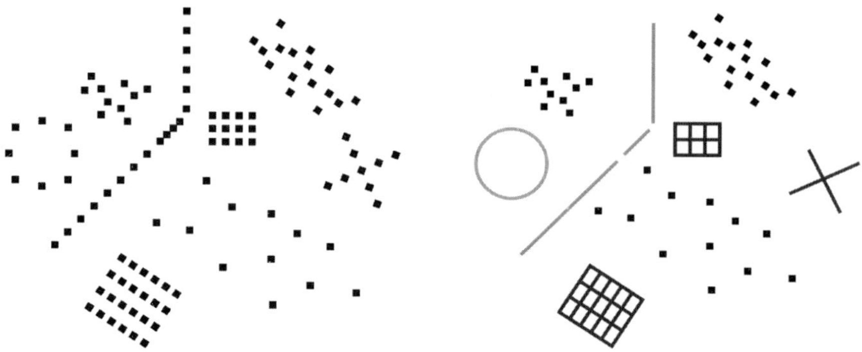

Fig. 3. Classification of a point set into linear, circular, grid, star and irregular shapes

2 Grid Detection

The following described approach is using structural knowledge in terms of grids. With grids we mean regular lattice-like layouts of buildings. More precisely the grid layout of the ground plan centroids (see Fig. 4a). In a grid structure every building belongs to two linear structures, which has to be preserved if possible. Like (Regnault 1996) we are using a neighborhood graph to detect the grid structures. In place of using a minimal spanning tree (MST) we are using the relative neighborhood graph (RNG) (Toussaint 1980). RNGs capture very well the inner structure of point sets, especially for the detection of grid like graph structures (see Fig. 4b). The MST is more useful for pure linear structures because in regular grid structures (equal distance between points) there is no unique MST. The MST for instance is a subset of the RNG. A general introduction to the subject of neighborhood graphs is given in Jaromczyk and Toussaint (1992) and (Anders 2004).

Built settlements based on development schemes frequently show a straight-line orientation. Groups of such linear structures can have a grid structure (see Fig. 4a). A grid is characterized by a set of mostly parallel lines, which are crossed by a second set of parallel lines. Frequently the sets of parallels intersect themselves approximately right-angled, this is however no compelling characteristic of a grid. Further characteristics of grid structures are rectangular or parallelogram-similar shape of the grid surfaces, straightness of the parallel line set, the convexity of the surfaces, a relatively constant side length relationship as well as a similar area of the rectangles.

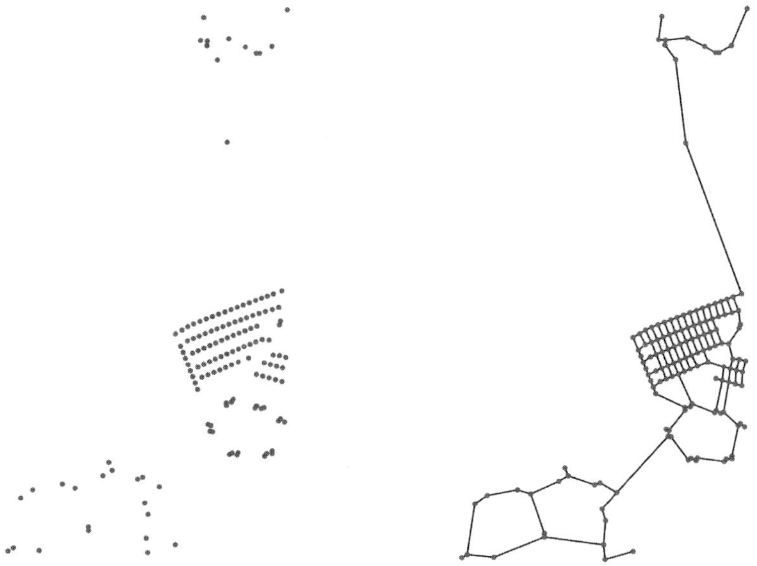

a) b)

Fig. 4. a) Building centroids **b)** The associated RNG

In practice the grid structures show rarely such "optimal" or ideal properties. Nevertheless the above-mentioned characteristics can be recovered in the data with appropriate deviations. According to the characteristics pointed out above an algorithm described in Heinzle, Anders, and Sester (2005) is used. Starting point for the algorithm are so called cross-nodes, at which four edges are crossing almost at right angle. Polygons belonging to such cross-nodes are potentially candidates for a grid if they meet the above-mentioned different criteria. Briefly described the similarity of a grid polygon candidate to its neighbor polygons is regarded, the surface size of the grid polygons are compared, and the convexity of the surfaces is examined. When we consider the example RNG shown in figure 4b this algorithm finds two grid structures (see Fig. 5a).

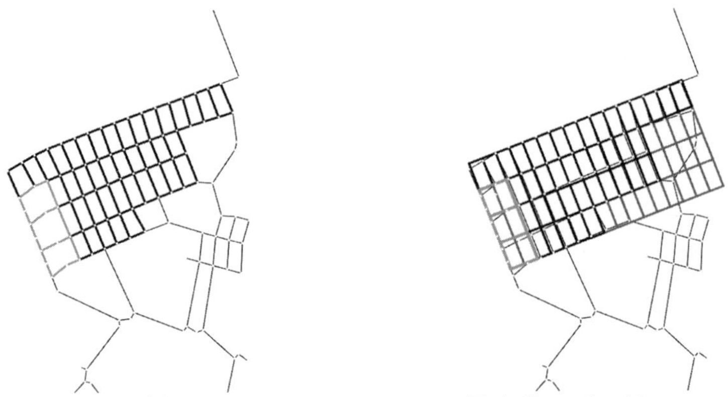

a) Detected grid structures b) Adjusted grid structures

Fig. 5. Detected and adjusted grid structures

3 Grid Adjustment

The above-detected grids are of course not perfect regular (see Fig. 5a). However our grid model describes perfect regular and complete grids, which means in detail that we consider an oriented (m x n)-matrix with m rows and n columns. The origin of the grid is the lower left node of the oriented matrix. We think that a complete regular grid is a simple useful model for generalization purposes. The number of rows and columns of our grid model can be determined from the detected grid graph and every grid point (x, y) can be assigned to a matrix entry (i, j). In general not all matrix entries will have an assigned grid point (like the grey lines in Fig. 5b). The position, orientation and size of the grid cells have to be computed by an adjustment model. Together with the matrix information we have a complete formal model of building blocks with grid structure, which can be used for further generalization processes. To compute an adjusted grid geometry we assume that the detected grid structure is the result of a transformation process which has distorted an optimal axis parallel grid (see Fig. 6).

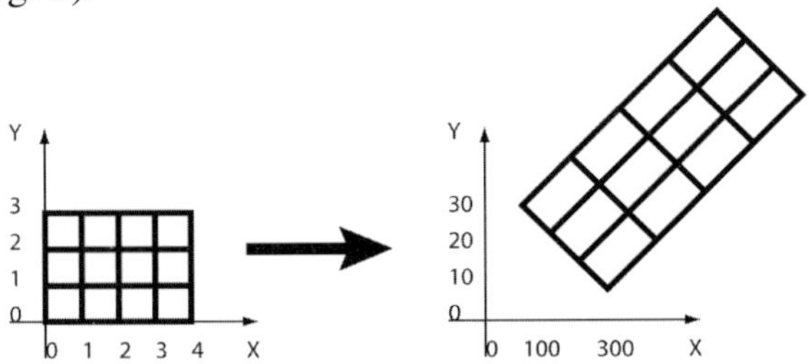

Fig. 6. Distortion process of an optimal axis parallel grid

The transformation parameters are computed by a least square adjustment. To achieve a linear adjustment model the transformation process is modeled as an affine mapping of the grid points (i, j):

$$\vec{p}_{i,j} = \begin{pmatrix} x_{i,j} \\ y_{i,j} \end{pmatrix} = A\vec{g}_{i,j} + \vec{t} = \begin{pmatrix} a & b \\ c & d \end{pmatrix}\begin{pmatrix} i \\ j \end{pmatrix} + \begin{pmatrix} e \\ f \end{pmatrix}, \qquad (1)$$

where i denotes the column and j the row of a grid point. From the detected grid structure we get N transformed grid points as observations and we have to solve the standard least square problem:

$$\sum_{k=1}^{N} \left\| \vec{o}_{i,j}^{\,k} - \vec{p}_{i,j}^{\,k} \right\|^2 \rightarrow Min! \qquad (2)$$

The affine mapping is able to model grid cells with different width and height (see Fig. 5b), but in general the grid cells are not perpendicular (see Fig. 7a). In certain cases it may be useful to model perpendicular grids. This can be achieved by a so-called Helmert transformation

$$\vec{p}_{i,j} = \begin{pmatrix} x_{i,j} \\ y_{i,j} \end{pmatrix} = sR\vec{g}_{i,j} + \vec{t} = \begin{pmatrix} a & -b \\ b & a \end{pmatrix}\begin{pmatrix} i \\ j \end{pmatrix} + \begin{pmatrix} c \\ d \end{pmatrix}, \qquad (3)$$

which describes one scaling, one rotation and one translation. The limitation of the Helmert-Transformation is that one can model only square grid cells, because there is only one scaling factor (see Fig. 7b).

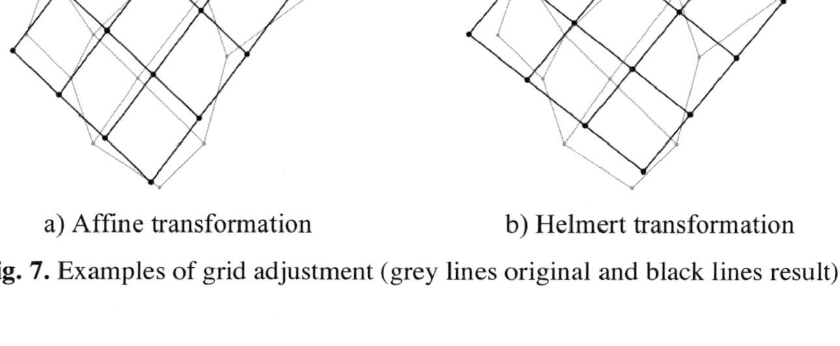

a) Affine transformation b) Helmert transformation

Fig. 7. Examples of grid adjustment (grey lines original and black lines result)

4 Grid Reduction

The adjusted grid structures are the input to the last grid typification process – the grid reduction (simplification). The reduction of elements in a grid is a discrete problem where we cannot remove simply single centroids. If we would do that the grid structure will be destroyed. In the case of grid reduction we can only reduce the number of rows and/or columns. The reduction of the rows and columns is not solvable in a unique way. Figure 8b shows an example with unchanged centroid positions.

a) 25% reduction of rows and columns by symmetrical sampling

b) 60% reduction of rows and columns by simple sampling

c) 60% reduction of rows and columns by symmetrical sampling

Fig. 8. Sampling results of the detected grid structures

If the enclosing area of the grid should remain one needs a symmetrical sampling process, which starts from both sides of the rows and the columns. The result of this approach is shown in Figure 8c. Figure 8a shows a reduction about 25 percent of the rows and columns where both grid structures from Figure 5 are still visible. After the sampling process the determined structure can be moved and scaled to fit the desired generalization constraints. In general after the sampling the distance between the centroids in row and column direction should be adjusted equally. To avoid gabs in the reduced grid and to get equal distances between the simplified grid points one can use the same adjustment process described in the section before. Figure 9 shows an example for the reduction of a given grid (see Fig. 9a) by removing one row of the grid (see Fig. 9b).

After removing the row there is an unbalanced (distorted) grid structure. Using the reduced grid points as observations for a new adjustment process yields to the result shown in Figure 10. It should be mentioned that it is not necessary to use the original grid points for the reduction process. It is also possible to use directly the adjusted grid points for the reduction process.

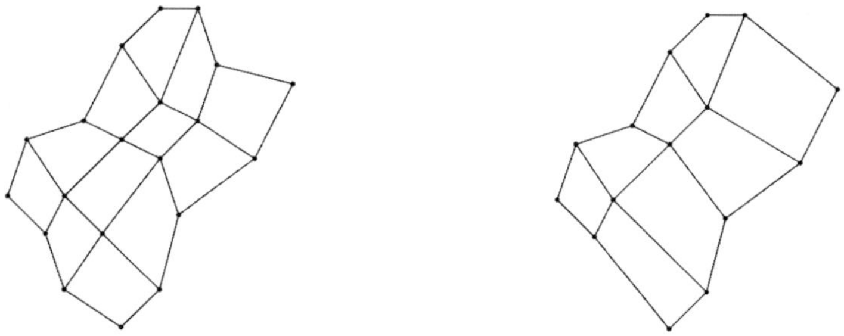

a) before reduction b) after reduction

Fig. 9. Example of grid reduction by removing one grid row

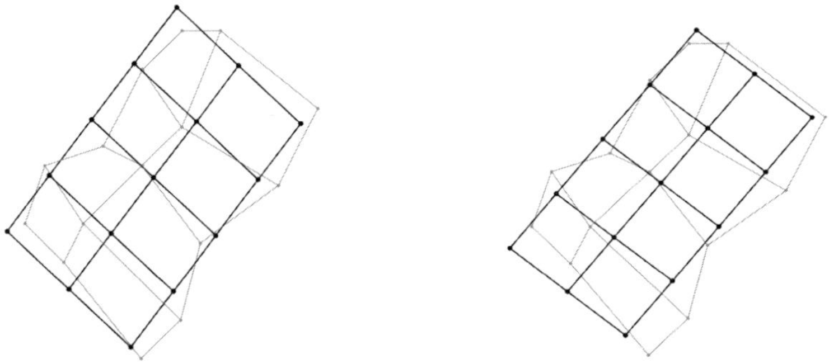

a) Affine transformation b) Helmert transformation

Fig. 10. Adjustment of a reduced grid structure

5 Conclusions and Outlook

In this paper a new approach for the typification of buildings is described. Existing approaches for linear structures did not consider the fact that in the case of building groups often several linear structures are coupled in a certain way and therefore this linear structures can not be processed independently from another without creating wrong results. The explicit modeling of grid structures enables the handling of several coupled linear structures in a consistent way. It has been found that the relative neighborhood graph is appropriate to detect grid structures in point sets. Further it was shown that the adjustment and simplification of such grid structures can be handled by a least square adjustment based on an affine or Helmert datum transformation, which is a well-known geodetic application.

Future work has to be done on the sampling (reduction) process of a grid structure. For now no unique formula exists, which takes into account a map scale as well as related map symbol sizes and distances. It is also interesting if there are useful non-linear adjustment models, which allows better fittings of the grid structure to the detected grid points. In the described linear adjustment models the distance between the detected grid points and the transformed optimal grid points is minimized. It may be better to minimize the perpendicular distance of the observed grid points and the adjusted grid edges.

References

Anders KH (2004) Parameterfreies hierarchisches Graph-Clustering-Verfahren zur Interpretation raumbezogener Daten. PhD Thesis, Institute for Photogrammetry, University of Stuttgart, http://elib.uni-stuttgart.de/opus/volltext/2004/2024/, (urn:nbn:de:bsz:93-opus-20249)

Anders KH, Sester M (2000) Parameter-Free Cluster Detection in Spatial Databases and its Application to Typification. In: Int Archives of Photogrammetry and Remote Sensing, Amsterdam, Netherlands, vol XXXIII, Amsterdam, Holland

Christophe S, Ruas A (2002) Detecting Building Alignments for Generalisation Purposes. In: 10th Int Symp on Spatial Data Handling, SDH 2002, Canada, Ottawa, pp 419–432

Heinzle F, Anders KH, Sester M (2005) Graph Based Approach for Recognition of Patterns and Implicit Information in Road Networks. XXII Int Cartographic Conf (ICC2005), ISBN: 0-958-46093-0

Jaromczyk J, Toussaint G (1992) Relative neighborhood graphs and their relatives. In: Proc IEEE, vol 80(9), pp 1502–1517

Müller J, Wang Z (1992) Area-patch generalization: a competitive approach. The Cartographic J 29:137–144

Regnault N (1996) Recognition of Building Clusters for Generalization. In: Kraak M, Molenaar M (eds) (1996), Advances in GIS Research, Proc of 7th Int Symp on Spatial Data Handling (SDH), vol 1, Faculty of Geod Engineering, Delft, The Netherlands, pp 4B.1–4B.14

Sester M (2000) Generalization Based on Least Squares Adjustment. In: Int Archives of Photogrammetry and Remote Sensing, Amsterdam, Netherlands, vol XXXIII, Part B4, pp 931–938

Sester M, Brenner C (2000) Kohonen Features Maps for Typification. In: Proc at the first GIScience Conf, Savannah, Georgia, USA

Toussaint G (1980) The relative neighborhood graph of a finite planar set. Pattern Recognition 12:261–268

Skeleton Based Contour Line Generalization

Krzysztof Matuk[1], Christopher Gold[2], Zhilin Li[1]

[1] Department of Land Surveying Geo-Informatics
Hong Kong Polytechnic University, Hong Kong
email: kmatuk@op.pl, lszlli@polyu.edu.hk

[2] GIS Research Centre, School of Computing, University of Glamorgan
Pontypridd CF37 1DL, Wales, UK
email: christophergold@voronoi.com

Abstract

Contour lines are a widely utilized representation of terrain models in both cartography and Geographical Information Systems (GIS). Since they are often presented at different scales there is a need for generalization techniques. In this paper an algorithm for the generalization of contour lines based on skeleton pruning is presented. The algorithm is based on the boundary residual function and retraction of the skeleton of contour lines. The novelty of this method relies on pruning not only the internal skeleton branches, but also those skeleton branches placed outside the closed contour polygon. This approach, in contrast to original method which was designed for closed shapes is capable of handling also open polygonal chains.

A simplified version of the skeleton is extracted in the first step of the algorithm and in the next a simpler boundary is computed. The simpler boundary as shown in this paper, can be found using three different ways: detection of stable vertices, computation of an average vertex and approximation of the boundary by Bezier splines.

1 Introduction

Generalization is one of the most important processes in cartography and GIS. As stated by Keates (1989), generalization is a process of adjusting the representation of a phenomenon for adaptation to a map scale. A more complex description of generalization has been presented by McMaster and Shea (1992). The authors divide the process of digital generalization into three parts: the analysis of philosophical objectives (why to generalize), cartometric evaluation (when to generalize) and spatial and attribute transformations (how to generalize). The philosophical objectives, besides adapting the map to viewer needs and reducing complexity, also minimize the use of resources (storage, processing time) and are also important to spatial accuracy and aesthetics.

Contour lines are the most widely used model for the representation of the terrain models in Cartography and GIS. Starting from the Douglas-Peucker algorithm (Douglas and Peucker (1973)), trough ϵ-band by Perkal (1958), natural principle by Li and Openshaw (1993) and line generalization based on the analysis of shape characteristics by Wang and Muller (1998), contour lines can be generalized in a number of ways.

In this work yet another answer to the question: "how to generalize?" is given in the form of a new algorithm for generalization of contour lines. The algorithm utilizes a medial axis of a two dimensional object as described by Blum (1967). In particular the algorithm makes use of a medial axis extracted from samples taken from the boundary of a shape or placed along an open curve. Extraction of the medial axis (or skeleton) from scattered points is described by several authors Attali (1997), Amenta et al. (1998) and Gold and Snoeyink (2001). The algorithm by Gold (1999) and Gold and Snoeyink (2001) appears to be the least complicated and is used in this study.

A simplification of a shape is possible by pruning its skeleton. Two interesting, from the point of view of this study, algorithms have been proposed by Gold and Thibault (2001) and Ogniewicz and Kübler (1995). The first is an iterative process, based on pruning the leaves of the skeleton only. The algorithm moves a leaf vertex of the skeleton to its parent. Simultaneously with pruning the skeleton, the outlying vertex of the boundary is moved onto the circle centered on the parent skeleton vertex.

The second algorithm is based on the boundary potential functions. The value of the boundary potential makes possible extraction of significant parts of a shape and its skeleton. This method has been proven to be a controllable and stable method for shape simplification. Unfortunately, due to taking into consideration internal skeleton branches only, the method does not apply to open polygonal chains.

In the algorithm presented in this paper the potential residual function ΔR_p is utilized as a measure of a feature size. The boundary potential residual as shown in Figure 1 is the distance along a boundary, between the two

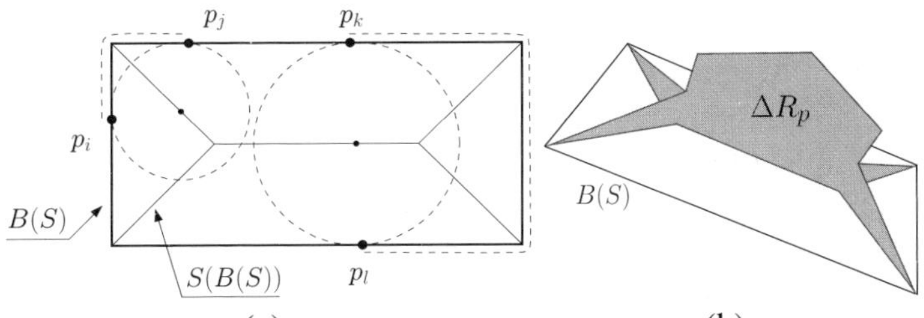

Fig. 1. Boundary distance between two points **(a)** and, a perspective view of the boundary potential function **(b)**

samples connected by a Delaunay edge (eg. $\overline{p_i p_j}$). The value of the distance is assigned to the Voronoi edge dual to $\overline{p_i p_j}$.

In order to find a useful cut off threshold, the idea of defining the smallest visible object (SVO) offered by Li and Openshaw (1992, 1993) is borrowed. The main idea of their algorithm relies on the removal of some of the features which are invisible at the desired scale of the map. The method is known as the Li-Openshaw algorithm and is based on the definition of the smallest visible object (SVO) given by equation:

$$SVO = S_t * D * (1 - \frac{S_f}{S_t}) \quad (1)$$

where:

S_t – scale factor of the source map
D – diameter of the SVO at the map scale. In the range of this diameter all information about the shape of the curve can be neglected
S_f – desired map scale factor

The organization of this paper is as follows: Section 2 gives definitions of symbols and definitions used in further sections. The new algorithm and results obtained of its application to some real data is described in Sections 3 and 4. Finally Section 5 presents conclusions and plans for future development of the algorithm.

2 Definitions

The algorithm presented in this work utilizes a few different techniques. In order to make the argument consistent, the introduction of some common symbols and definitions may be useful. Most are slight modifications of the terms described by Amenta et al. (1998), Ogniewicz and Ilg (1992), Ogniewicz et al. (1993), and Ogniewicz and Kübler (1995). Others are widely used in computational geometry.

The input of the algorithm is a set of samples taken from the boundary of a two dimensional shape. The shape is denoted as F and the samples obtained from its boundary as S ($S \subset F$). In order not to limit considerations to closed shapes, the samples can be also taken from an open curve.

The two dimensional Delaunay triangulation of the sample boundary points is used as a starting point. It is denoted as $\mathcal{DT}(S)$, the dual graph of the Delaunay triangulation, the Voronoi diagram is denoted as $\mathcal{V}(S)$. At the same time, for any edge e, $e \in \mathcal{DT}(S)$, the symbol $\mathcal{D}(e)$ denotes the Voronoi edge dual to e; which means that $\mathcal{D}(e) \in \mathcal{V}(S)$.

The boundary of the closed shape F can be approximated by a polygon. At the same time, a polyline can be used to approximate the boundary of an open curve. Since a distinction is not made between open and closed objects at this stage, both are denoted as $\hat{B}(S)$. The input set is not restricted to one boundary, which means that $\hat{B}(S)$ may be composed of many open curves and/or many closed shapes. This leads to the following statement: $\hat{B}(S) = \{\breve{B}_1(S), \ldots, \breve{B}_N(S)\}$, where $\breve{B}_k$ is any shape of the input set.

The medial axis of some shape F is a set of points having at least two nearest neighbors on the boundary of F. It has been proven by Brandt (1994) that when the sampling density approaches infinity, the Voronoi diagram of the boundary samples of a two dimensional shape converges to the medial axis of F. Sampling density in real data is never infinite, hence the medial axis of F can only be approximated. The approximation of the medial axis or the skeleton is denoted as $\mathcal{S}(\breve{B}_k(S))$.

3 Polygon and Polyline Simplification

As mentioned in the previous sections, the existing algorithms based on skeleton pruning have some disadvantages. These shortcomings prevent them from being applied for controllable generalization of contour line models of terrains. The method proposed by Ogniewicz et al. needs information about which skeleton part is internal for each polygon. While the algorithm by Gold and Thibault (2001) is able to prune the leaves of the skeleton only. It can not perform more extensive pruning. The algorithm presented here combines the advantages of both methods. It is able to perform a simplification by retraction of the skeleton branches on both sides of a curve. The simplification is not limited to skeleton leaves and is driven by the boundary potential function.

3.1 Simplification of a Skeleton

The simplification process starts from the extraction of the skeleton and computation of the boundary potential values for each Voronoi edge which is part of the skeleton. This is followed by extraction of the skeleton edges which have boundary potential greater than some threshold t ($t \in R^+$) (see Fig. 2). This leads to the following definitions:

Definition 1. *For some threshold* $t \in R^+$, $\mathcal{S}(\breve{B}_k(S),t)$ *denotes the skeleton of* $\breve{B}_k(S)$ *simplified for threshold* t.

Definition 2. *For some threshold* $t \in R^+$, $\breve{B}_k(S,t)$ *denotes the boundary* $\breve{B}_k(S)$ *simplified for threshold* t.

The pruning of the skeleton causes elimination of two dimensional features of a contour line. This operation, if performed in a coordinated way on all the contour lines, should prevent them from intersecting one another. The coordination must assure that if one polygon is placed inside the other before simplification, it will remain enclosed by the external polygon after simplification as well. The same condition should be fulfilled in the case of the variations on the contour lines or in other words two dimensional features. This means that not only must the direction of a simplification be the same for all corresponding two dimensional features but also the amount of simplification has to remain on a similar level.

A simplification parameter chosen for driving the simplification has to be both independent of the planar position of a contour line and also an elevation of the contour line. It should also be invariant under horizontal and vertical translations. Another restriction is the complexity of its computation. Cumulation of a terrain model size and the computation complexity

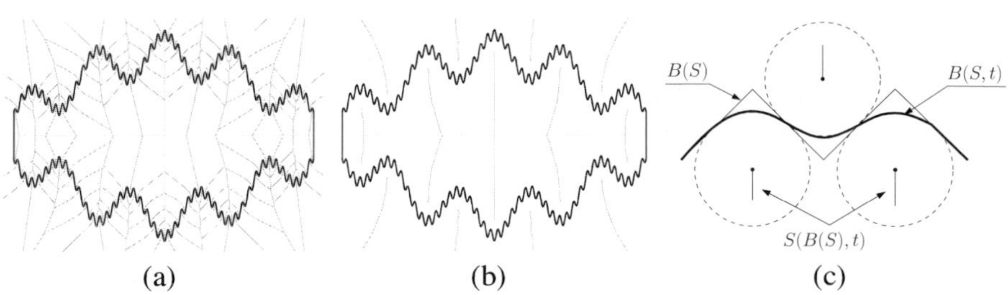

Fig. 2. Simplification of the skeleton. Source shape and its skeleton **(a)** and the same shape with the skeleton simplified for some threshold **(b)**. Interpolation of the shape between medial circles placed on the opposite sides of the boundary **(c)** (bold line)

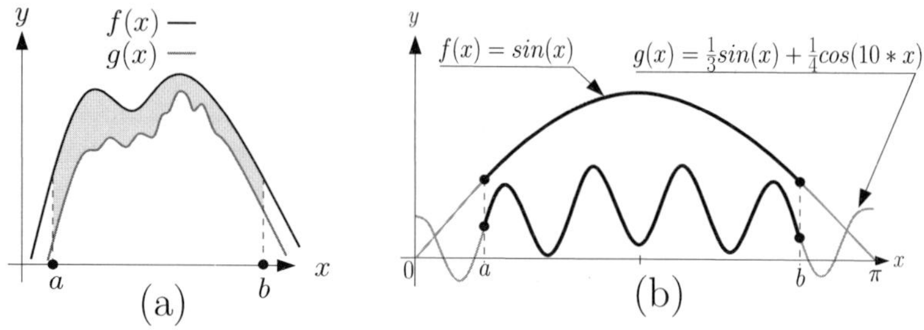

Fig. 3. Area enclosed by $f(x)$ is greater than area enclosed by $g(x)$ **(a)** the same is not always true in the case of the perimeter **(b)**

of the simplification parameter may seriously affect the performance of the simplification algorithm.

Two main quantitative parameters are related to the size of a closed planar figures. One is the area and the other the perimeter. Both are invariant under translation and rotation and both can be computed in linear time ($O(n)$). As can be seen in Figure 3a the area enclosed by an object ($g(x)$) placed inside another object ($f(x)$) is always smaller. Unfortunately this is not always true in the case of the perimeter. In fact it is quite easy to show a counterexample (see Fig. 3b). However, the situation as presented in Figure 3b is very unlikely to occur in the case of contour line data. Additionally the perimeter has been well described in the literature as the parameter for the skeleton based shape simplification. Owing to these reasons, the perimeter is used in the later parts of this study.

3.2 Extraction of the Simpler Shape from the Simplified Skeleton

In order to perform the skeleton based shape simplification, the problem of interpolation of the shape of medial circles, centered in the leaves of the simplified version of a skeleton (see Fig. 2c), must be solved. This problem is approached in this study in three different ways:

- detection of the stable vertices on the boundary
- computation of the average vertex
- approximation of the new shape by Bezier splines

The first method is based on the leaves of the simplified skeleton. Every leaf of the simplified skeleton is a Voronoi edge. This Voronoi edge is generated as a dual to some Delaunay edge between two sample points (see Fig. 4a). The two samples are generators of an important skeleton branch, which means that they are also important for the whole shape and may be selected as a part of its representation.

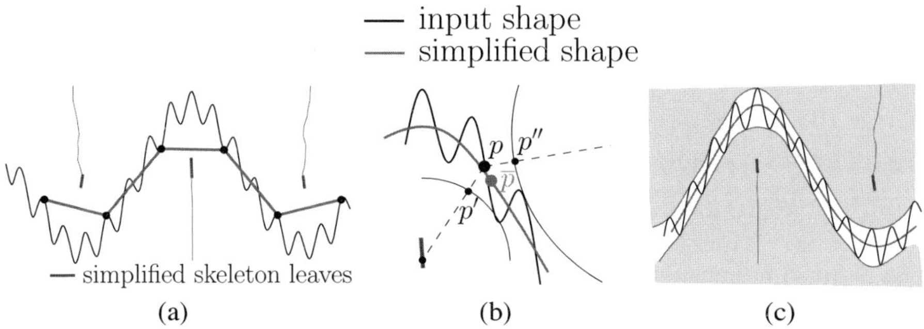

Fig. 4. Stable vertices detection **(a)**, an average vertex computation **(b)** and approximation by the Bezier splines **(c)**

The method, based on computation of the average vertex, treats both sides of the curve, being simplified, as two separated entities. Since the algorithm does not make a distinction as to which side is internal or more privileged, every sample is pruned to circles on two sides of the curve. This causes the sample to take more than one position after simplification (points p' and p'' in Fig. 4b). Since none of the positions is more privileged, average coordinates of the sample are computed and returned as its position after the simplification (point $\overline{p}$ in Fig. 4b).

The third method is similar to the computation of the average vertex. It also treats two sides of the curve separately. This results in two shapes bounding possible simplified curves on both booth sides (gray regions in Fig. 4c). The new, simplified shape can be obtained by creating a curve which is an interpolation between the two bounding shapes.

3.3 Generalization by Detection of the Stable Vertices

The algorithm presented here performs the skeleton retraction on both sides of the input shape. Figure 5a shows simplification of a shape for a small pruning threshold. The endpoints of the Delaunay edges ($\overline{s_1 s_2}$, $\overline{s_3 s_4}$, $\overline{s_5 s_6}$) dual to the leaves of the simplified skeleton are not influenced by retraction of the skeleton branches on the opposite side of $B(S)$. The threshold is small enough and does not cause conflicts on the neighboring features. With regard to the selected pruning threshold a new shape is obtained. The new shape is represented by vertices s_i, $i \in \{1 \dots 6\}$ and potentially endpoints of the input, if $B(S)$ is a polyline.

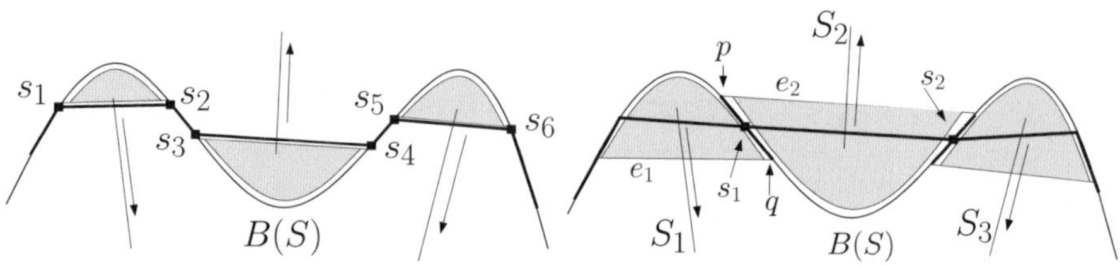

(a) (b)

Fig. 5. Small pruning threshold **(a)**, no conflicts occur and a bigger pruning threshold **(b)** conflicts on boundary occur

A more complicated situation occurs when the pruning threshold is greater than in the previous case. Figure 5b shows the situation, when some of the parts of $B(S)$ are pulled in opposite directions by adjacent skeleton branches. In order to obtain simplification of the boundary a vertex in which retraction balances must be found.

Let's consider the Delaunay edges e_1 and e_2 presented in Figure 5. They are dual to the skeleton leaves, being the result of pruning the skeleton branches s_1 and s_2 placed on the opposite sides of the boundary $B(S)$. If a Delaunay edge $e \in \mathcal{DT}(S)$ and $\mathcal{D}(e) \in \mathcal{S}(\breve{B}_k(S)$,t), e is called an importance marker of its endpoints.

In the case of retraction of S_1 the process can not retract the boundary $B(S)$ further than the boundary vertex q. While in the case of S_2, it will retract the boundary $B(S)$ to at most vertex p. For simplicity let us assume that

Algorithm 3.1: Shape generalization by detection of stable vertices

Data: Samples ($p_i \in \mathcal{P}$) from a shape ($\mathcal{S}$) , simplification threshold $t \in \mathbb{R}$
Result: Samples from a simpler shape, $p'_i \in P'$

```
for each B̆_k(S) ∈ B̂(S) do
    if B̆_k(S) is polyline then Add endpoints to ImpB(S) and SimB(S);
    for s ∈ B̆_k do
        n ← number of importance markers of s;
        if n > 0 then Add s to ImpB(S);
    end
    DetectStableVertices(ImpB(S));
end
```

the point of equilibrium is in the middle of the boundary between p and q. The case when $p = q$ may also occur, but in this case it can be also assumed that this vertex is the point of balance. The pseudocode of the algorithm can be found in Algorithm 3.1 and Algorithm 3.2 sections.

Algorithm 3.2: DetectStableVertices()

Data: Points important to a shape $ImpB(S)$
Result: Stable vertices from $ImpB(S)$
for *each* $q \in ImpB(S)$ **do**
  if q *is leaf* **then**
    $p \leftarrow$ predecessor of q (previous leaf) ;
    $n \leftarrow$ number of importance markers of q;
    $i_q, i_p \leftarrow$ importance markers of q and q;
    if *two of importance markers of* q *are on opposite sides of* $\breve{B}_k$ **then**
      Add q to $SimB(S)$;
    end
    else
      if i_q *and* i_p *are on the same side of* $\breve{B}_k \cap \overline{pq} \in \mathcal{DT}(S)$ **then**
        Add p and q to $SimB(S)$;
      end
      if i_q *and* i_p *are not on the same side of* $\breve{B}_k$ **then**
        Add $\lfloor (p+q)/2 \rfloor$ to $SimB(S)$;
      end
      $p \leftarrow q$;
    end
  end
end
return $SimB(S)$;

3.4 Generalization by Computation of an Average Vertex

The simplification process based on the simplification of the skeleton can be seen as a retraction of a boundary sample (p) to its parent circle (q). The parent circle is a first circle meet during the traversal from a skeleton vertex (C_0) (adjacent to the boundary sample) to the leaf node (C_t) of a simplified skeleton (see Fig. 6).

On a plane, every non self intersecting curve has two sides. The retraction can process in both directions. Additionally there can be more than one adjacent skeleton branch on each side of the curve. This means that every boundary sample can be retracted to at least two circles. In the case presented in

Figure 6b the vertex p_i is a sample from a sinusoidal curve. A simplification of the curve for some threshold causes p_i to be retracted to the circles c_{i_1} and c_{i_2}, centered in C_{i_1} and C_{i_2} respectively.

The vertex p_i after a retraction for some threshold to c_{i_1} takes the position denoted as p_{i_1} and after the retraction to c_{i_2} takes position denoted as p_{i_2}. Since there is no additional information about which of the circles is more important or which is internal, both positions are equally privileged. The new position of p_i after simplification is computed by averaging coordinates of p_{i_1} and p_{i_2} and denoted as p_i'.

The simplification threshold in the case of the average vertex method must be selected very carefully ownig to a possibility of generation of cracks on a simplified curve (Figure 6c). A solution for this problem can be execution of the average vertex algorithm in a few iterations, starting from a very small simplification threshold and gradually increasing it in every iteration. The same method will solve the problem presented in Figure 3b. The first iterations remove all high frequency components from the longer internal curve and cause very little effect on the shorter, external one.

3.5 Generalization by Approximation by the Bezier Splines

As mentioned above, the extraction of a simpler shape can be based on interpolation between two bounding regions (gray areas in Figure 7a). The regions are created as the result of independent retraction of the skeleton in two directions. In order to compute the control points for the splines the simplified skeleton is triangulated. Some triangles are constructed on skeleton vertices which belong to the skeleton branches placed on opposite sides of

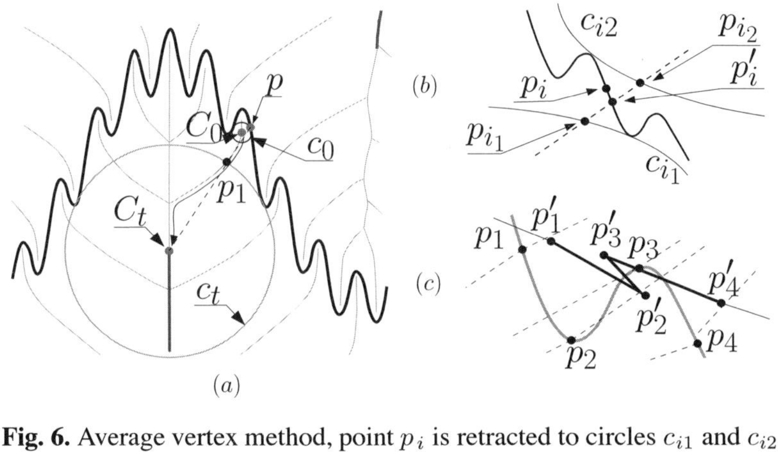

Fig. 6. Average vertex method, point p_i is retracted to circles c_{i1} and c_{i2}

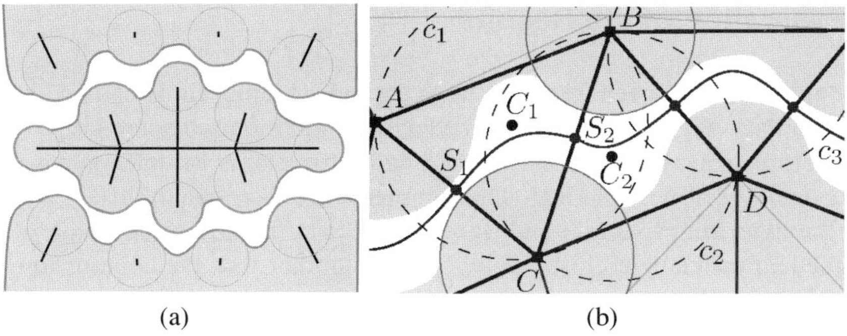

Fig. 7. The Bezier splines method. Simplified skeleton **(a)** and triangulated. Triangles constructed on different skeleton branches are taken into consideration (e.g., ΔABC and ΔBCD) **(b)** and the splines approximating the new shape are constructed

the shape being simplified. In the next step, a spline is constructed passing through points S_1 and S_2 (see Fig. 7b). The points are placed half way between the boundaries of the medial balls centered in points A and B (point S_1) and B and C (point S_2). The control point for the spline between points S_1 and S_2 is the circumcenter (C_1) of the circle c_1 constructed on the triangle ΔABC. The circumcenter C_2 of the circle c_2 is a control point for the next part of the spline. The circle c_2 is constructed on the triangle ΔBCD.

3.6 Generalization, Adaptation of the SVO

According to the definition presented in Li and Openshaw (1992), the smallest visible object (SVO) is equal to a circle with a radius related to the scales of both the target and the source map. In order to adapt it for use with residual functions, a slight modification of the condition used by Li and Openshaw (1992) has been made:

$$\Delta R_H(e) \leqslant \frac{d\pi}{2} \tag{2}$$

where d is the diameter of the SVO, as can be obtained using Equation 1. The meaning of the threshold obtained from Equation 2 is that the algorithm will remove all features for which the boundary length is smaller than half the perimeter of the circle representing SVO.

4 Evaluation of the Results

The algorithm has been tested on three datasets: Maple, Jin Island and Layer 6. Each have slightly different characteristics and required some preprocessing before generalization. The maple dataset contains samples taken from the boundary of a maple leaf. The samples were distributed closely enough for the algorithm by Gold and Snoeyink (2001) to extract the shape of the maple leaf as a single closed polygon. For another test, the contour lines of Jin Island in Hong Kong at the scale 1:10,000 were used. Data points from each layer of the input were triangulated on the first pass. On the next pass, each edge of the input was found in the corresponding triangulation and labeled as a part of the crust. All other Delaunay edges were considered as dual to the skeleton edges. The last dataset, Layer 6, was prepared as a set of points from one of the layers of Jin Island data. The crust and skeleton were extracted using the one step crust and skeleton extraction algorithm. Tests were conducted on a Pentium(R) M 1500 MHz laptop with 512 MB of RAM memory. For computation of the Delaunay triangulation, the Linux based CGAL library was used with lazy exact arithmetic.

The Maple leaf dataset was simplified for three different values of the smallest visible object: 5, 10, 25. They were substituted to Equation 2 and yield in following values of a cut off threshold ($\Delta R_H(e)$): 7.85, 15.71, 39.27 receptively. Figure 8 presents the results of the simplification.

The contour lines representing Jin Island in Hong Kong are in the scale 1:10,000. Tests have been conducted for reduction to 1:20,000 and 1:50,000 and 1:100,000, using different values of the SVO in output scale. Figure 9a presents the source contour lines of Jin Island at a scale of 1:10 000. Figures 9b...c present results of the generalization to scales 1:20 000 and 1:50,000, and the diameter of the SVO in the map scale set to 0.8 mm. As can be noticed in the images for scales 1:20,000 (see Fig. 9b) and 1:50,000

(a) (b) (c)

Fig. 8. Maple leaf simplified for SVO=5 **(a)**, simplified for SVO=10 **(b)**, simplified for SVO=25 **(c)**

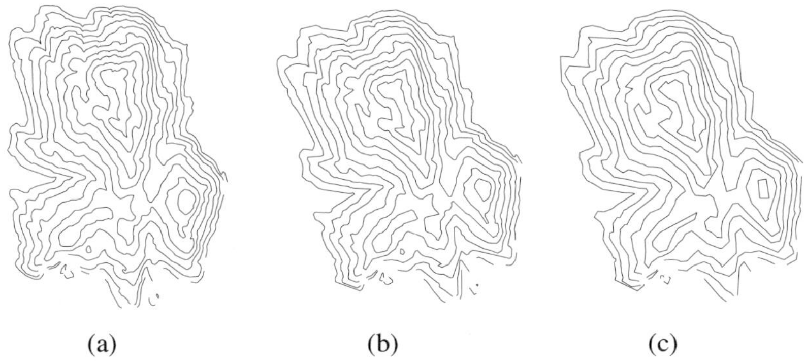

Fig. 9. Jin Island, (a) source data, (b) simplified for d=8, (c) simplified for d=32

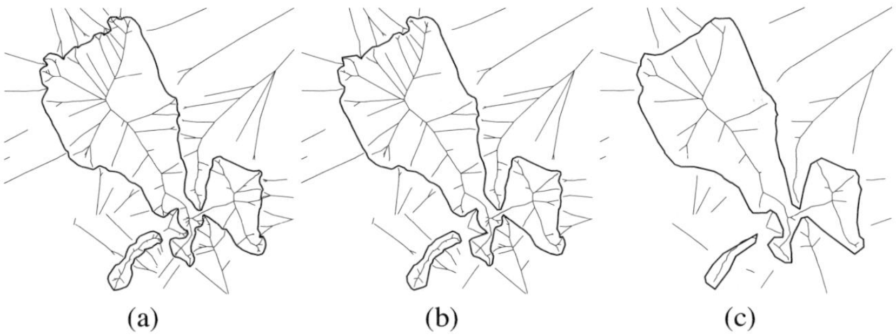

Fig. 10. Layer 6 of the contour lines of Jin Island, (a) source data, (b) simplified for d=8, (c) simplified for d=32

(see Fig. 9c) the algorithm behaves quite well and performs fully automatic generalization.

The preprocessing speed in case of the extraction of the crust and skeleton information, using the one step algorithm, was on average around 1,900 points/s. A noticeable drop in performance occurred in the case of extraction of the crust/skeleton information from coverage files (1,100 points/s). Extraction of a simplified boundary was performed from more than 1,000 vertices per second up to over 10,000 vertices per second.

4.1 Skeleton Pruning vs. Manual Generalization.

Figure 11 shows results of applying the algorithms based on the stable vertex method and average vertex to contour lines of Jin Island given in scale 1:1,000. The results are compared to contours simplified manually with the help of the Douglas-Peucker algorithm to scale 1:10,000.

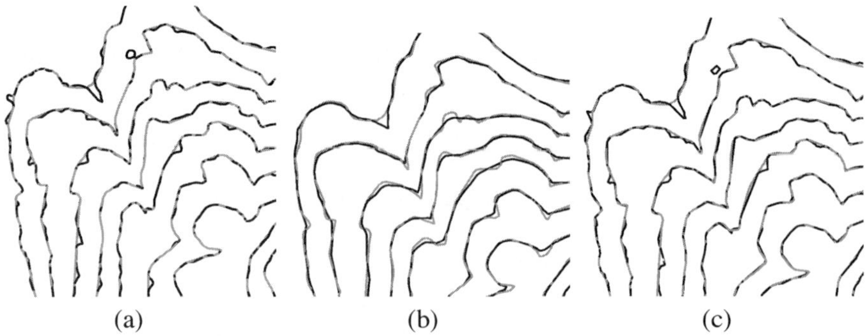

Fig. 11. Simplification of Jin Island **(a)** average vertex, SVO=4.5 m, **(b)** average vertex, SVO=45 m, **(c)** stable vertex, SVO=4.5 m. Gray line represent contour lines generalized manually, and black results of skeleton based generalization

5 Conclusions and Future Work

The proposed algorithms give good results in reasonable time. Their main advantage is that pruning can be parametrized using different types of parameters. Not only the boundary length, as shown in this paper but also the area covered by features and the parameters presented in work by Ogniewicz et al. may be used. According to tests performed so far the average vertex algorithm does not cause intersections between neighbouring contour lines even when applied with a big simplification threshold. However some cracks may occur when the pruning threshold is too big. In this case simplification should be performed for a few iterations with smaller thresholds, before applying the desired threshold.

Some problems which appear during simplification of bent, featureless objects (see Fig. 12) using the stable vertex method are the subject of ongoing research. The stable vertex algorithm also does not take into consideration the size of neighbouring features and computes the stable vertex exactly in

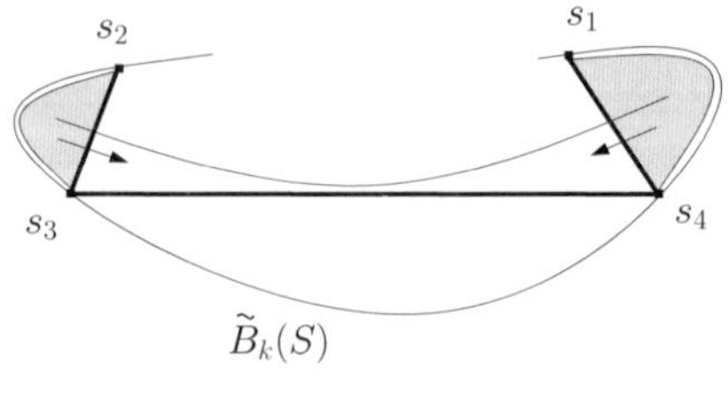

Fig. 12. Simplification of the bent, feature less shape

the middle of two leaf vertices, generated by skeleton branches placed on the opposite sides of the shape. Hopefully this issue can be solved in the future.

Storage of each layer of the contour line model in separate triangulations makes possible parallelization of the process. This should result in better performance on multiprocessors machines.

The work presented here is a part of a bigger project. The aim of this project is to remove three dimensional terrain features by simplification of 2D cross sections. All algorithms presented here were in fact designed to achieve this goal. In the future it is hoped to give theoretical guarantees for the algorithms. The guarantees should show that skeleton based simplification prevents contour lines from intersecting one another. Utilization of the area as the parameter driving the generalization process seems to be promising. Future research focuses on utilization of the area as well as on further development of the Bezier splines approximation algorithm.

Acknowledgments

The work described in this paper was fully supported by The Hong Kong Polytechnic University.

Terrain data where obtained thanks to courtesy of the Survey and Mapping Office, Lands Dept., HK SAR Government.

The code was implemented in C++ and uses CGAL library CGAL (2005) for the Delaunay triangulation.

References

Amenta N, Bern M, Eppstein D (1998) The Crust and the B-Skeleton: Combinatorial Curve Reconstruction. Graphical models and image processing: GMIP 60(2):125–

Attali D (1997) r-Regular Shape Reconstruction from Unorganized Points. In: Proc of the 13th ACM Symposium on Computational Geometry, pp 248–253

Blum H (1967) A Transformation for Extracting New Descriptors of Shape. In: Models for the Perception of Speech and Visual Form. MIT Press, pp 362–380

Brandt JW (1994) Convergence and continuity criteria for discrete approximations of the continuous planar skeleton. CVGIP: Image Understanding 59(1):116–124

CGAL (2005) http://www.cgal.org

Douglas DH, Peucker TK (1973) Algorithms for the reduction of the number of points required to present a digitized line or its caricature. The Canadian Cartographer 10(2):112–122

Gold C (1999) Crust and Anticrust: A One Step Boundary and Skeleton Extraction Algorithm. In: SCG '99: Proc of the Fifteenth Annual Symposium on Computational Geometry. ACM Press, New York, NY, USA, pp 189-196. ISBN 1-58113-068-6

Gold CM, Snoeyink J (2001) A One-Step Crust and Skeleton Extraction Algorithm. Algorithmica 30(2):144–163

Gold CM, Thibault D (2001) Map generalization by skeleton retraction. In: Proc 20th Int Cartographic Conf (ICC 2001). International Cartographic Association, pp 2072–2081

Keates JS (1989) Cartographic Design and Production, 2nd ed. Longman Scientific and Technical, New York

Li Z, Openshaw S (1992) Algorithms for automated line generalization based on a natural principle of objective generalization. Int J of Geographical Information Systems 6(5):373–389

Li Z, Openshaw S (1993) A natural principle for the Objective Generalization of Digital maps. Cartography and Geographic Information Systems 20(1):19–29

McMaster RB, Shea KS (1992) Generalization in Digital Cartography. Association of American Geographers, Washington, D.C.

Ogniewicz R, Ilg M (1992) Voronoi Skeletons: Theory and Applications. In: Proc IEEE Conf Computer Vision and Pattern Recognition, CVPR. IEEE Computer Society, Los Alamitos, California, pp 63-69. ISBN 0-8186-2855-3

Ogniewicz R, Szekely G, Naf M (1993) Medial manifolds and hierarchical description of 2d and 3d objects with applications to MRI data of the human brain. In: Proc 8th Scandinavian Conf on Image Analysis, pp 875–883

Ogniewicz RL, Kübler O (1995) Hierarchic Voronoi skeletons. Pattern Recognition 28(3):343–359

Perkal J (1958) An attempt at objective generalization, Translated by Jackowski W from Julian Perkal, Proba obiektywnej generalizacji. Geodezja I Kartografia VII(2):130–142

Wang Z, Muller JC (1998) Line Generalization Based on Analysis of Shape. Cartography and Geographic Information Systems 25(1):3–15

Conflict Identification and Representation for Roads Based on a Skeleton

Stuart Thom

Ordnance Survey, Research and Innovation, Romsey Road, Southampton, UK; email: stuart.thom@ordnancesurvey.co.uk

Abstract

This paper presents a method to detect, represent and classify conflicts between roads. The dataset used in the study is OS Integrated Transport Network™ (ITN). A partial skeleton is created from a constrained Delaunay triangulation of the road network. The skeleton is used to interrogate the space between the roads, creating 'conflict region' features. Nine types of 'conflict region' are characterized and examples of each are given, created from ITN data. A discussion is presented of possible uses for these features showing how they help to orchestrate the removal of conflicts from a road network as part of an automatic generalization process.

Key words: map generalization, road network, conflict detection, Delaunay triangulation, skeleton

1 Introduction

This paper presents a method to detect, represent and classify conflicts between roads. The aim is to provide the information required by a generalization process to solve these conflicts. Conflicts occur when the symbolization of the roads at the target scale overlap with neighboring roads.

There have been a number of studies aiming to resolve this problem. (Nickerson 1988) developed algorithms to detect road section interference and perform subsequent displacement. Mackaness and Mackechnie (1999) used graph theory, the graph being used to determine connectivity of junctions, to remove road sections in order to simplify road junctions.

The use of spatial structures to determine distance relationships was reported by Jones, Bundy, and Ware (1995) who used a constrained triangulation to detect conflicts on narrow sections of polygon boundaries. Narrow sections were detected by an analysis of the width of triangles. Conflict regions were then formed by sets of (entire) triangles, which were identified as too narrow. Van der Poorten and Jones (2002) have used constrained Delaunay triangulations and classification of triangles for line simplification. More recently (van Kreveld and Peschier 1998) used the Voronoi diagram to determine whether two roads were in conflict.

Work at IGN (the French National Mapping Agency) by Lemarie (2003) used the Beams displacement algorithm (Bader 2001) to resolve conflicts in their road network when depicted at 1:100,000 scale.

In our work the target map was at 1:50,000 (1:50k) and the level of conflicts, especially in urban areas is high. Displacement using spatial structures has been successful in building generalization (see Ai and van Oosterom 2002; Højholt 2000) where conflict levels are also high. It seemed a sensible extension to develop methods to utilize spatial structures built from linear features, as they had been so useful when generalizing polygon objects. For a thorough survey on the use of spatial structures in generalization see Regnauld (2005).

The software platform chosen for developing this generalization prototype is Clarity (Neuffer, Hopewell, and Woodsford 2004) developed by Laser-Scan(UK).

The paper is structured as follows: in Section 2 the reasons for using the partial skeleton are given. Creating a partial skeleton from the 'constrained' Delaunay triangulation is described in Section 3, the generation of the conflict regions is explained in Section 4 and their characterization forms Section 5. Finally some suggestions for their utilization are given in Section 6 followed by the conclusion.

2 Why Use the Skeleton to Represent Inter Feature Distances?

The approach adopted was to create conflict information using a skeleton derived from a 'constrained' Delaunay triangulation. There are important advantages of using the triangulation and its skeleton over traditional distance measuring techniques:

2.1 Property of the Skeleton

The most important property of the skeleton is that by definition it's a line that represents the equidistance to the sides of a polygon. As used here the two sides are two separate roads.

2.2 Exclusion Zone at Road Junctions

In the work by van Kreveld and Peschier (1998) they say that two roads are in *conflict* if their minimum distance is less than ***d***. Two roads that start (end) at the same intersection always have distance 0, but we do not want the algorithm to detect a conflict in this case. They redefined the smallest distance between two roads as the smallest distance while excluding both parts of the road closer than ***d*** from the two ends. The method used here utilizes the triangulation skeleton to give a more subtle end zone.

Figure 1 shows how the skeleton, created from a triangulation of the network of road centerlines, is used to filter distance measurement. The displayed roads with thick black pecks are depicted at their respective widths. In this study the radius of the exclusion zone (black circle) at a junction has been set to the largest of its road's widths. Measurements of the distance between the roads are taken at reference points along the skeleton starting from the junction but are ignored until the reference point is outside the zone i.e. transition occurs at the points indicated by arrows.

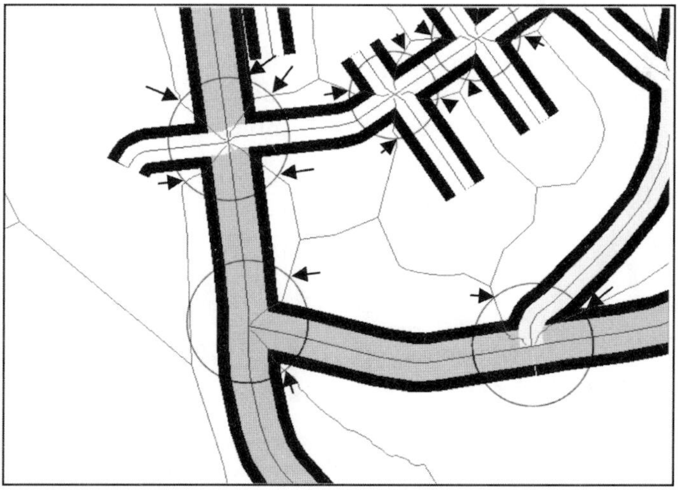

Fig. 1. Showing exclusion zones along the triangulation skeleton

2.3 Cul-de-sac End Gap

In urban areas culs-de-sac are plentiful. Their generalization has previously been limited to strategies such as delete all those whose length is below some threshold, or trim all culs-de-sac by some global length. This is treating them without reference to their context. It is thought that displacement algorithms, such as Beams, use the half width as a buffer to prevent conflict along their length *and* at their end, as illustrated in Figure 2.

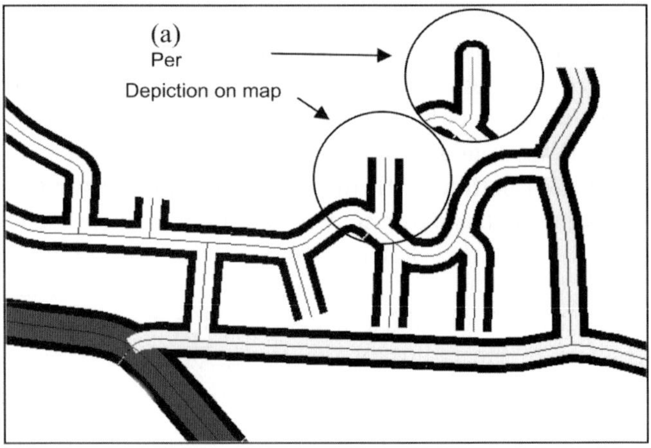

Fig. 2. Cul-de-sac widths

A more context sensitive approach is adopted here since, through the triangulation, the software can tell whether the end point of the cul-de-sac, or some interior point on the line, is in conflict. This allows the possibility of a separate threshold distance for cul-de-sac end point conflicts and would mean that culs-de-sac could be trimmed back to allow a user supplied constant gap between them and the other road that might be more pleasing to the eye (see Fig. 3).

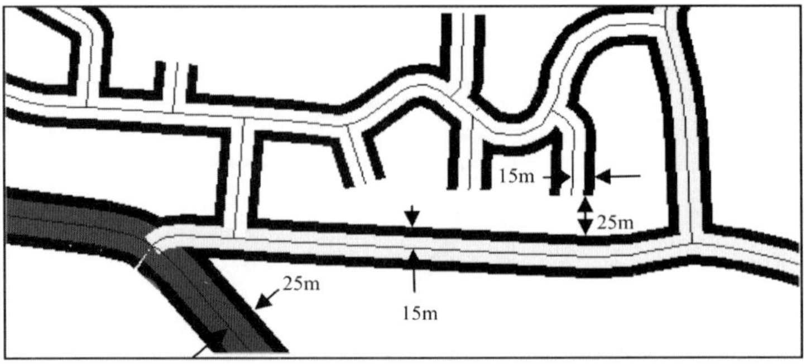

Fig. 3. Display after conflict resolution

3 Creating Partial Skeletons

Here we describe the initial generation of road strokes and triangulation. There follows the method used to grow the skeleton from the road junctions into the spaces between the strokes, to form the limbs. Then the method to connect limbs by tracking the backbones is detailed.

3.1 Creating the Strokes

The following processes rely on the road sections having been chained together to produce 'strokes'. In this study we have used a limiting angle of 120° in stroke production. Sections joined at two-armed nodes (i.e. two and *only* two road sections join at this point) are always placed in the same stroke whatever the continuity. Strokes have no branches but they may intersect one another. Sections can belong to only one stroke.

3.2 Creating the Triangulation

The road network needs to be divided into manageable areas to aid computation. The dataset can be partitioned using the major roads (M-ways, A-roads and B-roads) as boundaries to give a set of non-overlapping polygons, which are known as the partition areas. All the road sections within and comprising a partition area are used to generate a triangulation, e.g. that illustrated in Figure 4.

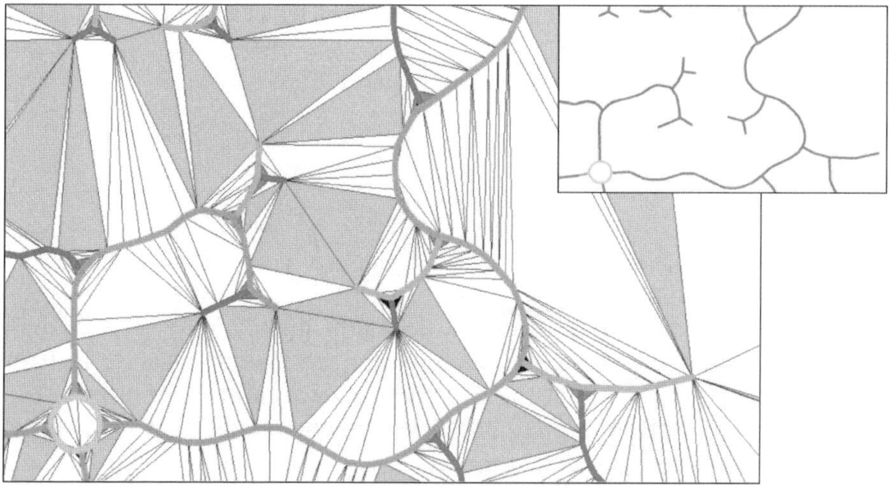

Fig. 4. Triangulation, various grayscales show separate strokes (inset shows just roads)

All the vertices of these features are used to populate the nodes of the triangulation. The triangulation was a 'constrained' type: – all the edges present in the road sections are forced into the triangulation. Additionally the triangulation was 'densified', adding vertices until all pairs of consecutive vertices are closer than some "densify interval". The interval used in this study was 10 meters. Where road sections were ***shorter*** than 10 meters, an interval of half their length was applied; therefore all road sections are represented by ***at least*** two triangle edges.

3.3 Construction of the Skeleton

The triangulation shown in Figure 4 has three distinct types of triangles:

- *seed triangles* (dark gray), those produced where two constrained edges meet at a *road junction point*, i.e. a point on the road network where two or more strokes join or cross.
- *branching triangles* (light gray), whose three nodes sit on road sections from three different strokes.
- *measurable triangles* (white), those produced with only one constrained edge.

In this study the skeleton refers to a graph, which consists of

- *nodes* situated at the centroid of seed and branching triangles, and
- *edges* which connect the skeleton nodes together.

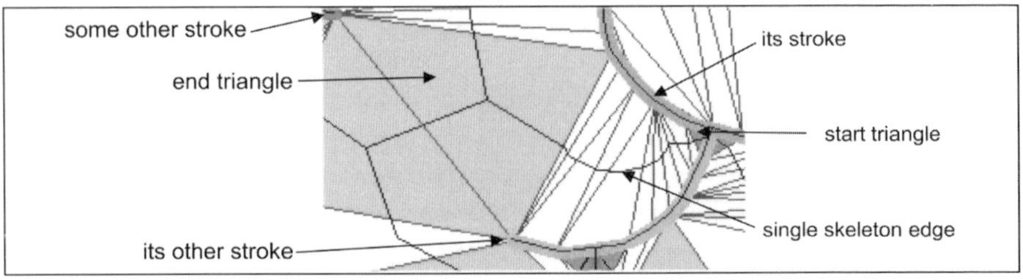

Fig. 5. The structure of a limb skeleton edge and its associated triangles

The skeleton edge steps from the skeleton node of the start triangle through the midpoint of the *midline* of each measurable triangle and finishes at the skeleton node of the end triangle. The *midline* of a triangle is defined as the line from the midpoint of the constrained edge to its opposite node. Figure 6 shows one of the triangle's *midline* as a dashed line.

Two types of skeleton edges can be defined depending on their terminating triangles:

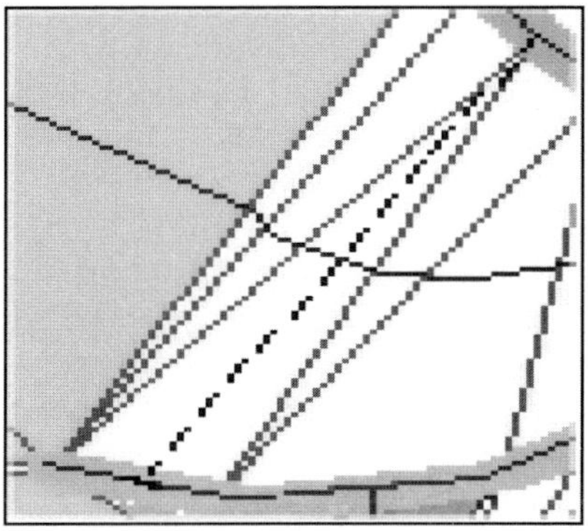

Fig. 6. The triangle midline

- a *limb* has a seed triangle at its start, its end can be a seed or a branching one.
- a *backbone* has branching triangles at its start and end.

3.3.1 Creating the Skeleton Limbs

For each seed triangle we track adjacent triangles on the basis that they span the same two strokes. Tracking stops when the adjacent triangle's test node, i.e. the node not in the pairs common edge, either:

1. holds neither of the strokes we've been tracking against, i.e. we have reached a branching triangle, or
2. holds both the strokes we've been tracking against, i.e. we have reached a seed triangle.

3.3.2 Creating the Skeleton Backbones

After we have created our limb skeleton edges, we begin tracking from the skeleton nodes that are situated at branching triangles. We act as follows:

- if a branching triangle has three skeleton edges, it is deemed fully tracked, otherwise
- a branching triangle's untracked triangle edge is tested to see if it has different strokes at its ends. If it does we use the triangle and its edge to spawn tracking, which proceeds until it reaches another branching triangle (see end condition 1 in the previous section).

In Figure 7 all the branching triangles are three-way branches, however there can be two-way branches, for example the case described in the next section. The resulting graph is termed a partial skeleton, since some parts of the triangulation are not visited by the skeleton, as can be seen (arrowed and elsewhere) in Figure 7. The triangles in these parts are attached to a single stroke, and are ignored.

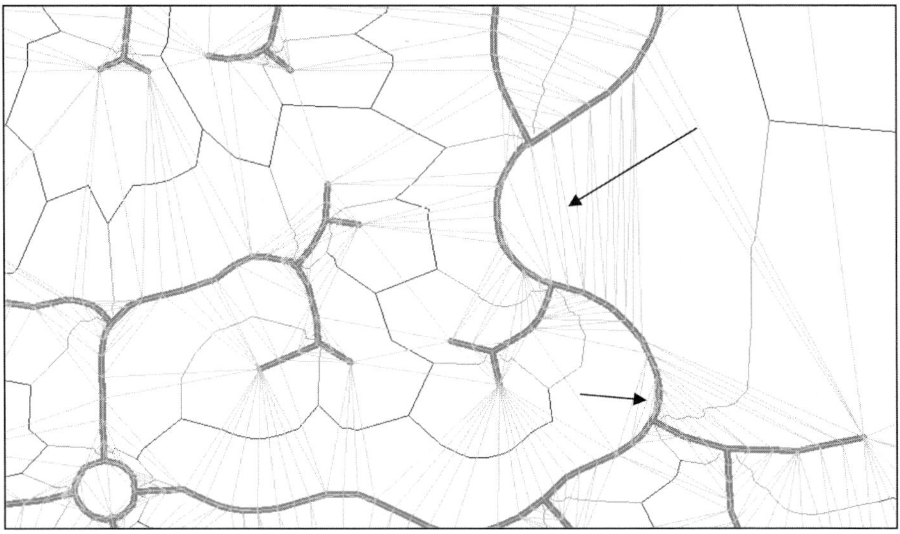

Fig. 7. Showing triangulation in light gray, skeleton limbs in medium gray, skeleton backbones in black

3.3.3 Special Case: Missing Seed Triangles

There are cases where the triangulation doesn't give a seed triangle where two strokes meet. Instead a pair of branching triangles with a shared edge is sited here, where the edge has one end node at the *road junction point* and the other node on a separate stroke. This is shown in Figure 8, and has resulted in skeleton edges A and B overlapping.

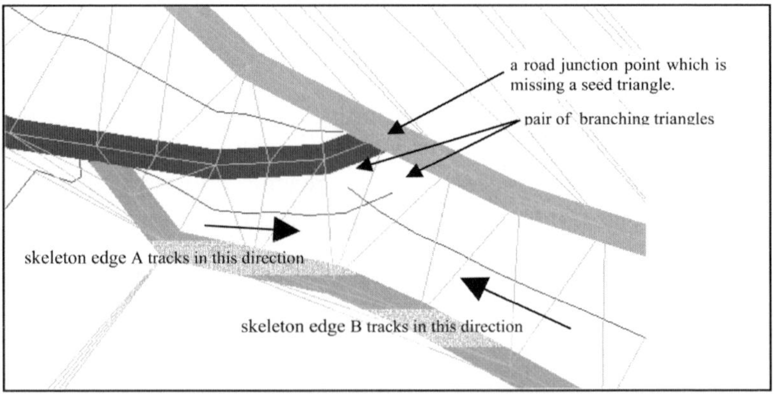

Fig. 8. End condition overlap, limb skeleton edges in red

To solve this, a list of terminating triangles is maintained during tracking. Each time we complete a skeleton edge its start and end triangles are added to the list. When tracking the next skeleton edge, the list is checked at each tracking step to ensure we do not pass over a terminating triangle. This is how the overlap is prevented.

4 Creating Conflict Lines

Before describing the analysis of the skeleton graph we need to understand how each skeleton edge is going to be processed. With certain limitations the software aims to place a *conflict line* feature on top of any part of the skeleton edge where it spots a conflict. A conflict line describes the extent and the intensity of a conflict.

4.1 Detecting the Conflict Extent

The software looks for conflicts along the skeleton edge's length; the test points are the edge's vertices excepting the start and end ones. The *inter stroke distance* is taken from the length of the *midline* (as characterized in Fig. 6) at each vertex.

The software uses the following parameters to determine a *threshold distance* for conflicts:

- the half widths of the two strokes of this skeleton edge. These are obtained by halving the external width for the various types of strokes (e.g. A road, B road etc.).
- the value for cul-de-sac end gap.

Generally the sum of half widths of the two strokes is the active *threshold distance* used when creating conflict lines.

For the *limbs* there is an extra distance termed the *exclusion distance,* which operates from the road junction point and precludes any conflict lines on the skeleton edge within that radius. The value of the exclusion distance is calculated from the larger of the stroke half widths.

A *short-leg limb* is one where none of the limb test points are outside the exclusion zone.

When outside the exclusion zone *limbs* are treated in the same way as *backbones*. The software creates conflict lines for the parts of the skeleton edge where the *inter stroke distances* are below *the threshold distance.*

For culs-de-sac, creating conflict lines is complicated by the fact that although, like all roads, they have width they do not usually cause conflicts beyond their end point. The triangulation allows us to detect when the *inter stroke distance* is being measured against a cul-de-sac end point. Local adjustment of the *threshold distance* incorporating the cul-de-sac end gap value allows more subtle control of end point conflicts.

4.2 Recording the Features Involved in the Conflict

Each conflict line has an associated *adjacency list*, which is a list of road sections which are involved in this conflict, the four possible directions for the conflicts are right, left, start and end.

The skeleton limbs and backbones are tracked looking for conflicts. Once we are storing a conflict line, we initially gather just two *adjacency lists*, the right and the left one. The list a road section is placed in is worked out using a vector from the previous conflict vertex to the current one.

Part of the analysis, described in next section, involves creating a conflict line, which represents several skeleton edges. At this stage the start and end list are populated. This usually occurs where *short-leg* limbs are involved.

In the case shown in Figure 9, a *reference triangle node* is used to place the limb's road sections in the correct conflict line's *adjacency list.* This ensures the road section lying between the two *road junction points* (circled) is placed in the start list for this conflict line.

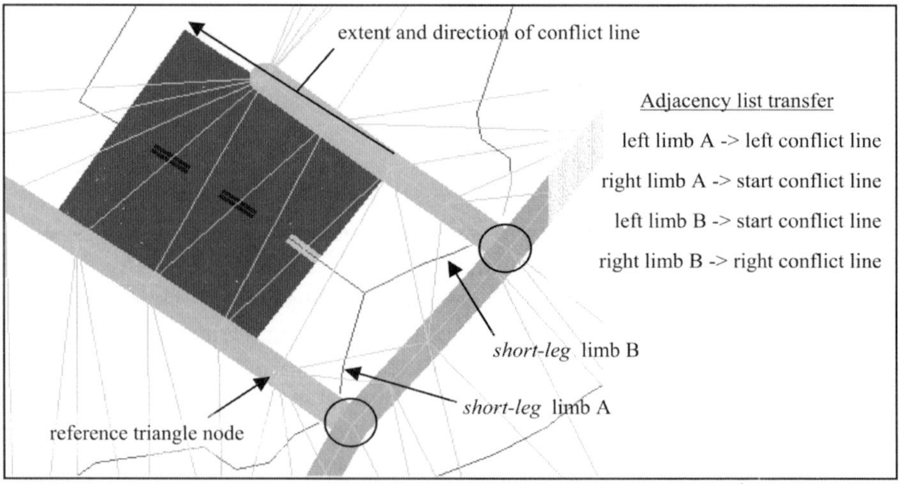

Fig. 9. Showing the combination of adjacency lists for a complex conflict line

5 Classifying the Skeleton Graph

The analysis process currently distinguishes nine different types of conflict scenarios. They can be distinguished by their display as detailed in Table 1. Conflict lines are displayed at the width of their closest separation (except the two small ones).

Table 1. Showing the display color and adjacency lists of different categories of conflict lines

description	left	right	start	end
general conflict	√	√	X	X
general conflict (at a junction)	√	√	X	X
open parallel conflict	√	√	√ or X	X or √
closed parallel conflict	√	√	√	√
cul-de-sac conflict	√	√	X	X
cul-de-sac / cul-de-sac conflict	√	√	X	X
staple conflict	√	√	X	X
small conflicting staple	√	√	X	X
small conflicting triangle	√	√	√	X

General Conflict

The majority of conflicts fall into the first two groups. In Figure 10 the conflict line is independent of the junction. Its adjacency lists contain the two road sections arrowed.

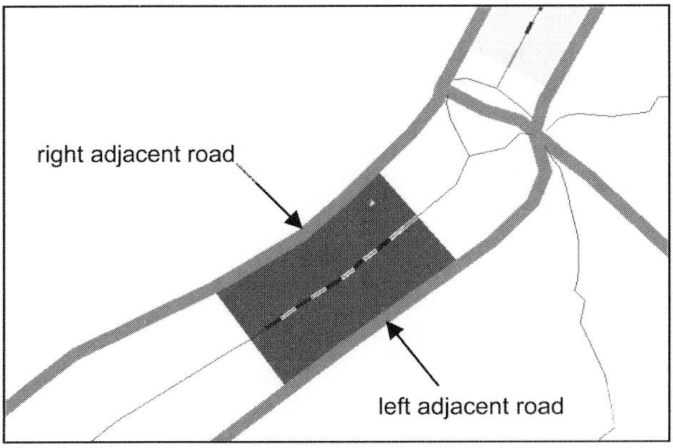

Fig. 10. General conflict

General Conflict (at a junction)

When we have acute angles where roads join at a junction the depiction of the junction is sensitive to the angle of join. The length of the conflict line gives extra information as to the urgency for resolution. Because the two conflict lines in Figure 11 sit upon limbs and touch the exclusion zone these are both junction conflicts. Their adjacency lists hold roads encountered when inside the zone. The exclusion zone is shown in gray in this and the following figures.

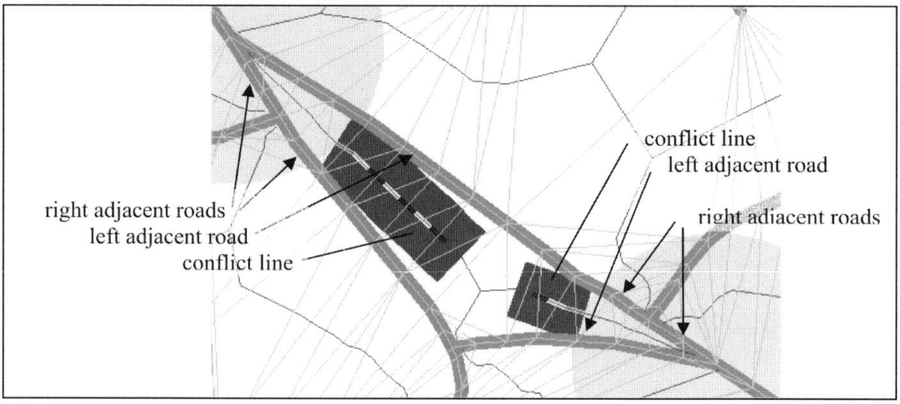

Fig. 11. General conflicts – extend to junction

Open Parallel Conflict

The software creates this type of conflict line if a skeleton edge has a conflict line that abuts to two *short-leg* limbs (see Fig. 12).

The start and/or end lists for this and the next parallel *conflict* lines are set up when the matching of the limbs occurs. The short-leg limbs left and right lists are apportioned to the correct start or end lists of these types of conflict lines, as illustrated in Figure 9.

Closed Parallel Conflict

Created when two short-leg limbs are found at both ends of a backbone that has a conflict line that extends its whole length (see Fig. 13).

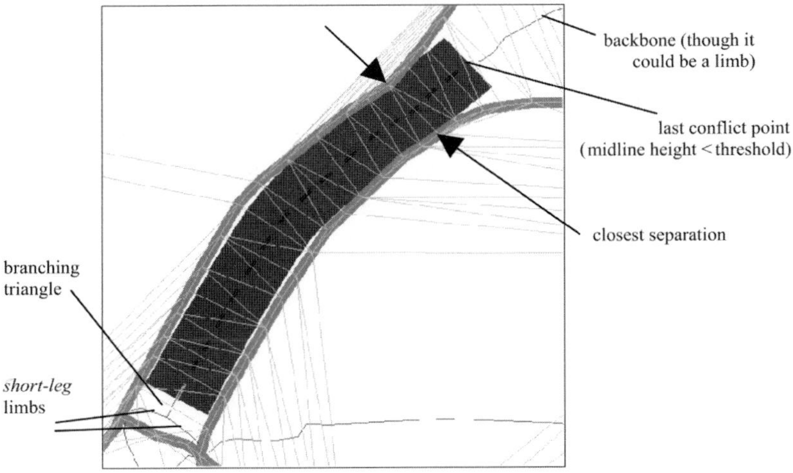

Fig. 12. Open parallel conflict

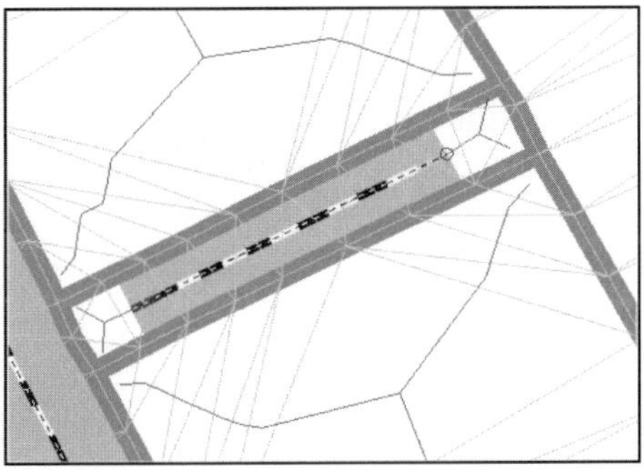

Fig. 13. Closed parallel conflict

Cul-de-sac Conflict

This is where the end point of a cul-de-sac is the nearest point to the conflicting road section. In this case we must distinguish between side-on and end-on conflicts, only conflicts, which are end-on will be termed cul-de-sac conflicts.

The first test is to check the triangle nodes on the conflict line's left and right side to see if any is at the end point of a cul-de-sac. A triangle node can be queried for its attached triangle edges. If a node has only one constrained edge, it must be at the end point of a cul-de-sac.

If only one of the sides has such a node, the software takes the triangle edge at the end of the cul-de-sac and the midline of the triangle where the conflict is and computes the angle between the two. If this is above say 45° the conflict is end-on, otherwise its side-on.

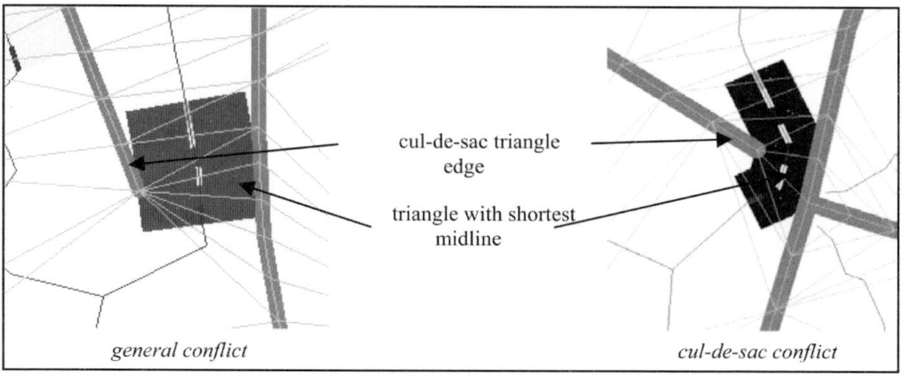

Fig. 14. Distinguishing cul-de-sac conflicts

Cul-de-sac / Cul-de-sac Conflict

This is created as follows:

Only skeleton edges with *two measurable triangles or less* (the one shown has one) qualify.

If the triangle nodes on the conflict line's left and right sides both contain a cul-de-sac end point then check whether the distance between the two cul-de-sac end points is below the *threshold* (in these conflicts the *midline* height of the triangle is not an accurate measurement of the closest separation).

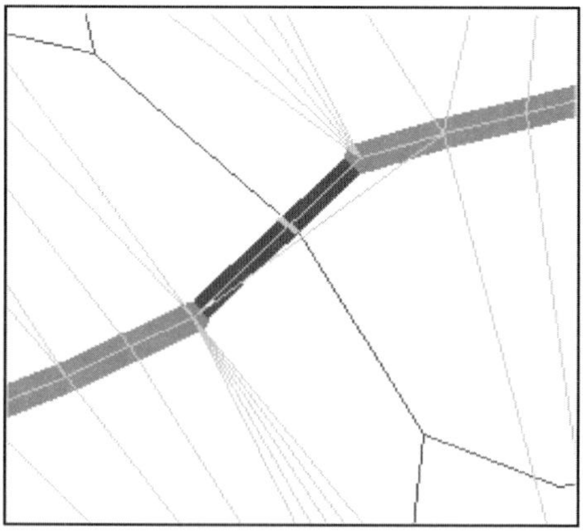

Fig. 15. Culs-de-sac / culs-de-sac conflict

Staple Conflict

Created when a *limb* is a staple, i.e. having seed triangles at both its ends, and tracking produced a conflict line, which begins at the start exclusion zone and continues in conflict all the way to the end exclusion zone.

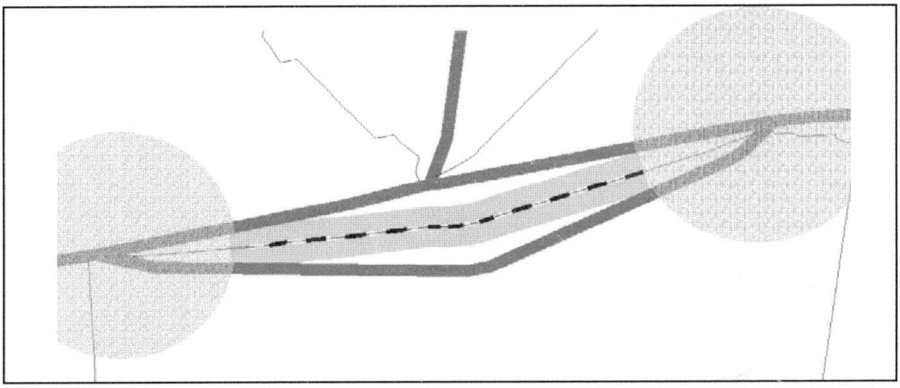

Fig. 16. Staple conflict – the conflict extends the length of the staple road structure

Small Conflicting Staple

Created where a *limb* (arrowed) is a staple, and tracking remains inside the two exclusion zones. The resulting conflict line is symbolic, depicted as a two-point line connecting the two seed triangle centroids.

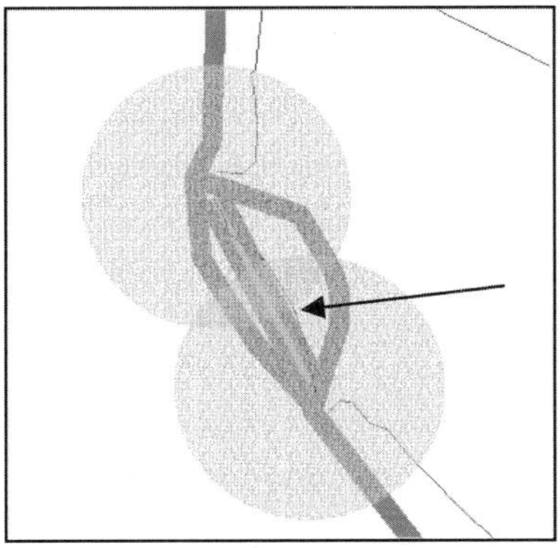

Fig. 17. Small conflicting staple

Small Conflicting Triangle

Created where all three *limbs* are *short-leg* ones. The resulting conflict line is symbolic, depicted as a three-point line connecting the three seed triangle centroids. Only the exclusion zone for the longest limb is shown here. The adjacency lists left, right and start are loaded with road sections from the three sides of the road triangle.

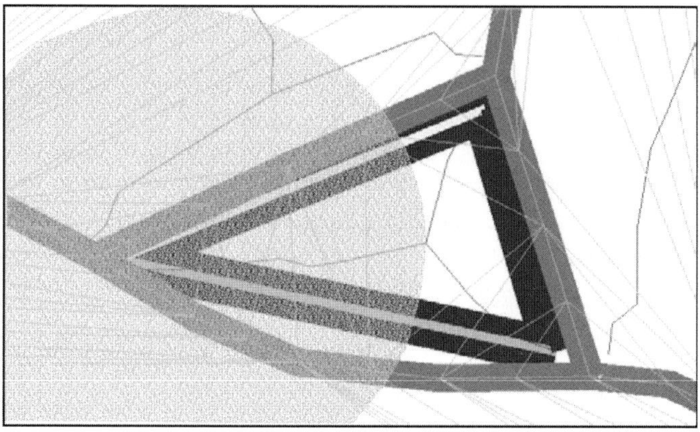

Fig. 18. Small conflicting triangle – showing edges and triangulation

6 Utilizing the Conflict Line Features

6.1 Identifying Road Structures with Poor Displacement Characteristics

The *small conflicting staples* may be best dealt with by removing one side of the staple completely by selecting which is the minor side and deleting its road section(s). Clearly if there were more than one section to delete this may be a more complicated road feature and some other actions may be necessary.

Some of the *small conflicting triangles* can be dealt with by setting their road section attributes as unitary features such as the *traffic island links* found in ITN i.e. possibly they were overlooked when the original attribution was applied. After manual intervention they can be collapsed using separate methods as described by Thom (2005). For the rest, measurements of the area within the junction can be taken. Triangle junctions are often enlarged at 1:50k as they are good way markers. Once they have been detected an automatic enlargement process can be applied. This ensures that the resulting generalized junction is represented correctly.

6.2 Criteria for Resolving Conflicting Staples

Once we can identify a staple in conflict, further measures can be obtained e.g. its length, from the length of the conflict line, and its area, from the triangulation it encloses. Comparison of these values against user set limits would allow us to simplify these road primitives selectively.

6.3 Trimming the Culs-de-sac

When presented with a group of conflicts its best to resolve the simplest ones first. Trimming conflicting cul-de-sac road sections is a sensible first move. The conflict line stores both the closest separation and the threshold distance, so the amounts to prune to achieve the required separation can be calculated. If the necessary prune is greater than or close to the cul-de-sac length we could delete the section.

6.4 Removing Multi Conflicting Culs-de-sac or Lollipops

Additionally one might ask if a cul-de-sac or lollipop is involved in a number of conflicts. The adjacency lists are stored as a two-way reference so it is easy to query each of the culs-de-sac/lollipops in the partition area and discover how many conflict lines they are listed by. Removing these multi conflicting culs-de-sac/lollipop road sections could be a good strategy.

6.5 Applying Displacement to Simple Right / Left Conflicts

Using the adjacency lists of the *general conflicts* it is easy to check if a conflict line's adjacent road sections reference other conflict lines. If there are no other conflicts this is a simple one with hopefully some room to move. Displacement using the Beams algorithm (Bader 2001) could be applied to these road sections with success.

6.6 Applying Displacement to Grouped Conflict Lines

Using the adjacency lists it is easy to trigger a process where a conflict line's adjacent road sections are searched for their involvement in other conflict lines etc. etc. In this way we can aggregate conflicts producing a list of conflicting road sections, which might be best displaced as a group.

Such a list may contain sections from more than one partition area, allowing displacement algorithms to act to resolve the local conflict scenario as a whole.

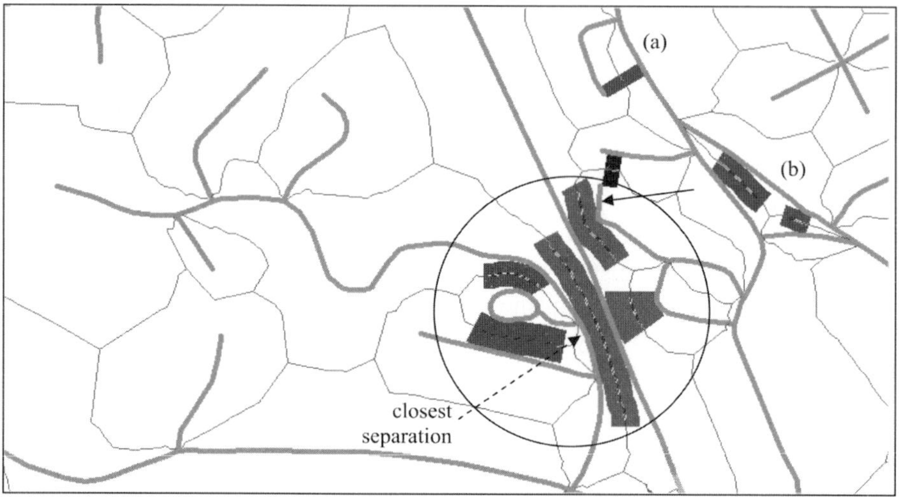

Fig. 19. Showing some conflict lines

The conflicts shown in Figure 19 might be best solved by the following sequence

1. trim cul-de-sac (arrowed)
2. remove lollipop as causes multiple conflicts.
3. do displacement on isolated conflict line at (a) – adjacency lists contain 2 road sections.
4. do displacement on two aggregated conflict lines at (b) – combined lists contain 5 road sections.
5. do displacement on larger aggregate (circled) – combined lists contain seven road sections.

Parallel Conflicts are a Special Case

Figure 20 shows two examples where recognizing stretches of parallel road sections would help their automatic generalization. In the top example the parallel service road shares its pecks with the major road it is parallel too. In the lower example two long parallel roads have been removed from the map.

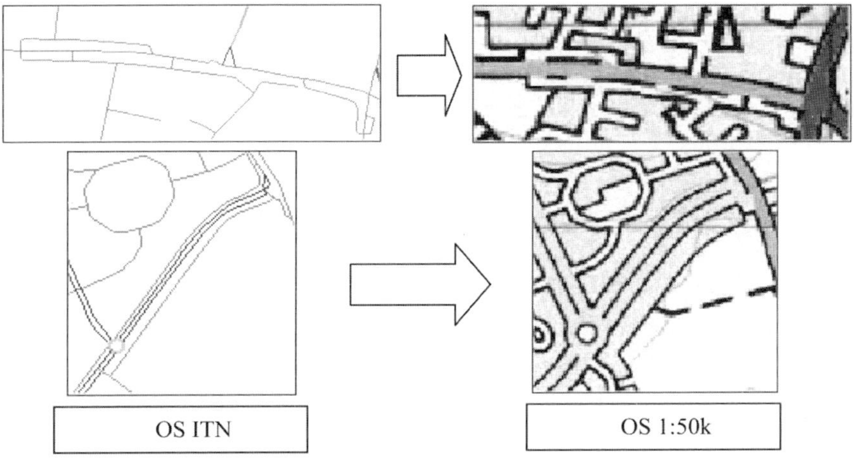

Fig. 20. Showing cartographic generalization of parallel road sections

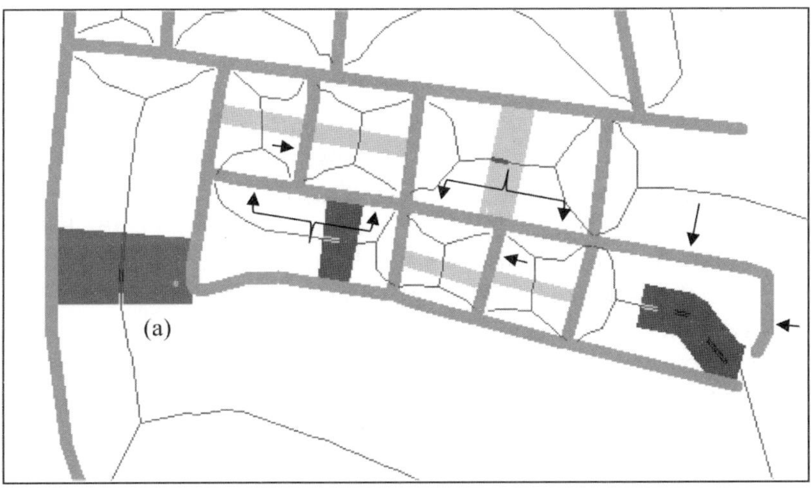

Fig. 21. Showing various parallel conflict lines

The conflicts shown in Figure 21 might be best solved in the following sequence:

1. trim cul-de-sac (arrowed)
2. remove two vertical roads (arrowed) – sideways (right/left, **not** start/end) matches from both sides.
3. regenerate triangulation and create conflict lines for these two changed blocks.
4. remove five horizontal roads (arrowed) – side matches on both sides plus continuity for cul-de-sac.
5. do displacement on general conflict (a).

6.8 Monitoring Conflict Resolution

After the various resolution strategies have been carried out a second set of conflict lines called conflict remnants is created. They are matched with the conflict lines. The adjacency lists are excellent for match making, since displacement solutions may move road sections considerable distances. Therefore proximity is not always going to be enough to secure the correct match.
There are a number of possible ways such matching can be interpreted:

- a conflict line without any matches is assumed resolved.
- a conflict line with its matched conflict remnant(s) can be compared as to their closest separation and conflict length to see if the conflict has been resolved at all.
- a conflict remnant without any matching conflict lines is assumed to be a new conflict caused by the displacements performed to resolve the first set of conflicts.

6.9 Possible Future Usage

There is a further possible use for the skeleton. In Figure 19 let us consider the road junction point below the circle. Two limbs attached to this junction are visible in the figure. If we can go from one limb, via the backbone skeleton edges, and connect to the second limb, this means that there is an isolated branch of the network attached to the road section between the limbs. This is possible in our example, though the tracking leaves the figure, later to return. Such procedures could return the distance between the roads in the branch and those in the adjacent network. In our case this would turn out to be the closest separation of the general conflict line (dashed arrow). On the basis of this, a displacement strategy could be chosen which simply displaced the whole branch *without distortion*. The required distance can be calculated from the conflict line's threshold distance and closest separation. An approximate direction can be obtained from the triangulation at the closest point.

7 Conclusions

A road network can be categorized by its road primitives. From simple structures like the lay-by (staple), triangle junction, and cul-de-sac to complex ones such as the dual carriageway and the elevated motorway junction, they are known to all road users. Some such structures are unitary, for example a dual carriageway and some are not, for example a service road providing access to shops from a main road. Unitary structures should be collapsed first. When generalizing at 1:50k it is necessary in the region of conflicts to extract as much information as to the category of the road primitives involved so as to resolve the conflict as sensibly as possible. The success of automatic generalization of road networks may well lie primarily on how successfully we can recognize, delimit and categorize the conflicts. This study uses a spatial structure based on the constrained Delaunay triangulation to create "conflict region" features, which have been shown to divide the various conflicts into groups. This will help to establish which road primitive a conflict belongs to. The conflict lines have a variety of other uses, which include orchestrating the removal of conflicts from a road network.

References

Ai T, van Oosterom P (2002) Displacement Methods Based on Field Analysis. In: Proc of the ISPRS Technical Commission II Symp

Bader M (2001) Energy Minimization Methods for Feature Displacement in Map Generalization. PhD Thesis, Department of Geography, University of Zurich, Switzerland

Højholt P (2000) Solving Space Conflicts in Map Generalization: Using a Finite Element Method. Cartography and Geographic Information Systems 27(1): 65–73

Jones CB, Geraint LB, Ware JM (1995) Map Generalization with a Triangulated Data Structure. Cartography and Geographic Information Systems 22(4): 317–31

Lemarie C (2003) Generalisation Process for Top100: Research in Generalisation Brought to Fruition. Fifth Workshop on Progress in Automated Map Generalisation

Mackaness WA, Mackechnie GA (1999) Automating the detection and simplification of junctions in road networks. GeoInformatica 3(2):185–200

Neuffer D, Hopewell T, Woodsford P (2004) Integration of Agent-based Generalisation with Mainstream Technologies and other System Components

Nickerson BG (1988) Automated Cartographic Generalization for Linear Features. Cartographica 25(3):15–66

Regnauld N (2005) Spatial Structures to Support Automatic Generalisation. 22nd Int Cartographic Conf, A Coruna

Thom S (2005) A Strategy for Collapsing OS Integrated Transport Network(tm) dual carriageways. In: 8th ICA Workshop on Generalisation and Multiple Representation, A Coruna

van der Poorten PM, Jones CB (2002) Characterisation and generalisation of cartographic lines using Delaunay Triangulation. Int J of Geographical Information Science 16(8):773–794
van Kreveld M, Peschier J (1998) On the Automated Generalization of Road Network Maps. In: Proc of the 3rd Int Conf in GeoComputation

This article has been prepared for information purposes only. It is not designed to constitute definitive advice on the topics covered and any reliance placed on the contents of this article is at the sole risk of the reader.

The 'Stroke' Concept in Geographic Network Generalization and Analysis

Robert C. Thomson

The Robert Gordon University, School of Computing,
St. Andrew St., Aberdeen, UK; email: rcthomson@yahoo.com

Abstract

Strokes are relatively simple linear elements readily perceived in a network. Apart from their role as graphical elements, strokes reflect lines of flow or movement within the network itself and so constitute natural functional units. Since the functional importance of a stroke is reflected in its perceived salience this makes strokes a suitable basis for network generalization, through the preferential preservation of salient strokes during data reduction. In this paper an exploration of the dual functional-graphical nature of strokes is approached via a look at perceptual grouping in generalization. The identification and use of strokes are then described. The strengths and limitations of stroke-based generalization are discussed; how the technique may be developed is also considered. Finally, the functional role of strokes in networks is highlighted by a look at recent developments in space syntax and related studies.

Key words: network, generalization, perceptual grouping, continuity, space syntax

1 Introduction

The automatic generalization of geographic networks can be viewed, at its simplest, as the progressive removal of segments in a principled fashion. For the generalization to be effective there are two goals to be achieved.

First, the reduction should preserve as far as possible the important features of the network – where the definition of importance will be context dependent, and may involve linkages with other networks and various non-network contextual data. Second, the visual character of the network map should be preserved where possible.

In some situations the two goals may be essentially the same. This paper argues that when no semantic information about a network map is available then the perceived importance of elements will correlate strongly with their perceptual salience. It follows that network features that combine both functional importance within the network and perceptual salience in the map representation should be preferentially retained in generalization.

Strokes are relatively simple network elements of this type. Strokes are computationally simple to derive from network data, and generalization based on strokes has been found to effective in several contexts. This paper reviews the use of strokes in generalization with the broad aim of clarifying their dual functional/graphical nature and how these properties support their role in effective generalization.

Aims and Structure. First perceptual grouping and its role in generalization are discussed. The procedures implemented at the Atlas of Canada for extracting strokes and using them in generalization will then be briefly described, noting how it is sometimes possible to adjust their functional or graphical roles in a generalization. Stroke-based network generalization and other analyses are then described. The strengths and the limitations of the stroke-based approach are discussed, and how the latter are being addressed. A final section considers the relationship between space syntax and network generalization. It will be shown that the functional role of strokes in networks is highlighted by recent developments in space syntax, and that space syntactic measures can be used to support stroke-based generalization.

The material presented here deepens, extends and updates the discussion of strokes, which formed one part of a recent, broader review of geographic network generalization (Thomson and Brooks 2006). In particular, the discussion of perceptual grouping has been revised to clarify the dual nature of the stroke and to place the use of strokes in a wider context of the search for patterns in networks that should be preserved in generalization; review material has been updated, and extended to consider applications of strokes outside generalization; a new section has been added examining links between strokes, generalization and space syntax analysis of road and street networks.

2 Perceptual Grouping

When looking at maps of road or river networks, natural linear elements will be seen which extend through junctions. These elements were termed 'strokes' (Thomson and Richardson 1999), prompted by the idea of a curvilinear segment that can be drawn in one smooth movement and without a dramatic change in line style. Strokes are paths of good continuation: they move through the network with no abrupt change in direction or character at junctions. Any geographic network can be completely decomposed into strokes. According to map type, the longer strokes present could be expected to represent the main courses of rivers or major routes; shorter strokes could be expected to represent tributary streams or minor roads.

Implications about fundamental relations among these elements follow from their perceived saliences. A longer and smoother stroke will naturally appear more important than one shorter and more meandering. Also, one stroke may terminate against another with implications of occlusion, tribute and less importance. The resulting perceived salience is a useful indication of the relative functional importance in the network of the features represented by these elements. Simply put, roads or rivers that look important in a network map usually *are* important. And saying that a stroke looks important is in effect saying it is a salient perceptual group.

Perceptual grouping is fundamental to human vision. Even with no high level or semantic knowledge available, the human visual system spontaneously organizes elements of the visual field, resulting in the perception of groups: Some arrangements of picture elements will tend to be seen as 'belonging together' in natural groups, which often appear to stand out from the surrounding elements, i.e. as 'figures' against 'grounds'. Many perceptual grouping principles have been identified, such as proximity, similarity, symmetry, uniform density, closure, parallelism, collinearity, cotermination and continuity (Wertheimer 1938).

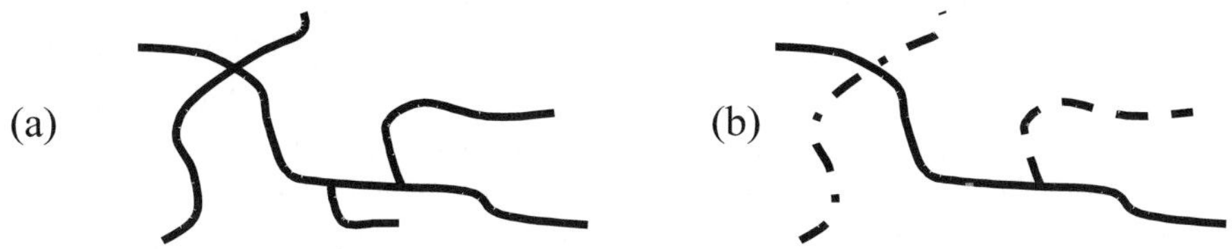

Fig. 1. A simple network with 8 arcs and 9 nodes resolves **(a)** into 4 strokes **(b)**

Figure 1 illustrates the principle of continuity or good continuation: the figure is naturally perceived as four smooth curves – the longest curve is crossed by one curve, and with two shorter curves incident on it. The four curves perceived are the strokes of this small network. In the arc-node data model this network would be represented by eight arcs and nine nodes.

Figure 2 shows a scatter of line segments in which several grouping principles can be seen to operate: the parallelism of four segments, the collinearity of others, the co-termination of four segments and some less well defined closed loops of segments. These latter groups can arise from simple proximity of their constituent line segments without touching.

Fig. 2. A scatter of line segments showing perceptual groups of collinearity, co-termination and parallelism

Perceptual grouping is a fundamental component of perceptual organization – the ability of a visual system to spontaneously organize detected features in images, even in the absence of high level or semantic knowledge (Palmer 1983; Witkin and Tenenbaum 1983; Lowe 1987). These principles are recognized as the basis for parsing the visual world into surfaces and objects according to relatively simple visual characteristics, in a process that operates independently of the domain being represented. Their importance in map interpretation (MacEachren 1995) and generalization (DeLucia and Black 1987) has also long been recognized.

The assumption that line-drawing recognition is a learned or cultural phenomenon is not supported by the evidence (Lowe 1987). The mechanisms being used for line drawing or map understanding have presumably developed from their use in recognizing three-dimensional scenes. The role of perceptual organization in vision is to detect those image groupings (such as those of Fig. 2) that are unlikely to have arisen by accident of viewpoint or position. These image groupings therefore indicate probable similar groupings in the scene (Lowe 1987). An important part of scene understanding is the recognition of occlusion, where one object is partly obscured by another. This is achieved using the same grouping principles, with the principle of good continuation applied to object/region boundaries. This was demonstrated in a study of a domain presenting analogous problems of interpretation (Thomson and Claridge 1989).

The idea that perceptual groups in maps indicate important phenomena in the world being represented is backed up by the findings of Zhang (2004a), who sought characteristic higher-level network patterns, and of Heinzle et al. (2005), who sought patterns in road networks that could indicate useful implicit or more complex information about the network. All the patterns they identified are instances of one or more perceptual grouping principles. The patterns (and main principles) are: star-like hubs (co-termination), parallel structures (parallelisms), grid-like patterns (parallelism, symmetry), loops (closure), density differences (uniform density).

Thus strokes, like other perceptual groups, are not simply graphical entities but carry implications of the probable presence of some notable phenomenon. Strokes are lines of good continuation in a network map, reflecting likely good continuation in the physical network – paths of natural movement of traffic or water, as applicable.

Yet strokes, like other perceptual groups, are important graphical elements. It has been suggested (Thomson and Brooks 2002) that the character of a map is a function of the perceptual groups it contains, and that consequently, during generalization, the preservation of salient perceptual groups – although necessarily in some attenuated form – should help retain that character. In a stroke-based generalization of a geographic network the more perceptually salient strokes are retained preferentially, and so may be expected to produce results that preserve important aspects of network character. If the generalization does not consider higher-order perceptual groups, however, these patterns may be lost from the map, with consequent loss of character.

3 Stroke-based Generalization

The methods of stroke-based generalization used at the Atlas of Canada have previously been described and illustrated (Thomson and Brooks 2000, 2002; Brooks 2003). The key steps are stroke building and stroke ordering.

3.1 Stroke Building

In the implementation used at the Atlas of Canada, stroke building is a local process, considering each node in turn and dependent completely on the properties of the network at that node neighborhood. This implies that strokes can be locally adjusted, with the update requiring only small processing time. The stroke-building algorithm is very efficient. Measured computational efficiency in practice appears to be approximately linear in the number of nodes.

At each node, a decision is made as to which (if any) incident pairs of arcs should be connected together. The simplest criterion to use is the angle of deflection that the join would imply. However, if other attribute data for the road or river segments are available then much more sophisticated processing is possible, with a rule base used to facilitate control of the concatenation. The rule set is the key to tuning strokes for a given application. The rules can be arbitrarily complex, with fallback rules to deal with cases unresolved by the primary rules.

Thomson and Brooks (2000) and, in greater detail, Brooks (2003) describe procedures for river networks where strokes equivalent to the main streams required by Horton ordering are derived – but with a greater flexibility that allows subsequent generalization results to be close to that of a human cartographer.

When processing road networks many road attributes may be available for use in stroke building. Road junctions themselves may also have associated relevant information. For example it may be possible to classify junctions as urban or rural, and then different rule sets can be applied for the different cases – for example, different limits may be set on acceptable angles of deflection. When road class information is available for arcs then generally better continuation is achieved by letting continuity of class override continuity of direction. Linking arc pairs with suitable continuity of direction can be restricted to those pairs with identical or similar road class. This eliminates problems such as a major road bending where a minor road joins it at an angle that would give a smoother continuation of direction between major and minor roads than between major and major.

Introducing a rule where continuation of road/street name at a junction was given precedence over continuity of direction was also found to yield good results. It was felt that street name continuity was a safer indicator of a natural functional network unit. Where such information is missing or incomplete, the strokes will be built according to geometric considerations.

Once decisions have been made at each node the strokes are assembled as sets of arcs that connect. The strokes that result will be the same regardless of the order in which the nodes are processed. The process is robust, in that there are no degenerate cases that cause the algorithm to fail, but to achieve meaningful results it is essential that the data have correct connectivity and, in the case of hydrologic networks, correct flow direction. (According to the context for generalization, one-way restrictions on traffic flow could be ignored or taken into account by additional rules on arc linking at junctions.)

The resulting strokes are concatenations of arcs each representing a path through the network from one terminal node to another. There are no restrictions imposed on the overall shape of the stroke that emerges. No consideration is taken of the total curvature of the stroke [c.f. Chaudhry and Mackaness' (2005) implementation], and the resulting stroke could in certain cases self-intersect or form a complete loop – thus allowing some orbital routes or ring roads, say, to be represented by single strokes.

3.2 Stroke Ordering

Once the strokes are constructed for a network they must be ordered appropriately, effectively assigning a ranking value or weight to each. The generalization can then proceed through the removal of some suitable proportion of strokes, in what can be viewed as a thresholding process. The data can be represented to the cartographer in an interactive display with slider control of the threshold percentage. Percentage reductions based on criteria such as Töpfer's radical law (Töpfer and Pillewizer 1966) could be applied, but the interactive method was preferred.

The success of the generalization will depend both on the suitability of the strokes found and the method used for their ordering. It is to be expected that the ordering method will depend on factors such as the criteria used in stroke building, the availability of arc attribute data, and the purpose of the generalization.

Stroke attributes on which to base the ordering will be derived from the attributes of their constituent arcs and nodes. Two important stroke attributes are ratio measures that do not depend on thematic attributes of constituent arcs: stroke length, which is readily found, and a measure of connectivity that can be derived easily from the degrees of the nodes within the stroke. Other stroke attributes depend on the available arc attribute data. Road class/category is an important road attribute with values that are normally nominal, e.g. "motorway" or "single track". However workable ordinal values usually follow from consideration of the relative road qualities. It may even be possible to derive workable ratio values to represent road class attributes. The problem is similar to that of deriving 'friction values' for roads, i.e. dimensionless multiplicative factors, which estimate the effective length of a road segment given its attributes (Richardson and Thomson 1996).

It may be noted that this method of ordering strokes does not use the implications about relative stroke importance that may be drawn from observing how one stroke terminates at a junction when another passes through it.

Network Connectivity. In the implementation at the Atlas of Canada strokes are ordered for generalization in two stages. In the first stage a stroke salience order is produced through sorting first by road category/class data (if available), and then by stroke length. There is no consideration of strokes' role in connectivity and hence no guarantee that the network would not become disconnected if generalization were applied directly to these strokes. Nevertheless, these strokes have been found to be useful for several applications (Section 3.3) and good generalization results can often be produced without further consideration of connectivity.

A second stage revision of stroke order eliminates the possible problem of roads becoming disconnected during stroke removal that has sometimes been reported (Zhang 2004b; Elias 2002). The reordering procedure used at the Atlas of Canada has been published (Thomson and Brooks 2000), albeit with a typing error. The algorithm ensures that if a stroke is removed then all pairs of its stroke neighbors remain connected by paths through the strokes that remain. The search for alternative paths is not exhaustive, but this did not adversely affect the results.

This reordering of strokes is effectively subordinating the perceptual salience of strokes to their structural role in maintaining network connectivity. Thus generalization on the basis of the revised order can lead to relatively salient strokes being removed while less salient strokes survive because of their greater structural importance.

3.3 Using Strokes

Use in Generalization. The most successful application of stroke-based generalization has been to hydrographic networks. The system developed at the Atlas of Canada was used to generalize the hydrology for a published 1:4M scale map of the three northern territories from source material in the GeoBase Level 0 hydrology dataset. The source data had been cleaned, attributes had been added, connectivity corrected and directionality computed (Brooks 2000, 2003). The results were comparable in quality with previous methods, and the new approach brought additional advantages in production (see below).

Misund et al. (2003), working in hierarchical GML modeling, proposed stroke extent (a measure roughly equivalent to length but faster to compute) as the best semantic generalization criterion to use for on-the-fly generalization of transportation networks. Elias (2002) found stroke-based generalization to be a simple and fast method of producing acceptable way finding maps, capturing most characteristic structures and major streets.

In a current Ordnance Survey project on the model generalization of road networks the stroke concept is adapted by adding traffic direction constraints to support the creation of single line representations for dual carriageways (Thom 2005). Strokes are detected for the individual carriageways, then paired and collapsed to produce the strokes representing the dual carriageway, in a network with the same connectivities as the original.

The stroke-based technique has also been adapted for generalizing fault lines in geological maps (Downs and Mackaness 2002), and Brooks (personal communication) found good results when applying the method to pipeline maps.

Tests on road network data at the Atlas of Canada and elsewhere (Chaudhry and Mackaness 2005; Elias 2002) have produced good results, and showed the feasibility and potential of the strokes-based approach. However, limitations of the method for urban road networks have been recognized by its authors and others (Section 3.4). For example, as the method stands, its application to road networks often entails additional special handling for certain junction types, including roundabouts. This could take the form of pre-processing to recognize and simplify road junctions. Mackaness and McKechnie (1999) have developed techniques that address this problem. With such techniques in hand, strokes could be extracted from network data over a range of scales. The extraction of strokes over multiple scales invites further investigation.

Other Uses. The Atlas of Canada drainage basins / hydrology dataset provided a good example of how the stroke model can also be used for other non-generalization applications. For example, strokes were used in matching name attributes to river tributaries, greatly increasing the efficiency of that process, and also in defining upstream drainage basins (Brooks 2003).

Elias (personal communication) found stroke salience a useful parameter to guide the automatic selection of the street labels when zooming street map data. Similarly, stroke length gave a quick means of classifying urban roads as major and feeder roads and, in an analogous situation, could even be used to categorize gas supply-line maps into main lines and house connections. Heinzle et al. (2005) used strokes in detecting certain important patterns in road networks.

The use of strokes and its related concepts in space syntactic network analysis is considered below (Section 4.2).

3.4 Further Development of Stroke-Based Generalization

Although stroke-based generalization preferentially retains the more salient strokes in a network and this helps to preserve the network character, this may not be sufficient when 'higher level' structures are perceptible in the network – i.e. patterns formed from groupings of arcs or strokes. The network may lose aspects of its character during generalization if certain patterns are disrupted or lost by stroke removal. The problem is most likely to occur in dealing with urban road networks, where such patterns are commonly found, and unlikely to arise with river networks.

As noted (Section 2), characteristic higher-level network patterns were identified by Zhang (2004a) and Heinzle et al. (2005). All were instances of one or more perceptual grouping principles. Ideally, network generalization should preserve these patterns where possible, perhaps necessarily in attenuated form, in order to preserve the network character. Stroke based generalization as described above does not attempt to retain any higher-level patterns.

The use of strokes can be viewed as the first, important step in preserving perceptual groupings in network generalization. Strokes can comprise several network arcs and in that sense are intermediate-level structures, but they are relatively simple perceptual groups and often serve as a basis for higher-level groupings. Preserving the more salient strokes helps preserve one facet of the network character during generalization. The surviving strokes may then preserve some higher-level perceptual groupings that are based on them, but this cannot be guaranteed without more relatively sophisticated processing. The work of Edwardes and Mackaness (2000) was an important advance in this direction.

Using a Network's Areal Dual. Edwardes and Mackaness recognized that the stroke-based approach to generalization is concerned only with the set of linear road or street objects and so may not provide sufficient consideration of the network's areal properties. Their solution to this problem (Edwardes and Mackaness 2000) goes a significant way in addressing the shortcomings of the stroke-based approach described above.

Strokes are used as one tool for characterizing the network in order to provide global information to the generalization process, but their method adopts the areal dual of the network as a second structure for characterization, using the two structures simultaneously to perform the generalization. The urban spaces are partitioned into city blocks using minimum cycles of streets, and generalization proceeds by sequential fusion of adjacent blocks when a block area is below a scale dependent threshold. A partition always aggregates with its neighbor across the weakest boundary, hence the weakest strokes are removed from the network and the effect on the overall good continuation of the network is minimized. The area size threshold can be varied using a function relating it to district density, which helps preserve network density patterns. Network connectivity is handled implicitly, since block aggregation cannot disconnect the network, although some special cases need additional processing. The algorithm produces good results, identifying and retaining the essential areal, linear, semantic and density patterns of the network and its constituent roads.

4 Strokes and Space Syntax

Space syntax is a method for measuring the relative accessibility of different locations in spatial system, including street networks (Hillier and Hanson 1984). Usually analysis proceeds via a derived 'axial map' – being a minimal set of straight-line segments called 'axial lines' which passes through each convex space and makes all axial links. From the axial map a map dual graph is extracted whose vertices represent the axial lines and whose edges link vertices whose corresponding axial lines intersect. This graph is then analyzed to derive a range of quantitative attributes for each axial line, which aim to describe aspects of their functionality in the network. Space syntax has proven to be a valuable tool for modeling and analyzing urban patterns with respect to human activity (e.g. pedestrian movement, traffic flow, burglary).

4.1 Space Syntax in Generalization

Mackaness (1995) first suggested that space syntactic measures could be used to guide generalization of urban road maps, but the lack of a workable transformation between road network and axial map prevented implementation. Since strokes bring no such problems it was suggested (Thomson and Richardson 1999) that such measures could be derived from a basis of strokes rather than axial lines, and used to support generalization.

The question of whether space syntactic measures can be used directly in an effective stroke-based generalization of urban networks has been answered, positively, for one particular interpretation of strokes. Jiang and Claramunt (2004) experimented in the generalization of a street network in which named streets were used as the spatial descriptors. The authors dismiss strokes as simply graphical elements and fundamentally different in character from named streets, which they see as functional units of a network. However, their justification for the use of named streets as structuring units, namely 'the observed fact that named streets often denote a logical flow unit or commercial environment that is often perceived as a whole by people acting in the city', could quite well be applied to strokes. Also, as noted above, the Atlas of Canada implementation of stroke extraction uses available road/street name information to guide the concatenation of road segments, resulting in continuous, named streets being defined as strokes.

Thus the Jiang and Claramunt experiments in fact provide a useful example of how space syntactic measures can be derived for strokes in a road network, and used as parameters to guide generalization. They investigated three space syntactic measures of centrality: connectivity, betweenness and integration. Connectivity measures the number of streets incident on a given street. Betweenness evaluates the extent to which a given street is part of the shortest paths that connect any two other streets. Integration reflects how far a given street is from every other. Their conclusion was that these measures, possibly in combination with other geometric and semantic properties, could support useful generalization.

4.2 Strokes in Space Syntax

Spatial data generalization and space syntax share similar goals in the analysis of spatial structure and functionality. This prompted the question of whether experience in network generalization could contribute to space syntax. In particular, could space syntactic analysis using strokes in place of axial lines bring benefits?

Axial lines represent straight lines of sight "possible to follow on foot" (Klarqvist 1993). Their relevance for pedestrian movement is thus clear. However, their relevance for vehicular traffic flow is less clear: the extent and continuity of roads could be expected to be important factors influencing the route taken by a motorist. Strokes, by definition and construction, aim to represent the lines of best continuation, and consequently could be expected to be appropriate elements for modeling and analysis in such contexts. Supportive evidence for this view may be drawn from studies such as those of Conroy Dalton (2001), which suggest that travelers appear to try to conserve linearity throughout their journey, avoiding unforced deflections.

Thus, the suggestion was made (Thomson 2003) that there is good reason to believe that for analyses of road networks on the urban scale or wider, and for traffic movement studies in particular, strokes could be more suitable spatial descriptors than axial lines. (Concerns about the limitations of axial lines as spatial descriptors for road networks had already led to the development of approaches that take into account the angular deflection between intersecting axial lines (e.g. Turner 2001; Dalton 2001), with the later extension of these methods to road centerline data (Dalton et al. 2003; Turner 2005) – but these techniques did not use strokes.) The use of strokes or stroke-like elements in space syntactic analysis has since been implemented, with positive results. Two examples follow; both reinforce claims that strokes are important functional units of networks.

"Continuity Lines". Figueiredo (2004) was directly influenced by the above suggestion, and applied the stroke building idea to concatenate axial lines into direct analogues of strokes, termed continuity lines (to better emphasize the idea of movement through a network with minimal deflection at junctions). It was found for a range of cities that all correlations between vehicular flows and syntactic variables (length, connectivity, integration) improve when using continuity lines in place of axial lines (Figueiredo and Amorim 2004). It was concluded that they revealed the importance of curved or sinuous paths, which is not clearly brought out by axial lines. Continuity maps were also felt to be more appropriate to the handling of long or extended paths and hence a better foundation for studies of traffic patterns (Figueiredo and Amorim 2005). Figueiredo's software could in principle derive continuity lines from both axial lines and road centerline data; no tests with the latter have been reported, but would be expected to produce similar results.

"Intersection Continuity Negotiation". Porta et al. (2004) made a comparative study of some structural properties of networks, and sought to develop methods that improve the correlation between measures of the structure of the network and measures of the dynamics on the network.

The method they developed used a technique to recognize the continuity of street segments through intersections and this information was used in constructing the dual graph. To detect continuities, the 'named streets' approach (Jiang and Claramunt 2004) was rejected as restrictive and costly (due to problems in establishing the required data). Instead, a generalization approach termed intersection continuity negotiation

(ICN) was adopted. ICN is identical with stroke construction in all important respects. The vertices of the resulting dual graph thus represent strokes, the arcs indicate stroke intersections, and syntactic analyses are as for conventional graphs derived from axial maps.

Useful features of the ICN (stroke-building) model were that it allows complex chains like loops and tailed loops to be recognized, and captures most of the continuity of paths throughout urban networks. Also, being based on a pure spatial principle of continuity, it avoids problems of social interpretation. From the combined analyses of the dual graph and a primal graph (derived directly from the original network map) new and useful measures of network structure are being developed (Porta et al. 2005).

5 Conclusions

Strokes are simple elements of a network, readily perceived in its map, whose visual salience broadly reflects functional importance within the network. The perceived paths of good continuation in the map indicate natural lines of flow in the physical network. Because of the good general correlation between these two aspects of strokes they form a suitable basis for network generalization, through the preferential preservation of the more salient strokes in data reduction.

This paper approached the functional-graphical nature of strokes via a wider consideration of perceptual grouping. Perceptual groups in a representation of the world such as an image or map generally reflect important features in the world being represented. Retaining the more salient perceptual groups in a network during generalization should therefore help to preserve both its functionally important features and the visual character of the map.

Hence the use of strokes can be viewed as a first, important step in preserving perceptual groupings in generalization. Strokes are relatively simple groups and often serve as a basis for higher-level groupings. Preserving the more salient strokes helps preserve one facet of the network character during generalization; the surviving strokes may preserve some higher-level perceptual groupings that are based on them, but this cannot be guaranteed without more relatively sophisticated processing. For networks without such patterns, such as river networks, strokes provide a basis for effective and efficient generalization – as demonstrated in the production of commercial maps.

Methods for implementing network generalization on the basis of strokes were described. Recent applications of strokes in generalization and other analyses were reviewed. The limitations of stroke-based generalization were also discussed, and further development of the technique was considered.

Finally, some links between network generalization, space syntax analysis, and strokes were highlighted. Here strong support for the view that strokes represent important structural/functional units of networks was found in examples of space syntactic network analysis incorporating the use of strokes or closely similar elements.

Acknowledgments

The discussions with Rupert Brooks and William Mackaness concerning the topics of this paper are greatly appreciated.

References

Brooks R (2000) National Atlas of Canada producing first map using automated generalization of framework data. Cartouche 39

Brooks R (2003) Atlas of Canada open-source generalization tools. Online document available at http://www.cim.mcgill.ca/~rbrook/atlas_gen/

Chaudhry O, Mackaness M (2005) Rural and urban road network generalization deriving 1:250000 from OS MasterMap. In: Proc 22nd Int Cartographic Conf, La Coruña, Spain

Dalton N (2001) Fractional configurational analysis and a solution to the Manhattan problem. In: Peponis J (ed) Proc 3rd Int Space Syntax Symp, Atlanta, Georgia, pp 26.1–26.13

Dalton N, Peponis J, Conroy Dalton R (2003) To tame a TIGER one has to know its nature: extending weighted angular integration analysis to the description of GIS road-center line data for large scale urban analysis. In: Hanson J (ed) Proc 4th Int Space Syntax Symp, London, pp 65.1–65.10

DeLucia A, Black TA (1987) Comprehensive approach to automatic feature generalization. In: Proc 13th Int Cartographic Conf, pp 168–191

Downs TC, Mackaness WA (2002) Automating the generalization of geological maps: the need for an integrated approach. The Cartographic J 39(2):137–152

Edwardes AJ, Mackaness WA (2000) Intelligent generalization of urban road networks. In: Proc GIS Research UK 2000 Conf, York, pp 81–85

Elias B (2002) Automatic derivation of location maps. IAPRS 34(4), Geospatial Theory, Processing and Applications

Figueiredo L. (2004) Linhas de Continuidade no Sistema Axial. Unpublished MSc Dissertation, Federal University of Pernambuco Recife

Figueiredo L, Amorim L. (2004) Continuity lines: aggregating axial lines to predict vehicular movement patterns. In: Proc 3rd Great Asian Streets Symp, Singapore

Figueiredo L, Amorim L (2005) Continuity lines in the axial system. In: Van Nes A (ed) Proc 5th Int Space Syntax Symp, Delft

Heinzle F, Anders K-H, Sester M (2005) Graph based approaches for recognition of patterns and implicit information in road networks. In: Proc 22nd Int Cartographic Conf, La Coruña, Spain

Hillier B, Hanson J (1984) The Social Logic of Space. Cambridge University Press, Cambridge

Jiang B, Claramunt CA (2004) A structural approach to model generalization of an urban street network. GeoInformatica 8(2):157–171

Klarqvist B (1993) A space syntax glossary. Nordisk Arkitekturforskning 2:11–12

Lowe DG (1987) Three-dimensional object recognition from single two-dimensional images. Artificial Intelligence 31(3):355–395

MacEachren AM (1995) How Maps Work: Representation, Visualization, and Design. Guilford Press, New York

Mackaness WA (1995) Analysis of urban road networks to support cartographic generalization. Cartography and Geographic Information Systems 22:306–316

Mackaness WA, Mackechnie GA (1999) Automating the detection and simplification of junctions in road networks. GeoInformatica 3(2):185–200

Misund G, Johnsen KE, Lindh M (2003) Hierarchical GML modeling of transportation networks. Presented at the 2nd Annual GML Developers' Conf [http://www.ia.hiof.no /~gunnarmi/omd /gmldev_03 (accessed 24.02.06)]

Palmer S (1983) The psychology of perceptual organisation: a transformational approach. In: Beck J (ed) Human and Machine Vision. Academic Press, New York, pp 269–339

Porta S, Crucitti P, Latora V (2004) The network analysis of urban streets: a dual approach. arxivorg preprint cond-mat/0411241

Porta S, Crucitti P, Latora V (2005) The network analysis of urban streets: a primal approach. arxivorg preprint cond-mat/0506009

Richardson DE, Thomson RC (1996) Integrating thematic, geometric and topological information in the generalization of road networks. Cartographica 33:75–83

Thom S (2005) A strategy for collapsing OS Integrated Transport Network™ dual carriageways. In: 8th ICA Workshop on Generalization and Multiple Representation, La Coruña, Spain

Thomson RC (2003) Bending the axial line: smoothly continuous road centre-line segments as a basis for road network analysis. In: Hanson J (ed) Proc 4th Int Space Syntax Symp, London, pp 50.1–50.10

Thomson RC, Brooks R (2000) Efficient generalization and abstraction of network data using perceptual grouping. In: Proc 5th Int Conf on GeoComputation, Greenwich UK

Thomson RC, Brooks R (2002) Exploiting perceptual grouping for map analysis, understanding and generalization: the case of road and river networks. In:

Blostein D, Kwon YB (eds) Graphics Recognition: Algorithms and Applications (= LNCS 2390). Springer, Berlin, pp 141–150
Thomson RC, Brooks R (2006) Generalization of geographical networks. In: Mackaness WA, Ruas A, Sarjakoski T (eds) The Generalization of Geographic Information: Models and Applications. Elsevier, Amsterdam
Thomson RC, Claridge E (1989) A 'computer vision' approach to the analysis of crystal profiles in rock sections. In: Pietikainen M (ed) Proc 6th Scandinavian Conf on Image Analysis, Oulu, Finland, pp 1208–1215
Thomson RC, Richardson DE (1999) The 'good continuation' principle of perceptual organization applied to the generalization of road networks. In: Proc 19th Int Cartographic Conf, pp 1215–1223
Töpfer F, Pillewizer W (1966) The principles of selection: a means of cartographic generalization. The Cartographic J 3(1):10–16
Turner A (2001) Angular analysis. In: Peponis J (ed) Proc 3rd Int Space Syntax Symp, Atlanta, Georgia, pp 30.1–30.11
Turner A (2005) Could a road-centre line be an axial line in disguise? In: Van Nes A (ed) Proc 5th Int Space Syntax Symp, Delft
Wertheimer M (1938) Laws of organization in perceptual forms. In: Ellis W (ed) A Source Book of Gestalt Psychology. Harcourt Brace, New York, pp 71–88
Witkin AP Tenenbaum JM (1983) On the role of structure in vision In: Beck J, Hope B, Rosenfeld A (eds) Human and Machine Vision. Academic Press, New York, pp 481–583
Zhang Q (2004a) Modeling structure and patterns in road network generalization. ICA Workshop on Generalization and Multiple Representation, Leicester, UK
Zhang Q (2004b) Road network generalization based on connection analysis. In: Proc 11th Int Symp on Spatial Data Handling, Leicester, UK

An Integrated Cloud Model for Measurement Errors and Fuzziness

Tao Cheng [1,2], Zhilin Li[2], Deren Li[3], Deyi Li[4]

[1] School of Geography and Planning, Sun Yat-sen University, GuangZhou, P.R. China
email: {lstc@polyu.edu.hk; chengtao@mial.sysu.edu.cn}
[2] Department of Land Surveying and GeoInformatics, The Hong Kong Polytechnic University, Hung Kom, Kowloon, Hong Kong
email: {lszlli@polyu.edu.hk}
[3] State Key Laboratory of Information Engineering in Surveying, Mapping and Remote Sensing, Wuhan University; email: {drli@whu.edu.cn}
[4] Tsinghua University {ziqin@public2.bta.net.cn}

Abstract

Two kinds of uncertainties – measurement errors and concept (or classification) fuzziness, can be differentiated in GIS data. There are many tools to handle them separately. However, an integrated model is needed to assess their combined effect in GIS analysis (such as classification and overlay) and to assess the plausible effects on subsequent decision-making. The cloud model sheds lights on integrated modeling of uncertainties of fuzziness and randomness. But how to adopt the cloud model to GIS uncertainties needs to be investigated. Indeed, this paper proposes an integrated formal model for measurement errors and fuzziness based upon the cloud model. It addresses physical meaning of the parameters for the cloud model and provides the guideline of setting these values. Using this new model, via multi-criteria reasoning, the combined effect of uncertainty in data and classification on subsequent decision-making can be assessed through statistical indicators, which can be used for quality assurance.

Key words: error, fuzziness, uncertainty, cloud model

1 Introduction

Almost all kinds of spatial data contain uncertainties since they are imperfect description of the complex reality. It is apparent that the use of uncertainty-laden spatial data without considering the intrinsic uncertainty involved will lead to serious consequence in concepts and practices.

Uncertainty in spatial database generally refers to data errors in attributes and in locations of spatial objects. There are many aspects of errors that have been addressed, including accuracy, reliability, bias and precision, etc. A great deal of research has been carried out in this area, from the looking at both location and attribute accuracies in initial values to estimating consequences of errors in the final outcome based upon probability theories (Fisher 2003; Zhang and Goodchild 2002).

Recently, a general framework for measurement errors and their propagation in various interrelated GIS and spatial operations (such as polygon overlay) has been formulated in a consistent and effective manner (Leung et al. 2004). Although it is declared that the model is suitable for general error analysis in GIS, their discussion mainly deals with location errors rather than attribute errors. Also, like most of other researchers, this general measurement error model is developed based on the assumption that the spatial objects can be defined precisely and identified crisply.

However, most spatial objects are naturally imprecise (fuzzy). The continuity, heterogeneity, dynamics and scale-dependence of spatial objects cause indeterminacies of spatial objects (Cheng 2002). Such indeterminacy is considered as another kind of uncertainty and is usually modeled by fuzzy set theories. Fuzzy logic methods have received great attention in many GIS applications (Burrough and Frank 1998; Cheng at al. 2001; Robinson 2003; Petry et al. 2005).

These two kinds of uncertainties – measurement errors and concept (or classification) fuzziness – have been discussed for a long time, but mostly in a separate manner. Little attention has been paid to the interaction between them, let alone the propagation of these combined uncertainties in further analysis and decision-making. The exception can be found in Heuvelink and Burrough (1993), which discussed error propagation in cartographic modeling using crisp and fuzzy classifications. They illustrated several situations that might arise when measured thematic data with attribute errors are classified. When the distribution of the measurement is well within the class boundaries, the classification is certain. On the other hand, when the distribution straddles the boundaries, the classification result is uncertain. The possible class and the corresponding membership value for these uncertain cases were obtained by the Monte Carlo method.

However, the influence of errors on the classification cannot be calibrated in an explicitly analytical form by Monte Carlo method. Cheng et al. (1997) analyzed the error propagation in an analytical approach, by generalizing the possible situations into four cases, either with or without errors in the measurement, and, either with or without fuzziness in the classification. An abstract form for the expectation of the classification result is derived based upon the distribution of the measurement. However, it is difficult to derive an analytical formula for nonlinear classification functions based upon the abstract form proposed.

Therefore, a formal basis is needed for estimating how errors in measurement of attributes can be propagated through GIS operation (such as classification and overlay), and for assessing the plausible effects on subsequent decision-making. An integrated model for spatial object uncertainties, i.e. measurement errors and classification fuzziness, should be developed.

In order to achieve such an aim, the first thing is to find a mathematical tool that is able to accommodate errors and fuzziness. Recently, a novel concept for uncertainty mediating between the concept of a fuzzy set and that of a probability distribution has been proposed (Li et al. 1998; Neumaier 2004). This concept overcomes the weakness of fuzzy set theory by considering the randomness of fuzziness in fuzzy class definition. It also provides uncertainty-reasoning strategies that would propagate the randomness and fuzziness in decision-making. However, how to adopt this model to describe the errors and fuzziness in GIS data is not explored since the physical meaning and setting of the parameters in the cloud model is not clearly defined. How to represent the fuzzy spatial object by the cloud model has not been discussed. Further, how to use the uncertainty-reasoning strategy for overlay analysis (multi-criteria decision) has not been investigated, although the cloud model theory has been applied in spatial knowledge discovery (Li et al. 2000).

Therefore, the aim of this paper is to explore the application of the cloud model theory in modeling, representing and reasoning of GIS uncertainties. The specific objectives of this paper are:

- To explore the application of the cloud model theory in modeling and representing uncertainties of GIS by proposing an integrated model for measurement errors and classification fuzziness: first investigate the propagation of errors in fuzzy classification, then check the possibility to integrate errors and fuzziness under the cloud model; and
- To assess the effects of uncertainty on subsequent decision-making via multi-criteria reasoning based upon cloud theory, i.e. propagation of uncertainties with errors and fuzziness

After introducing the concepts and theories of cloud model in Section 2, we will try to achieve these two objectives subsequently in Section 3 and Section 4. We will present an experiment to illustrate the practical usefulness of the integrated model and the uncertainty reasoning strategies. The paper will conclude with a summary of major findings and suggestion for further research.

2 Fundamentals of Cloud Model Theory

The conception of cloud is due to the challenge that the fuzzy logic has been confronted. For a fuzzy classification, a membership function has to be assigned, which is impossible to be proved uniquely correct in a real word application. Furthermore, the membership function of a fuzzy set is a one-point to one-point mapping from a (attribute) space U to the unit interval [0,1] of a fuzzy class (fuzzy concept). After the mapping the uncertainty of an (attribute) element belonging to the fuzzy concept becomes certain to that degree, a precise number. The uncertain characteristics of the original concept are not passed on to the next step of processing at all. This is the intrinsic shortcoming of the fuzzy set theory, which is often criticized by probabilists and experts in relevant fields (Li et al. 1998).

To overcome the intrinsic shortcoming of fuzzy set theory, cloud model was proposed as a novel concept for uncertainty mediating between the concept of a fuzzy set and that of a probability distribution. Li et al. (1998) first proposed the concept of cloud with the application of uncertainty reasoning. Neumaier (2004) provided the mathematical foundations about the cloud model. These two cloud models are quite similar. Here we mainly adopt the cloud model proposed by Li et al. (1998) since it has been used for spatial data mining (Wang et al. 2003).

2.1 Cloud Model

Cloud model is a model of the uncertain transition between a linguistic term of a qualitative concept (such as a fuzzy class) and its numerical representation (thematic attribute). In short, it is a model of the uncertain transition between qualitative and quantitative. Let U be the set $U=\{x\}$, as the universe of discourse, and C is a linguistic concept associated with U. The membership degree of x in U to C, $\mu(x)$, is a random variable with a probability distribution, taking values in [0,1]. A membership cloud is a mapping from the universe of discourse U to the unit interval [0,1]. That is, $\mu: U \rightarrow [0,1] \quad \forall x \in U \quad x \rightarrow \mu(x)$.

A cloud has following properties:

- cloud integrates the fuzziness and randomness of a linguistic concept in a unified way. For any $x \in U$ to the interval [0,1] is a one-point to multipoint transition, producing a membership cloud, which is a probability distribution rather than a membership curve;
- any particular drop of the cloud may be paid little attention to. However, the total shape of the cloud, which is visible, elastic, boundless, and movable, is most important. That is why it is called "cloud".

The concept of membership clouds is often pictured as a two-dimensional graph with the universe of discourse represented as one dimension. The geometry of membership clouds is a great aid in understanding fuzziness, defining fuzzy concepts, and proving fuzzy theorems. Visualizing this geometry may by itself be the most powerful argument for fuzziness. It is important to see the properties of the clouds.

For example, the concept of "distance is about 30 kilometers" can be represented as a membership cloud in Figure 1.

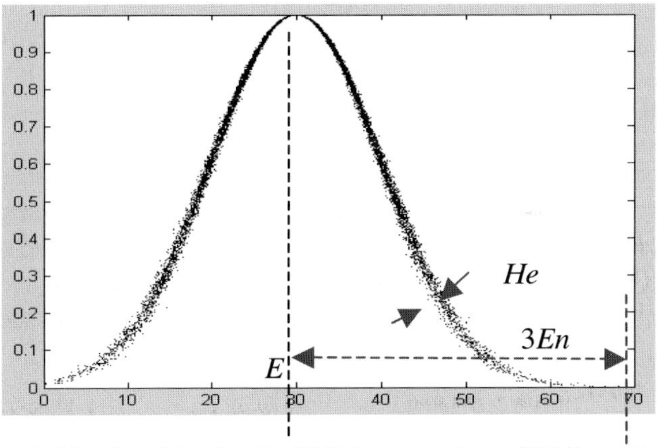

Fig. 1. Membership cloud of "distance is about 30 kilometers"

There are various ways to interpret the concept of clouds. The membership degrees at each x are all random numbers showing the deviation but obeying certain probability distributions. The thickness of the cloud is uneven; near the top and bottom of the cloud, the standard errors are smaller than that in the middle.

2.2 Mathematical Description of Clouds

A cloud can be characterized by three values (Li and Du 2005; also see Fig. 1):

- EXPECTED VALUE *Ex*. The expected value *Ex* of a membership cloud is the position at the universe of discourse, corresponding to the center of gravity of the cloud. In other words, the element *Ex* in the universe of discourse fully belongs to the concept represented by the cloud model.
- ENTROPY *En*. The entropy is a measure of uncertainty. In one aspect, *En* is a measure of randomness, showing the deviation of the cloud points which representing the concept; in another aspect, *En* is also a measure of fuzziness, showing how many elements in the universe of discourse could be accepted as the concept.
- SUPER ENTROPY *He*. It is a measure of the uncertainty of entropy, decided by the randomness and fuzziness of the entropy. *He* usually decides the thickness of the cloud.

Therefore, a cloud can be mathematically described as

$$\text{Cloud}=C(Ex, En, He). \tag{1}$$

2.3 Normal Cloud and Cloud Generator

There are different forms of clouds, symmetric, half-shaped or combined. The symmetric cloud represents the uncertainty of the linguistic concept in symmetry, which are usually bell-shaped (see Fig. 2a) or trapezoidal-shaped. The half-cloud represents the uncertainty of linguistic concepts in one side, either left (see Fig. 2b) or right. The combined cloud represents the uncertainty in asymmetry linguistic concepts (see Fig. 2c).

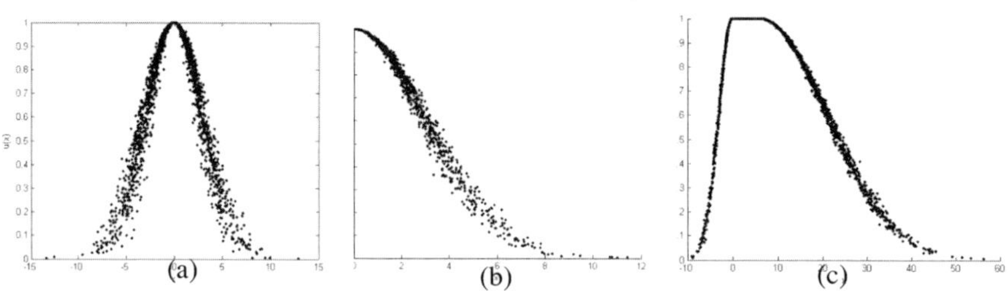

Fig. 2. Three types of cloud models
(a) Symmetric cloud – **(b)** Half-cloud – **(c)** Combined cloud

The normal cloud is developed based upon the bell-shaped membership function. The degrees of membership in the normal cloud obey a normal distribution. The bell-shaped membership functions $u(x) = exp[-(x-a)^2/2b^2]$ are mostly used in fuzzy set; and the normal distribution has been supported by results in every branch of both social and natural sciences, which is characterized by mean and deviation. Therefore, we will adopt normal cloud in our following discussion.

The mathematical expected curve (MEC) of the normal membership cloud with its expected value *Ex* and entropy *En* may be written as:

$$MEC(x) = \exp[-\frac{(x-Ex)^2}{2En^2}]$$

Let $x = Ex \pm 3 \times En$, then (2)

$$MEC(x) = \exp[-\frac{(x^{'}-Ex)^2}{2En^2}] = \exp[-\frac{3E_n^2}{2En^2}] = 0.011 \approx 0$$

That is to say that the elements beyond $Ex \pm 3En$ in the universe of discourse can be neglected for a linguistic concept.

Given three digital characteristics *Ex*, *En*, and *He*, to represent a linguistic concept, the normal cloud generator could produce as many drops of the cloud as you like (see Fig. 1). This kind of generator is called normal cloud generator (CG). All the drops obey the properties described previously. The algorithm of normal cloud generator is present in the Appendix.

3 An Integrated Model for Measurement Errors and Fuzziness

Like the fuzzy concept of "distance is about 30 Kilometers", if we think 30 km is the expected value for this concept and the fuzzy transition zone is 20 km (since we think the distance which is less than 10 and further than 50 km can not be considered as a "about 30 km" at all), then the concept is mapped as a fuzzy membership function in bell shape as $y = e^{-\frac{(x-30)^2}{2*10^2}}$. The membership curve of this function is shown in Figure 3a.

In case there are errors in the measurement x (distance), the error will be propagated to y. Figures 3b and 3c present the Monte Carlo stimulation when there are errors in the measurement of x with standard deviation σ_x=0.5 and σ_x=1, respectively.

The membership degrees at each x are all random numbers showing the deviation but obeying certain probability distributions. The deviation of y

forms an uncertainty band with uneven thickness. Near the top and the bottom of the uncertainty band, the deviations are smaller than that in the middle. The general thickness of the uncertainty band enlarges with the increase of σ_x.

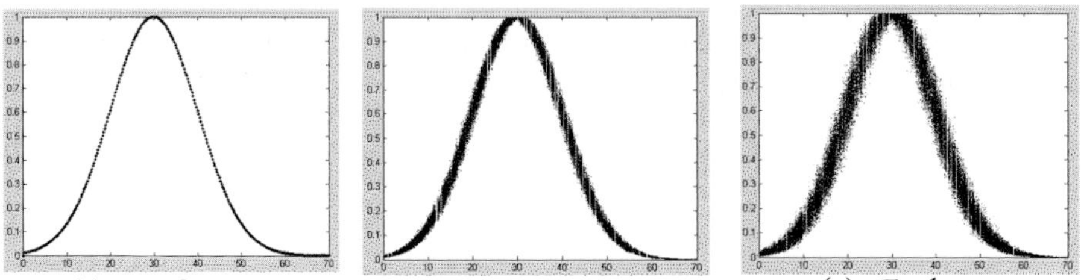

(a) σ_x=0 (b) σ_x=0.5 (c) σ_x=1

Fig. 3. Representing the concept of "distance is about 30 km" by a bell-shaped membership function as $y = e^{-\frac{(x-30)^2}{2*10^2}}$.

We can also derive the analytical form of the relationship of deviation of y (σ_y^2) with x. According to Leung et al (2004), the basic measurement error model for indirect measurement can simply be expressed as:

$$\begin{cases} Y = f(X) \\ X = \mu_x + \varepsilon_x, \varepsilon_x \sim (0, \Sigma_x), \end{cases} \tag{3}$$

where μ_x is the true-value vector, X is the random measurement-value vector, Y is the indirect measurement-value vector obtained by f, and ε_x is the random measurement error vector with zero mean *0* and the variance-covariance matrix Σ_x. According to (3), Y is random and its error variance-covariance matrix Σ_y is propagated from Σ_x.

Σ_y can be derived via the approximate law of error propagation (Leung et al. 2004), i.e.,

$$\sigma_y^{\ 2} == B\Sigma_x B^T, \tag{4}$$

$$\sigma_y^{\ 2} \approx \tilde{\Sigma}_y \equiv B_\mu \Sigma_x B_u^{\ T}, \tag{5}$$

Formula (4) is for the case when $f(x)$ is linear in x, i.e., $f(x) = a + BX$, where a is a constant vector and B is a constant matrix. Formula (5) is for the case when $f(x)$ is nonlinear in x, where B_u is the matrix of partial derivatives with respect to each of the components, which can be approximated by the Taylor series as

$$B_\mu \equiv \left(\frac{\partial f_i(x_i)}{\partial x_i} \right) \tag{6}$$

If a bell-shaped function is used for a fuzzy classification or a fuzzy concept, i.e., $y = e^{-\frac{(x-a)^2}{2*b^2}}$, and we assume each measurement is independent from others and the standard deviation of the measurement error is σ_x, then

$$B_\mu \equiv \left(\frac{\partial f_i(x_i^{'})}{\partial x_i} \right) = \frac{df(x)}{dx} = -\frac{(x-a)}{b^2} e^{-\frac{(x-a)^2}{2b^2}} \tag{7}$$

$\sum_x = \sigma_x \cdot \sigma_x = \sigma_x^2$

so

$$\sigma_y^2 \equiv \left(-\frac{(x-a)}{b^2} e^{-\frac{(x-a)^2}{2b^2}}\right)^2 \sigma_x^2 = \frac{(x-a)^2}{b^4} e^{-\frac{(x-a)^2}{b^2}} \sigma_x^2 \tag{8}$$

Formula (8) expresses that the fuzzy membership value *y* is random when there are errors in the measurement *x*. The dependence of σ_y^2 on *x* is nonlinear which means that the deviation of *y* is changing with *x* nonlinearly. We draw the relationship of σ_y^2 with *x* in Figure 4 with two cases. It shows that in both cases there are two peak points for σ_y^2 which are symmetric to the expectation of *x* on the left and the right, i.e., when $x \approx a \pm b$, $\sigma_y^2 = \max = \frac{1}{b^2}\sigma_x^2$. It also shows that in both cases the deviation of *y* is smallest when *x* is close to the mean, i.e. when *x*=a, σ_y^2=min=0. When *x* is beyond [*a*-3*b*, *a*+3*b*], $\sigma_y^2 \to 0$. It means that *y* scatters most when it is close to the waist of the uncertainty band, and at the top and the bottom of the band, it is more focused.

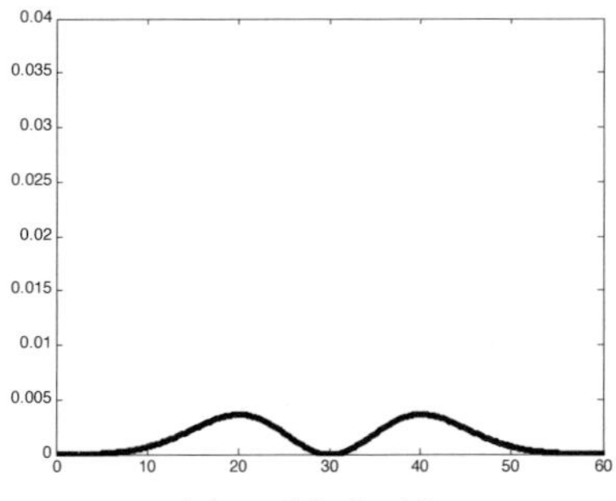

(a) *a*=30, *b*=10

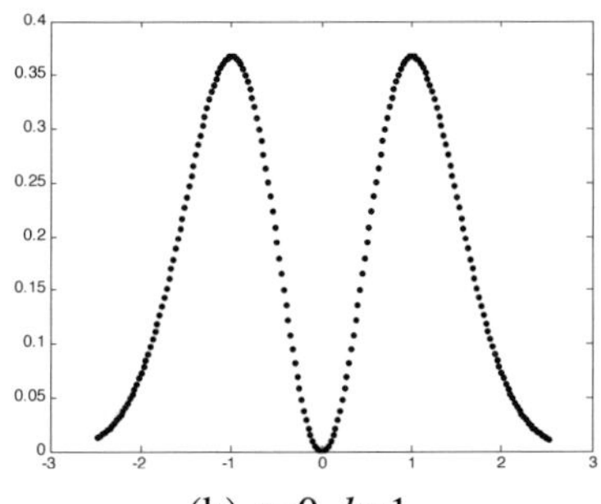

(b) *a*=0, *b*=1

Fig. 4. The relationship of σ_y^2 with *x*

All these properties fit quite well with the cloud model. In order to test our assumption that measurement errors and classification fuzziness can be integrated in the cloud model, we used the membership function to be the mathematical expectation curve of the cloud model (i.e. $Ex=a$, $En=b$), and used standard deviation of measurement errors (σ_x) to be the value of He in the cloud model (i.e. $He=\sigma_x$). The clouds for the concept "distance is about 30 km" created based upon two He are presented in Figure 5a and Figure 5b, corresponding to Figure 3b and Figure 3c, respectively. These clouds are created based on the algorithms of cloud generator (CG) presented in the Appendix.

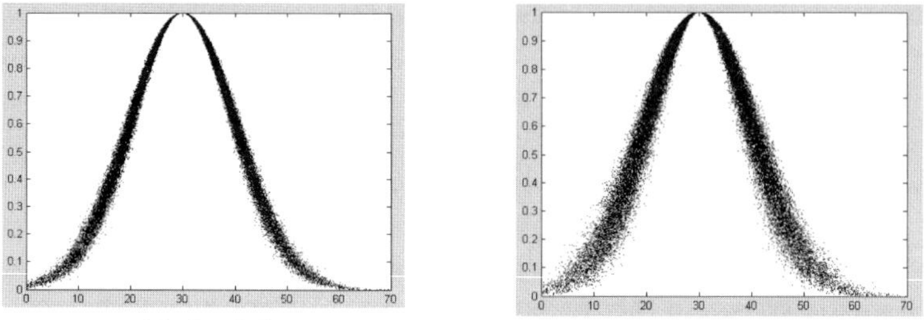

(a) He=0.5 (b) He=1

Fig. 5. Representing the concept – "distance is about 30 km" by cloud model with Ex=30, En=10

Comparing Figure 3 and 5, we found that

1. the general patterns of two figures are quite similar, i.e., the thickness of the uncertainty band and the thickness of the cloud change with x. Close to the waist, the degree of membership is most dispersed, while at the top and bottom the focusing is much better;
2. the deviation of the points in Figure 3 increases with the randomness of x (i.e. σ_x), while the deviation of the cloud drops in Figure 5 increases with the randomness of the degree of fuzziness (i.e. He);
3. the mathematic expected curve of the fuzzy classification with measurement errors (see Figs. 3b; c) and mathematic expected curve of the membership clouds (see Figs. 5a,b) are the same as the membership function (see Fig. 3a).

Therefore, it seems that we can use the cloud model to accommodate the errors and fuzziness. When there are measurement errors in fuzzy classification, we consider $He \approx \sigma_x$. This view provides the physical meaning of He for the cloud theory, which is not clearly explained in all the existing papers about cloud. It also provides a physical guideline to set up the value of He.

4 Uncertainty Reasoning Based Upon the Cloud Model

The representation of fuzzy spatial objects by the cloud model has be discussed in (Cheng et al. 2005). This section will illustrate the application of uncertainty reasoning based upon cloud model theory for decision-making.

Suppose there is a multi-criteria decision rule as follows (Burrough 1998, p 22)

IF SLOPE $\geq 10\%$ AND SOIL TEXTURE
=SAND AND VEGETATION COVER $\leq 25\%$
THEN EROSION HARZARD IS SEVERE

In case SL=9, ST=80 and VC=24, where SL, ST and VC are slope, percentage of sand in the soil and vegetation cover ratio, what is the possibility for the EROSION HARZARD?

4.1 Uncertainty Reasoning by Fuzzy Set Theory

If a fuzzy approach is taken, the three quantitative concepts in the conditions can be expressed as three fuzzy membership functions as follows. Here we just think they are all in triangle-shape for simplicity.

$$\mu_{SL}(x) = \begin{cases} 1 & x \geq 10 \\ \frac{1}{10}x & others \end{cases}$$

$$\mu_{ST}(x) = \frac{1}{100}x \tag{9}$$

$$\mu_{VC}(x) = \begin{cases} 1 & x \leq 25 \\ \frac{4}{3} - \frac{x}{75} & 25 < x \leq 100 \end{cases}$$

When SL=9, ST=80 and VC=24, $\mu_{SL} = 0.9$, $\mu_{ST} = 0.8$ and $\mu_{VC} = 1.0$.

As we know, logical operation with fuzzy sets are generalizations of the usual Boolean algebra applied to measurements that have partial membership of more than one set. The 'AND' in binary logic is replaced by a 'MIN' operation; and the 'OR' by a 'MAX' operation. Therefore, the reasoning rule for fuzzy set can be written as $SE_{F=}$ MIN($\mu_{SL}, \mu_{SL}, \mu_{SL}$)= MIN(0.9, 0.8, 1)=0.8 (Burrough 1998, p 22). This is a fix number. It means that the erosion hazard is severe with possibility 80%.

In case there are measurement errors in SL, ST, and VC with $\sigma_{SL} = 0.01$; $\sigma_{ST} = 0.02$ and $\sigma_{VC} = 0.1$, what decision can we make now? This situation cannot be handled by the fuzzy set theory.

4.2 Principle of Uncertainty Reasoning by Cloud Model Theory

We can use the cloud model to accommodate the errors in the fuzzy classification. We can also use the uncertainty reasoning strategies based on cloud model for decision-making. A number of algorithms for uncertainty reasoning are available (Li et al. 2000). Here we review two basic cases as examples, which will be used here.

❑ *Single Condition- Single Rule Generator*

If there is only one factor in a rule antecedent, it is called single condition-single rule (SCSR), which can be formalized as:

If *A* then *B*

where *A*, *B* are fuzzy concepts, such as "If the altitude is high, then the population density is low".

A and *B* can be mapped as two clouds, representing the mapping from the universes of discourses U_1 and U_2 to the concepts C_1 and C_2, respectively. If a numerical value *a* in the universe of discourse U_1 is given, a generator that produce drops representing the degree of *a* belonging to C_l is called an antecedent generator CG_A (on the left of Figure 6). Inversely, if a generator produces drops in the universe of discourse U_2 which belonging to C_2 with a given degree μ, $\mu \in [0,1]$, this generator is called a consequent generator CG_B (on the right of Fig. 6). The algorithms of the 1-D CG_A and CG_B are presented in the Appendix.

A single condition-single rule generator (SCSRG) can be made by connecting a CG_A with a CG_B as shown in Figure 6. The algorithm of the SCSRG is presented in the Appendix.

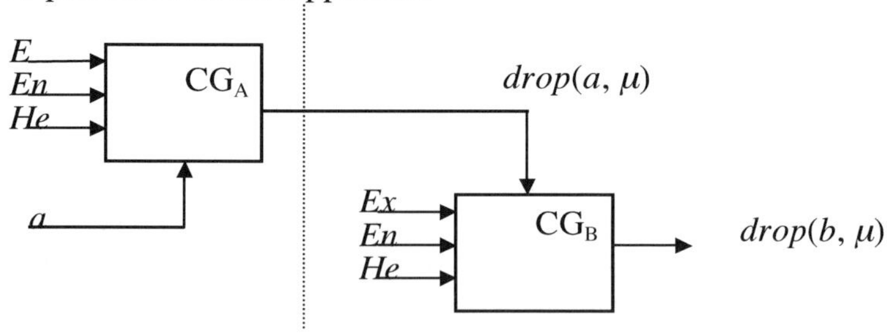

Fig. 6. Single condition-single rule generator (Li and Du 2005)

The joint distribution (*a*, *u*) is shown in Figure 7a, i.e. all the drops are on a horizontal line $x = a$. The joint distribution (*b*, *u*) is shown in Figure 7b, i.e. all the drops are on a vertical line $y = \mu$.

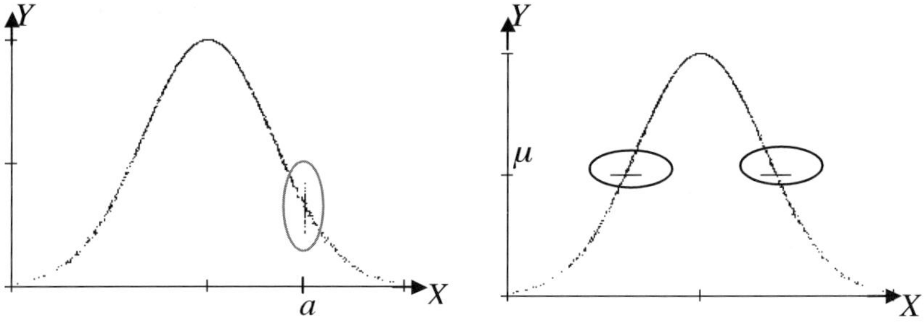

(a) Drops produced by CG_A (b) Drops produced by CG_B

Fig. 7. Joint distribution of drops (Li and Du 2005)

In a SCSRG, a given a in U_1 actives CG_A, which randomly generate μ. The value of μ reflect the degree of activation and it is used as the input for the CG_B, which randomly produce a cloud drop *drop*(b, μ). Therefore the SCSRG transfers the uncertainty implicitly. For a given a in U_1, the output b in U_2 is random. The uncertainty is transferred through μ from U_1 to U_2. Under such strategy, the rule generator guarantees the transfer of uncertainties in the reasoning.

❑ *Multiple Conditions-Single Rule Generator*

If a rule antecedent has two or more factors, it is called multiple conditions-single rule, which can be formalized as

If A_1, A_2, …, A_n then B

A multiple condition single rule generator (MCSRG) can be made by connecting multiple CG_A with a CG_B as shown in Figure 8. A_1, A_2, …, *An*, and B represent the concepts C_{A1}, C_{A2}, …, C_{An} and C_B corresponding to the universal of discourse U_{A1}, U_{A2}, …, U_{An} and U_B, respectively. A given x_1 in U_{A1} actives C_{A1}, producing μ_1; a given x_2 in U_{A2} actives C_{A2}, producing μ_2; till a given x_n in U_{An} active C_{An}, producing μ_n.

The relationship of multiple quantitative concepts in the antecedent of the rule is implicit. Logically it is difficult to use a "AND" operation to produce μ. Therefore, a new concept of "Soft-And" is proposed to produce μ from μ_1, μ_2, …, μ_n . Here "Soft-And" is considered as a quantitative concept and represented as a multi-dimensional cloud.

4.3 Uncertainty Reasoning by the Cloud Model Theory

In our case there are three factors in the rule antecedent, it is a three conditions single rule. The rule reflects the relationship of four quantitative concepts in *Slope, Soil Texture, Vegetation Cover* and *Erosion Hazard*, where A_1 =" $\geq 10\%$"; A_2 ="SAND"; A_3= " $\leq 25\%$"; B ="SEVERE".

The uncertainty reasoning based on the cloud model theory consists of three steps:

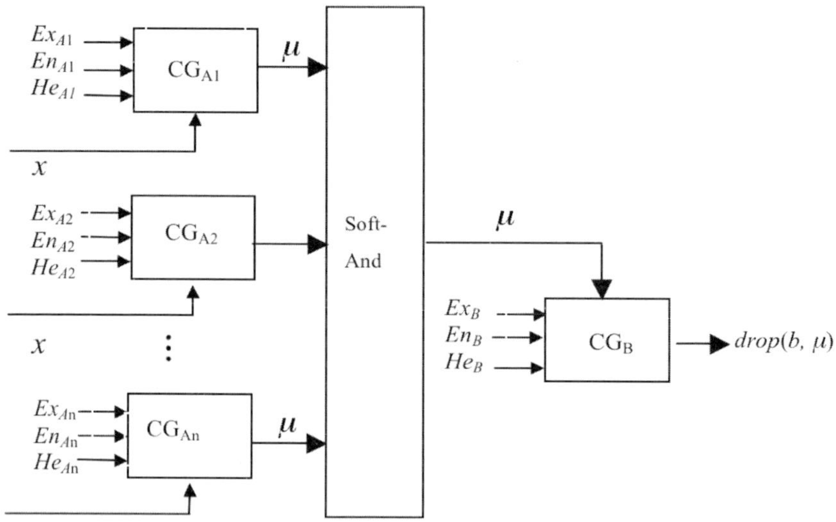

Fig. 8. Multiple conditions-single rule generator (MCSRG)

1) Active the Antecedent by CG_A

The three conditions in the rule antecedent can be represented as three quantitative clouds as follows:

$$C_{A1} = \begin{cases} 1 & x \geq 10 \\ C(10, 6/3, 0.01) & otherwise \end{cases}$$

$$C_{A2} = C(100, 15/3, 0.5) \tag{10}$$

$$C_{A3} = \begin{cases} C(25, 15, 0.1) & otherwise \\ 1 & x \leq 25 \end{cases}$$

The consequent of the rule can also be represented as a cloud as:

$$C_B = C(1, 0.2, 0.01) \tag{11}$$

When SL=9, ST=80 and VC=24, we use the CG_A to produce μ for each condition of the rule antecedent, i.e. SL=9 actives C_{A1} to produce μ_1; ST=80 actives C_{A2} to produce μ_2; VC=24 actives C_{A3} to produce μ_3.

2) Three Rule "Soft-And"

Three rule "Soft-And" can be considered as a three-dimensional (3-D) normal cloud $C(1, En_x, He_x, 1, En_y, He_y, 1, En_z, He_z)$. Each dimension corresponds to the value rang of certainty μ_1, μ_2 and μ_3, [0,1]. The statistical expectation of the cloud drops produced by "Soft-And" is (1,1,1). Further the cloud drop is from the expectation drop, smaller its certainty degree μ, reflecting the uncertainty of "AND". That's why it is called "Soft-And". The value μ produced via the "Soft-And" operation is different from "And" in the fuzzy set theory, which usually equals to $min\{\mu_1, \mu_2, \mu_3\}$.

The degree of "Soft-And" is adjusted by the values of (En_x, He_x; En_y, He_y; En_z, He_z). These values should be defined case by case (Li and Du, 2005). However, no guideline of setting these values has been provided. Here we think the uncertainties of three dimensions are the same and they contribute equally to μ, so $En_x=En_y= En_z=0.25$. Also, the randomness of three dimensions come from the randomness of the measurements, so $He_x=He$; $He_y=He_2$; $He_z=He_3$. Therefore, the 3-D cloud of the "Soft-And" is defined as C(1, 0.25,0.01;1,0.25,0.02;1,0.25,0.1) in our case.

We used μ_1 · μ_2 and μ_3 to be the input for the 3-D "Soft-And" cloud, and generated μ as follows:

$$u = \exp\left[-\frac{1}{2}\left[\frac{(\mu_1 - En_x)^2}{(En_x')^2} + \frac{(\mu_1 - En_y)^2}{(En_y')^2} + \frac{(\mu_1 - En_z)^2}{(En_z')^2}\right]\right]$$

$$\text{where } En_x' = En_x + He_x * randn(1) \tag{12}$$
$$En_y' = En_y + He_y * randn(1)$$
$$En_z' = En_z + He_z * randn(1)$$
$$\text{i.e. } (En_x', En_y', En_z') = G(En_x, He_x, En_y, He_y, En_z, He_z)$$

3) Produce Drop (μ, b) by CG_B

$$En_B' = En_B + He_B * randn(1)$$
$$b = Ex_B - En_B' * \sqrt{-2\ln(\mu)}. \tag{13}$$

Table 1 presents 30 random results of (b, μ) when x=[9, 80, 24]. It shows the final result of b is random. The average of thirty values of b is $0.7863 \approx 0.80$; the standard deviation is about 0.03. It means that the decision "the erosion hazard is severe" can be made based upon the observation with different possibilities, ranging between (0.69, 0.89) with expectation about 0.80.

Table 1. Results by "Soft-And" Reasoning

	μ_1	μ_2	μ_3	μ	b
	0.8807	0.7654	1	0.6031	0.7846
	0.8816	0.7759	1	0.6203	0.8135
	0.8801	0.7523	1	0.5714	0.7819
	0.8831	0.8141	1	0.7070	0.8388
	0.8829	0.7062	1	0.4702	0.7549
	0.8825	0.7408	1	0.5558	0.8035
	0.8830	0.7891	1	0.6361	0.8043
	0.8825	0.7814	1	0.6256	0.8088
	0.8821	0.7418	1	0.5523	0.7856
	0.8826	0.7677	1	0.6180	0.8072
	0.8832	0.7857	1	0.6540	0.8248
	0.8827	0.7467	1	0.6096	0.8084
	0.8837	0.7492	1	0.5565	0.7825
	0.8818	0.7294	1	0.5324	0.7861
	0.8834	0.7774	1	0.6284	0.8098
	0.8812	0.6297	1	0.3282	0.7092
	0.8829	0.7645	1	0.6026	0.8088
	0.8814	0.7382	1	0.5436	0.7908
	0.8841	0.7572	1	0.5996	0.7981
	0.8812	0.6867	1	0.4510	0.7356
	0.8826	0.7806	1	0.6106	0.8059
	0.8822	0.6947	1	0.4402	0.7319
	0.8825	0.7251	1	0.512	0.7659
	0.8814	0.8035	1	0.6798	0.8113
	0.8837	0.7231	1	0.5039	0.7667
	0.8821	0.7150	1	0.4953	0.7453
	0.8819	0.7148	1	0.4987	0.7563
	0.8815	0.6954	1	0.4647	0.7532
	0.8830	0.7276	1	0.5228	0.7772
	0.8824	0.8095	1	0.6877	0.8386
Average	0.8823	0.7463	1	0.5627	0.7863
Standard Deviation	0.0009	0.0402	0.000	0.0829	0.0311

Therefore, the result obtained by the fuzzy set theory is only a special case of that obtained by the cloud model theory. The uncertainties of errors and fuzziness are propagated together via "Soft-And" reasoning in the cloud model theory. The result obtained is a random variable, with a quality assurance indicated by the standard deviation. Caution has to be taken when the decision is made.

5 Conclusions

In this paper, an integrated cloud model has been proposed to represent combined uncertainties of measurement errors and fuzziness. The cloud model can be calibrated by simple mathematical characteristics (*Ex*, *En*, *He*), with *Ex* expressing the expectation value of the fuzzy concept, *En* expressing the acceptable range of the fuzzy concept, and *He* calibrates the errors in the measurement. When there are measurement errors in fuzzy classification, we consider $He \approx \sigma_x$. This view extends the existing cloud model by adding the physical meaning of *He* for the cloud model, which is not clearly explained in all the existing papers. It also provides the guideline of setting the value of *He* for the cloud model.

The rule generator guarantees the transfer of uncertainties in the reasoning. Both errors and fuzziness are considered in the propagation from source data to final output, which overcome the limits in fuzzy set theory. The concept of "Soft-And" provides a strategy of uncertain decision-making based upon multiple criteria. It also provides a tool for fuzzy overlay.

The experiment presented in Section 4 illustrates the practical usefulness of the model and the uncertainty reasoning. It also provides an example for setting the mathematical parameters for the cloud. It shows that when there are errors in the measurement, the final decision is not a constant but a random variable with deviation, which can be used as quality assurance for decision-making.

This integrated model only fits into the situations that the concepts or the classes could be modeled as a bell-shaped fuzzy membership function. However, it is found in literature that the bell-shaped fuzzy membership function is mostly used.

Acknowledgements

The research is supported is supported by the '985 Project – GIS and Remote Sensing Platform for Geo-Application' from the Chinese Ministry of Education (no. 105203200400006) and the Hong Kong Polytechnic University (no. G-YW92).

References

Burrough PA (1996) Natural Objects with Indeterminate Boundaries. In: Burrough P, Frank AU (eds) Geographic Objects with Indeterminate Boundaries. Taylor & Francis, pp 3–28

Burrough P, Frank AU (eds) (1996) Geographic Objects with Indeterminate Boundaries. Taylor & Francis, 345 pp

Cheng T, Molenaar M, Bouloucos T (1997) Identification of fuzzy objects from field observation data. In: Hirtle SC, Frank AU (eds) Proc of Conf of Spatial Information Theory (COSIT'97), Spatial Information Theory: A Theoretical Basis for GIS (= LNCS 1329). Springer-Verlag, Berlin, pp 241–259

Cheng T, Molenaar M, Lin H (2001) Formalizing fuzzy objects from uncertain classification results. Int J of Geographical Information Science 15:27–42

Cheng T, Li Z, Deng M, Xu Z (2005) Representing indeterminate spatial objects by cloud theory. In: Proc of the 4th Int Symp on Spatial Data Quality (ISSDQ 2005), Beijing, August 25-26, pp 70–77

Fisher P (2003) Data quality and uncertainty: ships passing in the night! In: Shi W, Goodchild M, Fisher P (eds) Proc of the 2nd Int Symp on Spatial Data Quality, pp 17–22

Heuvelink GBM, Burrough PA (1993) Error Propagation in Cartographic Modeling using Boolean and Continuous Classification. Int J Geographical Information Systems 7(3):231–246

Leung Y, Ma JH, Goodchild MF (2004) A general framework for error analysis in measurement-based GIS Part 1: The basic measurement-error model and related concepts. J of Geographical Systems 6:325–354

Li D, Cheung D, Shi X, Ng D (1998) Uncertainty reasoning based on cloud model in controllers. Computers Math Application 35(3):99–123

Li D, Di K, Li D (2000) Knowledge Representation and uncertainty reasoning in GIS Based on cloud theory. In: Forer P, Yeh AG, He J (eds) Proc of 9th Int Symp on Spatial Data Handling, pp 3a.3–14

Li D, Du Y (2005) Artificial Intelligence with Uncertainty. In press

Neumaier A (2004) Clouds, fuzzy sets and probability intervals. Reliable Computing 10:249–272

Petry F, Robinson V, Cobb M (2005) Fuzzy Modeling with Spatial Information for Geographic Problem. Springer-Verlag

Robinson VB (2003) Fuzzy sets in geographical information systems. Transactions in GIS 7:3–30
Wang S, Shi W, Li D, Li D, Wang X (2003) A Method of Spatial Data Mining Dealing with Randomness and Fuzziness. In: Shi W, Goodchild M, Fisher P (eds) Proc of the 2nd Int Symp on Spatial Data Quality, pp 370–383
Zhang JX, Goodchild MF (2002) Uncertainty in Geographical Information. Taylor and Francis, New York

Appendix (Li et al. 2000)

1. Cloud Generator (CG)

Input: the mathematical properties of the concept *C*: *Ex*, *En* and *He*; and number of cloud drops *n*
Output: cloud drops (x,μ)

BEGIN
  i=1:n
  $En' = \text{NORM}(En, He)$;
  $X=\text{NORM}(Ex, En)$;
  $\mu = e^{\frac{-(x-Ex)^2}{2(En')^2}}$;
  OUTPUT *drop* (x, μ);
  Next i;
END

2. 1-D Antecedent Generator (CG_A)

Input: the mathematical properties of the concept C_1: *Ex*, *En* and *He*; and *a*
Output: cloud drops (a,μ)

BEGIN
  $En' = \text{NORM}(En, He)$;
  $\mu = e^{\frac{-(a-Ex)^2}{2(En')^2}}$;
  OUTPUT *drop* (a, μ);
END

3. 1-D Consequent Generator (CG_B)

Input: the mathematical properties of the concept C_2: *Ex*, *En* and *He*; and μ
Output: drops(b, μ)

BEGIN
$En' = \mathrm{NORM}(En, He)$;
$b = Ex \pm En'\sqrt{-2\ln\mu}$;
OUTPUT *drop* (b, μ);
END

4. Single Condition Single Rule Generator (SCSRG)

Input: the mathematical properties of the concept C_1: *Ex*, *En* and *He*; and *a*
Output: drop(b, μ)

BEGIN
$En' = \mathrm{NORM}(En, He)$;
$\mu = e^{\frac{-(a-Ex)^2}{2(En')^2}}$;
$En' = \mathrm{NORM}(En, He)$;

If $a <= Ex_A$, $b = Ex - En'\sqrt{-2\ln\mu}$;
If $a > Ex_A$, $b = Ex + En'\sqrt{-2\ln\mu}$;
OUTPUT *drop* (b, μ);
END

The Influence of Uncertainty Visualization on Decision Making: An Empirical Evaluation

Stephanie Deitrick, Robert Edsall

Department of Geography, Arizona State University,
Tempe, AZ USA; 85282-0104

Abstract

Uncertainty visualization is a research area that integrates visualization with the study of uncertainty. Many techniques have been developed for representing uncertainty, and there have been many participant-based empirical studies evaluating the effectiveness of specific techniques. However, there is little empirical evidence to suggest that uncertainty visualization influences, or results in, different decisions. Through a human-subjects experiment, this research evaluates whether specific uncertainty visualization methods, including texture and value, influence decisions and a users confidence in their decisions. The results of this study indicate that uncertainty visualization may effect decisions, but the degree of influence is affected by how the uncertainty is expressed.

1 Introduction

Visualization has the power to increase the apparent quality of highly generalized or uncertain geographic data. Uncertainty visualization strives to bridge the gap between the imprecision of reality and the apparent precision of its digital representation to provide a more complete representation of data (Pang et al. 1997). Research has demonstrated that visualization of the uncertainty of complex spatial data can aid the process of decision making (MacEachren and Brewer 1995; Leitner and Buttenfield 2000; Cliburn et al. 2002). Although visualization environments do not necessarily constitute complete or all-inclusive views of every possible alternative, they can supply decision makers with a quick, and to a certain degree, reliable overview of reasonable solutions (Aerts et al. 2003).

The mere fact that a phenomenon is represented on a map may imply unwarranted authoritativeness in the data. Comprehensive analysis of a geographic dataset is facilitated by an integrated presentation of both the data and its uncertainty. Uncertainty in this context refers to uncertainty in input data, attribute data, model formulations, or graphical representation. Managing uncertainty for decision support involves quantifying the uncertainty present, and requires an understanding of how uncertainty propagates in the data, model, or simulation. Furthermore, it involves learning how to make decisions when uncertainty is present, and communicating that uncertainty to decision makers (Aerts et al. 2003). Researchers have responded to these challenges by developing concepts and techniques for the representation of uncertainty for use in decision support applications (Pang et al. 1997; Leitner and Buttenfield 2000; Cliburn et al. 2002).

This paper examines the results of a pilot study of decision-making based on maps with and without a representation of uncertainty, and reports results that suggest that uncertainty visualization has a significant influence on decisions. Specifically, our research focuses on the following:

- Does displaying uncertainty information result in different conclusions or decisions about the data?
- Does the inclusion of uncertainty result in a difference in expressed confidence about decisions or conclusions?

We examine relevant background literature, describe the methods and results of a human-subjects experiment we conducted, and discuss the experiment in the context of extending and generalizing its results. The results of this pilot study are preliminary findings, which will support an ongoing study of decision-making and uncertainty representation.

2 Background

2.1 Approaches to Uncertainty Visualization

Cartographic research offers a well-defined framework of guidelines for the display of different types of information. Recent research has theorized and demonstrated that some methods of depicting uncertainty may be superior to others; some suggested methods include supplementing thematic maps with added geometry or visual variables (on static or dynamic maps), or adding specific interactive capabilities on dynamic maps. Dynamic maps are a special type of map that includes maps that are interactive, animated or both. The pilot study discussed in this paper was designed to determine the degree to which the addition of uncertainty representation influences decision making, and, though interactive exploratory tools in a computer environment may prove important, we limited the scope of this study to the use of static maps. We look in particular, therefore, to previous studies of uncertainty visualization using paper maps.

MacEachren (1995) suggests three general methods for depicting uncertainty. The first two can be applied to static representations: first, representations can be compared with individual representations presented for an attribute and its associated uncertainty. Second, representations can be combined, where a single visualization presents both an attribute and its uncertainty – using appropriate visual variables, an attribute and its uncertainty are visualized by overlaying one on the other. The third method, possible in an interactive computer environment, is to utilize exploration tools that allow users to manipulate the display of both the data and their uncertainty (see also Howard and MacEachren 1996; Slocum et al. 2004). Fisher (1994) demonstrated that sound could be used in combination with animation for the representation of uncertain information.

Gershon (1998) proposed two general categories for techniques to represent uncertainty: intrinsic and extrinsic. *Intrinsic* representation techniques integrate uncertainty in the display by varying an object's appearance to show associated uncertainty. Such techniques include varying visual variables such as texture, brightness, hue, size, orientation, position, or shape (Gershon 1998). For example, finer texture or darker value could represent greater reliability and coarser texture and lighter values could represent unreliability (MacEachren 1992; Leitner and Buttenfield 2000; Slocum et al. 2004). MacEachren (1992) suggested that saturation is logical for depicting uncertainty, with pure hues representing reliable data and unsaturated hues representing unreliable data. Three years later, MacEachren (1995) proposed a new visual variable, "clarity," that would be particularly applicable to uncertain data representation. Cliburn et al. (2002) suggest that intrinsic methods provide a more general visualization of detailed uncertainty data, which non-technical users may prefer over extrinsic representations. *Extrinsic* techniques add geometric objects, including arrows, bars, and complex objects (such as pie charts), to represent uncertainty. This representation method implies that uncertainty is a variable separate from the data. Some of the more complex objects, such as error bars, may become confusing over large areas. Allowing the selection of specific regions or objects in an interactive exploratory environment may prevent complex objects from overwhelming the user (Cliburn et al. 2002).

2.2 Uncertainty Visualization for Decision Support

Decision support systems (DSS) have been defined as computer-based systems that integrate modeling and analytic tools with data sources, assist in the development, evaluation and ranking of potential alternative solutions, assist in the management and evaluation of uncertainty and enhance overall comprehension of problems and potential solutions (Mowrer 2000; Crossland et al. 1995). Although DSS incorporate inaccuracies (i.e. uncertainty), traditional (non-spatial) DSS do not provide a means for organizing and analyzing *spatial* data. Spatial decision support systems (SDSS) integrate the data and analytical models of traditional decision support systems with the spatial data organization and processing capabilities of GIS, remote sensing classification or spatial statistics, allowing decision makers to perform graphical analysis of spatial information (Cooke 1992; Sengupta and Bennett 2003; Mowrer 2000). As with the definition of a DSS, spatial decision support provide access to relevant information that otherwise might be inaccurate or unavailable. SDSS also provide detailed displays resulting in reduced decision time and enabling a better grasp of spatial problems due to better visualization of the problem to be solved (Crossland et al. 1995).

It has been argued that uncertainty information is a vital component in the use of spatial data for decision support (Hunter and Goodchild 1995; Aerts et al. 2003). Many techniques have been developed for communicating uncertainty in data and models for specific visualization applications, such as remote sensing, land allocation, water-balance models and volumetric data (Aspinall and Pearson 1995; Leitner and Buttenfield 2000; Bastin et al. 2002; Cliburn et al. 2002; Aerts et al. 2003; Lucieer and Kraak 2004; Newman and Lee 2004). Although cartography has a strong tradition of empirical research in map design and user comprehension, research into the effectiveness of uncertainty visualization as it relates to decision support is only beginning to emerge. Researchers have emphasized the need for empirical research to test the effectiveness of visual variables and their usefulness in depicting uncertainty (Evans 1997; MacEachren et al. 1998, Leitner and Buttenfield 2000; MacEachren et al. 2005).

In one study, MacEachren et al. (1998) developed and tested a pair of methods for depicting "reliability" of data on choropleth maps of cancer mortality information, and studied the effect of different visual depictions on accuracy of responses to tasks typical of epidemiological studies. They found that texture-overlay onto a choropleth map (a "visually separable" technique) was superior to a "visually integral" depiction (using color to represent both data and reliability) for decision-making using uncertainty. Additionally, Leitner and Buttenfield (2000) examined how the addition of attribute uncertainty information affects the decision-making process utilizing static maps, analyzing correctness and speed of responses to tasks relevant to urban planning. The addition of uncertainty with specific representation styles significantly increased the number of correct responses. They found that users identify the inclusion of uncertainty information as clarification and not as an addition of map detail.

3 Methods

At its most general, our study aimed to ascertain whether the inclusion of uncertainty information on a map (or set of maps) has a significant influence on decision-making. To do so, we conducted a human-subjects test consisting of a series of map reading and decision-making tasks that are representative of real-world tasks in water use policy and information dissemination. This test was administered using paper color maps of a variety of forms showing data sets from predictive models – and in some cases the uncertainty associated with those data sets – relevant to making decisions about water use. Specifically, we gave participants a series of different ranking tasks, identifying which regions were most vulnerable to water policy or water use changes, and which regions should be targeted in a marketing campaign for responsible water use. The survey instrument, including the maps, ranking tasks, and other questions, is described in detail below.

This study differs from similar studies in the past (MacEachren 1992; Leitner and Buttenfield 2000; Cliburn et al. 2002) in that we aimed to determine whether decision making *changes* as a result of incorporating uncertainty on maps, and not whether "correct" answers to specific questions were obtained or whether response times changed. To test for this, participants were randomly divided into two groups, one that was provided with uncertainty information present and one that was not. Participants in the first group (Group A) were asked questions related to maps of the data *and* their uncertainty, while those in the second group (Group B) were asked questions based on maps of the data alone. We assumed (with some caveats, discussed below) that, if we found any significant differences between average rankings from Group A and those from Group B, they could be attributed to the presence of uncertainty information on the maps.

Efforts were made to make the two groups otherwise similar. The questions for each group were identical, with the exception that surveys in Group A included questions and introductory statements referring to the uncertainty of the data. Base maps and color schemes were kept the same for both survey sets. In addition, the data on the maps were identical for each question, except, as mentioned above, the maps for the Group A surveys included representations of uncertainty. These measures were designed to help ensure that any variations between the groups were due to the inclusion of uncertainty.

In this study, we also sought to examine subtle differences in the connotations of the terms "certain" and "uncertain." In the survey and on the maps in Group A, some questions referred to the "certainty" of the data, while others refer to the "uncertainty" of the data. The concepts are, of course, similar in that they can both be either qualitative ("this value is uncertain") or quantitative ("this value is 80% certain," which is equivalent in this study to "this value is 20% uncertain"). Throughout the remainder of this document, we use the term "uncertainty visualization" to refer to both concepts.

The complete survey took approximately 15 to 20 minutes to complete. A majority of participants were students at Arizona State University, with some participants being from local planning and engineering companies.

3.1 Survey Overview

The first page of the survey identified the goal of the survey and provided a brief description of the participants' role in the survey. These introductory statements were purposely vague in order to avoid biasing Group B (our control group), stating that the goal of the project was simply to analyze the effects of specific visualization techniques on decision making (without making any mention of uncertainty). Several pages, discussed in turn below, followed, each with its own distinct map or maps and set of questions. The maps (an example of which is shown in Fig. 1) depicted predicted water consumption based on changes in households, population and/or income in a hypothetical region. Each set of questions was meant to simulate tasks typical of decision-making. A brief discussion of the purpose of each map follows.

3.1.1 Map 1

Map 1 was a simple choropleth map, as in Figure 1, showing water consumption in the hypothetical region. This map and its associated questions were intended to identify participants' basic ability to read and interpret maps. Responses to these questions from Groups A and B would be compared to ensure that the groups represented similar map-reading abilities. Participants were asked to identify areas (among four circled and labeled sub-regions) of greatest growth and to make decisions about where to start a water conservation awareness campaign. Because this section was meant to evaluate basic map reading abilities, neither survey represented or asked about uncertainty for these questions.

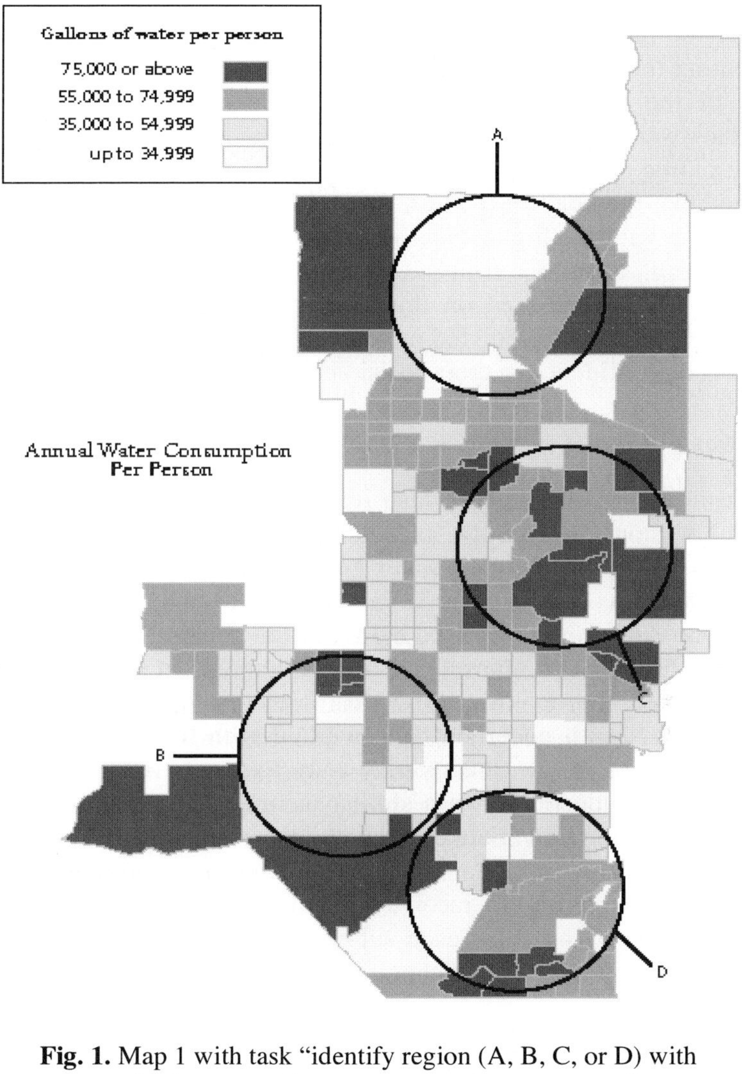

Fig. 1. Map 1 with task "identify region (A, B, C, or D) with the highest rate of water consumption per person"

3.1.2 Map 2

The questions for map 2 were intended to compare decisions made with (in Group A) and without (in Group B) representation of the relative *certainty* of the data. In both groups, we presented data with identical choropleth maps (but different from those in map 1). In Group A, we depicted the certainty of the data in a second choropleth map with lightness differences representing the certainty. Group B was not provided a second map in this section of the survey. We presented a "story problem" to participants, where they were in charge of determining where a media campaign to educate the public about water conservation should begin. The story problem explained that the goal for the task was to release the media campaign first in the region with the highest predicted increase, and then if the campaign was successful, to release it in the area with the next highest predicted increase and so forth. Based on this task, participants ranked the regions from highest to lowest predicted increase. They were also asked to identify their level of confidence in their decisions.

3.1.3 Map 3

The questions for map 3 were also intended to compare decisions made with and without uncertainty information. Once again, participants saw choropleth maps of predicted water consumption. Groups A and B saw identical thematic data in map 3, with identical color schemes and class breaks (map 3, however, depicted a different theme, in the same hypothetical region, than that in maps 1 and 2). Participants in Group A this time saw a bivariate choropleth map, with a depiction of uncertainty using a texture overlay, with uncertain data overlaid with hatch marks and more certain data with no texture overlay (MacEachren, Brewer, and Pickle (1998) depicted uncertainty in a similar manner in their study). Group B was shown only the univariate choropleth map. Once again, we presented a "story problem" to participants, where they were in charge of prioritizing four circled sub-regions to receive infrastructure improvements. Participants ranked the priority for each sub-region and identified their level of confidence in their decisions. Otherwise, the maps for the two groups were the same.

3.1.4 Map 4

Map 4 was only presented to participants in Group A. Map 4 was identical to map 1, except that certainty information was included as a texture overlay, with three levels of certainty represented with different texture densities. We asked the same question for map 4 as that for map 1. This allowed for within-subject comparison within Group A based on maps with and without certainty; with map 4, we would be able to determine if the certainty information resulted in different decisions made by the same person, and if the participants actually used the certainty information presented.

3.1.5 Exit Questions

Finally, we asked open-ended questions meant to determine if the maps were interpreted as effective decision-making tools (to both groups), whether uncertainty/certainty information was seen as negative or positive (to group A only), and whether the uncertainty information was viewed as useful for decision making (to group A only).

4 Pilot Study Analysis and Results

We collected the following data: rankings of sub-regions of the maps according to water-use decision making for each map and participants' confidence levels and opinions for each map. Responses to corresponding questions were compared between Groups A and B (and within Group A in the maps 1 and 4 comparison) to determine the significance of the variation between responses.

4.1 Analysis

Three types of information obtained during the study were examined to address the question of whether the representation of uncertainty affects decisions. We examined the differences in both rankings and confidence in those rankings between Groups A and B. Responses to open ended questions about the inclusion of certainty/uncertainty information were examined for Group A.

4.1.1 Ranking Comparison

Participant rankings for the questions for maps 1 through 3 were compared between the two Groups. For each map, participants ranked four regions (region A-D) on a scale of one to four, with one being the highest (priority, increase, consumption) and four being the lowest (priority, increase, consumption). To facilitate the analysis, we assigned each region a numerical value based on the ranking they received (for example, if the ranking was

ABCD, region A would have value 1, B would have a value of 2, etc.). Based on these values, we calculated the average ranking for each region for the Groups (region A had an average value for each Group A and Group B, as did regions B, C, and D). We then calculated the difference between the average values for each region (Group A minus Group B), and found the mean (absolute value) of all of these differences, for each map (1, 2, and 3). Our null hypothesis in each case was that there would be no difference between the rankings between Groups A and B. We evaluated this hypothesis by calculating a 95% confidence interval around the mean difference: if the participants in Groups A and B ranked the regions the same way, this confidence interval should include zero.

From Group A, we also compared the rankings for map 1 and those for map 4. The absolute value of the difference between the rankings for each region was calculated for each participant. A change in ranking from #4 to #1 would result in a score of three (four minus one), and a change from #2 to #4 would result in a score of two (four minus two). The minimum score between sets of rankings is, of course, zero, and the maximum is eight (a complete reversal: 4-3-2-1 to 1-2-3-4). Again, our null hypothesis was that there would be no difference between the rankings from map 1 (without uncertainty mapped) and map 4 (with uncertainty). We evaluated this hypothesis by calculating a 95% confidence interval around the mean difference for all participants: if the rankings from map 1 and map 4 were similar, this confidence interval should include zero.

4.1.2 Confidence in Rankings

For each ranking question, we also asked about the participant's confidence in the ranking decision. Participants identified their level of confidence on a five-point Likert scale from "not confident" (1) to "completely confident" (5). For each of the maps, we calculated the mean value for each Group and performed a two-sample t-test; our null hypothesis in this case was that there would be no difference between the mean confidence levels for each map between the two Groups.

4.1.3 Opinion Questions

The last four survey questions asked the uncertainty group participants to give their opinions about whether the inclusion of uncertainty information made them more/less confident in their decision, how the inclusion of uncertainty affected their decision and whether they viewed the inclusion of uncertainty/certainty information as negative. The analysis for these results consists of a summary of responses.

4.2 Results

We conducted the pilot study with volunteer participants sampled from among undergraduate and graduate students in the departments of Geography and Planning, as well as from professionals and decision makers in private planning and engineering firms. We collected 92 surveys in total, 48 in Group A and 44 in Group B (uneven because of incomplete responses and the randomized distribution between Groups). Of those 92, 87 were students and five were professionals[1].

4.2.1 Map 1

All participants were able to identify the area of greatest water consumption in map 1. In this instance, the correct answer is important in assessing whether participants understand the information presented in the map, and if we could reasonably compare Groups A and B. More than 83 percent of participants correctly ranked the regions from highest to lowest consumption.

When asked to identify the area where a water conservation awareness campaign should begin, over 90 percent recognized the region depicting the highest water consumption.

Table 1 identifies the results of the t-test for the average level of confidence in the rankings made for each subgroup. The results support the null hypothesis that there was no difference in rankings between the two groups. Based on these results we concluded that participants were drawn from the same population and that participants were able to recognize relevant spatial patterns represented in a choropleth map.

Table 1. Map 1 reading comparison

	Group A		Group B
Mean	3.73		3.86
Variance	0.75		0.86
N	48		44
hypothesized mean difference		0.00	
Df		88	
$t_{observed}$		0.72	
t_{crit}		1.66	
$p(t_{observed}<t_{crit})$		0.24	

[1] We sent 15 surveys to professionals; only 5 were returned completed. The rate of return for students (87 of 95) was higher, presumably because we remained in the classroom with the students – unlike the professionals – while they completed the survey.

4.2.2 Map 2

Figure 2 summarizes the rankings for map 2, and Table 2 summarizes the mean ranks for each sub-region, by Group, and the confidence interval for the overall mean difference. Our test showed evidence of a significant difference in rankings between those from Groups A and B (the 95% confidence interval does not include zero, and our null hypothesis is rejected).

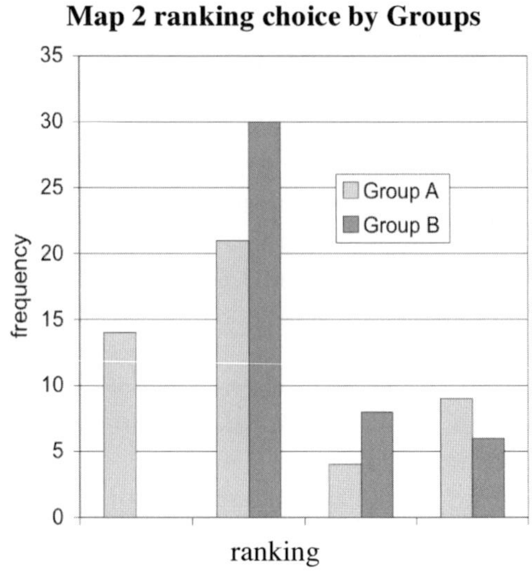

Fig. 2. Three common ranking orders for map 2: one popular ranking order for Group A was never chosen by Group B participants

Table 2. Map 2 ranking comparison

	Group A	Group B	Absolute difference
Mean rankings			
Sub-region A	3.04	3.77	0.73
Sub-region B	1.21	1.45	0.24
Sub-region C	2.29	1.84	0.45
Sub-region D	3.25	2.93	0.32
mean difference			0.44
σ			0.21
95% confidence interval (±0.30)		(0.14, 0.73)	

Table 3 identifies the results of the t-test for the average level of confidence in the rankings made for each subgroup. The average level of confidence expressed in both Groups was somewhat to almost completely confident (values of 3.46 and 3.61 out of 5.00). We cannot conclude that, in the case of map 2 and its associated ranking task, there was a statistically significant difference between the confidence ratings depending on the presence of certainty information

Table 3. Map 2 confidence comparison

	Group A		Group B
Mean	3.46		3.61
Variance	1.02		1.17
N	48		44
hypothesized mean difference		0.00	
Df		88	
$t_{observed}$		0.71	
t_{crit}		1.66	
$p(t_{observed}<t_{crit})$		0.24	

4.2.3 Map 3

Figure 3 summarizes the rankings for map 3, while Table 4 summarizes the mean ranking for each sub-region, by Group, and the confidence interval for the overall mean difference. Our test showed evidence of a significant difference in rankings between those from Groups A and B (the 95% confidence interval does not include zero, and our null hypothesis is rejected).

Table 4. Map 3 ranking comparison

	Group A	Group B	Absolute difference
Mean rankings			
Sub-region A	3.33	3.02	0.31
Sub-region B	2.19	2.98	0.79
Sub-region C	1.69	1.39	0.30
Sub-region D	2.58	2.39	0.20
Mean difference			0.40
σ			0.27
95% confidence interval (±0.27)		(0.03, 0.77)	

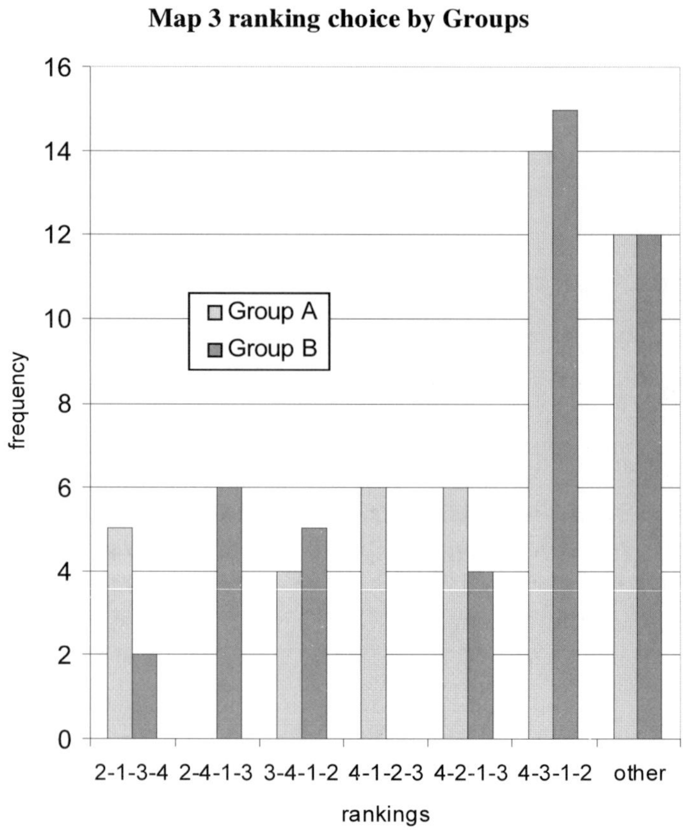

Fig. 3. Differences in rankings of sub-regions with and without uncertainty: the 2-4-1-3 ranking was popular among those who were not provided uncertainty (Group B), while none that were provided uncertainty (Group A) chose that ranking

Table 5 identifies the results of the t-test for the average level of confidence in the rankings made for each subgroup. The average level of confidence was somewhat confident (values near three). As can be seen from the results, it cannot be concluded that there is a statistically significant difference between the confidence in rankings for map 3 between those from Group A and Group B.

Table 5. Map 3 confidence comparison

	Group A		Group B
Mean	2.98		2.86
Variance	0.99		1.03
N	48		44
hypothesized mean difference		0.00	
Df		88	
$t_{observed}$		0.56	
t_{crit}		1.66	
$P(t_{observed}<t_{crit})$		0.29	

4.2.4 Map 1 vs. Map 4

Figure 4 summarizes the absolute difference in rankings for maps 1 versus 4 for each Group A participant. Over half (25 of 48) participants altered their ranking from map 1 to map 4 (shown on the graph as non-zero absolute ranking-difference scores), supporting the hypothesis that the presence of uncertainty on the maps influences decision-making. As discussed in section 4.1.1, a high raking-difference score indicates a significant change in ranking.

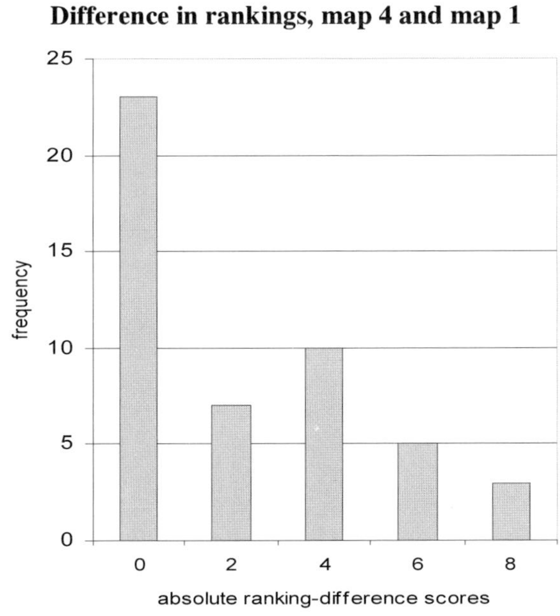

Fig. 4. Ranking-difference score frequency, comparing rankings with and without uncertainty information depicted. Non-zero scores indicate a change in ranking (the scoring is discussed in section 4.1.1)

Table 6. Map 1 v. Map 4 ranking-difference scoring

	ranking-difference scoring
Mean	2.27
σ	2.62
95% confidence interval (± 0.76)	(1.51, 3.03)

Table 6 summarizes the difference between and the confidence interval for the paired rankings. Our test showed evidence of a significant difference in rankings between Maps 1 and 4 (the 95% confidence interval does not include zero, and our null hypothesis is rejected). We can thus say that rankings changed when certainty was depicted. However, the average difference of 2.27 indicates that the change in rankings were subtle and not completely reversed (i.e. they may have reversed the middle values but kept the highest and lowest rankings the same).

Figure 5 summarizes the difference in confidence expressed for the rankings provided for map 1 and map 4. A negative value indicates that the participant had a higher degree of confidence for the decision made with the map without certainty information and a positive value indicates that they had a higher degree of confidence in the ranking made with the certainty map.

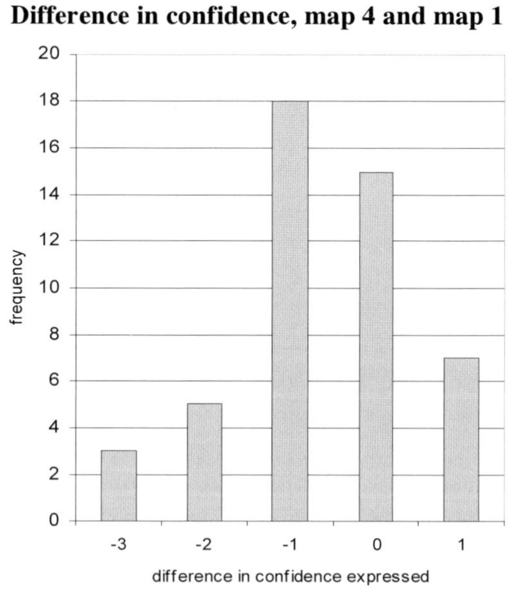

Fig. 5. The difference in confidence expressed between map 4 and map 1. A negative value indicates that confidence expressed for map 4 (with uncertainty presented) was lower than then expressed for map 1 (without uncertainty)

Table 7. Map 1 v. Map 4 difference in confidence self-scoring

	confidence scoring difference
Mean	-0.62
σ	1.13
95% confidence interval (± 0.31)	(-0.97, -0.32)

Table 7 summarizes the difference in expressed confidence for the paired results as well as the confidence interval for the mean difference. The null hypothesis that there is no difference between confidence scores with and without uncertainty depiction is rejected. The average difference in the level of confidence for the two questions was negative, indicating that confidence significantly decreased with the inclusion of certainty information.

4.2.5 Opinion Questions

Based on the opinion questions, most participants in Group A indicated that the inclusion of uncertainty information would influence their decisions, but that they would feel more confident if they had other data sources in addition to the uncertainty/certainty maps. Of Group A participants, 46 percent viewed uncertainty information as negative and certainty information as positive, 31 percent viewed neither uncertainty or certainty information as negative, 10 percent viewed both uncertainty and certainty as negative and the remaining 13 percent did not respond to the exit questions. When participants viewed the uncertainty or certainty information as positive, they also viewed the inclusion of the information as positive.

5 Summary of Results

The results for each maps ranking task showed a statistically significant difference in the rankings between Group A and Group B for maps 2 and 3, as well as within Group A for map 1 and map 4. The 95 percent confidence interval for each map comparison did not include zero, and we rejected the null hypothesis that there was no difference between responses based on maps with uncertainty and maps without uncertainty. The results for the confidence expressed identified no statistically significant difference between the confidence expressed for Group A and Group B responses for map 2 and map 3, however, there was participants expressed decreased confidence in their results for map 4 compared to their responses for map 1. These results and potential implications for future research are discussed in detail in the following section.

6 Discussion

The results of this study suggest that uncertainty visualization may influence decisions. The analysis suggested that there was a difference in rankings when both uncertainty and certainty information were included, although the differences were not extreme. Results for the expressed confidence in the decisions made found no statistical difference between confidence levels for map 2 and map 3; however, participants expressed decreased confidence when responses to map 1 and map 4 from Group A were compared.

This discrepancy in confidence level results suggests that factors other than inclusion or non-inclusion of uncertainty representations may be influencing confidence. For example, the ranking task in map 2 may have been more difficult than in map 4, or the addition of uncertainty in map 3 more obviously relevant in the ranking task than in map 2. Other factors that may have influenced expressed confidence include the complexity of the data or uncertainty classifications—map 3's general binary classification of data as certain or uncertain was less complex than map 4's representation of three degrees of certainty. The provision of more detailed information in map 4 may have contributed to the difference in confidence.

There are a number of factors about the administration of this survey that could be modified if it is to be repeated. The response rate among decision makers was below 50 percent. Increasing this response rate would allow comparisons between decision makers and others. Administering a paper-based survey to professionals in a variety of locations can be logistically prohibitive. The transition to a web based survey may lower the threshold for participation and increase the response rate with this group (Aerts et al. 2003). In addition, the survey instructions should more clearly identify that the region and data were hypothetical in nature and that the maps are not related (i.e. information in the first map should not influence responses to questions about map 2). In the exit questions, several participants noted that they attempted to utilize their knowledge of the region to interpret the data; however, since the regions geometry and size had been altered for the study maps and the data was created specifically for the study, this background knowledge made the maps confusing. Providing clearer instructions at the beginning of the survey would help to avoid this issue.

Extensions of the study could also include a third and fourth subgroup of participants. The maps should be divided into data only, data and uncertainty visualization, data and certainty visualization and data, and tabular or written description of data uncertainty. The methods of visualization should include varying levels of information detail, ranging from a simple classification of certain/uncertain to a range of saturation/value levels to represent a range of uncertainty/certainty values, as well as supporting information such as geographic reference data and detailed statistical information. These additions would identify whether uncertainty and certainty affect users differently and whether the inclusion of supporting information increases confidence.

7 Conclusions

The incorporation of uncertainty information into GIS applications and data sets is a vital component for the critical examination of spatial data for decision support. In this paper, we focused on the effect of a spatial representation of uncertainty on decision-making. We developed a pilot human-subjects experiment to evaluate the influence of uncertainty visualization in decision-making. Analysis of these tests suggests that the incorporation of a display of the spatial distribution of uncertainty information can significantly alter the decisions made by a map user. Our research, at this stage, is limited to tasks specific to water use and policy decision support, and is also limited to the use of static maps with specific uncertainty representation methods. There are many techniques that have been developed for communicating uncertainty in data and models for specific visualization applications, and research into the effect of these techniques on comprehension is ongoing. As shown in this study, uncertainty visualization may effect decisions, but the degree of influence is affected by how the uncertainty is expressed. We will use the results of this preliminary study to support future research into the effects of other uncertainty representations on user comprehension and decision-making.

References

Aerts JCJH, Clarke KC, Keuper AD (2003) Testing popular visualization techniques for representing model uncertainty. Cartography and Geographic Information Sciences 30:249–261

Aspinall RJ, Pearson DM (1995) Describing and managing uncertainty of categorical maps in GIS. In: Fisher P (ed) Innovations in GIS 2. Taylor & Francis, London, pp 71–83

Bastin L, Fisher PF, Wood J (2002) Visualizing uncertainty in multi-spectral remotely sensed imagery. Computers & Geosciences 28:337–350

Cliburn DC, Feddema JJ, Miller JR, Slocum TA (2002) Design and evaluation of a decision support system in a water balance application. Computers & Graphics 26:931–949

Cooke DF (1992) Spatial decision support systems: not just another GIS. Geo Info Systems 2:46–49

Crossland MD, Wynne BE, Perkins WC (1995) Spatial decision support systems: an overview of technology and a test of efficacy. Decision Support Systems 14:219–235

Evans BJ (1997) Dynamic display of spatial data reliability: does it benefit the map user? Computers & Geosciences 23:409–422

Fisher P (1994) Animation and sound for the visualization of uncertain spatial information. In: Hearnshaw HM, Unwin DJ (eds) Visualization in Geographic Information Systems. Wiley and Sons, London, pp 181–185

Gershon N (1998) Short Note: Visualization of an Imperfect World. IEEE Computer Graphics and Applications 18:43–45

Howard D, MacEachren, AM (1996) Interface design for geographic visualization: Tools for representing reliability. Cartography and Geographic Information Systems 23:59–77

Hunter GJ, Goodchild MF (1995) Dealing with error in spatial databases: A simple case study. Photogrammetric Engineering & Remote Sensing 61:529–537

Leitner M, Buttenfield BP (2000) Guidelines for display of attribute certainty. Cartography and Geographic Information Sciences 27:3–14

Lucieer A, Kraak MJ (2004) Interactive and visual fuzzy classification of remotely sensed imagery for exploration of uncertainty. Int J of Geographic Information Science 18:491–512

MacEachren AM (1992) Visualizing uncertain information. Cartographic Perspectives 13:10–19

MacEachren AM (1995) How Maps Work: Representation, Visualization and Design. Guilford Press, New York

MacEachren AM, Brewer CA (1995) Mapping health statistics: representing data reliability. In: Proc of the 17th Int Cartographic Conf, September 3-9, 1995, Barcelona

MacEachren AM, Brewer CA, Pickle LW (1998) Visualizing georeferenced data: representing reliability of health statistics. Environment and Planning A 30:1547–1561

MacEachren AM, Robinson A, Hopper S, Gardner S, Murray R, Gahegan M, Hetzler E (2005) Visualizing geographic information uncertainty: what we know and what we need to know. Cartography and Geographic Information Sciences 32:139–160

Mowrer HT (2000) Uncertainty in natural resource decision support systems: sources, interpretation, and importance. Computer and Electronics in Agriculture 27:139–154

Newman T, Lee W (2004) On visualizing uncertainty in volumetric data: techniques and their evaluation. J of Visual Languages & Computing 15: 463–491

Pang AT, Wittenbrink CM, Lodha SK (1997) Approaches to uncertainty visualization. The Visual Computer 13:370–390

Parikh M, Fazlollahi B, Verma S (2001) The effectiveness of decisional guidance: and empirical evaluation. Decision Sciences 32:303–331

Sengupta RR, Bennett DA (2003) Agent-based modelling environment for spatial decision support. Int J of Geographic Information Science 17:157–180

Slocum TA, McMaster RB, Kessler FC, Howard HH (2004) Thematic Cartography and Geographic Visualization, 2nd ed. Prentice Hall, Upper Saddle River, NJ

Modeling Uncertainty in Knowledge Discovery for Classifying Geographic Entities with Fuzzy Boundaries

Feng Qi[1], A-Xing Zhu[2, 3]

[1] Department of Political Science and Geography, University of Texas-San Antonio, 6900 N. Loop 1604 W., San Antonio, TX 78249, USA
[2] State Key Laboratory of Resources and Environmental Information System, Institute of Geographical Sciences and Natural Resources Research, Chinese Academy of Sciences, Building 917, Datun Rd., An Wai, Beijing 100101, China
[3] Department of Geography, University of Wisconsin-Madison, 550 North Park St., Madison, WI 53706, USA

Abstract

Boosting is a machine learning strategy originally designed to increase classification accuracies of classifiers through inductive learning. This paper argues that this strategy of learning and inference actually corresponds to a cognitive model that explains the uncertainty associated with class assignments for classifying geographic entities with fuzzy boundaries. This paper presents a study that adopts the boosting strategy in knowledge discovery, which allows for the modeling and mapping of such uncertainty when the discovered knowledge is used for classification. A case study of knowledge discovery for soil classification proves the effectiveness of this approach.

1 Introduction

Research on geographic knowledge discovery or spatial data mining has been receiving continuous attention in recent years. With the power to extract patterns from large volumes of spatial data, data mining enhances our traditional spatial data analysis abilities. Knowledge discovery from diverse data sources (e.g., image data, map data, etc.) also makes it possible to intelligently synthesize geographic knowledge that can be used with knowledge-based systems for geographic modeling and decision-making.

In previous research on geographic knowledge discovery, much effort has been made in geographic classification and category extraction (Zhang et al. 2005; Tadesse et al. 2005; and see Ester et al. 2001, Miller and Han 2001 for reviews of earlier studies). Although previous studies (Koperski and Han 1995; Castro and Murray 1998; Gopal et al. 2001; Qi and Zhu 2003; Zhang et al. 2005) have demonstrated the robustness of knowledge discovery methods in clustering or classifying geographic entities, the problem of uncertainty associated with class assignments in this process was rarely explored.

In fact, many geographic entities have fuzzy boundaries in both attribute space and geographic space (Burrough 1996). Fitting such entities into discrete categories with crisp boundaries induces uncertainties in the class assignments (Burrough 1996; Zhu 1997b). It has been demonstrated that two essential kinds of uncertainties are involved in this process: "ignorance uncertainty" and "exaggeration uncertainty" (Zhu 1997b). In order to provide information on the spatial distribution of such uncertainties in the classified products, using knowledge obtained through data mining, it is necessary to model the sources of such uncertainties during knowledge discovery.

This paper first examines the existence and nature of such uncertainties from the cognitive perspective, from which we then present an approach to modeling uncertainties through a machine learning strategy called boosting. We will then report a case study on knowledge discovery for soil classification to illustrate the approach. Conclusions are drawn and future work is outlined at the end.

2 Uncertainty in Classifying Geographic Entities with Fuzzy Boundaries

Cognitive psychology has long been concerned with categorization and classification. Before the 1970s, the classical category theory viewed all instances of a category as equal members of the category (Smith and Medin 1981). Members share common features that are singly necessary and jointly sufficient for defining the category (Smith and Medin 1981). Thus an instance that possesses all of the defining features is a category member with full membership, while an instance that lacks any of the defining features must be excluded from the category completely.

In the 1970s, psychologist Rosch (1973, 1978) presented an array of empirical studies that led to the establishment of a brand new category theory. The new theory, known as the "prototype theory", stresses the fact that category membership is not homogenous and that some members are better representatives of a category than others. This is noted as the "prototype effect" (Lakoff 1987). With prototype effects, internal structures of categories are graded from central to peripheral cases. The prototype of a category is the central concept of the category, and possible members are categorized on the basis of how similar they are to the prototype (Hampton 1995).

Our geographic environment is traditionally classified into various kinds of such categories, which exhibit prototype effects. Due to the complexity and continuous nature of most geographic phenomena, geographic classification is a complicated abstraction process that leads to prototype effects. One extreme approach for classifying geographic features is using an indefinite number of categories to approximate one-to-one correspondences for the modeling of spatial variations in full detail. However, such a representation has little practical value because the classification is not informative. What is usually done instead is to categorize in such a way that a few concepts capture rich situations, so the categories are as informative as possible. In this case, the continuous geographic features are often discretized to a limited number of distinct classes (categories). This leads to an irresolvable indefinite correspondence between the concepts and the represented world, which lends itself to the application of prototype theory, as discretization results in different degrees of memberships within each category. Such an inherent degree of category membership is one of the causes of prototype effects (Lakoff 1987).

Our traditional products of the classification practice, however, are rooted in the classical category theory and are often the so-called "area-class" maps (Mark and Csillag 1989). On an area-class map, geographic entities are delineated as polygons with crisp boundaries and no overlaps, where areas within each polygon share absolute membership to one of the prescribed classes. With such a representation, each class is actually reduced to its prototype: once an instance is assigned to a class, it is said to carry the typical properties of the class. In other words, the "individuality" of the instance is lost in this class assignment process.

This reduction of the heterogeneous classes to their prototypes introduces uncertainty in two aspects. Let's assume that two instances I_1 and I_2 are both classified as the same Class *B*. Figure 1 illustrates the positions of these instances with regard to the locations of three adjacent classes. The space shown in Figure 1 could be either the parameter space (where the x, y axes represent two properties that define the classes) or the physical space (where the x, y axes represent geographic coordinates).

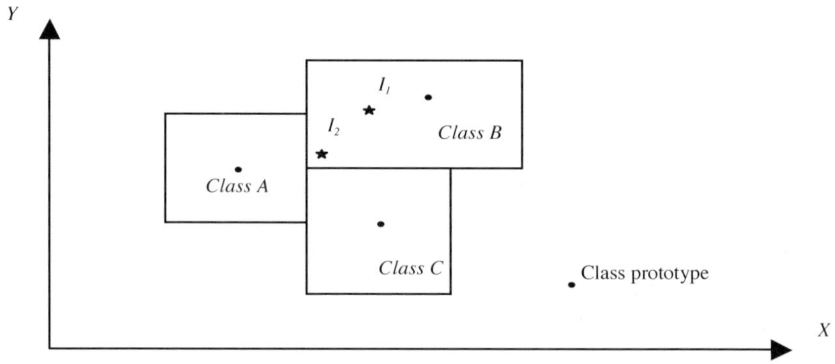

Fig. 1. Class assignments for instances I_1 and I_2. Source: Zhu (1997b)

As shown in Figure 1, it is apparent that neither I_1 nor I_2 is the actual prototype of Class *B*. By assigning these instances to Class *B* and having them bear the properties of the prototype, we ignore the differences between these instances and those of the class prototype. In this case, we committed a commission error, that is, we assigned a label to an instance, which does not fully "qualify" for it. During class assignment, we also ignore the fact that I_1 may bear some similarity to Classes *A* and *C*, as does I_2 to a different degree. By ignoring the similarities between an instance and other classes, we committed an omission error. Based on the nature of the errors, the uncertainties associated with the two types of errors are referred to as the ignorance uncertainty (associated with omission error) and exaggeration uncertainty (associated with commission error) (Zhu 1997b).

As discussed above, the source of the two types of uncertainty is the heterogeneity of category memberships. Reducing a category to its central concept overlooks its inherent prototype effects, which results in exaggeration of the similarity between an instance and the class prototype, and ignorance of the similarities between the instance and other classes. In order to model such uncertainty, it would be desirable to quantify the membership gradations within the geographic classes, and position an instance in relevance to its class prototype. The degrees of exaggeration and ignorance associated with the assignment of the instance to the class can then be estimated.

According to prototype theory, one possible source that leads to prototype effects is the employment of a so-called "cluster model" in categorization (Lakoff 1987). With such a model, the mental representation of a concept is not limited to a single structure, but consists of a composite set of cognitive models. The categorization of an instance is then based on the composite of categorization outputs from these cognitive structures. The membership of the instance to a certain category is determined through a mechanism similar to voting. When the outputs of all models coincide in the same category, the instance has the highest membership to the category and is deemed as no different from the category prototype. On the other hand, if the different models categorize the same instance differently, the instance will bear membership to multiple categories, with the membership degree to each category proportional to the number of votes favoring that category. This provides the cognitive basis for modeling membership gradations of geographic classes, and the associated uncertainty in class assignment through boosting.

3 Modeling Uncertainty Through Boosting

Boosting is one of a kind of machine learning strategies known as ensemble learning (Dietterich 1997). In ensemble learning, many classifiers (known as an ensemble) instead of one are trained from available examples and classification is done by letting the entire set of classifiers vote. Originally designed to increase classification accuracies of many well-established classifiers (such as neural networks and decision trees), this strategy of learning and inference actually corresponds to the cluster model that leads to prototype effects.

Various approaches to construct ensembles have been investigated by researchers in the Artificial Intelligence (AI) community; among which boosting has proved to be one of the most effective methods in many empirical studies (Dietterich 1997; Lawrence et al. 2004). With boosting, the training examples are manipulated to generate multiple classifiers in a number of iterations. At each iteration, the algorithm maintains a probability distribution over the training examples, and draws a training set by sampling with replacement according to the probability distribution.

Boosting has been best implemented to construct ensembles of decision trees. This study will use the *AdaBoosting* algorithm developed by Freud and Schapire (1996) to illustrate our approach to extracting knowledge for classifying geographic entities and modeling the uncertainties associated with classification. The algorithm integrates the sampling strategy with the information gain-based decision tree-training algorithm See 5 (Quinlan 2001) to generate multiple decision trees on the run. Specifically, a single decision tree is first constructed using all examples in the training data.

Pruning[1] is then conducted to avoid over fitting (Esposito et al. 1997). The pruned tree will assuredly make mistakes on some training examples. Such examples are given more attention in the follow-up iteration by assigning them higher weights than the correctly classified examples. Again, the second tree, after pruning, may still misclassify some examples. Such examples are then assigned even heavier weights during the next run. This process continues until a pre-determined number of trees are constructed or a pre-determined error rate is reached.

The output of *AdaBoosting* is a set of decision trees associated with their individual error rates. When using the boosted tree sets to classify a new example, votes from the tree sets are weighted and counted. The vote for an individual category y among all the possible categories is calculated as:

$$V_y = \sum_{i=1}^{M} \log\left(\frac{1-\varepsilon_i}{\varepsilon_i}\right) \times \delta(T_i(x) = y), \tag{1}$$

where x refers to the new example to be classified; $T_i(x)$ is its classification determined by a specific tree T_i in the tree set; ε_i is the error rate associated with this tree; M is the total number of trees; and δ takes on either 1 or 0 depending on whether T_i classifies x as y or not. Upon computing the votes for all potential categories, the votes can then be tallied and classification of the new example determined. For classification in the traditional manner, the category that gets the highest votes is determined as the classification:

$$Class = \frac{\arg\max}{y \in categories} V_y. \tag{2}$$

In order to model the uncertainty associated with this classification process, however, we need to know how typical the example is to its classified category and how similar it is to other categories. In this case, the votes secured by all potential categories are normalized, based on the total number of categories involved, to obtain the example's degrees of membership to these categories.

[1] A tree that classifies the training data perfectly may not be the tree with the best generalization performance when applied to real data since (1) there may be noise in the training data that the tree is fitting; and (2) the algorithm might be making some decisions toward the leaves of the tree that are based on very little data. This phenomenon is called over fitting, and an over fitted tree may not reflect reliable trends in the data. To avoid over fitting, various pruning algorithms have been developed to improve the decision tree performance on future examples (Esposito et al. 1997).

When classifying geographic entities using the boosted trees, the result is no longer a single piece of an area-class map. Rather, the spatial distribution of the geographic entity under concern can be represented using a similarity model (Zhu 1997a). The basic idea of the similarity model is using raster representation instead of the conventional area-class representation to depict the distribution of the entity in geographic space, and using a similarity representation of the entity in parameter space. The similarity representation is based on fuzzy logic under which a local instance can be assigned to more than one class with varying degrees of membership. Under this notion, an instance at pixel location (*i, j*) can be represented as an *n*-dimensional similarity vector, $S_{ij} = (S_{ij}^1, S_{ij}^2, \ldots, S_{ij}^k, \ldots, S_{ij}^n)$, where S_{ij}^k represents the similarity value or fuzzy membership of the instance to category *k*, and *n* is the total number of the prescribed categories. Under a raster representation scheme, a geographic entity over an area can be represented as a raster database of similarity vectors, with each vector corresponding to a pixel in the area. With this database, a traditional area-class map can be created through the process of "defuzzification" (Janikow 1999) whenever desired. Uncertainties associated with this classification process can also be computed.

As discussed in Section 2, the ignorance uncertainty associated with the omission error is caused by ignoring the similarities between an instance and classes other than the assigned one. Therefore, it is related to membership diffusion in the similarity vector. The more concentrated the membership in a particular class, the smaller is the ignorance uncertainty. And the more spreading out the membership values in the vector, the greater is the ignorance uncertainty. The ignorance uncertainty can thus be estimated using an entropy measure (Goodchild et al. 1994; Zhu 1997b):

$$U_{ij} = \frac{1}{\ln N} \sum_{k=1}^{N} (S_{ij}^k \cdot \ln S_{ij}^k), \tag{3}$$

where U_{ij} is the estimated ignorance uncertainty, S_{ij}^k is the similarity value of the instance at pixel *(i, j)* to category *k*, and *N* is the number of categories that the instance has similarity to. A value of 0 means that the instance has full membership to only one category, thus no ignorance uncertainty is in question. A value of 1, on the other hand, indicates that the instance is similar to all categories at the same degree, and that assigning the instance to any one of the categories would involve the greatest degree of ignorance uncertainty.

The exaggeration uncertainty associated with the commission error is caused by assigning an instance to a category that it does not fully "qualify" for. In other words, it is inversely related to the saturation of the membership to the assigned category. If an instance has full membership to a class, there is no exaggeration when the instance is categorized to that class. Also, the lower the membership of the instance to the assigned class, the greater is the exaggeration. The exaggeration uncertainty can thus be estimated as the following (Zhu 1997b):

$$E_{ij} = 1 - S_{ij}^{k}, \tag{4}$$

where E_{ij} is the estimated exaggeration uncertainty and S_{ij}^{k} is the similarity value of the instance at pixel *(i, j)* to its assigned category *k*. With the knowledge extracted as boosted decision trees and the classification results as similarity vectors, the two measures of uncertainty can be calculated to provide information on quality of both the classification results and the knowledge used for the classification.

4 Case Study

We report here a case study on knowledge discovery for soil classification to illustrate the approach. In the United States and many other countries of the world, the spatial distribution of soils is routinely collected, presented, and archived during soil surveys. Previous research has indicated that valuable knowledge was embedded in the archived soil maps and such knowledge could be revealed through knowledge discovery (Moran and Bui 2002; Qi and Zhu 2003). The knowledge concerns environmental conditions for each of the mapped soil classes to develop. Such knowledge can be used for soil classification and mapping during soil survey updates, when information on the environmental conditions of an area is available.

Our study area is the Raffelson watershed in southwestern Wisconsin, USA. It has remained free of direct impact from late Pleistocene era continental glaciers and has a mature topography with relatively flat, narrow ridges. Complex soils from many epochs of soil formation and movement can be found in the watershed. A recent soil survey indicates there are a total of 16 different soil series in the area (see Fig. 2). This study uses the soil series map in Figure 2 to extract the knowledge for classifying soils at the Raffelson watershed.

In order to extract knowledge of the sixteen soil classes in terms of their environmental configurations, a GIS database was first constructed to capture the soil-formative environment of the Raffelson watershed based on general soil pedology (McSweeney et al. 1994) and the local pedogenesis. The database contains five soil-formative environmental variables that are commonly used at the watershed scale and five spatial variables.

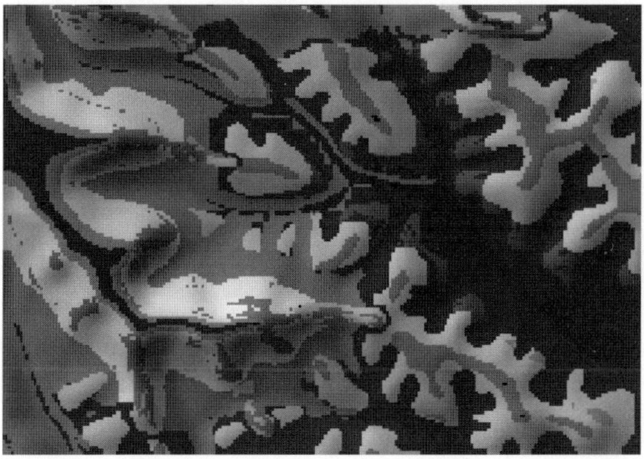

Fig. 2. Soil Series map of the Raffelson watershed

The primitive variables are elevation, slope gradient, planform curvature, profile curvature, and geology. Among the five spatial variables, three were included to capture the spatial relations of soil-formative environmental factors. They are distance to streams, topographic wetness index, and percentage of colluvium from competing bedrocks. The other two spatial variables capture topological and directional relations between soil classes: the upslope and downslope neighbors of each mapped location.

Next, soil pixels from the study area were stratified according to the different soil series assigned on the soil map. Each pixel was then labeled with the soil series name and associated with the values of environmental variables. The labeled examples were then fed to the boosting program to train ensembles of decision trees.

Once the knowledge was discovered and represented as an ensemble of decision trees, it was used for soil inference. Soil similarity vectors were computed for all pixels in the watershed, and a soil series map was created after defuzzification (see Fig. 3). Comparing the inferred map with the original map (see Fig. 2), we see they show similar patterns of the soil distribution. This indicates that the discovered knowledge captures well the spatial distribution of soils in relation to their environmental configurations in the study area. In an effort to further evaluate the inference results (and thus the discovered knowledge), we collected 99 field samples from the watershed and had them classified by experienced soil scientists from the local soil survey agency. Both the original map and the inferred map were then measured against the field samples to determine their accuracies. It turned out that both maps correctly mapped 83 of the 99 sites (with an accuracy of 83.8%). This confirmed the finding that the discovered knowledge captured well the soil distribution pattern in the original map.

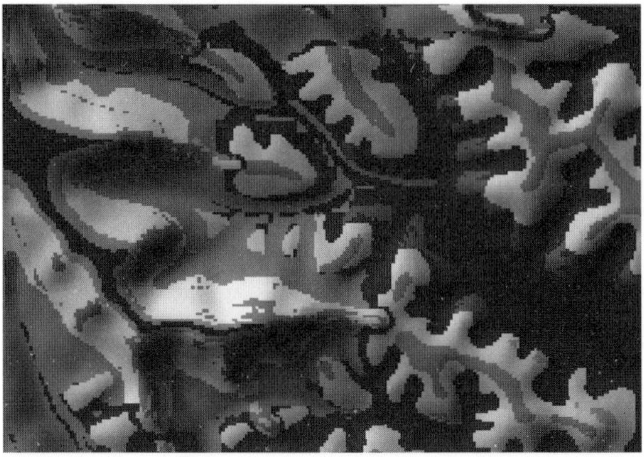

Fig. 3. Soil series map inferred from the derived ensemble

Uncertainty associated with the classification process was computed following Equations 3 and 4. Figure 4 shows the distribution of the ignorance uncertainty in the study area. Comparing Figure 4 to the soil series map (see Fig. 3), one can observe that uncertainty is high in areas at slope shoulder positions (area *A* in Fig. 4).

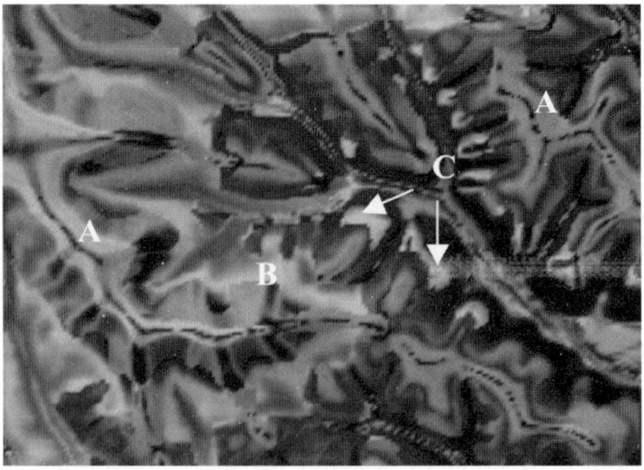

Fig. 4. Distribution of the ignorance uncertainty with light tones indicating high uncertainty values

This is because soils on the narrow shoulders are in transition from ridge soils to backslope soils. The soils are not similar to prototypes of any of the soil series. This is the same reason for the high uncertainty we see in transitional zones between bedrock-controlled soils and colluvium-based soils (area *B*). In fact, uncertainty always appears high at the fringe of a soil body, because soils at boundaries often bear some similarities to the adjacent soil types. Another observation that can be made from Figure 4 is the high uncertainty of small patches in the middle of the watershed (area *C*). It turns out that these patches are small areas of unique bedrock where three different soils develop. The crowding of three soil types in such a small area makes it difficult to separate the soil classes. Mixture of similarities to all three types is thus observed.

Figure 5 illustrates the distribution of the exaggeration uncertainty. It shows a different spatial pattern from that of the ignorance uncertainty, other than that high uncertainty is still seen at boundaries of soil bodies. The reason for high exaggeration near soil boundaries is that soils in transitional zones bear similarity to multiple soil classes but similarity is not high in any one of the classes.

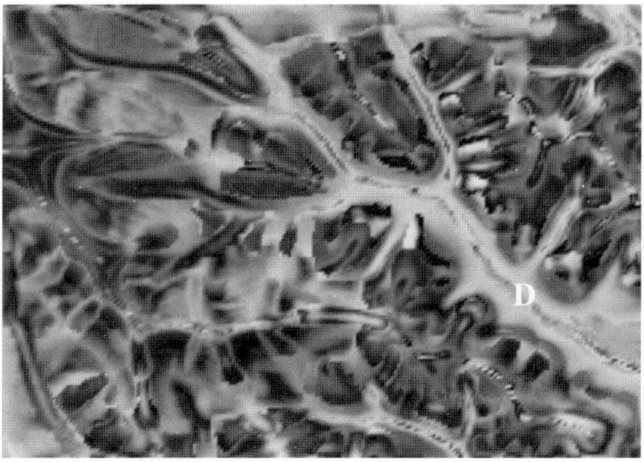

Fig. 5. Distribution of the exaggeration uncertainty with light tones indicating high uncertainty values

A major difference between the distribution of exaggeration uncertainty and that of ignorance uncertainty is the very high exaggeration uncertainty in the low valleys of the watershed (see area *D* in Fig. 5). This means that the soils at low elevation bear low similarities to even the soil class they are most similar to. This can happen in two circumstances. In the first, the soil is significantly different from any of the 16 soil classes; thus it bears low similarity values to all. In the second, the soil similarity vector is not an accurate representation of those local soils, and the low similarity values are the result of low confidence. It this study, the latter would be a more proper explanation, since an examination of the misclassified samples among the 99 field samples reveals that nearly half of the misclassifications occurred in the low elevation area.

As shown above, the uncertainty information provided through the two measures provides a sense of the quality of the knowledge obtained through knowledge discovery and the classification results (the inferred soil series map) using the knowledge. It helps identify areas where managerial decisions should be adjusted. For example, the transitional areas in the watershed are all classified as certain soil series. A direct implication is that these soils can be treated the same as the prototypes of the soil series. This could result in misuse of the soil resource in managerial practices, because of the high uncertainty associated with class assignments in these areas. The implication is that other managerial measures may have to be applied in using the soil resource in these areas. The example in the Raffelson watershed also shows that the knowledge extracted through knowledge discovery may not always capture well the soil distribution in valleys due to the high exaggeration uncertainty. Such information should offer a cautionary lesson in future use of the extracted knowledge in these areas.

The very reason that the uncertainty measures derived from the similarity vectors can be used to identify transitions between soil classes is that the similarity vectors contain information on the full range of membership of the local soil to all potential soil classes. These similarity vectors can thus be used to map soil properties in a fashion, which captures the gradual change between class prototypes. In the case study, continuous soil property maps of percentage of sand and silt in the *A* horizon were generated with the following formula following Zhu et al. (2001):

$$v_{i,j} = \frac{\sum_{k=1}^{n} m_{i,j}^{k} v^{k}}{\sum_{k=1}^{n} m_{i,j}^{k}}, \tag{5}$$

where v_{ij} is the property at site (i, j); v^k is the typical value of that property of soil class k; s_{ij}^{k} is the similarity value of soil class k at (i, j); n is the total number of soil classes in the area. For the sake of comparison, soil texture maps were also derived from the original map, by assigning each pixel the typical texture values of the labeled soil series.

The maps in Figures 6 and 7 show the distributions of the percentage of sand and silt in the surface soil, respectively. It is observable that the maps derived from the inference result and those based on the original soil map show comparable spatial details, given that both have the same spatial resolution. An apparent difference between the two sets of maps, however, is the smoothness of the look or the shown continuity of soil texture variations. The inferred texture maps tend to show more continuous changes of the texture values than those based on the original map.

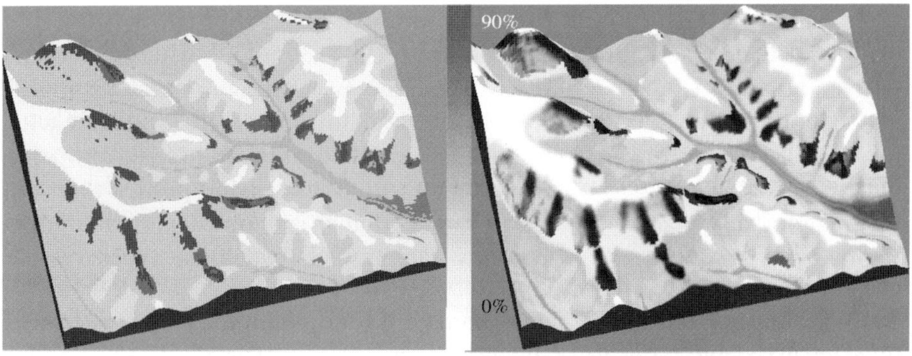

Fig. 6. (left) *A* horizon sand percentage derived from the original soil map (right) *A* horizon sand percentage inferred from data mining results

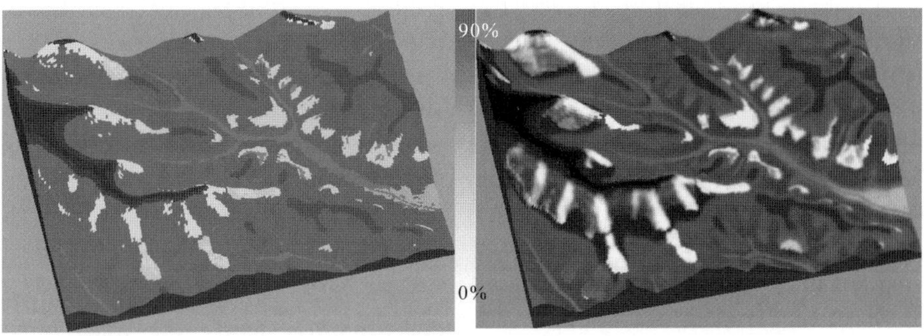

Fig. 7. (left) *A* horizon silt percentage derived from the original soil map (right) *A* horizon silt percentage inferred from data mining results

The two sets of maps were further compared using field measurements. Among the 99 field samples collected, 49 were given a texture analysis to determine the percentages of sand and silt in the *A* horizon. Three indices were computed to evaluate the performances of the property maps against the field samples: MAE, RMSE, and agreement coefficient (AC). The AC index is defined by Willmott (1984) as:

$$AC = 1 - \frac{n \cdot RMSE^2}{PE}, \tag{6}$$

where n is the number of observations and PE the potential error variance defined as:

$$PE = \sum_{j=1}^{n} (|P_i - \overline{O}| + |O_i - \overline{O}|)^2, \tag{7}$$

given that $\overline{O}$ is the observed mean, and P_i and O_i are the estimated and observed value, respectively. AC values vary between 0 and 1, where 1 indicates perfect agreement and 0 means complete disagreement between the estimated and observed values (Willmott 1984). Table 1 lists these computed statistics. The MAE and RMSE statistics for the inference result are consistently lower than those for the original soil map. This, together with the higher AC for the inference result, implies that the inferred property maps are better in terms of estimating continuous soil properties than the original soil map, and it stands to reason that this should be attributed to their ability to capture the continuous transitions between soil types.

Table 1. Accuracy of the derived *A* horizon texture in the Raffelson watershed: the inference result vs. the original map

	Percentage of sand			Percentage of silt		
	MAE	RMSE	AC	MAE	RMSE	AC
Inference result	8.74	13.92	0.82	7.41	12.06	0.82
Original map	10.66	16.63	0.67	9.51	14.31	0.67

5 Conclusions and Future Work

Previous studies on knowledge discovery using boosting usually focus on its ability to improve classification accuracy. This paper addressed the issue of uncertainty associated with classification using the discovered knowledge. We examined the nature of classification from the cognitive perspective, and found that boosting could be used to model the uncertainty associated with class assignment in the classification process. With boosting, an ensemble of classifiers could be constructed during knowledge discovery. When using the ensemble to classify new instances, a similarity vector is computed instead of the assigning of a single class label. The similarity vector can then be used to derive measures of two types of uncertainty: the ignorance uncertainty and the exaggeration uncertainty. The ignorance uncertainty is caused by membership ignorance, and exaggeration uncertainty measures the degree of exaggerating partial membership of an instance to full membership in the assigned class.

In a case study of knowledge discovery for soil classification, uncertainty images derived using the two measures helped to identify areas of potential problems on the inferred soil series map. It was also found that the similarity vectors inferred with the boosted decision trees can be used to model continuous soil properties. Soil property maps created with this approach captures the continuous gradations of soil properties between soil classes better than the maps based on the traditional "area-class" model.

The case study, however, employs a simple weighted average method to estimate the soil property at a location, based on the typical properties of all soil classes it is similar to. With the weights being the local soil's similarity values to all prescribed soil classes, this method takes into consideration membership diffusion and thus reduces the omission error. It does not, however, deal with the exaggeration of membership. Future study may address this problem and investigate better ways of utilizing the similarity vectors.

The boosting algorithm implemented in this study generates multiple classifiers to derive class memberships. Other algorithms that manage to maintain and iterate a population (such as genetic algorithms) could also be explored in the future. Furthermore, the spatial neighbor information (upslope neighbor and downslope neighbor) extracted through data mining was not used for soil inference in the current case study. Research should be conducted to utilize such information for inference in future works.

Reference

Burrough PA (1996) Natural objects with indeterminate boundaries. In: Burrough PA, Frank AU (eds) Geographic Objects with Indeterminate Boundaries. Francis and Taylor, London

Castro VE, Murray AT (1998) Discovering associations in spatial data – an efficient medoid based approach. In: Wu X, Kotagiri R, Korb KB (eds) Research and Development in Knowledge Discovery and Data Mining. Springer, Berlin

Dietterich TG (1997) Machine learning research: four current directions. American Association for Artificial Intelligence Publication:41

Esposito F, Malerba D, Semeraro G (1997) A comparative analysis of methods for pruning decision trees. IEEE Transactions on Pattern Analysis and Machine Intelligence 19:476–491

Ester M, Kriegel HP, Sander J (2001) Algorithms and applications for spatial data mining. In: Miller HJ, Han J (eds) Geographic Data Mining and Knowledge Discovery. Taylor & Francis, New York, pp 160–187

Freund Y, Schapire RE (1996) Experiments with a New Boosting Algorithm. In: Proc of the 13th Int Conf on Machine Learning, pp 148–156

Goodchild MF, Chin-Chang L, Leung Y (1994) Visualizing fuzzy maps. In: Hearnshaw HM, Unwin DJ (eds) Visualization in Geographical Information Systems. John Wiley & Sons, New York, pp 158–167

Gopal S, Liu W, Woodcock C (2001) Visualization based on fuzzy ARTMAP neural network for mining remotely sensed data. In: Miller HJ, Han J (eds) Geographic Data Mining and Knowledge Discovery. Taylor & Francis, NY

Hampton JA (1995) Testing the Prototype Theory of Concepts. J of Memory and Language 34:686–708

Janikow CZ (1998) Fuzzy decision trees: issues and methods. IEEE Trans Syst Man Cybern B: Cybern 28:1–14
Koperski K, Han J (1995) Discovery of spatial association rules in geographic information databases. In: Proc of 4th Int Symp on Large Spatial Databases, pp 47–66
Lakoff G (1987) Women, Fire, and Dangerous Things: What Categories Reveal about the Mind. University of Chicago Press, Chicago
Lawrence R, Bunn A, Powell S, Zambon M (2004) Classification of remotely sensed imagery using stochastic gradient boosting as a refinement of classification tree analysis. Remote Sensing of the Environment 90:331–336
Mark DM, Csillag F (1989) The nature of boundaries on 'area-class' maps. Cartographica 26:65–78
McSweeney K, Gessler PE, Slater BK, Petersen GW, Hammer RD, Bell JC (1994) Towards a new framework for modeling the soil-landscape continuum. Factors of Soil Formation: A Fiftieth Anniversary Retrospective. SSSA Special Publication 33:127–143
Miller HJ, Han J (2001) Geographic data mining and knowledge discovery: an overview. In: Miller HJ, Han J (eds) Geographic Data Mining and Knowledge Discovery. Taylor & Francis, New York, pp 3–32
Moran CJ, Bui EN (2002) Spatial data mining for enhanced soil map modeling. Int J of Geographical Information Science 16:533–549
Qi F, Zhu AX (2003) Knowledge discovery from soil maps using inductive learning. Int J of Geographical Information Science 17:771–795
Quinlan JR (2001) See5: An Informal Tutorial. URL: http://www.rulequest.com
Rosch EH (1973) Natural Categories. Cognitive Psychology 4:328–350
Rosch EH (1978) Principles of categorization. In: Rosch EH, Lloyd BB (eds) Cognition and Categorization. Lawrence Erlbaum Associates, Hillsdale, NJ
Smith EE, Medin DL (1981) Categories and Concepts. Harvard University Press, Cambridge, MA
Tadesse T, Brown JF, Hayes MJ (2005) A new approach for predicting drought-related vegetation stress: integrating satellite climate, and biophysical data over the U.S. central plains. ISPRS J of Photogrammetry and Remote Sensing 59:244–253
Willmott CJ (1984) On the evaluation of model performances in physical geography. In: Gaile GL, Willmott CJ (eds) Spatial statistics and models, D. Reidel Publi, the Netherlands, pp 43–460
Zhang B, Valentine I, Kemp P (2005) Modeling the productivity of naturalized pasture in the North Island, New Zealand: A decision tree approach. Ecological Modeling 186:299–311
Zhu AX (1997a) A similarity model for representing soil spatial information. Geoderma 77:217–242
Zhu AX (1997b) Measuring Uncertainty in Class Assignment for Natural resource Maps under Fuzzy Logic. PE & RS 63:1195–1202
Zhu AX, Hudson B, Burt JE, Lubich K, Simonson D (2001) Soil mapping using GIS, expert knowledge, and fuzzy logic. Soil Science Society of America Journal 65:1463–1472

Capturing and Representing Conceptualization Uncertainty Interactively using Object-Fields

Vlasios Voudouris[1], Peter F. Fisher[2], Jo Wood[2]

[1] School of Geography, Birkbeck, University of London
[1,2] The giCentre, Department of Information Science, City University, London, UK; email: {vv, pff1, jwo}@city.ac.uk

Abstract

We present a method for representing, recording and managing conceptualization uncertainty. We review components of uncertainty associated with semantics and metadata. We present a way of recording and visualizing uncertainty using sketching and suggest a framework for recording and managing uncertainty and associated semantics using Object-Fields. A case study is also used to demonstrate a software prototype that shows proof-of concept. We conclude by identifying future research challenges in terms of supporting dynamic exploration of uncertainty, semantics and field objects.

1 Introduction

Conceptualization uncertainty is regarded, in the present research, as the uncertainty that is introduced during the identification and conceptualization of geographic entities from data stored in digital field models. It is an element of uncertainty that is frequently overlooked, yet has a significant impact the way in which we use spatial data. Given that conceptualization can be a subjective process that varies between individuals, this form of uncertainty has particular importance in collaborative environments. In this paper, we show how the Object-Field model, specified by Voudouris, Wood and Fisher [29] who elaborated that of Cova and Goodchild [5], can be extended and visualized in order to represent conceptualization uncertainty. In other words, the model records the uncertainty of each collaborator simultaneously, which results in an overall uncertainty by combining the individual uncertainties (weighting or otherwise the individual uncertainties). This technique enables us to model the differences in uncertainties by using overlaid elementary units. Capturing conceptualization uncertainty using Object-Fields is a potentially useful approach in estimating uncertainty when appropriate data is not available, precision is not a key requirement or in critical situations where every second counts and a quick informed consideration of the uncertainty understood by collaborators is a requirement. In other words "a short term response to an uncertain future is to build in obsolescence" ([14], p 84).

In the following section we provide a brief overview of the different uncertainties outlined in the literature. We then introduce how the Object-Field model incorporates metadata. We demonstrate a sketching approach to represent uncertainty illustrating this with a collaborative visualization prototype. The paper concludes by identifying the limitations of the proposed approach and by suggesting future research challenges.

2 Conceptualizing Uncertainty

The presence of uncertainty in the geometric and attribute aspects of geographic information is well articulated in the literature ([6]; [30]; [27]). Molenaar [19] has identified the key issues of uncertainty as existential, extensional and geometric uncertainty. The last of these includes the issues of measurement error and the effect of mathematical processing (during acquisition, transformation). The first two are more concerned with the conceptualization of a phenomenon. Although these are fundamental in the subject domains, which are concerned with the creation of geographical information (soils, ecology, demography, etc.), they have made relatively little impact on Geographical Information Science until quite recently. Assessing the uncertainties associated with the existence of an object (conceptualization) and its potential spatial extent is particularly problematic because the interpretations are not culture-free. Of equal importance to the existence of an object is its semantic description ([11]; [12]). This means that the objects people identify are social constructs and their delineation is associated with the personal background and training of the individual operator or surveyor involved in the identification. Thus, uncertainty is also introduced by multiple conceptualizations. This source of uncertainty is emphasized here.

Many researchers point out that uncertainty spans a range of inconsistency. This is to say that inherent in the notion of uncertainty is an inconsistency of measurement, definition and conceptualization. It is, therefore, important to clearly define a sound theoretical understanding of uncertainty in designing data models ([6]). Hunter and Goodchild [13] and Gahegan and Ehlers [8] define error as objectively known inaccuracy/problems of information and unknown inaccuracy as uncertainty. Fisher [6] also proposes that ambiguity and vagueness are integral components of uncertainty. All of the components (higher and lower) of the uncertainty are influenced by two main processes, namely *conceptualization and measurement* based on that conceptualization ([23]). Fisher and Tate [7] describe the kind of uncertainties introduced in Digital Elevation Models (DEMs) during their construction, from the conceptual model to the final product (the digital model). From the conceptualization point of view, the informed selection of land surface and the relevant variables to measure is important as well as the selection of suitable methods for measurement and the application of suitable measurement and statistical processes. In other words, understanding of the processes tells us something about the associated uncertainties, including conceptualization uncertainty, even before we apply these processes.

Figure 1 aims to summarize the effect of the conceptualization and measurement processes on uncertainty in order to make a clear distinction how these processes influence uncertainty. These processes can influence uncertainty in a linear or loop manner. Longley et al. [15] outline three filters that influence uncertainty. The three filters distort the way in which the world is conceived, measures and represented, and analyzed.

Although mathematical methods for recording measurement uncertainties have been proposed, little attention has been given to recording uncertainties introduced by conceptualizations of a phenomenon. Yet his may be necessary when dealing with different conceptualizations of the same phenomenon by different people working together in a collaborative environment. Lowell [16] created a library of uncertainties by cataloging multiple interpretations of phenomena. This library is then used to estimate the uncertainty by using a distance measurement from certain points or by estimating variance among the interpretations. Estimating uncertainty from multiple interpretations is also suggested by this research but the method argued for here is not based on 'definite' objects. By recording interactively and visualizing how different people address these issues, a shift from data uncertainty to individual knowledge uncertainty is suggested.

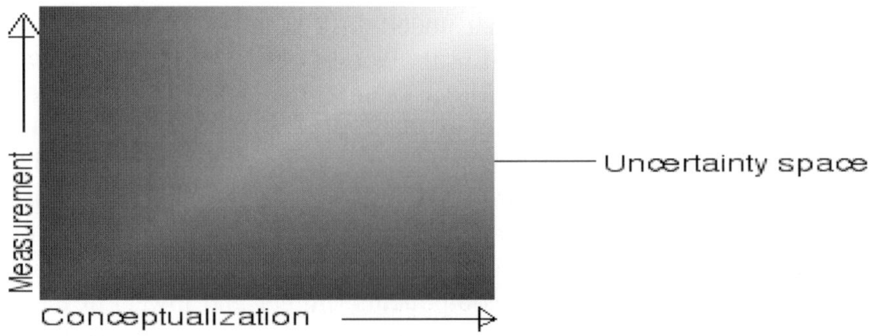

Fig. 1. Main impact of conceptualization and measurement processes on uncertainty. The darker the space, the higher the certainty

Uncertainty (defined in terms of data quality) is a component of metadata ([10]; [27]). Based on that we propose a three-way matrix including people, types of metadata and metadata levels (see Fig. 2). The 3D metadata matrix presented here enables us to capture conceptualization uncertainty at the object class, object and object element level for all personnel (especially the producers and intermediaries). Not all of the axes of the matrix are well catered for in existing metadata structures. For example, the 'object levels' axis manifests itself in the metadata plane, along with conceptualization uncertainty.

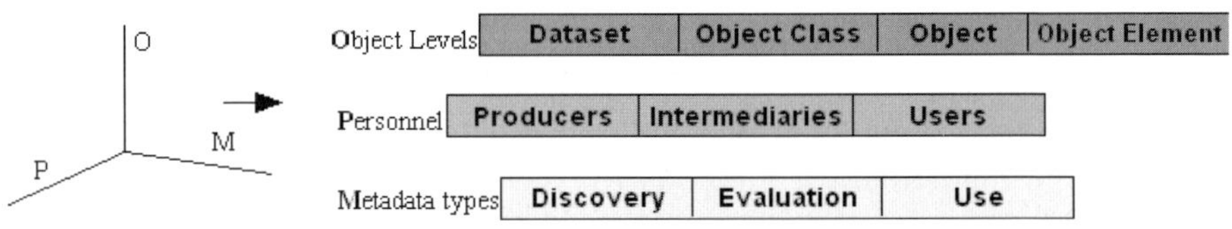

Fig. 2. Proposed axes of the 3D matrix for metadata

Not all of the tree axes are well catered for in existing metadata structures. 'Object levels ' axis manifests itself in the metadata plane. The kind of metadata that is recorded at the 'object levels' axis is influenced by the other two axes. For example, different metadata should be recorded at the tripletX [object levels-object. Personnel-users, metadata types-use] and at the tripletY [object levels-object. Personnel-users, metadata types-discovery]

The last part in the uncertainty of conceptualization is associated with semantics. Figure 3 presents semantic components that influence conceptualization uncertainty. We treat semantics in this context as being decomposed into selection criteria and definitions. Both of these components inform conceptualization uncertainty, as they allow a second user to see why a first user has identified a particular object in a field. As we shall see the selection criteria component is important in capturing conceptualization uncertainty using object-fields. It is, therefore, important to explicitly store and visualize these components along with a definition that justifies the selection of the object element. Both the selection criteria and definition elements facilitate the communication of geographic phenomena which are both well defined (crisp) concepts with uncertain spatial existence and uncertain concepts with well defined spatial existence.

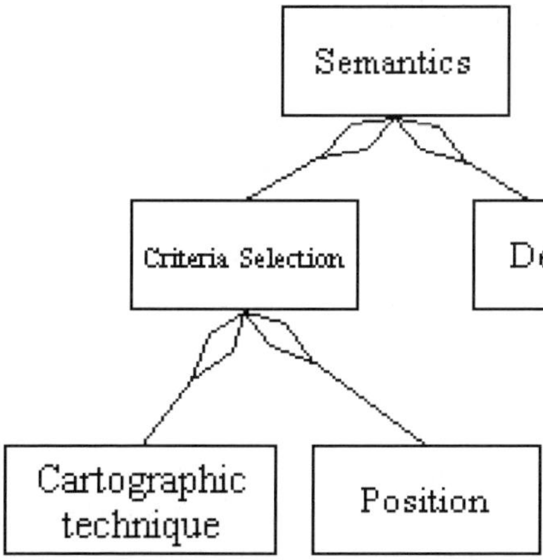

Fig. 3. Proposed semantic components associated with conceptualization uncertainty

One way of approaching the challenge of communicating uncertainty is by providing methods for depicting and interacting with data and uncertainty simultaneously. These methods should be able to merge two views of space: the personal and the scientific. The personal or cognitive view of space describes how we perceive space, both digital and physical, and is related to our knowledge, training and experience. In other words, we perceive the space and we add cognitive interpretations to make sense of what we perceive. A contrasting view is the classical scientific one that assumes an independently objective correspondence between geographic space and physics where the observer is assumed to be separate from that which is observed.

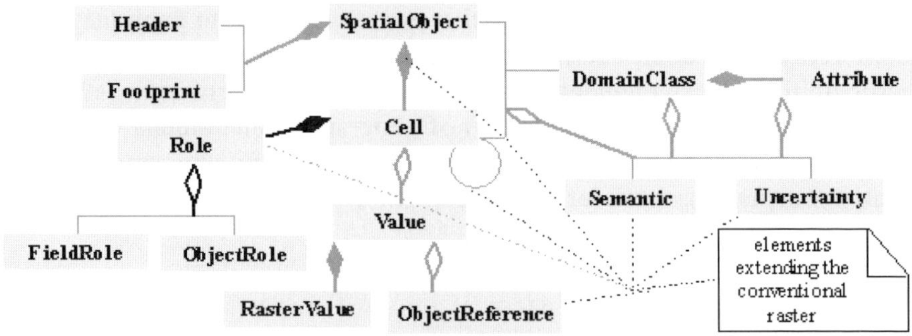

Fig. 4. Object-Field representation with semantics and uncertainty at different levels (source: Voudouris, Wood and Fisher [29])

3 Object-Field Representation and Metadata

The Object-Field model aims to link the field and object view of space by enabling the identification of objects from continuous phenomena with the retention of variable levels of properties such as uncertainty within the object and with semantic description of the user's understanding of the object. Variable levels of properties means that objects are associated with object element properties and with properties that are unique at the object level only. The model unifies the two models at the conceptual level and its implementation is grounded in the raster data structure.

The model is implemented using the object-oriented paradigm and the Java programming language, and is divided into two primary spaces, namely the object and the field space. The field space is closely associated with the scientific view of space and is used to represent observational data. The object space is associated with the human view of space and represents conceptual knowledge. This means that the object space augments the field space by adding knowledge through human-sensory input and interpretation to make sense of the field and communicate what is being represented in the field space. In other words, Figure 4 represents a formalized framework from which a coordinated field and object space can be systematically merged.

The current implementation of the Object-Field representation uses two *classes* (semantics and uncertainty) to store metadata. These metadata classes are used to handle interpretations and visualization of uncertainty and associated semantics. This is of particular importance in collaborative environments as different people bring different perspectives to bear on a situation. The need to explicitly integrate and use data and knowledge in data models has already been acknowledged by [18]. That knowledge can be derived automatically from, for example, knowledge discovery or defined externally by experts. Humans in general are recognized as knowing more than they can express in words [24], and so the uncertainty class here is designed to capture components of conceptualization uncertainty in a variety of data formats, including but not limited to natural language, numerical representations, sketching and images.

Representing uncertainty using sketching is a new approach proposed here. To our knowledge little experimental research has been done on the use of sketching in collaborative exploration and concept visualization. The idea of this work is to allow collaborators to draw over field representations to enable them to graphically represent their mental world/conceptualizations. For example, we can talk about a region like the City of London, but we usually do not know exactly the boundaries of that region. Nevertheless, we are able to reason about such a region. Therefore, sketching can provide us with a mechanism to visualize uncertain boundaries using rough drawings or sketches, which we might use to support our reasoning about the existence of the region. Thus we have a mechanism to account for Molenaar's [19] existential / extensional distinction. A similar approach is described by [16] who demonstrates how a human photo-interpreter "works from definite object or area – e.g., a lake, a clear-cut – and proceeds to less certain features" (p 2). Therefore, the aim of the sketching approach is to enable us to separate regions of different kinds for which the boundaries are uncertain. The *where* is represented using rough demarcation in the sketch (see Fig. 5) and the *what* using some degree of certain reasoning. In other words, by sketching an object, a user is asserting some degree of existential certainty, while recognizing its extensional uncertainty. The dialogue between collaborating users afforded by sharing such sketches allows both to be debated.

This *sketch representation* of conceptualization uncertainty enables the selective management of large amounts of information while holding several quick ways of looking at problematic situations at once, without disrupting the flow of inquiry and to enable participants to construct an understanding of the situation.

Although drawings may not be accurate, they are regarded as one of the most intuitive and innate methods of human communication of concepts because we draw before we can write or even speak. We emphasize that sketching is:(i)innately understood to be uncertain and imprecise; (ii) understood to be subjective;(ii)quick and easy to produce;(iii)easy to understand.

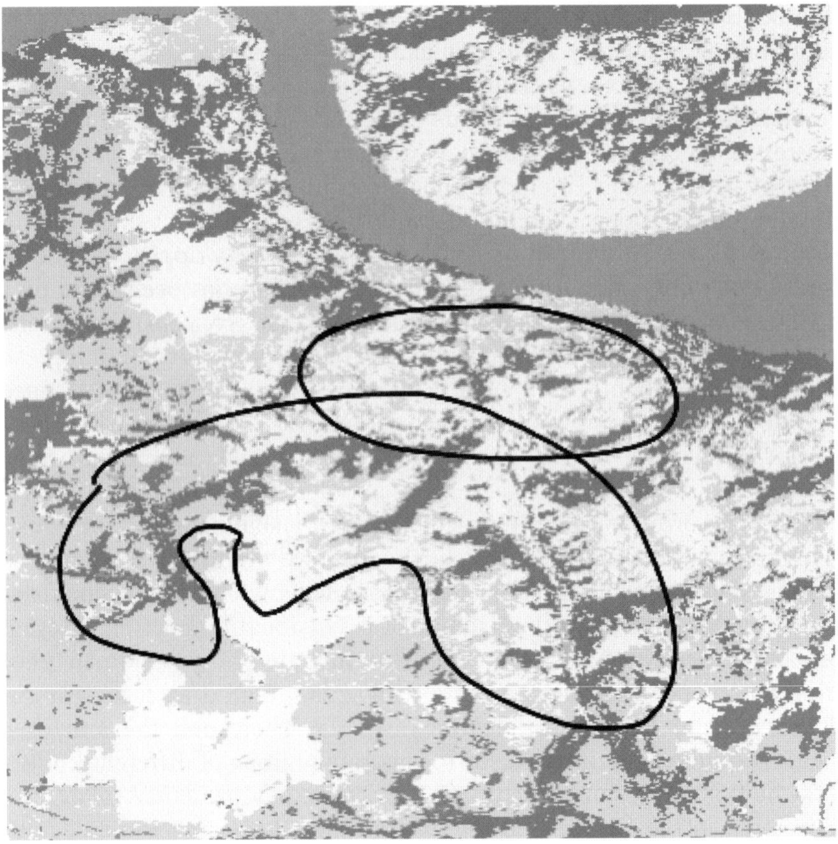

Fig. 5. Rough drawing/sketching to demarcate an uncertain region. Two different demarcations that demonstrate conceptual boundary uncertainty

In other words, sketching can be used, in a critical scenario for example, as a mechanism to demarcate uncertain phenomena and to process available descriptive information associated with these uncertain phenomena. This will enable us to take quick but informed consideration of the uncertainty involved. Thus sketching (rough drawing) can be used to represent obsolescence. This obsolescence, in our view, is particularly useful in collaborative environments as it enables the linking of people who are naturally good at getting ideas with those that are naturally good at elaborating ideas. Again, in a critical scenario, collaborators can intuitively and quickly represent and share their ideas while others, such as decision makers, can develop and evaluate these ideas to take informed decisions about the situation. These two processes, represented by humans, should be regarded as one mental process rather than separate ones.

Drawings are produced quickly and in a personal manner and in our case consist of a set of graphical objects drawn on a background (the field representations in the example presented above). Often sketched objects may be recalled from memory or constructed by imagination ([25]). Therefore, certain concepts can be better represented in a sketch. Shneiderman and Plaisant [26] argue that certain concepts can be more readily represented in graphical form than by using text. Furthermore, because "cognitivists viewed knowledge as abstract symbolic representations in the head of individuals" and the "constructivist school views knowledge as a constructed entity made by each and every learner through a learning process" (CSCL project 1999), we argue that sketches can:

- be a visual representation of an individual's abstract symbols held in mind
- be an effective mechanism for recording and manipulating ideas which is consistent with established ideas of learning, communication and development (e.g [20]; [22]; and
- provide users with a visual tool to specify and record their intention.

4 Case Study: Demonstrating the Application Using Land Cover Data

A central quality of the framework is the semantic and uncertainty information attached at the object class, object and object element level. These elements are increasingly important in communicating natural resource data such as land cover. Information to be attached to objects might include text descriptions using either free format text or using fixed format and predetermined dictionaries and photographs from the ground, or plan-views of the object in other geographic information types (aerial or satellite imagery or vector or other field data). All these can be readily stored and explored as semantics and uncertainty objects within the framework proposed (see Figs. 4, 6, 7 and 8) to communicate an understanding of land cover between collaborators. In particular, the object elements (raster cells in the current implementation) are treated explicitly as objects, which are used to define higher-level objects. This object-based approach enables us to retain the independence of the object elements while larger aggregates/compositions are possible. These object elements can store understandings that are expressed as semantic and uncertainty information. Objects, larger aggregates of object element, can be associated with additional or complementary semantic and uncertainty information. It is important to note that the current prototype stores understandings using free text format but due to the object-oriented approach used its extension is easily incorporated. By grouping object elements together, as we shall see, we manage to augment the spatial objects per se with relations that bind them together (spatial object in Fig. 4). An attention to spatial interactions/relations "strikes at the very heart of GI Science" ([1], p 129). These objects augment the field view with the concept of objects and this is particularly important when, for example, a feature represented as an area object is sufficiently large to be coded in the database as belonging to several object elements such as raster cells. Although conventional field structures such as the raster model can record the presence/absence of objects in adjacent cells of the field representation the notion of there being coherent objects at those places is lost. O'Sullivan and Unwin [21] have already acknowledged this by using a field type that is made up of a continuous pixel categorical values such as land cover classes.

These coherent objects can be used as a means for communicating the rationale of remapping of land cover data. This rationale can be driven by changes in the use or cover itself. These changes are also influenced either by the methodology employed or by policy initiatives ([3], 2003). Tracing these kinds of inconsistencies, due to a revised methodology or due to changes in the phenomena being measured, in a collaborative environment add additional sharing/communication difficulties. And uncertainty and semantics metadata can manage these difficulties (see Fig. 6).

Figure 6 demonstrates the idea of sharing knowledge, constructing knowledge and exploring data collaboratively using visual methods. These methods enable the building of conceptualizations /interpretations from observational land cover data and the connection of these conceptualizations to each other.

Figure 7 illustrates the construction of field objects with semantics and uncertainty from observational land cover data. These objects are composed of many cells. Which are constructed based on the field resolution and the footprint of a location selected by the user. In other words, observes click on the field representation and the footprint of this location is recorded. Then this footprint is used to identify the row and column numbers of the field data structure uniquely identifying the cell. This point footprint and the field resolution are used to visualize the cell. From this it is clear that any field representation can be used, as the resolution and the cells are identified dynamically.

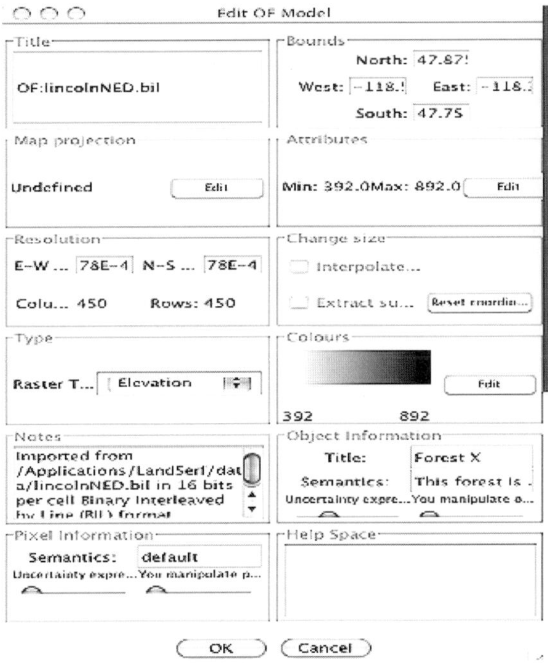

Fig. 6. Exploring land cover spatial and a spatial data

OF representation of land cover data. Participants are able to edit data produced by others. The sliders at the bottom of the figure enable retrieval of the field-objects and the associated elementary units (raster cell) along with attached uncertainties and semantics (metadata components).

Transparency is used to visualize uncertainty. Unlike MacEachren at al. [17] who depict areas of greater certainty using less transparency, we depict greater uncertainty using less transparency, as we wish to emphasize uncertainty as a metadata concept to be understood. The uncertainty can be specified both at the pixel level along with the associated semantics and at the object level. Recording uncertainty at the pixel level we can represent geographic features that exhibit well-defined and uncertain boundaries simultaneously. We also implement uncertainty at the object level for practical reasons. For example, it is impractical to request uncertainty input for a large number of pixels that make up an object. By associating uncertainty at the object level we introduce a mechanism that explicitly records at the object level, implicit user actions at the pixel level. This means that pixels are associated with object level uncertainty.

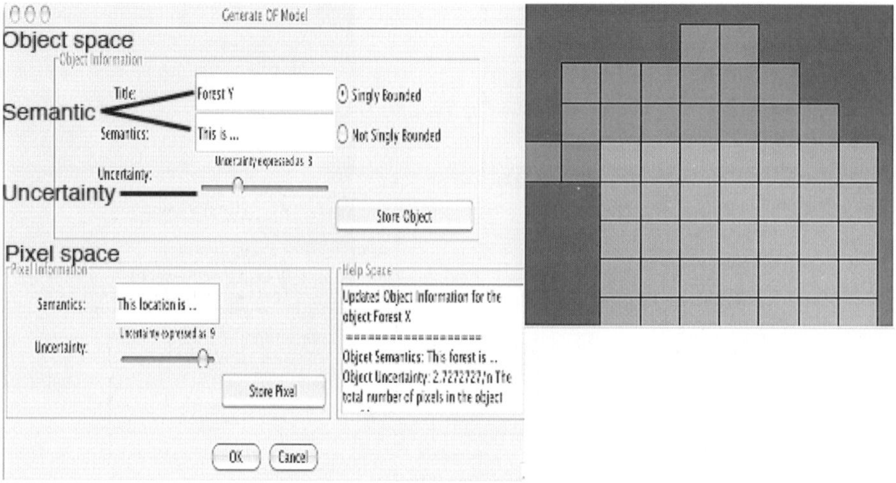

Fig. 7. Field objects with different Uncertainties and Semantics

Objects identified by different interpretations of the same visualized observational data, using the 'generate OF model' box. The model records not only the spatial extend but also uncertainty and semantics. Cell value represents conceptualization uncertainty defined by the number of users who have selected a cell as being part of the feature.

Furthermore, the uncertainty is also defined by the number of times a specific pixel/location is selected as being part of an object. The more users select a location, the less transparent the pixel becomes because the location acquires a crisp definition as a result of the number of people who conceptualize it as being part of the object. In other words, this approach is capable of representing crisp concepts represented as semantic objects and fuzzy spatial extents represented as uncertain objects (see Fig. 7). We can also refine this combined conceptualization uncertainty by giving different weightings to participants. For example, depending on the application requirements, people with local knowledge may influence the identification of locations as being part of an object more than remotely located people. Finally hue is used to represent field values.

Figure 8 is a UML diagram that represents the internal structure of Figure 7. It is also an instantiation of classes in Figure 4. The aim of this diagram is to emphasize the distinction between the objectives of the visual representation and the objectives of the database. The visual representation attempts to show its users something about the world whereas the objectives of the design of the database are to do with measurement, analysis and modeling of spatial characteristics of a phenomenon. In other words, it aims to separate the cartographic representation of a field object from its fundamental spatial characteristics.

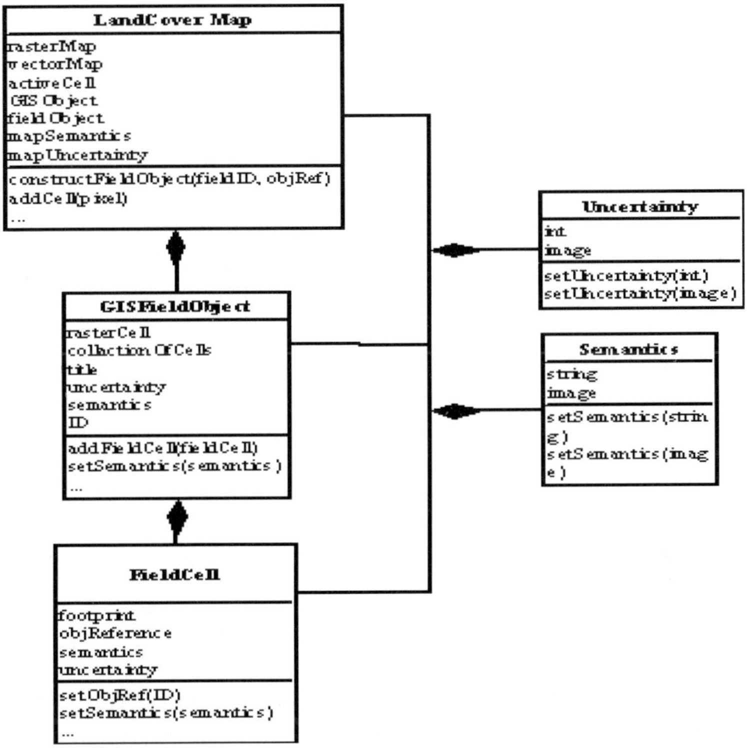

Fig. 8. UML overview diagram of the classes that represent the land cover taxonomy It is clear that the Object-Field model LandCoverMap is composed of GISFieldObjects and FieldCells but also a raster map as well as a vector map. The use of composition rather than inheritance avoids semantic conflicts introduced by inheritance and increases dynamic flexibility [2]

The quality of the model proposed here is that all the data (spatial and aspatial, including graphic and tabular data) are fully integrated and combined in a single database enabling reports and maps to be built, shared and explored together (see Figs. 7 and 8). Given the inherent time complexities and possibly conflicting working practices among collaborators it is particularly important to provide a mechanism that enables people to share observational data and derived knowledge simultaneously. By recording a number of GISFieldObjects it is possible to question, analyze and associate the changes in the semantic and uncertainty objects which will enable people to identify semantic and uncertainty inconsistencies. Not only the spatial extent of the region but also the cumulative uncertainty and semantic information is captured. This is accomplished by establishing an aggregation link between pixels and objects. This link also enables the recording of local and detailed information.

5 Discussion and Conclusions

Here we have proposed a mechanism from which conceptualization uncertainty can be recorded using Object-Field representations. Uncertainty can be recorded and visualized using sketching, natural language and numerical representations. It is also recorded explicitly using the number of times a location is selected to be part of a field object. It has also been suggested that a weighting approach can be adapted to accommodate expertise (local knowledge of the area under question).

Capturing conceptualization uncertainty is not a trivial task and the research presented here has only just investigated one way of capturing that uncertainty interactively using Object-Fields. There is a need to conduct a usability test to say with a degree of confidence whether the simultaneous visualization of observational data along with semantics and uncertainty can indeed enhance peoples' understanding of the data.

Object-Fields with semantics and uncertainty provide a potentially useful and usable geographical data model in collaborative modeling and decision support.

Although a sketching approach has been proposed (among others) in representing uncertainty, there is a clear need to explore this possibility further by extending the functionality of the application. Furthermore, the interaction styles required to interface with field objects need to be further explored. And finally the multimedia recording of the uncertainty metadata needs to be brought on board. Currently the prototype is being extended to record sketches, images and formatted data to enhance the kinds of information that are transferable. Furthermore, the generation of summary statistics at the object class, object and object element level is being researched.

It is important to verify that the architecture of the data model is robust and flexible. This means that new ways of recording uncertainty and semantic information associated with object class and object elements can be readily employed because of the object-oriented database paradigm employed here. What is required is to address the human-computer interaction (HCI) issues in terms of interactive exploration of field objects which is limited in the present implementation. For example, exploration of the spatial dependence between objects and their associated statistics is not supported in the current prototype. Furthermore, a more formal way of recording semantic information is required at all levels of the data model.

References

1. Batty M, Galton A, Llobera M (2005) Not Just Space: An Introduction. In: Fisher P, Unwin D (eds) Re-presenting. Wiley, Chichester, pp 127–135
2. Bloch J (2001) Effective Java Programming Language Guide. Addison-Wesley, London
3. Comber AJ, Fisher PF, Wadsworth RA (2003) Actor Network Theory: A suitable framework to understand how land cover mapping projects develop? Land Use Policy 20:299–309
4. Chrisman NR (1991) The error component in spatial data. In: Maguire DJ, Goodchild MF, Rhind DW (eds) Geographical information systems. Longman, London, UK, vol 1, Chapter 12, pp 165–174
5. Cova JT, Goodchild MF (2002) Extending geographical representations to include fields of spatial objects. Int J of Geographical Information Science 16(6):509–532
6. Fisher PF (1999) Models of Uncertainty in Spatial Data. In: Longley P, Goodchild M, Maguire M, Rhind D (eds), Geographical Information Systems: Principles, Techniques, Management and Applications. Wiley and Sons, New York, vol 1, pp 191–205
7. Fisher PF, Tate NJ. (in press) Causes and Consequences of Error in Digital Elevation Models. Progress in Physical Geography
8. Gahegan M, Ehlers M (2000) A framework for the modeling of uncertainty between remote sensing and geographical information systems. ISPRS J of Photogrammetry and Remote Sensing 55:176–188
9. Guesgen HW (2005) Fuzzy reasoning about geographic regions. In: Petry F, Robinson V, Cobb M (eds) Fuzzy Modeling with Spatial Information for Geographic Problems. Springer, New York, pp 1–14
10. Guptill SC, Morrision JL (eds) (1995) Elements of Spatial Data Quality. Elsevier, Oxford
11. Harvey F, Chrisman N (1998) Boundary Objects and the Social Construction of GIS Technology. Environment and Planning A 30(41):1683–1694
12. Harvey F (2005) The linguistic trading zones of semantic interoperability. In: Fisher P, Unwin D (eds) Re-presenting. Wiley, Chichester, pp 43–54
13. Hunter GJ, Goodchild MF (1993) Managing uncertainty in spatial databases: Putting theory into practice. In: Proc, URISA, Atlanta, July 25–29, 1993, pp 1–14
14. Lawson B (1980), How Designers Think: The design process demystified. The Architectural Press Ltd., London
15. Longley P, Goodchild M, Maguire D, Rhind D (2005) Geographic Information Systems and Science, 2nd ed. Wiley, Chichester
16. Lowell K (1997) Outside-in, inside-out: Two methods of generating spatial certainty maps. In: Proc of GeoComputation '97 & SIRC '97, pp 15–25
17. Maceachren AM, Robinson A, HopperS, Gardner S, Murray R, Gahegan M, Hetzler E (2005) Visualizing Geospatial Information Uncertainty: What We

Know and What We Need to Know. Cartography and Geographic Information Science 32:139–160
18. Mennis J, Peuquet DJ (2003) The role of knowledge representation in geographic knowledge discovery: a case study. Transactions in GIS 7(3):371–391
19. Molenaar M (1998) An Introduction to the Theory of Spatial Object Modelling. Taylor & Francis, London
20. Montessori M (1964) The Montessori Method. Schocken, New York
21. O'Sullivan D, Unwin D (2003) Geographic Information Analysis. Wiley
22. Papert S (1980) Mindstorms: Children, Computers and Powerful Ideas. Basic Books, New York Books, New York
23. Plewe B (2002) The nature of uncertainty in historical geographic information. Transactions in GIS 6(4):431–456
24. Schon AD (1991) The reflective practitioner: how professionals think in action. Biddles, Ltd., Guildford and King's Lynn
25. Scrivener SAR, Clark SM (1994) Sketching in Collaborative Design. Interacting With Virtual Environments. Wiley Professional Computing, England
26. Shneiderman B, Plaisant C (2005) Designing the User Interface: Strategies for Effective Human-Computer Interaction. Addison-Wesley
27. Shi W, Fisher PF, Goodchild MF (eds) (2002) Spatial Data Quality. Taylor & Francis, London
28. Sinton D (1978) The inherent structure of information as a constraint to analysis. In: Dutton D (ed) Harvard papers on geographic information systems. Addison Wesley, Reading, UK
29. Voudouris V, Wood J, Fisher PF (2005) Collaborative geoVisualization: Object-Field Representations with Semantic and Uncertainty Information. In: Meersman R, Tari Z, Herrero P et al. (eds) On the Move to Meaningful Internet Systems OTM 2005(= LNCS 3762), Springer, Berlin, pp 1056–1065
30. Zhang J, Goodchild MF (2002) Uncertainty in Geographical Information. Taylor & Francis, London

From Point Cloud to Grid DEM: A Scalable Approach

Pankaj K. Agarwal [*1], Lars Arge [**12], Andrew Danner [***1]

[1] Department of Computer Science, Duke University, Durham, NC 27708, USA; email: {pankaj, large, adanner}@cs.duke.edu
[2] Department of Computer Science, University of Aarhus, Aarhus, Denmark; email: large@daimi.au.dk

Abstract

Given a set $\mathcal{S}$ of points in $\mathbb{R}^3$ sampled from an *elevation* function $H : \mathbb{R}^2 \rightarrow \mathbb{R}$, we present a scalable algorithm for constructing a grid digital elevation model (DEM). Our algorithm consists of three stages: First, we construct a quad tree on $\mathcal{S}$ to partition the point set into a set of non-overlapping segments. Next, for each segment q, we compute the set of points in q and all segments neighboring q. Finally, we interpolate each segment independently using points within the segment and its neighboring segments.

Data sets acquired by LIDAR and other modern mapping technologies consist of hundreds of millions of points and are too large to fit in main memory. When processing such massive data sets, the transfer of data between disk and main memory (also called I/O), rather than the CPU time, becomes the performance bottleneck. We therefore present an *I/O-efficient* algorithm for constructing a grid DEM. Our experiments show that the algorithm scales to data sets much larger than the size of main memory, while existing algorithms do not scale. For example, using a machine with 1GB

* Supported by NSF under grants CCR-00-86013, EIA-01-31905, CCR-02-04118, and DEB-04-25465, by ARO grants W911NF-04-1-0278 and DAAD19-03-1-0352, and by a grant from the U.S.–Israel Binational Science Foundation.

** Supported in part by the US Army Research Office through grant W911NF-04-1-0278 and by an Ole Rømer Scholarship from the Danish National Science Research Council.

*** Supported by the US Army Research Office through grant W911NF-04-1-0278.

RAM, we were able to construct a grid DEM containing 1.3 billion cells (occupying 1.2GB) from a LIDAR data set of over 390 million points (occupying 20GB) in about 53 hours. Neither ArcGIS nor GRASS, two popular GIS products, were able to process this data set.

1 Introduction

One of the basic tasks of a geographic information system (GIS) is to store a representation of various physical properties of a terrain such as elevation, temperature, precipitation, or water depth, each of which can be viewed as a real-valued bivariate function. Because of simplicity and efficacy, one of the widely used representations is the so-called grid representation in which a functional value is stored in each cell of a two-dimensional uniform grid. However, many modern mapping technologies do no acquire data on a uniform grid. Hence the raw data is a set $\mathcal{S}$ of N (arbitrary) points in $\mathbb{R}^3$, sampled from a function $H : \mathbb{R}^2 \to \mathbb{R}$. An important task in GIS is thus to interpolate $\mathcal{S}$ on a uniform grid of a prescribed resolution.

In this paper, we present a scalable algorithm for this interpolation problem. Although our technique is general, we focus on constructing a grid digital elevation model (DEM) from a set $\mathcal{S}$ of N points in $\mathbb{R}^3$ acquired by modern mapping techniques such as LIDAR. These techniques generate huge amounts of high-resolution data. For example, LIDAR[1] acquires highly accurate elevation data at a resolution of one point per square meter or better and routinely generates hundreds of millions of points. It is not possible to store these massive data sets in the internal memory of even high-end machines, and the data must therefore reside on larger but considerably slower disks. When processing such huge data sets, the transfer of data between disk and main memory (also called *I/O*), rather than computation, becomes the performance bottleneck. An *I/O-efficient* algorithm that minimizes the number of disk accesses leads to tremendous runtime improvements in these cases. In this paper we develop an I/O-efficient algorithm for constructing a grid DEM of unprecedented size from massive LIDAR data sets.

Related Work. A variety of methods for interpolating a surface from a set of points have been proposed, including inverse distance weighting (IDW), kriging, spline interpolation and minimum curvature surfaces. Refer to [12] and the references therein for a survey of the different methods. However, the

[1] In this paper, we consider LIDAR data sets that represent the actual terrain and have been pre-processed by the data providers to remove spikes and errors due to noise.

computational complexity of these methods often make it infeasible to use them directly on even moderately large points sets. Therefore, many practical algorithms use a segmentation scheme that decomposes the plane (or rather the area of the plane containing the input points) into a set of *non-overlapping areas* (or *segments*), each containing a small number of input points. One then interpolates the points in each segment independently. Numerous segmentation schemes have been proposed, including simple regular decompositions and decompositions based on Voronoi diagrams [18] or quad trees [14, 11]. A few schemes using *overlapping* segments have also been proposed [19, 17].

As mentioned above, since I/O is typically the bottleneck when processing large data sets, I/O-efficient algorithms are designed to explicitly take advantage of the large main memory and disk block size [2]. These algorithms are designed in a model in which the computer consists of an internal (or main) memory of size M and an infinite external memory. Computation is considered free but can only occur on elements in main memory; in one *I/O-operation*, or simply *I/O*, B consecutive elements can be transfered between internal and external memory. The goal of an I/O-efficient algorithm is to minimize the number of I/Os.

Many $\Theta(N)$ time algorithms that do not explicitly consider I/O use $\Theta(N)$ I/Os when used in the I/O-model. However, the "linear" bound, the number of I/Os needed to read N elements, is only $\Theta(\mathrm{scan}(N)) = \Theta(\frac{N}{B})$ in the I/O model. The number of I/Os needed to sort N elements is $\Theta(\mathrm{sort}(N)) = \Theta(\frac{N}{B}\log_{M/B}\frac{N}{B})$ [2]. In practice, B is on the order of 10^3–10^5, so $\mathrm{scan}(N)$ and $\mathrm{sort}(N)$ are typically much smaller than N. Therefore tremendous speedups can often be obtained by developing algorithms that use $O(\mathrm{scan}(N))$ or $O(\mathrm{sort}(N))$ I/Os rather than $\Omega(N)$ I/Os. Numerous I/O-efficient algorithms and data structures have been developed in recent years, including several for fundamental GIS problems (refer to [4] and the references therein for a survey). Agarwal et al. [1] presented a general top-down layered framework for constructing a certain class of spatial data structures–including quad trees–I/O-efficiently. Hjaltason and Samet [9] also presented an I/O-efficient quad-tree construction algorithm. This optimal $O(\mathrm{sort}(N))$ I/O algorithm is based on assigning a Morton block index to each point in $\mathcal{S}$, encoding its location along a Morton-order (Z-order) space-filling curve, sorting the points by this index, and then constructing the structure in a bottom-up manner.

Our Approach. In this paper we describe an I/O-efficient algorithm for constructing a grid DEM from LIDAR points based on quad-tree segmentation. Most of the segmentation based algorithms for this problem can be considered as consisting of three separate phases; the *segmentation* phase, where the decomposition is computed based on $\mathcal{S}$; the *neighbor finding* phase, where for each segment in the decomposition the points in the segment and the relevant neighboring segments are computed; and the *interpolation* phase, where a surface is interpolated in each segment and the interpolated values of the grid cells in the segment are computed. In this paper, we are more interested in the segmentation and neighbor finding phases than the particular interpolation method used in the interpolation phase. We will focus on the quad tree based segmentation scheme because of its relative simplicity and because it has been used with several interpolation methods such as thin plate splines [14] and B-splines [11]. We believe that our techniques will apply to other segmentation schemes as well.

Our algorithm implements all three phases I/O-efficiently, while allowing the use of any given interpolation method in the interpolation phase. Given a set $\mathcal{S}$ of N points, a desired output grid specified by a bounding box and a cell resolution, as well as a threshold parameter $k_{\max}$, the algorithm uses $O(\frac{N}{B}\frac{h}{\log\frac{M}{B}} + \text{sort}(T))$ I/Os, where h is the height of a quad tree on $\mathcal{S}$ with at most $k_{\max}$ points in each leaf, and T is the number of cells in the desired grid DEM. Note that this is $O(\text{sort}(N) + \text{sort}(T))$ I/Os if $h = O(\log N)$, that is, if the points in $\mathcal{S}$ are distributed such that the quad tree is roughly balanced.

The three phases of our algorithm are described in Section 2, Section 3 and Section 4. In Section 2 we describe how to construct a quad tree on $\mathcal{S}$ with at most $k_{\max}$ points in each leaf using $O(\frac{N}{B}\frac{h}{\log\frac{M}{B}})$ I/Os. The algorithm is based on the framework of Agarwal et al. [1]. Although not as efficient as the algorithm by Hjaltason and Samet [9] in the worst case, we believe that it is simpler and potentially more practical; for example, it does not require computation of Morton block indices or sorting of the input points. Also in most practical cases where $\mathcal{S}$ is relatively nicely distributed, for example when working with LIDAR data, the two algorithms both use $O(\text{sort}(N))$ I/Os. In Section 3 we describe how to find the points in all neighbor leaves of each quad-tree leaf using $O(\frac{N}{B}\frac{h}{\log\frac{M}{B}})$ I/Os. The algorithm is simple and very similar to our quad-tree construction algorithm; it takes advantage of how the quad tree is naturally stored on disk during the segmentation phase. Note that while Hjaltason and Samet [9] do not describe a neighbor finding algorithm based on their Morton block approach, it seems possible to use their approach and an I/O-efficient priority queue [5] to obtain an $O(\text{sort}(N))$ I/O algorithm for the problem. However, this algorithm would be quite complex and therefore probably not of practical interest. Finally, in Section 4 we describe how to apply an interpolation scheme to the points collected for each quad-tree leaf, evaluate the computed function at the relevant grid cells within the segment corresponding to each leaf, and construct the final grid using $O(\text{scan}(N)) + O(\text{sort}(T))$ I/Os. As mentioned earlier, we can use any given interpolation method within each segment.

To investigate the practical efficiency of our algorithm we implemented it and experimentally compared it to other interpolation algorithms using LIDAR data. To summarize the results of our experiments, we show that, unlike existing algorithms, our algorithm scales to data sets much larger than the main memory. For example, using a 1GB machine we were able to construct a grid DEM containing 1.3 billion points (occupying 1.2 GB) from a LIDAR data set of over 390 million points (occupying 20 GB) in just 53 hours. This data set is an order of magnitude larger than what could be handled by two popular GIS products–ArcGIS and GRASS. In addition to supporting large input point sets, we were also able to construct very large high resolution grids; in one experiment we constructed a one meter resolution grid DEM containing more than 53 billion cells – storing just a single bit for each grid cell in this DEM requires 6 GB.

In Section 5 we describe the details of the implementation of our theoretically I/O-efficient algorithm that uses a regularized spline with tension interpolation method [13]. We also describe the details of an existing algorithm implemented in GRASS using the same interpolation method; this algorithm is similar to ours but it is not I/O-efficient. In Section 6 we describe the results of the experimental comparison of our algorithm to other existing implementations. As part of this comparison, we present a detailed comparison of the quality of the grid DEMs produced by our algorithm and the similar algorithm in GRASS that show the results are in good agreement.

2 Segmentation Phase: Quad-Tree Construction

Given a set $\mathcal{S}$ of N points contained in a bounding box $[x_1, x_2] \times [y_1, y_2]$ in the plane, and a threshold $k_{\max}$, we wish to construct a quad tree $\mathcal{T}$ [8] on $\mathcal{S}$ such that each quad-tree leaf contains at most $k_{\max}$ points. Note that the leaves of $\mathcal{T}$ partition the bounding box $[x_1, x_2] \times [y_1, y_2]$ into a set of disjoint areas, which we call *segments*.

Incremental Construction. $\mathcal{T}$ can be constructed incrementally simply by inserting the points of $\mathcal{S}$ one at a time into an initially empty tree. For each point p, we traverse a root-leaf path in $\mathcal{T}$ to find the leaf v containing p. If v contains less than $k_{\max}$ points, we simply insert p in v. Otherwise, we split v into four new leaves, each representing a quadrant of v, and re-distribute p and the points in v to the new leaves. If h is the height of $\mathcal{T}$, this algorithm uses $O(Nh)$ time. If the input points in $\mathcal{S}$ are relatively evenly distributed we have $h = O(\log N)$, and the algorithm uses $O(N \log N)$ time.

If $\mathcal{S}$ is so large that $\mathcal{T}$ must reside on disk, traversing a path of length h may require as many as h I/Os, leading to an I/O cost of $O(Nh)$ in the I/O-model. By storing (or *blocking*) the nodes of $\mathcal{T}$ on disk intelligently, we may be able to access a subtree of depth $\log B$ (size B) in a single I/O and thus reduce the cost to $O(N\frac{h}{\log B})$ I/Os. Caching the top-most levels of the tree in internal memory may also reduce the number of I/Os needed. However, since not all the levels fit in internal memory, it is hard to avoid spending an I/O to access a leaf during each insertion, or $\Omega(N)$ I/Os in total. Since $\mathrm{sort}(N) \ll N$ in almost all cases, the incremental approach is very inefficient when the input points do not fit in internal memory.

Level-by-level Construction. A simple I/O-efficient alternative to the incremental construction algorithm is to construct $\mathcal{T}$ level-by-level: We first construct the first level of $\mathcal{T}$, the root v, by scanning through $\mathcal{S}$ and, if $N > k_{\max}$, distributing each point p to one of four leaf lists on disk corresponding to the child of v containing p. Once we have scanned $\mathcal{S}$ and constructed one level, we construct the next level by loading each leaf list in turn and constructing leaf lists for the next level of $\mathcal{T}$. While processing one list we keep a buffer of size B in memory for each of the four new leaf lists (children of the constructed node) and write buffers to the leaf lists on disk as they run full. Since we in total scan $\mathcal{S}$ on each level of $\mathcal{T}$, the algorithm uses $O(Nh)$ time, the same as the incremental algorithm, but only $O(Nh/B)$ I/Os. However, even in the case of $h = \log_4 N$, this approach is still a factor of $\log_{\frac{M}{B}} \frac{N}{B} / \log_4 N$ from the optimal $O(\frac{N}{B} \log_{\frac{M}{B}} \frac{N}{B})$ I/O bound.

Hybrid Construction. Using the framework of Agarwal et al. [1], we design a hybrid algorithm that combines the incremental and level-by-level approaches. Instead of constructing a single level at a time, we can construct

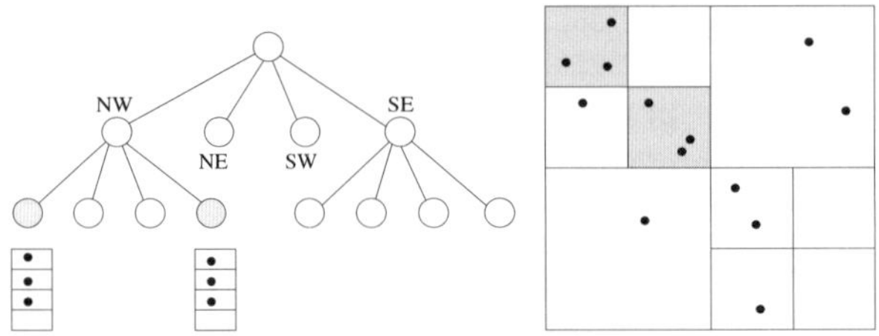

Fig. 1. Construction of a quad-tree layer of depth three with $k_{\max} = 2$. Once a leaf at depth three is created, no further splitting is done; instead additional points in the leaf are stored in leaf lists shown below shaded nodes. After processing all points the shaded leaves with more than two points are processed recursively

a *layer* of $\log_4 \frac{M}{B}$ levels. Because $4^{\log_4 \frac{M}{B}} = M/B < M$, we construct the layer entirely in internal memory using the incremental approach: We scan through $\mathcal{S}$, inserting points one at a time while splitting leaves and constructing new nodes, except if the path from the root of the layer to a leaf of the layer is of height $\log_4 \frac{M}{B}$. In this case, we write all points contained in such a leaf v to a list L_v on disk. After all points have been processed and the layer constructed, we write the layer to disk sequentially and recursively construct layers for each leaf list L_i. Refer to Figure 1.

Since a layer has at most M/B nodes, we can keep a internal memory buffer of size B for each leaf list and only write points to disk when a buffer runs full (for leaves that contain less than B points in total, we write the points in all such leaves to a single list after constructing the layer). In this way we can construct a layer on N points in $O(N/B) = \text{scan}(N)$ I/Os. Since a tree of height h has $h/\log_4 \frac{M}{B}$ layers, the total construction cost is $O(\frac{N}{B}\frac{h}{\log \frac{M}{B}})$ I/Os. This is $\text{sort}(N) = O(\frac{N}{B}\log_{\frac{M}{B}}\frac{N}{B})$ I/Os when $h = O(\log N)$.

3 Neighbor Finding Phase

Let $\mathcal{T}$ be a quad tree on $\mathcal{S}$. We say that two leaves are neighbors if their associated segments share part of an edge or a corner. Refer to Figure 2 for an example. If $\mathcal{L}$ is the set of segments associated with the leaves of $\mathcal{T}$, we want to find for each $q \in \mathcal{L}$ the set $\mathcal{S}_q$ of points contained in q and the neighbor leaves of q. As for the construction algorithm, we first describe an incremental algorithm and then improve its efficiency using a layered approach.

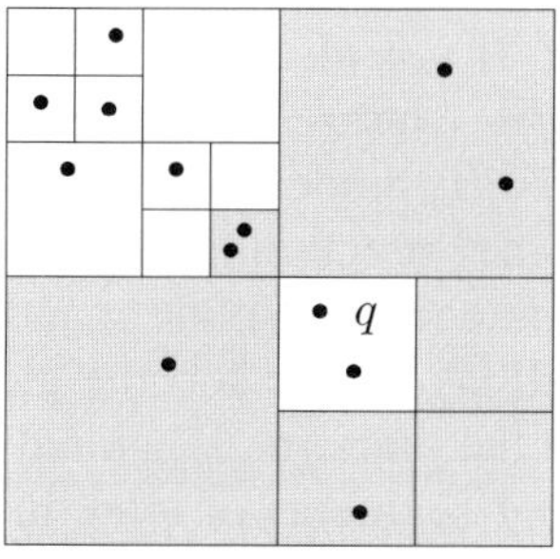

Fig. 2. The segment q associated with a leaf of a quad tree and its six shaded neighboring segments

Incremental Approach. For each segment $q \in \mathcal{L}$, we can find the points in the neighbors of q using a simple recursive procedure: Starting at the root v of $\mathcal{T}$, we compare q to the segments associated with the four children of v. If the bounding box of a child u shares a point or part of an edge with q, then q is a neighbor of at least one leaf in the tree rooted in u; we therefore recursively visit each child with an associated segment that either neighbors or contains q. When we reach a leaf we insert all points in the leaf in $\mathcal{S}_q$.

To analyze the algorithm, we first bound the total number of neighbor segments found over all segments $q \in \mathcal{L}$. Consider the number of neighbor segments that are at least the same size as a given segment q; at most one segment can share each of q's four edges, and at most four more segments can share the four corner points of q. Thus, there are at most eight such neighbor segments. Because the neighbor relation is symmetric, the total number of neighbor segments over all segments is at most twice the total number of neighbor segments which are at least the same size. Thus the total number of neighbor segments over all segments is at most 16 times the number of leaves of $\mathcal{T}$. Because the total number of leaves is at most $4N/k_{\max}$, and since the above algorithm traverses a path of height h for each neighbor, it visits $O(Nh)$ nodes in total. Furthermore, as each leaf contains at most $k_{\max}$ points, the algorithm reports $O(Nk_{\max})$ points in total. Thus the total running time of the algorithm is $O((h + k_{\max})N) = O(hN))$. This is also the worst case I/O cost.

Layered Approach. To find the points in the neighboring segments of each segment in $\mathcal{L}$ using a layered approach similar to the one used to construct $\mathcal{T}$, we first load the top $\log_4 \frac{M}{B}$ levels of $\mathcal{T}$ into memory. We then associate with each leaf u in the layer, a buffer B_u of size B in internal memory and a list L_u in external memory. For each segment $q \in \mathcal{L}$, we use the incremental algorithm described above to find the leaves of the layer with an associated segment that completely contains q or share part of a boundary with q. Suppose u is such a layer leaf. If u is also a leaf of the entire tree $\mathcal{T}$, we add the pair $(q, \mathcal{S}_u)$ to a global list Λ, where $\mathcal{S}_u$ is the set of points stored at u. Otherwise, we add q to the buffer B_u associated with u, which is written to L_u on disk when B_u runs full. After processing all segments in $\mathcal{L}$, we recursively process the layers rooted at each leaf node u and its corresponding list L_u. Finally, after processing all layers, we sort the global list of neighbor points Λ by the first element q in the pairs $(q, \mathcal{S}_u)$ stored in Λ. After this, the set $\mathcal{S}_q$ of points in the neighboring segments of q are in consecutive pairs of Λ, so we can construct all $\mathcal{S}_q$ sets in a simple scan of Λ.

Since we access nodes in $\mathcal{T}$ during the above algorithm in the same order they were produced in the construction of $\mathcal{T}$, we can process each layer of $\log_4 \frac{M}{B}$ levels of $\mathcal{T}$ in $\mathrm{scan}(N)$ I/Os. Furthermore, since $\sum_q |\mathcal{S}_q| = O(N)$, the total number of I/Os used to sort and scan A is $O(\text{sort}(N))$. Thus the algorithm uses $O(\frac{N}{B}\frac{h}{\log \frac{M}{B}})$ I/Os in total, which is $O(\text{sort}(N))$ when $h = O(\log N)$.

4 Interpolation Phase

Given the set $\mathcal{S}_q$ of points in each segment q (quad tree leaf area) and the neighboring segments of q, we can perform the interpolation phase for each segment q in turn simply by using any interpolation method we like on the points in $\mathcal{S}_q$, and evaluating the computed function to interpolate each of the grid cells in q. Since $\sum_q |\mathcal{S}_q| = O(N)$, and assuming that each $\mathcal{S}_q$ fits in memory (otherwise we maintain a internal memory priority queue to keep the $n_{\max} < M$ points in $\mathcal{S}_q$ that are closest to the center of q, and interpolate on this subset), we can read each $\mathcal{S}_q$ into main memory and perform the interpolation in $O(\text{scan}(N))$ I/Os in total. However, we cannot simply write the interpolated grid cells to an output grid DEM as they are computed, since this could result in an I/O per cell (or per segment q). Instead we write each interpolated grid cell to a list along with its position (i, j) in the grid; we buffer B cells at a time in memory and write the buffer to disk when it runs full. After processing each set $\mathcal{S}_q$, we sort the list of interpolated grid cells by position to obtain the output grid. If the output grid has size T, computing the T interpolated cells and writing them to the list takes $O(T/B)$ I/Os. Sorting the cells take $O(\text{sort}(T))$ I/Os. Thus the interpolation phase is performed in $O(\text{scan}(N) + \text{sort}(T))$ I/Os in total.

5 Implementation

We implemented our methods in C++ using TPIE [7, 6], a library that eases the implementation of I/O-efficient algorithms and data structures by providing a set of primitives for processing large data sets. Our algorithm takes as input a set $\mathcal{S}$ of points, a grid size, and a parameter $k_{\max}$ that specifies the maximum number of points per quad tree segment, and computes the interpolated surface for the grid using our segmentation algorithm and a regularized spline with tension interpolation method [13]. We chose this interpolation method because it is used in the open source GIS GRASS module `s.surf.rst` [14] – the only GRASS surface interpolation method that uses segmentation to handle larger input sizes – and provides a means to compare our I/O-efficient approach to an existing segmentation method. Below we discuss two implementation details of our approach: thinning the input point set, and supporting a *bit mask*. Additionally, we highlight the main differences between our implementation and `s.surf.rst`.

Thinning Point Sets. Because LIDAR point sets can be very dense, there are often several cells in the output grid that contain multiple input points, especially when the grid cell size is large. Since it is not necessary to interpolate at sub-pixel resolutions, computational efficiency improves if one only includes points that are sufficiently far from other points in a quad-tree segment. Our implementation only includes points in a segment that are at least a user-specified distance ε from all other points within the segment. By default, ε is half the size of a grid cell. We implement this feature with no additional I/O cost simply by checking the distance between a new point p and all other points within the quad-tree leaf containing p and discarding p if it is within a distance ε of another point.

Bit Mask. A common GIS feature is the ability to specify a bit mask that skips computation on certain grid cells. The bit mask is a grid of the same size as the output grid, where each cell has a zero or one bit value. We only interpolate grid cell values when the bit mask for the cell has the value one. Bit masks are particularly useful when the input data set consists of an irregularly shaped region where the input points are clustered and large areas of the grid are far from the input points. Skipping the interpolation of the surface in these places reduces computation time, especially when many of the bit mask values are zero.

For high resolution grids, the number of grid cells can be very large, and the bit mask may be larger than internal memory and must reside on disk. Randomly querying the bit mask for each output grid cell would be very expensive in terms of I/O cost. Using the same filtering idea described in Section 2 and Section 3, we filter the bit mask bits through the quad-tree layer by layer such that each quad-tree segment gets a copy of the bit mask bits it needs during interpolation. The algorithm uses $O(\frac{T}{B}\frac{h}{\log \frac{M}{B}})$ I/Os in total, where T is the number of cells in the output grid, which is $O(\text{sort}(T))$ when $h = O(\log N)$. The bits for a given segment can be accessed sequentially as we interpolate each quad-tree segment.

GRASS Implementation. The GRASS module `s.surf.rst` uses a quad-tree segmentation, but is not I/O-efficient in several key areas which we briefly discuss; constructing the quad tree, supporting a bit mask, finding neighbors, and evaluating grid cells. All data structures in the GRASS implementation with the exception of the output grid are stored in memory and must use considerably slower swap space on disk if internal memory is exhausted. During construction points are simply inserted into an internal memory quad tree using the incremental construction approach of Section 2.

Thinning of points using the parameter ε during construction is implemented exactly as our implementation. The bit mask in `s.surf.rst` is stored as a regular grid entirely in memory and is accessed randomly during interpolation of segments instead of sequentially in our approach.

Points from neighboring quad-tree segment are not found in advance as in our algorithm, but are found when interpolating a given quad-tree segment q; the algorithm creates a window w by expanding q in all directions by a width δ and querying the quad tree to find all points within w. The width δ is adjusted by binary search until the number of points within w is between a user specified range $[n_{\min}, n_{\max}]$. Once an appropriate number of points is found for a quad-tree segment q, the grid cells in q are interpolated and written directly to the proper location in the output grid by randomly seeking to the appropriate file offset and writing the interpolated results. When each segment has a small number of cells, writing the values of the T output grid cells uses $O(T) \gg \mathrm{sort}(T)$ I/Os. Our approach constructs the output grid using the significantly better $\mathrm{sort}(T)$ I/Os.

6 Experiments

We ran a set of experiments using our I/O-efficient implementation of our algorithm and compared our results to existing GIS tools. We begin by describing the data sets on which we ran the experiments, then compare the efficiency and accuracy of our algorithm with other methods. We show that our algorithm is scalable to over 395 million points and over 53 billion output grid cells–well beyond the limits of other GIS tools we tested.

Experimental Setup. We ran our experiments on an Intel 3.4 GHz Pentium 4 hyper-threaded machine with 1 GB of internal memory, over 4 GB of swap space, and running a Linux 2.6 kernel. The machine had a pair of 400 GB SATA disk drives in a non-RAID configuration. One disk stored the input and output data sets and the other disk was used for temporary scratch space.

For our experiments we used two large LIDAR data sets, freely available from online sources; one of the Neuse river basin from the North Carolina Floodmaps project [15] and one of the North Carolina Outer Banks from NOAA's Coastal Services Center [16].

Neuse river basin. This data set contains 500 million points, more than 20 GB of raw data (see Fig. 3a). The data have been pre-processed by the data providers to remove most points on buildings and vegetation. The average spacing between points is roughly 20ft.

Outer banks. This data set contains 212 million LIDAR points, 9 GB of raw data (see Fig. 3b). Data points are confined to a narrow strip (a zoom

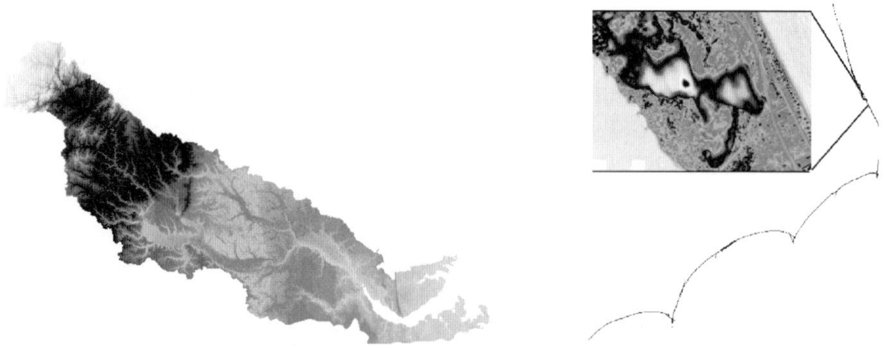

(a) (b)

Fig. 3. **(a)** Neuse river basin data set and **(b)** Outer Banks data set, with zoom to very small region

of a very small portion of the data set is shown in the figure). This data set has not been heavily pre-processed to remove buildings and vegetation. The average point spacing is roughly 3ft.

Scalability Results. We ran our algorithm on both the Neuse river and Outer Banks data sets at varying grid cell resolutions. Because we used the default value of ε (half the grid cell size) increasing the size of grid cells decreased the number of points in the quad tree and the number of points used for interpolation. Results are summarized in Table 1. In each test, the interpolation phase was the most time-consuming phase; interpolation consumed over 80% of the total running time on the Neuse river basin data set. For each test we used a bit mask to ignore cells more than 300ft from the input points. Because of the irregular shape of the Outer Banks data, this bit mask is very large, but relatively sparse (containing very few "1" bits). Therefore, filtering the bit mask and writing the output grid for the Outer Banks data were relatively time-consuming phases when compared to the Neuse river data. Note that the number of grid cells in the Outer Banks is roughly three orders of magnitude greater than the number of quad-tree points. As the grid cell size decreases and the total number of cells increases, bit mask and grid output operations consume a greater percentage of the total time. At a resolution of 5ft, the bit mask alone for the Outer Banks data set is over 6 GB. Even at such large grid sizes, interpolation – an internal memory procedure – was the most time-consuming phase, indicating that I/O was not a bottleneck in our algorithm. We also tried to test other available interpolation methods, including `s.surf.rst` in the open source GIS GRASS; kriging, IDW, spline, and topo-to-raster (based on ANUDEM [10]) tools in ArcGIS 9.1; and QTModeler 4 from Applied Imagery [3]. Only `s.surf.rst` supported the thinning

Table 1. Results from the Neuse river basin and the Outer Banks data sets

Dataset	Neuse		Outer Banks	
Resolution (ft)	20	40	5	10
Output grid cells ($\times 10^6$)	1360	340	53160	13402
quad-tree points ($\times 10^6$)	395	236	128	66
Total Time (hrs)	53.0	24.4	17.7	6.9
Time spent to... (%)				
Build tree	2.0	3.8	4.5	8.6
Find Neighbors	10.6	15.1	14.5	16.4
Filter Bit mask	0.2	0.3	13.1	8.0
Interpolate	86.4	80.4	52.6	57.8
Write Output	0.8	0.4	15.3	9.2

of data points based on cell size, so for the other programs we simply used a subset of the data points. None of the ArcGIS tools could process more than 25 million points from the Neuse river basin at 20ft resolution and every tool crashed on large input sizes. The topo-to-raster tool processed the largest set amongst the ArcGIS tools at 21 million points.

The `s.surf.rst` could not process more than 25 million points either. Using a resolution of 200ft, `s.surf.rst` could process the entire Neuse data set in six hours, but the quad tree only contained 17.4 million points. Our algorithm processed the same data set at 200ft resolution in 3.2 hours. On a small subset of the Outer Banks data set containing 48.8 million points, `s.surf.rst`, built a quad tree on 7.1 million points and computed the output grid DEM in three hours, compared to 49 minutes for our algorithm on the same data set.

The QTModeler program processed the largest data set amongst the other methods we tested, approximately 50 million points, using 1 GB of RAM. The documentation for QTModeler states that their approach is based on an internal memory quad tree and can process 200 million points with 4 GB of available RAM. We can process a data set almost twice as large using less than 1GB of RAM.

Overall, we have seen that our algorithm is scalable to very large point sets and very large grid sizes and we demonstrated that many of the commonly used GIS tools cannot process such large data sets. Our approach for building the quad tree and finding points in neighboring segments is efficient and never took more than 25% of the total time in any of our experiments. The interpolation phase, an internal step that reads points sequentially from disk and writes grid cells sequentially to disk, was the most time-consuming phase of the entire algorithm.

Comparison of Constructed Grids. To show that our method constructs correct output grids, we compared our output on the Neuse river basin to the original input points as well as to grid DEMs created by `s.surf.rst`, and DEMs freely available from NC Floodmaps. Because `s.surf.rst` cannot process very large data sets, we ran our tests on a small subset of the Neuse river data set containing 13 million points. The output resolution was 20ft, ε was set to the default 10ft, and the output grid had 3274 rows and 3537 columns for a total of 11.6 million cells. Approximately 11 million points were in the quad tree.

The interpolation function we tested used a smoothing parameter and allowed the input points to deviate slightly from the interpolated surface. We used the same default smoothing parameter used in the GRASS implementation and compared the distribution of deviations between the input points and the interpolated surface. The results were independent of $k_{\max}$, the maximum number of points per quad-tree segment. In all tests, at least 79% of the points had no deviation, and over 98% of the points had a deviation of less than one inch. Results for `s.surf.rst` were similar. Since the results were indistinguishable for various $k_{\max}$ parameters, we show only one of the cumulative distribution functions (CDF) for $k_{\max} = 35$ in Figure 4.

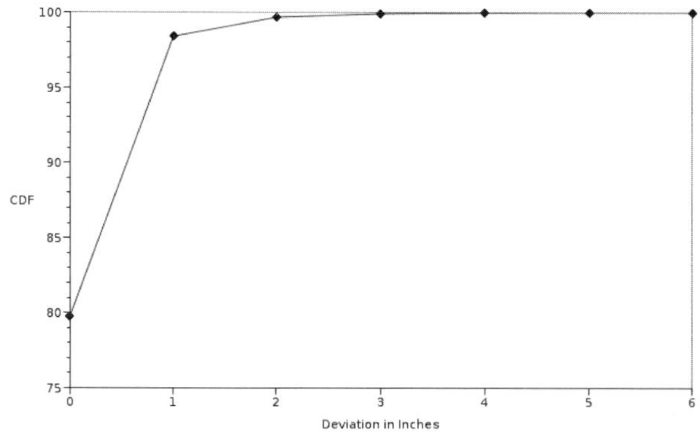

Fig. 4. Distribution of deviations between input points and interpolated surface ($k_{\max} = 35$)

Next, we computed the absolute deviation between grid values computed using `s.surf.rst` and our method. We found that over 98% of the cells agreed within 1 inch, independent of $k_{\max}$. The methods differ slightly because `s.surf.rst` uses a variable size window to find points in neighboring points of a quad-tree segment q and may not choose all points from immediate neighbors of q when the points are dense and may expand the

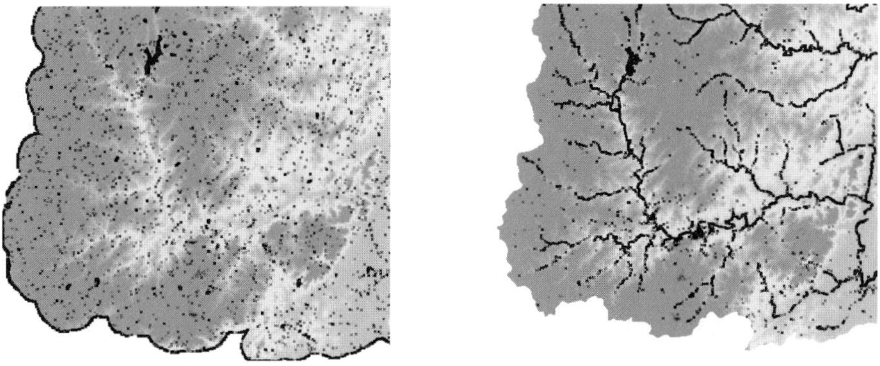

(a) (b)

Fig. 5. Interpolated surface generated by our method. Black dots indicate cells where the deviation between our method and **(a)** `s.surf.rst` is greater than three inches, **(b)** ncfloodmap data is greater than two feet

window to include points in segments that are not immediate neighbors of q when the points are sparse. In Figure 5a we show a plot of the interpolated surface along with an overlay of cells where the deviation exceeds 3 inches. Notice that most of the bad spots are along the border of the data set where our method is less likely to get many points from neighboring quadtree leaves and near the lake in the upper left corner of the image where LIDAR signals are absorbed by the water and there are no input data points.

Finally, we compared both our output and that of `s.surf.rst` to the 20ft DEM data available from the NC Floodmaps project. A CDF in Figure 6 of the absolute deviation between the interpolated grids and the "base" grid from NC Floodmaps shows that both implementations have an identical CDF curve. However, the agreement between the interpolated surfaces and the base grid is not as strong as the agreement between the algorithms when compared to each other. An overlay of regions with deviation greater than two feet on base map shown in Figure 5b reveals the source of the disagreement. A river network is clearly visible in the figure indicating that something is very different between the two data sets along the rivers. NC Floodmaps uses supplemental break-line data that is not part of the LIDAR point set to enforce drainage and provide better boundaries of lakes in areas where LIDAR has trouble collecting data. Aside from the rivers, the interpolated surface generated by either our method or the existing GRASS implementation agree reasonably well with the professionally produced and publicly available base map.

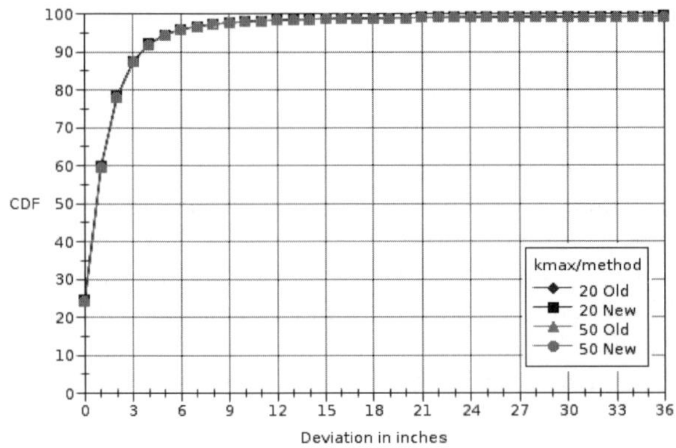

Fig. 6. Cumulative distribution of deviation between interpolated surface and data downloaded from ncfloodmaps.com. Deviation is similar for both our method and `s.surf.rst` for all values of $k_{\max}$

7 Conclusions

In this paper we describe an I/O-efficient algorithm for constructing a grid DEM from point cloud data. We implemented our algorithm and, using LIDAR data, experimentally compared it to other existing algorithms. The empirical results show that, unlike existing algorithms, our approach scales to data sets much larger than the size of main memory. Although we focused on elevation data, our technique is general and can be used to compute the grid representation of any bivariate function from irregularly sampled data points.

For future work, we would like to consider a number of related problems. Firstly, our solution is constructed in such a way that the interpolation phase can be executed in parallel. A parallel implementation should expedite the interpolation procedure. Secondly, as seen in Figure 5b, grid DEMs are often constructed from multiple sources, including LIDAR points and supplemental break-lines where feature preservation is important. Future work will examine methods of incorporating multiple data sources into DEM construction. Finally, the ability to create large scale DEMs efficiently from LIDAR data could lead to further improvements in topographic analysis including such problems as modelling surface water flow or detecting topographic change in time series data.

References

1. Agarwal PK, Arge L, Procopiuc O, Vitter JS (2001) A framework for index bulk loading and dynamization. In: Proc Int Colloquium on Automata, Languages, and Programming, pp 115–127
2. Aggarwal A, Vitter JS (1988) The Input/Output complexity of sorting and related problems. Communications of the ACM 31(9):1116–1127
3. Applied Imagery (5 March 2006) http://www.appliedimagery.com
4. Arge L (1997) External-memory algorithms with applications in geographic information systems. In: Kreveld M van, Nievergelt J, Roos T, Widmayer P (eds) Algorithmic Foundations of GIS (= LNCS 1340). Springer-Verlag, pp 213-254
5. Arge L (2003) The buffer tree: A technique for designing batched external data structures. Algorithmica 37(1):1–24
6. Arge L, Barve R, Procopiuc O, Toma L, Vengroff DE, Wickremesinghe R (1999) TPIE User Manual and Reference (ed 0.9.01a). Duke University (The manual and software distribution are available on the web at http://www.cs.duke.edu/TPIE/
7. Arge L, Procopiuc O, Vitter JS (2002) Implementing I/O-efficient data structures using TPIE. In: Proc European Symp on Algorithms, pp 88–100
8. Berg M de, Kreveld M van, Overmars M, Schwarzkopf O (1997) Computational Geometry – Algorithms and Applications. Springer Verlag, Berlin
9. Hjaltason GR, Samet H (2002) Speeding up construction of quadtrees for spatial indexing. VLDB 11(2):109–137
10. Hutchinson MF (1989) A new procedure for gridding elevation and stream line data with automatic removal of pits. J of Hydrology 106:211–232
11. Lee S, Wolberg G, Shin SY (1997) Scattered data interpolation with multilevel B-splines. IEEE Transactions on Visualization and Computer Graphics 3(3):228–244
12. Mitas L, Mitasova H (1999) Spatial interpolation. In: Longley P, Goodchild MF, Maguire DJ, Rhind DW (eds) Geographic Information Systems – Principles, Techniques, Management, and Applications. Wiley
13. Mitasova H, Mitas L (1993) Interpolation by regularized spline with tension: I. theory and implementation. Mathematical Geology 25:641–655
14. Mitasova H, Mitas L, Brown WM, Gerdes DP, Kosinovsky I, Baker T (1995) Modelling spatially and temporally distributed phenomena: new methods and tools for GRASS GIS. Int J Geographical Information Systems 9(4):433–446
15. NC-Floodmaps (5 March 2006) http://www.ncfloodmaps.com
16. NOAA-CSC (5 March 2006) LIDAR Data Retrieval Tool-LDART. http://www.csc.noaa.gov/crs/tcm/missions.html
17. Pouderoux J, Tobor I, Gonzato JC, Guitton P (2004) Adaptive hierarchical RBF interpolation for creating smooth digital elevation models. In: ACM-GIS Nov, pp 232–240
18. Sibson R (1982) A brief description of natural neighbor interpolation. In: Barnett V (ed) Interpreting Multivariate Data. John Wiley and Sons, pp 21–36
19. Wendland H (2002) Fast evaluation of radial basis functions: Methods based on partition of unity. In Chui CK, Schumaker LL, Stöckler J (eds) Approxi-

mation Theory X: Wavelets, Splines, and Applications. Vanderbilt University Press, Nashville, pp 473–483

Use of Plan Curvature Variations for the Identification of Ridges and Channels on DEM

Sanjay Rana

Department of Civil and Environmental Engineering, University College London, Gower Street, London WC1E 6BT, UK
email: s.rana@ucl.ac.uk

Abstract

This paper proposes novel improvements in the traditional algorithms for the identification of ridge and channel (also called ravines) topographic features on raster digital elevation models (DEMs). The overall methodology consists of two main steps: (1) smoothing the DEM by applying a mean filter, and (2) detection of ridge and channel features as cells with positive and negative plan curvature respectively, along with a decline and incline in plan curvature away from the cell in direction orthogonal to the feature axis respectively. The paper demonstrates a simple approach to visualize the multi-scale structure of terrains and utilize it for semi-automated topographic feature identification. Despite its simplicity, the revised algorithm produced markedly superior outputs than a comparatively sophisticated feature extraction algorithm based on conic-section analysis of terrain.

1 Introduction

Ridge and channel are two fundamental features of terrain morphology. Owing to their unique significance in the shape and structure of terrains, ridge and channel features are used in various terrain analyses ranging from drainage basin delineation (see e.g., Band 1986) to intervisibility computation (see e.g., Rana 2003). In addition, as a generic abstraction of the surface structure, their use also extends in the analysis and visualization of surfaces in socio-economic studies (see e.g., Okabe and Masuyama 2004), metrology (see e.g., Scott 2004), and computer graphics (see e.g., Belyaev and Anoshkina 2005). Naturally, an enormous amount of research has been done in the automated delineation of ridges and channels from various types of surface datasets such as triangulated meshes and raster DEMs. This paper proposes simple and novel improvements in the algorithms for ridge and channel extraction in raster DEMs, with significantly improved results.

As a background to the following discussion on various algorithms, a brief summary of the relevant aspects of feature extraction in raster DEMs is essential. In most cases of cell-based feature extraction algorithms, a terrain is studied in patches, which are *square windows* of m x m cells (also called kernel or filter) of DEM, centered on the cell of interest. The value of m is an odd integer greater than 2 and no more than the lesser amongst the number of rows and columns. In polynomial-fitting based algorithms, kernel is considered analogous to a set of regularly spaced points, each one typically derived from the geometric centre of the kernel cells. The result of the feature classification is assigned to a cell in the output raster, which is located at the same place as the kernel's central cell. Most feature extraction algorithms based on kernels above suffer from two fundamental limitations. As the kernel size remains fixed during a feature extraction, geographic features whose extents are not pronounced within the dimensions of the kernel (e.g., gently sloping features), could be incorrectly classified. Figure 1 shows the example of a 3 x 3 cells kernel on two ridges with different extents. In Figure 1a the point of maximum curvature lies within the kernel hence the central cell will be correctly classified as part of a ridge. However, in Figure 1b due to its shape, the top of the ridge is not so well defined within the kernel hence although the central cell probably belongs to a large ridge feature it will be identified as planar. This limitation is referred as the scale-dependency of the feature extraction algorithm. Another limitation relates to the odd number of cells in the *square window*, such that the cells on the edges of DEM remain unclassified. For further information on raster based spatial analysis, refer to an introductory text by DeMers (2002).

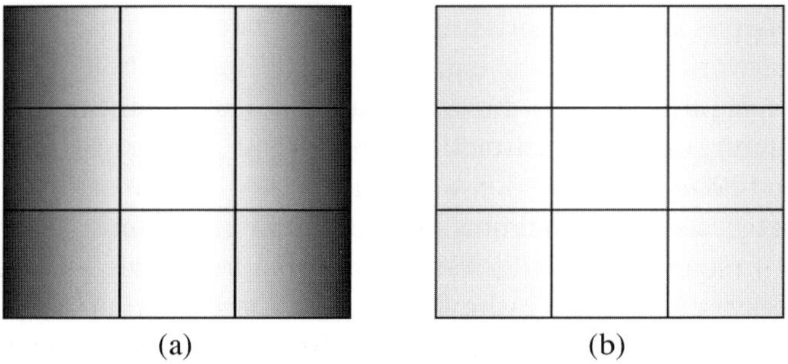

Fig. 1. Scale dependency in fixed kernel size based feature extraction. Darker color represents lower elevation and lighter color represents higher elevation
(**a**) Kernel can recognize the feature type as ridge and (**b**) Due to flattening at the ridge top, kernel recognizes it as a planar feature

The majority of existing ridge and channel feature extraction algorithms are primarily based on either the local elevation differences or the curvature (i.e., whether convex, concave and so on), over some terrain patch. Two popular examples of local elevation differences based algorithms include the Steepest Descent or D8 (Deterministic 8) algorithm (Peucker and Douglas 1975; O'Callaghan and Mark 1984) and, the comparatively advanced algorithm based on the conic section (Wood 1996). In the D8 algorithm, a cell is considered to be a candidate channel cell if it receives flow from an adjoining cell. A cell receives flow from another cell if its elevation is lower than the other cell (i.e., upslope cell). The total flow to each cell is then accumulated by following all the flow paths and the cells that receive flow from more than a certain number of upslope cells are classified as channel cells. Ridge cells are similarly derived but without a threshold condition. In the conic section algorithm the identification of ridge and channel cells is based on the sign of the quadratic determinant of a conic section polynomial, fitted over a kernel. A nil value for quadratic determinant indicates a parabolic type of conic section, which would occur at ridge and channel areas. In the D8 algorithm, the scale i.e., the geographic area, used in the classification is typically fixed to an equivalent area of 3 x 3 cells kernel. Thus, features whose extent is larger than 3 x 3 cells area could be missed, which makes the classification scale-dependent. On the other hand, the conic section algorithm allows the sampling of elevations over variable kernel sizes, hence it can be used for multi-scale feature visualization (Wood 1996). Some other local elevation differences based algorithms include the bilinear surface patches algorithm (Schneider and Wood 2004), various D8 algorithm variants (see e.g., O'Callaghan and Mark 1984; Band 1986; Skidmore 1990). Numerous other extensions of D8 algorithms have also been proposed that enforce drainage network consistency e.g., Random 8-node (Rho8) and DEMON stream tube algorithms. For more a detailed review of these drainage network modeling related extraction algorithms, refer to Gallant and Hutchinson (2000).

Curvature based ridge and channel extraction algorithms involve a combination of first and second derivatives of elevation, namely slope and curvature respectively. Figure 2 shows a part of channel around a point p and various relevant morphometric measures. Slope (d) is the maximum gradient at p. Aspect is the direction of maximum gradient. A number of curvature measures can be derived by intersecting the terrain surface along different planes (see Wood 1996; Gallant and Hutchinson 2000 for more details on types of curvature). Maximum (κ_{max}) and minimum (κ_{min}) curvatures are respectively the maximum and minimum curvature along any plane. Plan curvature (κ_{pl}) at the point p is the curvature of the line formed by the intersection of the terrain surface with a horizontal plane passing through p. Cross-section curvature (κ_{cr}) at the point p is the curvature of a line formed by the intersection of a plane tangential to the terrain surface (i.e., a plane that just touches the surface) with a plane passing through p, bounded by the normal and strike direction (direction orthogonal to aspect) of this tangential plane (similar to a tilted horizontal plane). It is a general convention to convert the plan and cross-section curvature values to negative and positive values to indicate convergent and divergent flows respectively. Therefore, ridges have positive κ_{pl} and κ_{cr} values and channels have negative κ_{pl} and κ_{cr} values. Numerous variants of this fundamental rule have been published widely in Computer Graphics (see e.g., Toriwaki and Fukumura 1978; Haralick 1983; Belyaev and Anoshkina 2005; Yoshizawa et al. 2005) and GIScience literature (e.g., Smith et al. 1990; Wood 1996). For instance, a simple algorithm by Wood (1996) proposes that a point is part of ridge if $d = 0$; $\kappa_{max} > 0$; $\kappa_{min} = 0$ or $d > 0$; $\kappa_{cr} > 0$ and a point is part of channel if $d = 0$; $\kappa_{max} = 0$; $\kappa_{min} < 0$ or $d > 0$; $\kappa_{cr} < 0$. Another variant, used more commonly in computer graphics discipline (see e.g., Fisher 1989), involves the maximum (κ_1) and minimum (κ_2) curvatures (called principal curvatures) derived by intersecting all the planes that contain the surface normal with the surface. A ridge (a convex cylinder) and a channel (a concave cylinder) both have $\kappa_2 = 0$ but $\kappa_1 > 0$ and $\kappa_1 < 0$ respectively. Like the conic section based algorithm, curvature based algorithms can be used over variable kernel sizes.

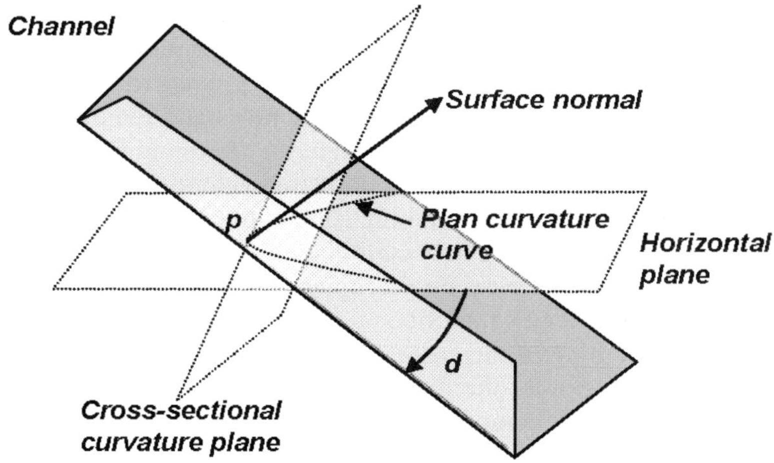

Fig. 2. A point *p* along a channel and the various morphometric measures, namely slope (*d*), plan curvature curve, surface normal and the cross-sectional curvature plane

In this we paper we propose a novel semi automated iterative algorithm, which employs the variation in plan curvature orthogonal to the ridge and channel axes as the basis for ridge and channel classification. In addition, we propose a statistical approach based on variations in the nominal counts of feature types during iterations, as a possible quantitative approach to evaluate and control the feature classification. The terms “features” and “topographic features” will be used for ridges and channels hence forth in the text.

2 Methodology

The overall algorithm consists of two main steps. Firstly, smoothing of the raster DEM and secondly, calculation of plan curvature and identification of topographic features. These steps are repeated until a desired feature classification has been achieved. The details on each of the steps above are given in the following sections.

2.1 Smoothing of Raster DEM

The removal of DEM noise is generally the first step in most feature extraction algorithms. In the case of basic steepest descent based algorithms, noise in DEM could lead to incorrect feature classification (see e.g., Wood

1996) and spurious pits (see e.g., Jenson and Domingue 1998). In the case of curvature-based algorithms, the dependence upon the surface derivatives necessitates a smoothing of DEM prior to feature extraction to avoid noise effects. A detailed discussion on the sources and removal of DEM noise is beyond the scope of this work. Refer to Martinoni (2002) for a recent review on DEM noise. Several smoothing techniques have been proposed in the literature ranging from the simple averaging to so-called feature preserving adaptive image averaging (Belyaev and Anoshkina 2005). This work uses a simple averaging technique built in the FOCALMEAN function of ArcInfo 9.0 GIS by ESRI.

Prolonged smoothing of the DEM also affects genuine topographic features by gradual erosion of the feature edges. This aspect of smoothing is commonly employed in computer graphics and vision, to study the multi-scale structure of surfaces (see e.g. Lindeberg 1994). This is unlike the DEM cell size resampling and variable kernel size approaches used in the GIScience (Wood 1996). To our knowledge, a comparison between these different types of techniques to study the multi-scale structure of terrains remains to be established. See section 3.1.1 for a hypothesis on the multi-scale structure revealed by prolonged smoothing.

2.2 Feature Classification

2.2.1 Plan Curvature

The computation of plan curvature is done within the ArcInfo 9.0. It is assumed that elevation, $z = f(x,y)$ and the plan curvature is derived by fitting the bi-variate quadratic polynomial[1] (see Eq. 1) over a 3x3 cells window with cell spacing l, shown below (ESRI 2005):

z_1	z_2	z_3
z_4	z_5	z_6
z_7	z_8	z_9

$$z = ax^2y^2 + bx^2y + cxy^2 + dx^2 + ey^2 + fxy + gx + hy + i, \quad (1)$$

$$a = [(z_1 + z_3 + z_7 + z_9) / 4 - (z_2 + z_4 + z_6 + z_8) / 2 + z_5] / l^4 \quad (2)$$

[1] It is based on the formula by Zevenbergen and Thorne (1987).

$$b = [(z_1 + z_3 - z_7 - z_9) / 4 - (z_2 - z_8) / 2] / l^3 \quad (3)$$

$$c = [(-z_1 + z_3 - z_7 + z_9) / 4 + (z_4 - z_6) / 2] / l^3 \quad (4)$$

$$d = [(z_4 + z_6) / 2 - z_5] / l^2 \quad (5)$$

$$e = [(z_2 + z_8) / 2 - z_5] / l^2 \quad (6)$$

$$f = (-z_1 + z_3 + z_7 - z_9) / 4l^2 \quad (7)$$

$$g = (-z_4 + z_6) / 2l \quad (8)$$

$$h = (z_2 - z_8) / 2l \quad (9)$$

$$i = z_5 \quad (10)$$

$$\kappa_{pl} = -2(dh^2 + eg^2 + fgh) / (g^2+h^2) * 100 \quad (11)$$

2.2.2 Classification Rules

The proposed algorithm extends the curvature-based algorithms with additional conditions. It is proposed here that a cell belongs to a ridge if $\kappa_{pl} > 0$; $e = \delta\kappa_{pl} / \delta t = 0$; $\delta e / \delta t > 0$; and a cell belongs to a channel if $\kappa_{pl} > 0$; $e = \delta\kappa_{pl} / \delta t = 0$; $\delta e / \delta t < 0$, where t is the direction orthogonal to feature axis. In other words, a ridge cell has a positive plan curvature and is at local maxima of plan curvature orthogonal to feature axis. In contrast, a channel cell has a negative plan curvature and is at local minima of plan curvature orthogonal to feature axis. Thus, the proposed algorithm evaluates a feature's entire extent, orthogonal to the feature axis. For simplicity in implementation and demonstration purposes, the current work assumes that ridge and channel feature axis is oriented along one of the four cardinal directions i.e., N-S, E-W, NW-SE, and NE-SW. Thus, the local maxima/ minima condition merely involves a comparison between the plan curvatures of the diagonally opposite cells with the plan curvature of the cell of interest. In summary, a cell is classified as a ridge/channel cell if it has positive/negative plan curvature and highest/lowest plan curvature value amongst any pair of diagonally opposite adjacent cells. Note that this definition will also classify the peak features as ridges and pit features as channels. Since saddle features have both ridge and channel morphology, they could be classified either as ridge or channel features depending upon the last conditional statement in the software which happens to evaluate the cell. The above anomalies are actually a useful side effect as they lead to better drainage network extraction. In the present demonstration, the proposed algorithm effectively combines the techniques of D8 and curvature based algorithms.

2.3 Experimental Setup

As mentioned earlier, the proposed algorithm has been developed in ArcInfo 9.0 using the Arc Macro Language (AML) script. A 3 x 3 cells size kernel has been used to classify DEM of three study areas, namely, Cairngorm area in Scotland (400 x 400 cells, 50 m resolution, Source: Ordnance Survey Landline, Fig. 3), Salisbury Hills in SW England (502 x 501 cells, 10 m resolution, Source: Ordnance Survey Landline, Fig. 4), and Round Mountain area in Nevada (451 x 485 cells, 30 m resolution, Source: USGS NED, Fig. 5). As can be seen from the figures, topography of these areas varies from gently rolling hills to incised cliffs. The robustness of the algorithm was particularly tested for the following factors:

- *Feature classification rule*: The feature classification of the proposed algorithm is compared with the relatively sophisticated conic section based algorithm available in the FEATURE NETWORK FUNCTION of LandSerf developed by Jo Wood (URL 1). For consistency, a 3 x 3 cells kernel is used in the LandSerf with no threshold criteria e.g., curvature- and slope- tolerance and distance decay. In addition, smoothed DEMs used for the proposed algorithm is also used for the LandSerf to avoid data bias. The assessment of the feature classification quality is based on a visual comparison with the morphological structure of the original DEMs. This is done so as to determine that a) feature classification was acceptable and b) effect of smoothing on feature preservation.
- *Cell size sensitivity*: It has been widely established that morphometric measures are dependent upon the DEM cell size (see e.g., Chang and Tsai 1991) hence three DEMS with different cell sizes (i.e., 10 m, 30 m, 50 m) are selected for the experiment.
- *Relief*: Due to the limited orientation and small size of the feature extraction kernel, it may suffer from scale-dependency issues. Hence the current study areas as shown in Figures 3–5 are chosen for their varied relief.

All the experiments are done on an IBM ThinkPad with 512 MB RAM and Pentium M 1.2GHz processor.

3 Results

Figures 3 to 5 show the feature classification derived from the proposed algorithm and LandSerf. The feature classifications based on original DEMs were significantly noisy. It was found out by trial and error that smoothing the DEMs 10 times produced the earliest most acceptable feature classification. As can be seen from figures the proposed algorithm accurately identifies most of the ridges and channels. The contrast between the outputs from the proposed algorithm and ones from LandSerf is self-evident. The proposed algorithm is able to identify several features unclassified by LandSerf. In addition, the proposed algorithm is able to localize the extent (i.e., thickness) of the feature more precisely despite being only limited to 4 cardinal directions. The computation time for each of the three study areas was approximately 1 minute.

3.1 Known Issues and Future Directions

Several interesting issues and aspirations arose during the experiments.

3.1.1 Effect of Smoothing on Feature Classification

In the proposed algorithm, smoothing of the DEM is one of the key factors that affect the spatial distribution of topographic features. As can be seen in Figures 3–5, each smoothing operation changes, and generally improves, the feature classification. A simple quantitative aspatial measure of such changes in feature classification is the ratio of number of feature cells to non-feature cells (henceforth referred as *feature content ratio*). A plot of the changes in *feature content ratio* between each consecutive smoothing would reveal how the feature classification evolves with varying smoothing (see Fig. 6). This plot reveals when the effect of smoothing appear to introduce (or stop having) marked effects on feature classification globally. Most notably,

- At the start of the smoothing, feature classification is noisy and *feature content ratio* shows a steep decline. However after around 5 iterations, the change in *feature content ratio* in all the curves varies gradually but fluctuates increasingly with more smoothing.

- The most striking feature of the curves is that all of them behave as power series with approximately the same R^2 values. This is quite intriguing and deserves further investigation as to whether this phenomenon is an artifact of the smoothing operator (since terrains themselves have dissimilar morphological structure) or whether it indirectly suggests something about the multi-scale structure of the terrain.

3.1.2 Limitations in feature classification

Admittedly, the present work leaves scope for future research, which would improve the appearance and applicability of the feature classification. Some of these desirable improvements, which are fairly straightforward to incorporate, include:

- Several of the spurious tiny ridges and channels seen in Figures 3–4 are partly due to the 3 x 3 cells kernel size, limited feature axis choices, and absence of any slope and curvature thresholds.
- The current demonstration doesn't include any post-processing to improve the presentation of the feature classification by thinning and smoothing the shape of the ridges and channels (see e.g., Yoshizawa et al. 2005) and making the classical interlocking ridge-channel network (Werner 1988) structure of drainage networks.

3.1.3 Using Feature Type Persistence for Classification

As seen in Figures 3–5, a DEM cell could be classified differently during a sequence of smoothing operations. However, DEM cells that do not carry noise may have the same feature type for several smoothing operations. Therefore, a possible method to derive a final feature classification could be to use the most *persistent* feature type assigned to each cell. Wood (1996) proposed a similar approach for feature classifications derived by varying kernel sizes.

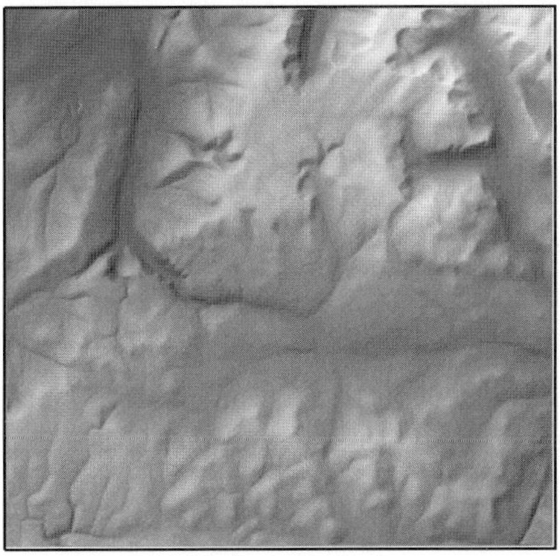

Hill shaded relief

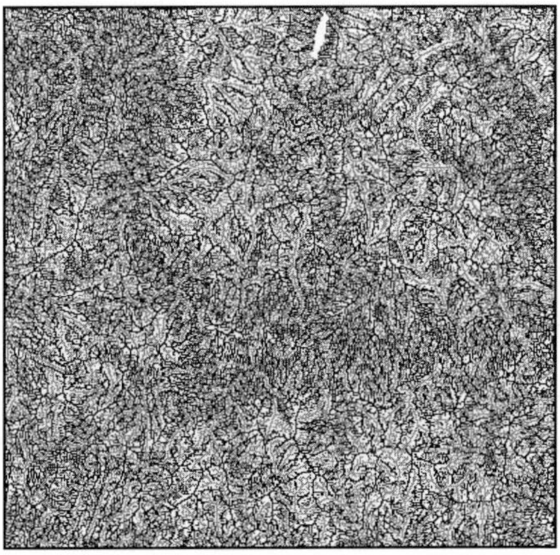

Feature classification without smoothing

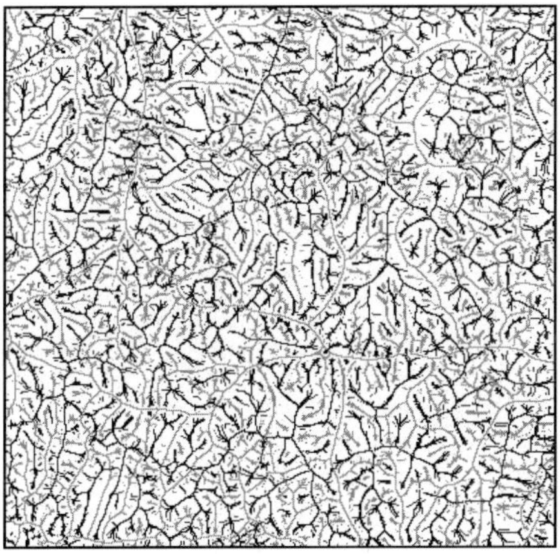

Feature classification after
10 smoothing iterations

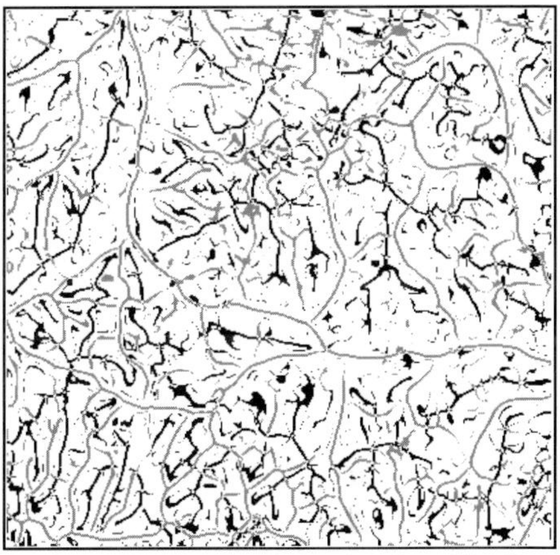

Feature classification using LandSerff,
after 10 smoothing iterations

Fig. 3. Hill shaded relief of the original Cairngorm DEM and the feature classifications. Black colored cells are ridges and gray colored cells are channels

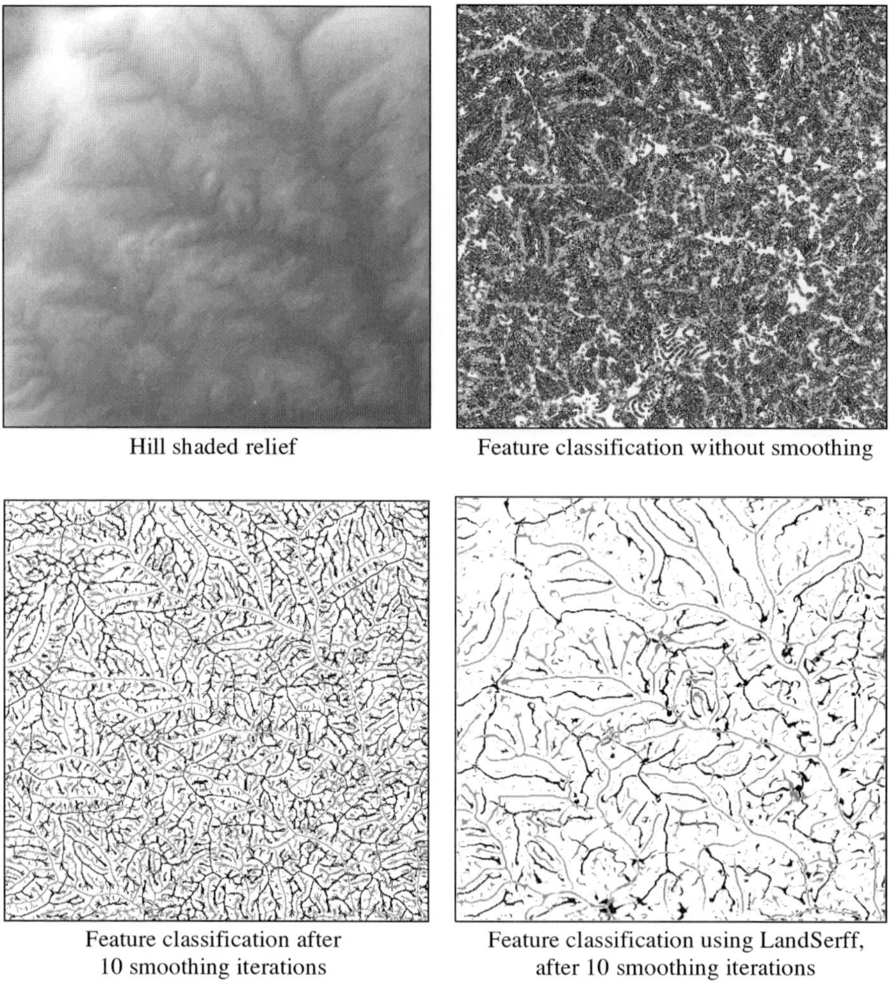

Fig. 4. Hill shaded relief of the original Salisbury DEM and the feature classifications. Black colored cells are ridges and gray colored cells are channels

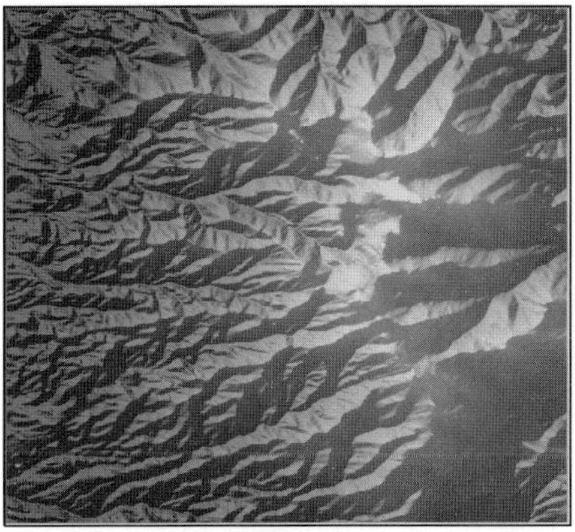

Hill shaded relief

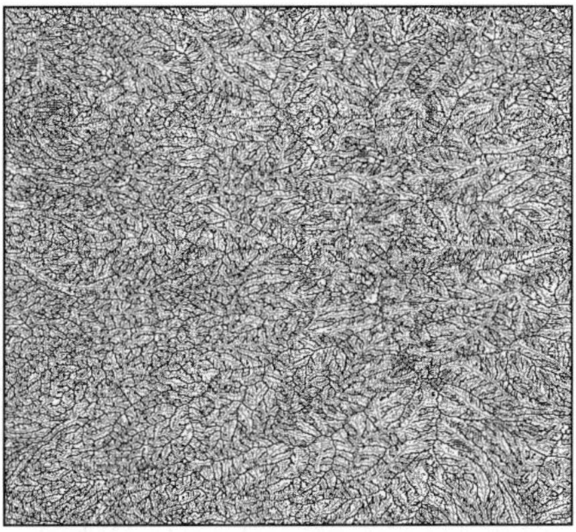

Feature classification without smoothing

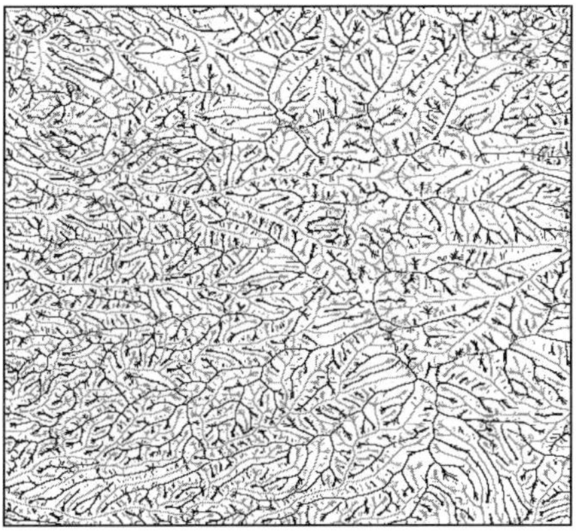

Feature classification after
10 smoothing iterations

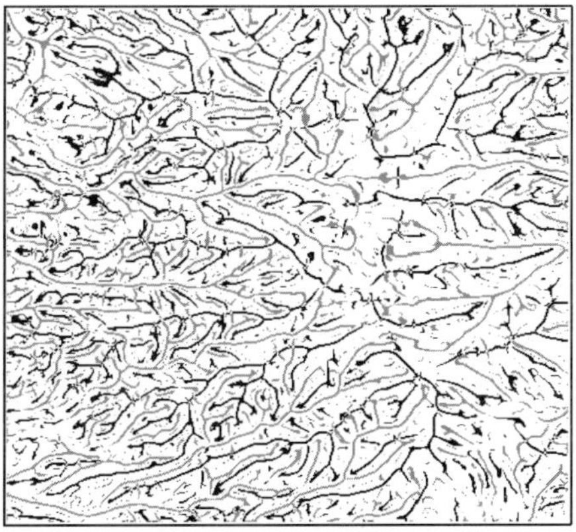

Feature classification using LandSerff,
after 10 smoothing iterations

Fig. 5. Hill shaded relief of the original Round Mountain DEM and the feature classifications. Black colored cells are ridges and gray colored cells are channels

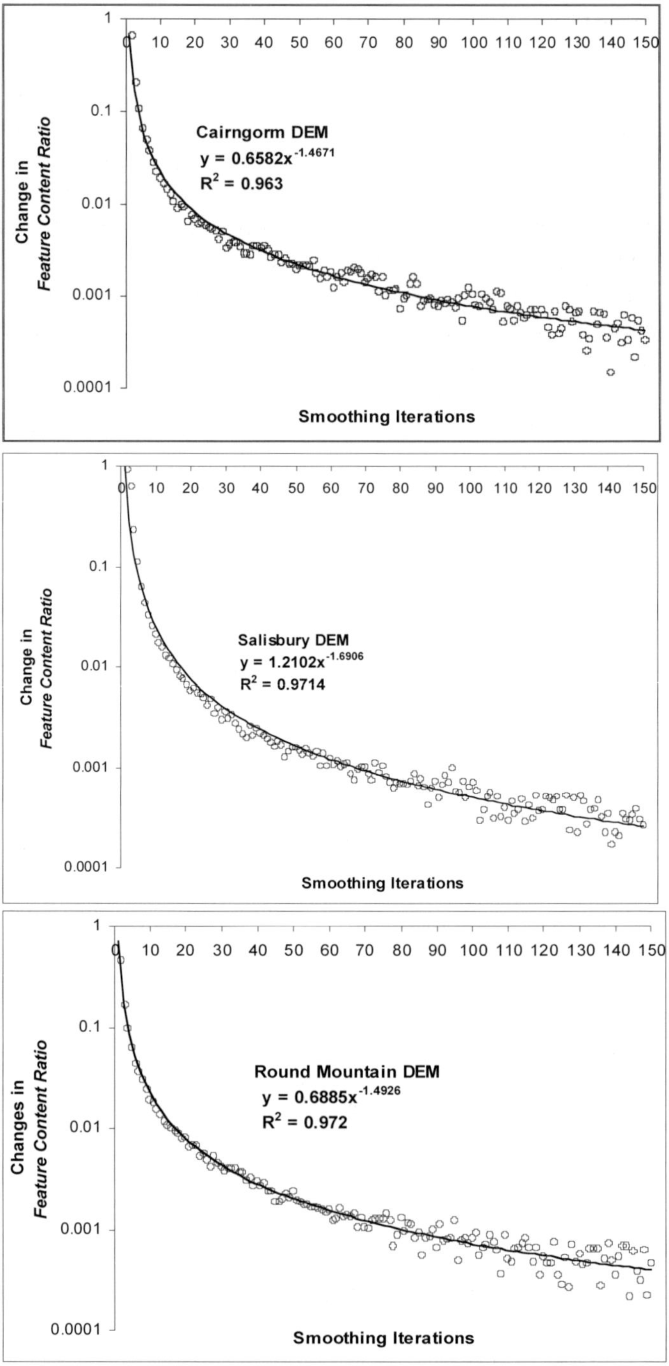

Fig. 6. Variation in the f*eature content ratio* with smoothing iterations

4 Conclusions

The paper presented an algorithm for detecting ridges and channels features in raster DEMs. It is proposed that ridge and channel features have positive and negative plan curvature respectively with descending plan curvature orthogonal to the feature axis from the higher to the lower reaches of the feature. The proposed algorithm has been demonstrated on three different areas with differing morphology and cell sizes. Despite its simplicity, the algorithm produced markedly superior outputs than a comparatively sophisticated feature extraction algorithm based on conic section modeling of terrain. The paper also evaluated the effect of smoothing on feature classification by plotting the change in the ratio of number of feature cells to non-feature cells. This *feature content ratio* interestingly behaves as a power series.

Acknowledgements

The Ordnance Survey funded part of the research work and also provided the Salisbury DEM. Cairngorm DEM was downloaded by Young-Hoon Kim from the EDINA service. Both Salisbury and Cairngorm DEMs are Crown Copyrighted and were provided by the Ordnance Survey for purposes of research. The Round Mountain DEM was downloaded from the Seamless Data Distribution System of the USGS.

References

Band LE (1986) Topographic partition of watersheds with digital elevation models. Water Resources Research 22:15–24

Belyaev A, Anoshkina E (2005) Detection of surface creases in range data. In: 11th IMA Conf on the mathematics of surfaces, Loughborough, UK

Chang K-T, Tsai BW (1991) The effect of DEM resolution on slope and aspect mapping. Cartography and Geographic Information Systems 18:69–77

DeMers MN (2002) GIS Modeling in Raster. John Wiley & Sons, New York

ESRI (2005) ArcInfo GIS help on CURVATURE function

Fisher RB (1989) From surface to objects: computer vision and three dimensional scene analysis. John Wiley & Sons, Chichester

Gallant JC, Wilson JP (2000) Primary topographic attributes. In: Wilson JP, Gallant JC (eds) Terrain analysis: Principles and Applications. John Wiley & Sons, New York, pp 51–85

Haralick RM (1983) Ridge and valley on digital images. Computer Vision, Graphics and Image Processing 22:28–38
Jenson, SK, Domingue JO (1998) Extracting topographic structure from digital elevation data for geographic information system analysis. Photogrammetric Engineering and Remote Sensing 54:1593–1600
Lindeberg T (1994) Scale-space theory in computer vision. Kluwer Academic Publishers, Dordrecht, The Netherlands
Martinoni D (2002) Models and experiments for quality handling in digital terrain modeling. Unpublished PhD Thesis, University of Zurich, Zurich, Switzerland
O'Callaghan JF, Mark DM (1984) The extraction of drainage networks from digital elevation data. Computer Vision, Graphics and Image Processing 28: 323–344
Okabe A, Masuyama M (2004) A method for measuring structural similarity among activity surfaces and its application to the analysis of urban population surfaces in Japan. In: Rana S (ed) Topological data structures for surfaces: an introduction to geographical information science. John Wiley & Sons, Chichester, pp 105–120
Peucker TK, Douglas DD (1975) Detection of surface-specific points by local parallel processing of discrete terrain elevation data. Computers Graphics and Image Processing 4:375–387
Rana S (2003) Fast approximation of visibility dominance using topographic features as targets and the associated uncertainty. Photogrammetric Engineering and Remote Sensing 69:881–888
Schneider B, Wood J (2004) Construction of metric surface networks from raster-based DEMs. In: Rana S (ed) Topological data structures for surfaces: an introduction to geographical information science. John Wiley & Sons, Chichester
Scott P (2004) An application of surface networks in surface texture. In: Rana S (ed) Topological data structures for surfaces: an introduction to geographical information science. John Wiley & Sons, Chichester, pp 157–166
Skidmore AK (1990) Terrain position as mapped from a gridded digital elevation model. Int J of Geographical Information Systems 4:33–49
Smith TR, Zhan C, Gao P (1990) A knowledge-based, two step procedure for extracting channel networks from noisy DEM data. Computers & Geosciences 16:777–786
Toriwaki J, Fukumura T (1978) Extraction of structural information from grey pictures. Computer Graphics and Image Processing 7:30–51
URL 1 (2005) LandSerf. http://www.landserf.org/ (accessed on 27 Dec 2005)
Werner C (1988) Formal analysis of ridge and channel patterns in maturely eroded terrain. Annals of the American Association of Geographers 78:253–270
Wood J (1996) The geomorphological characterization of digital elevation models. Unpublished PhD Thesis, University of Leicester, UK
Yoshizawa S, Belyaev A, Seide H-P (2005) Fast and robust detection of crest lines on meshes. In: ACM Symp on solid and physical modeling, Cambridge, USA
Zevenbergen LW, Thorne CR (1987) Quantitative analysis of land surface topography. Earth Surface Processes and Landforms 12:47–56

An Evaluation of Spatial Interpolation Accuracy of Elevation Data

Qihao Weng

Department of Geography, Geology, and Anthropology,
Indiana State University, Terre Haute, IN 47809, USA
email: qweng@indstate.edu

Abstract

This paper makes a general evaluation of the spatial interpolation accuracy of elevation data. Six common interpolators were examined, including Kriging, inverse distance to a power, minimum curvature, modified Shepard's method, radial basis functions, and triangulation with linear interpolation. The main properties and mathematical procedures of the interpolation algorithms were reviewed. In order to obtain full evaluation of the interpolations, both statistical (including root-mean-square-error, standard deviation, and mean) and spatial accuracy measures (including accuracy surface, and spatial autocorrelation) were employed. It is found that the accuracy of spatial interpolation of elevations was primarily subject to input data point density and distribution, grid size (resolution), terrain complexity, and interpolation algorithm used. The variations in interpolation parameters may significantly improve or worsen the accuracy. Further researches are needed to examine the impacts of terrain complexity in details and various data sampling strategies. The combined use of variogram models, accuracy surfaces, and spatial autocorrelation represents a promising direction in mapping spatial data accuracy.

Key words: spatial interpolation, grid digital elevation models, accuracy, statistical measures, spatial measures

1 Introduction

The methods of spatial interpolation have long been used in cartography and geography. In recent years, the topic of spatial interpolation has attracted great attention within the geographic information systems (GIS) community (Burrough 1986; Flowerdew and Green 1994). An assessment of accuracy of different interpolators is needed, because spatial data uncertainty introduced by spatial interpolation and the propagation of this uncertainty through GIS analyses will inevitably influence the quality of any decision-making supported by spatial data. Although various research endeavors have been devoted to studies of spatial data uncertainty (NCGIA 1989; Goodchild and Gopal 1989; Goodchild 1993; Hunter et al. 1995; Li 1998; Weng 2002), comprehensive evaluations of spatial interpolation accuracy of elevation data have not been conducted sufficiently.

Evaluation of spatial interpolation accuracy requests comparing the original elevation data with the interpolated elevation data. Such a comparison will result in height differences (or residuals) at the tested points. To quantify and analyze the pattern of deviation between the two sets of elevation data, conventional ways are to yield statistical expressions of the accuracy, such as in the form of root mean square error, standard deviation, and mean. Because the residuals vary spatially, a better understanding of spatial interpolation accuracy necessitates using spatial measures, such as accuracy surface. This paper attempts to undertake a comparative analysis of the performance of some widely used interpolation methods. This analysis will make use of both conventional statistical measures and spatial accuracy measures. A particular attention will be paid to the effects of variations in the interpolation parameters on spatial data accuracy.

2 The Spatial Interpolation Algorithms

Six commonly used spatial interpolation algorithms are examined and compared in this study. These interpolators can be used to convert randomly spaced data points into regularly gridded data points. They may be classified as either an exact or approximate algorithm, depending upon whether or not the original data points are preserved on the interpolated surface (Wren 1975; Lam 1983). The exact interpolators honor the data points on which the interpolation is based, such as ordinary Kriging, inverse distance, radial basis function, modified Shepard's method, and triangulation with linear interpolation. On other hand, an approximate interpolator is applied when there is some uncertainty about the original data points, such as inverse distance with smoothing, minimum curvature, and Shepard's smoothing method.

Kriging is a least-squares spatial interpolation method and a weighted-average estimator whose weights are functions of spatial covariance (Carr 1995). A notion of spatial covariance for a particular spatially distributed phenomenon is developed from a function known as a variogram (Carr 1995). Some commonly used variogram models are listed below:

Exponential:

$$\gamma(h) = C\,[1 - e^{-h}] \tag{1}$$

Linear:

$$\gamma(h) = C\,h \tag{2}$$

Quadratic:

$$\gamma(h) = \left\{ \frac{C\left[2h - h^2\right]}{C} \frac{0 \le \mathrm{h} \le 1}{\mathrm{h} > 1} \right\} \tag{3}$$

Spherical:

$$\gamma(h) = \left\{ \frac{C\left[1.5h - 0.5h^3\right]}{C} \frac{0 \le \mathrm{h} \le 1}{\mathrm{h} > 1} \right\} \tag{4}$$

where C is the scale for the structured component of the variogram; and h is the anisotropically rescaled, relative separation distance (Isaak and Srivastava 1989). In a typical variogram model, the length ("range" in the spherical and quadratic models) parameter defines how rapidly the variogram components change in value as distance of separation between observations increases. With the exception of the linear variogram model, which does not have a sill, the scale parameter (C) defines the sill for a variogram model. Therefore, the sill of a variogram model equals to the nugget effect plus the scale C. In most situations, the sill accounts for the covariance of the observed data (Carr 1995). A nugget effect is specified when there are potential errors in the collection of data.

Inverse distance algorithm is a weighted average interpolator. Unknown points are estimated in such a way that the influence of one data point relative to another declines with an increasing distance from the grid node. Weighting is assigned to data points by the use of a weighting power. The greater the weighting power, the less effect the points far from the grid node have during the interpolation. The mathematical expression of inverse distance algorithm is as follows (Davis 1986):

$$Z = \frac{\sum_{i=1}^{n}\left[Z_i/\left(h_{ij}+\delta\right)^{\beta}\right]}{\sum_{i=1}^{n}\left[1/\left(h_{ij}+\delta\right)^{\beta}\right]} \tag{5}$$

where Z is the interpolated point value, Z_i the neighboring data point, h_{ij} the distance between the grid node and data point, β the weighting power, and δ the smoothing parameter. The value of β may range from zero to infinity, but typically falls between 1 and 3. As β increases, the interpolated surface tends to be a polygonal one, in which polygons represent the nearest observation to the interpolated grid node. The use of a smoothing parameter δ aims to minimize the overwhelming influence of any particular data point, so that no single point is given a weighting factor equal to 1. Inverse distance without a smoothing parameter is fast but has the tendency to generate "bull's-eye" patterns of concentric contours around the data points.

Radial basis functions refer to a group of exact interpolators. Analogous to variograms in kriging, the functions define the optimal set of weights to apply to the data points in interpolating a grid node. Some of the most widely used functions are multiquadratic, inverse multiquadric, multilog, natural cubic spline, and thin plate spline. Because the multiquadratic function produces the most desirable result for most data sets, it is employed in this study. The formula for multiquadratic method is expressed as:

$$B(h) = \sqrt{(h^2 + R^2)} \tag{6}$$

Where h is the anisotropically rescaled, relative distance from the point to the node, and R^2 is the smoothing parameter. There is no optimal value for R^2. It is suggested that a reasonable trial value for R^2 should lie in between the average sample spacing and one-half the average sample spacing (Golden Software Inc. 1999). The larger the R^2 value, the smoother the terrain looks.

Shepard's interpolation algorithm uses an inverse distance weighted least squares method. As such, it is similar to the inverse distance interpolator. Because of its incorporation of local least squares, this method has the advantage of not generating a "bull's-eye" appearance over the inverse distance interpolator. The weighting is directional (Shepard 1968). In other words, when two points are closer, the weight for each point should be smaller (Wang 1990). The application of a smoothing parameter in the Shepard's method has the effect of generating a smoother terrain appearance. However, in order to get a practical interpolation result, the value of the smoothing parameter should be varied between zero and one (Golden Software Inc. 1999).

Triangulation with linear interpolation often usually uses Delaunay triangulation. The rules for creating Delaunay triangles are to connect each original point with its nearest neighboring points so that no triangle edges are intersected by other triangles. In other words, a triangulation of a set of points is a Delaunay triangulation if, and only if, the circumcircle of any of its triangles does not contain any other point in its interior (Weibel and Heller 1991). Each triangle is regarded as a local area, and an interpolation surface will then be fitted to each local area (Wang 1990). The triangulation method is an exact interpolator. The original data is honored closely. Linear interpolation is the simplest method for fitting a surface with Delaunay triangles. It works effectively with a moderate amount of evenly distributed data points, and is good at preserving break-lines.

Minimum curvature is an approximate interpolator. Minimum curvature algorithm uses the following equation in gridding (Smith and Wessel 1990):

$$C = \iint (\nabla^2 Z)^2 \, dx \, dy \tag{7}$$

The equation (7) is a valid approximation for the total curvature of Z when $|\nabla Z|$ is small. The Laplacian operator, $\nabla^2 Z$, is defined in multivariable calculus by:

$$\nabla^2 Z = \frac{\partial^2 Z}{\partial X^2} + \frac{\partial^2 Z}{\partial Y^2} \tag{8}$$

The implementation of the Laplacian operator in the SURFER program generates a grid using a standard five-point central difference formula. By repeatedly applying the equation over the grid, the minimum curvature algorithm aims to generate a smooth surface while honoring data points as closely as possible.

Each pass over the grid is counted as iteration. The grid node values are recalculated every time until successive changes in the values are less than the maximum residuals value set, or the maximum number of iterations has been reached. The maximum residuals parameter has the same unit as the original data, and the value (*MR*) may be set by the following formula:

$$MR = 0.001 * (Z_{max} - Z_{min}) \tag{9}$$

Where Z_{ma} and Z_{min} are the maximum and minimum value of original data points. The maximum number of iterations should be set at one to two times the number of grid nodes generated.

3 Data and Methods

3.1 Study Site Selection and Data Processing

To test the performance of the spatial interpolators, three test sites were selected from 7.5-minute (1:24,000) USGS topographic maps that have the contour interval of 20 feet. Each covered a 13,000 by 16,000 feet rectangular area. Site 1 is around Winterville, Georgia, where plain is the dominant type of topography. Site 2 centers in Stone Mountain, Georgia, a hill of moderate relief composed of granite bedrock. Site 3 covers Little River Canyon and the immediate proximities in Alabama. These sites represent different degree of terrain complexity from simple to most complicated, as measured by the standard deviation of elevations (see Table 1). The data capture process involved using the CAPTURE (R-WEL Inc. 1996) and Didger software (Golden Software Inc. 2001). A data set of 400 points was generated for each site, which were distributed randomly throughout the test areas. Another set of 400 randomly selected checkpoints were obtained and used for accuracy assessment. Surfer 7.0 (Golden Software Inc. 1999) was used for spatial interpolation. Both statistical and spatial accuracy measures were computed to reveal the closeness of interpolated surfaces to the reality.

Table 1. Topographic characteristics of the three test sites (Unit: feet)

Study Area	Terrain Type	Maximum Elevation	Minimum Elevation	Mean Elevation	Relative Relief	Standard Deviation of Elevation
Winterville, GA	plain	800	620	731	180	38
Stone Mountain, GA	hill	1683	800	976.79	883	205.93
Little River Canyon, AL	valley	1260	600	1016.2	660	270.69

3.2 Measuring the Accuracy of Interpolated Surfaces

To measure the statistical accuracy of interpolated surfaces, the mean, standard deviation, and root mean square error of the residuals were calculated for each interpolated surface. The mean of residuals is defined as the arithmetic average of residual values, mathematically expressed as:

$$\mathrm{d}z_{dr} = \frac{1}{n}\sum_{i=1}^{n}(z_{di} - z_{ri}) \tag{10}$$

The root mean square error (RMSE) measures the frequency distribution of deviations between an original elevation data and the interpolated data, and is mathematically expressed as:

$$RMSE_z = \sqrt{\frac{1}{n-1}\sum_{i=1}^{n}(z_{di} - z_{ri})^2} \tag{11}$$

The standard deviation is another commonly used aspatial measure for mapping accuracy, and is computed as follows:

$$\delta = \sqrt{\frac{1}{n-1}\sum_{i=1}^{n}[(z_{di} - z_{ri}) - dz_{dr}]^2} \tag{12}$$

In Equations (10) to (12), Z_{di} is defined as the ith elevation value measured on the interpolated surface, Z_{ri} the corresponding original elevation, dz_{dr} the mean deviation between the original and interpolated surface, and n the number of elevation points checked. Clearly, if $dz_{dr} = 0$, then $\delta = RMSE_z$, and the root mean square error can be replaced with the standard deviation of residuals.

In addition to statistical measures discussed above, it is often desirable to have a description and presentation showing the magnitude and spatial pattern of errors. Many authors have suggested that the distribution of errors in elevation models shows some forms of spatial patterns (Guth 1992; Li 1993; Wood and Fisher 1993; Monckton 1994). In this study, a spatial autocorrelation index, Moran's I statistic, was employed to quantify the magnitude of errors in each interpolated surface, which is mathematically expressed as:

$$I = \frac{n\sum_{u=1}^{n}\sum_{v=1}^{n} w_{uv}(dz_u - dz_{dr})(dz_v - dz_{dr})}{\sum_{u=1}^{n}(dz_u - dz_{dr})^2 \sum_{u=1}^{n}\sum_{v=1}^{n} w_{uv}} \tag{13}$$

Where dz_u is the deviation between the original and interpolated surface for each cell, dz_v the deviation between the original and interpolated surface for some neighboring cells, w_{uv} the weighting given to neighboring cells, and dz_{dr} the mean deviation between the original and interpolated surface. The Moran's index provides only a global description of spatial association, but fails to account for the impact of different spatial scales on the degree of clustering of errors (Monckton 1994; Wood 1996).

The best way to spatially observe and analyze the errors resulted from interpolation is to have a graphical representation – creating an accuracy surface. This representation has the advantage of clearly indicating where serious and perhaps anomalous errors occur. Comparison of such surfaces, for example, with a plot of the original input contours, can be extremely informative with respect to the occurrence and magnitude of errors in relation to such factors as the terrain slopes and distribution of input data (Shearer 1990). Scientific visualization of accuracy surfaces can further reveal characteristics and patterns that are not reflected in statistical measures (Ehlschlaeger and Goodchild 1994; Wood 1996).

4 Results

4.1 Accuracy of DEMs Generated by Various Interpolators

In assessing the accuracy of various interpolation algorithms, the same grid size input, i.e., 50 meters, was used. Table 2 shows the results of statistical analysis of residuals calculated for the three test sites. Among the six interpolators, Kriging (linear variogram, $C = 292$, A = 5.7) and Radial Basis Function (multiquadratic function, $R^2 = 0.01$) generated nearly identical accuracy in all test sites, as measured by RMSE, standard deviation, and mean. Triangulation with linear interpolation algorithm provided the worst interpolation, with the largest values of RMSE and standard deviation. Except for Test Site 2 (which has a modest mean value), this interpolator also generated the highest value of mean. The contour maps created by this algorithm exhibited distinct triangular faces, indicating that too few data points were used for the interpolation. Both Modified Shepard's Method (no smoothing) and Inverse Distance algorithm (ß = 3; δ = 0) generated a reasonably high statistical accuracy, comparable to Kriging and Radial Basis Function. If RMSE and standard deviation are the only measures, then the Modified Shepard's Method was the best interpolator for test sites 2 (RMSE: 3.9448; SD: 3.9338) and 3 (RMSE: 5.0999; SD: 5.0993). For test site 1, this method was a close second (RMSE: 5.7853; SD: 5.7849), next to the Inverse Distance algorithm (RMSE: 5.2385; SD: 5.2378). However, both Modified Shepard's Method and Inverse Distance algorithm tended to produce a "bull's eye" pattern. Minimum Curvature (max residual = 0.08, max iteration = 10,000) produced acceptably smooth interpolated surfaces, but its statistical accuracy was low. Indeed, it was the second worst interpolator in terms of RMSE and standard deviation in all test sites.

Three test sites had distinct statistical accuracy (with respect to RMSE and standard deviation), whatever the interpolation algorithm was selected. Site 1 always possessed the highest, Site 2 the second, and Site 3 the lowest accuracy. The only exception was observed with the Modified Shepard's Method, which produced the highest accuracy for Site 2. The difference in statistical accuracy in the three sites is mainly attributable to terrain complexity, because Site 1 has the lowest complexity while site 3 the highest. This finding is in agreement with Gao's (1997) work, in which he found that DEM accuracy was inversely associated with terrain complexity by a gridded DEM of the same resolution (grid size).

Table 2. Summary of the residuals calculated by different interpolation algorithms

Algorithmus	**Test Site 1**			**Test Site 2**			**Test Site 3**		
	RMSE	Standard Deviation	Mean	RMSE	Standard Deviation	Mean	RMSE	Standard Deviation	Mean
Kriging with linear variogram	6.23	6.23	-0.13	6.66	6.66	0.35	8.67	8.67	-0.22
Inverse distance squared	5.24	5.24	0.11	6.84	6.82	-0.06	7.16	7.16	-0.01
Minimum curvature	6.32	6.32	0.03	8.69	8.69	0.31	10.35	10.35	0.12
Modified Shepard's method	5.79	5.78	0.10	3.94	3.93	0.46	5.10	5.10	0.24
Radial basis functions	6.23	6.23	-0.13	6.66	6.66	0.35	8.67	8.67	-0.22
Triangulation with linear interpolation	6.92	6.92	-0.22	9.54	9.54	0.14	13.65	13.65	-1.17

4.2 Impact of Grid Size (GS)

The effect of grid size on the accuracy of interpolation was examined by comparing ten different resolution surfaces ranging from 10 to 100 meters using Kriging (with a linear variogram model) as the interpolator. Analysis of the residuals was first done by plotting RMSE vs. grid size (see Fig. 1). Clearly, as grid size became larger, the RMSE of the residuals in all test sites increased. In other words, the accuracy of interpolation was improved with reduction of grid size. The RMSE values were further regressed against grid size in order to reveal their apparently linear relationship. The linear models generated are all significant at 0.05 level. The coefficient of determination, r^2, was 0.9870, 0.9986, and 0.9973 respectively for the test sites. In terms of the distribution of regression residuals, there was no clear pattern of underestimates and overestimates. The residual becomes minimal around 70 meters. Moreover, the changing rate of RMSE was different in each site. A higher rate was accompanied with a more complex terrain. This finding confirms the proposition that densely observed points should be placed for rugged terrain in order to achieve the desired interpolation accuracy (Ackermann 1994, 1996).

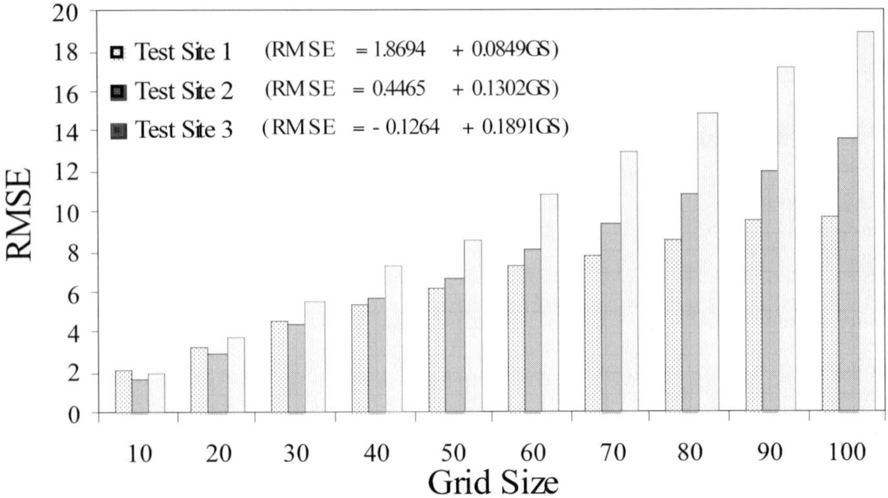

Fig. 1. Influence of grid size on interpolation accuracy

4.3 Smoothing Parameter (R^2) and Statistical Accuracy

Figure 2 shows the relationship between the RMSE of residual values and the smoothing parameter R^2 of the Radial Basis Function algorithm. It is found that the RMSE values increased as the value of R^2 increased with all types of terrains. In other words, a higher value of the smoothing parameter would result in lower interpolation accuracy. The increasing rate of RMSE differed, however, from one type of terrain to another. Test Site 3 (valley) displayed the highest rate, followed by Site 2 (hill) and Site 1 (plain). Therefore, the parameter should be used with caution if a high level of fidelity of representation is asked for, especially when a more complex terrain is mapped. Given the relationship between RMSE (Y) and the R^2 *(X)*, a linear regression model was built for each test site. These regression models are significant at 0.05 level. The coefficient of determination, r^2, is 0.9363, 0.9701, and 0.9737 respectively, suggesting that a closer correlation can be found with a more complex terrain. In terms of residual output, the linear models systematically overestimated the values of RMSE when the smoothing parameter was between 100 and 1000, but underestimated when it was less than 100 or larger than 1000. The regressed values emulated best the observed RMSE values at a R^2 value of 100. This optimal value related to average sampling spacing, and may be used as a reference for other projects that would use 1:24,000 USGS topographic maps and similar data point density.

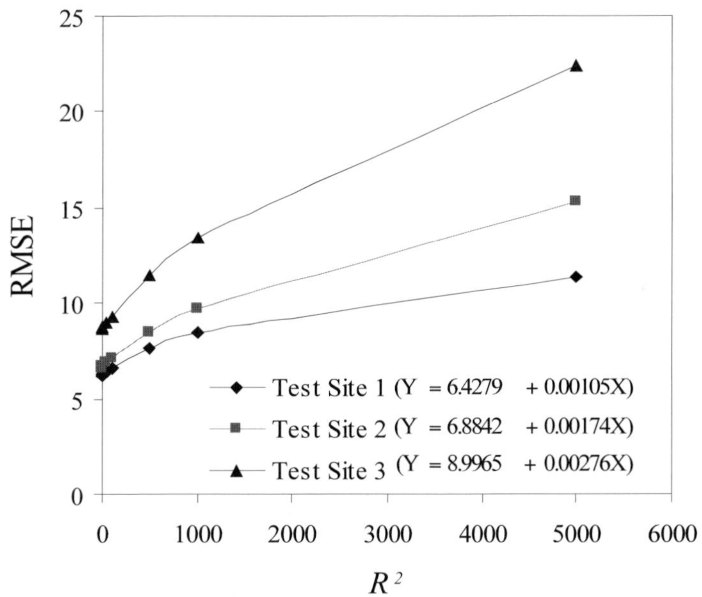

Fig. 2. Relationship between smoothing parameter R^2 and interpolation accuracy

4.4. Shepard's Smoothing Parameter (SSP) and Statistical Accuracy

In order to examine the impact of the SSP on interpolation accuracy, the RMSEs were calculated at ten different levels of smoothing. Figure 3 shows the relationship between the SSP and RMSE. Apparently, the values of RMSE increased dramatically with larger SSP. Using the RMSEs as the dependent variables and the SSPs as the independent variables, both linear and non-linear regression models were built for the three test sites. The non-linear models for all test sites provide greater explanation power in the light of the coefficient of determination, r^2. They improve the explanation power by 18.72%, 48.22%, and 38.55% respectively over the linear models. Three curves in Figure 3 are, though a little different, well fitted by a commonly observed logarithmic relationship: double-log functions. However, the non-linear models systematically underestimated the RMSEs with the SSP equal to 0.1 or from 0.8 to 1.0, but overestimated them between 0.2 and 0.7. A minimal disparity between the predicted and the observed RMSE was obtained at a SSP value of 0.7, which was the optimal smoothing value for all testing sites.

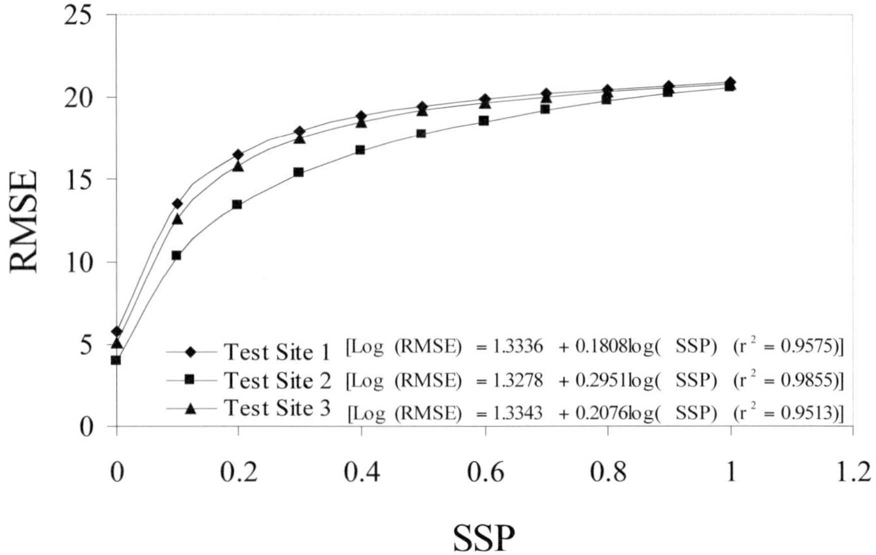

Fig. 3. Relationship between Shepard's smoothing parameter and interpolation accuracy

4.5 Weighting Power *(ß)* and Statistical Accuracy

Examination of the impact of weighting power on the accuracy of Inverse Distance interpolation focused on the β range from 1 to 3. Figure 4 show the results of this examination. It is clear that the RMSEs of residuals decreased as the weighting power β became larger. This is because larger weighting power generated a quicker falling off effect in weights with the distance from the grid node. In fact, as β continuously increased, the interpolation moved closer and closer to be a "nearest neighbor" interpolator, and the resulting surface would become polygonal, i.e., Thiessen polygons. In order to quantify the relationships, RMSE was regressed against β both linearly and non-linearly. The results indicate that the coefficient of determination, r^2, is consistently higher in the non-linear models, indicating logarithmic functions fit better the observed patterns of residuals. These regression models are all significant at 0.05 level, but the non-linear models improve the explanation power by 13.23%, 13.18%, and 14.93% respectively. Analysis of regression residuals indicates that these models underestimated the RMSEs when β ranges from 2.0 to 2.5, but overestimated them when β larger than 2.8. There was no optimal value of β found for all tested sites, but the range between 1.4 and 2.0 should give an intensive trial. Finally, the decreasing rate of RMSE with large β varied with the test sites. The more complex terrain, the larger the decreasing rate. In other words, more complex terrain was more sensitive to the changes in the weighting power.

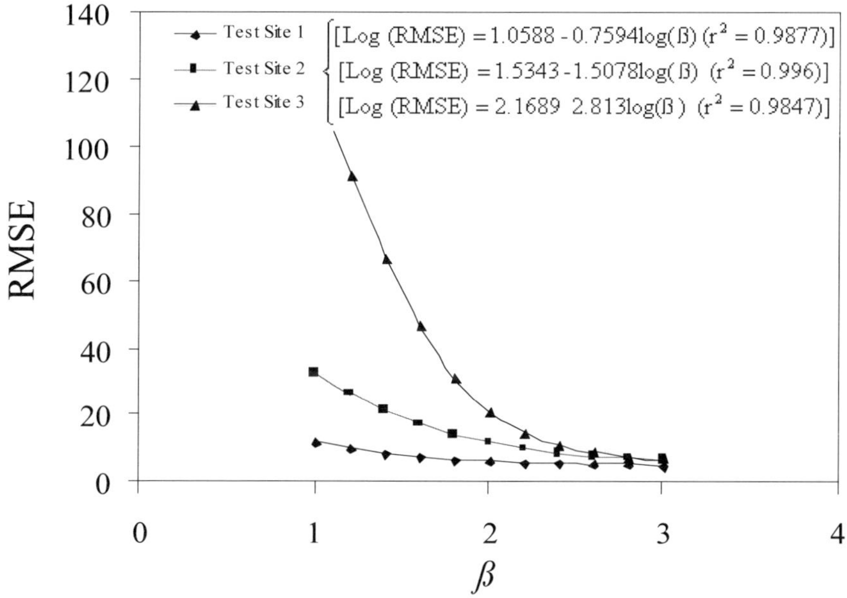

Fig. 4. Relationship between the weighting power (β) and interpolation accuracy

4.6 Spatial Distribution of the Residuals

An accuracy surface was created for every interpolation in the three test sites (see Figs. 5–7). The spatial patterns of error distribution can be visually examined from these maps, in which the magnitude of the residuals resulted from interpolation was represented by means of contours. It is clear from these maps that errors tended to cluster in the rugged areas where elevation changed rapidly. To understand the degree of error clustering in each accuracy surface, the Moran's index was computed. The resultant Table 3 shows that all surfaces had a value of the index larger than 0.8, indicating a reasonably high degree of clustering. A close look at this table, however, reveals many hidden differences in the pattern of error clustering in the surfaces. First, Test Site 1 had generally a lower value of the index (mean: 0.85), compared to Test Sites 2 (mean: 0.89) and 3 (mean: 0.88). This implies that the clustering of errors had something to do with the terrain complexity of a test site. The more complex the test site, the higher degree of error clustering it tended to be. Second, some interpolators did a better job than others in terms of revealing systematic errors, which were related to the rugged terrains. Both Kriging with Linear Variogram and Radial Basis Function stood out in this regard. Table 3 shows that the Moran's index had a larger value with a more complex test site (Site 1: 0.83; Site 2: 0.89; and Site 3: 0.90). Finally, a correlation analysis between the RMSE and the Moran's index gave a multiple r = 0.58, indicating that statistically better-interpolated surfaces did not necessarily result in less error clustering.

Table 3. Moran's index of the residuals of different interpolators

Interpolator	Test Site 1	Test site 2	Test Site 3
Kriging with linear variogram	0.83	0.89	0.90
Inverse distance squared	0.85	0.89	0.87
Minimum curvature	0.88	0.89	0.89
Shepard's method	0.86	0.84	0.83
Radial basis function	0.83	0.89	0.90
Triangulation with linear interpolation	0.88	0.90	0.88

5 Discussion and Conclusions

Spatial interpolation accuracy is an important topic in geographic information science. This paper has provided an examination on two fundamental issues that relate to it: (1) the performance of spatial interpolation algorithms, and (2) the methods for reporting the accuracy. The accuracy of spatial interpolation of elevations was found subject to many factors, but primarily to input data point density and distribution, grid size (resolution), terrain complexity, and interpolation algorithm used. The accuracy of interpolation improved with higher resolutions, but inversely correlated to terrain complexity. Denser input data points were needed to represent higher degree of terrain complexity. Errors tended to cluster in the rugged areas of a test site. There was no optimal interpolator for all test sites. The variations in interpolation parameters may significantly improve or worsen the accuracy.

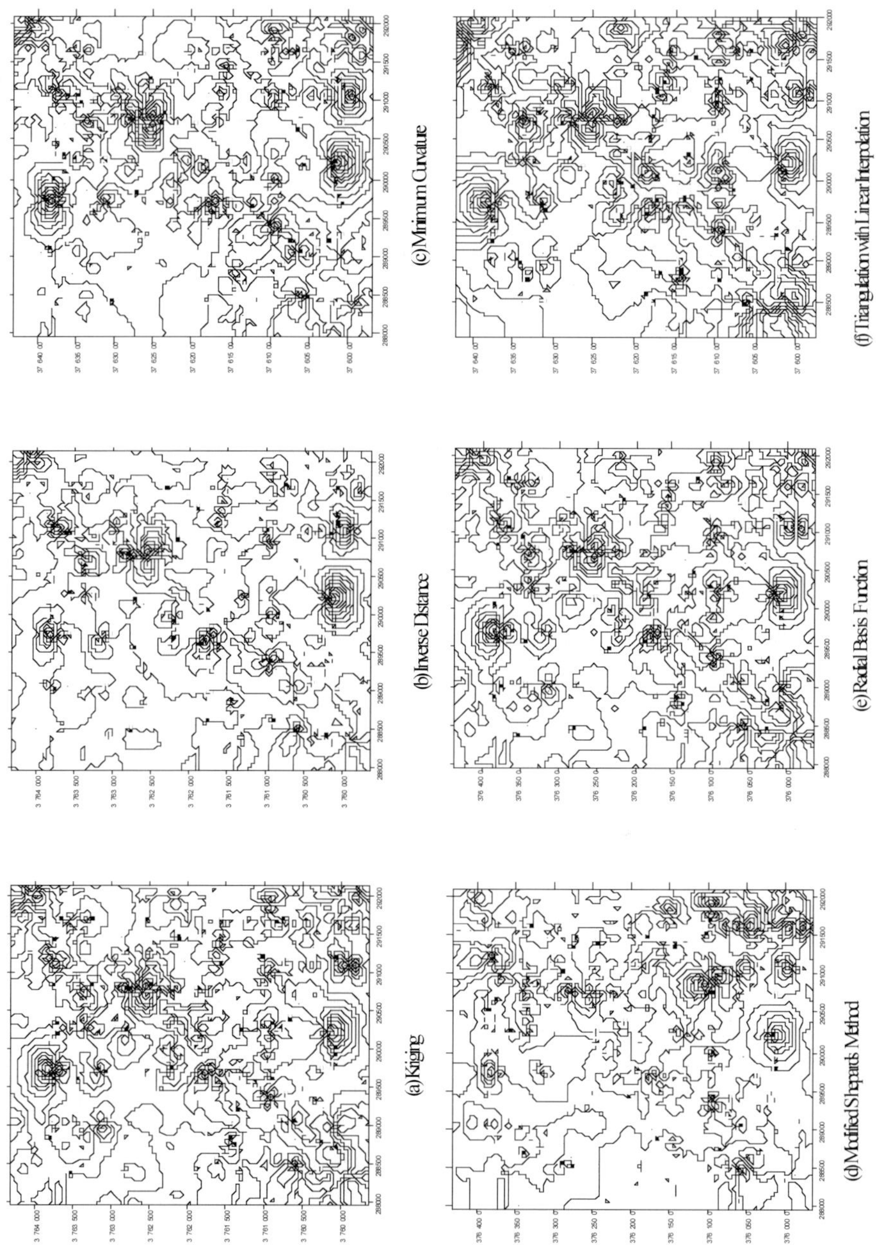

Fig. 5. FDEM accuracy surfaces of Winterville, Georgia (Test Site 1)

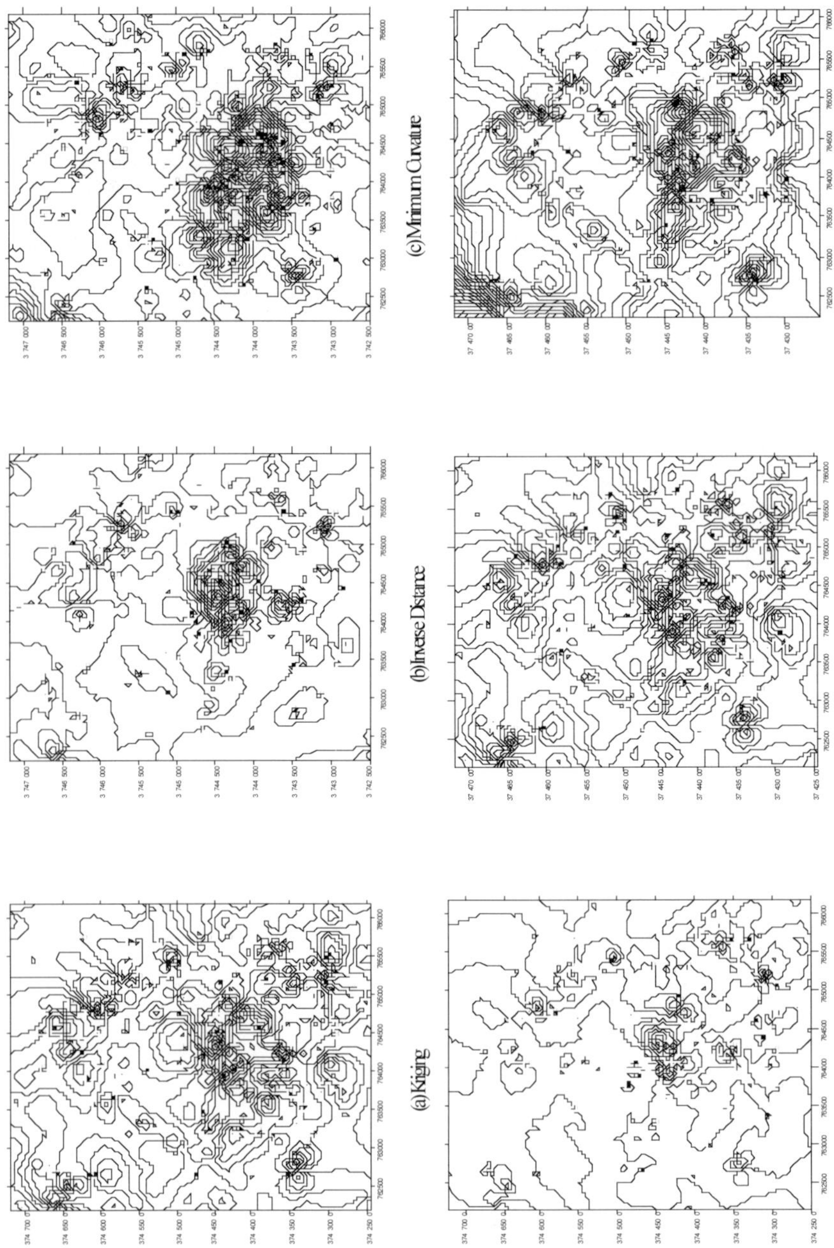

Fig. 6. DEM accuracy surfaces of Stone Mountain, Georgia (Test Site 2)

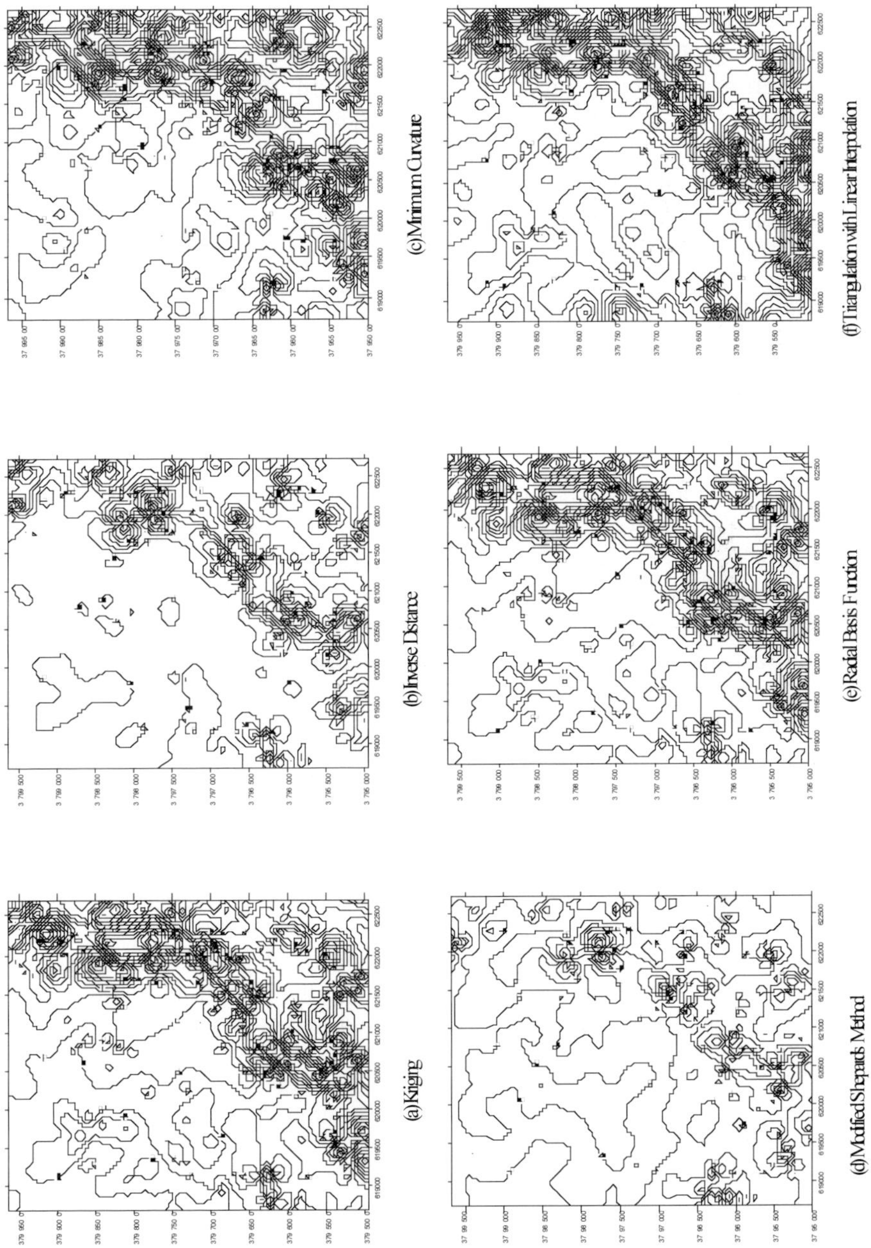

Fig. 7. DEM accuracy surfaces of Little River Canyon, Alabama (Test Site 3)

Statistical measures, such as root mean square error, standard deviation, and mean, cannot address the issue of spatial accuracy. To fully understand residual errors resulted from spatial interpolation, both statistical and spatial accuracy measures were needed. Accuracy surfaces were found valuable in identifying the magnitude and pattern of errors, from which spatial autocorrelation indices may be calculated.

Further researches are needed in the following issues in order to assess fully the spatial interpolation accuracy of elevation data: (1) Terrain complexity: The analysis should be extended into more complicated types of terrain in different geomorphologic settings. Test sites at different scales may result in varying interpolation accuracy. Comparative studies may also be conducted by using different input data density for different degrees of roughness in a surface or by taking breaklines. (2) Data sampling strategy: The analysis should also be extended into the impact of different data collection patterns (e.g., random vs. systematic; significant points vs. contouring). (3) Variogram modeling: Variogram models may be used in two stages of study: helping to know input data at the early stage and modeling the interpolation residuals at the later stage. The combined use of variogram models, accuracy surfaces, spatial autocorrelation, and even geovisualization represents a promising direction in accuracy mapping.

References

Ackermann F (1994) Digital Elevation Models: Techniques and Applications, Quality Standards, Development. In: Proc of the Symp on Mapping and Geographic Information Systems, ISPRS, vol 30, no 4. University of Georgia, Athens, GA, pp 421–432

Ackermann F (1996) Techniques and strategies for DEM generation. In: Greve C (ed) Digital Photogrammetry: An Addendum to the Manual of Photogrammetry. American Society of Photogrammetry and Remote Sensing, Falls Church, VA, pp 135–141

Burrough PH (1986) Methods of spatial interpolation. In: Principles of Geographic Information Systems for Land Resources Assessment. Clarendon Press, Oxford, England, pp 147–166

Carr JR (1995) Numerical Analysis for the Geological Sciences. Prentice Hall, Englewood Cliffs, NJ

Davis JC (1986) Statistics and Data Analysis in Geology, 2nd ed. John Wiley and Sons, New York

Ehlschlaeger CR, Goodchild MF (1994) Uncertainty in spatial data: defining, visualizing, and managing data error. In: GIS/LIS 1994 Proc, pp 246–253

Flowerdew R, Green M (1994) Areal interpolation and types of data. In: Fotheringham S, Rogerson P (eds) Spatial Analysis and GIS. Taylor & Francis, Bristol, pp 121–146

Gao J (1997) Resolution and accuracy of terrain representation by grid DEMs at a micro-scale. Int J of Geographic Information Science 11(2):199–212

Golden Software Inc. (1999) Surfer User's Guide. Golden, Colorado

Golden Software Inc. (2001) Didger User's Guide. Golden, Colorado

Goodchild MF (1993) Data models and data quality: problems and prospects. In: Goodchild MF, Parks BO, Steyaert LT (eds) Environmental Modeling with GIS. Oxford University Press, New York, pp 94–103

Goodchild MF, Gopal S (eds) (1989) Accuracy of Spatial Databases. Taylor and Francis, New York

Guth P (1992) Spatial analysis of DEM error. In: Proc ASPRS/ACSM Annual Meeting, pp 187–196

Hunter GJ, Caetano M, Goodchild MF (1995) A methodology for reporting uncertainty in spatial database products. J of the Urban and Regional Information Systems Association 7:11–21

Isaaks EH, Srivastava RM (1989) An Introduction to Applied Geostatistics. Oxford University Press, New York

Lam NS (1983) Spatial interpolation methods: a review. The American Cartographer 10(2):129–149

Li Z (1993) Theoretical models of the accuracy of digital terrain models: An evaluation and some observations. Photogrammetric Record 14(82):651–659

Li Z (1998) A comparative study of the accuracy of digital terrain models (DTMs) based on various data models. ISPRS J of Photogrammetry and Remote Sensing 49(1):2–11

Monckton C (1994) An investigation into the spatial structure of error in digital elevation data. In: Innovations in GIS 1. Taylor and Francis, London, pp 201–211

NCGIA (1989) The research plan for the National Center for Geographic Information and Analysis. Int J of Geographical Information Systems 3(2):117–136

R-WEL Inc. (1996) CAPTURE Digitizing Program, Version 3.1. Athens, Georgia. http://www.rwel.com

Shearer JW (1990) The accuracy of digital terrain models. In: Petrie G, Kennie TJM (eds) Terrain Modeling in Surveying and Engineering. Whittles Publishing Services, Caithness, pp 315–336

Shepard D (1968) A two dimensional interpolation function for irregularly spaced data. In: Proc 23rd National Conf ACM. Brandon/Systems Press, Princeton, pp 517–523

Smith WHF, Wessel P (1990) Gridding with continues curvature splines in tension. Geophysics 55(3):293–305

Wang L (1990) Comparative Studies of Spatial Interpolation Accuracy. Master Thesis, The University of Georgia, Athens, Georgia

Weibel R, Heller M (1991) Digital terrain modeling. In: Maguire DJ, Goodchild MF, Rhind DW (eds) Geographical Information Systems: Principles and Applications. Longman, London, pp 269–297

Weng Q (2002) Quantifying uncertainty of digital elevation models derived from topographic maps. In: Richardson D, van Oosterom P (eds) Advances in Spatial Data Handling. Springer-Verlag, New York, pp 403–418

Wood J (1996) The Geomorphological Charaterisation of Digital Elevation Models. PhD Thesis, Department of Geography, University of Leicester, Leicester, UK

Wood J, Fisher P (1993) Assessing interpolation accuracy in elevation models. IEEE Computer Graphics and Applications 13(2):48–56

Wren AE (1975) Contouring and the contour map: a new perspective. Geographical Prospecting 23:1–17

I/O-Efficient Hierarchical Watershed Decomposition of Grid Terrain Models

Lars Arge$^{\star 1}$, Andrew Danner$^{\star\star 2}$, Herman Haverkort$^{\star\star\star 3}$, Norbert Zeh$^{\dagger 4}$

[1] Department of Computer Science, University of Aarhus, Aarhus, Denmark; email: large@daimi.au.dk
[2] Department of Computer Science, Duke University, Durham, NC, USA email: adanner@cs.duke.edu
[3] Department of Computer Science, TU Eindhoven, Eindhoven, The Netherlands; email: cs.herman@haverkort.net
[4] Faculty of Computer Science, Dalhousie University, Halifax, Canada email: nzeh@cs.dal.ca

Abstract

Recent progress in remote sensing has made massive amounts of high resolution terrain data readily available. Often the data is distributed as regular grid terrain models where each grid cell is associated with a height. When terrain analysis applications process such massive terrain models, data movement between main memory and slow disk (*I/O*), rather than CPU time, often becomes the performance bottleneck. Thus it is important to consider I/O-efficient algorithms for fundamental terrain problems. One such problem

* Supported in part by the US Army Research Office through grant W911NF-04-1-0278, and by a Ole Roemer Scholarship from the Danish National Science Research Council. Part of this work was done while the author was at Duke University.

** Supported in part by the US National Science Foundation through grant CCR–9984099 and by the US Army Research Office through grant W911NF-04-1-0278.

*** This work was done while at University of Aarhus supported by a grant from the Danish National Science Research Council.

† Supported by the Natural Sciences and Engineering Research Council of Canada and the Canadian Foundation for Innovation.

is the hierarchical decomposition of a grid terrain model into *watersheds*– regions where all water flows towards a single common outlet. Several different hierarchical watershed decompositions schemes have been described in the hydrology literature. One important such scheme is the *Pfafstetter* label method where each watershed is assigned a unique label and each grid cell is assigned a sequence of labels corresponding to the (nested) watersheds to which it belongs.

In this paper we present an I/O-efficient algorithm for computing the Pfafstetter label of each cell of a grid terrain model. The algorithm uses $O(\mathrm{sort}(T))$ I/Os, the number of I/Os needed to sort T elements, where T is the total length of the cell labels. To our knowledge, our algorithm is the first efficient algorithm for the problem. We also present the results of a experimental study using massive real life terrain data that shows our algorithm is practically as well as theoretically efficient.

1 Introduction

Over millions of years, rainfall has been slowly etching networks of rivers into the terrain. Today, studying these river networks is important for managing drinking water supplies, tracking pollutants, creating flood maps, and more. Hydrologists can use large-scale digital elevation models, or DEMs, of the terrain along with a Geographic Information System, or GIS, to automate much of such studies. Often it is not necessary to study the entire terrain or river network at once; frequently one is only interested in regions that are downstream of a particular river, or the upstream areas that contribute flow to a particular river. By decomposing the terrain into a set of disjoint *hydrologic units* – regions where all water within the region flows towards a single, common outlet – one can quickly identify areas of interest without having to examine the entire terrain. The Pfafstetter labeling scheme described by Verdin and Verdin [16] defines a hierarchical decomposition of a terrain into arbitrarily small hydrological units, each with a unique label. These *Pfafstetter labels* also encode topological properties such as upstream and downstream neighbors, making it possible to automatically identify hydrological units of interest based on the Pfafstetter label alone.

In this paper, we describe an efficient algorithm for computing Pfafstetter labels efficiently on grid DEMs. Our algorithm is capable of handling massive high-resolution DEMs that are too large to fit in the main memory of even high-end machines. With recent progress in remote sensing technology, such as LIDAR, such DEMs are increasingly becoming available. Existing methods for determining hydrological units on grid DEMs use either manual methods [14], local filters [13, 10], or full terrain flow modeling [12] to identify terrain features and extract watersheds. While manual methods are often very ad-hoc, some of the main disadvantages of the current automatic methods is that they do not naturally define a hierarchical decomposition or a hierarchy that encode topological properties such as upstream and downstream neighbors. Furthermore, the existing algorithms cannot handle massive grid DEMs.

1.1 Pfafstetter Labels of Grid DEM

Conceptually, the definition of Pfafstetter labels [16] is independent of what DEM representation is used. However, for brevity we here only formally define Pfafstetter labels for grid DEMs.

Several different methods for modeling water flow on grid DEMs have been proposed; refer to [10, 8, 12, 15] for a discussion of the different methods. To model the direction water naturally flows from each cell s in the grid, most of these methods assign one or more *flow directions* from s to one or more of its (at most) eight neighboring cells. In the most common method [10], each cell s is assigned a single flow direction to the lowest of the lower neighboring cells. To model water flow off the terrain, cells on the boundary of the terrain (cells with less than eight neighbors) without any lower neighbors are assigned a flow direction to an imaginary cell ρ outside the terrain (the "outside sink"). The cells and flow directions naturally form a graph with an edge from cell s to cell t if s is assigned a flow direction to t. Assuming that the grid DEM does not contain any cells without lower neighbors other than the boundary cells, this graph is indeed a tree $\mathcal{T}$ since it contains $N-1$ edges (each cell except ρ has one downslope edge to a neighbor cell) and does not have cycles (flow directions go to lower cells). If we root $\mathcal{T}$ in ρ, each cell s with flow direction to t has t as its parent and is connected to ρ through a unique path of cells $s = s_1, t = s_2, s_3, \ldots, s_k = \rho$, where cell s_i is assigned a flow direction to s_{i+1}, i.e., water can flow from s to (the outside) ρ through $s_2, s_3, \ldots s_{k-1}$; water from cells in the subtree rooted in s drain through s on its way to (the outside) ρ. We call such a path in $\mathcal{T}$ a *river* $\mathcal{R}$ with *mouth* ρ (and *source* s). If the grid DEM *does* contain cells without lower neighbors other than the boundary cells, assigning flow directions as above to cells with a lower neighbor leads to a *forest* of trees where water in each tree can flow from a cell through parent cells to the root of a tree [5].

We define Pfafstetter labels of a grid DEM in terms of a forest of trees. For simplicity in this abstract, we only consider a single *binary flow tree* $\mathcal{T}$ with root ρ. Furthermore, we assume that each leaf l in $\mathcal{T}$ is augmented with a *drainage area* $d(l) \geq 1$, and that each internal node v in $\mathcal{T}$ is augmented with a drainage area $d(v)$ that is one plus the sum of the drainage areas of v's children. Note that if $d(l) = 1$ for every leaf l, then $d(v)$ is the size of the subtree rooted in v.

Pfafstetter labels of a binary flow tree $\mathcal{T}$ augmented with drainage areas are defined as follows. Let the *main river* $\mathcal{R}$ of $\mathcal{T}$ be the root-leaf path obtained by starting at the root ρ of $\mathcal{T}$ and in each node continuing to the child with the largest drainage area. The subtrees obtained if $\mathcal{R}$ is removed from $\mathcal{T}$ are called *tributary basins* or *tributary trees* and the root, t, of one of these subtrees, $\mathcal{T}^t$, is called a *tributary mouth*. First consider the case where at least four tributary mouths are obtained if $\mathcal{R}$ is removed. In this case, let v_2, v_4, v_6, v_8 be the four tributary mouths with largest drainage area, numbered in the order their parents are met when traversing $\mathcal{R}$ from ρ towards a leaf. Let p_i and s_i denote the parent and the sibling of v_i, respectively; both p_i and s_i are on $\mathcal{R}$. If we remove the eight edges incident to p_2, p_4, p_6 and p_8 (i.e. edges (v_i, p_i) and (s_i, p_i), for $i \in \{2, 4, 6, 8\}$), $\mathcal{T}$ is decomposed into four tributary basins rooted in v_2, v_4, v_6, and v_8, as well as five *interbasins* rooted at $s_0 = \rho$, s_2, s_4, s_6 and s_8. The Pfafstetter label of a node in the tributary basin rooted in v_i is i followed by the label obtained by recursively labeling the basin. The label of the nodes in the interbasin rooted in s_i (which includes nodes on $\mathcal{R}$) is $i + 1$ followed by the label obtained by recursively labeling the interbasin. In the case where $1 \leq k < 4$ tributary mouths are obtained when $\mathcal{R}$ is removed from $\mathcal{T}$, labels 1 through $2k + 1$ are assigned as above, while labels $2k + 2$ through 9 are not assigned. Finally, no labels are assigned when no tributary mouths are obtained, that is, when all nodes of $\mathcal{T}$ are on $\mathcal{R}$. Refer to Figure 1.

1.2 I/O-efficient Algorithms

When processing massive datasets that do not fit in main memory and must therefore reside on larger but considerably slower disks, transfer of data between disk and main memory (also called I/O) often becomes the performance bottleneck. In such cases the use of so-called *I/O-efficient* algorithms that minimize the number of disk accesses can lead to tremendous runtime improvements. I/O-efficient algorithms are algorithms designed in an *I/O-model* where the machine consists of an internal (or main) memory of limited size M and an infinite external memory. Computation is considered free but can only occur on data in main memory; in one *I/O-operation* (or simply *I/O*) B consecutive elements can be transfered between internal and external memory. The goal is to solve a given problem using as few I/Os as possible [1].

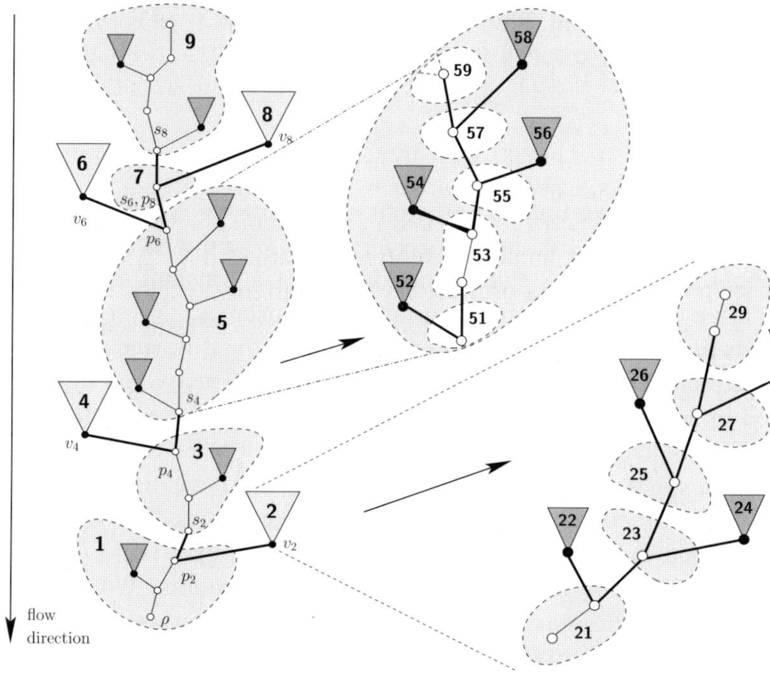

Fig. 1. *Left figure:* A flow tree $\mathcal{T}$ with the main river shown as white circles and tributary mouths as black circles (circle nodes constitute an augmented river). Removing the eight bold edges decomposes $\mathcal{T}$ into four tributary basins and five interbasins, each with the first digit in their Pfafstetter label shown in bold type. The remaining digits in the Pfafstetter label of the nodes in each basin (subtree) are computed recursively. *Two right figures:* First level of recursion for interbasin labeled 5 and tributary basin labeled 2

Trivially, the number of I/Os needed to scan through N elements in the I/O-model is $\Theta(\frac{N}{B}) = \Theta(\text{scan}(N))$. Aggarwal and Vitter showed that the number of I/Os needed to sort N elements is $\Theta(\frac{N}{B}\log_{M/B}\frac{N}{B}) = \Theta(\text{sort}(N))$. Note that $\text{sort}(N)$ is typically much smaller than N. Therefore tremendous speedups can often be obtained by developing algorithms that use $O(\text{scan}(N))$ or $O(\text{sort}(N))$ I/Os rather than $\Omega(N)$ I/Os; algorithms that are designed to work on data that fits in main memory often use $\Omega(N)$ I/Os when used in the I/O-model.

Numerous I/O-efficient algorithms and data structures have been developed, including many for GIS problems. Previous results that are particularly relevant for our work include $O(\text{sort}(N))$ I/O algorithms for various problems on trees [7] and various flow computation problems on large grid DEMs [5], as well as external stacks and priority queues on which N operations can be performed in $O(\text{scan}(N))$ and $O(\text{sort}(N))$ I/Os [3, 6], respectively. Refer to recent surveys for further results [17, 2].

1.3 Our Results

In this paper we present an I/O-efficient algorithm for computing the Pfafstetter labels of a flow tree in $O(\text{sort}(T))$ I/Os, where T is the total length of all labels. If each Pfafstetter label consists of a constant number of digits, e.g., if we truncate the labels, our algorithm uses only $O(\text{sort}(N))$ I/Os, where N is the number of nodes in the flow tree. If the flow tree and the labels fit in main memory, our algorithm uses $O(T)$ time. The overall algorithm is described in Section 2; it utilizes an algorithm for labeling a single river with tributary basins that consist of single nodes, described in Section 3. In Section 4 we investigate the practical use of the algorithm: we discuss how a flow tree that yields practically realistic watershed hierarchies (Pfafstetter labels) can be obtained in $O(\text{sort}(N))$ I/Os from a general grid DEM (with many cells without lower neighbors) using previous algorithms. We also present the results of a preliminary experimental study using massive real life terrain data that shows that our algorithm is practically as well as theoretically efficient.

2 Computing Pfafstetter Labels of Flow Tree

The recursive definition of Pfafstetter labels of a binary flow tree $\mathcal{T}$ naturally leads to a recursive algorithm to compute the labels: Compute the main river $\mathcal{R}$ and four largest tributary mouths, break the tree into nine subtrees, and recurse. Unfortunately, due to random data access patterns, it seems hard to make such a direct algorithm I/O-efficient. Instead our algorithm works by decomposing $\mathcal{T}$ into a set of rivers augmented with tributary mouths, Pfafstetter labeling them individually, and finally combining the labels of the individual augmented rivers to obtain the Pfafstetter labels for all nodes of $\mathcal{T}$.

Our decomposition of the flow tree $\mathcal{T}$ into augmented rivers is defined by a *tributary tree* $\mathcal{T}^t$, where each node l in $\mathcal{T}^t$ stores an augmented river $\mathcal{R}_l^t$ and where m is a child of l if and only if the parent of the mouth of $\mathcal{R}_m^t$ is on $\mathcal{R}_l^t$, that is, if $\mathcal{R}_m^t$ flows directly into $\mathcal{R}_l^t$. More precisely, the root r of $\mathcal{T}^t$ contains the path obtained by starting at the root ρ of $\mathcal{T}$ and in each node continue to the child with the largest drainage area; for each node v on the path we also include the (possible) child of v not on the path (called a *tributary mouth node*) in $\mathcal{R}_l^t$. Note that $\mathcal{R}_r^t$ is the main river $\mathcal{R}$ in the above definition of Pfafstetter labels of the flow tree $\mathcal{T}$ augmented with its tributary mouths. The root r has a child for each tributary basin of $\mathcal{R}$, that is, for each subtree of $\mathcal{T}$ obtained if $\mathcal{R}$ is removed from $\mathcal{T}$; the rivers in these children are obtained recursively. Note that this means that each tributary mouth is stored exactly twice, namely in $\mathcal{R}_r^t$ and as the mouth of the main river $\mathcal{R}_l^t$ in a child l of r. Refer to Figure 2.

Given a Pfafstetter labeling of each individual augmented river $\mathcal{R}_l^t$ in the tributary tree $\mathcal{T}^t$, we can combine these labels to obtain the Pfafstetter labeling of the whole flow tree $\mathcal{T}$ as follows. Consider the augmented river $\mathcal{R}_r^t$ stored in the root of r. As mentioned, $\mathcal{R}_r^t$ is the main river $\mathcal{R}$ in the definition of Pfafstetter labels of $\mathcal{T}$, augmented with its tributary mouths. Since in the definition of Pfafstetter labels of $\mathcal{T}$, the labeling of $\mathcal{R}$ only depends on the drainage area of its tributary mouths (first digit is determined by the four tributary mouths with largest drainage areas, and the rest recursively determined in each interbasin), the labels of the nodes in common between the main river $\mathcal{R}$ and the individually labeled augmented river $\mathcal{R}_r^t$ are indeed the same. Furthermore, the labels of the nodes in a tributary basin of $\mathcal{R}$ consists of some prefix determined by the labeling of the nodes on $\mathcal{R}$ (a digit for each recursive labeling step where the tributary basin is part of one of the four interbasins, followed by a digit determined in the recursive call where the tributary mouth has one of the four largest drainage areas), followed by the label obtained by recursively labeling the basin. The prefix is exactly the label assigned to the mouth of the tributary basin in the augmented river $\mathcal{R}_r^t$. Thus we can obtain the Pfafstetter labels for all nodes in $\mathcal{T}$ from a labeling of the augmented rivers in $\mathcal{T}^t$, simply by assigning the nodes in the main river $\mathcal{R}$ the labels of the corresponding nodes in $\mathcal{R}_r^t$ in the root r of $\mathcal{T}^t$, and recursively labeling the nodes in each subtree of r while prefixing the labels in the subtree rooted in child l with the label of the tributary mouth node in $\mathcal{R}_r^t$ corresponding to the mouth of the main river $\mathcal{R}_l^t$.

Intuitively, computing the tributary tree $\mathcal{T}^t$ from flow tree $\mathcal{T}$ is easier than computing Pfafstetter labels directly on $\mathcal{T}$. The definition of $\mathcal{T}^t$ suggest a

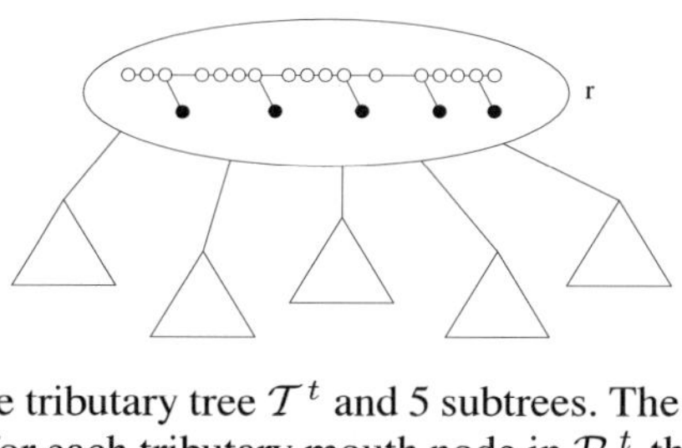

Fig. 2. The root r of the tributary tree $\mathcal{T}^t$ and 5 subtrees. The augmented river $\mathcal{R}_r^t$ is stored in the root and for each tributary mouth node in $\mathcal{R}_r^t$ there is one subtree of r

natural algorithm based on a DFS-traversal of $\mathcal{T}$, where in each step the child with largest drainage area is chosen. By modifying the known $O(\text{sort}(N))$ I/O algorithm for DFS-numbering nodes in a tree [7], it is possible to obtain a $O(\text{sort}(N))$ I/O algorithm for our special DFS-traversal problem. However, while the know general DFS-numbering algorithm is quite complicated (and therefore not of practical interest), the special structure of flow trees (decreasing drainage area along root-leaf paths) allows us to develop a simple and practical $O(\text{sort}(N))$ I/O algorithm. We describe this algorithm (which utilizes an I/O-efficient priority queue) in the full version of this paper. Similarly, once each individual augmented river in the tributary tree $\mathcal{T}^t$ has been labeled, an algorithm based on DFS-traversal (or a BFS-traversal) can be used to combine the labels from the augmented rivers to obtain the Pfafstetter labels of $\mathcal{T}$ in $O(\text{scan}(T))$ I/Os, where T is the total length of the labels. We also describe such a simple and practical algorithm in the full paper. We describe the remaining part of our algorithm, an $O(\text{scan}(T))$ I/O algorithm for computing the Pfafstetter labels of a single augmented river, in Section 3. This leads the following main result.

Theorem 1. *The Pfafstetter labels of a flow tree $\mathcal{T}$ can be constructed in $O(\text{sort}(N) + \text{scan}(T))$ I/Os, where T is the total size of the labels of all nodes in $\mathcal{T}$.*

Remarks. In the full paper we discuss the following properties of our algorithm. (i) It can easily be modified to handle non-binary flow trees in the same I/O-bound. (ii) It can easily be modified to handle forests rather than trees in the same I/O-bound. (iii) If each Pfafstetter label consists of a constant number of digits (elements), e.g. if we truncate the labels, it only uses $O(\text{sort}(N))$ I/Os. (iv) If $\mathcal{T}$, $\mathcal{T}^t$ and all labels fit in memory, we can easily design a Pfafstetter labeling algorithm that uses $O(T)$ time.

3 Labeling a Single River

In this section we describe a simple and I/O-efficient algorithm for computing the Pfafstetter labels of a single augmented river $\mathcal{R}_i^t$–a simple flow tree consisting of one path (river) where each node (possibly) has a tributary mouth node child. Our algorithm is described in Section 3.2; in Section 3.1 we first discuss a data structure, the Cartesian tree, used in the algorithm.

3.1 Cartesian Tree

Let $A = (a_1, a_2, \ldots, a_N)$ be a sequence of N elements, each with an associated weight, and let A_i denote the prefix $(a_1, a_2, \ldots, a_i)$ of A. The Cartesian tree $\mathcal{C}(A)$ of A is a binary tree defined as follows [9]: If A is empty, $\mathcal{C}(A)$ is empty. Otherwise, let a_i be the element with the largest weight in A; if there is more than one occurrence of the largest weight, a_i is the element that appears first in A. $\mathcal{C}(A)$ consists of a root v containing an element with weight $a(v) = a_i$, with a left subtree $\mathcal{C}((a_1, ..., a_{i-1}))$ (a Cartesian tree on the elements before a_i in A) and a right subtree $\mathcal{C}((a_{i+1}, ..., a_N))$ (a Cartesian tree on the elements after a_i in A). Note that the weights of elements on a root-leaf path in $\mathcal{C}(A)$ are nondecreasing.

The Cartesian tree $\mathcal{C}(A)$ of a sequence A can be constructed in $O(N)$ time using an algorithm that iteratively constructs $\mathcal{C}(A_i)$ from $\mathcal{C}(A_{i-1})$ as follows [9]: Let the rightmost path P of $\mathcal{C}(A_{i-1})$ be the path traversed by starting at the root r and repeatedly continuing to the right child until a node l without a right child is reached; note that this is not necessarily the path from the root to the rightmost leaf of $\mathcal{C}(A_{i-1})$. We construct $\mathcal{C}(A_i)$ by first traversing P from l towards r, until two adjacent nodes u and v are located such that $a(u) \geq a_i > a(v)$; if $a(l) \geq a_i$, $u = l$ and v is non-existing, and if $a(r) < a_i$, $v = r$ and u is non-existing. Then we construct a new node w containing an element with weight $a(w) = a_i$, and make w the right child of u and v the left child of w. Refer to Figure 3. The correctness of the algorithm follows from the fact that the weights of the elements along P are non-decreasing and that w is inserted as a right child without a left child; Refer to [9]. The linear time bound follows from the fact that all nodes on P traversed to find u and v (except u) are removed from P by the insertion of w (that is, they are not on the rightmost path of $\mathcal{C}(A_i)$) and therefore they are not traversed in later iterations; thus we traverse $O(N)$ nodes in total.

Given the sequence A stored as a list in external memory, we can implement the above algorithm such that we compute $\mathcal{C}(A)$ and store it as a sorted list C of post-order numbered nodes in external memory using $O(\text{scan}(N))$ I/Os; a post-order numbering of the nodes in $\mathcal{C}(A)$ is the numbering con-

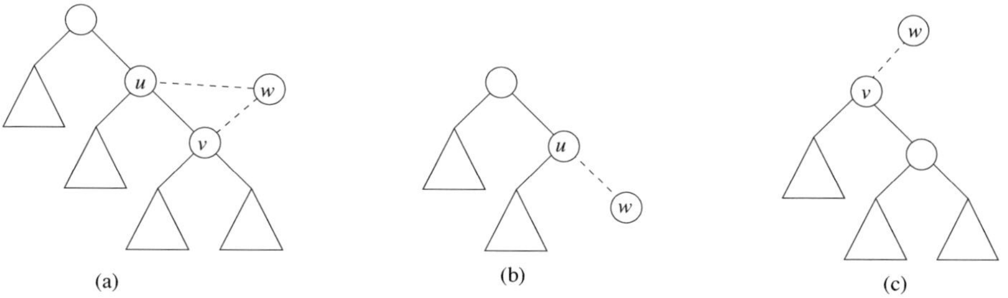

Fig. 3. Inserting w to obtain $\mathcal{C}(A_i)$ from $\mathcal{C}(A_{i-1})$; dotted lines indicate inserted edges. (a) $a(u) \geq a(w) > a(v)$ (b) $a(l) \geq a(w)$ (c) $a(r) = a(v) < a(w)$

sisting of a recursive numbering of nodes in the left subtree of the root r, followed by a recursive numbering of nodes in the right subtree of r, followed by the numbering of r, and where each node stores the numbers of each of its children. Note that the nodes on the rightmost path of $\mathcal{C}(A)$ have the highest post-order numbers.

To implement the algorithm I/O-efficiently, we maintain the following two invariants for $\mathcal{C}(A_{i-1})$: (1) Except for the nodes on the rightmost path P of $\mathcal{C}(A_{i-1})$, all nodes have been post-order numbered and stored in sorted order in a list C in external memory; (2) Nodes on P are stored on a stack S in the order they appear on P (with the leaf l on top of S), and each node stores the correct number of its left child (stored in C, if existing).

Initially C and S are empty. To compute $\mathcal{C}(A_i)$ from $\mathcal{C}(A_{i-1})$ while maintaining the invariants, we implement the traversal of P from l towards r used to find u and v as follows. Until u is on the top of S (or S is empty), we repeatedly pop a node s from S and insert it after the last element t in C; we number s with the number following the number of t and (except for l) we set its right child number equal to the number of t. Then we set the left child number of the new node w equal to the number of the last element v inserted in C (if existing), and push w on S. After computing $\mathcal{C}(A_N) = \mathcal{C}(A)$, we pop each node s from S in turn and insert it in C, while updating numbers and right child numbers as above.

That the above procedure maintains the first invariant can be seen as follows. Before the procedure, the nodes on the rightmost path of $\mathcal{C}(A_{i-1})$ stored on S have the largest numbers in the post-order numbering of $\mathcal{C}(A_{i-1})$, and by the first invariant the remaining nodes of $\mathcal{C}(A_{i-1})$ are stored in post-order number order in C. Since nodes are popped from S and inserted in C in post-order, the nodes of $\mathcal{C}(A_i)$ in C are also in post-order number order. The left and right child numbers of each node s inserted in C are also correct, since by the second invariant the left child number was already correct before the insertion, and the right child number is explicitly set to the last inserted node t (or left empty in the case of the first inserted node l), which also by the second invariant is the right child of s. That the procedure also maintains the second invariant can be seen as follows. By the second invariant the nodes on P are stored in order on S before the procedure. Since the nodes that are not on P in $\mathcal{C}(A_i)$ are popped from S, and since the only node pushed on S is the new leaf w on P in $\mathcal{C}(A_i)$, the nodes on P are also stored in order on S after the procedure; each node store the correct left child number, since the left child number of the only new node w is explicitly set to v. After computing $\mathcal{C}(A_N) = \mathcal{C}(A)$, invariant one implies that all but the nodes on P have been correctly numbered and stored in C. Since by invariant two, the nodes on P are stored in post-order number order on S, the list C correctly contains all nodes in $\mathcal{C}(A)$ in post-order number order after popping each element from S and inserting it in C.

Overall, the algorithm performs one scan of A and one scan of C, as well as $O(N)$ stack operations. Since a stack can easily be implemented such that each operation takes $O(1/B)$ I/Os (by keeping the top B elements in an internal memory buffer and only reading/writing to disk when the buffer is empty/full), the algorithm uses $O(\text{scan}(N))$ I/Os in total.

Augmented Cartesian Tree. In our augmented river labeling algorithm we will use a slightly modified version of the Cartesian tree, called an augmented Cartesian tree. An augmented Cartesian tree $\mathcal{C}_a(A)$ of a sequence $A = (a_1, a_2, \ldots, a_N)$ of N elements is simply a Cartesian tree $\mathcal{C}(A)$ of A, where each node v has been augmented with copies of the four nodes (post-order number, drainage area, and children post- order numbers) with largest weight in the subtree rooted in v; if two nodes have the same weight, the node with the weight that appear first in A is chosen. Note that one of these largest weight nodes is v itself. In the full version of this paper we show that we can easily modify our I/O-efficient Cartesian tree construction algorithm to construct an augmented Cartesian tree without performing any extra I/Os.

Lemma 1. *Given a sequence A of N elements, each with a weight, the augmented Cartesian tree $\mathcal{C}_a(A)$ can be computed and stored as a sorted list of post-order numbered nodes using $O(\text{scan}(N))$ I/Os.*

Observation 3.1 *The four largest weight nodes stored in the root r of an augmented Cartesian tree $\mathcal{C}_a(A)$ constitute a connected subtree of $\mathcal{C}_a(A)$ rooted in r.*

Proof. The four nodes containing the elements with largest weights trivially include r. Assume they do not form a connected subtree. Then one of them is a node v, other than r, whose parent u is not one of the four nodes; therefore the weight of u is smaller than the weight of v. This contradicts that the weights of nodes on any root-leaf path in $\mathcal{C}_a(A)$ are nondecreasing. □

3.2 Labeling a River

We are now ready to describe how to compute the Pfafstetter labels of an augmented river $\mathcal{R}_l^t$ with mouth (root) s_0 and source t. Recall that by the definition of Pfafstetter labels, the labels of $\mathcal{R}_l^t$ are obtained by first identifying the four tributary mouth nodes v_2, v_4, v_6 and v_8 with largest drainage area, numbered in the order they appear along $\mathcal{R}_l^t$, and label them $2, 4, 6, 8$. Then all edges incident to their parents p_2, p_4, p_6 and p_8 are removed, decomposing $\mathcal{R}_i^t$ into five interbasins rooted in s_0 and the siblings s_2, s_4, s_6

and s_8 of v_2, v_4, v_6 and v_8. Finally, each interbasin is labeled recursively, and the label of each node in the interbasin rooted in s_i is prefixed by $i+1$. In the case where $\mathcal{R}_l^t$ only has $1 \le k < 4$ tributary mouth nodes, labels $2k+2$ through 9 are not assigned; when there are no tributary mouth nodes (when $k=0$) no label (other than the possible prefix) is assigned.

The augmented Cartesian tree provides us with an easy way of computing the Pfafstetter labels of $\mathcal{R}_l^t$. Consider constructing an augmented Cartesian tree $\mathcal{C}_a(L)$ on the sequence L consisting of the nodes along $\mathcal{R}_l^t$ ordered from mouth to source, where each tributary mouth node v is stored between its parent p and sibling s, and where each river node has weight zero and each tributary mouth node v has weight equal to its drainage area $d(v)$. Refer to Figure 4. Note that if $\mathcal{R}_l^t$ has at least one tributary mouth node, then the root r of $\mathcal{C}_a(L)$ corresponds to the tributary mouth node v with largest drainage area. Splitting L at v (while removing v) corresponds to removing the two edges incident to the parent p of v, and results in two sequences $L_l = (s_0, \ldots, p)$ and $L_r = (s, \ldots, t)$ corresponding to two interbasins rooted in s_0 and the sibling s of v. The augmented Cartesian trees rooted in the children of r are exactly $\mathcal{C}_a(L_l)$ and $\mathcal{C}_a(L_r)$. Similarly, if the weights of the four largest weight nodes in L stored in r are all non-zero, they correspond to the four tributary mouth nodes v_2, v_4, v_6 and v_8 of $\mathcal{R}_l^t$ with largest drainage areas. Splitting L at v_2, v_4, v_6 and v_8 (while removing these nodes) corresponds to removing the edges incident to their parents p_2, p_4, p_6 and p_8, and results in five sequences $L_0 = (s_0, \ldots, p_2), L_2 =$

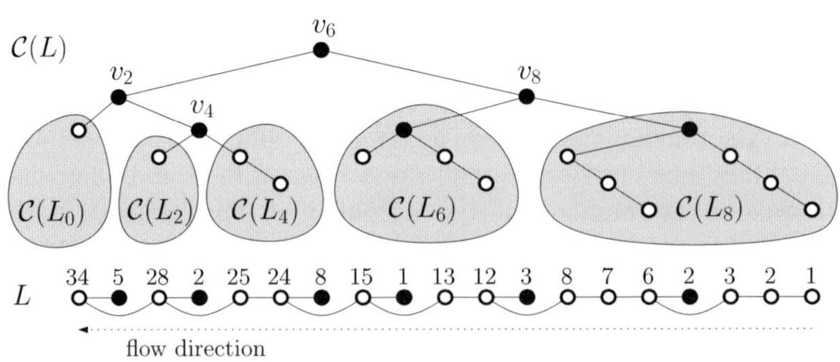

Fig. 4. *Bottom figure:* An augmented river with drainage areas (as it is stored in L); the weight of river nodes (white circles) is zero and the weight of tributary mouth nodes (black circles) is equal to their drainage area. *Top figure:* Cartesian tree $\mathcal{C}(L)$ with the four tributary mouth nodes v_2, v_4, v_6 and v_8 with largest drainage areas (weight), and the five Cartesian trees $\mathcal{C}(L_0), \mathcal{C}(L_2), \mathcal{C}(L_4), \mathcal{C}(L_6)$ and $\mathcal{C}(L_8)$ for the five interbasins obtained when removing edges incident to their parents p_2, p_4, p_6 and p_8 in L (removing v_2, v_4, v_6 and v_8 from $\mathcal{C}(L)$)

$(s_2, \ldots, p_4), L_4 = (s_4, \ldots, p_6), L_6 = (s_6, \ldots, p_8)$ and $L_8 = (s_8, \ldots, t)$ corresponding to the five interbasins rooted in siblings s_0, s_2, s_4, s_6 and s_8. By Observation 3.1, the nodes in $\mathcal{C}_a(L)$ corresponding to v_2, v_4, v_6 and v_8 form a connected subtree rooted in r, and if this subtree is removed, $\mathcal{C}_a(L)$ is decomposed into five subtrees (since it is binary) that are augmented Cartesian trees $\mathcal{C}_a(L_0), \mathcal{C}_a(L_2), \mathcal{C}_a(L_4), \mathcal{C}_a(L_6)$ and $\mathcal{C}_a(L_8)$ for the five interbasins. Thus the Pfafstetter labels of $\mathcal{R}_l^t$ can be obtained by labeling v_2, v_4, v_6 and v_8 with $2, 4, 6$ and 8, respectively, and recursively labeling $\mathcal{C}_a(L_0), \mathcal{C}_a(L_2), \mathcal{C}_a(L_4), \mathcal{C}_a(L_6)$ and $\mathcal{C}_a(L_8)$ while prefixing all labels in $\mathcal{C}_a(L_i)$ with $i + 1$. In the case where only $1 \leq k < 4$ of the weights of the largest weight nodes in L stored in r are non-zero, that is, if $\mathcal{R}_l^t$ only has k tributary mouth nodes $v_2, \ldots, v_{2k}$, removal of the subtree corresponding to $v_2, \ldots, v_{2k}$ decomposes $\mathcal{C}_a(L)$ into $k + 1$ augmented Cartesian trees $\mathcal{C}_a(L_0), \ldots, \mathcal{C}_a(L_{2k})$ that can be labeled recursively (that is, labels $2k + 2$ through 9 are not assigned). Finally, if the weights of all nodes stored in r are zero, $\mathcal{R}_l^t$ does not have any tributary mouth nodes and no labels (other than the possible prefix) should be assigned to $\mathcal{C}_a(L)$. Based on the above observations, we can design an I/O-efficient algorithm for Pfafstetter labeling an augmented river $\mathcal{R}_l^t$ given as a list L consisting of the nodes along $\mathcal{R}_l^t$ ordered from mouth to source, where each tributary mouth node v is stored between its parent p and sibling s, and where each river node has weight zero and each tributary mouth node v has weight equal to its drainage area $d(v)$.

We first construct an augmented Cartesian tree $\mathcal{C}_a(L)$ on L, stored as a sorted list C of post-order numbered nodes. Next we label each node in C, storing all labels in a list C_p, using a recursive traversal of $\mathcal{C}_a(L)$ as outlined above, where we always recursively visit the right subtree of a node v before recursively visiting the left subtree of v, and where we explicitly implement the recursion stack S. The stack S can contain two types of elements, namely *label* and *recursion* elements, both consisting of (the number of) a node v of $\mathcal{C}_a(L)$ and a Pfafstetter label (prefix) P. Initially, S contains a recursion element for the root r of $\mathcal{C}_a(L)$ (that is, an element with number N) and an empty label. We repeatedly pop an element from S and access the corresponding node v in C. If the element is a label element, we simply label v with P and insert it at the end of C_p. If it is a recursion element, we want to label the subtree of $\mathcal{C}_a(L)$ rooted in v, while prefixing all labels with P. To do so, we consider the four largest weight nodes v_2, v_4, v_6 and v_8 stored with v in C. Assume first that their weights are all non-zero. In this case we label v_2, v_4, v_6 and v_8 by pushing a label element for each v_i on S with the label P followed by i; we also recursively label $\mathcal{C}_a(L_0), \mathcal{C}_a(L_2), \mathcal{C}_a(L_4), \mathcal{C}_a(L_6)$ and $\mathcal{C}_a(L_8)$ by pushing a recursion element for each of their roots (obtained from v_2, v_4, v_6 and v_8) with labels P followed by $1, 3, 5, 7$ and 9, respectively, on S. We push the elements in the order they appear in a post-order traversal of the subtree rooted in v, where left subtrees are visited before right subtrees; note that this means that they appear in reverse post-order traversal order on S. In the case where only $1 \leq k < 4$ of the largest weight nodes stored with v in C are non-zero, we only push label elements corresponding to these nodes $v_2, \ldots, v_{2k}$ and recursion elements corresponding to $\mathcal{C}_a(L_0), \ldots, \mathcal{C}_a(L_{2k})$. Finally, if the weights of all the largest weight nodes stored with v in C are zero, we simply label v with P and insert it at the end of C_p, and push two recursion elements with label P on S; first an elements for the left child of v and then an elements for the right child of v (note that this will eventually label the whole subtree rooted in v with P).

That the above algorithm correctly computes the Pfafstetter label of $\mathcal{R}_l^t$ follows from the above discussion. The list C is constructed from L in $O(\text{scan}(N))$ I/Os (Lemma 1). Since we visit the nodes in $\mathcal{C}_a(L)$ in reverse post-order, the N accesses to C correspond to a backwards scan of C, and are therefore performed in $O(\text{scan}(N))$ I/Os. If T is the total size of the computed Pfafstetter labels, the labels are written to C_p in $O(\text{scan}(T))$ I/Os, and the $O(N)$ stack operations can also be performed in $O(\text{scan}(T))$ I/Os (since the combined size of the labels pushed on S is $O((T))$. After computing the labels of the nodes in C, stored in C_p, we can easily label the corresponding nodes in L in a single sorting step. However, by essentially reversing the way C was produced from L, we can also easily do so in $O(\text{scan}(T))$ I/Os. Thus $\mathcal{R}_l^t$ is labeled in $O(\text{scan}(T))$ I/Os in total.

Lemma 2. *Given an augmented river $\mathcal{R}_l^t$ as a ordered list L of N nodes along $\mathcal{R}_l^t$, where each tributary mouth node is stored between its parent and sibling, the Pfafstetter labels of $\mathcal{R}_l^t$ can be computed and stored with the nodes in L in $O(\text{scan}(T))$ I/Os, where T is the total size of the labels of all nodes in $\mathcal{R}_l^t$.*

4 Implementation and Experimental Results

In this section, we present the results of an experimental study of our Pfafstetter labeling algorithm. We first in Section 4.1 discuss how we implemented our algorithm to handle general grid DEMs (as opposed to the simplified case considered in the previous sections). In Section 4.2 and Section 4.3 we then discuss the data and experimental results, respectively.

4.1 Implementation

In the introduction we discussed how we can obtain a flow tree $\mathcal{T}$ from a grid DEM that (other than the boundary cells) *does not* contain any cells without a lower neighbor, simply by assigning each cell a flow direction to the lowest of its lower neighbors and from each boundary cell without a lower neighbor to a special cell ρ (the outside sink). Given the grid DEM with N cells in row (or column) major order, we can easily in $O(\text{scan}(N))$ I/Os construct a representation of $\mathcal{T}$ consisting of an unordered list of numbered nodes, where each node contains the numbers of its children, simply by scanning through the grid three rows at a time, while for each cell looking at cells in a 3×3 neighborhood.

In the, most common, case where the grid DEM *does* contain cells other than boundary cells without lower neighbors, often called *flat cells*, the above procedure leads to a forest of trees, since each cell without a lower neighbor becomes the root of a separate flow tree. Simply computing Pfafstetter labels for such a forest does not lead to realistic watersheds, because treating each flat cell as a sink does not model global water flow very well. Often flat cells appear together and form larger *flat areas*. These flat areas can be divided into *plateaus* that contain at least one *spill point* – a flat cell with at least one lower neighbor – and *sinks* that do not. Intuitively, a single plateau should not yield separate flow trees. Instead flow trees with a cell in the plateau should be connected by assigning flow directions such that water flows across each plateau to spill points. On the other hand, its often natural to regard each sink as giving rise to one separate watershed or flow tree. This can be accomplished by connecting all flow trees with a root in the sink, for example by assigning flow directions such that each cell in the sink has a flow path to one specific cell in the sink.

Using known algorithms, we can compute flat areas of a grid DEM and assign directions to plateaus and sinks as discussed above in $O(\text{sort}(N))$ I/Os [5]. We can also use known algorithms to compute the drainage area of each node in the resulting forest (with each leaf l having drainage area $d(l) = 1$) in $O(\text{sort}(N))$ I/Os [5]. After that we can compute Pfafstetter labels in $O(\text{sort}(T))$ I/Os using our algorithm described in the previous sections, modified to work on a flow forest rather than a flow tree and to handle flow trees that are not binary.

Often a grid DEM contains many small sinks that should intuitively not lead to separate watersheds. Therefore a common practice in flow modeling is to *flood* the DEM in order to remove all sinks, by simulating uniformly pouring water onto the DEM (while viewing the outside as a giant sink) until a steady-state is reached and all sinks are filled by accumulating water [10, 11]. Thus flooding produces a terrain in which all flat areas are plateaus, and assigning flow directions towards spill points then lead to a single flow tree for the grid DEM. I/O-efficient $O(\text{sort}(N))$ algorithms for flooding a grid DEM and for plateau flow direction assignment have been developed and implemented in the TERRAFLOW software package [5]. In fact, this software also computes the drainage area of each cell in $O(\mathrm{sort}(N))$ I/Os.

Our Pfafstetter implementation takes two input grids corresponding to a DEM, namely the corresponding flow directions and the corresponding drainage areas. To obtain a realistic watershed hierarchy (Pfafstetter labels), we used flooded grid DEM models, where all cells, including flat cells on plateaus, have already been assigned a flow direction, as well as had their drainage area computed by TERRAFLOW From these input grids we obtain the unordered list representation of $\mathcal{T}$ used in our Pfafstetter algorithm by a simple simultaneous scan of the two grids using $O(\mathrm{scan}(N))$ I/Os; in the same scan we also augment each node with the grid position of the corresponding cell. After that our implementation follows the algorithm described in the previous sections (modified to handle a non-binary flow tree), and after computing Pfafstetter labels of all nodes, we sort the nodes by grid position using $O(\mathrm{sort}(T))$ I/Os to obtain an output Pfafstetter label grid. (Optionally, we allow the user to truncate labels to a maximum length, so that each label fits in $O(1)$ $\log N$-bit words and the sorting of labels can be done in $O(\mathrm{sort}(N))$ I/Os). We implemented our algorithm in C++ using TPIE [4], a library that provides support for implementing I/O-efficient algorithms and data structures. The implementation work was greatly simplified by the fact that all main primitives of our algorithm – scanning, sorting, stacks and priority queues – are already implemented I/O-efficiently in TPIE or TERRAFLOW.

4.2 Datasets

To investigate the practical performance of our algorithms, as well as the realism of the computed watersheds, we conducted a set of experiments with five grid DEMs of varying size. The largest DEM covered the Neuse river basin in North Carolina at a resolution of 20 feet. It contained 396.5 million cells (such that the flow directions and drainage areas occupied 5.8Gbytes), and is publicly available from `ncfloodmaps.com`. The other four DEMs covered sub-basins of the upper Tennessee river basin at a resolution of one arc second (approximately 100 feet) and contained 2.7, 21.7, 30.8 and 147 million cells, respectively; these datasets are from the National Elevation Dataset (NED) from the United States Geological Survey, publicly available at seamless.usgs.gov.

4.3 Experimental Results

For each of the five input DEMs we used TERRAFLOW to compute filled DEMs with flow directions and drainage area, and then we used our implementation to compute Pfafstetter labels, truncated to nine digits. The experiments were run on a Dell Precision Server 370 (Pentium 4 3.40 GHz processor) with hyperthreading enabled and running Linux 2.6.11. The machine had 1 GB of physical memory, but we made sure that our implementation never used more than 256 MB by setting a kernel flag to limit memory to 256 MB and instructing TPIE to abort if more memory than this limit was allocated. All data was stored on a single 400 GB SATA disk drive.

Table 1 shows the time used to label each of the five input DEMs, not counting the the time used by TERRAFLOW. In all cases, the time taken by TERRAFLOW was more than five times the time taken by the Pfafstetter labeling routine.

Table 1. Size and Pfafstetter labeling time for the five DEMs

Dataset	Ten 1	Ten 2	Ten 3	Ten 4	Neuse
Input size (MB)	17	116	150	713	5,819
Size (mln cells)	2.7	21.7	30.8	147.0	396.5
Running time	0m30	6m51	10m29	58m10	187m43

Table 2 shows how much time is spent in the various phases of the algorithm, as a percentage of total time. Constructing $\mathcal{T}^t$ is the most time consuming phase of the algorithm. This is not unexpected, since this phase is the most complicated (it utilizes a priority queue) and performs $O(\text{sort}(N))$ I/Os. Interestingly, labeling $\mathcal{T}$ and $\mathcal{T}^t$ (using the augmented Cartesian tree) is a small fraction of the total time (this is somewhat expected, since $O(\text{scan}(N)) < O(\text{sort}(N))$). It is also interesting to note that reading and importing the initial grids (constructing $\mathcal{T}$) and exporting the final results is not an insignificant portion of the total time. Overall, we conclude that our algorithm is practically, as well as theoretically, efficient.

Table 2. Breakdown of labeling time for each of the five DEMs

Dataset	Ten 1	Ten 2	Ten 3	Ten 4	Neuse
Constructing $\mathcal{T}$	16%	9%	8%	7%	16%
Constructing $\mathcal{T}^t$	64%	65%	66%	69%	62%
Labeling $\mathcal{T}^t$ and $\mathcal{T}$	5%	8%	7%	6%	6%
Sorting labeled cells	8%	13%	14%	13%	12%
Exporting data	6%	4%	5%	4%	5%

The HUC (Hydrologic Unit Code) scheme developed by the Water Resources Division of the United States Geological Survey (USGS) [14] is an example of a manual hierarchical watershed decomposition scheme different from the Pfafstetter method; it is a (up to) twelve level hierarchical decomposition of the terrain in the United States. Maps with eight-digit HUC labels are currently available and ten to twelve digit HUC maps are in development. However, as discussed in the full version of this paper, Pfafstetter labels have several advantages over HUC labels.

To investigate how Pfafstetter label watersheds computed using our algorithm compare to the published digital USGS 8-digit HUCs, we compared the two for a portion of the French Broad–Holston river basin (Ten 3 in the Tables and USGS HUC 060101). As can be seen on Figure 5, the watershed boundaries agree well. The Pfafstetter method divides the basin into nine sub-basins, whereas the USGS HUC only has four sub-basins in the area; however Pfafstetter basins can easily be combined to form basins that are of approximately the same extent as the USGS basins (e.g., Pfafstetter basins 7, 8, and 9 can be combined to approximate USGS sub-basin 05). A close inspection of the overlay of the two watershed decompositions show minor

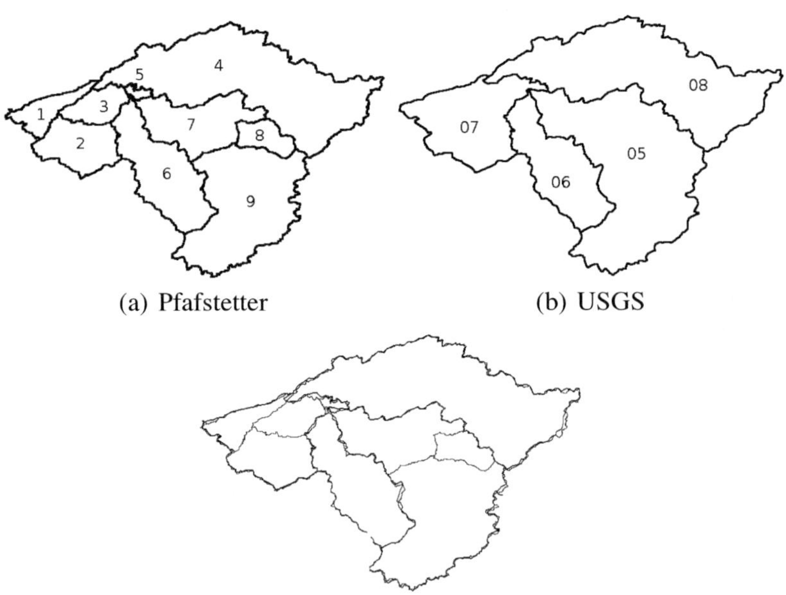

(a) Pfafstetter (b) USGS

(c) Overlay

Fig. 5. Comparison of Pfafstetter label watersheds to USGS HUCs in the French Broad–Holston river basin (HUC 060101). Common boundaries are generally in good agreement discrepancies between Pfafstetter and USGS HUC watersheds, but our Pfafstetter labels are consistent with the underlying elevation, flow direction and flow accumulation data. This consistency across multiple data layers is desirable in many GIS applications and avoids the need to rely on multiple heterogeneous data sets.

5 Conclusions

In this paper we presented an I/O-efficient algorithm for computing the Pfafstetter label of each cell of a grid terrain model. We also presented the results of a preliminary experimental study that showed that our algorithm is practically as well as theoretically efficient.

References

1. Aggarwal A, Vitter JS (1988) The Input/Output complexity of sorting and related problems. Communications of the ACM 31(9):1116–1127
2. Arge L (2002) External memory data structures. In: Abello J, Pardalos PM, Resende MGC (eds) Handbook of Massive Data Sets. Kluwer Academic Publishers,pp 313–358
3. Arge L (2003) The buffer tree: A technique for designing batched external data structures. Algorithmica 37(1):1–24
4. Arge L, Barve R, Hutchinson D, Procopiuc O, Toma L, Vengroff DE, Wickremesinghe R (2002) TPIE User Manual and Reference (ed 082902). Duke University. The manual and software distribution are available on the web at http://www.cs.duke.edu/TPIE/
5. Arge L, Chase J, Halpin P, Toma L, Urban D, Vitter JS, Wickremesinghe R (2003) Flow computation on massive grid terrains. GeoInformatica 7(4):283–313
6. Brodal GS, Katajainen J (1998) Worst-case efficient external-memory priority queues. In: Proc Scandinavian Workshop on Algorithms Theory (= LNCS 1432), pp 107–118
7. Chiang YJ, Goodrich MT, Grove EF, Tamassia R, Vengroff DE, Vitter JS (1995) External-memory graph algorithms. In: Proc ACM-SIAM Symp on Discrete Algorithms, pp 139–149
8. Freeman T (1991) Calculating catchment area with divergent flow based on a regular grid. Computers and Geosciences 17:413–422
9. Gabow HN, Bentley JL, Tarjan RE (1984) Scaling and related techniques for geometry problems. In: Proc of 16th ACM Symp on Theory of Computing, pp 135–143
10. Jenson S, Domingue J (1988) Extracting topographic structure from digital elevation data for geographic information system analysis. Photogrammetric Engineering and Remote Sensing 54(11):1593–1600

11. Morris D, Heerdegen R (1988) Automatically derived catchment boundary and channel networks and their hydrological applications. Geomorphology 1:131–141
12. O'Callaghan JF, Mark DM (1984) The extraction of drainage networks from digital elevation data. Computer Vision, Graphics and Image Processing 28
13. Peucker TK (1975) Detection of surface specific points by local parallel processing of discrete terrain elevation data. Computer Graphics and Image Processing 4:375–387
14. Seaber P, Kapinos F, Knapp G (1987) Hydrologic unit maps (= USGS water supply 2294). 63 p
15. Tarboton D (1997) A new method for the determination of flow directions and contributing areas in grid digital elevation models. Water Resources Research 33:309–31
16. Verdin KL, Verdin JP (1999) A topological system for delineation and codification of the Earth's river basins. J of Hydrology 218:1–12
17. Vitter JS (2001) External memory algorithms and data structures: Dealing with MASSIVE data. ACM Computing Surveys 33(2):209–271

Tradeoffs when Multiple Observer Siting on Large Terrain Cells

W. Randolph Franklin[1], Christian Vogt[2]

[1] Rensselaer Polytechnic Institute, Troy, New York, 12180–3590, USA
email: mail@wrfranklin.org
[2] email: chvogt@gmail.com

Abstract

This paper demonstrates a toolkit for multiple observer siting to maximize their joint viewshed, on high-resolution gridded terrains, up to 2402×2402, with the viewsheds' radii of up to 1000. It shows that approximate (rather than exact) visibility indexes of observers are sufficient for siting multiple observers. It also shows that, when selecting potential observers, geographic dispersion is more important than maximum estimated visibility, and it quantifies this. Applications of optimal multiple observer siting include radio towers, terrain observation, and mitigation of environmental visual nuisances.

Key words: terrain visibility, viewshed, line of sight, siting, multiple observers, intervisibility

1 Introduction

Consider a terrain elevation database, and an observer, $\mathcal{O}$. Define the *viewshed* as the terrain visible from $\mathcal{O}$ within some radius of interest, R, of $\mathcal{O}$. The observer might be situated at a certain height, $\mathcal{H}$, above ground level, and might also be looking for targets also at height $\mathcal{H}$ above the local ground. Also, define the *visibility index* of $\mathcal{O}$ as the fraction of the points within R of $\mathcal{O}$ that are visible from $\mathcal{O}$. This paper goes beyond merely computing viewsheds of individual observers. It combines a fast viewshed algorithm with an approximate visibility index algorithm, to site multiple observers so as to jointly cover as much terrain as possible.

The multiple observers case is particularly interesting and complex, and has many applications. A cell phone provider wishes to install multiple towers so that at least one tower is visible (in a radio sense) from every place a customer's cellphone might be. Here, the identities of the observers of highest visibility index are of more interest than their exact visibility indices, or than the visibility indices of all observers. One novel future application of siting radio transmitters will occur when the moon is settled. The moon has no ionosphere to reflect signals, and no stable satellite orbits. The choices for long-range communication would seem to include either a lot of fiber optic cable or many relay towers. That solution is the multiple observer visibility problem.

As another example, a military planner needs to put observers so that there is nowhere to hide that is not visible from at least one. This leads to a corollary application, where the other side's planner may want to analyze the first side's observers to find places to hide. In this case, the problem is to optimize the targets' locations, instead of the observers'.

Again, a planner for a scenic area may consider each place where a tourist might be to be an observer, and then want to locate ugly infrastructure, such as work yards, at relatively hidden sites. S/he may wish site a forest clearcut to be invisible to observers driving on a highway sited to give a good view. Finally, an architect may be trying to site a new house while following the planning board's instruction that, "You can have a view, but you can't be the view."

Our programs may easily produce a set of observers with *intervisibility*, i.e., their views of each other form a connected graph, but we do not impose that constraint in the experiments reported here.

In contrast to many other researchers, we consider that speed of execution on large datasets is important. Many prototype implementations, demonstrated on small datasets, do not scale up well. That may happen either because of the size and complexity of the data structures used, or because of the asymptotic time behavior. For instance, even an execution time proportional to $N \log(N)$, where N is the size of the input, is problematic for $N = 10^6$. In that case, the $\log(N)$ increases the time by a factor of 20. Some preliminary published algorithms may even be exponential if performing a naive search. Therefore, we strive for the best time possible.

In addition, large datasets may contain cases, which did not occur in the small test sets, that require tedious special programming by the designer. In a perfect software development process, all such cases would have been theoretically analyzed *a priori*, and treated. However, in the real world, testing on the largest available datasets increases our confidence in the program's correctness.

Next, a large enough quantitative increase in execution speed leads to a qualitative increase in what we can do. Only if visibility can be computed efficiently, can it be used in a subroutine that is called many times, perhaps as as part of a search, to optimize the number of observers. This becomes more important when a more realistic function is being optimized, such as the total cost. E.g., for radio towers, there may be a tradeoff between a few tall and expensive towers, and many short and cheap ones. Alternatively, certain tower locations may be more expensive because of the need to build a road. We may even wish to add redundancy so that every possible target is visible from at least two observers. In all these cases, where a massive search of the solution space is required, success depends on each query being as fast as possible.

Finally, although the size of available data is growing quickly, it is not necessarily true that available computing power is keeping pace. There is a military need to offload computations to small portable devices, such as a Personal Digital Assistant (PDA). A PDA's computation power is limited by its battery, since, approximately, for a given silicon technology, each elemental computation consumes a fixed amount of energy. Batteries are not getting better very quickly; increasing the processor's cycle speed just runs down the battery faster.

There is also a compounding effect between efficient time and efficient space. Smaller data structures fit into cache better, and so page less, which reduces time. The point of all this is that efficient software is at least as important now as ever. The terrain data structure used here is either a 1201×1201 matrix of elevations, such as from a USGS level-1 Digital Elevation Model cell, or a 2402×2402 extract from the National Elevation Data Set. The relative advantages and disadvantages of this data structure versus a triangulation are well known, and still debated; the competition improves both alternatives. This current paper utilizes the simplicity of the elevation matrix, which leads to greater speed and small size, which allows larger data sets to be processed.

For distances much smaller than the earth's radius, the terrain elevation array can be corrected for the earth's curvature, as follows. For each target at a distance D from the observer, subtract $D^2/(2E)$ from its elevation, where E is the earth's radius. The relative error of this approximation is $(D/(2E))^2$. It is sufficient to process any cell once, with an observer in the center. The correction need not changed for different observers in the cell, unless a neighboring cell is being adjoined. Therefore, since it can be easily corrected for in a preprocessing step, our visibility determination programs ignores the earth's curvature.

The radius of interest, R, out to which we calculate visibility, has no relation to the distance to the horizon, but is determined by the technology used by the observer. E.g., if the observer is a radio communications transmitter, doubling R causes the required transmitter power to quadruple. If the observer is a searchlight, then its required power is proportional to R^4.

In order to simplify the problem under study enough to make some progress, this work also ignores factors such as vegetation that need to be handled in the real world. The assumption is that it's possible, and a better strategy, to incorporate them only later.

This paper extends the earlier visibility work in [9] and [11], which also survey the terrain visibility literature. The terrain siting problem was identified as far back as 1982 by Nagy [2]. Other notable pioneer work on visibility includes [5, 18, 23]. [24] studied visibility, and provided the Lake Champlain W data used in this paper. [22] presented new algorithms and implementations of the visibility index, and devised the efficient viewshed algorithm that we use. One application of visibility is a more sophisticated evaluation of lossy compression methods [1]. [3, 4, 19] analyze the effect of terrain errors on the computed viewshed. [6] proposes modified definitions of visibility for certain applications. [17] explores several heuristics for siting multiple observers, and reports on the experimental tradeoffs that were observed. [25] discusses many line-of-sight issues. For more details on the results in this paper, see [13, 26]. An extended abstract was published in [7]. Lack of space here prevents the presentation of our experiments on the effect of lowered resolution on the quality of the siting.

The results reported here are part of a long project that may be called *Geospatial Mathematics*. Our aim is to understand and to represent the earth's terrain elevation. Previous results have included these:

1. a Triangulated Irregular Network (TIN) program that can completely tin a 10801×10801 block of 3×3 level-2 DTEDs [8, 10, 21],
2. Lossy and lossless compression of gridded elevation databases [12], and
3. Interpolation from contours to an elevation grid [15, 14, 16].

2 Siting Toolkit

This toolkit, whose purpose is to select a set of observers to cover a terrain cell, consists of four core C++ programs, supplemented with zsh shell scripts, Makefiles, and assorted auxiliary programs, all running in SuSE

Linux. The impact of this toolkit resides in its efficient processing of large datasets.

1. VIX calculates approximate visibility indices of every point in a cell. VIX takes several user parameters: R, the radius of interest, H, the observer and target height, and T, a sample size. VIX reads an elevation cell. For each point in the cell in turn, VIX considers that point as an observer, picks T random targets uniformly and independently randomly distributed within *R* of the point, and computes what fraction are visible. That fraction is this point's estimated visibility index.
2. FINDMAX selects a manageable subset, called the top observers, of the most visible tentative observers from VIX's output. This is somewhat subtle since there may be a small region containing all points of very high visibility. A lake surrounded by mountains would be such a case. Since multiple close observers are redundant, we force the tentative observers to be spread out as follows.
   a) Divide the cell into smaller blocks of points. If necessary, first perturb the given block size so that all the blocks are the same size, ± 1.
   b) In each block, find the K points of highest approximate visibility index, for some reasonable K, e.g., 3. If there were more than K points with equally high visibility index, then select K at random, to prevent a bias towards selecting points all on one side of the block.
3. VIEWSHED finds the viewshed of a given observer at height *H* out to radius, R. The procedure, which is an improvement over [11], goes as follows.
   a) Define a square of side $2R$ centered on the observer.
   b) Consider, in turn, each point around the perimeter of the square to be a target.
   c) Run a sight line out from the observer to each target calculating which points adjacent to the line, along its length, are visible, while remembering that both the observer and target are probably above ground level.
   d) If the target is outside the cell, because R is large or the observer is close to the edge, then stop processing the sight line at the edge of the cell.

   Various nastily subtle implementation details are omitted. The above procedure, due to [22], is an approximation, but so is representing the data as an elevation grid, and this method probably extracts most of the information inherent in the data. There are combinatorial concepts, such as Davenport-Schintzel sequences, which present asymptotic worst-case theoretical methods.

4. SITE takes a list of viewsheds and finds a quasi-minimal set that covers the terrain cell as thoroughly as possible. The method is a simple greedy algorithm. At each step, the new tentative observer whose viewshed will increase the cumulative viewshed by the largest area is included, as follows.
   a) Calculate the viewshed, $\mathcal{V}_i$, of each tentative observer $\mathcal{O}_i$. $\mathcal{V}_i$ is a bitmap.
   b) Let $\mathcal{C}$ be the cumulative viewshed, or set of points visible by at least one selected observer. Initially, $\mathcal{C}$ is empty.
   c) Repeat the following until it is not possible to increase $area(\mathcal{C})$, either because all the tentative observers have been included, or (more likely) because none of the unused tentative observers would increase $area(\mathcal{C})$.
      i. For each $\mathcal{O}_i$, calculate $area(\mathcal{C} \cup \mathcal{V}_i)$.
      ii. Select the tentative observer that increases the cumulative area the most, and update $\mathcal{C}$. Not all the tentative observers need be tested every time, since a tentative observer cannot add more area this time than it would have added last time, had it been selected. Indeed, suppose that the best new observer found so far in this step would add new area A. However we haven't checked all the tentative new observers yet in this loop, so we continue. For each further tentative observer in this execution of the loop, if it would have added less than A last time, then do not even try it this time.

In all the experiments described in the following sections, all the programs listed above are run in sequence. In each experiment, the parameters affecting one program are varied, and the results observed.

3 Vix and Findmax Experiments

Our goal here was to optimize VIX and FINDMAX, and to achieve a good balance between speed and quality. We used six test maps. Five of those maps were level-1 DEM maps, with 1201×1201 postings and a vertical resolution of 1 meter. The maps were chosen to represent different types of terrain, from flat planes to rough mountainous areas. Table 1 describes them, and Figure 1 shows them.

The sixth map is a National Elevation Data Set (NED) downloaded from the USGS "Seamless Data Distribution System". From the original 7.5-

Table 1. Elevation Statistical Values for the Level-1 DEM Maps

Name	Mean	Min	Max	Range	St dev
Aberdeen east	420.5	379	683	304	36.5
Baker east	1260.9	546	2521	1975	376.9
Gadsden east	257.6	118	549	431	73.7
Hailey east	1974.1	954	3600	2646	516.3
Lake Champlain west	272.5	15	1591	1576	247.8

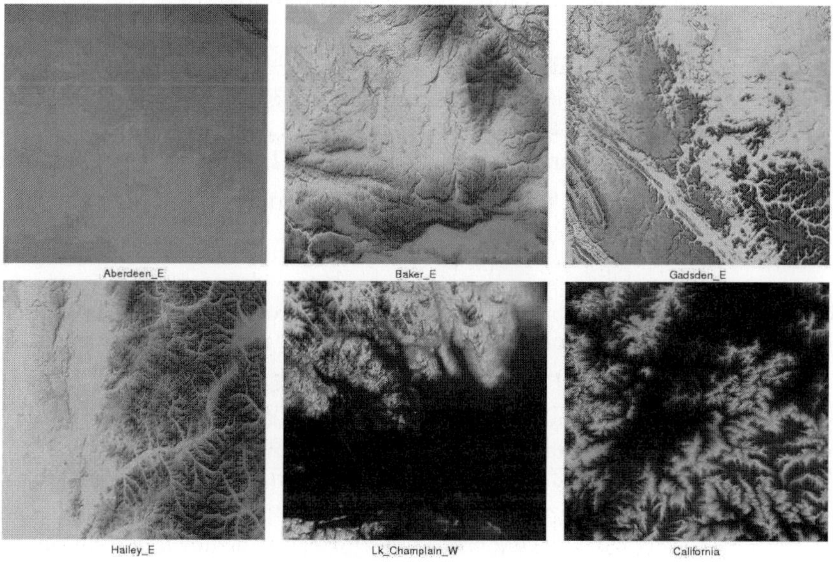

Fig. 1. The Test Cells

minute map with bounds $(41.2822, 42.4899)$, $(-123.8700, -122.6882)$, the first 2402 rows and columns were extracted. This map is from a rough mountainous region, and was chosen to test our programs on a larger higher resolution map, since some siting programs might have difficulties here. Table 2 gives its statistics.

Table 2. Elevation Statistics of the Large NED Map

Name	Mean	Min	Max	Range	St dev
California	706.9	205.9	2211.3	2005.4	2946.8

3.1 Testing VIX

These experiments tested the effect of varying *T*, the number of random targets used by VIX to estimate the visibility index of each observer. A higher *T* produces more accurate estimates but takes longer. Note that precise estimates of visibility indexes are unnecessary since they are used only to produce an initial set of potential observers, called the top observers. Actual observers are selected from this set according by how much they increase the cumulative viewshed.

We performed these tests with various values of *R* and *H*, on various datasets, The experiment consisted of five different test runs for all maps and an additional sixth test run for the larger map, as shown in Table 3. Each test run contained 10 different test cases, listed in Table 4. $T = 0$ gives a random selection of observers since all observers have an equal visibility index of zero.

Table 3. Parameter Values for the Different Test Runs of the Experiment (Italicized Case Only for the California Dataset)

Parameter	Test runs					
Radius of interest R	100	100	100	80	300	*1000*
Observer and target height H	5	10	50	10	10	*10*

Table 4. Parameter Values for the Different Test Cases of the Experiment

Parameter	Test cases									
Sample size T	0	2	5	8	12	15	20	30	50	200

Each test case was executed 20 times for the 1201×1201 maps and 5 times for the 2402×2402 map. Each time enough observers were selected to cover 80% of the terrain. (FINDMAX used a block size of 100 and 1008 top observers.) The mean number of observers over the 20 runs was reported.

Figure 2 shows results for $R = 300$ and $H = 10$. The results were normalized to make the output from the experiments with no random tests to be 1. That is, 1 is the result that can be achieved by randomly choosing top observers for SITE. Every value higher than one is worse than random, while every value lower then one is better. Figure 3 shows the Baker test case in more detail.

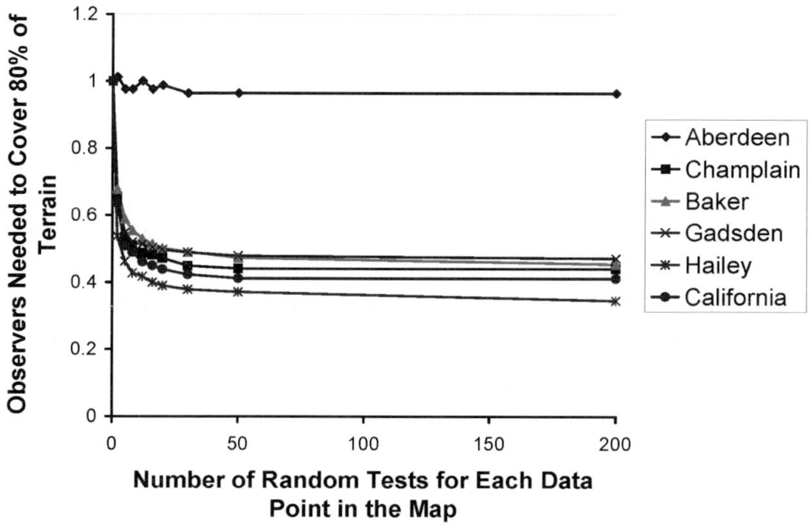

Fig. 2. Effect of Varying the Number of Tests per Observer on the Number of Observers Needed to Cover 80% of the Cell, for $R = 300$, $H = 10$

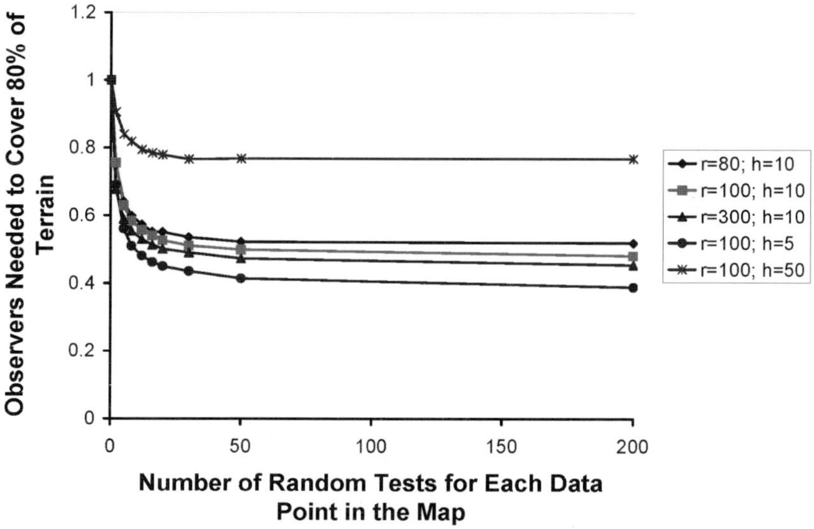

Fig. 3. Effect of Varying the Number of Tests per Observer on the Number of Observers Needed to Cover 80% of the Baker East Cell, for Various R and H

3.2 Testing FINDMAX

The purpose of the FINDMAX experiment was to evaluate the influence of FINDMAX on the final result of the siting observers problem. The two parameters evaluated were the number of top observers and the block size. The number of top observers specifies how many observers should be returned by FINDMAX. A larger number slows SITE because there are more observers to choose from, but may lead to SITE finally needing fewer observers. Therefore we want to keep this number as low as possible. It is computationally cheaper to increase the sample set in VIX than to increase the number of top observers. The block size specifies how much the top observers returned by FINDMAX are forced to spread out. A smaller number increases the number of blocks on a map and therefore reduces the number of top observers from a given block. This parameter has no influence on the computational speed.

Test Procedure

The experiment for the number of top observers consisted of 9 different test cases. It was only conducted on the level-1 DEM maps. During the experiment the values for the number of top observers ranged from 576 to 10080. In all the test runs a block size of 100 was chosen, resulting in 144 blocks. 576 top observers produced 4 observers per block; 10080 top observers produced 70 observers per block. All different values for the number of top observers are given in Table 5 together with their resulting number of observers per block.

The experiment for the block size was different for level-1 DEM maps than for the larger map. In the case of the level-1 DEM maps there were 9 different test cases with values for block size ranging from 36 to 300. This resulted in having between 1 and 1089 blocks per map. The number of top observers was chosen to be 1000.

The actual number depends on the number of blocks since each block needs the same number of top observers. In case of the larger maps there are 8 different test cases with values for block size ranging from 80 to 2402. This results in having between 1 to 900 blocks per map. The number of top observers was chosen to be 2000. The actual number depends on the number of blocks since each block needs the same number of top observers. All the different settings are given in Table 5.

Evaluation

In the sample size experiment, each test case was executed 20 times, with the entire application run each time until the site program was able to cover

Table 5. The parameters for block size and top observers are given for the different test cases. The values in the "Blocks" column represent the actual number of blocks used by FINDMAX given the size of the map and the parameters for block size and top observers. The values in the "obs/block" column represent the number of top observers that FINDMAX calculates for each block

Experiment	**Parameters & Numbers**	**Test Cases**								
Top Observers	Block Size	100	100	100	100	100	100	100	100	100
	Top Observers	576	864	1008	1296	1584	2016	3024	5040	10080
	Blocks	144	144	144	144	144	144	144	144	144
	Obs/Block	4	6	7	9	11	14	21	35	70
Block Size	Block Size	36	50	63	75	80	100	150	200	300
	Top Observers	1089	1152	1083	1024	1125	1008	1024	1008	1008
	Blocks	1089	576	361	256	225	144	64	36	16
	Obs/Block	1	2	3	4	5	7	16	28	63
Block Size	Block Size	80	100	150	200	300	500	1201	2402	
	Top Observers	2700	2304	2048	2016	2048	2000	2000	2000	
	Blocks	900	576	256	144	64	25	4	1	
	Obs/Block	3	4	8	14	32	80	500	2000	

80% of the terrain. VIX used $R = 100$, $H = 10$, and $T = 20$. The resulting number of observers needed to cover the 80% was noted, and the arithmetic mean from the results of the same test case calculated.

In the block size experiment, each test case was executed 20 times for the level-1 DEM maps and 5 times for the larger test. The evaluation of the results is slightly different. The site program ran until 100 (400 for the larger map) observers were sited. The parameters used for VIX were $R = 100$, $H = 10$, and $T = 20$. The amount of terrain visible by the final observers was then noted. The reason for changing the evaluation method was due to the problem that in some test cases we were not able to cover 80% of the cell.

Figure 4 shows for different maps how much terrain can be seen by 100 observers. For all data sets the parameters used were $R = 100$ and $H = 10$. The results are normalized by 1. For each map the best result achieved by any value for the block size was considered to be 1. The results of the experiments using different values for the block size were scaled accordingly. Therefore the highest value that can be achieved is 1. Everything below one is worse.

Figure 5 shows for the larger map how much terrain can be seen by 100 observers. For the data sets the parameters used were $R = 100$ and $H = 10$. The results are normalized by 1. The best result achieved by any value for the block size was considered to be 1. The results of the experiments using

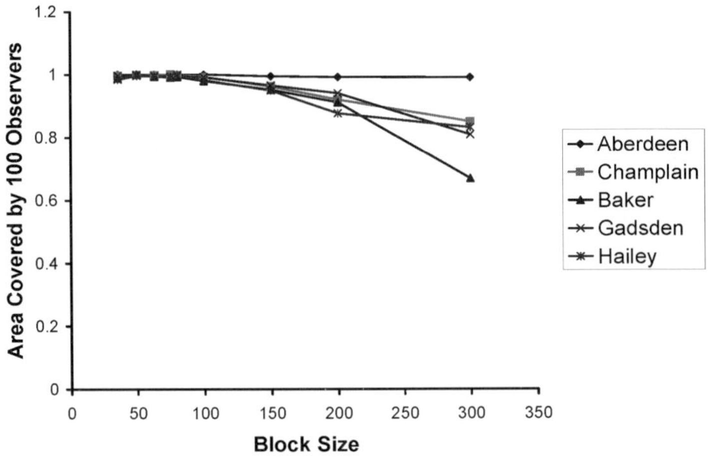

Fig. 4. Effect of Block Size on the Area Covered by 100 Observers, for Various 1201 × 1201 Cells

different values for the block size were scaled accordingly. Therefore the highest value that can be achieved is 1. Everything below 1 is worse.

Figure 6 shows for different maps how many observers are needed to cover 80% of the data. For all data sets the parameters used were 100 for the radius of interest and 10 for the observer and target height. The results are normalized by 1.The results of the experiments that were achieved by computing 576 top observers was considered to be 1. Lower values are worse.

4 Conclusions

4.1 VIX Experiment

- A sample size of 20 to 30 random tests for VIX is a good balance between the quality of the result and the computational speed. Surprisingly this value is good for a wide range of parameters and terrain types.
- VIX improved the result on the level-1 DEM maps in the best case by reducing the amount of observers needed to 39% compared to randomly

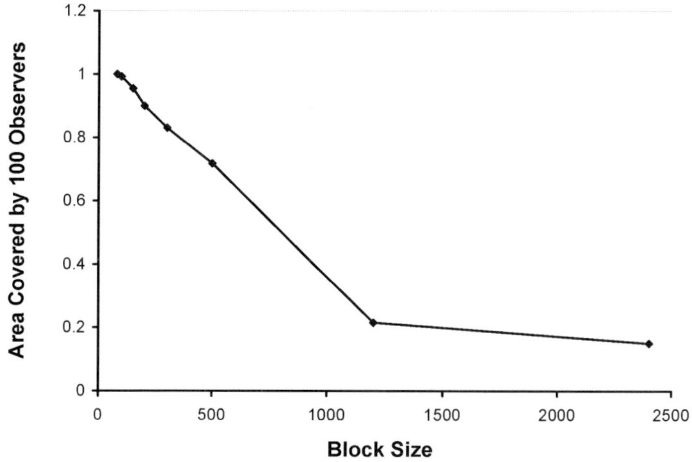

Fig. 5. Effect of Block Size on the Area Covered by 100 Observers, for the Large Cell

selecting top observers. The largest improvements were achieved for large or rough terrain for large R or low H. The smallest improvement was achieved on flat terrain.

- On the larger map the improvement of VIX was even bigger. Possible explanations are that this terrain is the roughest, and that there were fewer top observers per data point than in the smaller maps.

4.2 FINDMAX Experiment

- The block size should be chosen to be small, i.e., 2 to 5 observers per block. When covering a larger fraction of the terrain, more blocks with a smaller number of observers per block is important.
- Increasing the number of top observers in FINDMAX increases the quality of the result, but requires much more time. It is cheaper to increase the number of random tests in VIX, but there is a limitation for what can be achieved by increasing the number of random tests. The best results in the entire experiment were achieved with 10,000 top observers. This might not be obvious when comparing the graph of the results from the VIX

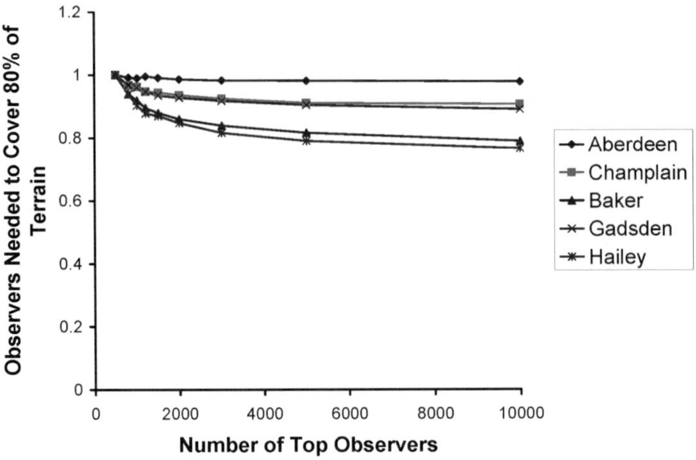

Fig. 6. Effect of Varying the Number of Top Observers Returned by FINDMAX on the Number of Observers Needed to Cover 80% of the Cell, for Various 1201×1201 Cells

experiments with the results from the FINDMAX experiments. However, during the FINDMAX experiments a relatively large number of random tests was chosen. Therefore the visibility index for FINDMAX was of a high resolution.

5 The Future

The various tradeoffs mentioned above and the above experiments illuminate a great opportunity. They tell us that shortcuts are possible in siting observers, which will produce just as good results in much less time.

Another area for investigation is the connectivity of either the viewshed, or its complement. Indeed, it may be sufficient for us to divide the cell into many separated small hidden regions, which could be identified using the fast connected component program described in [20].

There is also the perennial question of how much information content there is in the output, since the input dataset is imprecise, and is sampled only at certain points. A most useful, but quite difficult, problem is to determine what, if anything, we know with certainty about the viewsheds and observers for some cell. For example, given a set of observers, are there some regions in the cell that we know are definitely visible, or definitely hidden? We have earlier demonstrated an example where the choice of interpolation algorithms for the elevation between adjacent posts affected the visibility of one half of all the targets in the cell.

This problem of inadequate data is also told by soldiers undergoing training in the field. Someone working with only maps of the training site will lose to someone with actual experience on the ground there.

Finally, the proper theoretical approach to this problem would start with a formal model of random terrain. Then we could at least start to ask questions about the number of observers theoretically needed, as a function of the parameters. Until that happens, continued experiments will be needed.

Acknowledgements

This paper was supported by the National Science Foundation grant CCR 03-06502, and by DARPA/DSO under the Geo* program.

References

1. Ben-Moshe B, Mitchell JSB, Katz MJ, Nir Y (2002) Visibility preserving terrain simplification: an experimental study. In: Symp on Computational Geometry (ACM), pp 303–311
2. De Floriani L, Falcidieno B, Pienovi C, Allen D, Nagy G (1986) A visibility-based model for terrain features. In: Proc Second Int Symp on Spatial Data Handling, 5–10 July 1986, Seattle, Washington, pp 235–250
3. Fisher PF (1991) 1st experiments in viewshed uncertainty — the accuracy of the viewshed area. Photogrammetric Engineering and Remote Sensing 57(10):1321–1327
4. Fisher PF (1992) 1st experiments in viewshed uncertainty — simulating fuzzy viewsheds. Photogrammetric Engineering and Remote Sensing 58(3):345–352
5. Fisher PF (1993) Algorithm and implementation uncertainty in viewshed analysis. Int J Geographical Information Systems 7:331–347
6. Fisher PF (1996) Extending the applicability of viewsheds in landscape planning. Photogrammetric Engineering and Remote Sensing 62(11):1297–1302
7. Franklin WR, Vogt Chr (2004) Efficient observer siting on large terrain cells (extended abstract). In: GIScience 2004: Third Int Conf on Geographic Information Science, U Maryland College Park, 20–23 Oct 2004
8. Franklin WR (1973) Triangulated irregular network program. ftp://ftp.cs.rpi.edu/pub/franklin/tin73.tar.gz

9. Franklin WR (2000) Applications of analytical cartography. Cartography and Geographic Information Systems 27(3):225–237
10. Franklin WR (2001) Triangulated irregular network computation. http://www.ecse.rpi.edu/Homepages/wrf/sw.html#tin
11. Franklin WR, Ray C (1994) Higher isn't necessarily better: Visibility algorithms and experiments. In: Waugh TC, Healey RG (eds) Advances in GIS Research: Sixth Int Symp on Spatial Data Handling, Edinburgh, 5–9 Sept 1994. Taylor & Francis, pp 751–770
12. Franklin WR, Said A (1996) Lossy compression of elevation data. In: Seventh Int Symp on Spatial Data Handling, Delft, August
13. Franklin WR, Vogt Chr (2004) Multiple observer siting on terrain with intervisibility or lo-res data. In: XXth Congress, Int Society for Photogrammetry and Remote Sensing, Istanbul, 12-23 July
14. Gousie M, Franklin WR (1998) Converting elevation contours to a grid. In: Eighth Int Symp on Spatial Data Handling, Vancouver BC Canada, July, Dept of Geography, Simon Fraser University, Burnaby, BC, Canada, pp 647–656
15. Gousie M, Franklin WR (2003) Constructing a DEM from grid-based data by computing intermediate contours. In: Hoel E, Rigaux P (eds) GIS 2003: Proc of the Eleventh ACM Int Symp on Advances in Geographic Information Systems, New Orleans, pp 71–77
16. Gousie MB (1998) Contours to digital elevation models: grid-based surface reconstruction methods. PhD thesis, Electrical, Computer, and Systems Engineering Dept, Rensselaer Polytechnic Institute
17. Kim YH, Rana S, Wise S (2004) Exploring multiple viewshed analysis using terrain features and optimisation techniques. Computers and Geosciences 30:1019–1032
18. Lee J (1992) Visibility dominance and topographic features on digital elevation models. In: Bresnahan P, Corwin E, Cowen D (eds) Proc 5th Int Symp on Spatial Data Handling, vol 2. International Geographical Union, Commission on GIS, Humanities and Social Sciences Computing Lab, U South Carolina, Columbia, South Carolina, USA, August, pp 622–631
19. Nackaerts K, Govers G, Orshoven JV (1999) Accuracy assessment of probabilistic visibilities. Int J of Geographical Information Science 13(7):709–721
20. Nagy G, Zhang T, Franklin WR, Landis E, Nagy E, Keane D (2001) Volume and surface area distributions of cracks in concrete. In: Arcelli C, Cordella LP, Sanniti di Baja G (eds) Visual Form 2001: 4th Int Workshop on Visual Form IWVF4, (= LNCS 2051) Capri, Italy, 28-30 May. Springer-Verlag, Heidelberg
21. Pedrini H (2000) An Adaptive Method for Terrain Approximation based on Triangular Meshes. PhD Thesis, Rensselaer Polytechnic Institute, Electrical, Computer, and Systems Engineering Dept
22. Ray CK (1994) Representing Visibility for Siting Problems. PhD Thesis, Rensselaer Polytechnic Institute
23. Shannon RE, Ignizio JP (1971) Minimum altitude visibility diagram – MAVD. Simulation:256–260
24. Shapira A (1990) Visibility and terrain labeling. Master's Thesis, Rensselaer Polytechnic Institute

25. US Army Topographic Engineering Center (ed) (2004) Line of Sight Technical Working Group. http://www.tec.army.mil/operations/programs/LOS/
26. Vogt Chr (2004) Siting multiple observers on digital elevation maps of various resolutions. Master's Thesis, ECSE Dept, Rensselaer Polytechnic Institute

Scale-Dependent Definitions of Gradient and Aspect and their Computation*

Iris Reinbacher[1], Marc van Kreveld[1], Marc Benkert[2]

[1] Institute of Information and Computing Sciences, Utrecht University, P.O.box 80 089, 3508 TB Utrecht, The Netherlands
email: iris@cs.uu.nl, marc@cs.uu.nl

[2] Department of Computer Science, Karlsruhe University, Postfach 6980, 76128 Karlsruhe, Germany
email: mbenkert@ira.uka.de

Abstract

In order to compute lines of constant gradient and areas of constant aspect on a terrain, we introduce the notion of scale dependent local gradient and aspect for a neighborhood around each point of a terrain. We present three definitions for local gradient and aspect, and give efficient algorithms to compute them. We have implemented our algorithms for grid data and we compare the results for all methods.

Key words: geomorphology, multiscale, gradient, aspect, algorithms

1 Introduction

Geomorphometry is concerned with the precise measurement and quantitative description of the shape of landforms. Given a terrain T, the most important measures to classify landscapes are slope, as well as profile curvature and plan curvature. The value for slope at each point of the terrain is usually divided into *gradient*, i.e. the steepness of the slope, and *aspect*, the cardinal

* This research was partially supported by the Netherlands Organisation for Scientific Research (NWO) through the BRICKS project GADGET.

direction in which the slope faces. Using such measures and classifications, the goal is for example to derive drainage maps, specify areas in mountains that have high danger of avalanches, or study how a certain area has been formed.

Using some numerical value for gradient, and the classification convex or concave for plan and profile curvature, it is possible to identify landforms like convergent and divergent shoulders, footslopes, or crests, swales, and plains (see e.g. [5, 7, 9]). Slope can also be used to compute shaded relief maps and for parametric terrain classification.

Contour maps of terrains where each curve represents constant height are very common. Similar maps with curves representing constant gradient values – for simplicity we will call them *isogradients* – are useful in geomorphometry.

An important influencing factor of geomorphometry is at which scale we are studying the terrain. Large scale maps provide the most detail, the smaller the scale gets, the more information is lost due to generalization, and generating small scale maps from large scale maps is an important issue for automated map generalization [8]. Another problem for the classification of landforms with respect to scale is that the morphometric class may change at different scales. Fisher et al. show an example, where the same point is classified as ridge, planar slope, and channel as the scale increases [6]. The scale of the terrain model also influences for example the area of a lake or the length of a seashore. This influence can be especially crucial when the investigated spatial object already has fuzzy boundaries [3, 4]. The book [11] edited by Tate and Atkinson presents research on scale related issues in GIS.

As a new method that yields isogradient and isoaspect maps, we introduce *local gradient* and *local aspect* for each point of a terrain. The basic idea is to define the local gradient for a point based on the gradient value of the other points in some neighborhood. The size of the neighborhood can be chosen, which makes the definition scale dependent. This way, points that are close to each other are likely to have similar gradient values and from there, we can derive the desired isogradients. Our definitions yield a continuously changing value of the local gradient value on for example TINs, whereas the standard definition on a TIN does not yield continuity.

For parametric terrain classification, it is important to determine generalized isogradients. These can be computed in various ways. Firstly, we can generalize the terrain and then derive isogradients from this. Alternatively, we can first derive the isogradients from the terrain and then simplify them using line simplification. Our definitions provide a third method.

This paper is structured as follows: In Section 2 we will introduce three different definitions for local slope on a terrain. All but one will give rise to continuous isogradient lines and isoaspect areas. We have implemented our different methods for grid data and we compare the results for different sizes of the neighborhood in Section 3. In Section 4 we will present efficient algorithms to compute isogradient lines and isoaspect areas on a TIN terrain according to those definitions. For simplicity reasons, we will there deal with square neighborhoods only. Conclusions and an outlook to possible future work can be found in Section 5.

2 Definitions for Local Slope

We are given a terrain T, for example represented by a TIN (Triangular Irregular Network) or a DEM (Digital Elevation Model), where every point (except for the points and edges of a TIN) has constant values for slope. By default, slope is defined by a plane tangent to the surface at any given point. It consists of two components, the gradient, which is the maximum rate of change of altitude, and the aspect, the compass direction of the maximum rate of change. We will refer to this definitions as the *standard definitions* of gradient and aspect. Usually, gradient is measured in percent or degrees and aspect in degrees, which are converted to a compass bearing. We want to derive from these values a map with scale dependent curves or areas representing constant values of gradient or aspect respectively. Throughout the paper, we will call them *isogradients* and *isoaspects* respectively. We introduce the notion of *local slope*, i.e. *local gradient* and *local aspect*, for each point p of the terrain T and some neighborhood around p. A natural choice for such a neighborhood is a disc with radius r, centered at p, which we will denote by D_r.

It is obvious that the choice of the size of the radius influences the resulting isogradients and isoaspects and is therefore very important. If we choose r small, such that we mainly average over points close to p, we will get many, detailed isolines. If we choose a large value for r, we will get few, more smooth isolines.

The basic idea is to compute for every point p that is the center of a disc D_r the local gradient and aspect depending on the points that lie inside the radius r. We can do this in the following three ways:

1. Uniform weighing over the neighborhood D_r
2. Non-uniform weighing over the neighborhood D_r
3. Maximum value in the neighborhood D_r

Note that standard gradient and aspect values need not be defined everywhere, e.g. on the edges and vertices of a TIN. In this case we can either exclude these points from the computation, or assign values from a neighboring point. We will first give the basic definitions and properties of local gradient and aspect for a circular neighborhood around p.

2.1 Uniform Weighing over Neighborhood

Gradient

In the uniform weighing over the given neighborhood D_r, we compute the weighted gradient sum over all points in the neighborhood D_r. More formally, we can state this by the following equation:

$$gradient(p) = \frac{1}{area(D_r)} \int_{p' \in D_r} gradient(p')\, dxdy \tag{1}$$

Here p' is a point in the neighborhood D_r.

For the local gradient, we can choose whether we treat it as a simple scalar value or as a vector. The following example shows that this makes a difference. If we have two equally sized, adjacent cells with the same gradient value, say 10, and their outer normals pointing in opposite directions, we will get as average local gradient the value 10, when treating gradient as a scalar. Another way to look at it is to treat gradient as a vector at each point on the terrain T, and to compute the vector sum of all gradient vectors inside D_r. We take as local gradient at p the value indicated by the resulting vector, which gives as local gradient the value zero in this example.

When treating gradient as vector, and the neighborhood cuts off the terrain at the same height everywhere, the resulting gradient vector should indicate gradient zero (see Fig. 1). We can achieve this by normalizing the z-component of the gradient vector to 1, before weighing with the area covered by D_r. The local gradient at p is computed as the vector sum of the gradient values of all points inside D_r, divided by the total area of D_r. The resulting vector gives the local gradient value at p.

In both cases, whether we treat the given gradient value as a scalar or as a vector, the uniform weighing leads to a continuous gradient function over the whole terrain.

Aspect

The value for the aspect at a point p is always given as a vector which is projected onto a unit circle centered at p in the xy-plane. As the aspect gives the compass direction of the maximum gradient at a point, it can have the following nine discrete values: N, NE, E, etc. for the eight cardinal directions, and an additional value F (for Flat) in case the point has gradient 0.

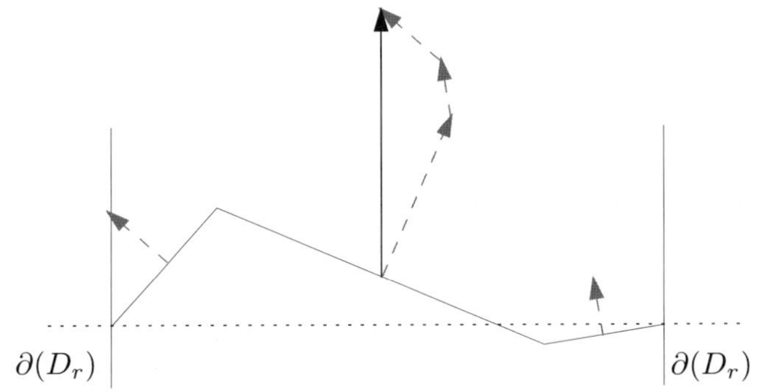

Fig. 1. When the terrain is cut off by D_r at the same height everywhere, the resulting gradient vector should point upwards, indicating value zero

As we want the local aspect to correspond to the value of local gradient at each point p, we use the same uniform weighing as for the gradient given above, and use the same vector sum. The resulting vector at p needs to be projected to the xy-plane and normalized onto the unit circle.

It is easy to see that on a TIN, the aspect values given for each point may jump from one of the possible values to any other one as we pass an edge of the terrain. However, with our method of uniform weighing, i.e. averaging over the neighborhood D_r, the local aspect value becomes continuous. Here we mean by continuous that the aspect value can change from one value to another only by moving through adjacent values or a flat zone. That means that aspect value North may change to aspect value East only by passing through the values Northeast or Flat.

2.2 Non-uniform Weighing over Neighborhood

Gradient

In the non-uniform weighing we give a higher importance to areas that are closer to p and a lower importance to areas that are closer to the rim of the disc D_r. We do this by a weight that decreases linearly with the distance to p. The weight values themselves form a cone with its tip at p. Outside of the neighborhood D_r the weight is zero.

Computing the weighted average for each point p' inside the disc D_r is equivalent to computing the height of the point on the cone C directly above p' times the gradient value at the point p'. See Figure 2 for an illustration of this definition on a TIN. Stated more formally, we get the following integral representing the local gradient:

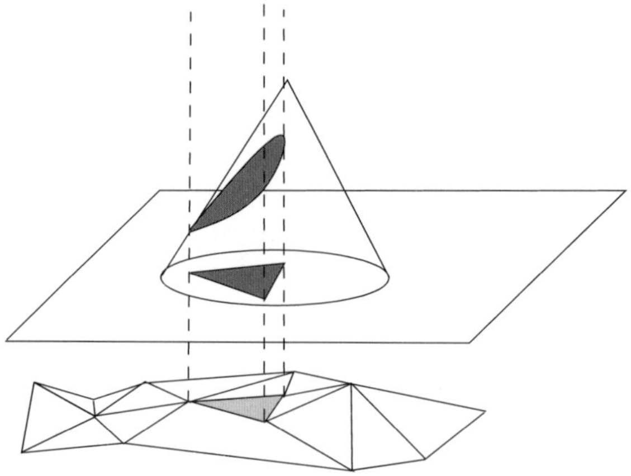

Fig. 2. Computing the non-uniform weight for one triangle of a TIN is equivalent to computing the intersection of the weight cone with the prism erected on the triangle

$$gradient(p) = \frac{1}{vol(C)} \int_{p' \in D_r} (h - \frac{h}{r}\sqrt{x^2 + y^2}) \cdot gradient(p')\, dxdy \quad (2)$$

Here, h denotes the height of the cone and r the radius of the disc, x and y are the coordinates of the point p' with respect to $p = (0, 0)$, the center of D_r.

Again, we can treat the gradient value as either a scalar or a vector. In the first case we compute the local gradient value by summing up all weighted values and dividing by the total volume of the cone. In the second case, when treating the gradient value as a vector, we again need to normalize the z-component before the weighing, and the local gradient value is a weighted vector sum as before. Also the non-uniform weighing method yields a continuous gradient function over the whole terrain.

Aspect

For the local aspect vector at a point p, weighted non-uniformly over the disc, we can use the same model with the linearly decreasing weight that corresponds to the local gradient. The local aspect is the weighted vector sum with the resulting vector at p, projected into the xy-plane and normalized onto the unit circle centered at p.

The model of non-uniform weighing also gives local aspect values that are continuous, which means their values can only change to adjacent values.

2.3 Maximum Value in Neighborhood

Gradient

In the maximum value method, we set the local gradient to be the absolute maximum gradient value from p to any other point p' inside D_r. The gradient between two points p and p' in space is defined as their z-distance divided by the Euclidean distance in the xy-plane, which leads to the following equation for the local gradient:

$$gradient(p) = \sup_{p' \in D_r \setminus \{p\}} \frac{|p_z - p'_z|}{\sqrt{(p_x - p'_x)^2 + (p_y - p'_y)^2}} \tag{3}$$

As the local gradient is not averaged but depends only on one single value, there is no distinction between scalar or vector value. Furthermore, note that the point p' giving the maximum gradient may not be unique. This model also leads to a continuous gradient function.

Aspect

As local aspect at p we choose the vector between p and the point that gives the maximum gradient and project it into the xy-plane and onto the unit circle centered at p. When there is more than one point p' giving the maximum gradient, the aspect is generally not well defined. In this case, we could choose the point p' that has the largest z-distance to p.

Note that this is the only definition that does not lead to continuous local aspect values, as whenever the point of maximum gradient changes, the local aspect jumps directly to the new maximum, without having to pass through adjacent values (or Flat) first.

3 Experimental Results for Grid Data

We have implemented our methods for DEM data in Java. We downloaded the data from [1], it is a 358 by 468 pixel grid representing an area of 11.3 by 14.5 kilometers northwest of Denver, USA, with a grid spacing of approximately 30 meters.

In all figures that we will discuss in the following, we have used a circular neighborhood around p of given radius r. We approximate the circle D_r on the grid as follows: Every grid cell whose center is closer to p than r is part of D_r, all other grid cells are not. The chosen values for the isogradients were 0.3 (light gray), 0.6, 0.9, 1.2 and 1.5 (black). The aspect maps correspond to the gradient maps. There, white represents a flat area, black has aspect facing South and light gray is aspect North. To increase readability of the aspect maps, no distinction was made between the coloring of East and West.

Note that, when the chosen neighborhood did not fully cover the terrain, i.e. at the edges and corners of the data set, we do not have a proper neighborhood to define local gradient and aspect, so we set the value for local gradient and aspect to zero. This is the reason for having a frame of width r around each of the figures.

Due to space restrictions, we can only present a small subset of our derived maps. The full set of figures for the comparison of all methods for three different radii can be found in [10].

3.1 Comparison of Different Radii

Here we discuss the influence of different radii on the outcome of isogradients and isoaspects. In order to investigate this influence separated from the different methods, we only look at the uniformly weighted scalar gradient and aspect method for the four different radii $r = 0, 5, 10, 15$.

We can see in Figure 3 that the smaller the radius and therefore the area of influence, the more detailed the gradient map becomes. We get many isolines with highly detailed boundaries that are relatively short. Also the spacing between isolines with different values is small. When gradually increasing the radius, we get less detailed contour maps, the isolines get longer and more smooth and also the distance between isolines with different values gets larger.

The situation for aspect is similar (see Fig. 4). Again, the aspect map with smallest radius of the neighborhood shows the highest detail, and the map with largest radius has the largest and smoothest areas. Note especially the gradual decrease of size of the flat white area representing a lake in the upper right corner of the aspect map. As an area is flat only if the aspect points exactly in z-direction, even one cell inside the neighborhood that is outside the flat area will change the whole outcome of the averaging and cause the local aspect to be non-flat.

3.2 Comparison of Different Methods

Here we compare the outcome of all five different methods for gradient for constant radius $r = 10$. We see in Figure 5 that the maximum value method shows the most detailed map. This is as expected, as the maximum value

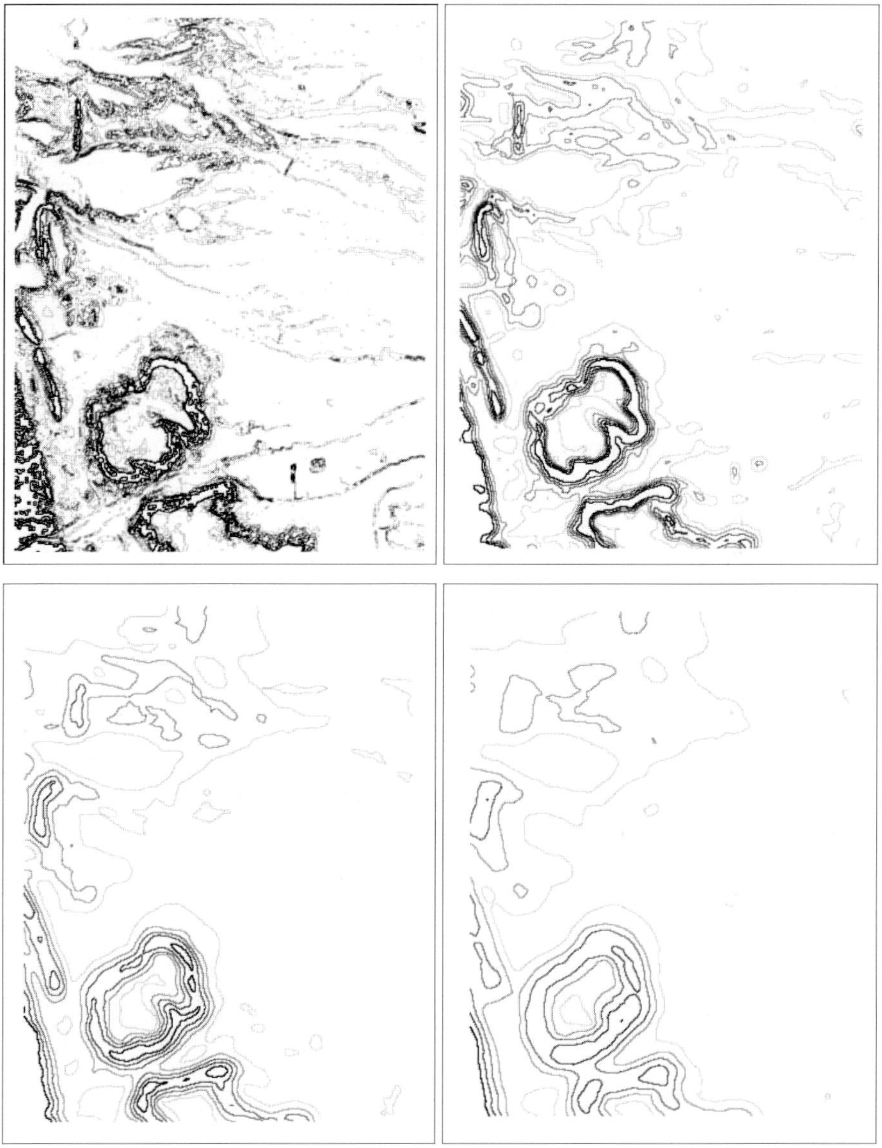

Fig. 3. Comparison of the four different radii for the gradient, uniformly weighted scalar method. From top to bottom and left to right: $r = 0$ (standard), $5, 10$ and 15

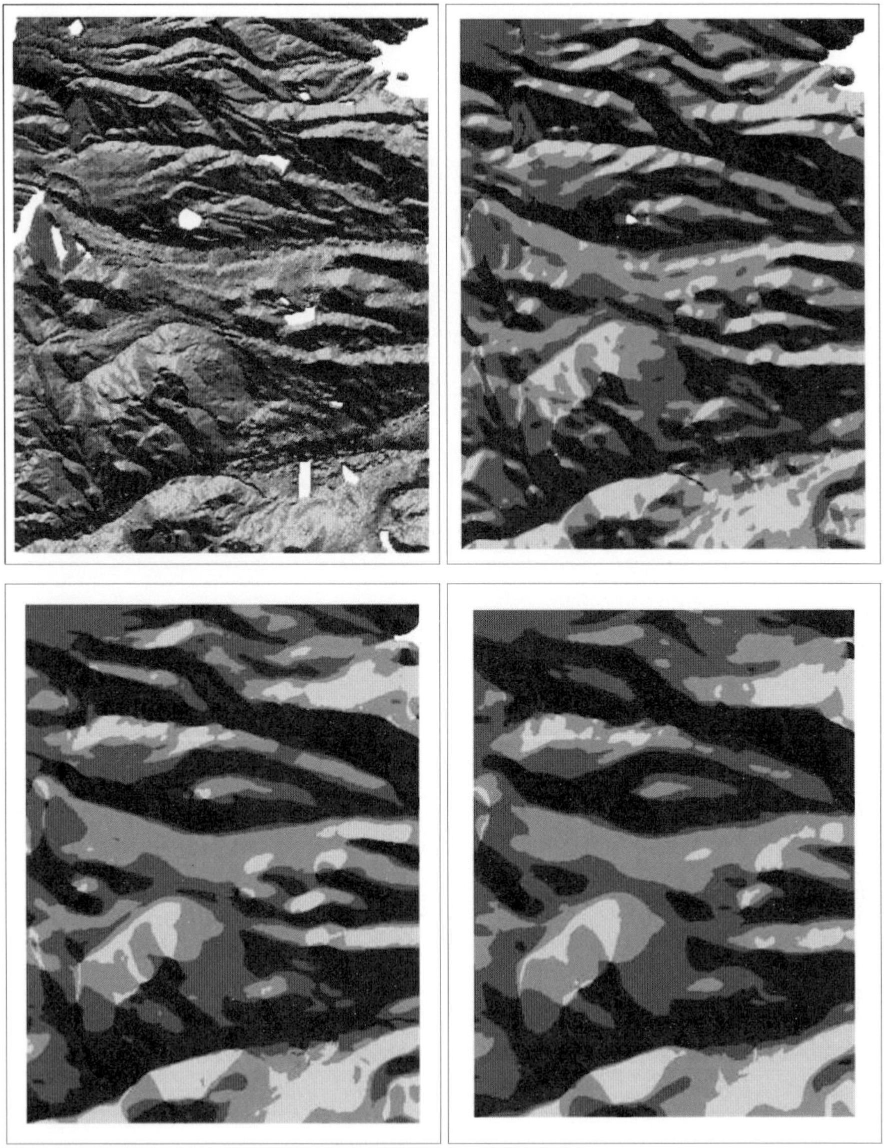

Fig. 4. Comparison of the four different radii for the aspect, uniformly weighted method. From top to bottom and left to right: $r = 0$ (standard), $5, 10$ and 15

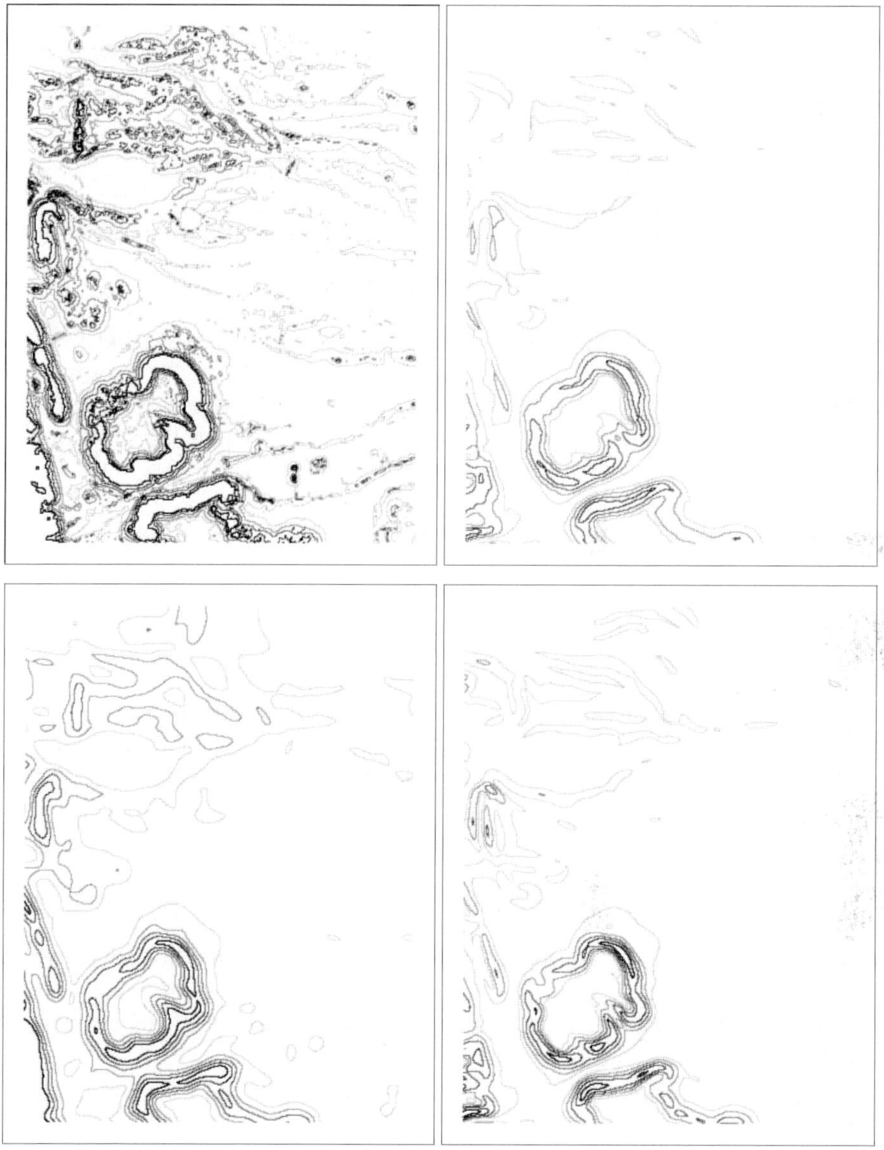

Fig. 5. Comparison of four different methods for gradient with radius $r = 10$. From top to bottom and left to right: Maximum value, uniform weighing (vector method), non-uniform weighing (scalar), non-uniform weighing (vector)

may change rapidly inside the neighborhood even for adjacent points. Furthermore we see that the scalar methods, both for uniform (see Fig. 3, bottom left) and non-uniform weighing, produce more detailed maps than the vector methods. This is especially visible in the upper half of the figures. Also this was expected, as the vector method provides additional averaging. There seems to be more detail in the non-uniform weighing methods than in the uniform weighing methods. However, this difference is small and which method should be chosen depends on the application.

In general, it seems that the uniform weighing method, either scalar or vector, provide the best output when looking for a generalization of the isogradients. When high detail is desired, the maximum value method is to be preferred. It is also the only method that preserves the highest isogradient class well. It gives a smoothing of different character than the other methods.

Due to space limitations, we cannot include the outcome of the different methods to compute isoaspect maps. They can be found in [10]. However, the general observations made in case of gradient hold here as well.

4 Algorithms for Local Slope and Isolines

In this section we will describe efficient algorithms to compute contour maps of constant local gradient and constant local slope. For reasons given in the next paragraph, we will limit ourselves to present the algorithms for square neighborhoods D_r around a point p on a TIN terrain. Note that all algorithms can easily be adapted to the use ofregular polygons, if a better approximation to a circular neighborhood is desired. Furthermore, our focus will lie on computing the local scalar gradient for each method; however, computing the local vector gradient and local aspect is very similar for all three methods.

As all methods can be implemented in a straightforward way for a gridded DEM by using windows with radius r around p, we will focus on presenting the algorithms for a TIN. In most cases, the neighborhood around p will intersect more than one triangle of the TIN. For the first two methods, where we compute the (weighted) average, we need to know the area of each triangle that is intersected by the neighborhood. This area of intersection is given by a function in the coordinates of p. In case of a circular neighborhood this function may consist of up to a linear number of terms, all involving square roots. Such functions generally cannot be simplified, and hence, the equations describing curves of constant gradient are too complex to be used. Therefore we describe only the case of a square neighborhood with side length $2r$, denoted by D_r. In this case, the area of intersection is given by a simple quadratic function, where the problems mentioned above do not occur.

Note that gradient and aspect are not well-defined on the edges and vertices of a TIN. We overcome this problem by assigning the values of one of the neighboring triangles to the edges and vertices of the TIN.

As mentioned in the last paragraph, we want the local gradient (aspect) at each point p to be determined by a function $f(x, y)$, depending only on the coordinates of p. It therefore makes sense to subdivide each triangle of the TIN into cells such that for each point inside a cell the function $f(x, y)$ is determined by the same features, i.e. the same edges or vertices of the TIN. Whenever we cross the boundary of such a cell, this list of features influencing $f(x, y)$ changes. In each of these cells, the function given there can be evaluated in constant time to compute the local value for gradient or aspect at each point p. Also, for every chosen gradient value we can compute the isogradient in one cell in constant time. We have the following lemma for the number of cells and a corollary on the time bound to compute this subdivision.

Lemma 1. *(From [12]) Let E be the set of n edges of a planar subdivision, and let Q be a square of fixed size and orientation. There are $O(n^2)$ distinct subsets of edges of E intersected by Q for the placements of Q.*

Corollary 1. *(From [12]) Given a connected subdivision S with n edges and a square Q, the arrangement representing all distinct placements of Q with respect to S can be constructed in $O(n \log n + k)$ time, where $k = O(n^2)$ is the number of distinct placements.*

4.1 Uniform Weighing Method

In the uniform weighing method, we construct the subdivision as explained above. For every center point p of D_r in each cell we get a quadratic function $f(x, y)$ in the coordinates of p, which gives the local gradient. We can evaluate this function for every point in one cell in constant time. Also, for every chosen gradient value we can compute the isogradient in one cell in constant time.

The subdivision of the given terrain into cells is a representation of the gradient (aspect) of every point of the terrain. The image of the gradient function is continuous and constists of a quadratic number of patches. It is not differentiable at the boundaries of the cells. We can summarize:

Theorem 1. *Given a TIN terrain T with n triangles, we can compute a subdivision into cells, such that in each cell the uniform weighted local gradient (aspect) is a fixed, quadratic function $f(x, y)$ in the coordinates of* *each point p and a square neighborhood D_r with side length $2r$ around it in $O(n \log n + k)$ time, where $k = O(n^2)$ denotes the number of cells overall.*

4.2 Non-uniform Weighing Method

In the non-uniform weighing method, we give a higher importance to points that lie close to p and a lower importance to points that lie at the boundary of the neighborhood. We can do this by applying a weight that decreases linearly with the distance to p. As our neighborhood D_r is now a square, it is natural to use the Manhattan distance. This way, the weight values form a pyramid with its tip at p.

Computing the weight for each point inside a triangle is equivalent to computing the intersection volume of the pyramid P centered at p with a prism A, which has a triangular base and edges parallel to the z-axis. The volume of one such prism can be computed as

$$V_A = A_{xy} \cdot \frac{a+b+c}{3} \tag{4}$$

where A_{xy} denotes the area of the projection of the terrain triangle into the xy-plane and a, b, c denote the side lengths of the three sides of the prism. Note that there are resulting solids that need not be such prisms, e.g. for the triangle containing p. However, all these solids can be split up in triangular-based prisms, such that we can use Equation (4) given above.

The algorithm is as follows: Again, we subdivide each triangle into cells as above. Here, we can further subdivide these cells. First, such that each prism A has a triangular top; we do this by projecting the slanted sides of the pyramid onto the terrain. Second, such that each prism A has a triangular base as well, by triangulating each of the cells. Now each cell has linear boundaries and is the base of exactly one triangular prism A. For each cell we can directly apply Equation (4) to compute the weight for each point. The asymptotic bound on the number of cells remains the same: Projecting the slanted sides of the pyramid can at most double the number of cells, which can now have at most six sides. Therefore, the number of cells that have a fixed, cubic function $f(x, y)$ for local gradient (aspect) remains quadratic.

Again, the subdivision of the given TIN into cells is a representation of the gradient (aspect) of every point of the terrain. We get a gradient function that is continuous, consists of a quadratic number of patches, and is not differentiable at the boundaries of the cells. We summarize our findings in the following theorem.

Theorem 2. *Given a TIN terrain T with n triangles, we can compute a subdivision into cells, such that in each cell the non-uniform weighted local* *gradient (aspect) is a fixed, cubic function $f(x, y)$ in the coordinates of each point p and a square neighborhood D_r with side length $2r$ around it in $O(n \log n + k)$ time, where $k = O(n^2)$ denotes the number of cells overall.*

4.3 Maximum Value Method

In the maximum value method, the maximum absolute gradient inside D_r with respect to p is taken as local gradient. We observe that on a TIN, the maximum gradient can only occur at points or edges of the terrain or the boundary of D_r.

Computing the gradient from p to any other point p' inside D_r is straightforward. We can show that for an arbitrary line ℓ and any point p that is not on ℓ, the gradient between p and ℓ can have only one maximum. That means that the maximum gradient between a point p and an edge e of the terrain T either lies in the interior of e or at one of its endpoints. Note that when only part of an edge e lies inside D_r, we only consider this part for the computation of the maximum gradient. The point of maximum gradient on ℓ can be determined by using simple analytical methods starting from Equation (3) and the equation for ℓ and the coordinates of p. This takes constant time for each line and thus for each edge e.

The algorithm is as follows: We generate a subdivision of each triangle as given above. We can further subdivide each cell such that in each cell of this refined subdivision there is exactly one vertex or edge of the TIN or the boundary of D_r that defines the maximum gradient. That means that the boundary of each cell is determined by exactly two different features, and the gradient to both of them is the same.

We can determine the boundaries between the features as follows: For every point in each cell of the subdivision of the triangle, we determine the gradient function to each of the features inside the cell. For each feature, this will be some surface in three dimensions that is the image of a function in x and y. To find the one feature that determines the maximum for a given point, we need to find the pointwise maximum of all surface patches inside the cell. The pointwise maximum of m surfaces is called their upper envelope, it has complexity $O(m^{2+\epsilon})$ and can be computed in $O(m^{2+\epsilon})$ time [2], where $\epsilon > 0$ is an arbitrarily small constant. In each cell of the refined subdivision that we determine this way, there is only one feature such that the gradient from each point in this cell to that feature is maximal.

It is easy to see that the pointwise maximum of all surface patches over the whole terrain is the representation of the gradient of every point of the terrain.

The number of patches that are continuous everywhere, but not differentiable at the boundaries of the cells, is $O(n^{4+\epsilon})$.

For the local aspect, we take the vector from p to the point with maximal gradient and convert it to the aspect value. It is easy to see that the subdivision we get, has the same boundaries as the subdivision for local gradient. Note that now, the representation of the local aspect is not continuous. This is easy to see, as whenever we cross a boundary of a cell, the point of maximum gradient may change abruptly, and so will the vector that points to it. Therefore, this method does not lead to continuous local aspect values.

We denote by m the maximum number of edges of the terrain that are intersected by any square of fixed side length $2r$ and summarize:

Theorem 3. *Given a TIN terrain T with n triangles, we can compute a subdivision into cells, such that in each cell the maximum value local gradient (aspect) is a fixed function $f(x, y)$ in the coordinates of each point p and a square neighborhood D_r with side length $2r$ around it in $O(n^2 \cdot m^{2+\epsilon})$ time.*

Note that in theory, m can be as large as $\Theta(n)$, but typically it is much smaller.

5 Conclusions and Future Work

In this paper we have introduced the notion of scale dependent local slope, i.e. local gradient and local aspect, for each point of a given terrain. We suggested and analyzed three different methods to compute local slope inside a neighborhood with radius r around a point. From there, it is straightforward to compute contour maps with lines of constant gradient or areas of constant aspect.

The results of the implementation on a gridded DEM show the expected smoothing behaviour. The uniform weighing scalar method shows good output, as it is also the easiest to implement, it may be the method of choice. However, which method to prefer for a certain application, depends on the desired level of detail of the result. The same holds for the choice of radius.

Future work includes more extended experiments, for example to compare the length of isogradients and the areas of isoaspects for different methods and radii. Furthermore, it is interesting to develop similar methods as investigated in this paper to compute scale dependent contour maps for plan and profile curvature and other measures used in geomorphometry.

Acknowledgements

We would like to thank Peter Lennartz for valuable discussions and Tim Adelaar for implementing our algorithms and conducting the experiments.

References

1. Mapmart homepage – http://www.mapmart.com/samples/samples.htm
2. Agarwal PK, Schwarzkopf O, Sharir M (1996) The overlay of lower envelopes and its applications. Discrete Comput Geom 15:1–13
3. Burrough PA, Frank AU (eds) (1996) Geographic Objects with Indeterminate Boundaries. GISDATA II. Taylor & Francis, London
4. Cheng T, Fisher P, Li Z (2004) Double vagueness: Effect of scale on the modelling of fuzzy spatial objects. In: Developments in Spatial Data Handling, Proc 11th Int Symp on Spatial Data Handling, pp 299–314
5. Evans IS (1986) The morphometry of specific landforms. Int Geomorphology, Part II:105–124
6. Fischer P, Wood J, Cheng T (2004) Where is Helvellyn? Fuzziness of multi-scale landscape morphometry. In: Transactions of the Institute of British Geographers 29(1):106–128
7. Mitchell CW (1991), Terrain Evaluation, 2nd ed. Longman, London
8. Müller JC, Lagrange JP, Weibel R (eds) (1995) GIS and generalization. GISDATA I. Taylor & Francis, London
9. Pennock DJ, Zebarth BJ, Jong E de (1987) Landform classification and soil distribution in hummocky terrain. Saskatchewan, Canada Geoderma 40:297–315
10. Reinbacher I, Kreveld M van, Adelaar T, Benkert M (2006) Scale-dependent definitions of gradient and aspect and their computation. Technical Report UU–CS–2006–011, Utrecht University
11. Tate N, Atkinson P (eds) (2001) Modelling Scale in Geographical Information Science. John Wiley and Sons, Chichester
12. Kreveld M van, Schramm E, Wolff A (2004) Algorithms for the placement of diagrams on maps. In: Proc 12th Int Symp ACM GIS (GIS'04), pp 222–231

Development Density-Based Optimization Modeling of Sustainable Land Use Patterns

Arika Ligmann-Zielinska[1], Richard Church[2], Piotr Jankowski[3]

[1] Departments of Geography, San Diego State University, Univ. of California, Santa Barbara; email: ligmannz@rohan.sdsu.edu
[2] Geography Department, University of California Santa Barbara, CA 93106-4060, USA; email: church@geog.ucsb.edu
[3] Department of Geography, San Diego State University, CA 92182-4493, USA; email: piotr@geography.sdsu.edu

Abstract

Current land use patterns with low-density, single-use, and leapfrogging urban growth on city outskirts call for more efficient land use development strategies balancing economy, environmental protection, and social equity. In this paper, we present a new spatial multiobjective optimization model with a constraint based on the level of neighborhood development density. The constraint encourages infill development and land use compatibility by requiring compact and contiguous land use allocation. The multiobjective optimization model presented in this paper minimizes the conflicting objectives of open space development, infill and redevelopment, land use neighborhood compatibility, and cost distance to already urbanized areas.

1 Introduction: Sustainable Land Use Patterns

Current urban land uses exhibit inefficient patterns that are of major concern for sustainable development (Leccese et al. 2000; Silberstein and Maser 2000; Ward et al. 2003; Williams et al. 2000). Low residential densities, sprawl and leapfrog fragmentation of urbanization, rapid open space development at the urban edge without considering the redevelopment of declining inner cities, and patches of single land use all dominate the current urban form (Galster et al. 2001; Grimshaw 2000; Silberstein and Maser 2000; Williams 2000). Such trends lead to an increasing ethnic and economic separation, deterioration of the environment, loss of agricultural land and wilderness, and the erosion of society's architectural heritage (Leccese et al. 2000, preamble). Research suggests that up to 70% of the consumed energy is dependent on land use arrangements (Barton 1990). In consequence, the importance of sustainable land use allocation cannot be underestimated.

In this paper, we report findings from experimenting with a new spatial optimization model for sustainable land use planning. Motivated by a conflict-laden nature of urban activity allocation, we have developed a multiobjective sustainable land use allocation model that supports infill development, balances conflicts of neighboring land uses, encourages accessibility to existing urban areas, and analyzes tradeoff between the conversion of undeveloped land and redevelopment. The novelty of the model is a constraint promoting neighborhood land use contiguity and compactness. The constraint imposes a user specified minimum value of neighborhood development density for a given location. By testing the model in the land use-planning context of Chelan, a small town located on the eastern slopes of Cascade Mountains in Washington State, we generate a set of compromise spatial alternatives representing conflicting development goals.

Williams et al. (2000) described a list of building blocks for sustainable urban form. These include urban layout and size, housing type, open space distribution, mix of uses, and various growth options like intensification, extensification, or decentralization. Many sustainable urban forms may stem from these postulates (Guy and Marvin 2000). Moreover, these principles are spatially explicit in their majority, and therefore GIS-coupled spatial analysis and modeling may provide a potentially useful methodology serving sustainable land use planning. Consequently, we propose to define *sustainable urban land use allocation* as a normative model that recognizes and evaluates current land use pattern and introduces changes that promote compatibility of adjacent land uses, neighborhood compactness, infill development, and politically defensible redevelopment.

The remainder of this article is structured as follows. Section two provides a brief synopsis of existing spatial optimization constraints and models for land use allocation with the focus on their usefulness in guiding our modeling effort. In section three, we formulate and describe the model. We develop a density based design constraint (DBDC) for compact neighborhood development, which promotes infill and counteracts a fuzzy urban-rural fringe. Section four reports about land use planning for Chelan, which served as the context for model evaluation. We verify model assumptions through a qualitative and quantitative assessment of trade-offs among model objectives. The final section summarizes the research presented here and outlines future model refinements.

2 Optimization Techniques for Land Use Allocation

The utility of optimization as a normative tool for spatial problems is widely recognized (Arthur and Nalle 1997; Church 1999; Church 2002; Chuvieco 1993; Malczewski 1999). These generative techniques allow for multiple scenario analysis, where the outcomes obtained are non-inferior or Pareto optimal to the objectives contained in the model (Cohon 1978). Land use allocation problems comprise a subset of spatial optimization models, and involve efficient distribution of activities over feasible sites in order to meet demand and maintain physical, economic, environmental, or social constraints. Models involving allocation of spatial activities are not unique and span over such domains as urban and regional planning, forest management, reserve design, site restoration, facility location, land acquisition, or waste landfill siting (Aerts and Heuvelink 2002; Aerts et al. 2003, Benabdallah and Wright 1992; Brookes 2001; Brotchie et al. 1980; Chang et al. 1983; Cova and Church 2000; Dökmeci et al. 1993; Gilbert et al. 1985; Minor and Jacobs 1994; Nalle et al. 2002a, 2002b; Ward et al. 2003; Williams 2002; Williams and ReVelle 1996; Wright et al. 1983). The majority of land use allocation models involve integer programming, where the variables are often binary, and represent two-choice decisions of whether or not to allocate a particular activity to a specific site (Malczewski 1999).

The major shortcoming of most allocation models is the absence of existing land use patterns in model initialization (Church 1999, 2002). The models usually convert completely undeveloped (green-field) areas, where every allocation of activity is new to the land under consideration (a revolutionary approach). This is a particularly flimsy assumption in urban planning, which by and large involves a modification of an existing situation (an evolutionary approach) and not building from scratch. "In brown-field planning (i.e. adding to, taking away, or transforming an existing configuration) there must be the capability to solve for a new configuration which maintains much of what currently exists and which adds or moves specific facilities to better locations" (Church 1999, p 302). A valuable exception to green-field development is a reserve network design model by Nalle et al. (2002), which extends an existing reserve pattern and also evaluates spatial efficiency of this scenario against an open space conversion case. The sustainable urban land use allocation model, which we present in this paper, builds upon the brown-field planning premise.

To the authors' knowledge, the only spatial optimization model that explicitly addresses urban sustainability is the regional scale model by Ward et al. (2003). Their model allocates over time zoning options such as rural residential, urban residential, commercial, industrial, recreational, and special use to aggregate planning units based on regional population projections. Sustainability is addressed through incorporation of economic, social, or environmental requirements to the model, and minimization of deviations from these targets. The model produces fractions of residential use allocated to aggregate spatial units, and is further integrated with a local Cellular Automata (CA) model that assigns these zoning proportions to finer-grained spatial units based on several local scale suitability measures. While the Ward et al. (2003) model presents a significant step towards modeling sustainable land use allocation, it is still inadequate in terms of addressing the variety of spatially explicit sustainability aspects (like contiguity, compactness, or infill development) mainly due to the fact that it is a regional model and these spatial characteristics may be obtained only indirectly through the integration with the CA model.

A variety of sprawl development metrics could be utilized in spatial optimization for sustainable urban activity allocation like contiguity, which represents the degree to which a specific use has been allocated to land in an unbroken fashion (Aerts et al. 2003; Galster et al. 2001; Williams 2002; Wright et al. 1983), or compactness, defined as an allocation of like land uses next to or in direct proximity of each other (Aerts et al. 2003).

We have classified the existing contiguity and connectedness land use allocation constraints into three methodological categories: network based contiguity, edge based compactness, and adjacency based clustering. Network based contiguity constraints use various concepts of graph theory in the quest of ensuring contiguous land patterns, where locations are nodes and their adjacency is represented with arcs. These include network flow (Shirabe 2005), dual graph (Williams 2002), and ordered closeness (Cova and Church 2000). Edge based compactness optimizes the ratio or product of development perimeter length to a certain measure of total development area (Benabdallah and Wright 1992; Gilbert et al. 1985; Minor and Jacobs 1994). A variation of this approach is a core/buffer constraint that produces clusters of land uses surrounded by buffer zones (Aerts et al. 2003; Williams and ReVelle 1996). Finally, clustering based on adjacency emerges from formulations that use concepts of direct topological touching of spatial units (Aerts et al. 2003; Fischer and Church 2003; Wright et al. 1983). In the sections that follow, we propose an alternative approach to encourage local clustering of land uses, based on the idea of core/buffer connectedness, and called here a density based design constraint (DBDC).

3 A Multiobjective Model for Sustainable Land Allocation

We chose four objectives as suitable for model formulation:

1. Minimization of new development. This encourages redevelopment and efficient urban land utilization
2. Minimization of redevelopment. This encourages only the economically defensible spatial change. By varying the importance between objective one and two, we allow for tradeoff between new development and redevelopment
3. Minimization of the incompatibility of adjacent allocated land uses. This helps to promote a quality of environment
4. Minimization of distance to already developed areas, which acts as a coarse-equivalent to accessibility.

3.1 Model Notation

The model was developed for a regular grid of cells. As already mentioned, the land use of each cell is homogenous. Also, we do not allow for urban land redevelopment leading back to open space, which is in practice very unlikely since once urbanized, land "typically stays that way" (Nalle et al. 2002b, p 60; Silberstein and Maser 2000). Our model utilizes the concept of neighborhood, which we define as a Moore neighborhood of range r = 1 (Weisstein 2005).

Given these assumptions, consider the following notation:

i,j	$1,2,\ldots,n$; where n is the total number of cells in the study area
l, m	$1,2,\ldots k$; types of urban land uses
u	undeveloped land use type
D_l	set of cells that already have land use l
D	set of developed cells, all subsets of D are mutually disjoint
U	set of cells of undeveloped land
B_j	set of j's neighbors that are undeveloped
e_j	existing land use of cell j
t_l	number of cells that initially have land use l, where $l = 1,2,\ldots k$
c_{lm}	estimated compatibility index between land use l and m (the higher the more compatible the land uses), if $l = m$, then $c_{lm} = 1$, in the model l is represented by d_j
s_j	number of initially developed cells within j's neighborhood
r_j	resistance to change for already developed j, the higher the coefficient, the less probable that redevelopment occurs
$dist_j$	distance to the nearest developed area (in cells)
v_l	estimated demand for land use l (in cells)

d_j dominant urban land use type within the neighborhood of j[1], The dominant land use type is the preferred (most compatible) land use to be allocated at j; $d_j = 1,2,\ldots k$ or $d_j = u$, if the neighborhood is undeveloped[2]

b minimum required number of neighboring cells that are developed after allocation

Variables

x_{jum} 1, if undeveloped land at location j is changed to m; 0, otherwise

x_{je_jm} 1, if current land use e_j at location j is changed to m, $m \neq e_j$; 0, otherwise

3.2 Model Formulation

Minimize

$$\sum_{j \in U} \sum_{m} x_{jum} \tag{1}$$

$$\sum_{j \in D} \sum_{m \neq e_j} r_j x_{je_jm} \tag{2}$$

$$\sum_{j \in U} \sum_{m} (1 - c_{d_jm}) x_{jum} + \sum_{j \in D} \sum_{m \neq e_j} (1 - c_{d_jm}) x_{je_jm} \tag{3}$$

$$\sum_{j \in U} \sum_{m} dist_j x_{jum} \tag{4}$$

Subject to

$$\sum_{m \neq e_j} x_{je_jm} \leq 1; \forall j \in D \tag{5}$$

$$\sum_{m} x_{jum} \leq 1; \forall j \in U \tag{6}$$

$$t_l - \sum_{j \in D_l} \sum_{m \neq l} x_{jlm} + \sum_{j \in (D - D_l)} x_{je_jl} + \sum_{j \in U} x_{jul} \geq v_l; \forall l \tag{7}$$

$$s_j + \sum_{i \in B_j} \sum_{m} x_{ium} \geq b \sum_{m} x_{jum}; \forall j \in U \tag{8}$$

$$x_{jlm} \in \{0,1\}; x_{jum} \in \{0,1\}; x_{je_jm} \in \{0,1\} \tag{9}$$

---

1 The dominant land use type within neighborhood is the one that covers the maximum neighborhood area. For regular grid, where area of each cell equals some constant value, we can determine d_j as the one having maximum number of neighboring cells (including self); for ties, the dominant land use is chosen randomly

2 The dominant land use type is set to ‘undeveloped’ if and only if for a regular grid, all neighbors (including self) are undeveloped

3.3 Description of the Model

With this land use allocation model, we seek to promote compactness, contiguity, and infill of urban development. The model supports redevelopment whenever it is politically defensible and economically reasonable. For each land use, its new location should be as close as possible to other, compatible land uses. Thus, compactness and mixed uses may be achieved simultaneously.

The first two objectives (see Eqs. 1, 2) allow for tradeoff evaluation between the minimization of the conversion of undeveloped land, and the minimization of redevelopment. Thus, assigning variable importance between these objectives, we can encourage either compact or diffuse growth (Ward et al. 2003). Through minimizing redevelopment, we seek to minimize the change of current urban land use and therefore we encourage only reasonable redevelopment. For example, we could assign low resistance to change (r_j) values to derelict inner city sites, and hence increase the probability of redevelopment for these areas. Since undeveloped lands do not have the r_j value, their resistance to change implicitly equals 1 (which is the highest). Consequently, in our model any urban area with a lower r_j value ($r_j < 1$), has higher probability of land use change than the undeveloped land. Thus, giving open space areas the highest resistance to change, we 'penalize' allocation of urban uses to these areas and therefore encourage protection of undeveloped land. Objective (3), adopted from Gilbert's et al. (1985) concept of amenity and detractor cells, minimizes incompatibilities of development or redevelopment between site j and its neighborhood and thus addresses adjacent conflicts of land. The level of incompatibility is estimated between the candidate land use and the existing dominant land use d_j within j's neighborhood. By assigning equal compatibilities for different land uses (e.g. both residential-residential and residential-commercial have compatibility of 1), this objective promotes mixing of adjacent uses (see Table 1). The last objective (see Eq. 4) minimizes the distance of new development to already developed sites.

Table 1. The matrix of land use compatibility

Land use	PSV	SFR	ORE	COM	MFR	IND
Undeveloped	1.0	1.0	1.0	1.0	1.0	1.0
Preserved (PSV)	1.0	1.0	1.0	0.5	1.0	0.0
Single-family residential (SFR)	1.0	1.0	1.0	0.7	1.0	0.0
Other Residential (ORE)	1.0	1.0	1.0	0.7	1.0	0.0
Commercial (COM)	0.5	0.7	0.7	1.0	0.7	0.2
Multi-family residential (MFR)	1.0	1.0	1.0	0.7	1.0	0.0
Industrial (IND)	0.0	0.0	0.0	0.2	0.0	1.0

Constraints (5) and (6) ensure that we can allocate maximally one land use to each cell j. Equation (7) guarantees that the demand for land use l is satisfied. This constraint not only permits allocation of undeveloped land but also relocation of already urbanized areas (Lemberg and Church 2000). Conditions (9) guarantee that the decision variables are binary.

3.4 Density Based Design Constraint

Equation (8) represents the density based design constraint (DBDC). It ensures that we will allocate to a given cell j if and only if the sum of the cell's initially and newly developed neighbors is at least equal to a threshold development density value b (see Fig. 1). Therefore, the higher the value of b in this constraint, the more compact and contiguous is the pattern obtained and thus leapfrog development is prevented. DBDC, combined with the compatibility objective (see Eq. 3), is a surrogate to traditional zoning in urban planning.

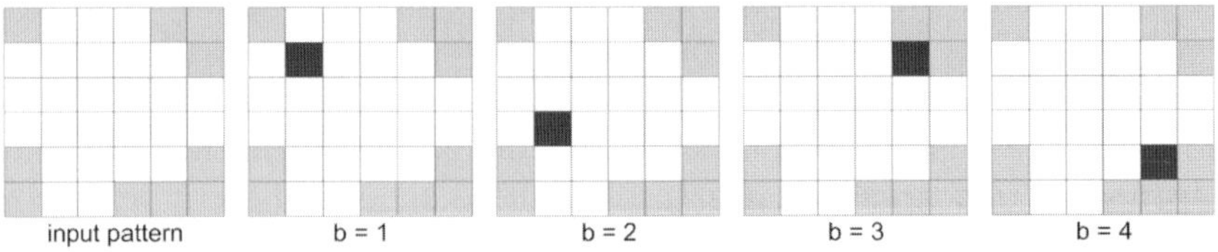

Fig. 1. Density based design constraint (see Eq. 8). With all other objectives and constraints unchanged, the new land use (black) is allocated to the neighborhood that meets b value

The maximum value of b depends on the size of the neighborhood considered. For example, if the neighborhood is a depth of one cell about a focus cell j - Moore neighborhood with $r = 1$ (Weisstein 2005), there are eight neighbors. Thus, an absolute maximum size of b cannot be larger than 9, if you include the focus cell as well. There will always be a set of cells on the perimeter of a given allocated land use (boundary cells). Observe the situation when the focus cell is on the perimeter of a given cluster of cells allocated to a given land use. The worst case or the case in which the focus cell has the fewest developed neighbors is when the cell is in the corner of the perimeter. If it is the corner cell then at most three other cells of urban land use can be allocated within its neighborhood. Thus, counting the focus cell as well, the value of b could be no larger than 4, which is the worst case for high connectivity. Hence, if $r = 1$, then the neighborhood will be a $2r+1$ by $2r+1$, which is 3x3, in size. The maximum possible size of b for a corner cell is then $r+1$ by $r+1$. Thus, if we had a neighborhood size of $r = 2$ or two cells deep about the focus cell j, then the neighborhood would be a 5x5 (i.e. $2r+1 = 5$) and the maximum size of b is 9, which represents 3x3 ($r+1 = 3$). The model could be executed with larger b values, but there is no guarantee that a feasible solution exists, since in such cases the constraint forces inward development and forbids development within the neighborhood of the boundary cells. This condition is especially true when the problem assumes no existing developed land.

4 Model Evaluation

In order to assess the robustness of the model we considered the following questions:

1. Under what conditions is the pattern compact and contiguous?
2. What values of b in DBDC intensify the level of infill development?
3. What conditions promote maintaining the compatibilities of newly allocated land uses to already existing land uses?
4. What is the degree of redevelopment prescribed by the model?

4.1 Initialization and Solution

We used Chelan – a small town in Chelan County, Washington, US (see Fig. 2) for an experimental application of the model.

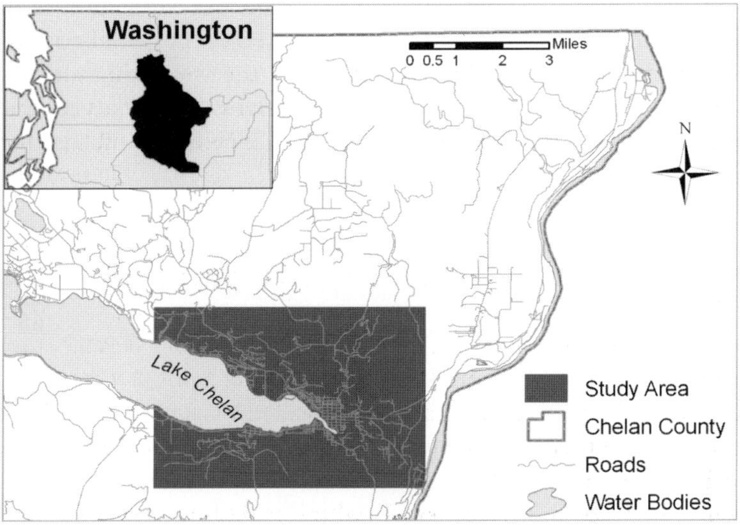

Fig. 2. Study area – Chelan City, Washington State, USA

Elsewhere (Ligmann-Zielinska et al. 2005), we tested the model on a smaller hypothetical example to examine a wider range of model parameters. Chelan suffers from two conflicting growth-related pressures: Pacific salmon habitat protection on one side, and rapid inward migration of population from the nearby Puget Sound metropolitan area on the other (Ligmann-Zielinska and Jankowski 2006). Therefore, a model to design various land use allocation scenarios is indispensable to evaluate impacts that these two opposing forces may have on future development of Chelan.

The cell resolution of 2 acres was selected as the basic unit of analysis, which resulted in 7384 raster cells representing the area of 14768 acres (61.5 km^2), about 24300 binary variables and 9100 structural constraints. The set of land uses under consideration comprises single-family residential, multi-family residential, other residential, commercial, industrial, and undeveloped. Water, evergreen forest, beaches, orchards, and institutional land were set to 'preserved' and excluded from the analysis (CCCLUP 1998). We used the commercial GIS software ArcGIS 9.0 (ESRI http://www.esri.com/) and Python scripting to generate additional input layers including: 'resistance to change', 'number of developed neighbors', 'dominant neighborhood land use', and 'distance to developed cells'. The 'Resistance to change' for developed cells was obtained based on slope, elevation, and parcel value (CCCLUP 1998).

The demand for land (see Table 2) was linearly extrapolated from the year 2000 ratios of population to land use area for each land use category. We used population projections from the Office of Financial Management, Washington State (CCCLUP 1998, Appendix H), to estimate the demand for land in the year 2025, the final year of land use allocation. Land use compatibility (see Table 1) was established based on personal communication with the Chelan county planners. We used the weighting method (Cohon 1978) to vary the importance of objectives and analyze tradeoffs among them, where 1 means the lowest importance and 9 means the highest importance.

Table 2. Demand for allocation

Land use	Current number of cells (2000)	Demand (2025)
Commercial	34	65
Industrial	4	8
Multi-family Residential	2	4
Other Residential	137	261
Single-family Residential	879	1673

For instance, a scenario with weights 9151, has the importance of objective (1) set to 9, the importance of objective (2) set to 1, the importance of objective (3) set to 5, and the importance of objective (4) set to 1. In the experiments, the DBDC *b* value was set to 1, 2, and 3. For higher values of b (like $b = 4$), we could not generate any results in a reasonable time, which we arbitrarily set to 10 minutes based on the rationale that in a meeting to evaluate land use allocation alternatives stakeholders will be unlikely to tolerate longer computation times. This confirms our hypothesis about the possible values of *b* for DBDC (Sect. 3.4).

The model formulations were written using the MPS file format (standard format for Linear and Integer Programming). We used the exact branch-and-bound technique of the Linear Integer solver in Lingo 9.0 extended version, by LINDO Systems, Inc. (http://www.lindo.com/ 2004), on a Mobile Intel Pentium(R) CPU 3.06GHz and 448 MB RAM. All solutions obtained were globally optimal, and the maximum solution time was 32 seconds. We observed an increased solution time and increased number of solver iterations for higher values of *b* in DBDC.

4.2 Results and Discussion

Figure 3b shows an example solution for equal objective weights (all set to 1) and maximum value of *b* (set to 3) in DBDC.

In general, the obtained land allocation patterns are plausible (see Fig. 3). A rough qualitative exploration of the patterns reveals that low values of *b* in DBDC may result in a more erratic pattern regardless of the preferences for objectives (see Fig. 4). Higher level of infill occurs for high *b*.

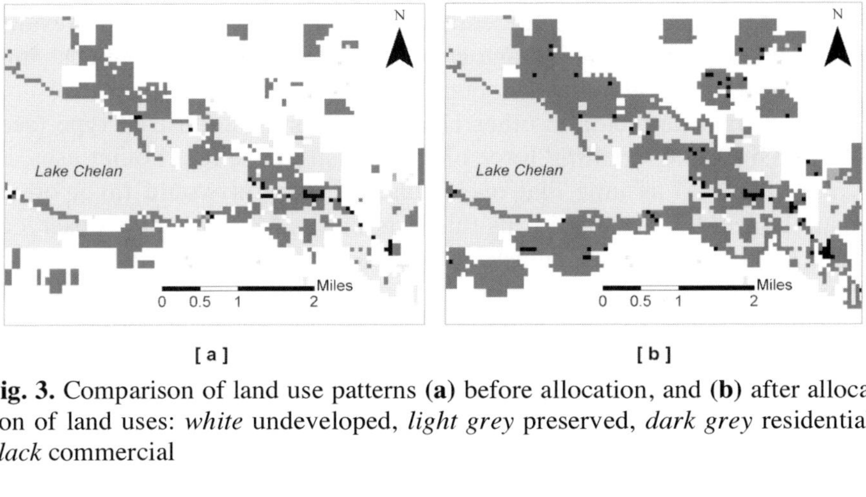

Fig. 3. Comparison of land use patterns **(a)** before allocation, and **(b)** after allocation of land uses: *white* undeveloped, *light grey* preserved, *dark grey* residential, *black* commercial

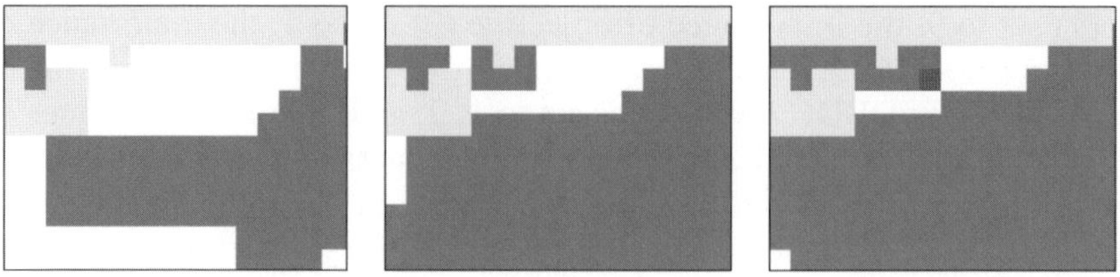

Fig. 4. Infill development with increasing *b* in DBDC for equal objective weights (see Eq. 8)

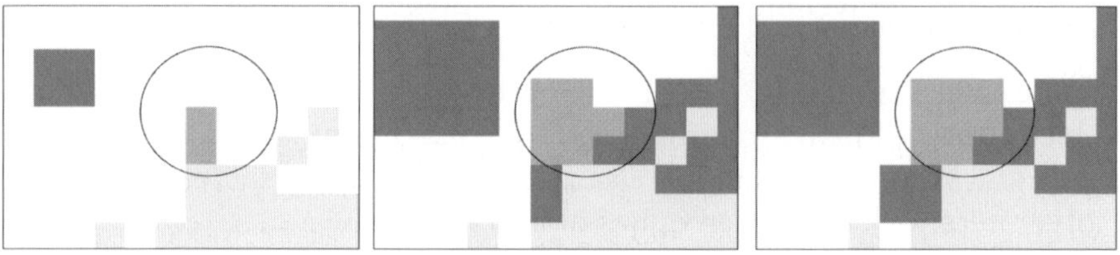

Fig. 5. Compatibility for increasing *b*-value in DBDC with equal objective weights (see Eq. 8); the circled area is an industrial cluster

We calculated the fragmentation of urbanization before and after allocation using a ratio of outer perimeter to development area as a metrics of fragmentation. For all cases, the fragmentation considerably decreased as compared to the initial land pattern (see Fig. 6a). On average, the improvement was about 34% of decrease in fragmentation value. Although the differences in improving the fragmentation are minor (32%–36%), we do observe some correspondence between the value of *b* in DBDC and the decrease in fragmentation value. The value of $b = 1$ results in the least improvement in fragmentation. Interestingly, the highest improvement in fragmentation was obtained for $b = 2$.

The only real detractor to other land uses was the industrial type (see Table 1). Given the repelling force of the industrial use to all other uses, it was reasonable to assume that one industrial location would favor other industrial sites. Figure 5 confirms this hypothesis. Moreover, we observed that for $b = 1$ the only two industrial clusters within the study area were of approximately equal size. However, with the increasing *b* value, one of these clusters began to grow and the other began to shrink (see Fig. 5). No other industrial cluster emerged, which was consistent with the Chelan comprehensive plan assumptions (CCCLUP 1998). Additionally, more mixing of uses occurred with the increase of *b*.

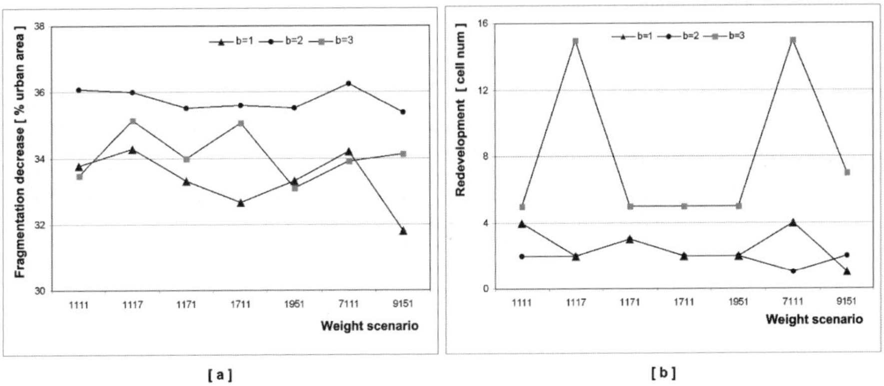

Fig. 6. The influence of increasing *b* in DBDC: **(a)** on fragmentation decrease, and **(b)** on redevelopment increase

The final aspect of model results is the degree of redevelopment. We recognize that obtaining sustainable land use patterns is in practice highly constrained by the economic and political viability of such decisions. We believe that minimizing the negative impacts of growth should be considered in any normative model for planning; otherwise these models will not gain any widespread use (Lemberg and Church 2000). In case of redevelopment minimizing the negative impacts of growth means encouraging redevelopment of already developed land. The redevelopment levels computed for Chelan, ranging from 4 to 30 acres (see Fig. 6b), might be considered satisfactory, given the size of the study area (14768 acres). Interestingly, the increasing value of *b* also has a positive impact on the level of redevelopment. Finally, there exists more redevelopment variability with higher value of *b* in DBDC (see Fig. 6b).

In general, the Chelan application of the model produced very plausible outcomes. All of the sustainability postulates represented in the model objectives were satisfied and we observed a considerable improvement between the current state of landscape and the proposed plan.

5 Future Work

Although the results of land use allocation were acceptable, the model revealed a slight drawback. The sensitivity analysis, implemented by varying the importance of objectives, proved that the obtained pattern is stable. This may be considered either a benefit, since the model is robust, or a drawback. Using generative techniques like spatial optimization should allow the stakeholders to choose from a set of scenarios that are both good and different from each other, since not every planning objective of interest may be introduced into the model in the form of mathematical formulation (Brill et al. 1990; Chang et al. 1983). Therefore, the decision makers should be provided with alternatives that allow for consideration of other – less defined – planning goals.

The current DBDC only considers whether a neighbor is developed or not. It neglects the neighbor's use compatibility. In the future version of the model, the constraint will be extended to ascertain some level of neighborhood compatibility. For example, for $b = 3$ (minimum 3 neighbors developed) at least 2 neighbors must be compatible. Such constraint would force some predefined level of compatibility for the allocated land use, instead of just minimizing the incompatibility objective, which does not guarantee that the adjacent uses are compatible (Arthur and Nalle 1997). Moreover, the varying extent of neighborhood considered in DBDC may significantly influence the generated patterns. Thus, in the future version of the model, we would like to analyze different shapes and widths of neighborhoods.

6 Conclusions

The complexity of sustainability in urban land use patterns has been widely debated in planning literature. In this paper, we have presented a mathematical programming model that addresses the problem by simultaneously considering such sustainability objectives like new development, redevelopment, land use compatibility, and accessibility. Additionally, the model may encourage infill development thanks to the density based design constraint. The land use patterns generated by the model present some idealized frames of reference, which allow for the exploratory analysis of current dissatisfactory land use arrangements, and reveal the possibilities for improving urban environments we live in.

References

Aerts JCJH, Heuvelink GBM (2002) Using simulated annealing for resource allocation. Int J of Geographic Information Science 16(6):571–587

Aerts JCJH, Eisinger E, Heuvelink GBM, Stewart TJ (2003) Using Linear Integer Programming for Multi-Site Land-Use Allocation. Geographical Analysis 35(2):148–169

Arthur JL, Nalle DJ (1997) Clarification on the use of linear programming and GIS for land-use modeling. Int J of Geographical Information Science 11(4): 397–402

Barton H (1990) Local global planning. The Planner 26, October:12–15
Benabdallah S, Wright JR (1992) Multiple Subregion Allocation Models. J of Urban Planning and Development 118(1):24–40
Brill ED, Flach JM, Hopkins LD, Ranjithan S (1990) MGA – A decision support system for complex, incompletely defined problems. IEEE Transactions on Systems Man and Cybernetics 20(4):745–757
Brookes CJ (2001) A genetic algorithm for designing optimal patch configurations in GIS. Int J of Geographical Information Science 15(6):539–559
Brotchie JF, Dickey JW, Sharpe R (1980) TOPAZ – General Planning Technique and its Applications at the Regional, Urban, and Facility Planning Levels (= Lecture Notes in Economics and Mathematical Systems 180). Springer-Verlag
CCCLUP (1998) City of Chelan Comprehensive Land Use Plan (1998) 2002 updates, copy available from the City of Chelan, http://www.cityofchelan.com/ (November 2005)
Chang SY, Brill ED Jr, Hopkins LD (1983) Efficient random generation of feasible alternatives: A land use example. J of Regional Science 22(3):303–314
Church RL (1999) Location modelling and GIS. In: Longley P et al. (eds), Geographical Information Systems, 2nd ed. John Wiley & Sons, New York, pp 293–303
Church RL (2002) Geographical information systems and location science. Computers & Operations Research 29:541–562
Chuvieco E (1993) Integration of linear programming and GIS for land-use modeling. Int J of Geographic Information Systems 7 (1):77–83
Cohon JL (1978) Techniques for Generating Noninferior Solutions. In: Multiobjective Programming and Planning. Academic Press, New York, pp 98–162
Cova TJ, Church RL (2000) Contiguity constraints for single-region site search problems. Geographical Analysis 32:306–329
Dökmeci VF, Cagdas G, Tokcan S (1993) Multiobjective Land-Use Planning Model. J of Urban Planning and Development 119(1):15–22
Fischer DT, Church R (2003) Clustering and Compactness in Reserve Site Selection: An Extension of the Biodiversity Management Area Selection Model. Forest Science 49(4):1–11
Galster G, Hanson R, Ratcliffe MR, Wolman H, Coleman S, Freihage J (2001) Wrestling Sprawl to the Ground: Defining and Measuring an Elusive Concept. Housing Policy Debate 12(4):681–717
Gilbert KC, Holmes DD, Rosenthal RE (1985) A multiobjective discrete optimization model for land allocation. Management Science 31(12):1509–1521
Grimshaw J (2000) Chapter Four. In: Leccese M, McCormick K (eds) Charter of the New Urbanism. McGraw – Hill, pp 35–38
Guy S, Marvin S (2000) Models and Pathways: The Diversity of Sustainable Urban Futures. In: Williams et al. (eds) Achieving Sustainable Urban Form, E & FN Spon, Taylor & Francis, London, pp 9–18
Leccese M, McCormick K (eds) (2000) Charter of the New Urbanism. McGraw – Hill
Lemberg DS, Church RL (2000) The school boundary stability problem over time. Socio-Economic Planning Sciences 34:159–176

Ligmann-Zielinska A, Church RL, Jankowski P (2005) Sustainable Urban Land Use Allocation with Spatial Optimization. Proc of the 8th Int Conf on Geocomputation, University of Michigan, Eastern Michigan University, USA, August 1–3, 2005

Ligmann-Zielinska A, Jankowski P (2006) Agent-Based Models As Laboratories For Spatially Explicit Planning Policies. Environment and Planning B: Planning and Design, forthcoming

Malczewski J (1999) Introduction to GIS. In: GIS and Multicriteria Decision Analysis. John Wiley & Sons, pp 15–80

Minor SD, Jacobs TL (1994) Optimal Land Allocation for Solid- and Hazardous-Waste Landfill Siting. J of Environmental Engineering 120(5):1095–1108

Nalle DJ, Arthur JL, Montgomery CA, Sessions J (2002a) Economic and spatial impacts of an existing reserve network on future augmentation. Environmental Modeling and Assessment 7:99–105

Nalle DJ, Arthur JL, Sessions J, (2002b) Designing Compact and Contiguous Reserve Networks with a Hybrid Heuristic Algorithm. Forest Science 48(1): 99–105

Shirabe T (2005) A model of contiguity for spatial unit allocation. Geographical Analysis 37:2–16

Silberstein J, Maser C (2000) Land-Use Planning for Sustainable Development. Sustainable Community Development Series, CRC Press LLC

Ward DP, Murray AT, Phinn SR (2003) Integrating spatial optimization and cellular automata for evaluating urban change. The Annals of Reg Sci 37:131–148

Weisstein EW (2005) Moore Neighborhood. Math World – A Wolfram Web Resource (http://mathworld.wolfram.com/MooreNeighborhood.html May, 2005)

Williams JC (2002) A Zero-One Programming Model for Contiguous Land Acquisition. Geographical Analysis 34(4):330–349

Williams JC, ReVelle CS (1996) A 0-1 programming approach to delineating protected reserves. Environment and Planning B: Planning and Design 23:607–624

Williams K (2000) Does Intensifying Cities Make them More Sustainable? In: Williams et al. (eds) Achieving Sustainable Urban Form, E & FN Spon, Taylor & Francis, London, pp 30–45

Williams K, Burton E, Jenks M (eds) (2000) Achieving Sustainable Urban Form, E & FN Spon, Taylor & Francis, London

Wright J, Revelle C, Cohon J (1983) A multiobjective Integer Programming Model for the land acquisition problem. Regional Science and Urban Economics 13:31–53

Building an Integrated Cadastral Fabric for Higher Resolution Socioeconomic Spatial Data Analysis

Nadine Schuurman, Agnieszka Leszczynski, Rob Fiedler, Darrin Grund, Nathaniel Bell

Department of Geography, Simon Fraser University,
8888 University Drive, Burnaby, British Columbia, V5A 1S6, Canada

1 Introduction

In 2003, Paul Longley called for a new urban geography based on disaggregate, high resolution socioeconomic data [36]. In so doing, he made a compelling argument for human geographers to engage with GIScience (GIS) to better understand a range of socioeconomic activities, from general population characteristics to the evolution of cityscapes. Longley speculated that the impediment to a greater embrace of GIS by human geographers was the narrow scope of extant data sets – limited in both scale and extent – which did not adequately capture the range of human activity. Analysis has been further stymied by measurement difficulties, which have compounded the inadequacy of socioeconomic data. The effect has been a theoretically focused urban geography in which spatial pattern is frequently ignored.

While Longley may have underestimated the cultural divide between human and applied geographers [31,40], he is correct in arguing that urban geography might have been better served by exploratory data analysis complemented by intuitive forms of visualization. These techniques better explain the new geographic manifestations of consumption and citizenship, create new possibilities for the integration of data from multiple sources, and allow for the emergence of more realistic neighborhood 'types' and geodemographic 'lifestyles' profiles [36]. Longley suggests that we draw on sources outside the familiar census and other publicly available data, such as the right marketing data accumulated at the individual and neighborhood level by private firms. Such data could be integrated with existing data sources to broaden the world of socioeconomic analysis.

Recently there has been a focus on designing 'meaningful' or 'optimal' census geographies [3]. The sensitivity of census data to areal unit definition has been overlooked by many users in the past because the ability to test alternative geographies was not yet possible [2,51]. Unfortunately, there are many ways to partition space into zones – it is therefore unlikely that one definitive 'optimal' solution exists [22]. The (potential) ability for users to create custom geographies, tailored to their specific analysis needs, may end up replacing problems associated with the arbitrary nature of zone design (known as the modifiable area unit problem, MAUP), with a new problem – the *user* modifiable area unit problem. One way of circumventing the propagation of scaling errors in analysis is to start at the rooftop level and aggregate as needed. This creates the possibility of more nuanced socioeconomic analysis.

In the following pages, we describe a novel methodology comprised of a series of protocols developed to enhance the resolution of socioeconomic data and demonstrate their value for socioeconomic analysis. First, we describe a protocol for integrating cadastral data from multiple socioeconomic attributes from multiple jurisdictions. A seamless spatial fabric of property data carries with it the ability to assign socioeconomic attributes at the rooftop (household) level. Second, we detail how land tenure (rental and ownership) data reported at the Dissemination Area level (DA)[1] can be disaggregated to the individual cadastre, allowing socioeconomic neighborhoods to be characterized based on the household. We will illustrate this with the example of lone parent families and housing tenure. Third, we describe the utility of survey data to estimate the income of individual households. This is complemented by the fourth and final component, which involves the assignation of property assessment data – from real estate boards or government agencies – to individual and rental properties. In combination, these four protocols yield an invaluable high-resolution fabric for spatial socio-economic analysis.

As few efforts have been made to date by human geographers to increase the resolution of socioeconomic data – and indeed none have proposed the use of cadastral data as a means of doing so – this is entirely new territory. Longley and Harris [37] argue that rather than using archaic theories of the city, *new* models, based on high-resolution socioeconomic data, are needed. In using cadastral data as a framework for socioeconomic

---

[1] Dissemination Areas (DAs) are the smallest unit for which Canada Census data are released. They comprise 700-800 people and are conceptually and spatially congruous with the US Census Block Face.

analysis, we address the call for a "data-rich GIS-based model building" [36, 37:871].

2 Looking for High Resolution Data in the Literature

Geographical analyses of socioeconomic welfare and population health are closely tied to notions of deprivation, and have long been dependent on aggregate statistics derived from national censuses. The census constitutes a coarse assessment of the population, derived from the quantification of social and housing indicators. These are aggregated prior to dissemination in order to preserve the anonymity of individuals enumerated [11,38,46]. Although researchers continue to draw upon the census because it is the only concise, reliable source of data from which to determine the spatial distribution of welfare [11,22,29,36,46], scholars have identified the census as an inherently impoverished source of data whose use for socioeconomic study is fundamentally limited by the coarse and arbitrary zonal boundaries of data collection [8,11,16,22,29,32,36,37,41,46,49,58]. A consequent effort has been made to expose and address what Mennis [41] refers to as the analytical and cartographical pitfalls of deriving conclusions about the well-being of populations from discretionarily aggregated data and their conventional thematic display.

The strongest objections to the use of census data for socioeconomic analysis concern the arbitrary nature of spatial census partitioning. Census geography has not been designed for the purpose of conveying understanding of the spatial patterning of socioeconomic phenomena; rather, boundaries, often adhering to street networks and other features of the built environment, are superimposed on the landscape so as to facilitate efficiency in data collection [8,10,11,22,36,42,49,58]. Raper [49] argues that the census collection rationale is guided by an understanding that temporal and spatial arrangements of populations and housing stock can be quantified and, furthermore, that the measurement of corresponding characteristics 'naturally' differentiates an area from its surroundings.

The data collection logic, which guides the drawing of census boundaries results in the discretization of socioeconomic continuity [10,11,32,36, 37,49,58]. The issue of continuous variation gives rise to a plethora of problems. Principal amongst these is the assumption that change occurs abruptly at the crisp boundaries of internally homogeneous zones. While all researchers working with object models for spatial analysis must contend with how to represent geographic continuity, scholars argue that the use of such data has unique and significant implications for studies of deprivation and social welfare. Because the city is a highly heterogeneous landscape inhabited by mobile and fluid populations, this is particularly problematic for urban analyses [22,36,49]. Consequently, differences *within* an enumeration unit are often greater than differences between zones. However, while this has long been recognized, Harris and Longley [22] contend that the nature of such aggregate surrogate data necessitates that researchers presume – for the purposes of analysis – that all individuals residing within the same census zone live under the same economic circumstances.

This flawed assumption and the crisp zonal boundaries of census geographies are the sources of two errors that plague socioeconomic analysis: the modifiable area unit problem (MAUP) and its sidekick the ecological fallacy (EF). The EF, illustrated by Openshaw [45], reveals that individual and areal correlations frequently do not correspond. Because the enumeration of spaces and people reduces individuals to statistics aggregated to represent populations, individuals are conflated with the areas in which they live. The EF precludes qualifying socioeconomic disparities between individuals as a function of the areas in which they reside [11,16,29]. Yet this methodology persists, and is compounded by the MAUP, another scaling problem. Privacy requirements force the aggregation of individual census responses into areal units prior to dissemination [51]. Areal units, however, are recognized as arbitrary and modifiable [44]. Because census boundaries are arbitrary, there is no concrete relationship between socioeconomic phenomena and the areal units over which they are tabulated [8]. Accordingly, analytical results from areal data are sensitive to how the units are defined [44] – the examination of data at different levels of abstraction yields different and often competing results [8,32,37,41,49,59]. As Openshaw suggests, areal units are "neither neutral nor meaningful entities, [but] are exogenous to all subsequent uses of the data" [45: p 18].

The aforementioned analytical problems are only exacerbated by the conventions of visualizing census data using choropleth maps. Thematic mapping grossly misrepresents the distribution of aggregate socioeconomic statistics [8], conflates the individual with their area of residence, and further perpetuates an understanding of socioeconomic phenomena as internally homogenous to the zone [37]. Eicher and Brewer [13] contend that because such areal representations of data convey abrupt changes in values at enumeration boundaries, they reinforce an understanding of social reality as occurring within discrete space. Tate [58] explains that it is precisely because the census enumeration model is based on the discretization of socioeconomic continuity that aggregate data lends itself so well to choropleth mapping. A subsequent problem concerns the averaging of socioeconomic characteristics, such as population density, over the entire area of a census unit. Because large non-residential areas are included in the tabulation, with population characteristics calculated over these lands, indicator statistics of distribution are significantly skewed and non-representative of spatial – and social – reality [8]. Hence the "surface shadowing effects" of thematic mapping obscure the true underlying geographic distribution of these variables [8,32:25].

The use of aggregate data, and reliance on choropleth cartography to convey statistics, has serious implications for the formation and delivery of socioeconomic policy by the state. Census geography and its visualization are a way of "politically understanding" social welfare, and they engender an understanding of socioeconomic phenomena as "regularities" that can be altered via policy to produce both desirable and predictable results [11:41–42]. Census geographies are imbued with power and represent particular political interests, articulated onto the landscape via an elitist political language used to label areas and people as deprived [10,11,22,49]. The results of census geometries – tied to notions of normality/abnormality – are often either the overestimation or underestimation of deprivation in terms of both intensity and extent [11,22].

Because they do not reflect actual social activity, Longley and Harris [37] assert that aggregate statistics can only be used as very *indirect* indicators of behavior and welfare. The formal geographies used to inform and administer resources and services often bear no relevance to lifestyles or socioeconomic trends. This arises because many continuous socioeconomic phenomena are *not* quantifiable and are therefore left out of the enumeration; indeed, only some things are measured, often to the exclusion of others [39]. Moreover relevant information is further obscured when data are aggregated to spatially normalized zones [22]. The data lost is not peripheral, but indeed has much to say about how zones are determined and how the data itself is generalized [22,36]. The census of Great Britain, for example, does *not* collect data on income [9], even though income has been identified as a primary indicator of health [33,62,63,64]. This has led Holloway et al. to observe that "when population density, or any socioeconomic variable, is mapped by choropleth techniques, the results often tell us more about the size and shape of the enumeration unit, than about the people actually living and working within them" [26:285].

Furthermore, as articulated by Bracken and Martin, areas where socioeconomic extremes exist "tend to be those for which the spatial basis of enumeration is least appropriate" [8]. Inequality does not exist exclusively at the scale of census polygons [22]. Harris and Longley [22] assert that because of the problem of continuous variation identified above, census geography serves to obscure where deprivation actually exists. Moreover using census units to qualify the socioeconomic welfare of subsections of the population, and subsequently basing the delivery of services on this geography, is based on the fundamentally flawed assumption that census units are suitable for these practices. According to Kirby [29], policy is consequently often guided by the belief that resources should be targeted where concentrations of deprivation exist. However, he postulates that such a policy approach is based on a distorted understanding of the spatial manifestations of poverty. In his review of the literature, he points to contentions in urban theory that "easily located pockets of hardship, which account for the bulk of such families, do not exist" [29:179]; rather, deprivation is widespread. While there are indeed areas that can be identified as deprived, deprived people live in all types of neighborhoods, as do people who enjoy a high level of welfare [29]. To this extent, Kirby argues that:

> Census and census-type variables, which relate strongly to housing and income, can only provide information on deprivation that stems from wider social inequalities, and which cannot be addressed by spatial remedies. [29:178]

This is similarly expressed by Fieldhouse and Tye [16]. Accordingly, "policy initiatives directed to the inner areas can simply act as palliatives to problems originating elsewhere" [29:179]. Fieldhouse and Tye [16] identify the results of policy based on aggregate areal statistics as the delivery of assistance to only those deprived persons living in areas defined as deprived; similarly, deprived individuals outside of deprived zones become *doubly* deprived because they are denied access to the services and resources they require. Indeed, Minot [43] contends that the effectiveness of service delivery, such as poverty alleviation programs, hinges on the geographic unit of analysis and targeting.

3 Generating Socioeconomic Surfaces: Review of Previous Methodologies

To circumvent the descent into the ecological fallacy (EF) and modifiable area unit problem (MAUP), and to better inform policy, scholars have identified the need for detailed, disaggregate socioeconomic data from which to draw valid inferences about the spatial distribution of social welfare [8,13,16,22,26,32,36,37,39,41,58]. However, because the census is the only source of reliable socioeconomic data and individual privacy must be preserved, one of the only means for garnering higher-resolution data is to disaggregate the census [8,13,16,22,26,32,36,37,39,41,42]. Not only does disaggregation of data provide a much better indication of where deprivation is concentrated [16], but Holloway et al. [26] indicate that breaking down enumeration units begets more relevant statistics, allowing for more *meaningful* analysis.

Generating maps of socioeconomic characteristics using dasymetric mapping is a means of modeling population demographics on a continuous scale [11,13,22,32,41]. This is a technique, which allows for the isolation of data into areas that appear internally homogeneous *independent* of the boundaries of original data collection [13]. An extension of areal interpolation methods, dasymetric mapping employs ancillary sources – either land-use data or spectral imagery – to identify trends in the data [41]. This approach is a great improvement over the choropleth model principally because it liberates data from the arbitrary zones of data collection, and allows data to be re-aggregated at multiple – and *meaningful* – scales [13,39,41]. Dasymetric mapping may be conducted using both polygon and raster based methods, although the latter technique is preferred because surfaces provide a much better representation of the continuous variation of socioeconomic spatial reality [58]. While some analysts approach socioeconomic surface modeling using a search radius [21,32], most socioeconomic surfaces are built using some form of what Tate [58] refers to as "kernel density estimation" [58:301]. These include models built by Bracken and Martin [8], Martin [38], Mesev [42], and Thurstain-Goodwin and Unwin [59]. Bracken and Martin [8] used a simple inverse distance weighting function on population-weighted centroids of census units to generate a population density surface. Martin's [38] technique, referred to by Bailey and Gatrell [4] as *adaptive kernel estimation*, improves upon the former simple distance-decay model by adjusting the size of the kernel to reflect the density of census population-weighted centroids.

More current approaches to census data disaggregation use ancillary data to exclude non-residential areas from analysis [13,26,41,42]. Eicher and Brewer [13] evaluated a number of both polygon and raster-oriented dasymetric mapping techniques. Of particular interest is the "grid three-class method," a weighting technique for redistributing population values to land use for each county on the basis of a set ratio of 70% urban, 20% agricultural, and 10% forested [13:129]. The immediate disadvantage of this approach is the assumption that the distribution of land use is uniform in each county. Mennis' [41] model addresses this limitation by accounting for urban population density. Urbanization classes indicative of the degree of urban development – *high density urban, low density urban, and non-urban* – were derived from spectral imagery to generate a population surface. Using urban development as a proxy for residential density, Mennis assigns the population data to the grid cell via a two-step process: first, relative differences in population among the three urbanization classes are accounted for from census block group data, and subsequently, the proportion of area represented by each urban class for each block group is calculated. This technique improves upon simple dasymetric mapping techniques by accounting for the proportion of the data collection unit (in this case, the census block group) that each ancillary class represents, rather than distributing population to pixels on the basis of the class alone (e.g. urban vs. rural).

Each of these techniques points to the value of working at appropriate resolutions rather than those assigned by administrative fiat. In the following section, we demonstrate a method for building an integrated cadastral fabric that can subsequently be populated with socioeconomic data at the household level.

4 The Rise of the Household: Why We Should Use Cadastral Data for Socioeconomic Analysis

Residential property values reflect consumption, which has been identified as a sound measure of welfare [25]. House value as a consumption indicator is unique because unlike the consumption of goods such as automobiles – the patterns of which say a lot about access to public transportation and lifestyle choices [9] – longitudinal trends in housing prices are more direct indicators of socioeconomic inequality [35]. In their study of the influence of the forces of social change on the urban real estate markets in Toronto and Vancouver, Ley et al. [35] correlate a growing social gap with the concentration of inflation in the prices of single-family homes. As evidenced by price increases experienced outside of the traditionally wealthy neighborhoods, variables such as social status had a minimal impact on housing market inflation in Vancouver. The researchers instead attribute inflation in single-family home prices to dwelling *types*, which, because they are closely correlated with *family type* and confer information about the number of adults earning a wage income in a household. This informs the analysis of household income because it suggests that households with greater collective or higher individual incomes are likely to concentrate in a particular housing type. Ley et al. [35] also conclude that the concentration of real estate inflation in this particular housing type is indicative of a growing socioeconomic gap in the city of Vancouver.

The Greater Vancouver Regional District (GVRD), located within the province of British Columbia (BC), houses over two million people residing within twenty-one municipalities. It constitutes one metropolitan area consisting of 424 census tracts (CTs), which further break down into 3369 dissemination areas (DAs). By mapping distribution of the income in the GVRD at the cadastral level, minute trends in the surface are manifest and signal socioeconomic transition. The reasons for this are two-fold. First, urban cadastres represent a high-resolution spatial fabric. The primitive spatial unit of the cadastre is the legal parcel. Thus cadastral level data creates the possibility of conducting socioeconomic analysis at the household level. Second, the cadastre is a framework for which information is typically already available. Many of the variables used for socioeconomic analysis, such as income, are not associated with the cadastre. However, information regarding dwelling values and tenure type is typically assigned to each legal parcel. In the Greater Vancouver metropolitan area, this information is available from the British Columbia Assessment Authority (BCAA), which collects this data for purposes of taxation. In this case we can use residential property values as a proxy for household income. The value of a cadastral fabric as a basis for socioeconomic analysis has been under-explored in urban and population health geography chiefly due to 'lack of data'. Our first protocol is the development of cadastral data sets.

5 Methodology, Protocol by Protocol

5.1 Protocol One: Building an Urban Data Set House by House

In some regions, integrated cadastral systems are in place. In such circumstances the creation of a seamless fabric is an unnecessary step. In this instance, we investigated the feasibility of using the Province of British Columbia's seamless digital cadastre being developed by the Integrated Cadastral Initiative (ICI). The ICI program is one of the numerous complementary initiatives in Canada charged with integrating the private land cadastral systems – presently managed by municipalities – into a standardized data set. However, the Vancouver portion of the ICI project is not complete. Private utility companies and other agencies also maintain cadastral data but it is unavailable or cost prohibitive to acquire. Our use of the cadastre is as an 'empty' digital spatial framework which can be subsequently populated with attributes relevant to socioeconomic analysis. The goal was therefore to ensure consistency across the fabric.

A number of challenges preclude the simple merging of cadastres across municipal boundaries. Each municipality in the GVRD test region keeps its own cadastral data set with unique attributes and spatial units. In an attempt to build a seamless cadastral fabric, problems include differences in attributes, spatial boundary definition, scale of collection, software specific file formats, as well as administrative and cultural resistance. Ironically problems of a geometric or spatial nature were generally considered computationally trivial [7,23,24,61] in the context of combining data from multiple contiguous municipalities. Preservation of attribute integrity is more complex. Retention of land use attributes associated with the cadastre is key as it allows use of land use/land cover as an ancillary data source to differentiate between residential and non-residential areas. Land use refers to the activities occurring on a particular piece of property, including both the intent and the reality of how a given parcel within a metropolitan boundary is altered by human decisions [20]. Although some attribute integration problems – such as standardization of common fields – may also be described as trivial, land use is variably categorized and understood.

However, because it is complicated by semantic heterogeneity, integration of land use taxonomies is non-trivial and therefore belongs to the purview of "application semantics," which has been recognized as the most difficult aspect of data interoperability [7]. Land use classification is a municipal government responsibility in the GVRD. There is no regional land use classification standard, and resulting taxonomies of land use are disparate and thereby incommensurable. For example, one community may identify *single family* as a land use designation; this designation may include two or three different types of single-family types in a competing classification (see Table 1).

Table 1. Summary of zoning designations based upon industrial, low-rise residential, high-rise residential and single-family residential land use categories for the identified communities

City	Industrial	Res Low	Res High	Single Family
Coquitlam	9	2	2	5
North Vancouver	7	1	1	3
Port Coquitlam	5	2	2	1
Richmond	7	1	1	1
Surrey	5	2	2	3

Indeed, in the same way that municipalities adhered to different survey control standards in their data collection, they have all devised internally unique land use classification schemes. Encountering incompatibilities between non-spatial attributes is routine. However, while survey control and projection incompatibilities can be resolved via algorithmic transformations, semantics cannot be normalized in this way. This is contrary to proposed GIScience solutions oriented towards the automation of the semantic integration process [1,6,12,14,17,18,28,30,34,48,50,52,55,57]. Automated solutions furthermore do not overcome institutional resistance to vertical integration and the difficulty of gaining consensus on standards for automation across jurisdictions [24].

We addressed issues of semantic incommensurability using a methodology from the social sciences that we have termed *database ethnographies* [54]. An ethnographic approach was used to interview data stewards in order to glean important contextual information about categorization, rationale for collection and other factors necessary to understand the semantic terms – or attributes – used in the respective municipal databases [54]. While database ethnographies do not resolve the lack of agreement on naming conventions, they are a means of identifying semantic differences. This is a first step before any talk of consensus or standardization – if at all possible – can begin.

Database ethnographies were the basis for resolution of land use integration [54]. They permitted the identification of attributes common to each data set as a basis for establishing linkage with the British Columbia Assessment Authority (BCAA) land use code system. The BCAA assigns each legal parcel in the province a code, which describes the dominant use of the property; this code can be accessed via the BCAA jurisdiction/roll number attribute encoded with each legal parcel. However there is no standard governing how the jurisdiction/roll number is assigned to each cadastre. Moreover, not all municipalities use BCAA data as the basis for determining land use – some smaller jurisdictions ground truth and record observed land use; others yet view municipal zoning as synonymous with land use. The implication is that although a BCAA land use code *exists* for every parcel of each municipal cadastre, it is not necessarily included as an attribute with the digital data, or it is not disseminated to data users. For municipalities that do not rely on BCAA codes for land use determination, this data was unavailable. By using database ethnographies, sufficient information was gleaned from database stewards to enable a rational integration of non-spatial attributes for the integrated cadastre. Though this methodology was arguably more time-consuming than automated integration, it is the basis for defensible very high-resolution urban data.

Figure 1 captures the non-triviality of attempting either automated integration or direct comparison of land use data sets across jurisdictional boundaries. Lack of taxonomic standardization meant that in order to employ the land use data, it was necessary to aggregate to lower-resolution GVRD land use classes, which are generalized from the aggregation of land use information from all of the local governments. The GVRD generalized land use classes served as a 'lowest common denominator' allowing cross-jurisdictional comparisons of land use within the region.

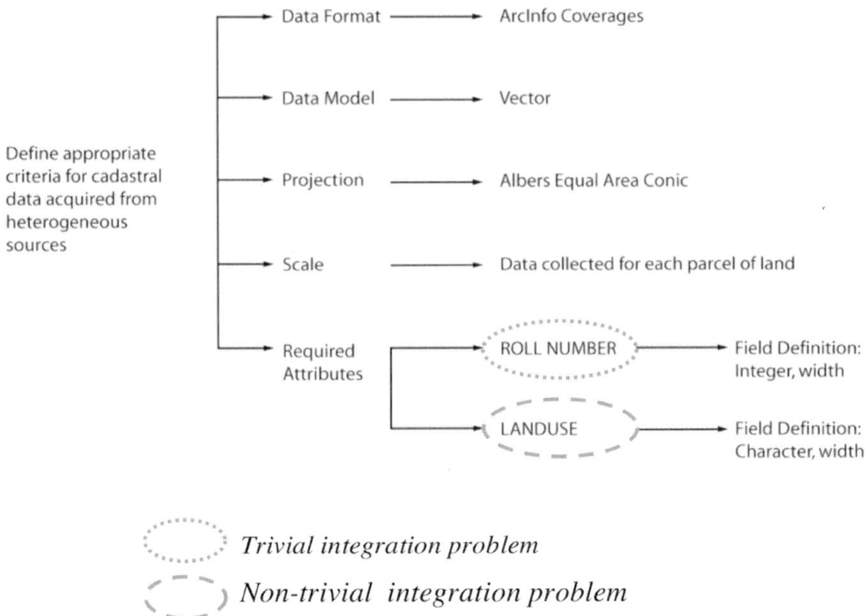

Fig. 1. Differentiating between trivial (Roll Number) and non-trivial (Landuse) attribute integration problems in generating a high-resolution cadastral fabric

These integration protocols are somewhat generic in that similar issues dog the development of integrated data sets in every jurisdiction. However, our approach is distinguished by 1) the differentiation between trivial and non-trivial semantic problems, and 2) the detailed interviews conducted with data stewards to determine how they derived land use classifications. This strategy is described in Figures 2 and 3.

This series of steps permitted the creation of a seamless integrated cadastre for the Vancouver region. By differentiating between trivial and non-trivial semantic attributes, different methodologies could be employed to solve each set of problems.

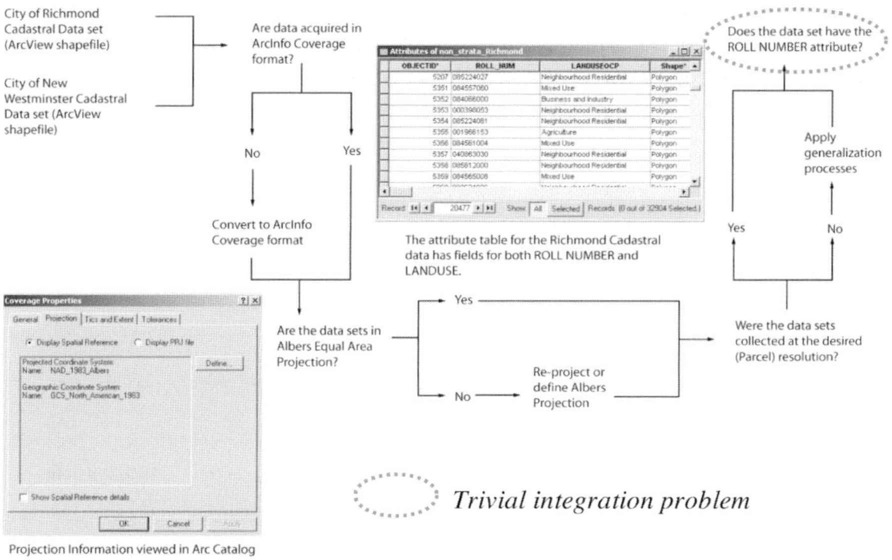

Fig. 2. Solving trivial integration problems for cadastral data. Roll Number integration is easily solved by using a numeric identifier for all municipalities

5.2 Protocol Two: Own or Rent Designating Tenureship at the Cadastral Level

Once the integrated cadastral fabric was created, the challenge was to populate individual residential properties with socioeconomic data. Canadian Census data can be procured in cross-tabulated tables that preserve – in a predetermined and therefore limited way – the multidimensionality of individual census responses otherwise lost in areal census data. This allows researchers to create custom population counts (for areal units) from differing configurations of the available data dimensions. When information on dwelling type is included as a data dimension, cross-tabulated census tables can be used in conjunction with land use (or cadastral) polygons to assign demographic/socioeconomic characteristics spatially within census areal units. This is a substantial benefit to researchers who rely on census data to provide a broad perspective of residential life, as dwelling type – along with tenure status – are good indicators of socioeconomic status. Although not available for all census categories, cross-tabulated data have the potential to provide researchers with a more granular depiction of how certain attributes are distributed within areal aggregation units.

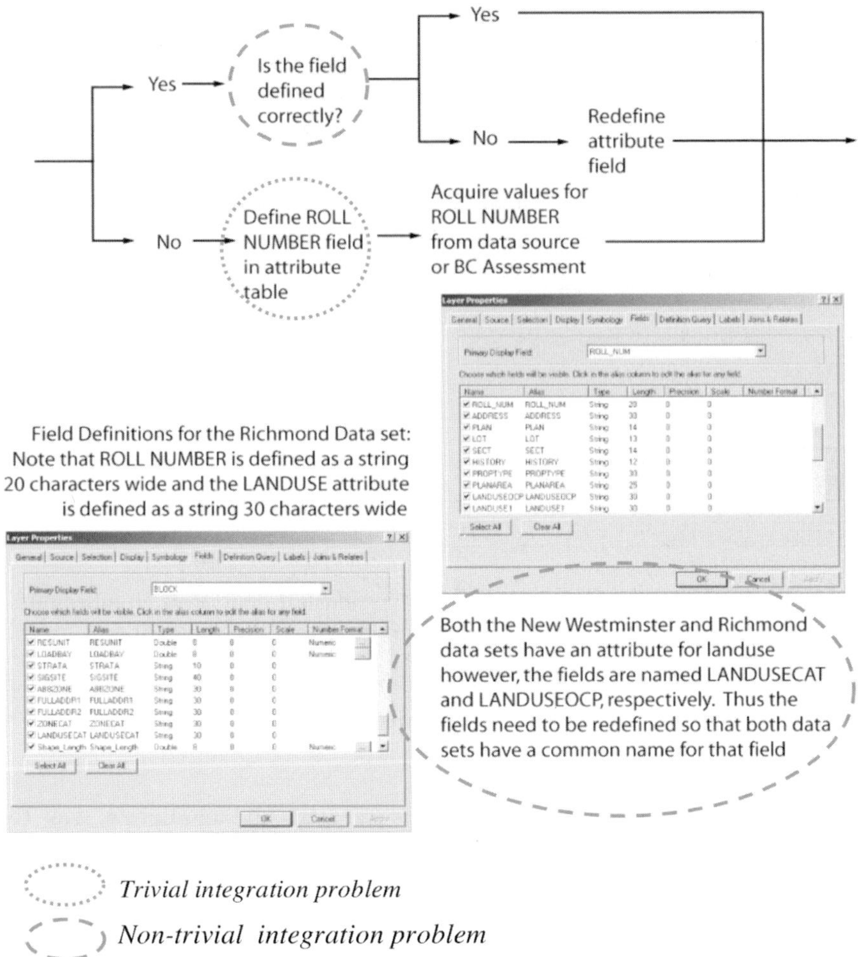

Fig. 3. Non-trivial integration problems require semantic interpretation. In this instance, database ethnographies were introduced as a methodology to get the semantic context from database stewards. Individual database managers were interviewed and institutional context and usage were clarified for semantic terms

Figure 4 illustrates the potential utility of linking cross-tabulated census data with cadastral level land use polygons. In areas where housing characteristics (dwelling type and tenure) are heterogeneous, such data enable researchers to identify meaningful socioeconomic variation otherwise masked by areal counts and/or averages even when fine-scale small areas data is used. Figures 6 and 7, produced using a cross-tabulation that included dwelling type, housing tenure status, and household type, illustrates the value of this technique in practice. Examining tenure status first, Figure 5 shows that when the 260 renter households living in DA identification number 59152888 are mapped according to their proportions in each dwelling type, most renters live in the northeast portion of the DA in high-rise buildings. Similarly, Figure 6 maps the distribution of single parent families by dwelling type providing a more accurate indication of where lone parent families reside within the DA being examined. Even though DA 59152848 contains more lone parent families than its surrounding DAs, cross-tabulated data reveals they are primarily located within the southeast quadrant of the DA. The two examples provided both link census data with cadastral (property parcels) polygons rendering variation within DAs visible.

DAUID	TtlStr	SDH	AptG5	AptL5
59152886	160	10	160	0
59152887	210	20	185	0
59152888	260	10	205	15
59152889	10	0	0	0
59152921	35	10	0	0

* numbers can vary +/- due to Census rounding errors

TtlStr = Total - Structural Type of Dwelling
SDH = Single Detached Housing
AptG5 = Apartment building greater than 5 stories
AptL5 = Apartment building less than 5 stories

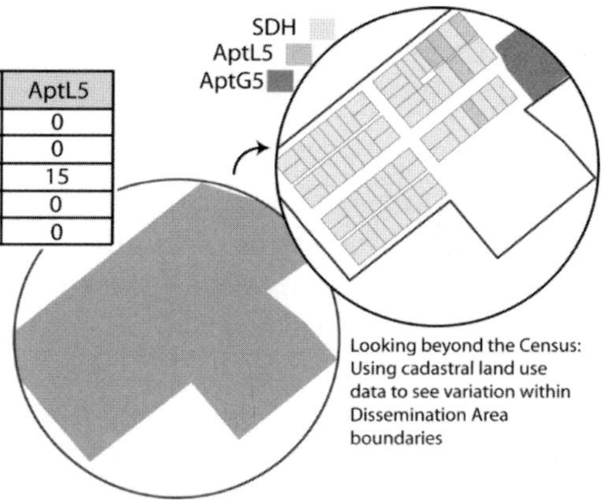

Fig. 4. Linking cross-tabulated census data to cadastral land use on a *DAUID* (Dissemination Area Unique Identifier) basis

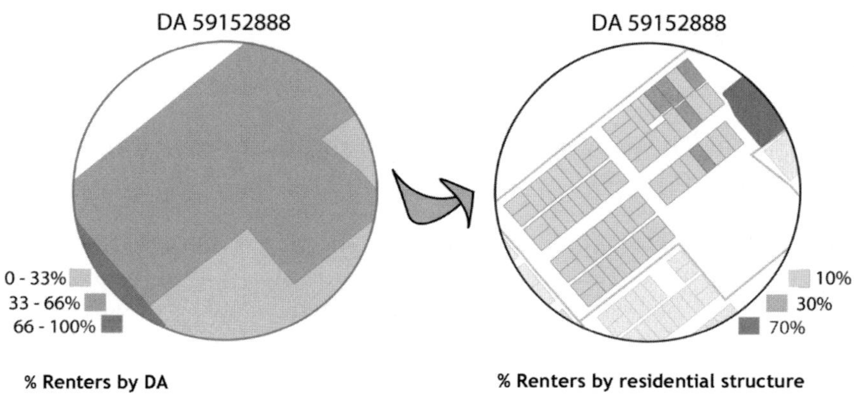

Fig. 5. Mapping the percentage of renters per legal parcel provides a more precise indication of where renters reside at the cadastral level compared to indicators of rental tenure aggregated at the *DA* (Dissemination Area) level

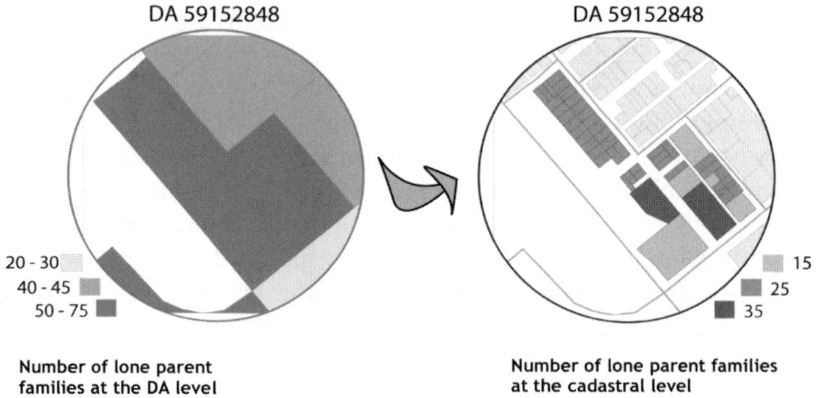

Fig. 6. The number of lone parent families as an aggregate statistic at the *DA* (Dissemination Area) level versus the number of lone parent families at the cadastral level within the *DA*

Understanding the distribution of land tenure and household type (in this case the marital status of families with children) is an anecdote to the generalization of census data and could be further complemented by understanding the distribution of income at the household level. Perhaps most importantly this approach identifies possible spatial mismatches between census boundaries and underlying social geographies. In the absence of local knowledge spatial mismatches of this nature can produce misleading representations and conclusions. The technique described here could be used to improve the resolution of spatial analysis in urban research generally, but in particular it offers a means to improve population health research, as both housing tenure and the incidence of lone parent families are commonly used to characterize relative health patterns by a number of socioeconomic deprivation indices [19,27,47,60].

5.3 Protocol Three: Visualizing Household Survey Data at the Rooftop Level

In an ideal GIScience scenario, one would be able to disaggregate household income from the census or other source to the rooftop level. Such an exercise would be fraught with uncertainty and dogged by issues of privacy – especially given the tentative nature of the resultant model. As an alternative, we chose to use survey data acquired by postal questionnaire as part of a study on new immigrants and hidden homelessness [15]. The questionnaire was sent to all households living in rental apartments in two DAs selected based on their percentage of recent immigrants and low household income was reported in the previous (2001) Canadian census.

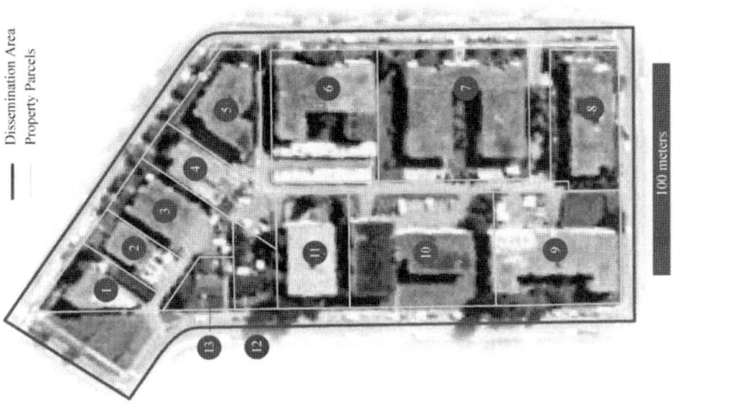

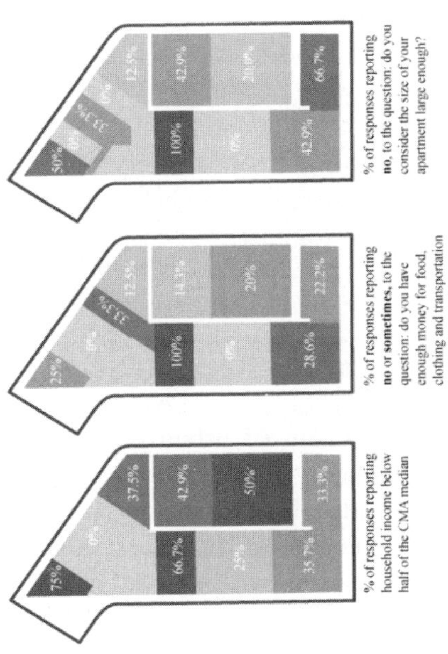

Map ID	Land Value (per Sqm)	Num Units	Gross Land $	Gross Improvements $
1	$516.60	11	$494,000.00	$612,000.00
2	$516.12	10	$374,000.00	$494,000.00
3	$516.54	18	$812,000.00	$601,000.00
4	$516.01	10	$635,000.00	$324,000.00
5	$508.77	26	$845,000.00	$1,223,000.00
6	$516.40	36	$1,564,000.00	$1,329,000.00
7	$508.88	62	$2,224,000.00	$2,971,000.00
8	$516.47	36	$1,159,000.00	$1,442,000.00
9	$520.59	48	$1,718,000.00	$2,510,000.00
10	$515.97	45	$1,668,000.00	$1,964,000.00
11	$516.11	27	$840,000.00	$1,488,000.00
12	$436.09	1	$268,000.00	$4,500.00
13	$411.93	1	$244,000.00	$0.00

Source: BC Assessment (2005)

Fig. 7. Survey data from individual rental households was aggregated at the cadastral (rooftop) level. Despite the limitations of an incomplete census, these data – when mapped using small multiples – illustrate the micro-scale variation in social conditions present within the *DA* surveyed

A response rate of 20.7% was achieved; this is marginally higher than the 20% sampling rate of Census Canada. Asked to indicate their income from a set list of household revenues, 23.8% of respondents from one DA (Edmonds) said they earned less than $1,000 per month, while 15.1% of responses from the second DA (Metrotown) were under this level. The majority of respondents from both areas, however, reported annual household incomes that fall below $24,000, which is less than half the Vancouver Census Metropolitan Area (CMA) median household income. Results for the Metrotown DA are illustrated in Figure 7.

Figure 7 demonstrates the degree of detail that becomes available using survey data in conjunction with cadastral data. In this case, the survey data from individual households within rental properties was aggregated to the building (rooftop) level. Despite the limitations of an incomplete census, these data – when mapped using small multiples – illustrate the micro-scale variation in social conditions present within the DA surveyed.

This survey was designed primarily to confirm the presence or absence of poverty in the area and shed light on the experiences of residents in ways not addressed by census data. It was not a random sample – as all households within the chosen areas were surveyed – and there was no control for who would respond. By contrast, polling firms actively pursue a random sample by targeting subgroups when surveying to ensure proper matching between respondents in the sample and the general population. As a result, non-response bias and sampling error is likely present in the results. The power of this data is magnified, however, when combined with property and land tenure data, which together suggest possible future research directions and questions.

5.4 Protocol Four: Assigned Value at the Rooftop Level

The fourth protocol for increasing resolution of urban datasets involves taking real estate assessment data and assigning them to individual cadastres. In this instance, data was provided by the British Columbia Assessment Authority (BCAA) in spreadsheet form. BCAA assigns a real estate value to each property parcel in the province split into land and building value.

Total assessed values for residential property allow a glimpse into a dimension of net worth other than pay-cheque; moreover, disparities in property values are indicative of the degree of socioeconomic inequality within a metropolitan area [35]. Total assessed value for a property is sufficient information to determine what level of household income would normally be required to support a particular property.[2] Clearly, there are

[2] Minimum household income required was calculated assuming a 25% down-payment, 25 year amortization period, monthly payments, a 6.0% rate of interest and a maximum mortgage payment-to-income ratio of 28%. The interest rate re-

limitations, but as a modeling exercise it allows us to leverage property parcel data to determine the differences in household income required to own property across an area. Given that individual households buy property at different times and at different mortgage rates, term length, and down-payment size, the assumptions in the model are not universally applicable. This is especially true in the GVRD where we frequently find recent immigrants who are asset rich but income poor. This approach does, however, yield interesting insight into housing costs for an assumed average household and the required income levels needed for such a household to support itself. Figure 8 illustrates that household values are affected by a clustering effect. Visualizing this clustering at a very high resolution, we are able to avoid the problem of losing small pockets of high and low valued homes. Such pockets are eradicated when data are aggregated – even to minimum census units. It is why, by examining household value at the cadastral level, that pattern emerges.

This approach to using alternative (non-census) high-resolution data to generate a model of a non-census socioeconomic variable is very much a part of what Longley and Harris [37] urged geographers to do. The business community has long engaged in modeling in this manner for exploratory purposes or strategic decision-making (even if it is less than perfectly accurate). We are reminded that all models are wrong; some are useful [53].

6 Conclusion: Moving Beyond the Census

Smith [56: p 32] notes greater specificity and the ability to move "beyond the comfort zone of census tract analysis" is required to understand neighborhoods and the individuals that populate them. In this paper we have described a series of strategies to develop higher resolution data as a means of breaking out of traditional administrative units in order to better understand social phenomena.

---

flected the average rate (for a 5 year fixed-rate closed mortgage) posted by Canada's five major banks (Bank of Montreal, Bank of Nova Scotia, CIBC, Royal Bank and TD Canada Trust) on December 13th, 2005. The maximum mortgage payment-to – income ratio was set to 28% in order to leave room for other shelter related expenses like property tax, insurance, utilities, and maintenance costs, and still meet the threshold typically used by mortgage lenders (shelter cost-to-income ratio of 32%).

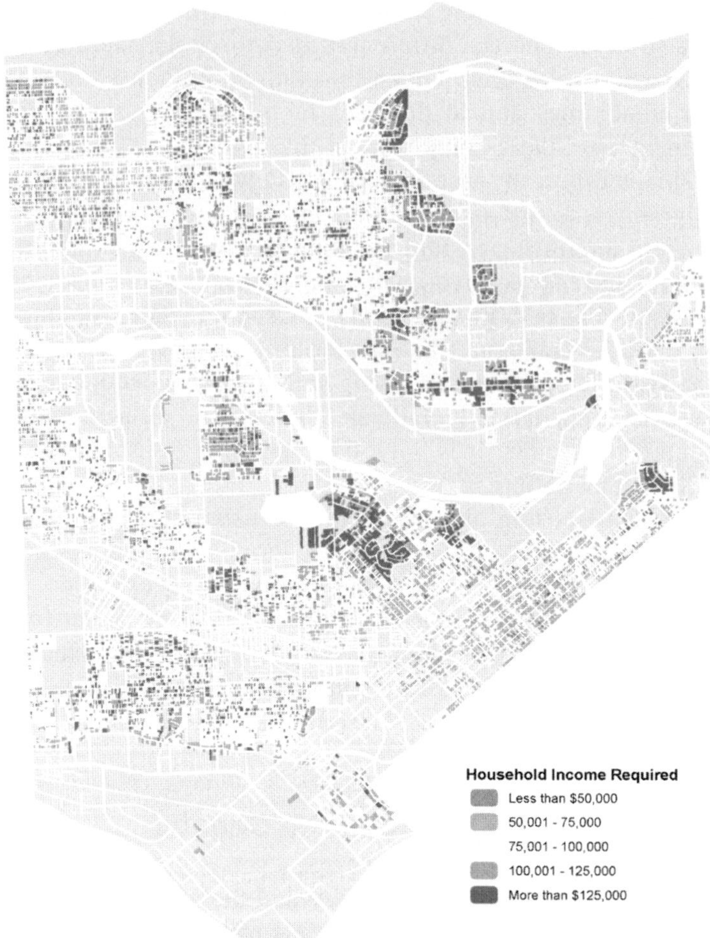

Fig. 8. The minimum household income required to purchase – via mortgage – a single-family detached home in Burnaby (suburb of Vancouver, Canada) assuming a 25% down-payment, 6.0% interest rate, 25 year amortization, and maximum income to mortgage payment ration of 28%

Producing more granular representations requires moving beyond the accepted scales and categories embedded in census data analysis and employing research approaches that can illuminate the 'blind-spots' of the census. The census remains the familiar stand-by for socioeconomic analysis for three main reasons: 1) use of census geographies enables use of the suite of non-spatial attributes collected at regular intervals by national governments; 2) the geography of its tidy boundaries are freely available to researchers; and 3) its reporting structures are acceptable among reviewers of all socioeconomic analysis. We argue, however, against the comfort zone of census geographies and its suite of attributes as a strategy for potentially revealing pattern and characteristics that are not visible in the aggregate and that are revealed using attributes typically collected at the individual level – such as tax assessment values.

This strategy is not at first glance appealing. It promises to involve many hours of data acquisition and attribute integration of disparate cadastral data sets. It certainly involves the introduction of uncertainty with the disaggregation of data from other administrative units. In short, this approach is messy yet it promises a means of better understanding human geography. We have pursued it in the belief that is better to have an untidy high-resolution analysis that reveals new and unpredictable patterns than clean aggregated data that reinforces what we know already.

References

1. Abel DJ, Ooi BC, Tan K-L, Tan SH (1998) Towards integrated geographical information processing. IJGIS 12:353–371
2. Alvanides S, Openshaw S (1999) Zone Design for Planning and Policy Analysis. In: Stillwell J, Geertman S, Openshaw S (eds) Geographical Information and Planning: European Perspectives. Springer-Verlag, New York, pp 299–315
3. Alvanides S, Openshaw S, Rees P (2001) Designing your own geographies. In: Rees P, Martin D, Williamson P (eds) The Census Data System. John Wiley & Sons, Chicester
4. Bailey TC, Gatrell AG (1995) Interactive spatial data analysis. John Wiley & Sons, New York
5. Ballantyne B (1997) The research agenda for cadastral surveying: against sectarian struggles. Geomatica 51:427–436
6. Bittner T, Edwards G (2001) Towards an Ontology for Geomatics. Geomatica 55:475–490
7. Bishr Y (1996) Overcoming the semantic and other barriers to GIS interoperability. IJGIS 12:538–543
8. Bracken I, Martin D (1989) The generation of spatial population distributions from census centroid data. Env Plan A 21:538–543
9. Bramley G, Smart G (1995) Modelling Local Income Distributions in Britain. Reg Stud 23:239–255
10. Campari I (1996) Uncertain Boundaries in Urban Space. In: Burrough PA, Frank A (eds) Geographic objects with indeterminate boundaries. Taylor and Francis, London, pp 57–69
11. Crampton JW (2004) GIS and Geographic Governance: Reconstructing the Choropleth Map. Cartographica 39:41–53

12. Devogele T, Parent C, Spaccapietra S (1998) On spatial database integration. IJGIS 12:335–352
13. Eicher CL, Brewer CA (2001) Dasymetric Mapping and Areal Interpolation: Interpretation and Evaluation. CAGIS 28:125–138
14. Fabricant SI, Buttenfield B (2001) Formalizing Semantic Spaces for Information Access. Ann Amer Assoc Geog 91:263–280
15. Fielder R, Schuurman N, Hyndman J (Forthcoming) Improving Census-based Socioeconomic GIS for Public Policy: Recent Immigrants, Spatially Concentrated Poverty and Housing Need in Vancouver. ACME
16. Fieldhouse EA, Tye R (1996) Deprived people or deprived places? Exploring the ecological fallacy in studies of deprivation with the Samples of Anonymised Records. Env Plan A 28:237–259
17. Fonseca FT, Egenhofer MJ, Agouris P, Camara G (2002) Using ontologies for Integrated Information Systems. TGIS 6:231–257
18. Fonseca FT, Egenhofer MJ, Davis CA Jr., Borges KAV (2000) Ontologies and knowledge sharing in urban GIS. Compt Env Urb Sys 24:251–272
19. Frohlich N, Mustard C (1996) A regional comparison of socioeconomic and health indices in a Canadian province. Soc Sci Med 42:1273–1281
20. Grimm N, Grove J, Pickett S, Redman C (2000) Integrated approaches to long-term studies of urban ecological systems. BioScience 50:585–671
21. Harris R, Frost M (2003) Indicators of urban deprivation for policy analysis GIS: going beyond wards. In: Kidner D, Higgs G, White S (eds) Socioeconomic Applications of Geographic Information Science, Innovations in GIScience 9. Taylor and Francis, New York, pp 231–242
22. Harris R, Longley P (2003) Targeting Clusters of Deprivation within Cities. In: Stillwell J, Clarke G (eds) Applied GIS and Spatial Analysis. John Wiley & Sons, Chichester, pp 89–110
23. Harvey F (1999) Designing for interoperability: Overcoming semantic differences. In: Egenhofer MJ, Fegeas R, Kottman CA (eds) Interoperating Geographic Information Systems. Kluwer Academic Publishing, Boston, pp 58–98
24. Harvey F, Buttenfield BP, Lambert SC (1999) Integrating geodata infrastructures from the ground up. Phot Eng Rem Sens 65:1287–1292
25. Hentschel J, Lanjouw JO, Lanjouw P, Poggy J (2000) Combining Census and Survey Data to Trace the Spatial Dimensions of Poverty: A Case of Ecuador. World Bank Econ Rev 14:147–165
26. Holloway SR, Schumacher J, Redmond RL (1999) People and Place: Dasymetric Mapping Using ARC/INFO In: Morain S (ed) GIS Solutions in Natural Resource Management: Balancing the Technical-Political Equation. OnWord Press, Santa Fe, pp 283–291
27. Jarman B (1983) Identification of underprivileged areas. BMJ 286:1705–1708
28. Kashyap V, Sheth AP (1996) Semantic and schematic similarities between database objects: a context-based approach. VLDB J 5:276–304
29. Kirby A (1981) Geographic contributions to the inner city deprivation debate: a critical assessment. Area 13:177–181

30. Kuhn W (2001) Ontologies in support of activities in geographical space. IJGIS 15:613–631
31. Lane SN (2001) Constructive comments on D. Massey 'Space-time,' 'science' and the relationship between physical geography and human geography. Trans Inst Brit Geog 26:243–256
32. Langford M, Unwin DJ (1994) Generating and mapping population density surfaces within a geographical information system. The Cart J 31:21–26
33. Laporte A (2002) A note on the use of a single inequality index in testing the effect of income distribution on morality. Soc Sci Med 55:1561–1570
34. Laurini R (1998) Spatial multi-database topological continuity and indexing: a step towards seamless GIS data interoperability. IJGIS 12:373–402
35. Ley D, Tutchener J, Cunningham G (2002) Immigration, Polarization, or Gentrification? Accounting for Changing Housing Prices and Dwelling Values in Gateway Cities. Urb Geog 23:703–727
36. Longley P (2003) Geographical Information Systems: developments in socio-economic data infrastructures. Prog Hum Geog 27:114–121
37. Longely PA, Harris RJ (1999) Towards a new digital data infrastructure for urban analysis and modeling. Env Plan B 26:855–878
38. Martin D (1989) Mapping population data from zone centroid locations. Trans Inst Brit Geog 14:90–97
39. Martin D (1998) Automatic Neighbourhood Identification from Population Surfaces. Compt Env Urb Sys 22:107–120
40. Massey D (1999) Space-time, 'science' and the relationship between human geography and human geography. Trans Inst Brit Geog 24:261–276
41. Mennis J (2003) Generating Surface Models of Population Using Dasymetric Mapping. Prof Geog 55:31–42
42. Mesev V (1998) The Use of Census Data in Urban Image Analysis. Phot Eng Rem Sens 64:431–438
43. Minot N (2000) Generating Disaggregated Poverty Maps: An Application to Vietnam. World Devl 28:319–331
44. Openshaw (1983) The Modifiable Area Unit Problem, Concepts and Techniques. In: Modern Geography 38. Norwich, Geobooks.
45. Openshaw (1984a) Ecological fallacies and the analysis of areal census data. Env Plan A 16:17–31
46. O'Sullivan D (2004) Too Much of the Wrong Kind of Data: Implications for the Practice of Micro-Scale Spatial Modeling. In: Goodchild MF, Janelle DG (eds) Spatially Integrated Social Science. Oxford University Press, New York, pp 95–107
47. Pampalon R, Raymond G (2000) A Deprivation Index for Health and Welfare Planning in Quebec. Chron Dis Can 21:104–113
48. Pundt H (2002) Field Data Collection with Mobile GIS: Dependences Between Semantics and Data Quality. GeoInformatica 6:363–380
49. Raper J (2001) Defining Spatial Socio-Economic Units: Retrospective and Prospective. In: Fran A, Raper J, Chelan J-P (eds) Life and Motion of Socio-economic Units. Taylor and Francis, New York, pp 13–20

50. Raubal M (2001) Ontology and epistemology for agent-based wayfinding simulation. IJGIS 15:653–655
51. Rees P, Martin D (2002) The debate about census geography. In: Rees P, Martin D, Williamson P (eds) The Census Data System. John Wiley & Sons, Chichester, pp 27–36
52. Rodriguez AM, Egenhofer MJ (2003) Determining semantic similarity among entity classes from different ontologies. IEEE Trans Knowl Data Eng 15: 442–456
53. Schuurman N (2002) Reconciling Social Constructivism and Realism in GIS. ACME 1:75–90
54. Schuurman N, Leszczynski A (Forthcoming) Ontology-based metadata. TGIS.
55. Sheth AP (1999) Changing Focus on Interoperability in Geographic Information Systems: From system, syntax, structure to semantics. In: Goodchild MF, Egenhofer MJ, Fegeas R, Kottman C (eds) Interoperating Geographic Information Systems. Kluwer Academic Publishers, Boston, pp 5–30
56. Smith HA (2004) The Evolving Relationship between Immigrant Settlement and Neighborhood Disadvantage in Canadian Cities, 1991-2001. RIIM, Vancouver. URL: http://rim.metropolis.net/Virtual%Library/2004/WP04-20.pdf
57. Stock K, Pullar D (1999) Identifying Semantically Similar Elements in Heterogeneous Spatial Databases Using Predicate Logic Expressions. In: Vckovski A, Brassel KE, Schek H-J (eds) Second Int Conf on Interoperating Geographic Information Systems (INTEROP'99), LNCS 1580. Springer-Verlag, Zurich, pp 231–252
58. Tate N (2000) Surfaces for GIScience. TGIS 4:301–303
59. Thurstain-Goodwin M, Unwin DJ (2000) Defining and Delineating the Central Areas of Towns for Statistical Monitoring Using Continuous Surface Representations. TGIS 4:305–317
60. Townsend P, Phillimore P, Beattie A (1988) Health and Deprivation. Croom Helm, London.
61. Vckovski A, Brassel KE, Schek H-J (eds) (1999) Interoperating Geographic Information Systems (= LNCS 1580). Springer-Verlag, Zurich
62. Wilkinson RG (1992) Income distribution and life expectancy. BMJ 304: 165–168
63. Wilkinson RG (1996) Unhealthy Societies: The Afflictions of Inequality. Routledge, New York.
64. Wilkinson RG (1999) Putting the picture together: prosperity, redistribution, health and welfare. In: Marmot M, Wilkinson RG, Social Determinants of Health. Oxford University Press, New York, pp 256–274

Analysis of Cross Country Trafficability

Åke Sivertun, Aleksander Gumos

GIS/HCS Department of Computer and Information Science
Linköpings Universitet, 581 83 Linköping, Sweden
email: akesiv@ida.liu.se; g-alegu@ida.liu.se

Abstract

Many decisions – not only in the field of Emergency Management or Military oriented actions - require nowadays in addition to reaching verdicts a large amount of spatial and geographical information data. If these data are handled in Geographical Information Systems – GIS, we are introducing new possibilities to handle and analyze this type of information in a way that divert substantially from traditional handling of the paper maps. A Geographical Information System is an IS with the capabilities not only to handle currently being produced digital maps in raster and vector formats but in addition analyze those for instance together with Remote Sensing techniques like GPS positioning and combining it with a real time intelligence reports. The development of the societies parallel to globalization and global dependencies trends, some symptoms of climate changes, ageing population, more complex societies and more complex systems lead also to a grater demand for more sophisticated information and information systems (Trnka 2003; Trnka et al. 2005a and 2005b; Quarantelli 1999; Rubin 1998; Rubin 2000; Kiranoudis et al. 2002; Mendonça et al. 2001; Beroggi 2001; Johnson 2002). The research teams at IDA/LiU have extensive experience with testing various forms of data capture, real time analyzes and diffusion of geographically registered data through, for example, mobile GIS technology.

However, we have experienced a necessity for development of both entirely GIS-based models and supportive to them data to be analyzed, in order to improve all crucial phases of the Emergency Management scenario reasoning during a preventive and as well an information provisions stages. In this way information could be regarded as a strategic infrastructure that is now being investigated on the Swedish national level as well as by the European Union, for example through the European Network of Excellence – the GMOSS. One goal for GMOSS is to investigate good procedures for Emergency Management and Crisis Response, and as a consequence to build standardized and harmonized geographical databases that can be used in decision support systems. What is still to be added to the agenda is an implementation of several GIS-based models that are making use of all those databases for prediction of potential hazards, for preventive works and action plans. Objectives in this article are to contribute to the development of such models and to investigate necessary data for use in rescue-, relief- and preventive works, and to stress obligations for data uptodateness for structuralizing better preparedness plans etc.

1 Cross Country Trafficability

This article is mostly based on a pilot study concerning methods to build GIS based models that can be included in a network based decision support system (Gumos 2005). As to reduce the amount of unprocessed and indexed data the user has to take into consideration in a stressing emergency – like a forest fire – we suggest that more stable and already existing geographical data are pre-processed into easy to adopt information that can be used together with dynamic data concerning development of scenarios – threats and resources. The model developed and tested here is aimed to predict trafficability through terrain – a task as relevant for military logistics as for emergency management and relief. The modelling of trafficability is a further development of a suggested development approach for Geographical Information (GI) where the data are collected for many different purposes (like in forestry for management reasons) but further shared and used in planning as to predict vulnerable areas from fires and other emergency events. Doing so and developing models that aid plan actions in case of such emergency situations can save lives, environment and properties. Traditional model approaches are often just oriented to optimizing or predicting for one task (like transports) and it is almost impossible to build a single (mathematical) model that can handle all the different factors that ought to be considered in a real emergency or military situation. However by harmonization of the models in a way that they are able to integrate an information system and later present the result of their computing outcome as a interoperable input into other models, it is at least in theory possible to see where the white spots are on the map – where we don't have enough information – or identify the black spots where we can agree on the situation and have enough information to make a judgement. These definitely require several additional programs to handle the time aspect, the status of all vehicles and other resources that are involved in the operation. Ideally, if the GIS can act as a core information system combining the results, it will have huge benefits at the final cross-country trafficability modelling. In this project we have tried to evaluate the data sources needed and some models to evaluate feasible routes through the terrain. We perceive this study as one of several components in a much more complex system for building information and knowledge databases that are possible to include and share data into models to achieve a broader overview of the situation. Accomplished evaluations can in their turn be shared for creating a common picture of understanding – a base for strategic and tactic decisions.

2 Previous Studies and Theoretical Framework

Having in mind some very first in-field research that has been made for US Army purpose (e.g. Terrain trafficability studies by Wood and Snell 1960); one can observe a strict military aspect of trafficability issue, that was done for two general vehicle types: having tracks (tanks) and having wheels (trucks, jeeps etc). Later on, the idea was developed and nowadays the argument for trafficability models are focused as much towards the civil applications as armed conflicts because the attention is more directed towards geo-hazards (e.g. flood, earthquake, tsunami, volcano eruptions etc.), unwilling forest fires, ecological catastrophes or rescue services. The knowledge about the accessibility to the Nature may save human life, reduce expenditure costs and also lessen the time needed to apply the remedy. The civil vehicles mainly are wheeled vehicles, and before going cross country there must be made as much reconnaissance as possible in order to find in detailed map the possible or preferable route, or corridor of movements, in the specific region. In Forestry and Agriculture nowadays, various set of machines and techniques exist, dealing more and more with the problem of analyzing an acceleration of soil erosion due to the extensive harvesting and particularly machine-soil interaction, that takes place in different weather conditions and topographical locations. According to the directives for general practices on the sensitive sites, reduction of the erosion risk (and loss of production) is under investigation, and some initial proposal exist, where vehicle operators are guided to chose an optimal route and asset e.g. proper tires pressure to prevent soil compaction. One of the simplest equations for the incorporation of the factors regarding the path distance for a moving vehicle used in these studies is (DeMers 2002):

$$\text{Fuel used} = SD \times F \times HF \times VF \tag{1}$$

Where:
SD – is the total slope distance
F – is the surface friction factor
HF – is the horizontal factor
VF – is the vertical factor relating to down or up slope movement.

The military SHAKEN project (Gil et al. 2003) was an attempt to let domain experts enter and verify logical rules describing features of a military scenario. The experts used two methods to verify their rules – in a declarative inference engine where rules describing trafficability of terrain used to control whether legal actions are performed by the user. In Birkel (2003) are four Non-GIS Military Models for trafficability presented:

1. Mod-SAF/SIMNET Trafficability (The Modular Semi-Automated Forces environment),
2. CCTT Trafficability (The Close Combat Tactical Trainer),
3. WARSIM Trafficability (The Warfightering Simulation),
4. NRMM II (The NATO Reference Mobility Model II)

However, in most of the literature they have not implemented their trafficability models into a GIS environment so most of the models had to be transferred to the GIS environment. In this study we have used a more traditional GIS approach with classification and combination of different features using the weights from existing models adding cost functions to different suggested paths.

The terrain factors are of greatest interest for the trafficability characteristics. Further on in the Birkel's report, there are mentioned following terrain factors that contribute for performing a drivability weights /ranges/ evaluation: Slope, Obstacle description, Surface material, Soil type, Soil strength, Freeze/thaw depths, Surface roughness, Surface slipperiness/ wetness/ice, Snow, Non-woody vegetation, Woody vegetation, Hydrology (refer to Table 1).

Similarly, The Canadian Space Agency hyperspectral satellite mission plans were to automatically derive specific information regarding the terrain parameters for every of each scanned pixel of the investigating terrain. Some of them are summarized in Table 2.

Table 1. General cross-country trafficability parameters (*stated briefly by Edlund 2004)

Trafficability parameters		
Terrain factors		Vehicle factors*
Relatively static	Dynamic	Width, length, height, override diameter, maximum gap to traverse, ground clearance, maximum step, maximum gradient, maximum tilt, specific ground pressure, maximum straddle, etc
Geology, Soils, Topography, Hydrography, etc	Weather Conditions, Human activities, etc	

Table 2. Surface features of military interest, after Canadian Defense Department J2 Geomatics (1999)

Surface Feature	Example
Surface material	Quarries, Surface roughness, Boulder fields, Rock out-crops, Soil type Disturbed soil
Surface drainage	Linear features (rivers, ditches, shorelines) Area features (lakes, flood areas) Point features (dams, locks)
Vegetation	Deciduous, coniferous, Canopy closure, Hedgerows, Grasslands, Swamps, marsh, bogs
Transportation network	Roads (type, bridges, tunnels), Railways, Airstrips, Ports, Ferry sites
Obstacles	Walls, Fences, Towers
Near-shore bathymetry	

Some of the above military non- GIS models are taking into account different classes of ground vehicles. Various parameters of the mobile units performance were calculated, taking into consideration seasonal weather changes. Some of such general driveability parameters are: velocity, resistance force, acceleration, maximum speed, deceleration, turning rate, and climb angle, load class etc. Several army vehicles either on wheels or on tracks have been taken to the terrain tests in the different landscapes and different weather conditions (e.g. Shoop et al. 2004). Those data are very suitable to involve into GIS based models, however, that was not possible to do within the frames of this report. We were at the first stage trying to implement GIS based models including as many as necessary of the huge number of factors of stable factors (and some dynamic) that are involved in judgments of trafficability and capable to aid in as well strategic as tactical decisions. Edlund (2004) has in her report "Driveability analysis – using a Digital Terrain Model and Map Data, briefly indicated those components that affect the trafficability through terrain. However she was limiting herself to use only the digital elevation in the thesis. Digital terrain is of importance and several experiments have been done to establish a better spatial resolution than the 50x50 m grid that is generally available from the Swedish Land Survey. This was also our approach. Sharing the opinion that these elevation data were too rough, we wanted to combine these 50x50 m elevation grid data with the height curves. However these vector features were without an absolute height data and our task was to connect them with the digital elevation model map layer. To perform an analysis based on all these factors using all the necessary data is a long lasting and difficult task. However if it were possible to automate at least some of the analyses it would be a great assistance for the decision makers by introducing several alternative approaches for the analysis and the results in the form of a rough Go/No Go area classification (in stressful situations or when there are doubts about the quality or reliability in data) and more sophisticated analyses with classes as Go, Restricted, Slow, Very Slow, alternatively No Go, Restricted, Several Restricted, Unrestricted etc.

The models and data investigated to be involved in the models were as follows

- Geomorphology (elevation, slope, natural landform obstacles)
- Hydrology (rivers, lakes, catchments areas, wetness index, hydrogeology)
- Land Use and Land Coverage (agriculture practices, vegetation, build-up areas etc)
- Soil types (soil geotechnical parameters)
- Human made structures (roads, channels, bridges, restricted areas etc); In the future also other human introduced factors like contaminated areas, fuel spill areas, minefields and similar man-made obstacles are possible to be added to the models to complete the picture.

GIS models are based on several methods to use vector and raster data. As a role – man-made data are most suitable to be represented as vector and natural phenomena as raster data. In this case both types of data representations have been used as to perform the different tasks required. By for example using elevation points in a grid divided by 50x50 m and the different tools to describe topography from that, we were able to calculate gradients and slopes and several factors of use in evaluating not only direct obstacles to traffic but as input in models to estimate soil wetness classes and further combinations of classes that together form obstacles or costs for movements. However, the height curves from the digital vector topographic maps (but without elevation attribute data attached) were added to the digital terrain model as to improve the estimation of elevation to approximately 5x5 m (see Fig. 1).

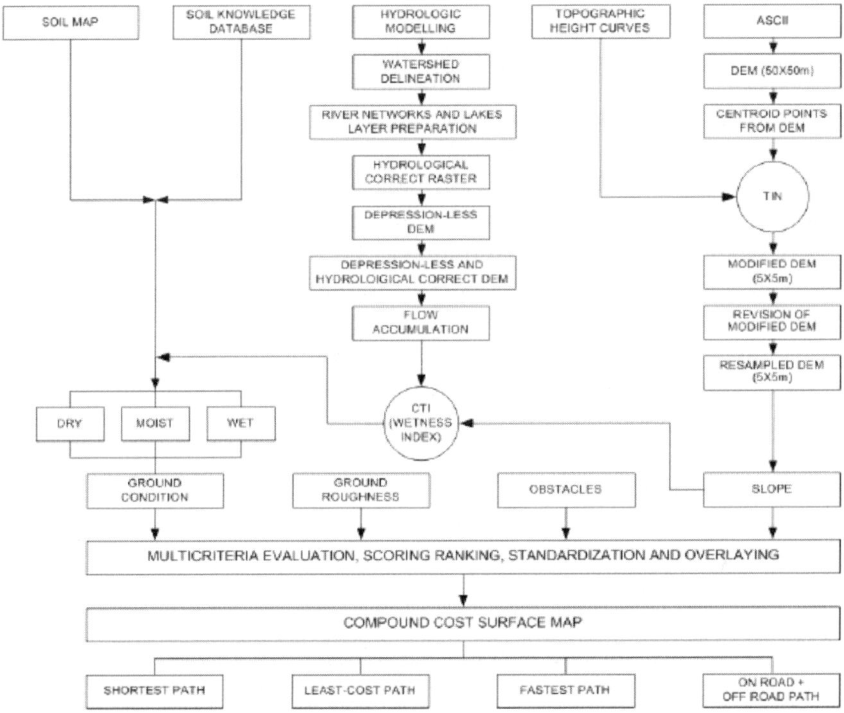

Fig. 1. Schematic presentation of methodological approach for the Trafficability Modeling in GIS (Gumos 2005)

In the same way it is often reasonable to be able to convert between raster and vector formats to perform different tasks and for presentation in different media and other program modules.

Soil type is another class of factors that cannot be used in a naive way. Here a soil particles, base mineral and genesis are all of importance to evaluate the impact on trafficability. In our modeling we have developed a special module to estimate the impact from the different soil properties.

Hydrological models were investigated for the purpose of to find out soil wetness classes. Those models originate from the results of using the topographic models, the soil models, the vegetation, land use etc.

With the base in the several models investigated we have also found several factors that are not connected – in a simple way – to the physical conditions. Different soils in different locations responding to the weather, the precipitation/snow, temperature etc in a way that we have not been able to investigate but that is still possible to be included in a parallel model or be manually added by and experienced user. By adding intelligence reports it is possible to classify e.g. all wetlands (except the ones with several big trees or other obstacles) as suitable for certain transports – if they have been found deeply frozen after several days with cold weather. In this way it is possible in a GIS to mix the use of automated analyses with manual classification and interpretation in cases when the models fail to handle all the complexity of reality or other data capture process fail. These models for trafficability are further easy to share in briefing conclusions (or if wanted in details) with members of staff and other involved forces. One may consider a situation where the basic command unit commander after being able to see the report supported by the basic assumptions (that e.g. the wetlands are frozen) is taking an action of certain level of critique, knowing basic facts about the terrain which conditions are validated by GIS modelling. It becomes than truly a decision base reasoning.

3 Methods and Techniques

3.1 Data Acquisition

In the created test application we were using only open and available data resources (refer to the Table 3). However, there are possible other sources of data that could be included in the future – in practice perhaps after being acquaintance with guidance from preliminary studies in a more general level.

Table 3. Datasets used for the Cross-Country Trafficability study

No	Datasets	Format	Scale/Resolution	Source
1	The topographic map	Vector	1:50K	GSD*
2	Terrain Elevation Databank	Binary file	50m x 50m	GSD
3	Quaternary Deposits Map	Raster	1m x 1m	SGU**
4	General Map of the Östergötland County	Vector	1:250K	GSD
5	Bedrock Geology Map of the Östergötland County	Vector	1:250K	SGU
6	Watersheds of the Östergötland County	Vector	Minimum 200 km^2 area size	SMHI***
7	Detailed Watersheds of the Östergötland County	Vector	Average 35 km^2 area size	SMHI

*GSD – the Geographical Sweden Data produced by The National Land Survey of Sweden
**SGU – the Geological Survey of Sweden
***SMHI - the Swedish Meteorological and Hydrological Institute

The test application was performed to investigate the quality, possibilities and problems that might occur in data processing and present models appraisal. Validation, which is needed to be performed, should take into de-liberation a dependency of various factors on each other and requirement from the data to be in a certain format and resolution to be acceptable for employed models.

The factors that have been investigated were tested over the location of the area equivalent to the topographic map in the scale of 1:50K, sheet 8FNO Linköping. The area with it's northern slopes of a tectonic rift is divided by the southern "Östgöta" plain with the shallow lake Roxen + 33,3 m a.s.l. , partly forested and with "mountainous area character" north from the lake (elevation round 100 m a.s.l) were mainly studied. There is not a dramatic difference in elevation but the landscape is divided with narrow valleys in the crystalline bedrock and in the investigation area a sub peneplane overlaid with clay that in its turn have been eroded by a creak forming gullies, e.g. in Stjärnorp. In non-clay areas bare rock is mixed with glacio-fluvial deposits and partly with areas of huge amounts of boulders. Vegetation is a mix of different-aged coniferous forest and also partly with deciduous forest including several hundred-year-old oak trees impossible or hard to pass over also with very powerful vehicles. The trafficability analysis in the area corresponds to 625 km^2.

3.2 Constructing the Soil Knowledge Database

When trying to construct a soil database we have to investigate the ontological base for existing soil maps (see Figs. 2, 3 and 4) and comparing them with the needs when building a trafficability application.

Soils with its complex physical, chemical and hydrological characteristics have been converted into a knowledge database including soil texture, soil grain-size distribution classification, soil strength and permeability proper-ties, soil capillarity, frost activity of the soils, soil consistence and as a result of this – Soils trafficability. Soil strength parameter is used as an index of suitability of the ground for the off-road mobility of vehicles (Mitchel 1991). The soil strength parameters have been prepared also to be a driving force for discreteness the Atterberg's Consistency Limits for the three different soil moisture classification levels (states): Dry, Moist and Wet (saturated).

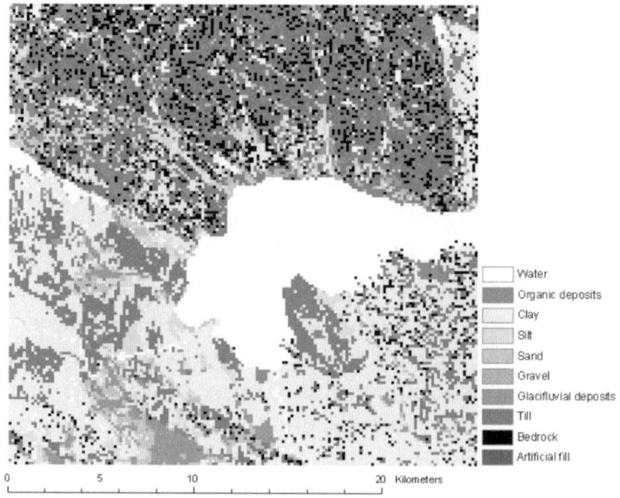

Fig. 2. Classification Quaternary deposits

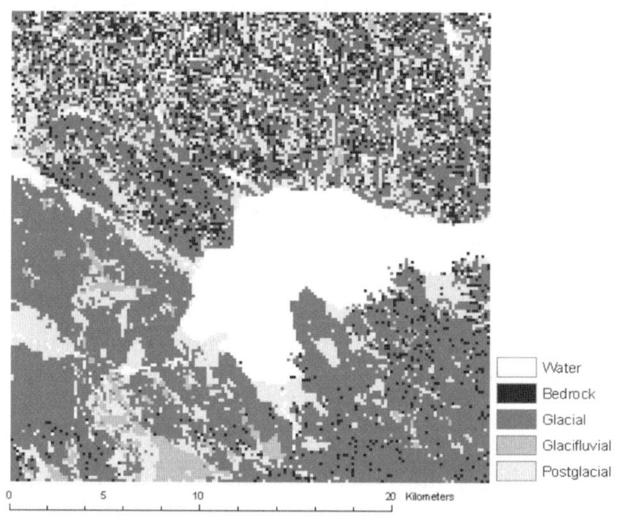

Fig. 3. Genesis of the Quaternary deposits

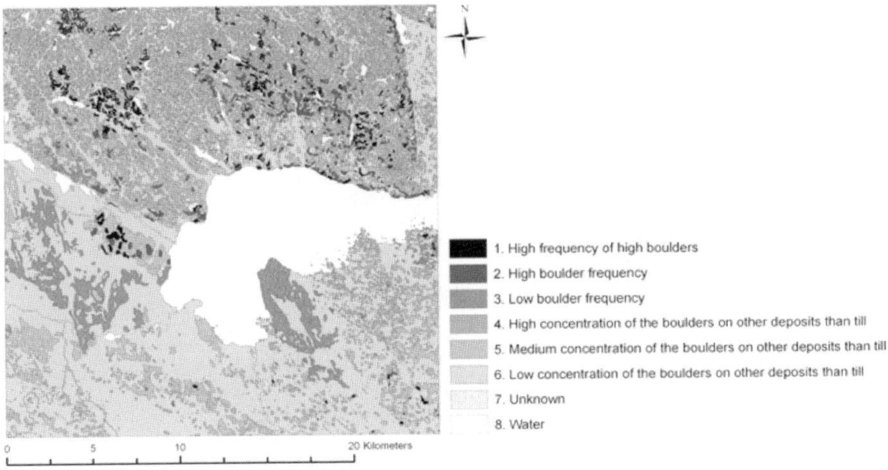

Fig. 4. The Boulder Frequency of the till Surface Map (raster 5x5m)

3.3 Hydrologic Modeling

There are several GIS based hydrological models for estimating the soil moisture like Andersson & Sivertun (1989) and Burrough & McDonnell (1998). Watersheds /drainage basins, the river network layer, the lake layer forming together a hydrological raster that has been prepared together with the enhanced Digital Terrain Model (DTM) and the Compound Topographic Index (CTI). CTI also knows as Steady State Wetness Index, is an equation adjustable for computing the topographic moisture accumulation. The formula is defined as:

$$CTI = \ln (As / \tan\beta) \qquad (2)$$

where: **As**-stands as the contributing catchment area per unit extent, orthogonal to the flow direction, and **tanβ** -is the Slope measured in degrees with **β > 0**

CTI equation formula originates from the studies that were conducted in order to calculate the changes of the moisture character in soils, in relation to their catenal hill slope position with significant relationship to both soil- and hydrology sciences (see Fig. 5).

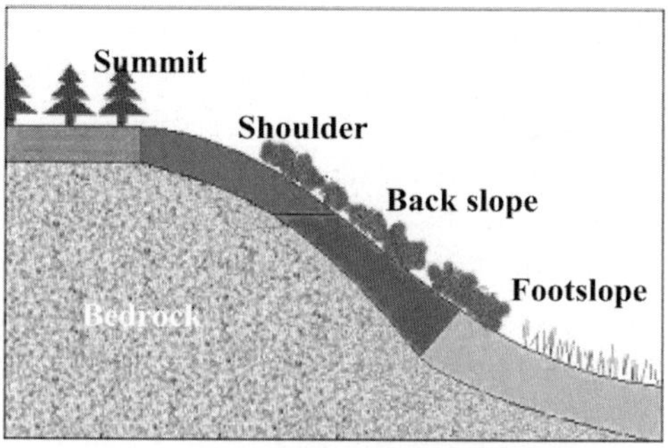

Summit – well-drained and developed soils
Shoulder – eroded and moderate developed soils
Back slope – deposited soils from summit and shoulder
Foot slope – moisture soils

Fig. 5. The principle for CTI classification

The soil and topography interrelations are significant especially when a spatial thinking is utilized during environmental analysis. The recalled role of topography in soil formations and soil genesis as a result of differences in relief is a well-known factor, however not isolated from the other factors. The catena concept originates from these primary studies and is being further developed by e.g., Pauw and Pertziger (2000); Gessler et al. (2000), Brown et al. (2000a) and Brown et al. (2000b).

3.4 Estimating Terrain Roughness Parameters

In this procedure, firstly, based on the Boolean statements (Boolean overlay), criteria were taken from the following maps datasets (see Table 4).

Next, they were converted (reclassified) into 0 or 1 assertion, marking 0 as suitable ('Go') and 1 as not suitable ('NoGo') for off-road drivability.

Table 4. List of Factor maps used in Boolean method

No	Factor map name	Format	Pre-processing 1st	Pre-processing 2nd
1	The Land Use Map	Area	Rasterization	Reclassification
2	The Area with Large Boulders Map	Area		
3	The Bog Map	Area		
4	The Human Made Objects Map	Point		
5	The Road Network Map	Line		
6	The Watercourses Map*	Line		

The second modeling stage was so called Saaty's Analytical Hierarchy Process (AHP). AHP has been chosen because it allows user to assign weights to a set of factors (maps), in created matrix of pair wise comparisons (ratios) between those factors (Jones 1997). This quantification of criteria is fully continuous and certain degree of suitability expresses the importance of each factor for decision makers (ed. Longley et al. 1999). The process of assigning the weights inside every each factor is called standardization. Three factor maps were analyzed using AHP (see Table 5), trying to answer to the question: if it is possible to drive cross-country than how good (convenience) is the possible route?

Table 5. Assignment of weight between the factor maps

No	Factor map name	The Slope	The Ground Condition	The Ground Roughness
1	The Slope	Equal importance between the factors*		
2	The Ground Condition			
3	The Ground Roughness			

* To simplify a modeling, AHP factors were weighted arbitrary equally.

4 Results

The study has shown that available data sources are possible to use in a proposed model for trafficability analysis in terrain. We have obtained in following order The Ground Condition Map (see Fig. 6), and later Go and No Go cost surface maps (see Figs. 7 and 8).

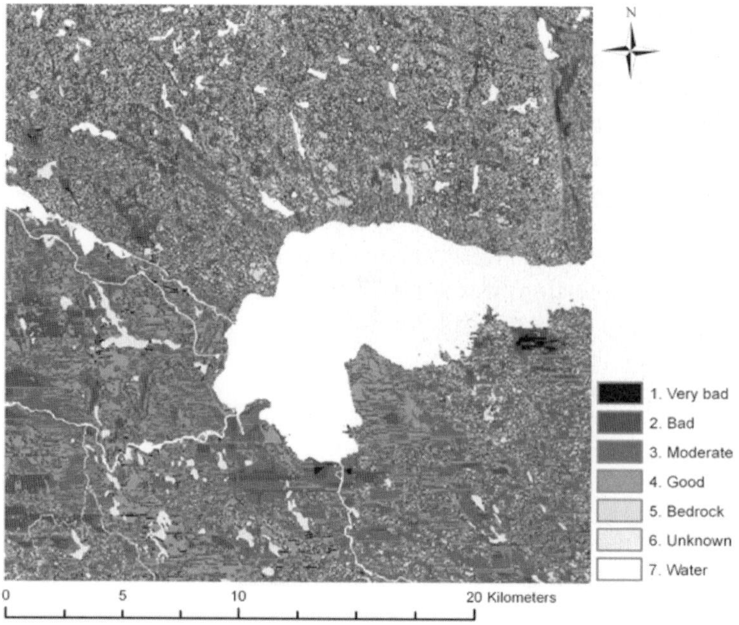

Fig. 6. The Ground Condition Map

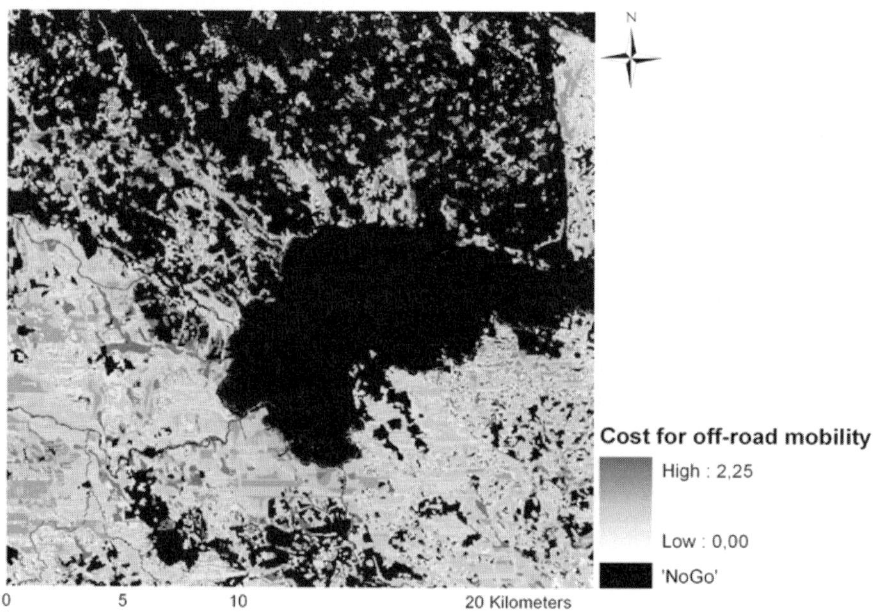

Fig. 7. Go – No Go and cost surface for off road trafficability

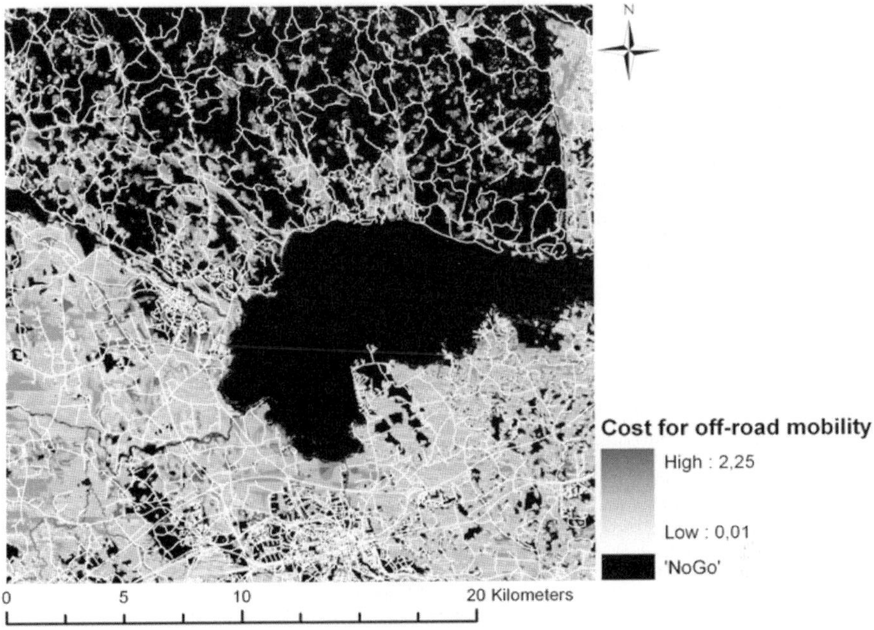

Fig. 8. On and Off-road Map with the Road Network
(Roads are given minimum – 0.01 cost to travel by, other terrain types have higher values)

However, considerable work has to be done to prepare the data sets for the analysis and elevation data was found to be partly both unreliable and insufficient in the area of investigation. As data are in a GIS database it is further possible to add other factors as precipitation, temperature and other dynamic factors of relevance. It is also easy to share the result with other users to add intelligence reports both concerning own resources and development of situation. Further in a GIS it is possible to analyze the situation with the scale or zoom factor desired as to see general overview (previous Figs. 7 and 8) and in the other hand specific details (see Fig. 9).

It is also possible to show the result of the off road network analyze in a 3D View (see Fig. 10) for control of the reliability of the simulation.

In the last example existing road network was added to demonstrate the possibility to combine traditional network analysis with the off road trafficability models (see Fig. 11).

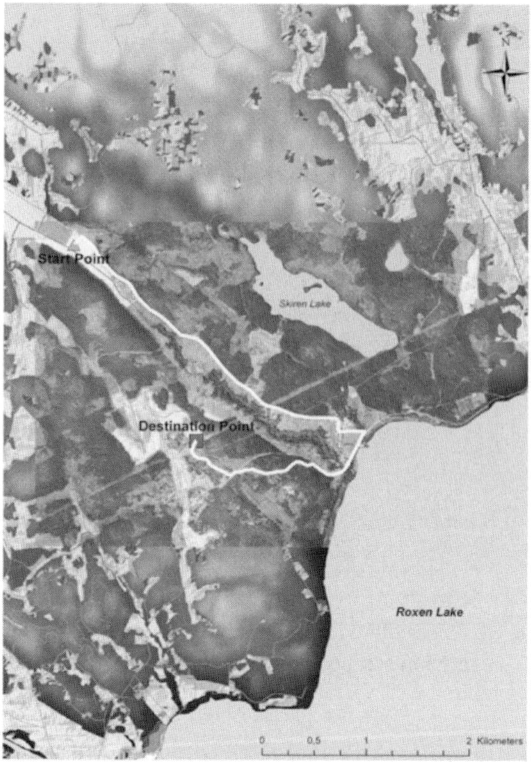

Fig. 9. An experimental field ground in the designated area next to the Stjärnorpravinen (Stjärnorp's gully)

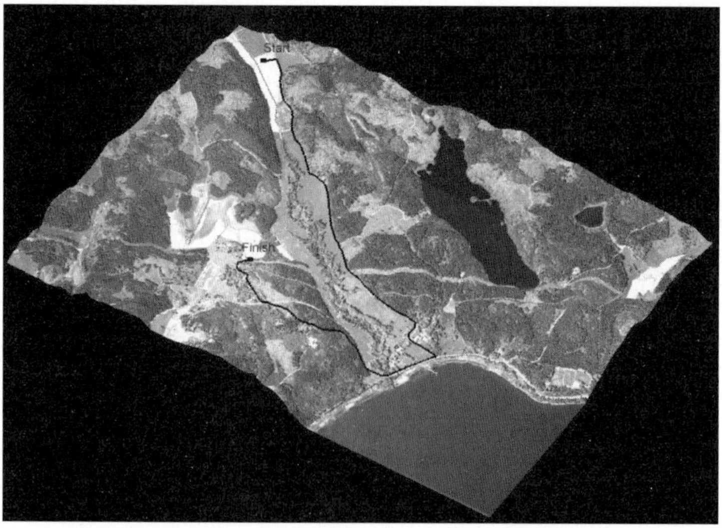

Fig. 10. 3D Scene of the experimental ground field

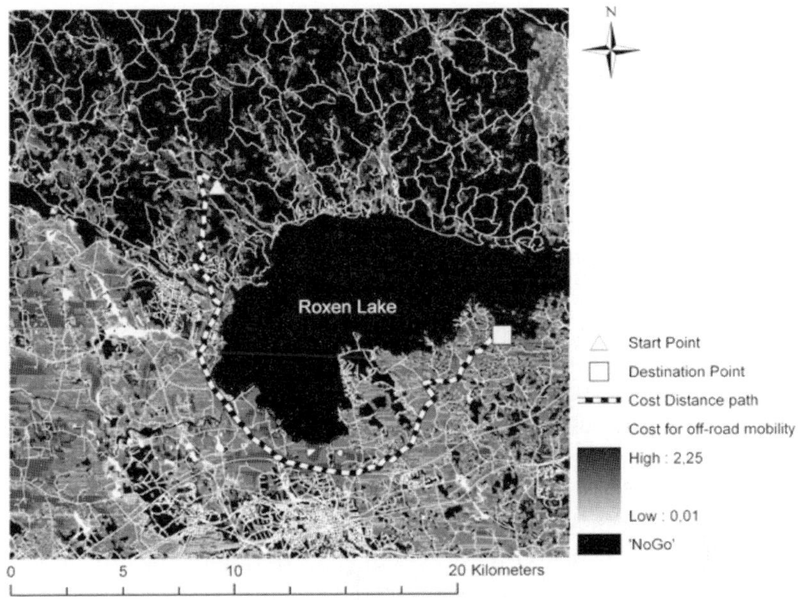

Fig. 11. Secondary randomly chosen destination path for the cross-country vehicle performance

5 Discussion

In this work we have shown that we were able to get necessary data stored, processed and analyzed to run more sophisticated GIS based simulation for trafficability in terrain, being employed by stable factors.

What is further needed are metrological data, data and information concerning present situation and dynamic man made obstacles as well as particular knowledge about the own forces. Such detailed knowledge about the own forces will be available in the near future as a part of the development of Network based organizations. The support systems will be able to share detailed information about every person with sensors and every vehicle through the built in computer and data busses in the same way as the civil mechanic makes the diagnosis on your car in the workshop. By making this information available for those that are asking for it (inside your own organization) it will – if needed – be possible to direct those units that are suited for that through the most difficult paths but redirect the other for tasks that they still have capacity for. In this way it will be possible to make also better tactical and strategic decisions. In an earlier work Trnka (2003) was concluded that GIS analyses used in emergency management could help decision makers and politicians to assess protection related problems, to support their decisions and to help them prioritize the available resources and protection measures. This is supported by Uran and Janssen (2003) and Crossland et al. (1995) pointing on that the better quality was obtained in decisions supported by analyses in GIS environments with, at the same time, less time needed for the decision-making process. These authors are however, pointing out the late adoption of the GIS technique in organizations. Perhaps the explanation is that these organizations are relatively conservative and not actually changing their organizational behavior so it would support the use of new technology! An analogue example is how the new telecom system for the rescue forces in Sweden has been designed – mainly to support a spoken interaction and present command structures!

At the moment the development of organization and the technological development of information systems are not harmonized as to take advantage of the progress in each other domain.

The Network based C3 structures will require other types of interaction supporting the members to share ready made analyses – if time is restricted – or digging deeper in the data sources as to find out even more delicate solutions to the problems if time and pre-knowledge allows that. Here the staff is in full charge – able to choose the sources and methods they want depending on the situation and – to share that with all members with few restrictions!

What Information Science can provide are the tools of highest perfection leaving to the persons that has to make the decisions to do so.

Also usability in this trafficability module is depending of how accurate the model describes the situation and how valid and reliable the data are. By introducing a soil wetness index we contribute to this estimated value however depending on access to the soil and other databases we have used here. In the future new sensors like multi-spectral LIDARs (Airborne Laser Scanning) and a mix of sensors and databases can provide the user with all the necessary stable as well as dynamic variables.

LIDAR and InSAR (Interferometric Radar) techniques have been tested and compared in order to depict the forest canopy dimensions and terrain elevation models (Andersen et al. 2004). For the concept of trafficability, the outcomes are of greatest importance in terms of accuracy improvement of the recorded earth surface features.

Acknowledgements

This work was supported by the Swedish National Defence College (FHS) and GMOSS NoE.

References

Andersen H-E et al (2004) A Comparision of Forest Canopy Models Derived From LIDAR and InSAR Data in a Pacific Northwest Conifer Forest. Int Archives of Photogrammetry and Remote Sensing 34 (Part 3/W13).211–217

Andersson L, Sivertun Å (1989) A GIS-supported Method for Detecting the Hydrological Mosaic and the Role of Man as a Hydrological Factor. Diss. Linköping Studies in Arts and Science no. 33, Linköping, p 1-27 and in: Landscape Ecology 5(2) 1991:107–124. SPB Academic Publishing bv, The Hague

Artman H, Persson M (2000) Old practices – new technology: Observations of how established practices meet new technology. In: Designing Cooperative Systems – The Use of Theories and Models. Proc of the 5th Int Conf on the Design of Cooperative Systems (COOP'2000), Sophia-Antipolis, France, pp 35–49

Attneave F (1959) Applications of information theory to psychology: A summary of basic concepts, methods, and results. Holt, Rinehart & Winston, New York

Bach C, Scapin DL (2004) Obstacles and Perspectives for Evaluating Mixed Reality Systems Usability. In: Workshop MIXER – Exploring the Design and Engineering of MR system, IUI-CADUI, Funchal, Portugal, no 13–16, pp 72–79

Brehmer B (2002a) Beslutsstöd i ROLF-staben. Teknisk rapport, Krigsvetenskapliga in-stitutionen, Försvarshögskolan

Brehmer B (2002b) Nästa steg: ROLF 2010 i det nätverksbaserade försvaret – bilaga 1 till milstolpsrapport 2002-12-16 för FHS' projekt Network Warfare. Teknisk rapport, Krigsvetenskapliga institutionen, Försvarshögskolan

Crossland MD, Wynne BE, Perkins WC (1995) Spatial decision support systems: An overview of technology and a test of efficacy. Decision Support Systems 14:219–235

DARPA. Rapid knowledge formtion. http://www.ksl.stanford.edu/projects/RKF/

Dörner D (1980) On the difficulties people have in dealing with complexity. Simulation & Games 11(1):87–106

Edlund S (2004) Driveability analysis – using a Digital Terrrain Model and Map Data. Master Thesis paper, Linköping universitet, LITH-IDA-EX-04/031-SE

Edwards W (1954) The theory of decision making. Psychological Bulletin 51(4): 380–417

Ferguson RW, Forbus KD (2000) GeoRep: A flexible tool for spatial representation of line drawings. I AAAI/IAAI:510–516

Gumos KA (2005) Modelling the Cross-Country Trafficability with Geographical Information Systems. Master Thesis in Geoinformatics ISRN LIU-IDA-D20-05/012-SE

Gustafsson T, Carleberg P, Nilsson S, Svensson P, Sivertun Å, Le Duc M (2004) Mixed reality för tekniskt stöd. Mixed Reality for technical support. Linköping, FOI, 57 p (FOI-R-1198-SE)

Hiltz SR, Turoff M (1985) Structuring computermediated communication systems to avoid information overload. Communications of the ACM 28(7):680–689

Hollnagel E, Woods DD (2005) Joint cognitive systems: Foundations of cognitive systems engineering. Taylor & Francis Books Inc., Boca Raton, FL

Jones B (1997) Geographical Information Systems and Computer Cartography. Addison Wesley Longman Limited. ISBN: 0 582 04439 1

Klein GA, Oramasu J, Calderwood R, Zsambok CE (eds.) (1993) Decision making in action: Models and methods. Ablex, Norwood, NJ

Kraemer KL, King JL (1988) Computer-based systems for cooperative work and group decision making. ACM Computing Surveys 20(2):115–146

Langston J (2002) Toward practical knowledge-based tools for battle planning and scheduling. In: Eighteenth National Conf on Artificial Intelligence. American Association for Artificial Intelligence, pp 894–899

Leifler O, Eriksson H (2004) A research agenda for critiquing in military decision-making. In: The Second Swedish-American Workshop on Modeling and Simulation

Leifler O, Eriksson H (2005) A Research Agenda for Critiquing in Military Decision-Making Report to the Swedish National Defence College (FHS)

Leifler O, Jenvald J (2005) Critique and visualization as decision support for masscasualty emergency management. In: Van de Walle B, Carlé B (eds) Proc of the Second Int Conf on Information Systems for Crisis Response and Management, Brussels, Belgium, pp 155–160

Leifler O, Johansson B, Persson M, Rigas G (2004) Developing critiquing systems for network organizations. In: Proc of IFIP 13.5 Working Conf on Human Error, Safety and Systems Development

Le Duc M (2000) Elements of Innovation Management in Computer Software and Services. In: Proc of The Ninth Int Conf on Management of Technology (IAMOT 2000), February 20–25, 2000, Miami, Florida, USA

Longley PA et al. (1999) Geographical Information System Vol. 1 – Principles and Technical Issues, pp 493–502, ISBN: 0471-33132, New York

Miller JG (1960) Information input overload and psychopathology. American J of Psychiatry 116:695–704

Mitchel C (1991) Tarrain Evaluation – An introductory handbook to the history, principles, and methods of practical terrain assessment: Longman Scientific& Technical, UK, ISBN 0-582-30122-X

Moray N (1967) Where is capacity limited? A survey and a model. In: Sanders AF (ed) Attention and performance I. NorthHolland Publ Comp, Amsterdam

Open CYC, http://www.opencyc.org/ (last access Oct 2005)

Persson M (2000) Visualization of information spaces for command and control. In: ROLF 2010 – The Way Ahead and The First Step. Gotab Erlanders, Stockholm

Silverman BG (1992) Critiquing human error. A knowledge based human-computer collaboration approach. Academic Press, UK, London

Slocum KR et al. (2003) Trafficability Analysis Engine, Cross Talk. The J of Defence Software Engineering, June 2003 issue

Sundin C, Friman H (eds) (2000) ROLF 2010 – The Way Ahead and The First Step. Gotab Erlanders, Stockholm

Talbot D (2004) How tech failed in Iraq. Technology Review Nov:36–44

Turing AM (1936) On computable numbers, with an application to the Entscheidungsproblem. Proc of the London Mathematical Society 2(42):230–265

Trnka J (2003) GIS in Protection Work: A Case Study – Sörmlandskustens Fire and Rescue Service – Analysis of Fire Safety and Fire & Rescue Service Performance IDA, LiU

Trnka J, Le Duc M, Sivertun Å (2005a) Inter-organizational Issues in ICT, GIS and GSD – Mapping Swedish Emergency Management at the Local and Regional Level (= Proc of the 2nd ISCRAM Conf), Brussels, Belgium.

Trnka L, Le Duc M, Sivertun Å (2005b) Utilization and Exchange of Geo-spatial Data in Swedish Emergency Management (= Proc of the 10th ScanGIS Conf), Stockholm, Sweden

Uran O, Janssen R (2003) Why are spatial decision support systems not used? Some experiences from the Netherlands. Computers, Environment and Urban Systems 27:511–526

Printing: Krips bv, Meppel
Binding: Stürtz, Würzburg